WASTES IN THE OCEAN

Volume 1:   Industrial and Sewage Wastes in the Ocean
Iver W. Duedall, Bostwick H. Ketchum, P. Kilho Park, and
Dana R. Kester, Editors
Volume 2:   Dredged Material Disposal in the Ocean
Dana R. Kester, Bostwick H. Ketchum, Iver W. Duedall, and
P. Kilho Park, Editors
Volume 3:   Radioactive Wastes and the Ocean
P. Kilho Park, Dana R. Kester, Iver W. Duedall, and Bostwick H.
Ketchum, Editors
Volume 4:   Energy Wastes in the Ocean
Iver W. Duedall, Bostwick H. Ketchum, P. Kilho Park,
and Dana R. Kester, Editors
Volume 5:   Deep Sea Waste Disposal
Dana R. Kester, Bostwick H. Ketchum, Iver W. Duedall,
and P. Kilho Park, Editors
Volume 6:   Near-Shore Waste Disposal
Bostwick H. Ketchum, P. Kilho Park, Dana R. Kester,
and Iver W. Duedall, Editors

LEAD AND LEAD POISONING IN ANTIQUITY
Jerome O. Nriagu

INTEGRATED MANAGEMENT OF INSECT PESTS OF POME AND STONE FRUITS
B. A. Croft and S. C. Hoyt, Editors

VIBRIOS IN THE ENVIRONMENT
Rita R. Colwell, Editor

WATER RESOURCES: Distribution, Use and Management
John R. Mather

COGENETICS: Genetic Variation in Susceptibility to Environmental Agents
Edward J. Calabrese

GROUNDWATER POLLUTION MICROBIOLOGY
Gabriel Bitton and Charles P. Gerba, Editors

CHEMISTRY AND ECOTOXICOLOGY OF POLLUTION
Des W. Connell and Gregory J. Miller

SALINITY TOLERANCE IN PLANTS: Strategies for Crop Improvement
Richard C. Staples and Gary H. Toenniessen, Editors

ECOLOGY, IMPACT ASSESSMENT, AND ENVIRONMENTAL PLANNING
Walter E. Westman

CHEMICAL PROCESSES IN LAKES
Werner Stumm, Editor

INTEGRATED PEST MANAGEMENT IN PINE-BARK BEETLE ECOSYSTEMS
William E. Waters, Ronald W. Stark, and David L. Wood, Editors

PALEOCLIMATE ANALYSIS AND MODELING
Alan D. Hecht, Editor

BLACK CARBON IN THE ENVIRONMENT: Properties and Distribution
E. D. Goldberg

GROUND WATER QUALITY
C. H. Ward, W. Giger, and P. L. McCarty, Editors

TOXIC SUSCEPTIBILITY: Male/Female Differences
Edward J. Calabrese

ENERGY AND RESOURCE QUALITY: The Ecology of the Economic Process
Charles A. S. Hall, Cutler J. Cleveland, and Robert Kaufmann

AGE AND SUSCEPTIBILITY TO TOXIC SUBSTANCES
Edward J. Calabrese

ECOLOGICAL THEORY AND INTEGRATED PEST MANAGEMENT PRACTICE
Marcos Kogan, Editor

ENERGY AND
RESOURCE QUALITY

ENERGY AND RESOURCE QUALITY

The Ecology of the Economic Process

CHARLES A. S. HALL

Cornell University
Current Affiliation: University of Montana

CUTLER J. CLEVELAND

University of Illinois

ROBERT KAUFMANN

University of New Hampshire
Current Affiliation: University of Pennsylvania

A Wiley-Interscience Publication

JOHN WILEY & SONS

New York · Chichester · Brisbane · Toronto · Singapore

Library of Congress Cataloging in Publication Data:

Hall, Charles A. S.
  Energy and resource quality.

  (Environmental science and technology)
  ''A Wiley-Interscience publication.''
  Bibliography: p.
  Includes index.
  1. Power resources.   2. Industry—Power supply.
3. Human ecology.   I. Cleveland, Cutler J.
II. Kaufmann, Robert (Robert K.)   III. Title.
IV. Series.

TJ163.2.H344   1985   333.79   85-16869
ISBN 0-471-08790-4

Printed in the United States of America

10 9 8 7 6 5 4 3 2 1

To all our parents,
and especially to Addison S. Hall,
who passed away while this book was being completed.
Addison Hall was an engineer and a wonderful father
who encouraged his son from the beginning to ask
"How does it work?"

We miss him.

CONTRIBUTORS

DAVID BEHLER
Institute of Environmental Studies
University of Wisconsin
Madison, Wisconsin

JONATHAN CHAPMAN
Institute of Tropical Forestry
Rio Piedras, Puerto Rico

MELINDA DOWER
Division of Water Resources
Department of Environmental Protection
State of New Jersey
Trenton, New Jersey

PAUL JACOBSON
Laboratory of Limnology
University of Wisconsin
Madison, Wisconsin

CARRIE KOPLINKA-LOEHR
Department of Conservation Education
Cornell University
Ithaca, New York

CHRIS NEILL
Center for Wetlands
Louisiana State University
Baton Rouge, Louisiana

CHARLES STAVER
Department of Vegetable Crops
Cornell University
Ithaca, New York

BEVERLY I. STRASSMANN
Museum of Zoology
University of Michigan
Ann Arbor, Michigan

GARY WAYNE
San Rafael, California

FOREWORD

This is a book written in the tradition of the Physiocrats. It is a book about economics from a physical point of view. Physiocrats believe that the ultimate source of economic value resides in the natural resources—in direct proportion to the degree of order stored in those resources and in inverse proportion to the physical cost of finding and extracting that order. The authors are much more sophisticated than their eighteenth century physiocratic counterparts in Europe. Land—and its ability to harness the sun—was the source of value in those distant times. In this book solar energy is studied but the emphasis is on the economic importance of the fossil and nuclear fuels.

While economists may have once concerned themselves with the importance of natural resources, two important points made in the last 25 years encouraged them to divert their attention. The first was made with the book *Scarcity and Growth,* by Barnett and Morse, which showed that resource harvesting costs, as measured in terms of direct capital and labor requirements, generally declined from the late nineteenth century up to the late 1950s (except for forestry). The second was developed in a theoretical paper by Robert Solow. He argued that increased use of capital goods could offset any decline in natural resource scarcity so that a constant real income could be maintained, providing certain seemingly reasonable assumptions were true. Both arguments have fallen on the same sword; substitution, a keystone principle in economics, does not apply well to finite resources, whether the finiteness arises from a rate-limited process (solar energy use) or stock-limited processes (the fossil and nuclear fuel use). Barnett and Morse ignored the massive quantities of natural resources poured into the resource-harvesting process, substituting in an unmeasured way for the declining use of capital and labor. Solow did not account for the demand his increasing need for capital would have for natural resources—that is, his theory also requires a hidden, indirect subsidy of natural resources. It was not an economist but the geologist M. K. Hubbert who predicted accurately that the decline in U.S. oil production would begin in 1970, and his prediction was based strictly on physical measures.

Economics is the study of how in detail to distribute the effects of natural resource surpluses or shortages—it is not a process for predicting when they will occur and how large they will be. Furthermore, economics is almost silent on the human health and ecosystem impacts of resource gathering and use. Such impacts are regarded as external to the economic system (externalities cannot be priced) and are largely ignored by the profession. The modern Physiocrats have no such constraints, as our authors clearly demonstrate. They see resource scarcity as causing an increased substitution of low-waged labor for high-waged labor and low-return rate for high-return rate capital. Therefore, natural resource scarcity reduces the material content of the lifestyles of wage earner and stockholder alike.

In the 1930s and 40s we bore a great deal of environmental and human health insults in ignorance. As the nation became wealthier we learned to recognize and manage these problems reasonably well. But as rising natural resource scarcity increases the level of these insults, and simultaneously reduces material wealth, an increasingly difficult choice will be whether or not we will continue to choose environmental protection over other goods and services.

The authors obviously want readers to understand the physical meaning of resource scarcity and its ultimate impact. It then becomes the responsibility of the readers to translate this knowledge, by a process not at all understood by economist or physiocrat, into mutual restraint for their present and future good. Consequently, this book is very appropriate for undergraduates in both science and liberal arts. It is clearly written and argues effectively from a large empirical base. It fills a much overlooked niche in their current education and it aids the rare kind of informed judgment desperately needed as our democracy faces increasing natural resource scarcity.

Bruce Hannon

Department of Geography
University of Illinois

ix

SERIES PREFACE

Environmental Science and Technology

The Environmental Science and Technology Series of Monographs, Textbooks, and Advances is devoted to the study of the quality of the environment and to the technology of its conservation. Environmental science therefore relates to the chemical, physical, and biological changes in the environment through contamination or modification, to the physical nature and biological behavior of air, water, soil, food, and waste as they are affected by man's agricultural, industrial, and social activities, and to the application of science and technology to the control and improvement of environmental quality.

The deterioration of environmental quality, which began when man first collected into villages and utilized fire, has existed as a serious problem under the ever-increasing impacts of exponentially increasing population and of industrializing society. Environmental contamination of air, water, soil, and food has become a threat to the continued existence of many plant and animal communities of the ecosystem and may ultimately threaten the very survival of the human race.

It seems clear that if we are to preserve for future generations some semblance of the biological order of the world of the past and hope to improve on the deteriorating standards of urban public health, environmental science and technology must quickly come to play a dominant role in designing our social and industrial structure for tomorrow. Scientifically rigorous criteria of environmental quality must be developed. Based in part on these criteria, realistic standards must be established and our technological progress must be tailored to meet them. It is obvious that civilization will continue to require increasing amounts of fuel, transportation, industrial chemicals, fertilizers, pesticides, and countless other products; and that it will continue to produce waste products of all descriptions. What is urgently needed is a total systems approach to modern civilization through which the pooled talents of scientists and engineers, in cooperation with social scientists and the medical profession, can be focused on the development of order and equilibrium in the presently disparate segments of the human environment. Most of the skills and tools that are needed are already in existence. We surely have a right to hope a technology that has created such manifold environmental problems is also capable of solving them. It is our hope that this Series in Environmental Sciences and Technology will not only serve to make this challenge more explicit to the established professionals, but that it also will help to stimulate the student toward the career opportunities in this vital area.

Robert L. Metcalf
Werner Stumm

PREFACE

This book is about the basic resources of natural and industrial systems and the ways that these resources interact with energy. Although the authors' original interests and training were in the energy and material relations of natural ecosystems, we have expanded our interests into what we call the energy and material flows of industrial ecosystems, a subject traditionally considered under the aegis of economics. Today, ecology and economics are normally considered, and taught, as quite different disciplines and from quite different perspectives, but we find that the two disciplines have much in common. One important reason is that both words, ecology and economics, are derived from the Greek *oikos*, meaning "pertaining to the household." Another is the central position of production in at least large segments of each discipline.

The very different approaches and subject matter that presently characterize the two disciplines have not always been so different. Much of the earliest work in the formal discipline of ecology (which is, incidentally, not related especially closely to the concept of "ecology as environmental concern" that permeates the contemporary media) was oriented toward the practical economic problems of plant and fish production. George Woodwell, past president of the Ecological Society of America and one of the leading spokespersons of contemporary ecology, laments the fact that most modern academic ecology has turned its back on these more practical problems and has dealt increasingly with quite esoteric, although often fascinating, problems. Similarly, many early economists, including the physiocrats of the 18th century, focused their analysis on how human society interacted with, and was influenced by, natural systems and the attributes of the physical resources found in those systems. But just as much of modern ecology has turned its back on the interactions between natural and modern industrial systems, so has much of modern neoclassical economics ignored the physical and biotic underpinnings of economic production. Some economists are concerned by this trend, and, for example, Wassily Leontief, a Nobel laureate in economics, laments the fact that modern economics has emphasized increasingly highly theoretical analyses that often have little bearing on what is going on in the real world. This book is an attempt to reunite again, to some degree, the two disciplines in order to understand better the interdependencies between human and natural systems, using energy and natural resource quality as common denominators for measuring activity in both types of system.

This book uses two approaches toward providing that knowledge of energy and resources: the first, developed in Chapters 1–6, is a statement and demonstration of our own philosophy that energy, other natural resources, and economic systems comprise a fundamental interacting ecosystem whose mechanisms cannot be understood by considering those components in isolation. From this perspective, economics is defined as the study of how energy is used to transform natural resources into goods and services to meet society's material needs and how these goods and services are allocated. We use the concept of *return on energy investment* as our principal conceptual tool, one that links together the often diverse phenomena that must consititute a comprehensive text on energy, society, and the environment.

Our second basic approach is an overview of what we perceive to be the essential properties of society's basic resource systems, including an assessment of the energy and environmental costs of those resources (Chapters 7–22). Editorials follow each part. Our newest and most important arguments are found in the first and second parts, especially in Chapters 2 and 4. Much of the rest of the book is a more detailed documentation of what we present in these chapters.

We believe, as do our editors at Wiley, that understanding the interdependencies between natural and economic systems, and their ramifications in our society, is as fundamental to a basic liberal or technical education these days as are the more traditional disciplines of biology and physics. We also believe that an understanding of energy and its role in human affairs is so inextricably tied in with virtually all questions of environment and economics that it provides a focal point for any study of environmental sciences or economics. We emphasize U.S. data here, but we believe the ideas are applicable essentially to any country.

The format and level of presentation of this book are designed both for general readership and for use as an upper division or graduate textbook for a course on energy itself, environmental sciences, or as an alternative text for an economics course. The book assumes a general knowledge of basic biology, physics, and economics, but we have made an attempt to make the major points stand by themselves.

Our motivation for writing this book is twofold. First, as ecologists we have found the modern human- and fossil-fuel-dominated landscapes that increasingly constitute the world to be fascinating entities for inquiry, every bit as interesting as the natural systems with which they compete for space. We call these relatively new landscapes of fossil-fuel-powered agriculture, fishing, manufacturing, transport, and commerce *industrial ecosystems*, and in a sense this book is very much a text in ecology. The conceptual tools that ecologists have developed for analyzing rivers, estuaries, and forests have exciting new applications in the analysis of industrial ecosystems. On the other hand, many concepts from natural systems are not transferable to human behavior and economic systems, and vice versa, because of some very large differences made by human culture.

Our second motivation for writing this book stems in large part from our considerable dissatisfaction with much of modern economic theory, and with the policies that are based on that body of theory. In our opinion, there has been little or no sophisticated treatment of the physical attributes of the resource base in much of modern economic theory or in the standard economics textbooks even though those resources are the basis for economic productivity and virtually all wealth. For example, Samuelson, in his influential text in economics, says: "In social science there is no law like the conservation of energy to prevent the creation of purchasing power." In contrast, the study of the economies of natural ecosystems has always included an assessment of the constraints provided by the limited energy available from the sun. Human

economic systems have escaped those constraints temporarily through the use of fossil fuels and the types of technologies made possible by these fuels. Since most modern economic theory was derived during times of expanding availability of high-grade energy resources, much of that theory could ignore the fundamental constraints imposed ultimately by the depletion of high-quality energy and other resources. Already the results of the depletion of high-quality fossil fuels has become increasingly a constraining factor for the United States and the world economies, causing many of the economic problems of the past decade. Clearly, old economic models of both the left and the right no longer seem adequate to explain and regulate economic behavior as they did before 1970. While different economic theorists and different political parties are attempting to patch together new variants of old theories with bits of political dogma in an attempt to solve the problems, many of the problems deepen and prove increasingly intractable to resolve in traditional ways. We have no easy solutions to these economic problems, nor, on the other hand, are we necessarily pessimistic about the future. But to us it is clear that any successful solutions must be made within the context of the changing resource conditions that we present in this text even though they are rarely mentioned in most contemporary economic analyses.

Although this book has an explicit *energy* perspective for the examination of economic systems, we are not suggesting that an energy approach to exchange or value should be implemented in society, for there are many good reasons to keep the money-based system presently in place. Instead, the approach found in this book should be viewed as a supplement to, rather than a replacement for, standard economic and environmental analyses, one that can give important information that is not, and cannot be, reflected in the market. We leave the assessment of the importance and validity of our approach to economics based on the basic laws of physics and principles of biology to the judgment of the reader and future generations. We do believe, however, that some concepts and methodological tools presented in this book will be of use to all readers.

Finally, we do not and cannot offer an explicit political philosophy for dealing with the information we present, for we believe that the virtues and vices of a liberal versus a conservative, or a centralized versus decentralized, political regime should be argued independently of, although with cognizance of, the information and approaches that we present. We believe that any given analysis will have different and often new constraints when viewed from our perspective. We examine and critique a number of existing philosophies from such a perspective, both in the editorial sections at the end of each part and especially in Part Five. As teachers and colleagues we have been fascinated by watching the changes that have occurred as students of initially disparate political philosophies have wrestled with, and changed their opinions because of, some of the facts and perspectives presented in this book. We hope that you too will be stimulated in a similar manner.

Since 1978 we have seen the nation's interest in energy wax and wane several times. As of this writing, energy is not a widely discussed topic in the media or the nation's Capitol due in large part to the decline in world oil prices. Analysts are gleefully predicting the demise of OPEC and there is talk of further decline in the price of oil. Some people say that the national economy is doing well and that our energy problems are solved.

We have been asked numerous times whether our book is still relevant. Our answer is that the book is *at least* as pertinent as it was in the 1970s for at least three reasons. The first is that the basic principles laid out in the book are just as appropriate when energy opportunity is expanding as when it diminishes, because the basic energy–economy links that we demonstrate operate whether energy resources are increasing or declining. Second, in international markets energy is still at least three times as expensive in real terms as it was in 1970, and this price increase has had, and will continue to have, severe impacts on developing countries. Third, although many attribute recent modest economic success to the unfettering of the market and imply that this proves the market can solve any economic or resource scarcity problem, we believe the contrary to be true. According to *The New York Times* (April 28, 1985), real family income, which had been increasing regularly since at least 1940, peaked in 1973 but has declined somewhat since then. For another example, allowing the price of oil and gas to increase nominally provides incentives for conservation and substitution while creating incentives for the petroleum industry to find new sources of petroleum. Conservation and substitution of petroleum have indeed occurred, but a closer examination shows that a large portion of the

"conservation" can be attributed to reduced and even negative economic growth compared to before 1973. Increased exploration and development also occurred; unfortunately, as we detail later, that increased exploration did not significantly alter the rate at which we *found* petroleum. The economic incentives mostly meant that we increased the rate at which we extracted oil from previously discovered fields.

One would clearly be foolish to try to predict international petroleum prices in today's volatile world, but whether prices go up or down, we believe the information contained in this book is valuable. Yet, sooner or later—perhaps in much less than a decade and certainly well within the lifetimes of most who read this book—our major domestic resources of high-quality fossil fuels will be depleted. At that time increasing resource scarcity, which underlies many of the economic problems of the 1970s and made those of the 1980s increasingly difficult to solve, almost certainly again will figure prominently in our economic and social lives. We will then need to modify existing economic and political mind sets in order to solve our national problems. We will also be saddled with the repayment of the enormous national and international debts generated in the 1980s when we as a nation tried to continue to live beyond our energy means.

To our eyes the major cause for concern is not that we may find ourselves as a nation somewhat poorer and less powerful in international competition but rather whether our democratic way of governing, which we personally value very highly and which has served us well in times of plentiful resources, can continue to ensure that a slower growing or even shrinking pie can be shared equitably. We also worry about whether a government can retain public support should austerity be the only thing that legitimately can be promised. The real challenge, then, is political in nature, and will require politicians very different from those we now encounter.

CHARLES A. S. HALL
CUTLER J. CLEVELAND, JR.
ROBERT KAUFMANN

Ithaca, New York
Urbana, Illinois
Durham, New Hampshire
September 1985

ACKNOWLEDGMENTS

The concepts, examples, and philosophy presented in this book represent the ideas, arguments, criticisms, and hard work of many people. The concepts originated in Howard Odum's classes and discussions in Chapel Hill more than a decade ago, although most of what is presented here represents rather large transformations of those ideas. Sandra Brown, Robert Costanza, Bruce Hannon, Robert Herendeen, Robert Howarth, and Ariel Lugo shared many of the initial ideas with us. Much of what appears here has been strongly influenced by our students and colleagues at Cornell in the past decade, and we wish to acknowledge the assistance of all those with whom we have worked closely. Support from our editors at Wiley, Don Deneck, Trev Leger and Mary Conway, as well as moral support and continued interest from Georgia Cohen of Macmillan and Robert Lakemacher at Merrill were important to us personally. Parts of the book were prepared principally by students and friends under our direction, and their names are listed in their respective chapters. We thank Colin Baker, Paul Colinvaux, Earl Cook, Jon Conrad, Robert Costanza, Herman Daly, Dick Hennemuth, Richard Houghton, Robert Howarth, Lewis Incze, Buzz Lavine, Karin Limburg, John Machta, Barrien Moore, William Odum, David Pimentel, Robert Pohl, F. Harvey Pough, Alan Roger, K. V. Sarkanen, Tomás Schlichter, Steven Schilizzi, Mike Sigler, Tomás Viatoris, and Bruce Wunder for critical reviews of specific material, and Laurie Badger, Juliet Bakker, Lauren Hovi, Carrie Koplinka-Loehr, Diana Voyles, and Robin Wakeman for editorial assistance. F. Robert Wesley did much of the photography, Ed Brothers did a number of figures, and Ken Bechman, Phil Bogdonoff and Sheila Underhill aided in the preparation of the bibliography. The book would not be a reality if it were not for the skilled, dedicated, and cheerful efforts of Terry Purino who organized and typed the various drafts and, in general, made things happen. Some of the original ideas for the book were initiated with a small grant from the New York College of Agriculture and Life Sciences. Continuing support was provided by the Center for Wetland Resources, Louisiana State University, the Geography Department, University of Illinois, and The Complex Systems Research Center, University of New Hampshire, whose director, Barrien Moore, gave a great deal of encouragement.

C. A. S. H.
C. J. C.
R. K.

CONTENTS

PART ONE INTRODUCTION 1

1. Energy and Living Systems, 3

PART TWO ENERGY AND ECONOMICS 25

Overview, 27

Historical Perspective, 31

2. Economics From an Energy
    Perspective, 33

3. Value, 69

4. Resource Quality, 77

5. Energy and Manufacturing, 105

6. Food, Energy, and Agriculture, 115

Editorials, 143

**PART THREE CHARACTERISTICS
AND MAGNITUDE OF U.S. ENERGY
RESOURCE SYSTEMS** 151

The Formation of Fossil Fuels, 153

7. Petroleum, 161

8. Imported Petroleum, 189

9. Natural Gas, 201

10. Shale Oil, 221
    by Chris Neill and Charles A. S. Hall

11. Coal, 229

12. Nuclear Energy, 259
    *by Paul Jacobson and Charles
    A. S. Hall*

13. Solar Energy, 285
    *by Gary Wayne, Charles A. S. Hall,
    and David Behler*

14. Conservation, 321

Editorial, 343

**PART FOUR ENVIRONMENTAL
AND HUMAN HEALTH IMPACTS
OF ENERGY EXTRACTION AND USE** 351

15. Impacts from Extraction and
    Processing, 357

16. General Impacts of Burning
    Fossil Fuel, 383

17. Human Health Impacts, 413
    *by Carrie Koplinka-Loehr and
    Charles A. S. Hall*

Editorial, 431

**PART FIVE ENERGY AND THE
MANAGEMENT OF RENEWABLE
NATURAL RESOURCES** 435

18. Fisheries, 437

19. Forest Resources and Energy Use
    in the Forest Products Industry, 461
    *by Jonathan Chapman and Charles
    A. S. Hall*

20. Cropland, 481
    by Charles Staver and Melinda Dower

21. Rangelands, 497
    by Beverly I. Strassmann

22. Water from an Energy Perspective, 515

Editorial, 525

Bibliography, 535

Index, 569

Metric–English Equivalence

Metric unit		English equivalent
Length		
millimetre (mm)	= 0.03937	inch (in)
metre (m)	= 3.28	feet (ft)
kilometre (km)	= .62	mile (mi)
Area		
square metre (m^2)	= 10.76	square feet (ft^2)
square kilometre (km^2)	= .386	square mile (mi^2)
hectare (ha)	= 2.47	acres
Volume		
cubic centimetre (cm^3)	= 0.061	cubic inch (in^3)
litre (l)	= 61.03	cubic inches
cubic metre (m^3)	= 35.31	cubic feet (ft^3)
cubic metre	= .00081	acre-foot (acre-ft)
cubic hectometre (hm^3)	= 810.7	acre-feet
litre	= 2.113	pints (pt)
litre	= 1.06	quarts (qt)
litre	= .26	gallon (gal)
cubic metre	= .00026	million gallons (Mgal or 10^6 gal)
cubic metre	= 6.290	barrels (bbl) (1 bbl = 42 gal)
Weight		
gram (g)	= 0.035	ounce, avoirdupois (oz avdp)
gram	= .0022	pound, avoirdupois (lb avdp)
tonne (t)	= 1.1	tons, short (2,000 lb)
tonne	= .98	ton, long (2,240 lb)
Specific combinations		
kilogram per square centimetre (kg/cm^2)	= 0.96	atmosphere (atm)
kilogram per square centimetre	= .98	bar (0.9869 atm)
cubic metre per second (m^3/s)	= 35.3	cubic feet per second (ft^3/s)
litre per second (l/s)	= .0353	cubic foot per second

Metric unit		English equivalent
Specific combinations—Continued		
cubic metre per second per square kilometre [(m^3/s)/km^2]	= 91.47	cubic feet per second per square mile [(ft^3/s)/mi^2]
metre per day (m/d)	= 3.28	feet per day (hydraulic conductivity) (ft/d)
metre per kilometre (m/km)	= 5.28	feet per mile (ft/mi)
kilometre per hour (km/h)	= .9113	foot per second (ft/s)
metre per second (m/s)	= 3.28	feet per second
metre squared per day (m^2/d)	= 10.764	feet squared per day (ft^2/d) (transmissivity)
cubic metre per second (m^3/s)	= 22.826	million gallons per day (Mgal/d)
cubic metre per minute (m^3/min)	= 264.2	gallons per minute (gal/min)
litre per second (l/s)	= 15.85	gallons per minute
litre per second per metre [(l/s)/m]	= 4.83	gallons per minute per foot [(gal/min)/ft]
kilometre per hour (km/h)	= .62	mile per hour (mi/h)
metre per second (m/s)	= 2.237	miles per hour
gram per cubic centimetre (g/cm^3)	= 62.43	pounds per cubic foot (lb/ft^3)
gram per square centimetre (g/cm^2)	= 2.048	pounds per square foot (lb/ft^2)
gram per square centimetre	= .0142	pound per square inch (lb/in^2)
Temperature		
degree Celsius (°C)	= 1.8	degrees Fahrenheit (°F)
degrees Celsius (temperature)	= [(1.8 × °C) + 32]	degrees Fahrenheit

(Quadrillion Btu)

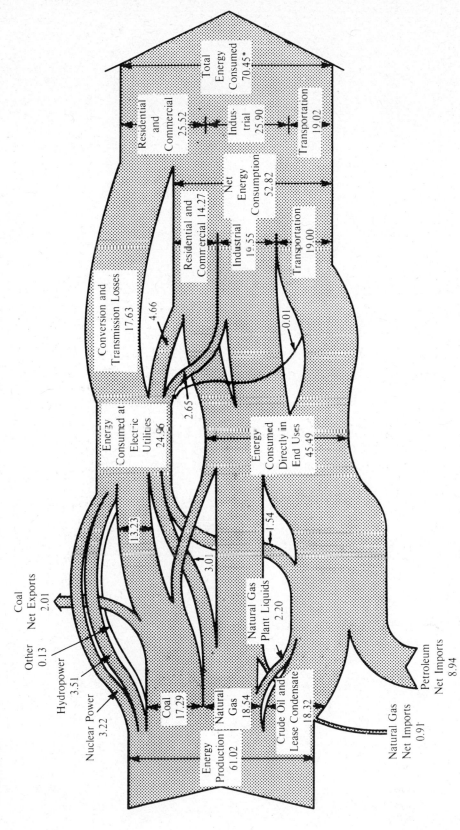

*Total Energy Consumption with conversion and transmission losses allocated to end-use sectors in proportion to the sectors' use of electricity.

Sum of components does not equal total due to independent rounding; the use of preliminary conversion factors; and the exclusion of changes in stocks, miscellaneous supply and disposition, and unaccounted for quantities.

Energy Flow Diagram for the United States, 1983

ENERGY AND
RESOURCE QUALITY

PART ONE

INTRODUCTION

It is our purpose in Part One to synthesize a number of seemingly unrelated concepts in order to build a more comprehensive understanding of some important forces in nature and, especially, in human society. These forces have a single common denominator: energy. If we are successful, the reader will see that many contemporary events are not isolated activities, responding to different causative factors, but are rather parts of a clear—if not always desirable—pattern. Ideally the reader will come away with a more informed basis for making difficult future decisions and a greater appreciation of the interrelatedness of events. At a minimum our goal is to leave the reader with a series of tools that will assist in understanding contemporary economic events, weighing alternatives while acknowledging constraints, evaluating government and private projects, and, ultimately, achieving a sensitivity toward investing our remaining high-quality energy capital as wisely as possible.

1

ENERGY AND LIVING SYSTEMS

In this book we illustrate the pervasive role of energy in society and the manner in which social patterns of energy use, geologic constraints, and the quality of various natural resources are fundamental to the economic process. To develop our premise, we first consider the physical characteristics of energy as a resource and the way it is used in nature and in society, its relation to economic activity and the environment, and the ways by which laws of mass balance, thermodynamics, and their corollaries constrain human activities.

The terms *energy, power,* and *work* have a variety of meanings depending on the frame of reference of the observer. Power to a political scientist has quite a different meaning than it does to a physical chemist. Because these concepts, and particularly the term energy, are used extensively throughout this book, it is necessary to define them as precisely as possible. For such definitions we turn to physics, where the most precise definitions exist, and our aim is to stay as close to such definitions as possible except as explicitly modified. The next several pages are a bit dull, but essential. For those reasonably well versed in the principles of physics we suggest that you turn to page 8, or even to Chapter 2, if you wish to get to the heart of our arguments.

Stated simply, energy is the capacity to do work. Work (W) is done when a force (F) moves a material body over a distance (d) according to the following relation:

$$W = Fd \qquad (1.1)$$

In this book we measure energy in kilocalories (kcal = kilogram-calories = 1000 gram-calories). One kilocalorie is the quantity of energy that, when completely converted to heat, will warm 1 kilogram of water 1 degree Celsius. A variety of other units are available for measuring energy and work, but all are interconvertible (Table 1.1). For example, 1 kcal = 3.968 British thermal units (Btu) = 0.0012 kilowatt-hour (kWh) = 4184 joules (J).

QUALITIES AND FORMS OF ENERGY

By using heat (i.e., kcal) as a measurement standard, we ignore some specific properties of other forms of energy that are very important for specific cases. Clearly there is more to sound or motion energy than its thermal equivalent, and electricity used to run a computer has a special property that makes it more valuable than its ability to merely heat water. We say that these forms of energy have a special *quality*. The quality or form of energy is especially important from an economic perspective (we talk more about energy quality on p. 55).

Power is the time rate of doing work, or, more generally, the time rate that energy is converted from one form to another. The power rating of an electric motor, for example, is the rate at which the motor converts electrical to mechanical energy. Kilowatts (kW) and horsepower (hp) are common units of power. For example, 2 kWh = 1 kW applied for 2 hr.

Energy occurs in many different forms: heat energy (kinetic energy of random motions of matter), mechanical energy (energy of organized motion of matter), electrical energy (forces of charged particles acting on one another), chemical energy (energy of chemical bonds due to atomic-level electrical forces), nuclear energy (energy of nuclear binding), gravitational energy (energy due to gravitational attractions), and light energy (energy of electromagnetic radiation).

Mechanical energy exists in two forms: *potential* energy and *kinetic* energy. Kinetic energy (E_k) is associated with the physical motion of a body, given by

$$E_k = \tfrac{1}{2}\,mv^2 \qquad (1.2)$$

where m and v are the mass and velocity of the body having the kinetic energy. Potential energy (E_p) is stored energy or energy of position. Gravitational potential energy of a body of weight (w) raised to a

Table 1.1. Energy Conversion Factors

Small-Scale Units

1 kilocalorie (kcal or Cal) =		1 British thermal unit (Btu) =		1 joule (J) =		1 kilowatt-hour (kWh) =	
4184	J	1054	J	1	J	3.6×10^6	J
1000	cal	252	cal	0.239	cal	8.604×10^5	cal
3.968	Btu	1	Btu	9.485×10^{-4}	Btu	3413	Btu
1	kcal	0.252	kcal	2.39×10^{-4}	kcal	860.4	kcal
1.162×10^{-3}	kWh	2.929×10^{-4}	kWh	2.778×10^{-7}	kWh	1	kWh
			10^{-6}	MBtu			3.413×10^{-3}	MBtu
4.843×10^{-8}	MW-day	1.22×10^{-8}	MW-day	1.157×10^{-11}	MW-day	4.167×10^{-5}	MW-day
1.326×10^{-10}	MW-yr	3.341×10^{-11}	MW-yr	3.169×10^{-14}	MW-yr	1.141×10^{-7}	MW-yr
		10^{-9}	mQ			3.413×10^{-12}	mQ
3.968×10^{-15}	Q	10^{-15}	Q	9.485×10^{-19}	Q	3.413×10^{-12}	Q

Medium-Scale Units

1 million kcal =		1 million Btu (MBtu) =		1 megawatt-day (MW-day) =	
4.184×10^9	J	1.054×10^9	J	8.64×10^{10}	J
		2.52×10^8	cal	2.065×10^{10}	cal
3.968×10^6	Btu	10^6	Btu	8.195×10^7	Btu
10^9	cal	2.52×10^5	kcal	2.065×10^7	kcal
1.162	kWh	292.9×10^{22}	kWh	2.4×10^4	kWh
4.843×10^{-2}	MW-day	0.0122	MW-day	1.0	MW-day
1.326×10^{-4}	MW-yr	3.341×10^{-5}	MW-yr	2.738×10^{-3}	MW-yr
3.968×10^{-9}	Q			8.195×10^{-8}	Q

Large-Scale Units

1 quadrillion kcal =		1 quadrillion Btu (1 quad or 1 Q) =		1 megawatt-year (MW-yr) =	
4.184×10^{18}	J	1.054×10^{18}	J	3.156×10^{13}	J
10^{18}	cal	2.52×10^{17}	cal	7.542×10^{12}	cal
3.968×10^{15}	Btu	10^{15}	Btu	2.993×10^{10}	Btu
10^{15}	kcal	2.52×10^{14}	kcal	7.542×10^9	kcal
1.162×10^{12}	kWh	2.929×10^{11}	kWh	8.766×10^6	kWh
4.843×10^7	MW-day	1.22×10^7	MW-day	365.2	MW-day
1.326×10^5	MW-yr	3.341×10^4	MW-yr	1.0	MW-yr
3.968	Q	1.0	Q	2.993×10^{-5}	Q

Source: R. H. Romer, *Energy: An Introduction to Physics.* Copyright © 1976 by W. H. Freeman and Company.

Note: Other units of energy are:

1 erg = 10^{-7} J

1 foot-pound (ft-lb) = 3.241×10^{-4} kcal = 356 J

1 therm = $2.52 \times 10^2 = 10^5$ Btu, often used in reporting sales of natural gas (1 therm is almost exactly the energy content of 100 ft^3 of natural gas).

1 kiloton = 10^{12} cal = 4.184×10^{12} J, approximately the energy released in the explosion of 1000 tons of TNT. (The kiloton, and the megaton, are frequently used in referring to A-bomb and H-bomb explosions.)

1 horsepower-hour (hp-hr) = 640 kcal = 0.746 kWh = 2.686×10^6 J, the energy delivered by 1 hp acting for 1 hr.

1 barrel of petroleum = 1.5×10^6 kcal = 0.1364 metric tons.

1 metric ton of coal = 7 million kcal

1 metric ton of wheat = 3.5 million kcal

height (h) above a reference level is

$$E_p = wh \qquad (1.3)$$

Water stored in a water tower is an example of gravitational potential energy. If that water is allowed to flow out of the tower, its potential energy is turned into kinetic energy in the form of falling water. If the storage of water is behind a hydroelectric dam, the kinetic energy of the flowing water can be used to do work on the blades of a turbine, which in turn can be converted into electrical energy. If the falling water is used to turn a waterwheel, the kinetic energy of the falling water is converted to mechanical energy (grinding grain, etc.). There are other types of potential energy in addition to gravitational: chemical (e.g., a disconnected battery), thermal, and so on.

Potential energy is converted to kinetic energy when it is used to set a body in motion. When the body comes to rest, the energy is not lost in a strict physical sense. Rather the energy associated with the motion of organized matter has become dissipated into random motion of single molecules and atoms. In the water example the kinetic energy of water flowing over a waterfall is not lost when the water reaches the pool below but is converted into heat energy. In other words, the ordered, directed motion of the falling water is converted into disordered motion. *Heat* is the kinetic energy associated with the random motion of atoms and molecules.

The mechanical equivalent of heat went unnoticed until it was observed in careful experiments by Joule and Mayer over 120 years ago. The reason was simple: the heat equivalent of mechanical energy is extremely small. The water at the bottom of Niagara Falls is only one-eighth of a degree Celsius warmer due to the conversion of mechanical to thermal energy than it was on top. To raise its temperature 1°C, the water would have to fall 425 meters (m) instead of about 50 m (Thirring, 1958).

LAWS OF THERMODYNAMICS

The *first law of thermodynamics,* or the law of the conservation of energy, states that energy can be changed in form but neither created nor destroyed. Stated another way, the energy of an isolated system is constant.* Accordingly, if one form of energy is changed to another form, the same total quantity, expressed as heat equivalents, remains after the

transformation—but not with the same quality or ability to do work. The quality of at least some of the energy will have been altered, and the energy may no longer be useful for doing a particular type of work. Returning to the water example, the loss of mechanical energy in the falling water is compensated exactly by an increase in heat energy even though such a change may be difficult to measure. As was noted earlier, only a small amount of heat is gained from the transformation of a large quantity of mechanical energy or, alternately, a kilocalorie of potential energy can do a lot of work. One kcal, for example, is equal to 427 kilogram-meters (kg-m), which is the energy that would be needed (at 100% efficiency) to lift a kilogram 427 m, or 100 pounds (lb) more than 30 feet (ft).

Thus mechanical work is converted readily to heat, but the reverse is not true. A brake applied to a rotating wheel or human physical exertion are processes that convert mechanical energy completely into heat energy. That heat, however, cannot be completely converted back into mechanical energy without addition of further energy. This condition is described by the *second law of thermodynamics* which places distinct restrictions on energy transformations. The two most important conditions are (1) heat cannot be transformed into work with 100% efficiency, and (2) heat spontaneously flows from a body of higher temperature to one of lower temperature. The second condition is readily observed in everyday life.

The second law defines *quality* differences between types of energy. This distinction restricts energy conversions. Although we can always convert 100% of high-quality work (mechanical energy) into low-quality heat, we can never convert 100% of our heat into work. Furthermore, the second law states that every transfer of high-quality energy will be less than 100% efficient; some of the high-quality energy will be degraded into heat energy. An example is that an electric motor warms up when running, so not all the electricity it uses is transformed to shaft horsepower.

The second law pervades science, technology, and even everyday life. It sets an upper limit on the efficiency of power plants, and it plays a role in the

* In physics, a closed system refers to one that has energy but no matter crossing its boundaries. An open system has both matter and energy crossing its boundaries, whereas an isolated system has neither matter nor energy entering or leaving. Such systems are idealizations since there is always some exchange of energy and matter between a system and its environment.

transmission of satellite broadcasts. It is manifest as friction between moving parts in machines and in the direction of chemical reactions. The net effect of every energy transformation is the degradation of higher-quality energy to lower-quality energy. Although high-quality energy can be produced from intermediate-quality source energy, a side effect is the production of large quantities of low-quality energy so that the net effect is a decrease in the total quality of the energy products relative to the original source.

In all its forms the second law governs the irreversibility of macroscopic processes. This is often described as increasing the randomness of the universe. Conversely, it can be viewed as decreasing the capability of the universe for performance of useful work. The concept of irreversibility can be understood easily in terms of food. An uncooked egg can be cooked at any time. Once cooked, however, it cannot possibly be coaxed into being uncooked. Further, once eaten, it cannot become uneaten (at least not in the form of a cooked egg). Each egg transformation is completely irreversible. The degradation of energy quality is irreversible as well.

Within nature and human society energy changes are involved in all physical and chemical processes. As matter changes state, such as water from a liquid to a gas, or as chemical bonds are formed and broken, energy is absorbed from, or released to, the environment. Chemical reactions that release energy are called *exothermic,* and those that require energy are called *endothermic.*

At any point in time matter will have thermal energy stored as a function of mass, temperature and the heat capacity of the material, inertial energy stored as a function of mass and velocity, and gravitational energy as a function of mass and position. Thermal, inertial, and gravitational energy are *physical energies.* Matter also will have *chemical energy* which can be released when chemical bonds are rearranged.

Efficiency

Efficiency is the ratio of useful work delivered to the total energy input, where both are measured in the same energy units. Sadi Carnot (1824), a French engineer, showed that the maximum efficiency of work that could be obtained from a thermal gradient is proportional only to the ratio between the temperatures. The greater the temperature difference,

the larger the gradient and the more efficient an energy transformation can be. Considering an idealized heat engine, Carnot showed that the theoretical maximum efficiency (E_{max}) of a heat engine can be calculated from the heat source T_1 (i.e., boiler temperature) and the heat sink T_2 (i.e., the temperature of river or lake water used for cooling; T_1 and T_2 in degrees kelvin) by the following equation:

$$E_{max} = \frac{T_1 - T_2}{T_1} \qquad (1.4)$$

E_{max} is known as the Carnot efficiency of an energy transformation process. Equation (1.4) indicates that for any finite and positive heat sink temperature, E_{max} will always be less than 100%. The Carnot efficiency for a fossil fuel electric power plant is 65–80%, although, in practice, efficiencies of only 35–40% are attained because machines normally are used more rapidly than the rate that gives the maximum Carnot efficiency.* An additional consideration is that in real machines additional energy is converted to heat by friction in bearings, linkages, and so on. Once energy lost as heat is dispersed into the aquatic or atmospheric environment as low-grade waste heat it is unavailable to do useful work.

The most important chemical gradient for most energy processes is the *redox gradient.* As a rule chemical fuels are reduced compounds rich in carbon–hydrogen bonds. Energy is released by oxidation of the fuels, and the oxidized compounds produced, such as H_2O or CO_2, are relatively energy poor. This basic chemical gradient between reduced fuels and oxidized by-products is what life operates in and on. Both biotic and fossil energy supplies are, or were, produced principally by green plants reducing atmospheric carbon using solar energy in photosynthesis. That energy is released in living organisms by a whole host of complex biotic reactions that sequentially oxidize the reduced carbon in small steps along small gradients, meanwhile using that energy for life processes while losing some useful energy to heat. In fossil fuel combustion, fuels of various combinations of carbon and hydrogen are more or less completely oxidized to CO_2 and H_2O in essentially one step.

Slesser (1978) cites an example that demonstrates the importance of gradients in energy transformations. Slesser observed that North Sea oil re-

*Technically it is more appropriate to use the Rankin cycle efficiency for an electric power plant but that need not concern us here.

serves were about 6 billion barrels (bbl) of oil equivalent, equal to about 9×10^{15} kcal of heat. A small amount of that oil can be burned to create a large thermal gradient—1000°C or more. That gradient can be used to do many kinds of useful work with a Carnot efficiency of up to about 80%. The heat content of the water in the North Sea is about 4.5×10^{18} kcal, or about 500 times as much energy as the oil buried beneath it. To utilize the heat in the North Sea, however, would require utilizing the temperature gradient between the warm upper waters and cold water at depth. The maximum temperature difference between the surface (10°C) and deep (6°C) waters occurs in the summer months. Evaluating this gradient in Equation (1.4) produces a Carnot theoretical maximum efficiency of about 1%. Actual operating efficiencies would be much less, even discounting the enormous energy capital costs of tapping the very large areas of the North Sea that would be required.

The efficiency of an energy conversion process can be improved by increasing the temperature gradient between source and sink, but it normally is inconvenient and uneconomic to use anything but the surrounding environment as a heat sink, which limits a modern power plant, for example, to a Carnot efficiency of about 80%. In practice, most processes operate at efficiencies even less than their Carnot efficiency—35–40% for our power plant example. This is because Carnot efficiencies can be achieved only under *ideal* conditions, one of which is that the transformation must occur infinitely slowly. Humans, for obvious reasons, are concerned with processes that operate on a finite time scale. The faster a given process is carried out, the lower its efficiency normally becomes. The trade-off between the rate of energy conversion and the theoretical maximum efficiency with which it is converted to work has important implications for all biological and economic processes (Odum and Pinkerton, 1955; Curzon and Ahlborn, 1975).

Entropy

The concept of entropy makes the implications of the second law easier to understand. The entropy of a system is a measure of the randomness of the system. Entropy is therefore an inverse measure of the order of a system—the greater the degree of order, the lower the entropy of a system. A barrel of crude oil represents a storage of ordered, low-entropy material. When it is burned in an engine, some of its energy is converted into useful work while the rest of it is converted into low-grade, high entropy heat. That low-grade heat no longer possesses the potential to do useful work because its temperature is similar to the environment. In other words, the gradient in energy potential has been decreased. When this occurs, we say that the entropy of the system has increased. Entropy can also be used to describe the work potential of nonfuel materials. Formal analysis of the relationship between energy and information has been undertaken by Shannon and Weaver (1963), but that is of little direct concern to this book.

To create order in one part of a system requires that energy (work) be expended there or in another part of the system. So an increase in order (decrease in entropy) in one part of the system implies a decrease in order (increase in entropy) in another part of the system or, as it is normally said, in the universe as a whole. To assemble books in a library, for example, energy was used not only to construct and maintain the library itself but also to support the poets, scientists, and other people who wrote the books. Similarly, large quantities of solar energy were used to drive the photosynthetic and geological processes that grew the trees from which the books were made and that formed the fossil fuels that supported the work of the scientists and, to some degree, most recent poets.

The second law represents the fact that in a closed system the only processes that can occur spontaneously are those that increase the total entropy (disorder) of the system. When a barrel of oil is burned, it no longer has potential to perform work. Therefore, an increase in entropy is associated with a decrease in potential to do useful work. For an open system at fixed pressure and temperature the relation between entropy (S) and ability to do work is represented by

$$G = H - T S \qquad (1.5)$$

where G is the *Gibbs free energy*, T is the temperature, and H is the enthalpy, or heat content, of the system. The Gibbs free energy is often used as a measure of the ability of a system to perform useful work. *Available* or free energy has potential for work, at least for human purposes. *Unavailable* energy has no potential to do work relative to human economic purposes.

The importance of the concept of entropy to human economic systems has been developed by

Georgescu-Roegen (1971) and Odum (1971). The concept, however, is often ignored or misunderstood by many analysts. For example, it is commonly argued that although the entropy of the universe is increasing—as the second law says it must—the ultimate heat death of our sun is billions of years away, a time scale much too large for humans to worry about, much less plan for. Thus entropy is said to be of no concern to human affairs. This argument would be valid if our existence depended solely on solar energy. Modern industrial economies, however, have evolved not only around solar energy but around sources of fossil fuel, whose rate of formation is negligible on a time scale of interest to humans. The fact that a kcal of petroleum can be burned only once for human economic purposes is just one example of how thermodynamic principles set very definite limits on human activity. It takes work to organize matter into useful economic artifacts, and anything once organized is subject to forces that tend to disorder matter as time passes (e.g., grapefruits in a pyramid tend to become spread out, mechanical structures rust and break, organic matter decays). A continual flow of new, low-entropy energy is therefore necessary to maintain any organized structure, be it an organism, an automobile, or a city. Thus it can be said somewhat loosely that entropy destroys that which humans create and that such natural destruction can be countered only by a continual investment of useful work.

A BIOLOGIST'S PERSPECTIVE

Although this book is principally about contemporary industrial systems, the procedures that we use for this examination were developed initially in the biological sciences, and much of our original training and research interests were in organismal and ecosystem biology. A principal tool for examining both the relations of organisms to their environment and the biotic processes of landscapes is the measurement and synthesis of energy budgets, for energy is the most essential currency of biology. Examining energy budgets yields wonderfully useful formalisms and constraints that add rigor to biological analyses, help assess life strategies, and that synthesize seemingly disparate information (e.g., see Odum, 1971; Warren, 1971; Pyke et al., 1977; Schoener, 1982; Aspey and Lustick, 1983). We began many years ago to consider human economic

systems from the same energy perspective and have found it curious that few others also do so for it seems a logical undertaking. A few wonderful books exist from earlier years, especially Fred Cottrell's 1955 *Energy and Society,* but they largely have been ignored—while there has been a plethora of books on economics that virtually do not mention energy. The following is an introduction to how energy flows are studied in the biological sciences that sets the stage for later, more comprehensive assessments of energy in industrial ecosystems.

Energy is the Ultimate Limiting Resource

The most important point about energy and its relation to other resources in both biological and economic systems is not that "everything can be reduced to energy" (which is false) but rather that *every* material (and most nonmaterial) resource has an associated *energy cost,* so that every potentially limiting resource is limiting in part because its energy cost is too high. It follows that *high-quality resources* are those that require less energy per unit of resource obtained. Another critical point about the unique importance of energy is its generality as a resource, for with sufficient energy reserves an organism (or human society) can divert energy—within the present genetic or technical range of possibilities—to the acquisition or mobilization of, essentially, whatever other resources are in short supply. Thus a plant is often considered nitrogen limited, for the addition of nitrogen increases that plant's growth. But even a plant in a nitrogen-limited environment normally is surrounded by many billions of atoms of nitrogen. The limitation is equally the energy required to pump nitrogen against the steep nitrogen gradient that exists between the plant's cells and the soil. Likewise, if water is limiting, energy could be used to gain water by greater mechanical pumping, by growing longer roots and so on. Each of these materials is certainly limiting, but each is also limited by energy, for if energy is available, it can be diverted to reducing the effects of most limitations. If the resource availability or quality is low the gains to an organism from using that low-quality resource may not be worth the very large energy investment to get it, for each use of energy diverts energy from the plant's or animal's reserves, and hence from other possible uses. Thus each use has what is called an *energy opportunity cost.* Over evolutionary time various different diversions of the plant's energy resources can be favored or not depending upon relative evolutionary advantages.

Much of the modern science of biology examines the ways in which electromagnetic energy from the sun is captured by green plants, stored by various biochemical processes as chemical potential energy, and used by the plant or animal to do work (i.e., rearrange molecules, concentrate chemicals, maintain cellular structure, produce locomotion, etc.). Some biologists view the process of Darwinian selection as a law of energy, the "fourth law of thermodynamics." This "law" states that life is made up of unlikely, and hence nonrandom, combinations of molecules that superficially appear to be an exception to the second law of thermodynamics. Such combinations are unlikely from a strictly physical perspective but are very likely due to laws of natural selection, which have produced organisms that counter the natural processes of degradation in their own cells (e.g., see Morowitz, 1968). Such local exceptions to the otherwise relative randomness of chemicals in the universe occur only at the expense of free energy elsewhere so that the second law is in no way violated if the correct system boundaries are used (Figure 1.1). In a sense, living organisms and human societies *borrow* a little free energy from the sun through the capture of direct solar radiation, the consumption of other organisms, or by burning fossil fuels, and use this energy to produce structures appropriate for their needs, while degrading some of that borrowed energy to heat. Living systems have very little to do with controlling the rate at which available short-wave solar radiation is changed to long-wave, low-temperature heat energy for the universe as a whole. That process is controlled by the rate at which the sun burns and the reflectance of the Earth's surface and the other objects that the sun strikes. What little a given biotic system does intercept and store, however, is of critical importance to that system, to ourselves as a species and to our society, and each living system controls the production of entropy within its own boundaries very carefully through homeostatic mechanisms.

All organisms and human societies require an almost continuous supply of energy from the sun, foodstuffs, and/or fuels to counteract the entropic forces of nature. Relatively few large organisms (seeds, hibernating woodchucks, cold-blooded animals at low temperatures) can live for more than a month or two without energy inputs in the form of

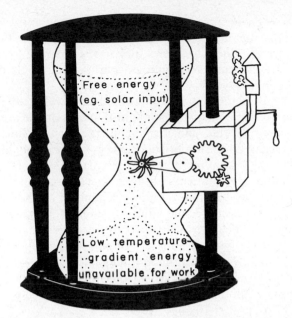

Figure 1.1. A mechanical analogy to thermodynamic processes of life. The hourglass represents the continued passage of free energy (i.e., the sun's energy) to unavailable low-grade thermal energy; the paddle wheel represents living creature's interception of some of this energy to maintain otherwise unlikely biotic structure. (Adapted from Georgescu-Roegen, *Bioscience* Vol. **27**(4):266–270. Copyright © 1977 by the American Institute of Biological Sciences.)

food or sunshine. Animals depend on continual sunshine and photosynthesis by plants as well as their own feeding efforts for their survival—processes that have so far been predictable enough for most species to have lasted for millions of years. Some animals and microbes are less dependent on recent photosynthesis. For example, bacteria and worms live in the soil using energy fixed by plants in years past, and fish in the bottom of the sea consume organic material produced near the surface in earlier years or even decades. There are a few organisms that ingest or use energy that came from chemical processes not directly related to the sun, but they are unimportant on a global scale. Most modern human societies require both energy from the sun *and* old sunlight stored in the form of fossil fuels.

All the creatures of the Earth face a common constraint: the total solar energy income is relatively fixed, changing little from year to year or century to century. Populations that become more abundant get a little larger share of the available energy for themselves, but other less successful populations *lose* energy. To our knowledge there has been no large-scale change in the quantity of energy that arrives on the Earth every day from the sun over billions of years, nor in the quantity fixed by green plants for millions, perhaps hundreds of millions, of years.

Thus plants have a restricted and well-defined source of energy—the sunlight that falls on the area covered by the plant leaves. In the United States the mean daily rate of solar energy input is roughly 3000–4000 kcal/m^2 per day, from which plants in suitable environments are able to capture about 1–50 kcal/day for their own use depending on the fertility of the site. Some animals, such as grazing antelope, walk from plant to plant, enabling them to exploit many square meters of sun-derived energy. Studies have shown that animals may eat from 5 to 40% of the energy that is fixed per unit area by plants, although this varies widely from one ecosystem to another (Golley, 1968).

In some special cases animal and even plant groups are what we call *energy subsidized;* they are able to exploit the solar energy that has been captured over a region larger than the one in which they live—without investing their own energy for that exploitation—because of additional energy inputs beyond the sun that strikes the space they live on. An example is an oyster living in an oyster reef, where the energy *subsidy* is in the form of the tidal currents that carry the plant material to the oysters. These reefs are found only in areas of swift tidal currents, and the oysters filter the passing water for plant material that was produced over a much wider area. Where such energy subsidies exist, as in an oyster reef, a stockyard, a manure pile, or a city, much denser animal populations can exist, and the organisms can exploit the productivity of a large area without investing large amounts of their own energy in transport.

An interesting parallel exists between a subsidized animal community, such as the oyster reef, and modern industrial society in that both depend on energy subsidies in amounts much greater than the direct solar energy available to them. In the case of the reef, tidal energy subsidies allow the oysters to exploit the plant production of an area many times that of the reef. In our modern society fossil fuels allow humans both to harvest food from a very large and often distant area and to increase the agricultural production of that land. One net effect is that oysters and people are often much denser in number than other animals. It is debatable whether the technologies of our society are good or bad, but it is clear that they are extremely energy intensive

and energy subsidized by fossil fuels and therefore could not exist in anything like their present form if they were dependent on solar energy inputs alone.

The Use of Energy by Organisms

All organisms need to capture energy in the form of sunlight or food to grow and reproduce. Additionally organisms must interact with, and adapt to, other energy flows of the environment (wind, temperature, the forces of waves) that tend to alter, both positively and negatively, the physiological well-being of the individual. We next describe the consumption and transfer of energy by organisms for several individual species in order to give an introduction to our view of the role of energy in living systems. The interested reader is referred to Warren (1971), Brett and Groves (1979), and Townsend and Calow (1981) for other developments of similar ideas.

Until the beginning of this century, many biologists believed that living organisms violated the laws of thermodynamics, for they appeared to be counter examples of the tendency for entropy to increase (Adams, 1920). We now know that living organisms do not violate the laws of thermodynamics if the correct boundaries for examining the energy flows in and out of an organism are drawn. Living organisms maintain their organized states by capturing high-quality, low-entropy energy and matter from their environment, using it to grow, repair damage, and reproduce, and then releasing that energy back to the environment in the form of low-quality, high-entropy waste heat. Based on this simple thermodynamic model, the functioning of biological organisms can be characterized both quantitatively and qualitatively by analyzing their use of energy.

Living organisms must use energy in order to synthesize new biomass from the raw materials of CO_2 or their food. Yet not all the energy available to an individual can be used for growth. An organism must invest energy to maintain a state of physiological and chemical equilibrium, a state known as *homeostasis*. Organisms are constantly subjected to normal catabolic (i.e., structure destroying or, as it is sometimes called, entropic) processes that tend to break apart the highly ordered structure of cells—as well as additional stresses such as adverse environments or disease that shift the homeostatic state toward disorder. Both processes require the organism to use energy for what biologists call *maintenance respiration* in order maintain or return to the homeostatic state. Organisms also normally grow for much of their life and attempt to reproduce before they die. Both growth and reproduction are normally energy-intensive processes that require considerable energy above the quantities used for maintenance respiration by a healthy adult. Later in this chapter we examine the energy budgets of several specific organisms to illustrate the allocation of energy between these three essential functions.

The allocation of energy resources to growth, maintenance, and reproduction has been studied thoroughly in many species. The energy content of various materials, including organisms and their foods, is measured using a *calorimeter,* and the rate of energy use by organisms is normally measured by analyzing oxygen changes in a device called a *respirometer* or, more rarely, by using radioactively labeled metabolic compounds. A general representation of energy flow through, and allocation within, an individual organism is represented in Figure 1.2. Specific values for the organism's allocation of these flows are shown for a plant, the wild strawberry, a fish (sockeye salmon), a snake, and a mammal—in this case our own species.

*Autotrophs,** or green plants, use chlorophyll to capture energy from solar-derived photons and store this energy by restructuring the carbon atoms of carbon dioxide derived from the surrounding atmosphere or water into complex organic compounds. *Heterotrophs* obtain metabolically useful energy from the consumption of the organic molecules in the food they obtain from other organisms, although many heterotrophs also have complex behavioral patterns to use thermal energy from the sun to reduce heat losses. Once within the digestive system, some portion of the ingested food is assimilated into the blood in the form of simple sugars, amino acids, and fatty acids. The remainder of the food is egested as feces. The energy contained in metabolic waste products (e.g., urea) which are excreted as urine or other nitrogenous wastes is included in calculations of excreted energy. The *assimilation efficiency* for food is the ratio of energy absorbed into the blood stream over the energy in the food ingested.

Notice that the assimilation efficiency of the autotroph and the three species of heterotrophs in Figures 1.2 differ. The strawberry plant captures and turns into new strawberry plants only about 1% of the solar energy that falls on its leaves. Because heterotrophs use higher-quality, preexisting organic molecules as sources of energy, their assimilation efficiency is greater than autotrophs, although the actual efficiency often depends on the food they eat. A toad, for example, eats insects that have a hard shell and are difficult to digest, and its assimilation efficiency is a relatively low 74%. A snake eats more digestible prey such as amphibians and mammals and has a greater assimilation efficiency of 87–89% (Smith, 1971). Hummingbirds represent one extreme because they eat extremely digestible sugar and therefore have a high assimilation ratio of 98% (Hainsworth, 1974). Humans eat both easily digestible animals and less digestible plant material and have intermediate assimilation efficiencies.

Assimilated food is partitioned between metabolism and growth. Each may be considered an *investment* of available energy for maintaining and/or increasing that organism's probability of survival and, ultimately, its ability to reproduce. The energy used for metabolism is essential to maintain the homeostasis of the individual and to carry out everyday functions. For example, most organisms maintain ionic balance by using energy to secrete excess salts taken in with their food, but they also use energy to move after prey or away from predators.

An organism can grow only if it assimilates more energy than needed for maintenance metabolism, prey capture, defense, and so on. The energy used for growth is subdivided between structural growth (e.g., depositing bone or muscle) and the production of fat, which is used for energy storage. The new tissue that an individual synthesizes is always less than the food assimilated because some energy is lost as waste heat during biosynthesis. Although it is readily apparent that food energy is essential for an organism's well-being, our perception of the ultimate significance of energy in biology goes far beyond the statement that an organism simply requires sufficient food energy to live. We, and many (but certainly not all) biologists, view energy as the ultimate arbitrator of biotic actions, including survival, reproduction, and patterns of natural selection. Although it is not possible at this time to say that natural selection always operates as we suggest next, we will attempt to explain how many very diverse biotic activities can be interpreted from an energy perspective.

We begin from an evolutionary perspective. Most biologists use the word *fitness* to describe the

---

*The more inclusive terms *phototrophs* or *primary producers* are preferred by some scientists.

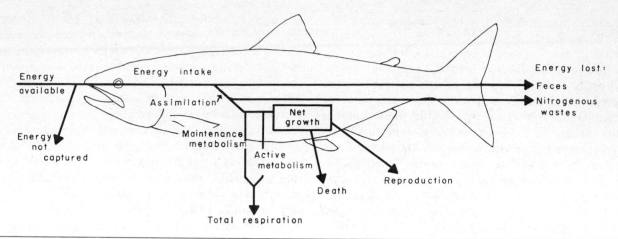

Energy	Wild Strawberry (%)	Salmon[d] (%)	Snake (%)	Adult Human Being (%)
Intake[a]	100 (~1% of solar)	100	100	100
Lost as feces	[b]	15	11	25
Assimilation, which includes:	85[c]	85	89	75
Energy lost in nitrogenous waste	[b]	5	?	5
Total respiration	54	40	~23	69
Maintenance metabolism	40	20	12	50
Active metabolism (moderate level of activity)	[b]	20	~11	19
Growth (including reproductive products)	31	40	66	0.9[e]
Reproductive products	9	2	~16	0.1[e]

[a] Gross photosynthesis for the plant, or food eaten by animal.
[b] Not applicable for plants.
[c] About 15% is reflected from, or passed through, the leaves (Chabot, personal communication).
[d] Juvenile silver salmon, at 17° and at a ration of 100 kcal/kg per day (Averitt, 1969, in Warren, 1971).
[e] Computed as one 3-kg baby per 60-kg adult per 50 years and weight gain of 0.5 kg/yr per 60-kg adult.

Figure 1.2. Overview of the use of energy by plants and animals, with some approximate values for (a) wild strawberries, (b) Pacific salmon, (c) the snake *Heterodon platyrhinos*, and (d) an adult human. (Adapted from Jurik, 1983; Warren, 1971; Smith, 1971; Brett and Groves, 1979; and Tuttle and Schottelius, 1965.)

relative contribution of a morphological, physiological, or behavioral property (or series of properties) to the future representation of an organism's genes in future populations. Fitness and those properties that contribute to fitness are rarely defined explicitly but rather in terms of future descendants of the organism. Thus a *fit* trait (or gene) is recognized only after the fact—that is, after survival and reproduction have or have not taken place. The circularity of such arguments has not gone unnoticed. Unfortunately it has not been easy to decrease that circularity because it is difficult to compare quantitatively different costs and benefits of an action or trait—the "apples and oranges" question—and because the contributing factors to an organism's fitness are many and diverse.

We and many other biologists believe that a less circular definition of fitness may be possible by using an energy-based approach, although at this time no complete and comprehensive theoretical or empirical development exists. Nevertheless, enough pieces exist to produce the beginnings of such a synthesis. After all, the concepts are basic and, for the most part, simple.

The basic tenet of an energy-based approach to selection is that organisms act so as to maximize their reproductive potential by selection for the largest difference between energy gains and energy losses over time. A large difference means a high-energy surplus relative to competitors. A large energy surplus contributes to fitness in many ways: by allowing larger numbers and/or size of eggs as well as through improving the probability of the offspring's survival through protection, nourishment, and teaching. An interesting trade-off exists between the number of eggs laid and quantity of energy an adult can invest in parenting the offspring so as to maximize the number of young produced and their probability of survival once they leave the nest. Lack (1954) observed that Swiss starlings lay between one and ten eggs in early spring when food is abundant. Those starlings that laid an intermediate number (four or five) of eggs were able to fledge the greatest number of nestlings. Adults that laid more than five eggs fledged fewer young, probably because each egg was endowed with less energy reserves and also because the parents could not bring enough food to the nest to provide a full diet for all the birds. Adults that laid one or two eggs fledged a greater percentage of their young than birds that laid four or five eggs but, because of the small number of eggs laid, fledged a lesser number of nestlings. The most common clutch size was indeed the number (four or five) that produced the largest number of fledged young.

Problems with a very simple energy analysis sometimes appear to arise when other factors are considered. For example, an organism may become larger than the size that best exploits the most abundant food resource in a given environment. For another example, many biologists think of predation as an independent cost unrelated to energy. But an inclusive energy theory has little difficulty in accommodating these problems. Predation may be viewed as the ultimate energy loss to an individual or a population. We view the energy cost of predation in much the same way that Lloyds of London views the possible loss of a commercial ship it insures: some probability of loss is essential if (economic) survival and profit are to be obtained. Behavior (of sea captains or organisms in nature) is meant to minimize the risk of loss while attempting to maximize gains. Obviously no losses would occur if the ships never left home port, but no gains would occur either. Captains who load too much sail in order to try to increase profitability are likely to lose more than they gain. In the same way an organism could avoid predation by spending its life under a rock, but it would eat little (unless it is a salamander) and probably not reproduce. Presumably there has been natural selection to balance the possibility of being eaten in a feeding foray versus the energy gained in that feeding foray. A 1000-kcal organism exposed to one chance in a thousand of

predation would have to gain, on average, well over 1 kcal of food (due to less than 100% assimilation of energy) per feeding foray to make it worthwhile to endure that risk of predation. Such calculations are best made for groups of organisms, rather than for single individuals where predation is an all or nothing event.

Of course particular characteristics of the environment—what we call various attributes of *resource quality*—influence the energy costs and gains of living. A most important attribute of this is the trophic productivity of the environment, that is, the fundamental rate of energy fixation by the green plants of that environment or the additions from neighboring environments. The efficiency of transfer of energy through the various components of food webs is also an important component of resource quality to a particular organism, as are the physical and chemical attributes of the organism's environment that raise or decrease the energy cost or efficiency of living. For example, the chemical composition of an organism's environment may include pollutants that increase physiological energy costs, or the presence or absence of shelters can increase or decrease the energy cost of living to an organism by shielding the organisms from predation or by decreasing the energy cost of adjusting to the physical environment. Various environments impose energy costs, as well as opportunities, upon organisms. Considerable literature exists (e.g., Syle, 1956; Odum, 1967b; Lugo 1978) that interprets and assesses stress, either natural or human caused, as an energy drain that decreases net production and that may make the system of interest less competitive. When energy drain to stress factors (to high level of pollutants, low number of hiding places so that loss to predation is excessive, etc.) is high, individuals or populations decline or go extinct.

In summary, we view the acquisition and loss of energy resources as fundamental to a comprehensive view of organismal (and other) biology. It does not answer yet all questions, but it is an extremely useful perspective that is increasingly unifying the diverse fields of biology and that offers the hope of genuine and quantifiable synthesis.

Energy and the Structure of Ecosystems

Just as the lives of individual organisms can be viewed from an energy perspective, so can populations of organisms and the whole units of landscape that we call ecosystems. One of the fundamental tools used to analyze ecosystems is the idea of *food chains,* which trace the transfer of energy originally derived from the sun through the biological system. The word *trophic* comes from the Greek word meaning food, and the study of feeding relations among organisms is called trophic analysis. More specifically, the study of *trophic dynamics* emphasizes the rates and quantities of transfer. Trophic studies are an important component of ecology because many of our management concepts and objectives are oriented toward understanding and directing the trophic relations of nature, especially those important to people. In addition, trophic processes determine energy availability and hence what is and what is not possible for a given organism within an ecosystem.

Autotrophs are found at the beginning of most food chains because they are able to capture the inorganic energy of solar-derived photons and store a small fraction of that energy by creating complex biotic molecules. The pathway of energy conversion and transfer (the eating of one organism by another) goes from the initial capture of solar energy by autotrophs to the *herbivores* that eat plants to the first level *carnivores* (that eat herbivores) on to the top carnivores. It is a biological convention to assign different species to different places—*trophic levels*—in the food chain by examining what they eat and their position relative to the initial conversion of inorganic energy by autotrophs. Based on this principle, a green plant is assigned to trophic level 1, a herbivore to trophic level 2, a carnivore that eats herbivores to trophic level 3, and so on. The power flow of an organism per unit area, or of a trophic level, is called *productivity* and normally is expressed in units of kilocalorie per square meter per time. *Primary production* is the fixation of solar energy by green plants; *secondary production* is the accumulation of animal or decomposer living tissue. Gross productivity is total energy captured, whereas net production subtracts the energy required for respiration.

A classic early study of trophic relations was done by Wells et al. (1939) who examined qualitatively the flow of energy from the sun through the food chain to the herring fisheries of the North Sea (Figure 1.3). The first study that attempted to examine the quantitative importance of the flows at each trophic level was by Lindeman (1942) who explicitly quantified the flow of energy from sun through primary producers to higher trophic levels in a bog in Minnesota. Another important study was that of Odum (1957) who developed new field techniques using oxygen production and consumption to measure explicitly the energy fixed or used by each trophic level and even of whole ecosystems, in this case Silver Springs in Florida. Both Lindeman and Odum found that by far the largest proportion of the energy captured at a given trophic level was utilized by that trophic level for maintenance respiration and was unavailable to higher trophic levels. Lindeman introduced the concept of *trophic efficiency,* defined as the ratio of production at one trophic level to production at the next. Trophic efficiency is commonly from 10 to 20% but may occasionally be very different (Kozlovsky, 1968). The concept is important and familiar in agriculture where beef or fish production per hectare is much less than the production of plants of that same area, due principally to the large maintenance respiration of the animals.

Recent research has emphasized that most trophic relations occur not as simple straight-line chains but as more complicated *food webs,* in which a given species and even different life stages of that species eat from different trophic levels. For example, a herring whose diet contained 50% algae and 50% herbivorous crustaceans would be assigned to trophic level 2.5. Many, perhaps most, organisms are omnivores rather than strictly herbivores or carnivores. The single most important attributes of food *quality,* other than its energy content, is its protein content, which is approximately proportional to the ratio of nitrogen to carbon in the food.

Another important conclusion of recent research is that for many ecosystems a great deal of the food energy that flows from one group of organisms to another does so only after microbial transformations. For example, an abundant plant in many coastal environments is the marsh grass *Spartina alterniflora.* Only a very small percent of the annual production of this grass is eaten directly by grasshoppers and other herbivores (Smolly, 1958), but many organisms eat *Spartina* indirectly. Microorganisms, especially fungi and bacteria, invade and colonize the dead *Spartina* as it lies on the marsh surface or as it drifts with the tides. The *Spartina* is a relatively poor source of nitrogen, so many of these microorganisms depend on their own ability to fix nitrogen from the atmosphere for protein synthesis. The combination of the dead *Spartina* and

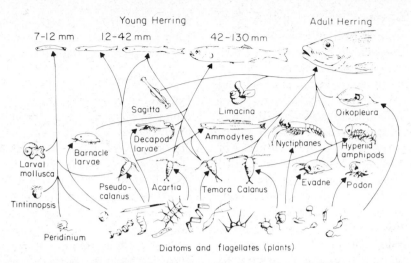

Figure 1.3. The food relations of the herring at different stages in its life. Sizes of herring indicated are 0.6 to 1.3 cm, 1.3 to 4.5 cm, 4.5 to 12.5 cm, and over 12.5 cm. (From Wells, Huxley, and Wells, 1939.)

the associated microorganisms is called *detritus*. This detritus has a greater nutritional value and hence is a higher quality resource for consumers than the dead grass alone because the microorganisms have made the grass more digestible and also have, through their own production and nitrogen fixation, increased the protein value of the detritus. Detritus food chains are now known to be the dominant trophic pathway in many environments, including many or most salt marshes, grasslands, small streams, and even open oceans. Cows and certain other herbivorous animals carry the microorganisms with them, for they live symbiotically in the cow's stomach.

Based on the capture, transfer, and use of energy by biological systems, we can represent the functioning of either an individual organism or even whole ecosystems by the paddle wheel and associated apparatus in Figure 1.1. The flux of solar energy along a gradient from short-wave, high-energy photons to long-wave, low-energy heat is represented in our diagram by the flow of sand in the hourglass from elevated, relatively high-energy sand to less elevated, relatively low-energy sand. Autotrophs can be represented as the paddle wheel that intercepts and converts a small percentage of this free energy to organic forms. Without the input of free energy and its capture by autotrophs, living systems would eventually run down and die.

The free energy captured by the plants is used by heterotrophs, which are represented by the machinery that is attached to the paddle wheel. Energy

is stored for a short time in the ecosystem in the biomass of the organisms, soil, and so on, and some of this energy is transferred from one trophic level to the next. At each of these transfers most of the energy is lost from the system as waste heat. Eventually, all the energy originally captured by autotrophs is degraded and returned to the ecosystem as waste heat. Ecosystems do not recycle energy; instead, they capture a small percentage of the energy that falls on them, concentrate it, use it to do work, and then return it to the environment in the form of low-grade heat that is no longer available to do work.

Adaptation from an Energy Perspective: Energy Investments

Most biologists believe that the only purpose of nonhuman life is to continue the processes of life, perhaps with small, fine-tuned changes over evolutionary time. Natural selection has been perceived as "an existential game the object of which is to keep playing" (Slobodkin, 1964). It is strange, perhaps, that such a seemingly uncaring process has produced the diversity, beauty, and richness of the biotic world. Most biologists believe that Darwinian selective pressures in response to changing environments have been the most important factors shaping the nature of the biotic world as we know it. Many examples exist of organisms being fine-tuned through natural selection to the particular environment in which they are found. Food is often scarce, resulting in intense competition and selective pressure for sequestering available food resources better than competing individuals or species.

We view the course of natural selection over the past three and one-half billion years as a series of *energy investments* into different life possibilities. Organisms, through the process of natural selection, are continually faced with the "choice" of how to invest their energy resources. The word choice is put in quotes because organisms themselves do not choose. Rather, different patterns of energy investment are represented in different genetic codes. These codes, and the resultant suite of morphological, physiological, and behavioral characteristics, ultimately do or do not contribute to surplus energy and hence reproductive potential. The codes are encouraged or not according to differential survival of offspring. The poorer investments failed, often because the environment changed, whereas the better ones are still with us. Examining the capture and use of energy by organisms allows the quantification of the costs and benefits of different processes and actions so that the effectiveness of different responses to environmental conditions can be compared (e.g., the costs of migration and physiological adaptations vs. the benefits of exploiting a food resource in a distant and physiologically hostile environment). Such studies often, but not always, show that behaviors or adaptations of organisms that initially appear strange or counterproductive to our eyes make a great deal of sense from an energy perspective.

One of the important criteria for these energy investments is that they be favorable: a favorable investment is defined as one where more than 1 kcal is returned per kilocalorie invested *and* where a greater return on investment is achieved relative to alternative available choices, thus favoring the survival of the organism and, ultimately, that investment pattern. This important principle and its relation to the quality of energy will be emphasized as we analyze the use of energy in economic systems in Chapters 2–6. Another important aspect of these energy investments is that it appears that organisms, like the industrial machinery discussed at the start of this chapter, are *deliberately* inefficient; highest efficiencies are normally achievable only at low process rates that would be disadvantagous in a competitive environment. (This interesting and general principle will be developed more fully on p. 63.) In the following sections we examine several

adaptations from an energy perspective and analyze their energy balance in terms of contributions to the survival of the individual.

Food Capture

For an organism to have energy available for maintenance, growth, and reproduction, it must obtain more energy from its food than the amount of energy it uses to capture it. In other words, the *energy return on energy invested* for capture must be greater than 1. The concept of energy return on investment is one of the unifying concepts of this book and will be discussed later in relation to economic systems. Prey capture—recovering energy from the environment—is one important example. Here we examine both prey capture by predators and foraging behavior from an energy perspective.

Our first example is from evolutionary ecology. Current interpretation of the evolution of life is in large part based on integrating the changes that we observe in the fossil record from the perspective of the ways in which these organisms evolved different, and often more complex, procedures for exploiting new (food) energy resources. For example, McFarland et al. (1979) summarize our knowledge of the importance of the development of a muscular pumping (vs. ciliary) mode of filter feeding as an important evolutionary development that helped encourage vertebrate evolution—the energy investments in the new filtering apparatus allowed the capture of more phytoplankton and was, presumably, an energy investment with a favorable return. Likewise the later development of jaws and fins allowed the exploitation of other energy resources, such as smaller fish. The energy costs of various types of activities in animals has been studied extensively (e.g., Warren, 1971; McNab, 1980; Bartholomew, 1982). Such reviews suggest that, in general, smaller organisms use more energy per gram than larger ones and, of particular interest here, more recently evolved groups of animals (especially warm-blooded birds and mammals) tend to use more energy per gram than older ones. This suggests a general, but by no means universal, tendency of at least some major groups to evolve toward forms that have higher-energy investments that provide greater energy returns to the organisms. McNab (1980) reviews the literature on mammalian evolution and concludes that as a rule organisms tend to be selected for as high a reproductive potential as the environment's food resources allow (i.e., to be high-energy investment, high-energy return species). Nevertheless, as we shall see, many opportunities still exist for less energy-intensive species life-styles. This is somewhat analogous to Cottrell's (1955) discussion of modern industrial societies—the development of a new technology (be that jaws or steam engines) allows the accelerated exploitations of new energy resources and gives large selective advantages to those with the technology. Yet many environments remain where lower investments and lower returns are favored.

Some modern species have evolved complex foraging patterns in response to energy return on investment criteria. An interesting example concerns the way bumblebees, *Bombus occidentalis,* obtain their energy, nector, and pollen from flowers that contain different quantities of energy. Because the bees must use energy to maintain their thorax temperature between 30 and 44°C in order to fly, the number of flowers that they must visit per hour to obtain an energy *profit* is related to air temperature and the energy content of the flowers they visit (see Figure 1.4). Since the colony is the unit of reproduction (only queen bees can reproduce), it is necessary for the rate of energy profit of the hive to remain high even after workers have made individual energy profits. To keep the rate of energy return (power) high to the hive, workers must forage for an energy profit as long as possible. They visit high-energy flowers in the morning when it is cold, so that they can offset the energy costs of thermoregulation, and visit low-energy flowers during warm periods when the energy cost of thermoregulation are reduced and when the nectar of the high-energy flowers (and hence the quality of that resource) has been depleted by other bumblebees (Heinrich, 1979).

An interesting division between herbivory and carnivory exists in lizards and may be related to the size of the lizard and the size and quality of the prey items available to them (Pough, 1973). All small lizards of the families iguanidae and agamidae are carnivorous, which provides them with a high quality diet, whereas most larger lizards are herbivorous (see Figure 1.5). This pattern sometimes is reflected in the ontogeny of individual species, for many young lizards are carnivorous but turn into herbivores as they grow larger. Although the specific metabolic rate (kcal/kg) decreases with increasing size, the absolute amount of energy required by a lizard increases with its body size as does the proportion of the energy gained from prey capture necessary to capture that prey. As a result the energy

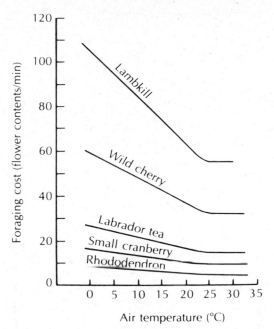

Figure 1.4. Calculated foraging cost, in terms of the number of flowers that queen bees (weighing 0.5 g) have to visit per minute to realize a positive energy balance. For these calculations it is assumed that the bees spend half their time in flight and regulate their thoracic temperature at 35°C. The calculations, based on nectar available in flowers, show that this feeding strategy is not economically feasible, particularly at low temperature, unless the bees make physiological and behavioral adjustments. Lambkill, although visited by queens, was exploited primarily by workers. (From Heinrich, "The Role of Energetics in Bumblebee-Flower-Interactions." In *Coevolution of Animals and Plants,* ed. Gilbert and Raven, pp. 141–158. University of Texas Press, 1975.)

return on "investment in predation" for carnivorous lizards decreases as the body size of the lizards increases, so that apparently it is not energetically feasible, at least in the ecological circumstances of the present geologic period, for large lizards to be carnivorous.

Endothermy versus Ectothermy

Two strategies for maintaining body temperature are endothermy and ectothermy. *Endothermic* animals like birds and mammals generate most of their body heat internally from food energy and maintain a relatively constant body temperature, whereas *ectothermic* animals such as most fish, amphibians, and reptiles depend on environmental energy flows for most of their body heat (e.g., ambient environmental temperature or direct sunlight). Generally, the body temperature of ectotherms fluctuates in response to environmental conditions. These two *evolutionary strategies* have an important effect on

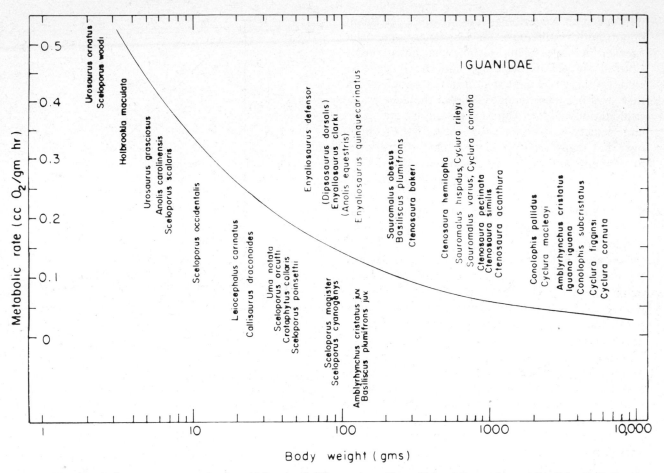

Figure 1.5. Metabolic rate, per gram, for Iguanid lizards of different sizes. (From ''Lizard Energetics and Diet'' by F. H. Pough. *Ecology* **54**(4):837–844. Copyright © 1973 by Ecological Society of America. Reprinted by permission.)

the need for, and partitioning of, energy within organisms, which in turn affects the food webs of ecosystems.

Ectotherms were once thought to be *cold-blooded,* but research has shown that many ectotherms attempt to maintain their preferred body temperatures—which often are comparable to that of the endotherm's—by both physiological and behavioral processes such as sunning (Cowles and Bogart, 1944; Dewitt, 1967). When ectotherms are unable to maintain body temperatures near their preferred temperature, their physiological abilities (e.g., muscular abilities, coordination, digestive efficiency) decrease. During these periods individuals are most vulnerable to predation. Ectotherms therefore are active only when environmental conditions permit them to maintain their preferred body temperature. As a result, the time ectotherms are able to carry out their biological activities is limited to specific periods of the day and year. For ex-

ample, Porter et al. (1973) found that the Mojave Desert lizard *Dipsosaurus dorsalis* was active only at temperatures between about 38 and 43°C. The lizards extend the time period that they are at that temperature by moving into the sun or shade, or by climbing shrubs to intercept breezes, yet they must spend over 80% of the total time in a year in their borrow where temperatures are more moderate. Thus their ability to forage is greatly limited. Endotherms, such as humans, are much less constrained by environmental conditions. Because they are able to maintain their body temperature somewhat independently of environmental conditions, they may remain active at temperatures that would incapacitate ectotherms.

Yet, if we examine ectotherms from an energy perspective, advantages of this seemingly limited life-style do exist. Because ectotherms do not generate their own body heat, they generally require less food energy per unit of body mass than endotherms; therefore, they are less likely to starve when food is scarce. For example, the resting metabolic rate for a 20 g lizard (ectotherm) is about 1.2 cal/g per hour, whereas a 20 g mammal (endotherm) consumes about 9 cal/g per hour (Pough, 1980). As a result ectotherms are able to survive in areas where the energy flow through the ecosystem is small, sporadic (e.g., deserts) or unpredictable, since they use less energy while waiting for the next source of food.

Because endotherms must generate their own body heat, a greater percentage of the food assimilated by an endotherm must be used for maintenance respiration. As a result endotherms have a smaller percentage of the energy they assimilate left over for growth or reproduction. The greater percentage of assimilated food available for growth and reproduction in ectotherms allows them greater efficiency in converting food into new biomass. The net efficiency of turning food into new living tissue for mammals and birds ranges between 0.5 and 2.3%, whereas the value for amphibians and reptiles ranges between 6.3 and 49% (Pough, 1980).

These differences in efficiency can have important effects in a food chain. Burton and Likens (1975) showed that although the ectothermic salamanders in the Hubbard Brook ecosystem consumed a smaller percentage of net primary productivity, they had a higher annual rate of net biomass production than birds and mammals in analogous trophic positions. It would seem that the higher energy requirements and lower efficiency of warm-blooded birds and mammals would put them at a disadvantage. The higher and more constant temperatures of endotherms, however, allow them higher and more sustained rates of muscular activity over a much wider temperature range, and a greater ability to sequester food when it is abundant. Endothermy is an energy investment that allows a larger capture of available energy flows in the organism's environment. Neither evolutionary strategy, ectothermy or endothermy, is better, since obviously both birds and reptiles exist. But each is favored in specific environments or at specific times (Pough, 1980).

Migration

Many different species undertake extensive, energy-intensive migrations. We will analyze salmon migration in the ocean from an energy perspective.

Although the analysis is specific to salmon, the principles probably hold for migrations undertaken by many other species (Day et al., in press).

Salmon have extremely complex life histories. They are born in fresh water, spend a few weeks to several years in rivers or lakes, and then spend the bulk of their adult life at sea where they undertake extensive migrations. At first glance this life history may seem counterproductive, for it is difficult physiologically and expensive energetically to change from a fresh water environment to a salt water environment and also to undertake these extensive migrations. Since salmon can survive and grow readily in either fresh or salt water, as well as over a wide range of temperatures, the fish clearly do not have to migrate. Some individuals do not, for example a sockeye salmon that does not migrate is called a kokanee. So why do most salmon migrate?

Many studies of migration have indicated that there are considerable food energy resources that can be exploited by migration, especially food resources that vary over space and time. The kokanee salmon does not grow as large, nor is it nearly as abundant, as the sockeye salmon that migrates. A preliminary study by Hall and English (1984) indicated that salmon that migrate in the ocean should grow at least as large and probably larger than salmon that don't migrate. Likewise, Hall (1972) calculated that fish migrating in a small North Carolina stream gained at least 3 kcal for each kilocalorie invested in migration. In the case of salmon, the fish appear to be "programmed" to be in about the right place at the right time to take advantage of the plankton blooms (areas of high phytoplankton productivity) that travel northward and westward along the Northwest coast of North America every year.

We view salmon migration, like other complicated adaptations, as an energy investment with a return to the individual or the population greater than the investment and at least as great as competitive alternative investments. It is not possible, however, to say that fish migrate *because* they have a favorable energy return on investment—for there are additional environmental and physiological criteria and, in addition, science is not set up to answer that type of question. We can say only that those fish which we have examined that do migrate gain at least as much energy due to the migrations as the energy they invest. The favorable return on investment appears to give a selective advantage to those that migrate because they will have more energy available to invest in reproduction. In the case of salmon the migratory sockeye return at age 3 or 4 much larger, and with many more eggs per female, than the nonmigratory kokanee of the same age. The larger number of eggs confers a clear selective advantage to the sockeye which consequently are much more abundant than the kokanee.

Territorial Defense

One of the most appealing and characteristic attributes of nature during spring is the sound of birds singing. In earlier times it was thought that birds were doing so for their own enjoyment, to attract mates or perhaps as a manifestation of a benevolent creator who had placed birds on this planet for the enjoyment of people. Although such rationales were aesthetically appealing, it was not clear why birds should invest so much of their time and energy in such a seemingly nonproductive activity.

The modern view of bird songs was elucidated by Eliot Howard (1920), who studied carefully the behavior of singing birds in his own garden. He concluded that birds sing to advertise the boundaries of their territories and to indicate that any other male bird of their species (and sometimes other species) would be challenged and chased, if possible, from their territory. Such an activity might at first seem to be rather wasteful for the singing bird, because the time spent singing was time not feeding. Additionally the singing males, which often are brightly colored, are more susceptible to predation.

But recent studies by Gill and Wolf (1975, 1979) and other investigators have found that the energy invested in singing and otherwise proclaiming and defending territories was time well spent from the perspective of that particular bird. By eliminating competition for the preferred and readily available food sources, more food is available to the bird holding the territory. This energy gain remains positive even after the energy costs of defending the territory are subtracted. As a result more food is available for feeding the young of the birds that hold territories and, according to the principles of natural selection, genes that produce the territorial behavior in birds are favored in future generations.

A direct relation between the availability of food energy and the need for and size of territories was found experimentally by Simon (1975) and Ferguson et al. (1983), who observed that the size of territory defended, and the intensity of the territorial defense by a lizard, depended on the quantity of food available. When the investigators introduced extra quantities of food to lizards defending a territory, the size of the territory defended by individual lizards diminished. In addition some individuals stopped defending territories altogether. When the extra quantities of food were withdrawn, the lizards resumed their previous territorial behavior.

Defense from Predation

An important example of the trade-offs that occur in natural selection is the way in which large green plants protect themselves from insects. Many herbivorous insects are capable of very large reproductive rates, and many insect populations could easily increase much more rapidly than their predators. Yet the food of herbivorous insects (green plants) is abundant, as a look out of most of our windows substantiates. As a rule the insects have not increased to levels that would seriously decrease the abundance of green plants even though they are physiologically capable of such an increase.

It was once thought that the principal reason that the insects had not defoliated the world was that the predators of insects kept the populations in check (Hairston, Smith, and Slobodkin, 1960). But the growing field of chemical ecology has another explanation that appears more important, although it does not negate the importance of predation. It has been found that one way green plants defend themselves is by filling their tissues with unpalatable and toxic materials. These materials discourage most insects from consuming the tissues, reduce the assimilation efficiency of the insects, and, ultimately, reduce their reproductive rate (Whittaker and Feeny, 1971; Feeny, 1970).

Examples of these kinds of chemical defenses are everywhere in our everyday life. Oak leaves, for example, are about 8% tannins, compounds that might seem to be of little direct use to the oak trees since they contain no chlorophyll and do not contribute to the structure of the plant. Tannins, however, are very unpalatable substances for insects, and greatly retard the efficiency with which the insects can turn oak leaf biomass into insect biomass and hence their growth and reproduction rates. This serves to discourage the insects from eating too high a percentage of oak leaves. Thus the tannins serve as a type of natural insect inhibitor, as do the chemicals that give flavor to onions, mustard, tea and many other foods that add, in moderation, interest to our diet. Some of these *secondary compounds* are found to increase following severe insect attack, which may help explain why gypsy moth infestations do not last more than a year or two.

The tannins are not without cost to the plants. Not only do they absorb sunlight and fill spaces that might otherwise contain chlorophyll-bearing structures, but they also must be synthesized by the plant. The energy and carbon that goes into synthesizing the tannins cannot be used by the plant to grow more and shade out competitors, repair damage from insects and other herbivores, or produce seeds. These trade-offs are presumably optimized over evolutionary time.

In summary, the complex world of nature can be viewed as a series of evolutionary adaptations to the energy available to organisms, resulting in patterns of energy investments that, when successful, produce positive returns—generally measured as biotic production and reproduction—that are encouraged through the process of natural selection. Although we have emphasized individual organisms here, these energy relations interact at the level of the entire ecosystem (see e.g. Odum 1977; 1983). Since the basic energy resource, solar radiation, is essentially constant, living systems are constrained to exist within the limits of this energy source. Human-controlled industrial systems are not yet constrained in the same way.

ENERGY AND ECONOMIC SYSTEMS

Ecological analysis includes the study of the production and patterns of allocation of organic materials and other resources in natural communities. Economics is the study of the production and patterns of allocation of goods and services in human societies. Although the operation of these two systems may seem very different, many parallels in the use, allocation, and regulation of energy flows exist because the two systems are constrained by the same physical laws. Goods and services can be produced from raw materials in either type of system only when solar, human, or fossil energy is applied to those raw materials. Our approach to economic analysis emphasizes this perspective. The two principal sources of energy for natural and human economic systems are the sun, operating through biotic systems, and solar-derived fossil fuels, operating through human-controlled and catalyzed industrial systems. These energy flows and their products interact in complicated and interdependent ways in what we call our national economies. The singular

importance of energy in our economic system can be seen from the following scenarios. If we were to remove the dollars from our economy, it could continue to function, although in very different ways. If we were to remove any of the principal commodities, say, iron ore, fertilizers, or trees, the economy would continue to function, although, again, very differently. If energy were removed from the economy, however, it would cease to exist once storages were used up. The highly ordered, low-entropy aggregates of elements that we call economic goods, and the motions we call services, would dissipate into functionless, more random patterns without a steady input of high-quality energy.

Wealthy economic systems require large quantities of energy to develop and maintain that wealth. For example, in the United States a quantity of energy equal to that obtainable from burning nearly 70 lb of coal, or 10 gal of oil, is used per person each day to maintain our standard of living. Thus energy has a very fundamental role in economic systems, one that is different from other commodities: although there are substitutes among energy sources, there is no substitute for energy, and unlike most other commodities energy can be used only once. All commodities require energy for their production, including all the inputs required for their production (capital, labor, technology). The analysis can be carried one step further: the study of economics may properly be considered the study of how human beings take solar and fossil energy and apply it to natural resources to produce and distribute goods and services.

Such an analysis would be only an interesting academic exercise if the availability of high-quality energy resources were not changing. But in reality energy resources are becoming strained in many different ways, and it is no longer possible to assume that growth in energy inputs will simply accompany any ongoing economic or population expansion, as generally has been the case in the past. More important, since it takes energy to run the energy-producing sections of our economy, which in turn run the rest of our economy, it is essential to ask some basic questions about how energy is invested to recover energy. Specifically, since petroleum is our principal high quality fuel, and since petroleum reserves are increasingly depleted, it becomes a question of critical national importance as to how the remaining petroleum should be used or invested. On a personal level, our best estimates of known and projected remaining liquid and gaseous petroleum reserves in the United States divided by the number of people amounts to about 20,000 gal/person. In a sense that petroleum is our "money" in the bank, our national prime capital, and we are using it at about 1000 gal/person per year. Thus one critical question for both ourselves as individuals and our economy as a whole is, "How will you spend your last 20,000 gal of petroleum?" This book is meant to give much of the information necessary to make intelligent use of petroleum and other critical resources, and to assist in investing any energy capital in our economy most wisely. It also is designed to examine how the uses of energy resources interact in complicated ways to produce personal wealth and aggregate national economic well-being, as well as environmental degradation.

APPENDIX 1.1: GETTING A FEEL FOR CALORIES

For some aspects of everyday life it is simple and interesting to calculate the energy costs of doing certain activities. For example, if a 150-lb (65 kg)

hiker walked to the top of a mountain that was 4000 ft (1219 m) higher than its base, 600,000 ft-lb (780,000 newton-meters) of work would be required, neglecting the much smaller energy required for overcoming friction during horizontal displacement. This is the same amount of work required to lift three 50-lb bags of cement to the top of a 4000-ft tower. Since a foot-pound is equal to 0.000324 kcal and human beings are about 20% efficient, about 1000 kcal would be needed for the hike, which is about two good-sized peanut butter sandwiches or, if the hiker didn't eat, about one-quarter of a pound (one-ninth of a kilogram) of his or her fat reserves, not including the energy used for maintenance metabolism. If the hiker went up on a ski lift, which also has an efficiency of roughly 20%, again one-quarter pound of petroleum, or one-thirty-sixth of a gallon, would be used, neglecting friction, since a gallon of gasoline has about 36,000 kcal. The weights of the two fuels (body fat and petroleum) required are similar because both are highly reduced hydrocarbons and have about the same amount of energy per pound. This simple example indicates the remarkable amount of work that can be done on a very small amount of energy. Even if

Table 1.2a.   Average Wattage and Estimated Energy Consumption of Various Appliances

Appliance	Average Power Required (W)	Estimated Electrical Energy Used per Year (kWh)	Estimated Electrical Energy Used per Year (thousand kcal)
Air conditioner (window, 5000 Btu/hr)	1565	1390	1196
Clock	2	17	15
Clothes dryer	4855	990	851
Dishwasher	1200	363	312
Fan (window)	200	170	146
Food blender	385	15	13
Food freezer (15 ft^3)	340	1200	1032
Food freezer (15 ft^3, frostless)	440	1760	1514
Hair dryer	380	14	12
Light bulb	100	100	86
Microwave oven	1450	190	163
Radio-phonograph	110	110	95
Refrigerator (12 ft^3)	240	730	628
Refrigerator (12 ft^3, frostless)	320	1215	1045
Stove	12,200	1175	1011
Television			
Black and white, solid state	55	120	103
Color, solid state	200	440	379
Toothbrush	7	0.5	0.4
Washing machine (automatic)	512	103	89

Table 1.2b.   Energy Requirements for Passenger Transportation

Mode of Transport	Maximum Capacity (passengers)[a]	Vehicle Mileage (miles/gallon)[b]	Passenger Mileage (passenger-miles/gallon)[b]	Energy Consumption kcal/passenger-mile
Bicycle	1	1560	1560	20
Walking	1	450	470	66
Intercity bus	45	5	225	140
Subway train (10 cars)	1000	0.15	150	210
Volkswagen sedan[c]	4	30	120	260
Local bus	35	3	105	300
Automobile[d]	4	12	48	650
747 jet plane	360	0.1	36	870
707 jet plane	125	0.25	31	1000
Concorde SST	110	0.12	13	2370
Ocean liner	2000	0.005	10	3125

Source: R. H. Romer, *Energy: An Introduction to Physics.* Copyright © 1976 by W. H. Freeman and Company.

[a] The relative effectiveness of various modes of transportation can be drastically altered if a smaller number of passengers is carried.

[b] Miles per gallon of gasoline or the equivalent in food or in other fuel; all values must be regarded as approximate.

[c] Long-distance intercity travel.

[d] Typical American automobile, used partly for local travel and partly for long-distance driving.

Table 1.2c.   Energy Requirements for Freight Transportation

Mode of Transport	Mileage (ton-miles/gallon)	Energy Consumption (kcal/ton mile)
Oil pipelines	275	113
Railroads	185	170
Waterways	182	170
Truck	44	700
Airplane	3	10600

Source: R. H. Romer, *Energy: An Introduction to Physics.* Copyright © 1976 by W. H. Freeman and Company.

we wish to drive an automobile that weighs 20 times our hiker, and that also has an efficiency of 20%, to the top of the mountain, it would still take less than a gallon of gasoline. But building that 2-ton automobile would require about 25 million kcal, or the energy in 712 gal of petroleum. Table 1.2a–c shows the energy cost of various everyday activities. It would take a very strong person working full time turning the handle of a generator just to keep a 100 W light bulb lit, but our fossil fuel slaves do enough work each day to keep several hundred such bulbs lit!

PART TWO

ENERGY AND ECONOMICS

OVERVIEW

What factors enabled the United States to achieve its high material standard of living and position of economic and military importance relative to the rest of the world? Most wealthy nations possess in abundance one or more of the following natural resources: fertile agricultural land, fossil fuels, or metal ores. (Japan is an exception, perhaps, that we consider later.) The United States is one of the few nations that has, or had, large quantities of almost all these important ingredients of economic growth. Despite their importance the supplies of natural resources have been given relatively little consideration in most analyses of U.S. prosperity. Instead, most people believe that U.S. economic, political, and other social institutions are the principal agents responsible for our affluence. In contrast to this popular view we develop the hypothesis that much of our wealth springs from our ability to empower human labor in the production process with our abundant endowment of high quality fossil fuels. From our perspective much of the success of the U.S. economy—founded on free-market principles and a political system based on democracy—were made possible by abundant supplies of high-quality fuels and other natural resources. This view does not negate the importance of other factors, for clearly political and economic factors encourage people to use natural resources at different rates. Nevertheless, these social factors cannot create material wealth from nothing. Most people are not especially aware of the relationship of energy and wealth simply because the production, and the energy used in production, occur elsewhere. Wealth seems to be much more directly related to store shelves and full wallets, but that is only a small part of the story.

Hence we assess U.S. prosperity from a thermodynamic perspective that emphasizes the production of goods and services from natural resources rather than from the utility-based perspective of neoclassical economics, which emphasizes the exchange and consumption of economic output according to subjective human preferences. Although human preferences and technology influence what commodities are considered useful and how they are distributed, we choose to focus on the supply of natural resources and human capabilities for converting them to useful goods and services. By emphasizing the production of goods and services, we do not ignore the importance of human preferences. Rather, we assume that the goods and services produced have value to some segment of society, an assumption that allows us to place relatively little emphasis on human tastes and preferences. On the other hand, most neoclassical economists assume that the mechanisms of a free market will automatically insure adequate supplies at reasonable cost. This assumption allows them to place relatively little emphasis on the supply and quality of fuels and other natural resources. This may have been a valid assumption in the past when unexploited fuel and other natural resources were abundant, but, as we shall see in Part Two, the importance of energy to economic production, coupled with the declining supply and quality of fuels and other natural resources, already limits our ability to extract and convert natural resources into useful goods and services regardless of the utility these commodities provide. The consequences are beginning to be seen as the plethora of economic problems since 1973.

We analyze the economic process from an energy perspective because both natural and human economic systems use energy to upgrade various chemical elements and simple compounds to more complex and economically useful forms. Ecological and geological systems use solar and radiation energies to drive photosynthetic and other biogeochemical processes that take biotic building blocks such as CO_2, nitrogen, phosphorus, and sulphur and produce natural resources such as plants, animals, soil, clean air, and water. Similarly, human economic systems use solar, fossil, and atomic energies to extract natural resources from the environment and turn these raw materials into finished goods and services that are useful to humans. In a broad sense, therefore, economics is a discipline that analyzes how human material wants and needs are satisfied by transforming natural resources into goods and services.

Economic production* is a work process, which, like any other process, requires energy. Thus energy powers economic production. Energy, of course, is not the only input to the economic process, since labor, capital, and other natural resources also are required. But capital, labor, and all other inputs require free energy for their production and maintenance. Because of this interdependence with energy, the cost of every input can be analyzed according to the energy required to realize that resource. As a result the availability of energy constrains our ability to locate, extract, and convert natural resources to useful goods and services.

Human labor is important in an economic sense because it provides energy to perform various economic tasks and because in combination with machines and other tools it can be used to direct and control other larger energy flows. But the work done by labor would not be available without inputs of solar and fossil fuel energies that provide the food, clothing, shelter, transportation, and other goods and services necessary to keep a laborer alive and functioning at a level that enables the laborer to perform his or her respective economic role.

Capital subsidizes human energies, increasing the rate and/or the efficiency of many economic processes, thereby increasing dramatically the material output that a worker can produce in a day. Like labor, however, capital structures require energy for their production, maintenance, and especially operation.

Energy is special because, unlike labor and capital, it can be used only once and the availability of many important fuels is declining. Energy is used in economic production in two general steps (see Figure 2.1). First, solar, geological, and other natural energies produce a wide variety of natural resources and environmental services that are essential for human existence and economic production. In the second step, human-directed production energies—including human muscle, draft animals, fossil fuel, and atomic energies—are used to locate, extract, refine, process, assemble, transport, and

*A glossary of basic economic terms defined from our energy perspective is included as Appendix 2.2.

otherwise upgrade the organizational state of natural resources to useful goods and services. We call all energies that humans control and direct for their own purposes *economic energies*. Economic energies are used to do *economic work:* the transformation of natural resources into economic output.

The amount of economic work possible depends on both the quantity and quality of energy directed to the task and the efficiency of the process. Although human technology plays an important role, the quantity and quality of fuels sets upper bounds on both of these aspects of economic production. *Higher quality* fuels exist in more concentrated forms, are more flexible, require less processing, and can be used with higher end-use efficiency. For example, 100 kcal of electricity can do more economic work, in general, than 100 kcal of coal, which can do more work than 100 kcal of sunlight, and so on. The idea of energy quality will be considered in more detail later.

Another important aspect of fuel quality is the amount of energy required to locate, extract, and refine fuels to a socially useful state. The process in which society invests some of its already extracted (surplus) energy to make available additional quantities of fuel is called an *energy transformation process*. This aspect of fuel quality is measured by energy return on investment (EROI),

$$EROI = \frac{\text{kcal of fuel extracted}}{\substack{\text{kcal of direct and indirect energy} \\ \text{required to locate, extract, and} \\ \text{refine that fuel}}}$$

EROI is the ratio of the *gross* amount of fuel extracted in the energy transformation process to the economic energy required to make that fuel available to society. The denominator of the EROI ratio includes both direct fuel use and other less obvious indirect economic energies. For example, in addition to the fuels—such as natural gas—used to pump oil out of the ground, fuel is used indirectly to produce capital equipment,* including oil rigs, drill bits, and refineries, to support workers who control this process, and to support government services such as road construction used by the energy industry. (See Figure 2.2 for a schematic represen-

*Throughout this text we use the term *capital* to refer to physical capital equipment (plants and equipment) that is used in the production process. This should not be confused with a more limited use of the term which means financial assets, and which is not our use.

tation of an energy transformation process that shows how energy inputs and fuel outputs are grouped to calculate a fuel's EROI.)

The EROI for a particular fuel is not constant. The energy cost of extracting fuels varies as the quality of the average deposit changes and as the extraction technology evolves. As nonrenewable fuels are depleted, lower-quality deposits are used. These lower-quality deposits often require new technologies that use more fuel directly and indirectly to extract a kilocalorie of fuel. Because these lower-quality deposits often require more energy to recover, their EROI declines (Part Three). We have found that the development of increasingly sophisticated technologies rarely offsets the increasing energy requirements for finding and extracting lower-quality fuel resources.

We use EROI and the related ideas of *economic work* and *output per unit energy invested* as the conceptual glue to bind the chapters of this book. We do so because these terms measure the importance of fuels and other natural resources to society. For example, net energy is a more relevant measure of a nation's energy supply than gross energy because net energy is the energy available to produce final goods and services. If the aggregate EROI for fuels used in the United States were 1 kcal returned for 1 kcal invested (1:1), no fuel would remain to drive our cars, heat our homes, or produce the other goods and services associated with our high standard of living—it all would have been used up simply to extract and process new amounts of fuel.

The aggregate EROI for fuels is important because it sets the upper bound on how much work an economic system can do, and ultimately on the standard of living possible. Other physical factors are important, such as the efficiency at which net energy supplies are converted to useful goods and services and a host of cultural and political factors that affect economic output and its distribution. Because of these human factors, a wide range of possibilities exists within the limits set by the supply of net energy. Thus natural resources do not determine what humans can and cannot do, but they do set important constraints. Human factors cannot override thermodynamic constraints so that, all other things equal, economies with access to fuels with higher EROI can do more useful economic work than those with lower EROI.

The *quality* of resources other than fuel also effects the level of economic output. The quality of individual sources of a nonfuel natural resource is determined in large part by the amount of solar or geologic energy used in the past to concentrate and upgrade them, as in the purification and transport of water in the hydrologic cycle, the production of fish and forests, or the concentration of mineral ores. Lower-quality sources of natural resources require more energy to recover than their higher-quality counterparts. Recovering lower-quality deposits of natural resources increasingly diverts economic energy from other uses. These diverted energies are called *energy opportunity costs* because these economic energies are no longer available to produce alternative goods and services. Although natural energies embodied in resources ultimately determine and limit the supply of resources available to humans, the economic energy cost of recovery is the critical energy flow to analyze because of these opportunity costs. We therefore define the *quality* of a given natural resource in terms of the amount of economic energy used to make a unit available to society. Natural resource quality is discussed in detail in Chapter 4.

The following five chapters form the core of this book. Chapter 2 develops our basic approach, Chapter 3 reviews the importance of the concept of *value* in both standard economic and energy analysis, Chapter 4 assesses natural resource quality from an energy perspective, Chapter 5 examines in detail energy use in manufacturing, and Chapter 6 looks at energy use in agriculture. In these chapters we develop our basic arguments about the relation of both energy and resource quality to economic activity. We compare our operating hypotheses and models of the production process with the theories and policies of modern neoclassical economics (see Chapter 3 for a short development of that body of theory). In doing so, we do not advocate replacing modern economic theory with an energy theory. Rather, we attempt to develop a different perspective that complements many economic analyses by including the physical principles that influence the formation, extraction, and processing of natural resources from which economic output is derived and that ultimately constrain any economic theory, plan, or activity. We believe that our approach complements the neoclassical analysis of how human preferences influence the allocation of goods and services in an economy while offering a framework that links fuels and other natural resources to contemporary economic, political, and social models of human welfare.

HISTORICAL PERSPECTIVE:
EARLIER ENERGY ANALYSTS

The importance of energy for human economic systems has been noted by a number of investigators in the past, although there were few empirical analyses until recently. Many classical economists realized that labor produced wealth, but they were not specifically interested in the fact that it was the energy used by labor that transformed less useful materials into more useful ones. The ideas of these and other classical economists are reviewed in Chapter 3. In England, Germany, France, and other European nations there was considerable interest in the use of science for the advancement of the human condition, including economic well-being. Some of the investigators appear specifically to have thought about and understood the degree to which energy use itself is an important component of economic and cultural change. For example, the British astronomer Sir John Herschel wrote in 1833 that "the sun's rays are the ultimate source of almost every motion that takes place on the surface of the earth." The German chemist Wilhelm Ostwald wrote in 1907 that the "progression of culture is characterized by an increase of man's control of the relations of energy, and the history of civilization becomes the history of man's advancing control over energy." G. H. MacCudry wrote in 1933 that "the degree of civilization in any epoch is measured by the ability to use energy for human advancement." The great German physicist Erwin Schrödinger was "convinced that (the second law of thermodynamics) governs all physical and chemical processes even if they result in the most tangled phenomena such as organic life, the genesis of a complicated world of organisms from primitive beginnings, and the rise and growth of human cultures" (Schrödinger, 1935). Social scientists interested in the importance of energy in human culture and economics included Grahamn Clarke (1946), and T. N. Carver (1935). Early biologists interested in the importance of energy, including its impor-

tance in social systems, include Herbert Spencer (1851) and Alfred Lotka (1936).

The first investigator to assess in any great depth the importance of energy in economic processes was Frederick Soddy: "The flow of energy should be the primary concern of economics" (Soddy, 1936). Soddy was a Nobel Laureate in chemistry who spent a considerable portion of his professional life examining the role of energy in economic production and in developing economic theories related to debt. The first detailed empirical investigation of the role of energy in the U.S. economy was by Ayers and Scarlatt (1952), who wrote an extremely interesting and detailed analysis of the energy situation at that time in a format that is not entirely different (although considerably less detailed) than this present volume.

But the most insightful and comprehensive assessment of the role of energy in various human societies was by Fred Cottrell, who was a professor of government in Miami (Ohio) University for many years after an earlier career as a railroad man. Cottrell's *Energy and Society* is an extremely perceptive, readable, and insightful analysis of the role of energy in human affairs. The most fascinating chapters detail the ways that ancient human civilizations were influenced by the development of new energy technologies and how that gave some societies economic advantage over others. Many of our ideas about EROI and resource quality, although developed independently, can be gleaned from these earlier volumes by Ayers and Scarlatt and Cottrell, although they did not have at that time the empirical record needed to understand and document the correlations and changes that we can now see.

Cottrell introduced the term *converter* as a general term for machines that turn one form of energy into another, more socially useful form. The first and most important such converters are humans themselves, which, according to Cottrell, turn solar-derived food energy into mechanical and other economic work at an efficiency of about 20 percent. Thus a person who eats about 3000 kcal/day can do about 600 kcal/day of work, a little less than 1 hph (horsepower-hour). The domestication of animals was probably the first direct supplement to human muscles and increased the work that could be done by one person (with a horse) to about 10 hph/day, about 6250 kcal. The average work that a horse does, of course, is much less because normally the horse is used only seasonally. But, according to Cottrell, the important thing about horses, and human-directed energy converters in general, is that they increase the *power* of a human worker, so that more work can be done when it is needed, for example, during the planting season. Planting crops at the best time increases the productivity of a farmer, since crops planted earlier or later are more likely to be affected by adverse weather, decreasing the food energy yield per kilocalorie of invested energy.

Cottrell was intrigued by the ways that the human harnessing of auxiliary energy sources, such as river flows and wind energy, in antiquity greatly increased the economic productivity of those who harnessed those resources. For example, the construction of even the earliest sailing ships by Phoenicians and Romans increased the power output of an individual by up to 250 times and allowed great concentrations of wealth and military power. Thus ancient Rome could exploit the granaries of Egypt, in turn allowing the employment of large numbers of artisans to build Rome and soldiers to expand the Empire. Cottrell even attempts to calculate the energy saved by the military conquests of the Mediterranean (an energy investment), which allowed Roman ships to sail unarmed so that the economic potential of the sailing ships was increased a great deal.

Cottrell elaborates in some detail on the ways in which the development of new energy converters (e.g., a sailing ship) and new energy resources (e.g., the wind) influenced the social development of cities and regions, and allowed a much greater concentration of people than could exist when the only energy resource was agriculture.

Another important concept introduced by Soddy and Cottrell and developed more fully by Cook (1980) is the changing role of lending and the concept of debt. In low-energy societies debt normally accrued only in response to special circumstances such as a bad harvest or a marriage. In general it was thought something to be avoided by both people and religious institutions. With the emergence of higher-energy societies the concepts of borrowing and investment became increasingly allowable, and even desired, as large investments were needed to build sailing ships and other expensive converters. Today, of course, investment and debt are the normal way of running personal and corporate economic affairs.

2

ECONOMICS FROM AN
ENERGY PERSPECTIVE

We have defined economics as the discipline concerned with the transformation of natural resources to goods and services designed to meet human needs and desires. Because material goods and services can be produced only by transforming natural resources, economics should have close ties to other disciplines relevant to this transformation. These include physics, because that discipline describes the thermodynamic limits on the availability of energy and matter for human purposes, geology, because it describes the geochemical processes that form the relatively rare fuel and mineral deposits that humans use as natural resources, and ecology, because it describes the interdependencies between the biotic and abiotic components of natural and human systems. A thorough analysis of the economic process must include each discipline's principles that describe how they constrain the economic system's ability to satisfy the material demands of its constituents.

Modern neoclassical economics, which has dominated most Western economic theory and politics since the mid-19th century, has ignored almost totally the physical, geological, and biotic underpinnings of economic production. We believe that much of the recent failure of standard economic theory to explain and remedy present economic problems as successfully as it once did stems from its failure to recognize and incorporate both recent and long-term changes in natural resource quality into its models and policy suggestions.

THE NEW ECONOMIC DIFFICULTIES

Stable consumer prices, full employment, and increased per capita production have been economic and political goals for most Western nations since at least the 1930s. A principal means for attempting to

realize these, and other, economic and social objectives has been the encouragement of aggregate growth of national economies through fiscal and monetary policies. These policies appear to have been at least partly successful from the late 1940s to the early 1970s. During that period, real economic production grew at an average annual rate of 4%, recessions were relatively short and mild, and inflation rates rarely exceeded 4%/yr. Since 1973, however, these same nations have experienced irregular and even negative economic growth rates, often coupled with high unemployment, unprecedented inflation and deficits, and stagnating or declining labor productivity.

These and other events often seem to defy explanation by, or even to contradict, some of the most fundamental models of neoclassical economics that have guided the economic prosperity of the preceding 50 years. Economists from many different camps are perplexed by "stagflation," for, according to neoclassical theory, inflation and economic stagnation should be mutually exclusive. Predictions of greatly increased oil discoveries and production due to deregulation of oil prices have not been borne out. The recent spate of inaccurate economic predictions reflect these contradictions; in one survey, forecasts of quarterly economic activity were wrong four times out of five with respect to not only magnitude but even sign (*Business Week*, July 27, 1981, p. 11).

Criticism has been leveled from within the discipline of economics itself. Drucker (1982) states that "both as economic theory and as economic policy Keynesian economics is in disarray." A Nobel Laureate in economics, Wassily Leontief (1982) described many current economic models as "unable to advance, in any perceptible way, a systematic understanding of a real economic system." Instead, they are based upon "sets of more or less plausible but entirely arbitrary assumptions" leading to "precisely stated but irrelevant theoretical conclusions." Bailey (1981) and Kuttner (1985) chronicle the failures, mutual conflicts, and frustrations with a number of economic models. The once highly advertised "supply side" economic approach to our current problems has not produced all, or even much, of what was promised, while many other economic problems have gotten worse.

Many authors have suggested various ideas and policies based on one or another variation of the neoclassical model to explain the current economic malaise of the country, and to recreate the favorable economic climate that once prevailed. For example, Lewis (1982) emphasizes the decline in technical innovation, Mansfield (1980) a decline in real research and development, Denison (1979) a decline in labor productivity, several prominent economists a failure of federal policy (*New York Times,* November 8, 1981, p. 41), and the Reagan administration the consequences of too high a level of taxation and too much governmental interference in the market. Remarkably, neither the above assessments, nor the economic models found in most textbooks consider changes in the natural resource base of our economy with any particular degree of sophistication.

Responding to Leontief's challenge, Glassman (1982) suggests that a greater diversity of theory is needed in economics to supplement the conditioned expectations of formal economic theory. We agree, and much of this chapter is devoted toward that end. The traditional policies described above have been more or less ineffective in combating current economic problems because they ignore the mechanisms by which human social and economic processes are constrained by the supply and quality of natural resources. Due to the interdependencies of economic and natural processes, human economies respond to changes in natural resource quality which affect the amount and rate of fuel throughput, and the amount of fuel and other economic energies required to convert natural resources to useful goods and services. These responses to natural resource changes go well beyond the impact of rising prices and have important implications for economic policy as well as theory, for they in large part determine what is and is not possible in the economy. As stated by F. G. Tryon (1927), author of one of the first detailed empirical analysis of fuel use in the United States:

Anything as important in industrial life as power deserves more attention than it has yet received from economists. A theory of production that will really explain how wealth is produced must analyze the contribution of this element energy.

EMPIRICAL TESTS OF ECONOMIC THEORY

The energy approach to economics is relatively young with some of its conceptual tools still in the developmental stage. The remainder of this chapter is devoted to analyzing, from an energy perspective, the process by which human needs are satisfied. We formalize our approach by presenting a series of hypotheses, each accompanied by an alternative example of a more traditional hypothesis that may or may not represent the dominant view of economists in the United States today depending on one's perspective.

We compare these hypotheses to the empirical record of natural resource use and economic output in the United States for the entire period for which data are available. Empirical testing of economic theories is a difficult but essential procedure too frequently ignored. Difficulties include simultaneous changes in variables that make controlled observation almost impossible. Empirical analyses of the time-series and cross-sectional data presented here cannot be used to prove hypotheses unequivocally, nor can they assure us that parameters will not change in new ways in the future. Empirical assessments, however, are a useful tool for identifying hypotheses consistent with reality and rejecting hypotheses that are not.

After reading and analyzing the discussions to follow, we believe the reader will see why it is useful to analyze the physical underpinnings of the economic process and to understand fully the relation between human well-being and natural resource availability. Energy-based analyses help assess which economic activities are more likely to succeed and which are less likely to succeed because of fuel and other natural resource constraints.

ENERGY LAWS AND THEIR RELATION TO THE PRODUCTION PROCESS

The relation of people to material goods and services normally is studied under the aegis of economics, wherein neoclassical economists measure value according to the utility these goods and services provide to people. Money is the traditional measure economists use to trace the flow of goods and services through the economy. But money has value only because it can be exchanged in the market for goods and services now or in the future. Because goods and services are derived ultimately from natural resources, these resources are the real source of material wealth for humans, not the money that represents them in market transactions. Unfortunately many economists appear to have lost sight of this truth and have resorted to manipulating money flows as a proxy for the physical flows of goods and services. This approach is not always effective because natural resources obey a different set of laws from money flows. Frederick Soddy (1926) was perhaps the first to note the irony of our preoccupation with money flows:

> . . . debts (money) are subject to the laws of mathematics rather than physics. Unlike (material) wealth which is subject to the laws of thermodynamics, debts do not rot with old age and are not consumed in the process of living.

Economic processes take place in the physical world and are subject to the same physical laws that operate on and constrain other physical, chemical, and biotic processes. As a result, the economic process ultimately is constrained by the laws of energy which set limits on the availability and rate of throughput of matter and energy through the economy. Within these limits economic, legal, political, and societal institutions set the rate of production and its distribution:

> The laws expressing the relations between energy and matter . . . necessarily come first . . . in the whole record of human experience, and they control, in the last resort, the rise and fall of political systems, the freedom or bondage of nations, the movements of commerce and industry, the origin of wealth and poverty, and the general physical welfare of the race. (Soddy, 1926.)

In his magnum opus *Entropy and the Economic Process,* Georgescu-Roegen (1971) eloquently describes how important the laws of energy and matter are to a truly comprehensive analysis of economics. Georgescu-Roegen observes that despite the inexorable bond between the production process and energy laws, economists have failed to pay attention to these laws, the "most economic of all physical laws."

The laws of thermodynamics and the law of conservation of mass form the basis of an energy perspective for economic production. Recall that the first law limits our supply of matter and energy, and hence the goods and services derived from them, because matter and energy can be neither created nor destroyed, only changed in form. Theoretically the ultimate availability of natural resources is set by the quantity and distribution of matter on Earth, and the availability of energy is limited to the quantity stored in the Earth's crust (fossil and atomic fuels) and the quantity of solar radiation intercepted by the Earth.

The second law of thermodynamics, in conjunction with the state of human technology, sets practical limits on the availability of matter and energy for human purposes. This law states that the randomness (or entropy) of the universe as a whole increases constantly and energy must be expended to reverse this tendency. Economic production usually increases the order of natural resources; therefore economic energy must be used to do the economic work associated with this ordering.

Most resources such as minerals are abundant in low concentration on the Earth's crust but rarely occur in an economically useful form. The mineral deposits humans mine are highly concentrated compared to average crustal rock and extremely rare, but even these concentrated deposits must be upgraded to make them useful. It takes economic energy to explore for and upgrade these minerals. For example, energy is used to increase the organizational state of iron from a socially useless, highly dispersed, and random ore to more concentrated, purified, and useful ingots, sheets, plates, and tubes of metal. These materials are upgraded even further to produce the final goods and services useful to consumers. As we shall see in Chapter 4 and in Chapters 7 through 14, the quality of a source of a given resource can be measured in terms of the amount of economic energy needed to upgrade it to an economically useful form.

Increasing the organizational state of matter is essential for producing goods and services, but this also has substantial energy costs. Although the process of upgrading the organizational state of matter produces a more economically useful product, it also converts the free energy of petroleum or other fuels into less useful, low-grade energy that is dissipated as waste heat into the environment. Such energy cannot do economic work and therefore cannot be considered as a resource.

We have already stated that one implication of the second law of thermodynamics is that matter in a low state of entropy (high degree of organization) tends to change back toward a greater state of entropy spontaneously over time. To maintain matter in its low entropy, organized state, energy must be expended. Both living beings and inanimate economic structures require continuous inputs of free energy to maintain their highly organized state. In other words, not only is economic energy required to produce a good or service, it also must be used to maintain a finished good or service against spontaneous processes of degradation such as rusting, breakage, and decay. As our industrial society matures, *maintenance energy costs* can become quite large. For example, in 1982 Congress and the president increased the federal tax on gasoline by $0.05/gal, in large part to generate revenues to repair the degraded condition of the nations' bridges and interstate highway system. Even a *steady-state* economic system, which has a constant amount of physical infrastructure, needs economic energy to pay its maintenance costs just as a living organism must invest substantial portions of its energy budget for maintenance respiration. Daly (1983) states:

> . . . a large part of (economic) production is really the maintenance cost of the capital stock, a measure of the regrettably necessary activities of depletion, pollution, and labor that are required to maintain the capital stock against the ravages of physical depreciation that inevitably occur as the capital is used to satisfy human wants.

THE PRODUCTION PROCESS

Economic goods and services are produced, distributed, and consumed in several steps (Figures 2.1 and 2.2). In the first step, natural energies are used to convert the basic matter of the Earth to natural resources and public service functions. These processes constitute the basic economy of nature. In the second step, economic energies are used by *extractive sectors* of the human economy to recover natural resources, such as fuels, ores, food, and fibers, from nature and convert them to raw materials. Extractive sectors include industries such as mining, agriculture, fisheries, and forestry. Some raw materials such as fresh fish are brought directly

to market but most are concentrated, refined, and otherwise processed before they are sent to *manufacturing sectors* which combine raw materials into intermediate goods. These goods are distributed and sold to final demand sectors through *commercial sectors* of society. Services also are provided principally through the commercial sector.

The Neoclassical View

The traditional neoclassical model of the production process is shown in Figure 2.3. The basic units of the production process are the firm and the household. Households own or control the factors of production (land, labor, and capital) that the firm buys or rents from households in return for factor payments (wages, rents, royalties, etc.). The firm employs these factors according to the current state of technology and their marginal costs to transform natural resources into goods and services. The firm sells these goods and services back to households in return for personal consumption expenditures. According to this model, economic production occurs in a *closed system* in which the factors of production (fuel, capital, labor, and natural resources) and finished goods and services cycle endlessly between firms and households.

Production from an Energy Perspective

Production from an energy perspective is shown in Figure 2.4. The human economy is an *open system* that depends on a net input of energy, natural resources, and other environmental services. In this open system economic production is powered by a *linear* flow of low entropy energy from *outside* the economic system. The stock of matter on which the economic system operates (natural resources) is produced initially by natural biogeochemical processes powered by solar and atomic energy. The ability of the human economy to convert natural resources to useful structures depends on the natural energies used in the past to upgrade these elements to natural resources and the economic energies available to convert these resources to useful goods and services. New, useful energy must be extracted from the environment for the process to continue, for if the throughput of new fuel and matter were cut off, the economy would cease to operate once the stores of fuel already accumulated

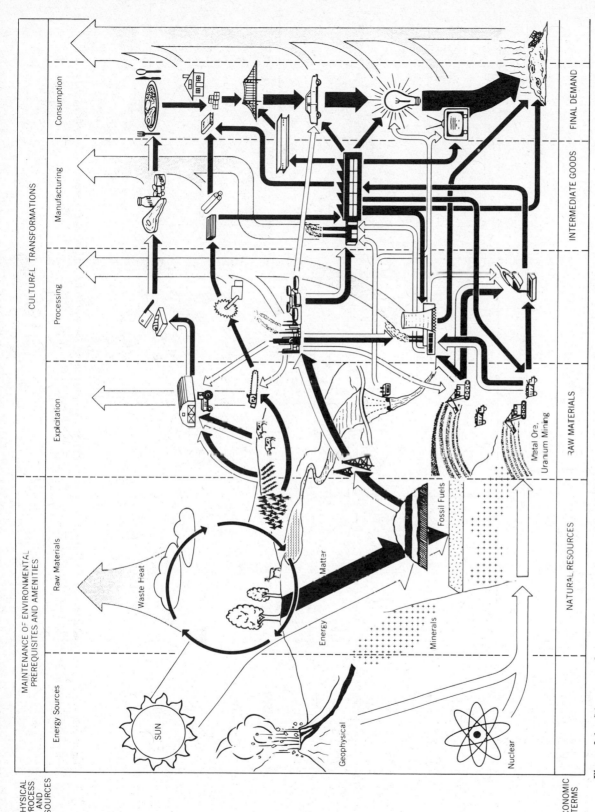

Figure 2.1. The steps of economic production. Natural energies drive geological, biological, and chemical cycles that produce natural resources and public service functions. Extractive sectors use economic energies to exploit natural resources and convert them to raw materials. Raw materials are used by manufacturing and other intermediate sectors to produce final goods and services. These final goods and services are distributed by the commercial sector to final demand. Eventually, the goods and energy return to the environment as waste products (original drawing by Doug Bogen).

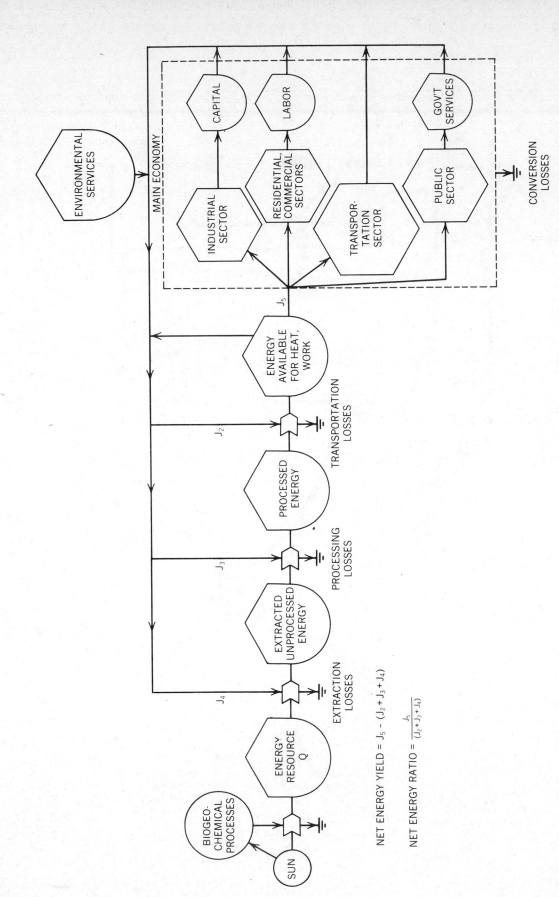

NET ENERGY YIELD $= J_5 - (J_2 + J_3 + J_4)$

NET ENERGY RATIO $= \dfrac{J_5}{(J_2 + J_3 + J_4)}$

Figure 2.2. Energy flows used to recover fuel from the environment, including the energy costs of labor and capital. Grouping of energy flows shows how EROI is calculated. J_2 and J_3, may be left out of many calculations for convenience.

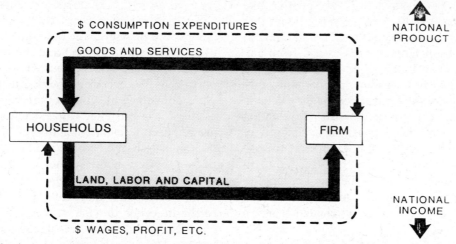

Figure 2.3. Neoclassical circular flow model of economic production. Households sell or rent land, natural resources, labor, and capital to firms in return for rent, wages, and profit (factor payments). Firms combine the factors of production and produce goods and services in return for consumption expenditures, investment, government expenditures, and net exports. (Modified from Heilbroner and Thurow, 1981.)

were used up. The dependence on a net input of low-entropy energy dictated by the laws of thermodynamics led Daly (1983) to conclude that "replenishment of the physical basis of economic life is not a circular affair."

Because the human economy is an open thermodynamic system, a circular, self-feeding, and self-renewing representation of economic production is fatally flawed. The circular flow model shown in Figure 2.3, which is found in most elementary economic texts, may be a valid representation of dollar flows, but the material quantities that give value to those dollars do not and cannot flow in circular fashion, nor can they replenish themselves. Therefore, neoclassical models of economic production do not give fuel the special consideration consistent with the laws of physics. Instead, neoclassical economists consider fuels to be just another input like capital or labor (Huettner, 1976; Berndt, 1978).

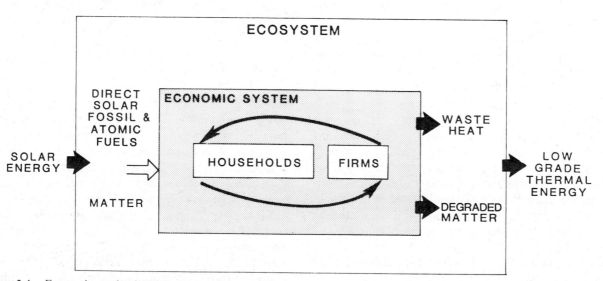

Figure 2.4. Economic production from an energy perspective. A continuous input of high-quality, low-entropy fuels enters the economic system, and the economy uses these fuels to upgrade natural resources, driving the circular flow between households and firms in the process. The fuel is degraded and put back into the environment as low-quality, high-entropy heat. (Modified from Daly, 1977.)

We disagree. From an energy perspective like the one represented in Figure 2.3, it is evident that energy is the only *primary* factor of production because it cannot be produced or recycled from any other factor—it must be supplied from outside the human economic system. Labor, capital, and technology are *intermediate* inputs because they depend on a net input of free energy for their production and maintenance. In other words, the availability of free energy is a necessary, but not sufficient, condition for the availability of labor, capital, and technology. With energy defined as the primary input to the production process, the intermediate factors of land, labor, and capital can be evaluated from an energy perspective.

The Energy Equivalents of the Factors of Production

Fuel has an obvious energy component represented by the heat released upon combustion, a quality that distinguishes it from all other resources. Fuel also has a less obvious, noncombustible energy component equivalent to the energy required to locate and extract it from nature, as well as the energy required to process it and deliver it to its final point of use.

Land is valuable in large part because it intercepts solar energy and because plants, which use solar energy as their fuel, produce economically useful crops and fibers, as well as other environmental services not sold in the market. Land also may be valuable if it has useful minerals, but mineral deposits have energy costs of recovery that increase as the quality of the deposit declines. This relation is exhibited by most mineral resources and is developed more fully in Chapter 4.

Capital structures represent energy that has been used in the past to transform natural resources into organized structures that can be used in the present to empower human labor in economic activity. This stock of organized structures is "used up" (worn out or becomes obsolete) as it is used in economic production. Energy analysts measure the contribution of capital in production by the portion of embodied energy in capital spread over the goods or services produced by the unit of capital over its life span.

Human labor might not seem to be closely related to energy beyond the metabolism of the worker, but, as we develop in Chapter 5, human labor in an industrial society requires not only the food energy that powers the work done by the laborer but also the large quantities of fossil energy used to produce the food and the other goods and services bought with the worker's paycheck. Some portion of the energy embodied in goods and services consumed by a laborer are beyond those necessary for subsistence (food, shelter, clothing) yet are required to maintain a laborer in his or her productive role in the economy. For example, a salesperson requires not only energy in the form of food, shelter, and clothing but also the energy embodied in an automobile, magazine subscriptions, and so forth, in order to visit customers, keep up with changes in the field, outwit competitors, and the like.

Embodied energy is the total amount of energy used to produce a good or service. This measure can be used as a common denominator for intercalibrating the "apples and oranges" of a complex economy and much of what follows illustrates why a physical measure, such as embodied energy, may complement dollar measures for many economic analyses. The methods by which embodied energy is calculated depends on estimating the energy used directly in manufacturing a product plus the energy used to manufacture the inputs to the final production process. Our purpose here is not to elaborate on these analytical techniques, for they are discussed in detail in Chapter 5. Rather, it should be noted that a variety of techniques are readily available for calculating the embodied energy of any good or service.

ECONOMIC GROWTH

Economic output of the U.S. economy generally has grown over the last century. Since 1890 the dollar value of real gross national product (GNP), one measure of economic activity, has grown 19-fold (Figure 2.5). This increase represents the product of a growing population and rising output, or production, per person. Until 1940 the increase in real GNP was split equally between a growing population and a rise in per capita production. Since 1940 per capita real GNP has risen more rapidly than population, so that by 1980 per capita real GNP had risen 5.2-fold while population had risen 3.6-fold.

The energy events of the 1970s heightened interest in the impacts that more expensive, less available fossil fuels might have on future economic

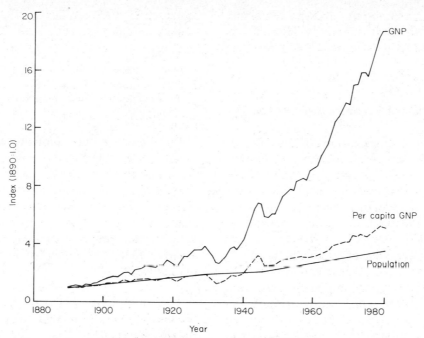

Figure 2.5. Components of economic growth relative to 1890 (1890 = 1.0). Real GNP is the product of a growing population and rising output per person. (Data from U.S. National Accounts and U.S. Bureau of Census.)

growth. Many analysts believe that the U.S. economy can continue its historic rates of growth while reducing its dependence on energy. Two factors have been suggested to decouple the link between energy use and economic production: (1) technological advances that increase the amount of economic output produced per unit of energy and (2) substitution of capital and labor for fuel. In this section we analyze technology and substitution from both a neoclassical and an energy perspective and evaluate their potential for weakening the energy-economic link.

Economic Growth from a Neoclassical Perspective

Neoclassical economists argue that in the long run economic growth is possible if the productivity of capital and/or labor increase (*total factor productivity*) or if capital stocks accumulate and population grows (*total factor input*). According to their model of economic growth, output increases as more people are employed or as each person becomes more productive, both of which are made possible by technological advance.

A technological improvement allows economic output to grow by increasing labor productivity. After the new technology is put in place, output grows

even if the same number of people remain employed and work the same number of hours per year. Technological improvements also can increase output by raising employment. By increasing the output produced by each laborer, this rise encourages firms to hire more workers.

Economic output also can grow if total factor input increases. Total factor input refers to the number of laborers or the stock of capital. With a given state of technology, total factor input can increase only if capital stocks grow in proportion to population. Assuming a given population and a wage determined by the marginal productivity of labor, employment can rise only if there are more people willing to work at this wage. This necessitates a growing population according to the model of neoclassical economists who assume that (at equilibrium) everyone who is willing to work at the present wage level already is employed. Furthermore, the capital needed to provide a workplace for this growing population also must be available.

Some of the most thorough analyses of economic growth in the United States from a neoclassical perspective are those by Denison (1974, 1979). Denison found aggregate economic growth (as measured by potential national income) in the United States from 1948 to 1969 grew at about 4%/yr. Denison calculated that about half of the observed growth was due to increasing factor inputs and the remaining half to increases in factor productivity. His analysis essentially did not test for the possible importance of energy.

The Neoclassical View of Technological Change

Firms employ varying combinations of fuel, capital, labor, and natural resources to produce output. Technology refers to the method by which these factors are combined to produce economic output based on the existing level of mechanical, engineering, or scientific sophistication. Firms choose, or at least are supposed to choose, combinations of inputs that minimize the cost of producing a unit of output, given the relative prices of the inputs. In economics the concept of technology is expressed analytically by the *production function,* which is the quantitative cause-and-effect relationship between material and human labor inputs to a process and the resulting output. In its simplest form the production function is

$$Q = f(x, y, \ldots) \qquad (2.1)$$

where Q is the quantity of output produced and $x, y, \ldots$ are the various inputs combined to produce Q. Simply put, the production function is a catalog of all recipes found in the cookbook of the prevailing technologies for obtaining a given output from a given set of inputs (Georgescu-Roegen, 1971). Technological changes and improvements that alter the inputs used to produce a unit of output are viewed as new combinations of labor, capital, and so forth, that reduce the dollar costs of producing a good or service.

It is important to note that total factor productivity and technological change are not measured directly in neoclassical analyses but rather are assigned the *residual* growth in output after increases attributable to the *tangible* factors of capital and labor have been accounted for. Based on the neoclassical view of the production process, it is not surprising that technological change has proved difficult to quantify precisely. Technological change usually is described as an exogenous driving force, resulting from advances in human knowledge that increase the productivity of capital and labor. Barnett and Morse (1963) would have us believe that "a strong case can be made for the view that technological progress is automatic and self-reproductive in modern economies, and obeys a law of increasing

returns." E. F. Schumacher (1973), a not so traditional economist, states:

> All history—as well as all current experience—points to the fact that it is man, not nature, who provides the primary resource: that the key factor of all economic development comes out of the mind of man.

But the recent stagnation and even decline in the productivity of U.S. labor and capital both absolutely and relative to other industrialized nations is inconsistent with the traditional view of technical change as an internally generated, constantly expanding process. This suggests that there may be other explanations of technological change that are based on factors not included in traditional neoclassical models.

Technology and Productivity from an Energy Perspective

An energy perspective does not downplay the importance of technology or its role in economic growth. Rather it recognizes that firms using natural resources to produce output make their production decisions based on costs of hiring more workers relative to the output they produce. But where neoclassical economists see technology as the creation of the human mind, an energy perspective focuses on the energy cost of the materials and tools required to implement technology. In the discussion that follows we explore the history of and explanations for technology from an energy perspective.

The cultural anthropologist Leslie White (1959) observed that arguing that ideas from the "mind of man" are the source of, and determine, the development of technology "explains nothing and shuts the door to further inquiry." If ideas determine the form and direction of technological change, what determines how ideas are generated and implemented or which ones are sufficiently useful to be adopted? Certainly the fuel and other natural resources available to a society influence strongly the technical means of production that a firm might have access to, consider using, or eventually adopt. As we have noted, energy is needed to do the economic work that transforms the natural resources into goods and services. Based on this need, we view technology as simply the specific methods by which energy is applied to upgrade and transform natural resources.

From an historical perspective, technological advance has taken the form of using increasing amounts of higher-quality fuels to empower human labor in production. The physical chemist Wilhelm Ostwald (1911) stated:

> Progress of technical science is characterized by the fact: first, that more and more energy is utilized for human purposes, and secondly, that the transformation of the raw energies into useful forms is attended by ever-increasing efficiency.

In a similar fashion Boretsky (1975) observed:

> ... historically, at least to date, the essence of most innovations in civilian technology has been the substitution, usually in all kinds of equipment and mechanical implements, of BTU's for human and animal energy.

In preindustrial agrarian societies the level and rate of economic work done were determined primarily by the quantity of human labor available, since labor was supplemented only by the diffuse natural energies of wind, water, and direct solar radiation. Large concentrations of economic wealth were possible only if one controlled large numbers of slaves or serfs. In these societies the quantity and rate of economic output (mostly agricultural commodities) were low because humans convert fuel to work at a relatively low rate. Humans eat about 3000 kcal/day and can transform only about 20% to physical work, equivalent to about 600 kcal, or a little less than 1 hph (horsepower-hour) per person-day. This is a small amount of work compared to the large amounts of work done by the prime movers of modern industrial societies, such as internal combustion engines, electric motors, and steam turbines. Indeed, the percentage of the total work done by labor and domestic animals in the United States has declined steadily since 1850 (Figure 2.6). Similarly, Slesser (1978) estimated that by the 1970s the United States was using 200 times more energy through its prime movers than its human population alone could supply. This shift from reliance on human labor power and diffuse natural energies with relatively low energy return on investment to highly concentrated fossil fuels enabled economic output to expand much more rapidly than the number of laborers. From this perspective the United States owes much of its material affluence to its rich endowment of natural resources, especially its high-quality fossil fuels.

An example of the subsidization of human labor

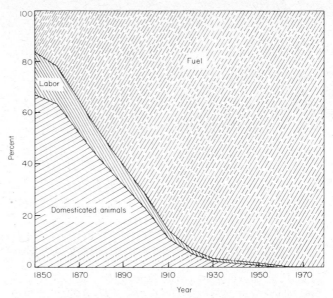

Figure 2.6.   Annual rate of work (in horsepower) done by fuel, labor, and domestic animals in the United States since 1850. (Data from Dewhurst and Associates, 1955, and updated with data from *Monthly Labor Review* and *U.S. Statistical Abstracts.*)

by energy, in both the form of fuel and energy embodied in capital structures, occurred in the bituminous coal industry between 1954 and 1967. During this period employment decreased 55% as extraction techniques based primarily on labor power were replaced by continuous mining machines powered by fossil fuels. As a result the amount of fuel used per coal miner increased almost 100%. Despite the sharp drop in employment coal production during the period increased over 40%, and labor productivity increased 102%.

Strong evidence is available to support the existence of a link between fuel use per laborer, labor productivity, and per capita GNP. Boretsky (1975) noted that higher labor productivity rates in the United States relative to Western Europe were associated with the fact that U.S. civilian employees use nearly double the quantity of fuel used by their European counterparts. Historical trends in labor productivity within the United States also are related to the amount of fuel used per laborer (Figure 2.7). For most of this century the amount of energy used per worker-hour rose rapidly, and labor productivity increased concurrent with that rise. Since the early 1970s, however, the amount of fuel used per worker-hour has leveled off, and labor productivity has stagnated. Indeed, the amount of fuel used per worker-hour accounts for 99% of the varia-

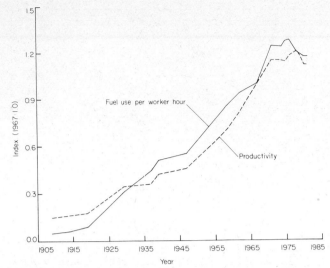

Figure 2.7. Labor productivity (dotted line), as measured by real value added per production worker-hour, and energy used per production worker-hour (solid line) relative to 1967 (1967 = 1.0). (Data from Census of Manufacturers, *Annual Survey of Manufacturers,* and *U.S. Historical Statistics.*)

tion in manufacturing labor productivity between 1909 and 1980.

In general, we may state that increased labor productivity has been made possible through technical advances that enable people to increase their use of energy, both directly (fuel) and indirectly (capital). In turn the technological advances that powered productivity increases were themselves made possible by using higher-quality fuels. If this relation continues to hold true, future technologies and their productivity rates will depend, in large part, on future supplies of high-quality fuels. This simple and fundamental relation between productivity, technology, and fuel resources apparently is not yet understood by most economists and politicians. It is often assumed that just the opposite is true—that there is no relation between fuel use, technological change, and labor productivity. For example, in their investigation of the possible impacts on economic output of reduced fuel use in the future, the National Academy of Sciences (1980) stated that "it is assumed that labor productivity growth will be independent of energy growth." Although this may be true in the future, it would be contrary to the experience of the past 70 years, the entire period for which adequate data are available.

Of course many factors besides energy affect labor productivity, many of which have been included in analyses like Denison's (1979). For ex-

ample, a more educated labor force increases labor productivity. But these other factors *also* tend to require energy for their implementations. Omitting the importance of fuel ignores one of the largest contributors to stagnating labor productivity and its attendant problems of stagnating production and inflation.

As the price of fuel has risen since 1973, reversing its historic downward trend, subsidizing workers with increasing amounts of fuel has become more expensive. This discourages further increases in fuel use per worker-hour that would boost productivity. The relation between fuel use and productivity supports those analysts who argue that rising fuel prices are a major contributor to stagnating labor productivity since 1973 (Uri and Hasein, 1982; Tatom, 1979). Thus, as the supply and quality of fuels declines, it will become increasingly important to evaluate the degree to which technology can reduce the energy equivalents in fuel, capital, and labor and the energy embodied in natural resources needed to produce a unit of output. The potential for such reductions is expressed by the economic concept of substitution.

Substitution

Many people believe that the U.S. economy can maintain its historic rates of growth by replacing or substituting fuel with labor and capital. This strategy assumes that fuel, capital, and labor are *independent* inputs to production, which is to say that the availability of capital and labor does not depend on the supply of fuel and other natural resources. From a physical perspective, however, these factors of production are not independent of each other but ultimately depend on the net amount of fuel recovered by energy transformation processes. Because of these interdependencies, substituting fuel with capital and/or labor does not automatically reduce the total amount of energy used to produce a good or service.

Neoclassical Model of Substitution

The physical interdependence between fuel, capital, and labor generally is not accounted for in neoclassical models of production. Instead, these factors are represented by autonomous dollar flows. According to neoclassical models, market forces select production processes that minimize the collective dollar costs of inputs. Simply put, firms try to produce goods and services at the lowest possible cost. The relative cost and/or availability of fuel, capital, and labor therefore influences how they will be combined in the production process.

In economics, substitution refers to the process by which changes in the relative price of a factor causes a relative increase in the amount used of another, less costly factor. For example, if the price of fuel decreases relative to labor, fuel is substituted for labor and labor productivity increases. Figure 2.8 shows changes in the real price of fuel, capital, and labor between 1958 and 1981. Until the early 1970s the real price of fuel and capital decreased while the real price of labor increased. The decreasing prices of fuel relative to labor favored the use of fuel relative to labor. Between 1958 and 1974 the use of fuel per unit of output in the manufacturing sector nearly doubled while the use of labor dropped by nearly 40% (Figure 2.9).

As the price of fuel rose, capital and labor were substituted for fuel, reversing the historic trend. Since 1974 the price of fuel has more than doubled while the price of labor has dropped slightly. This has discouraged further substitutions of fuel for labor, thereby decreasing the amount of fuel used per unit output, halting the decline of labor inputs, and increasing the quantity of capital used (Figure 2.9).

The degree to which fuel, labor, and capital can be substituted for one another is represented empirically by price *elasticities of substitution*. For fuel the elasticity of substitution measures the ease of replacing fuel with capital or labor in the production process. In terms of Equation (2.1) the elasticity of substitution between fuel and labor is simply the percentage change in the input ratio of fuel to labor brought about by a percentage change in the relative price of fuel to labor. If the elasticity of substitution between fuel and labor is greater than one, then output could be maintained and even increased with decreasing amounts of fuel by substituting labor for fuel. Some economists have estimated the elasticity of substitution between fuel and labor to be greater than one (Nordhaus and Tobin, 1972), while others believe it to be in the range of 0.2–0.6 (Hogan and Manne, 1977).

Representing the factors of production as physically independent of one another can be traced to the neoclassical production function. According to Equation (2.1), capital, labor, and fuel are interchangeable to a large extent and physically indepen-

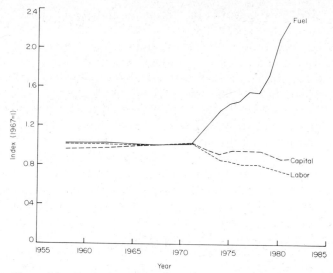

Figure 2.8. Real price for labor, capital, and fuel in the United States (1967 = 1.0). (Data from *U.S. Historical Statistics* and *U.S. Statistical Abstracts*.)

dent of one another. This implies that the use of natural resources can be reduced to essentially zero while economic output can be maintained by increasing capital and labor.

According to the neoclassical view of substitution, sharply rising fuel costs should not be cause for alarm because technological advances will devise less costly substitutes for fuel and other natural resources as they become relatively more expen-

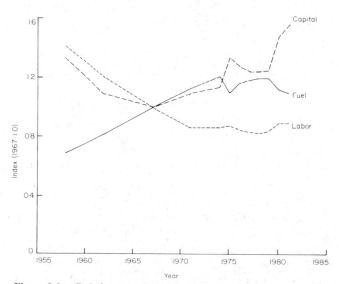

Figure 2.9. Relative quantity of fuel, capital, and labor used by U.S. manufacturers to produce a real dollar's worth of value added (1967 = 1.0). (Data from *Annual Survey of Manufacturers, Census of Manufacturers, Survey of Current Business*, and *U.S. Statistical Abstracts*.)

sive. In the words of Nordhaus and Tobin (1973), two well-respected economists, "there are no limits on the feasibility of expanding the supplies of nonhuman agents of production," and "reproducible capital is a near perfect substitute for land and other exhaustible resources." And Barnett and Morse (1963) state: "as differential resource scarcities lead to changes in relative costs, there are opportunities to make . . . substitutions that will ameliorate, and in some cases forestall, the appearance of diminishing returns." Barnett and Morse go on to say that substitution may even ameliorate resource extinction and that the list of these substitutions is "virtually endless in a scientifically and technologically advanced society."

Substitution from a Physical Perspective

From what we have already learned of the importance of energy and matter for economic production, the neoclassical view of substitution presents a physical impossibility. Fuel and other natural resources are not like any other factor of production. As Georgescu-Roegen (1979) notes, capital cannot "create the stuff out of which it is made." From a physical perspective substitution cannot replace energy completely (including the energy of labor) because each factor of production depends ultimately on an input of net energy for its own production and maintenance. Declining supplies of energy have a substantial impact on economic growth despite potential substitutions.

Substituting fuel with capital and labor replaces one type of energy with another, all of which were derived from fuel. Because of these interdependencies, the direct fuel savings of substitution are offset, to at least some degree, by increasing the indirect energy costs of capital and labor. As shown in Figure 2.9, the reduced fuel used per unit of output pointed out by several analyses (OTA, 1983; Ross and Williams, 1977) has been offset to some degree by the increased amount of capital used to produce a unit of output. In some cases such substitution can increase the total amount of energy used to produce a good or service. For example, the quantity of fuel used directly to produce a kilogram of corn in the United States decreased 105 kcal between 1959 and 1970, but the quantity of energy embodied in the capital and labor increased by 143 kcal (Pimentel et al., 1974). Despite the direct fuel savings, overall fuel use increased 38 kcal per kilogram of corn.

At the level of the firm, elasticities of substitution calculated by neoclassical economists may reflect accurately the degree to which fuel can be replaced with other factors (Field and Grebestein, 1980; Pindyck, 1979; Griffin and Gregory, 1976). A firm can save fuel by purchasing more fuel-efficient machines, installing insulation, and so on. The problem comes when one extrapolates from the level of the firm to the level of the economy as a whole. In order to calculate net fuel savings to society for a given substitution, the quantity of energy used in other sectors of the economy to produce the capital or support the labor that is substituted for the direct use of fuel must be subtracted from the gross fuel savings. When this is done, substitutions that save fuel do exist throughout the economy, but their number and magnitude will be less, perhaps much less than what is generally believed. Analyses, such as the much-heralded *Energy Future* by Stobaugh and Yergin (1979) and *Soft Energy Paths* by Lovins (1977), that look only at sector-by-sector potential for fuel conservation must therefore be viewed with caution.

This does not mean that fuel cannot be saved through appropriate investment of our resources. For example, Hall et al. (1979) found that in central New York, energy invested in the production and installation of building insulation would yield substantially more net thermal energy than would the construction of a new 800-MW coal-fired electric plant whose principal purpose was to supply resistance heat for buildings. But for many endeavors the savings is much less than anticipated. (See Chapter 14.)

THE EMPIRICAL RELATION BETWEEN FUEL USE AND ECONOMIC OUTPUT

So far we have shown that—from an energy perspective—fuel and other natural resources are the basis for economic production. In addition to goods and services, capital and labor are produced and maintained by continuous inputs of new low-entropy matter and energy. We defined technology as the process by which human muscle power is empowered with nonhuman energies. Finally, the ability of substitution to maintain or increase output with less energy is limited by the physical interdependence between fuel, capital, and labor.

If the physical perspective we have developed thus far is correct, we would expect to find a close relation between fuel inputs and economic production. Indeed, a number of investigators have used empirical analyses to show that a strong statistical correlation exists between fuel use and the production of goods and services, both between nations, states, or economic sectors at a given point in time (cross-sectional analyses) and over time (historical analysis). Interestingly some economists and politicians use these same cross-sectional and historical studies to argue the opposite: that the link between fuel use and economic activity is mostly a coincidence. Huettner (1981) and Lovins (1977), for example, argue that although a strong relation between fuel use and economic output existed in the past, the two can be decoupled in the future, allowing the economy to grow while using much less fuel. Other analysts, including Costanza (1980) and Cleveland et al. (1984) argue that the link is very strong and that a significant reduction in fuel use, whether by design or by fuel supply limits, probably would reduce the yearly production of goods and services.

The empirical relation between fuel use and economic output has been studied extensively from both neoclassical and physical perspectives (Schurr and Netschert, 1960; Berndt and Wood 1974; Darmstadter et al., 1977; Berndt and Wood, 1979; Starr 1979; Schurr, 1982). Resolving the nature of this relation is important for, as we shall see in Part Three, the quality and supply of domestic petroleum is declining, reducing net energy inputs to the U.S. economy. This trend would be of little economic or social concern if the United States could replace petroleum with alternate fuels that have EROI ratios equal to or greater than oil and gas, just as the United States switched from wood to coal to petroleum over the past 150 years (Figure 2.10). But all proposed alternative fuel systems to date have EROI significantly lower than historic fossil fuels (Table 2.1), although it is not yet clear what the EROI for new solar, breeder and fusion technologies will be. All of these fuel sources will be discussed in detail in Part Three.

The reduced supply of net energy implied by declining EROI will have a negative economic impact on the ability of our nation to maintain its current high standard of living if economic activity and energy use continue to be tightly linked. If the link between the two is weak or can be weakened—that is, if conservation and substitution can reduce substantially the amount of energy used per unit of economic production—historians will count the depletion of high quality fuels as simply another in a long line of resource exhaustions that did not alter the

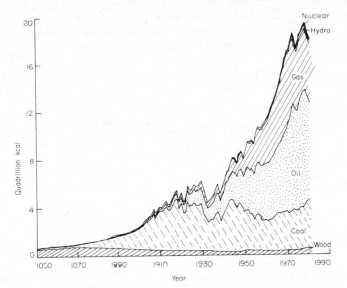

Figure 2.10. The contribution of various fuels to the U.S. energy budget from 1859 to 1981. Data are not corrected for quality. (Data from *U.S. Historical Statistics Monthly Energy Review,* and DOE Annual Report to Congress.)

fundamental course of economic growth over the last century. In this section we examine the empirical relation between fuel use and economic output in industrialized economies carefully in order to determine just how important fuel quality and availability are to the production of goods and services.

Indexes of Economic Output

A standard for measuring economic activity is essential for examining the relation between national fuel use and the production of material goods. Throughout this book we use the gross domestic product (GDP) and the gross national product (GNP) as indexes of aggregate national output. The *gross domestic product* measures the sum dollar value of all the goods produced and services rendered within a nation's economic system each year, regardless of the national origin of workers, corporations, or distribution of these goods and services between foreign or domestic markets. The *gross national product* measures the dollar value of all goods produced and services rendered by the legal citizens or corporations of a nation, regardless of where these goods or services have been produced or sold. In other words, the GDP is an index for economic production of a geographical area, whereas the GNP is an index for economic production by legal inhabitants of an economic system.

Economic output can be calculated two ways.

Table 2.1.   Estimates of Energy Return on Investment (EROI) Ratios for Some Existing and Proposed Fuel Supply Technologies

Process	EROI
Nonrenewable	
Oil and gas (wellhead)	
1940's	Discoveries > 100.0
1970's	Production ~ 20.0,
	Discoveries ~ 8.0
1980's	Production ~ 10.0
Coal (mine mouth)	
1950	80.0
1970	30.0
Oil shale	0.7–13.3
Coal liquefaction	0.5– 8.2
Geopressured gas	1.0– 5.0
Renewable	
Ethanol (sugarcane)	0.8– 1.7
Ethanol (corn)	1.3
Ethanol (corn residues)	0.7– 1.8
Methanol (wood)	2.6
Solar space heat (fossil backup)	
Flat-plate collector	1.9
Concentrating collector	1.6
Electricity Production	
Coal	
U.S. average	9.0 (27.0)
Western surface coal	
No scrubbers	6.0 (18.0)
Scrubbers	2.5 (7.5)
Hydropower	11.2 (33.6)
Nuclear (LWR)	4.0 (12.0)
Solar	
Power satellite	2.0 (6.0)
Power tower	4.2 (12.6)
Photovoltaics	1.7 (5.1)–10.0 (30.0)
Geothermal	
Liquid dominated	4.0 (12.0)
Hot dry rock	1.9 (5.7)–13.0 (39.0)

Source: C. J. Cleveland, R. Costanza, C. A. S. Hall, and R. K. Kaufmann, 1984. Energy and the U.S. economy: a biophysical perspective. *Science* **225**:890–897. See that paper, note 54, for specific references.

Note: Numbers in parentheses for electricity generation include a quality factor based on a heat rate of 2646 kcal/kWh. The energy cost of the fuel itself (e.g., the sun) is not included for these electricity generation number.

The GNP is comprised of four basic elements of final demand. *Personal consumption expenditures* (65% of GNP in 1982) represent the dollar value of all goods and services purchased by consumers to support their daily activities. The *gross private domestic investment* (13%) includes the dollar value of all investments made in new capital by private individuals or firms. Investment can be thought as the replacement of worn out capital (depreciation) plus the sacrifice of goods and services that could be consumed in the present to ensure that a greater amount of goods will be available in the future (growth in productive capacity). *Government expenditures* (20%) are the dollar value of all goods and services purchased or invested by the government to fulfill its functions—national defense, supporting research, highway construction, and so on. Finally, *net exports* (2%) refer to the difference between the dollar value of a nation's exports and imports and will be positive or negative depending on the nation's balance of trade. The *national product,* the upper line in Figure 2.3, includes all monies spent on final demand goods and services.

Economic output also can be calculated as the sum of the *value added* to natural resources and raw materials during their production and allocation in the market. Starting with the extraction of natural resources from the ground and ending with the purchase of a good by a final demand sector, each step of the production process upgrades the organizational form of the commodity thereby increasing the commodity's monetary (and embodied energy) value. Economists call this process *adding value* to the preexisting product. This process can be viewed as *adding order* in which the organizational state of matter is upgraded in each step of production (Roberts, 1982). In Figure 2.3 total value added is equivalent to the sum of all dollar flows in the bottom half (or *national income* portion) of the model. In theory, calculating the GNP by either the value-added or the final demand method should produce the same result.

The gross national product is used to measure the annual output of the economy. Economists often divide the GNP by the total population of a country to obtain per capita GNP and use this index to measure the *standard of living* or the *material quality of life* for a citizen within that country. Large per capita GDP and GNP reflect the availability of large quantities of goods and services to the consumer and are equated with a high material standard of living.

Despite its widespread use as an indicator of quality of life, the per capita GNP is a rather poor indicator of a society's *total* standard of living. Although an increase in the GNP normally is interpreted as a positive sign, Daly (1973) notes our obsession with increasing the GNP reinforces our view of consumption of material goods as an end in itself, rather than as a means of achieving some higher, more desirable end. As Daly states, ". . . the medical bills paid for cigarette-induced cancer and pollution induced emphysema are added to GNP when they should clearly be subtracted for detracting from human welfare."

Daly believes these additions should be labeled "swelling" not "growth." Similarly the satisfaction of wants created by "brainwashing" the public through advertising in the mass media also represents "swelling" not growth. Robert F. Kennedy gave a moving and ultimately poignant reminder of the inadequacy of the GNP as a measure of social welfare:

> . . . for the gross national product includes air pollution and advertising for cigarettes, and ambulances to clear our highways of carnage. It counts special locks for our doors, and jails for the people who break them. The gross national product includes the destruction of redwoods and the death of Lake Superior. It grows with the production of napalm and missiles and nuclear warheads, and even includes research on the improved dissemination of bubonic plague. The gross national product swells with the equipment for the police to put down the riots in our cities, and although it is not diminished by the damage these riots do, still it goes up as slums are rebuilt on their ashes. It includes Whitman's rifle and Speck's knife, and the broadcasting of television programs which glorify violence and sell goods to our children.

Nor does the GNP include the value of some natural services used in the production process because these services are not *paid for* by firms producing economic output. Lavine and Butler (1982) note that in addition to the traditionally defined inputs of labor, capital, and fuel, vast amounts of sunlight, oxygen, clean air, and water are used by the economy each year. Unlike the owners of land, labor, and capital, however, nature is not compensated by the users of its services, despite the fact that many of those services are degraded by human use and made unavailable for future users. Thus a policy of maximizing the GNP is practically equivalent to a policy of maximizing pollution and depletion of natural resources.

Another problem with using the per capita GDP or GNP as a measure of a nation's overall standard of living is that some extremely important aspects concerning the *distribution* of economic output are not reflected by the GNP. Although two countries may have a similar per capita GNP, the quality of life for its citizens may differ depending on the distribution of wealth. These and other contradictions have prompted many prominent economists including Nordhaus and Tobin (1972) to look for alternative ways to measure the quality of life.

Despite its shortcomings, analysts interested in examining the relation between fuel use and economic output use the GDP and the GNP as measures of output because these indexes are the most detailed, sophisticated, and generally appropriate data about economies gathered by national governments. Ideally one also would like data on production and consumption in inflation-proof physical units (i.e., bushels of corn or tons of cars sold) to compare with their monetary representation in the GNP calculations (i.e., dollar value of cars sold). Since these data are not available, nor always helpful, we too use the GNP and the GDP as measures of economic output.

Indexes of Fuel Use

A measure for annual fuel use in the economy also is needed to analyze the relation between energy and economic activity. Reliable data on annual fuel use were compiled by Schurr and Netschert (1960) in their volume *Energy in the American Economy, 1850–1975* published for Resources for the Future. Schurr and Netschert compiled a time series on U.S. fuel production and use from 1850 to 1955, including estimates in both physical and thermal units. They also included more aggregated estimates of wind, water, sailing vessel, and draft animal power used in the nation's early economic development. Interestingly the federal government did not keep detailed and centralized information on fuel use until after World War II. The lone exception was the Bureau of Mines' *Minerals Yearbook* which recorded data on mineral fuel production and use since the 1880s.

Since the mid-1940s a variety of federal government agencies have compiled weekly, monthly, and annual fuel use data to varying degrees of con-

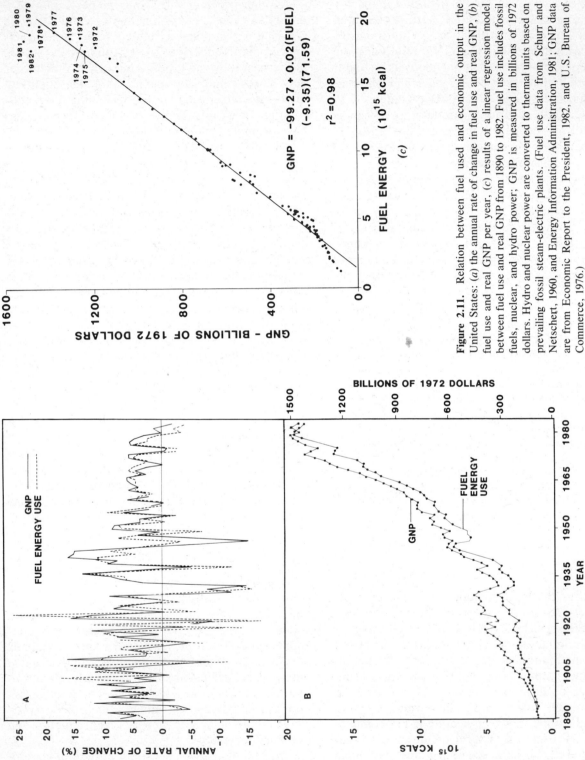

Figure 2.11. Relation between fuel used and economic output in the United States: (*a*) the annual rate of change in fuel use and real GNP, (*b*) fuel use and real GNP per year, (*c*) results of a linear regression model between fuel use and real GNP from 1890 to 1982. Fuel use includes fossil fuels, nuclear, and hydro power; GNP is measured in billions of 1972 dollars. Hydro and nuclear power are converted to thermal units based on prevailing fossil steam-electric plants. (Fuel use data from Schurr and Netschert, 1960, and Energy Information Administration, 1981; GNP data are from Economic Report to the President, 1982, and U.S. Bureau of Commerce, 1976.)

sistency and reliability. Since the mid-1970s, the Energy Information Administration within the Department of Energy has published comprehensive fuel-use data in their *Monthly Energy Review* and *Annual Report to Congress*.

Since the early 1800s the United States has relied on five major fuel sources: wood, coal (bituminous and anthracite), petroleum (crude oil and natural gas), hydropower, and nuclear power. In all analyses to follow, these five quantities are included in the calculation of fuel use. Fuel wood is sometimes excluded, for reasons explained later and noted as appropriate.

The Empirical Relation between Fuel Use and GNP

Economic work has been defined as the process in which fuel, capital, and labor are combined to transform natural resources into goods and services. The quantity of goods and services produced ultimately depends on the *quantity* of economic energy available and the *efficiency* with which that energy is used. The three variables are related in the following manner:

$$Q = E \times n \qquad (2.2)$$

where Q is the amount of goods and services produced per year (real GNP), E is the total amount of energy used per year, and n is the efficiency of fuel use expressed in number of goods per unit fuel used. The greater the quantity of economic energy that an economic system has available to it and the greater its efficiency in turning economic energy into goods and services, the greater the quantity of goods and services it can produce. Thus nations with large supplies of energy and fuel-efficient economies can produce large amounts of economic output.

The Historical Relation between Fuel Use and GNP in the United States

The close relation between fuel use and economic growth, as measured by inflation corrected or *real* GNP in the United States since 1860, is shown in various ways in Figure 2.11. A linear regression between fuel use and real GNP lends considerable support to the trend suggested in Figure 2.11 that from 1890 to 1982, fuel use and economic output have been closely related in the United States. Statistically, about 99% of changes in real GNP can be explained by changes in fuel use, and vice versa (Figure 2.11). The high coefficient of determination in Figure 2.11c is consistent with the hypothesis

that, at least in the past, economic output and fuel use have been tightly linked. Although a causal relation from fuel use to GNP, or vice versa, cannot be verified, a contemporaneous link between the two variables is supported strongly (see Akarca and Long, 1980).

Some have argued that statistical correlations like those presented in Figure 2.11 are less powerful than their high statistical correlation indicates because, in a continually growing economy, correlations could be found between economic output and any variable selected randomly. In other words, the high correlation in Figure 2.11 might reflect *time trends* in fuel use and the GNP in a growing economy rather than a close relation between fuel use and the GNP produced in a given year or set of years. This hypothesis can be evaluated using an advanced statistical procedure called the Box–Jenkins time-series analysis. For the interested reader the methods of this procedure are given in Cleveland et al. (1984). Here it is important to note only that Box–Jenkins procedures confirm a close relation between annual rates of change in fuel use and real GNP. This statistical procedure indicates that the GNP produced in a given year can be described well by a linear function of the fuel used in that year and the GNP produced in the previous year.

A cross-sectional analysis (i.e., comparing different sectors of the U.S. economy in the same year) of fuel use and economic output within the United States also supports the hypothesis that fuel use and economic output are tightly coupled. Costanza (1980, 1981) found a close relation between the sum of fossil fuel, hydro, and nuclear power used by 87 sectors of the U.S. economy and the dollar value of their output if the fuels used to support labor and government services are included (Figure 2.12). This relation is confirmed for all years in which the requisite input–output tables are available. Costanza's findings indicate that when both direct and indirect energy costs are included in estimates for embodied energy values, the energy cost of virtually all final demand goods and services are similar. Both the time-series and the cross-sectional analyses of fuel use and economic output by the U.S. economy provide strong statistical evidence for a very tight correlation between the two variables, at least in the past.

International Comparisons of Fuel Use and GDP

International comparisons of fuel use and economic output also support a close relation between these

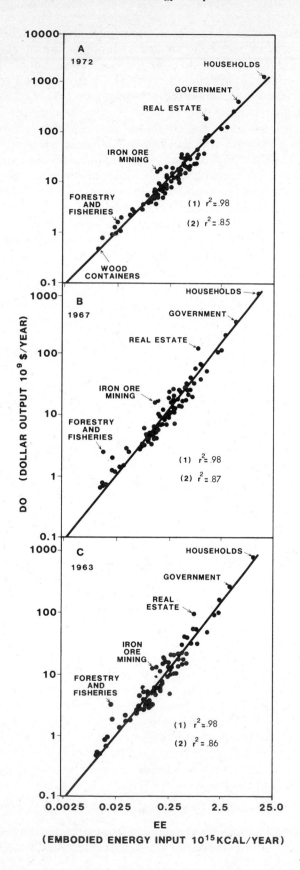

variables for most industrialized nations. Figure 2.13 depicts an international cross-sectional comparison of fuel use and real GDP for 35 nations in 1972. The high correlation coefficient for this regression suggests that regardless of the absolute amount of fuel used by a nation, a strong correlation exists between fuel use and the output of a nation's economy.

A statistical analysis of fuel use and the GNP for individual nations over time supports the strong correlation suggested by Figure 2.11. The International Institute for Applied Systems Analysis (IIASA), Hafele (1981) recently analyzed historical trends in fuel use and the GDP for 25 nations. Despite considerable variations in the primary energy/GDP ratio between nations, IIASA found that the *change* in fuel use relative to the *change* in GNP (energy output elasticity) over time for all nations studied was 0.99. This indicates a nearly 1:1 correspondence between fuel use and the GNP. Zucchetto and Walker (1981) reached similar conclusions in an independent analysis.

The importance of the general energy–economic relation has not been missed by lesser developed countries (LDCs), which often have large labor forces but are capital and energy poor. These countries have found that to increase their per capita economic output, they must increase their ability to do economic work. But the way in which economic energies are used is not fixed, and considerable debate continues over whether or not complete industrialization is required to increase production of material wealth (Schipper, 1977; Dunkerley et al., 1981). Obviously nations with especially high quality resources, such as rich agricultural land (e.g., Argentina), high-grade ore (e.g., Zimbabwe), or especially effective industry (e.g., Japan) have been able to increase economic production with a lower rate of fuel use. Nevertheless, even these nations use large quantities of fuel energy—there is an oil tanker every 100 km between Japan and the Near East bringing oil to fuel Japan's industries (United Nations, 1983).

Figure 2.12. Cross-sectional analysis of embodied energy inputs and dollar out for the U.S. economy in 1963, 1967, and 1972. Energy input includes direct fuel use, and fuel embodied in materials, and labor and government services purchased by each sector measured at the point of consumption. r^2 are the correlation coefficients. Item (1) refers to the correlation coefficient with all data points included. Item (2) refers to all points excluding households and governments. (From Cleveland et al., 1984.)

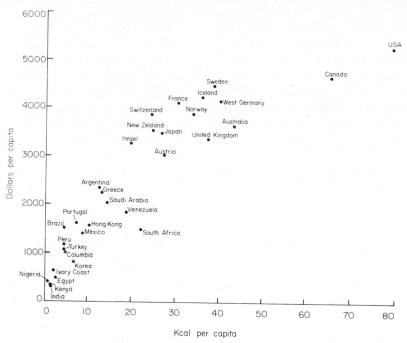

Figure 2.13. Fuel use per person in kilocalories versus GDP per person in 1970 dollars for 1974. (Data from Zucchetto et al., 1982.)

The historical and cross-sectional analyses presented here support a strong relation between economic activity and fuel use. A critical, and as yet largely unanswered, question hinges on our ability to increase substantially the efficiency with which we use fuel to upgrade natural resources and produce wealth. In the next section we discuss in detail the factors affecting the historical and cross-sectional differences in energy efficiency. Before we turn to this discussion, however, we conclude our discussion on the importance of fuel by noting that, although fuel use and the production of goods and services are strongly correlated, energy by and of itself is not valuable. Rather, it is people's ability to use fuel directly and indirectly through technology and capital investments to produce economic goods and services that makes fuel valuable. As stated by Sherry (1979),

> The customer does not want nuts and bolts, they want things fastened. We should not look at energy as so much BTU's or kilowatt-hours, or barrels of oil equivalency, but as to what energy provides us in terms of useful living.

Chemical energy stored in bonds of fossil fuels, which we currently value so greatly, has little intrinsic economic value. Instead, energy is valuable because people can use it to produce useful goods and services from natural resources which would otherwise be relatively useless to humans.

Energy Efficiency

The concept of *energy efficiency* has a variety of meanings depending on the perspective of the observer. In general, neoclassical economists define efficiency in relation to *Pareto optimality,* which is the pattern of allocation of a scarce resource so that no one can be made better off without making someone else worse off. Physicists define efficiency in relation to the laws of thermodynamics by comparing the useful work accomplished with the theoretical maximum amount of useful work possible. Energy analysts define efficiency in terms of the quantity of fuel used to produce a unit of economic output. The most commonly used index of an economic system's energy efficiency is the ratio of fuel used by that system in a given period to the real GNP produced during that period (the fuel/GNP ratio).

The efficiency with which fuel is used to convert natural resources into goods and services is important for several reasons, particularly as the supply and quality of fuel resources dwindle. Producing more economic output per unit of fuel is a desirable goal because it could alleviate the U.S. dependence on imported oil and conserve domestic fuel supplies. As Schurr et al. (1979) state, "If a given level and compositional content of national product can be economically accommodated with less energy rather than more, that possibility very likely spells a social benefit to the country." A further benefit of increased fuel efficiency would be reduced levels of thermal, chemical, and other types of pollution associated with fuel extraction and combustion for a given level of economic output (Steinhart and Steinhart, 1974).

Many of the historical and cross-sectional studies of fuel use and economic output described in the previous section have been used by those who argue that technology and substitution can decouple substantially the link between fuel use and production. Despite the strong correlation between the use of fuel and the production of economic wealth in Figures 2.11–2.13, there is no reason why the relation should be deterministic. For example, the wide spread in the GNP produced by nations that use about 40 million kcal per person per year has been used to argue that some nations are more efficient than others at using fuel to produce output than others (e.g., Schipper and Lichtenberg, 1976). Indeed, the fuel/GDP ratio varies greatly among industrialized nations (Figure 2.14). Similarly, the amount of fuel used to produce a real dollar's worth of GNP in the United States declined 42% between 1929 and 1983 (Figure 2.15). This ratio declined 19% since 1973 when real fuel prices rose sharply after a long and steady decline.

To many analysts these differences in the amount of fuel needed to produce a real dollar's worth of GNP (energy efficiency) suggest that the connection between fuel use and economic output is rather elastic—that is, the economy can continue its historic rate of growth while consuming less energy. For example, Ross and Williams (1977) state: ". . . there are substantial opportunities for uncoupling energy growth from economic growth in the present economy, so that a major reduction in projected future energy consumption levels can be achieved without major dislocations or sacrifice of economic growth." Ross and Williams conclude that the United States can maintain a "robust economy with greatly diminished and perhaps zero growth in aggregate energy consumption." These authors are not alone in their optimism concerning the potential for increased energy efficiency. Many other notable studies have concluded that the link between economic output and fuel is highly variable and that the economy will become more efficient in response to higher fuel prices and fuel scarcities (Lovins, 1977; Stobaugh and Yergin, 1979; Schurr et al., 1979; National Academy of Sciences, 1979).

We believe that the preceding interpretations of the changes in the fuel/real GNP ratio overestimate the degree to which fuel use and output can be decoupled. To date, several important parameters related only very indirectly to most people's perception of efficiency have been identified that account for a significant portion of the variation in the fuel/

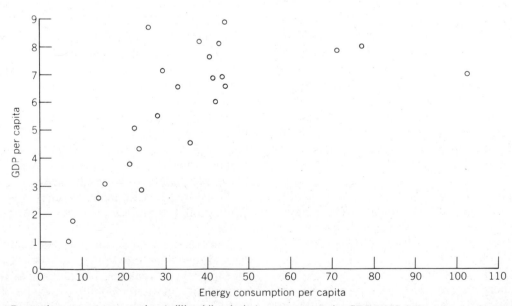

Figure 2.14.   Per capita energy consumption (million kilocalories) versus per capita GDP (1975 dollars) for the 24 OECD nations in 1978. The value to the far right is Luxembourg. (Data from *U.N. Statistical Yearbook*.)

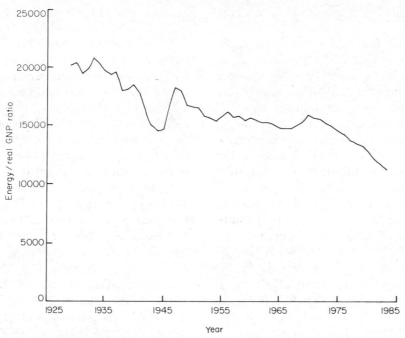

Figure 2.15. Fuel efficiency in the United States as measured by fuel used per 1972 dollar of GNP. (Data from U.S. National Accounts, *U.S. Historical Statistics,* and *Monthly Energy Review.*)

real GNP ratio. These parameters include the types of fuels consumed, the quantity of fuel consumed in the household sector, the product mix, fuel prices, and the uncertainties of record keeping. Although the patterns affecting these parameters are complex, they do suggest that waste is not necessarily responsible for much of the historic and cross-sectional differences in energy efficiency. In the following section we examine how these factors affect historical and cross-sectional comparisons of energy efficiency and evaluate the magnitude of their impact.

Quality of Fuel Used

Many studies have pointed out the effect of the types of fuels used on the fuel/real GNP ratio (Schurr and Netschert, 1960; Darmstadter, 1977; Schurr et al., 1979; Dunkerley, 1980). Standard energy units, such as kilocalories or British thermal units, measure the heat produced when a fuel is burned. According to the laws of thermodynamics, heat is the lowest common denominator among fuels. But measuring just the heat content of fuels misses important qualitative differences among fuels, and these differences affect energy efficiency.

Changes in the quality of fuel consumed can influence energy efficiency if one fuel type can do more useful work per heat equivalent than another.

Petroleum is a more efficient fuel than coal for most purposes, as measured by both thermal and economic criteria. For example, a coal locomotive uses five times the amount of thermal energy a diesel locomotive uses to pull the same train (Adams and Miovic, 1968). Economic research indicates that oil and gas generate more economic value per kilocalorie than coal by factors ranging from 1.3 to 2.45. (Slesser, 1978; Alexander et al., 1980).

Electricity is an even higher quality fuel than petroleum and coal. The higher quality of electricity is particularly useful in manufacturing where its conversion to mechanical work and heat at the point of application can be controlled exactly with respect to time and where power requirements could be matched to the task (Devine, 1982). As Schurr (1982) points out,

> with the use of electrical motors, which could be flexibly mounted or attached to individual machines, the sequence and layout of productive operations within the factory could be made to match the underlying logic of the productive process, as opposed to the more constrained organization implied by a system of shafts and belts linked to a single prime mover.

Thus a kilocalorie of electricity generated from hydro and nuclear sources can produce more output than a kilocalorie of petroleum or coal, which would be reduced to one-third of a kilocalorie when converted to electricity.

An analysis of the fuel/real GNP ratio in the United States indicates that 71.5% of the variation in this ratio between 1929 and 1983 can be explained by changes in the type of fuel consumed (Kaufmann, in preparation). As the relative share of petroleum, nuclear power, and hydropower increased, more economic work was done (more GNP produced) per kilocalorie of fuel used. Between 1929 and 1972 the fuel/real GNP ratio declined steadily. During the same time period the proportion of petroleum in the U.S. fuel budget increased from 29.8 to 79.3% (Figure 2.16a). Since 1972 the contribution of nuclear and hydro electricity increased from 1.57 to 3.28% while the importance of petroleum shrank from 79.3 to 74.2%. These post-1972 changes had roughly offsetting effects on U.S. energy efficiency that halted the decline in the fuel/real GNP ratio caused by the increasing quality of fuels used.

Household Fuel Consumption

Energy efficiency also responds to changes in the proportion of fuel used by intermediate sectors, such as manufacturing, versus final demand sectors, such as personal consumption expenditures.

Significantly less energy is required to produce a dollar's worth of nonfuel goods and services purchased by households (i.e., refrigerators or radios) compared to direct fuel purchases, such as gasoline or electricity, since the latter contain both embodied and direct fuel energy. For example, an average dollar's worth of fuel purchased by households in 1972 contained 140,000 kcal of embodied energy and heat energy, whereas a dollar's worth of most of the nonfuel goods or services purchased by households contained only 8900–11,000 kcal (Hannon, 1983).

The direct fuel purchases by households are one representation of the energy cost of labor (see Chapter 5). Although the energy used to operate a machine has a large impact on the output generated (Figure 2.7), the fuel consumed by households has only a relatively small effect on labor productivity. Thus the amount of fuel used directly by labor has a large effect on the total amount of energy used to produce a good or service, although it has only a small effect on the amount of output produced. Thus the way in which people spend their paychecks has a large effect on the energy costs of goods and services when that computation is made

at the national level or for specific analyses when including the full energy cost of labor.

Direct fuel use by the household sector can be measured by the percentage of GNP accounted for by personal consumption expenditures on fuel (gasoline, home-heating oil, etc.). The fuel/real GNP ratio is very sensitive to this factor, and changes in household fuel use account for 23.6% of the variation in this ratio between 1929 and 1983 (Kaufmann, in preparation). This ratio declined during periods marked by a relative and/or absolute drop in household fuel purchases (1941–1945 and 1973–1981; Figure 2.16b). Conversely, the ratio rose during periods marked by increased household fuel use (1945–1950 and 1966–1970; Figure 2.16b). These changes in household fuel purchases are caused by both market and nonmarket factors (Lareau and Darmstadter, 1982). Higher fuel prices in the 1970s certainly reduced the demand for fuel by households. Nonmarket factors such as the natural gas shortages in the mid-1970s and voluntary and involuntary reduction of fuel use by households during World War II also produced shifts in household fuel use.

Similarly, international differences in the relative portion of the fuel budget consumed by households explains much (52%) of the international variation in energy efficiency (Howarth et al., in preparation). Nations that consume a large percentage of their energy budget in household sectors, such as the United States and Canada, have a relatively high fuel/real GNP ratio, whereas nations that consume a relatively small percentage of their energy budget in the household sector, such as Japan, have a relatively low fuel/real GNP ratio.

International differences in the magnitude of the energy consumed in the final demand sector often is dictated by the physical environment, so cold countries often use more energy in the household sector (see Hafele, 1981; Darmstadter et al., 1977). For example, Canada, which is well to the right of the regression line for Figure 2.13, uses a disproportionally large percentage of its fuel to transport goods and heat homes because its population density is low and its winters are cold. In addition, cost differences in fuels and demographic patterns have a major impact on household fuel purchases. The price of gasoline and other fuels in the United States are much lower than in European nations, and this relative abundance has led to housing, working, and commuting patterns that encourage large personal fuel purchases (Dunkerley, 1980).

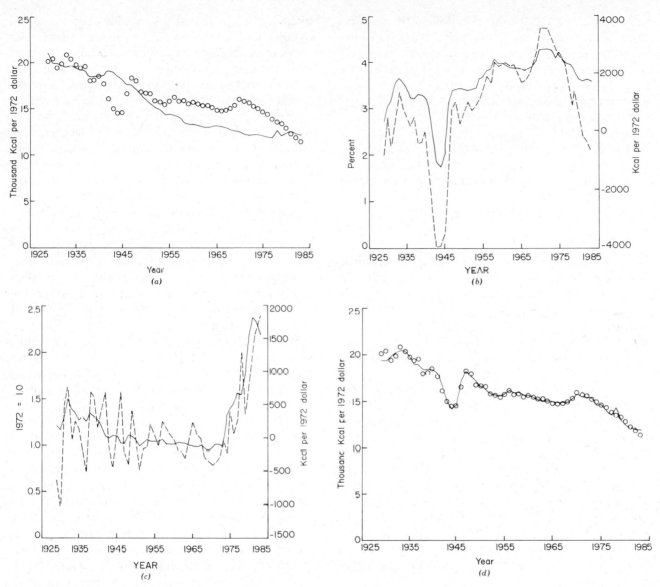

Figure 2.16. (*a*) Fuel/real GNP ratio predicted by the percentage of petroleum, nuclear, and hydro electricity in the U.S. fuel budget (solid line) and observed values (circles). (*b*) Differences between observed fuel efficiency and fuel efficiency as predicted by changes in the U.S. fuel budget (dotted line, residual from Figure 2.16) and the percent of real GNP (1972 dollars) accounted for by real Personal Consumption Expenditures on direct fuel purchases (solid line). (*c*) The difference between the actual energy/real GNP ratio and the energy/real GNP ratio predicted by changes in fuel quality and household fuel consumption (dashed line), and the price of fuel (solid line). (*d*) Prediction of energy/real GNP ratio after corrections for fuel quality, household fuel consumption and fuel prices (line) vs. actual ratio (circles).

Fuel Prices

According to neoclassical economic theory, rising fuel prices should reduce the quantity of fuel used to produce a unit of output. An examination of the empirical relation between fuel prices and the energy/real GNP ratio by Kaufmann (in preparation) confirmed the statistical significance of fuel prices but found that they accounted for less than 1% of the total variation in the energy/real GNP ratio. Prior to 1972, stable fuel prices had a relatively minor effect on the energy/real GNP ratio (Figure 2.16*c*). Rapidly rising fuel prices since 1972 have played an important role in the ratio's rapid decline during that period.

Based on this relation between fuel prices and

the energy/real GNP ratio, Kaufmann (in preparation) estimated energy's own price elasticity of demand to be −0.19—a 1% rise in energy prices reduced the amount of energy needed to produce a dollar's worth of output by 0.19%. While this estimate is lower than that made by many economists, it indicates that microeconomic fuel savings are not offset completely by other sectors increasing their fuel use. If fuel prices continue to rise, spurring conservation, it is possible that the amount of energy needed to produce a dollar's worth of output can be reduced significantly, but we believe that the degree has been overemphasized.

Product Mix

There exists a large range in the direct fuel required to produce a dollar's worth of various goods and services, so many argue that differences or changes in the composition of national output will lead to changes in fuel use per unit output. It is often assumed that the maturation of the U.S. economy and increased energy scarcity will facilitate substitution away from an energy-intensive product mix. Since the 1930s the U.S. economy has undergone several structural changes. In the early stages of industrialization the extractive and manufacturing sectors dominated the economy. As the economy developed, the importance of these basic industries declined while the importance of consumer services increased. Ginzberg and Vojta (1981) found that since 1930 the provision of services displaced the production of goods as the principal economic activity. Between 1929 and 1980 employment in the service sectors increased from 55 to 65% of the total work force, and from 1948 to 1978 the portion of GNP accounted for by services increased from 54 to 66%. A National Academy of Sciences (1979) report argues:

> Heavier energy consumption relative to national output was thus associated . . . with the stronger influence of heavy industry. The 1920–1945 period was one in which lighter manufacturing and the broad services component of national output grew rapidly, and the corresponding decline in energy consumption relative to national product was a result.

The impact of changes in the composition of national output probably has had far less of an impact on historical fuel efficiency than the preceding statement implies. The proposition that the evolution of the U.S. economy away from energy-intensive basic industries to one dominated by lighter, less energy-intensive manufacturing and service establishments will substantially increase fuel efficiency is not supported by empirical analyses, at least so far. As described earlier, Costanza (1981) has shown the seemingly wide variation in energy intensities per dollar output across all sectors of the economy is greatly diminished when all indirect energy costs are included. Consistent with these findings, the percentage of GNP coming from fuel-intensive industries or services does not account for the declining fuel/real GNP ratio in the United States in a statistically significant manner.

Cross-sectional analyses of the U.S. economy do show significant differences in the ratio of direct fuel use to output as a function of product mix. Comparisons of fuel use per gross state product (GSP) for different states indicates that the relative size of the extractive sector of a state's GSP is reflected in its fuel/GNP ratio. Starr (1979) found that states that have a very large extractive sector, such as Texas and Louisiana, have a much higher fuel/GSP ratio than do states such as Connecticut and Vermont, whose economies are dominated by manufacturing and commercial enterprises.

Likewise, international differences in the relative importance of extractive sectors are reflected by differences in fuel/real GDP ratios. Nations like Switzerland and Belgium are resource poor but import fuel and other natural resources embodied in intermediate goods from resource-rich nations. The energy embodied in these intermediate goods are not included in their accounts for national fuel use; thus their total fuel use appears low although the "consumption" of embodied energy is higher. Conversely, nations that have large energy-intensive sectors, such as Luxembourg with its large steel industry, use large amounts of fuel directly, so their energy efficiency appears low. The relative importance of industries that are directly fuel intensive account for 27% of the international differences in fuel/real GNP ratios. Consistent with these results, Darmstadter et al. (1977) found very little differences in international fuel use per unit output in a given industrial sector, except for the differences in the proportion of fuel-intensive products.

Accounting Irregularities

Problems with bookkeeping affect the reported size of fuel use and the GNP, and therefore the estimate of energy efficiency. For example, GNP calculations do not include the substantial but variable portion of many nations' economic activity that does not fit into traditional national accounts and/or is either illegal or not reported. Such economic activity is not included in either tax or official economic records. Housework is not included in the national accounts, yet the same work done by maids or butlers assigns monetary value to that labor and is included. Cherry (1980) argued that the changing role of women from unpaid housewives to paid workers may be responsible for a significant proportion of recent increases of the GNP. Illegal activities and income not reported for tax purposes represent black market activities, and this underground economy may be a significant portion of the GNP of the United States. David O'Neill of the U.S. Census Bureau (1983) estimated the underground economy in the United States to be roughly 7.5% of the GNP in 1982.

Data for fuel use also are not complete or always comparable among nations. The use of such household fuels as wood or cow dung is difficult to estimate because such fuels are rarely traded in the market. Since their importance varies among nations, less developed nations often underreport the quantity of energy used. Incomplete reporting of energy use also is a problem in the United States, where fuel wood for home heating reemerged as a significant contributor to the nation's total energy budget during the late 1970s. Fuel wood consumption was estimated to be over 0.5 quadrillion kcal in the early 1980s, nearly half of nuclear powers' recorded contribution of 1.1 quadrillion kcal in 1980 (DOE, 1982).

Data obtained from market transactions also do not incorporate the use of most renewable energies, such as solar, tidal, and wind power, which certainly contribute to economic production and well-being. Kemp et al. (1981) examined the relation between the GNP and the use of both storage energy (fossil fuels, uranium), which are usually included in monetary assessments, and solar-derived energy (e.g., agricultural products, timber expressed as embodied energy) plus indirect renewables (e.g., wind, elevated water). Many of the solar-derived sources normally are not included in economic statistics such as those compiled by the World Bank or United Nations but clearly contribute to economic well-being. Kemp et al. (1981) found that including renewable fuel resources along with stored energy resources in national energy accounts improved the correlation between fuel use and GNP for different nations.

The fuel/real GNP ratio also does not include social costs that are not included in market transactions. Examples of these hidden costs are the externalities associated with fuel use. These costs are called *externalities* because we consume some of these services every day but pay nothing for them since they are not represented by a dollar value in the marketplace. Despite the fact that many of nature's services are depicted by the market as being *free,* or more accurately, without cost to the consumer, human use of nature for natural resources or as a depository for many of the noxious by-products of the economic process often renders that system less useful for various future uses. The costs of some of these externalities are considered in Chapters 15 and 16.

The close relation between fuel use and economic activity reemerges once historical and cross-sectional differences in the types of fuels consumed, household fuel consumption, fuel prices and the relative importance of fuel-intensive sectors are accounted for. Changes in fuel use, and household fuel consumption and fuel prices account for 97% of the variation in the fuel/real GNP ratio in the United States since 1929 (Figure 2.16d). Similarly, international differences in fuel use, household fuel consumption, and the relative importance of fuel-intensive sectors account for 85% of international differences in fuel/real GDP ratios (Figure 2.17). These close relations indicate a tight coupling between fuel use and economic production. This implies that the diminishing supply and quality of fuel are almost certain to have a negative impact on the outputs of U.S. and other economies. The degree of this impact for both the United States and the world is as yet unresolved, for there may indeed be some improvement in the efficiency with which we turn fuel into wealth. On the other hand, the improvements to date have been far less than the uncritical use of energy/GNP ratios has led many people to believe.

INFLATION

In its simplest form inflation can be described as a rise in the general price level of goods and services. The fact that inflation has emerged as a major economic and political concern in most industrialized economies suggests that its roots go beyond any particular economic or political ideology. Particularly since the Depression, the U.S. federal govern-

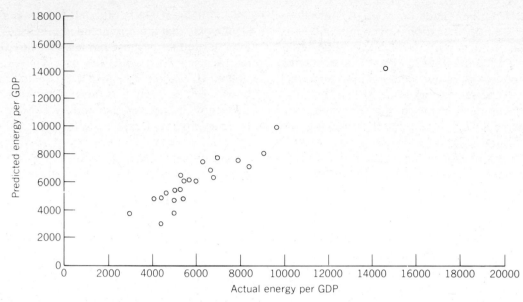

Figure 2.17.   Values for the energy/GDP ratio of the OECD nations after correction for personal energy consumption, the importance of fuel-intensive industries, and fuel quality. Compare with uncorrected values in Figure 2.14. (Data for 1978.)

ment has manipulated its spending (fiscal policies) and the money supply (monetary policy) to fine-tune the economy so as to avoid large oscillations in the business cycle. The resultant inflation rates of 2–4%/yr were viewed as a legitimate and acceptable price to pay for the stabilization of severe and unpredictable fluctuations in output and employment.

In the 1970s and early 1980s, however, periods of double-digit inflation and high levels of unemployment appeared, sometimes simultaneously, thus violating a basic neoclassical explanation of inflation which held that inflation and unemployment were negatively related. This contradiction is reflected by the relative impotency of fiscal and monetary policies to stimulate growth for long periods without producing greater inflation. In this section we describe the two predominant economic theories of inflation and offer an alternative based on an energy perspective.

Keynesian Theory and Fiscal Policy

According to Sir John Maynard Keynes and his followers (Keynesians), the Depression and subsequent recessions were crises of too little demand. Keynes felt that the government could offset this lack of demand directly by increasing its purchases of goods and services through its fiscal policy.

Franklin Roosevelt's New Deal policies repre-

sented a large increase in government spending and marked one of the first attempts by government to control aggregate economic production. These policies were somewhat successful even though many argue that it was not until large increases in defense spending during World War II that the economy was pushed out of the Depression. Following the war, many feared the slowdown in defense spending would plunge the economy back into the Depression. To avoid such a fall, the government increased its spending to stimulate demand during low periods of the business cycle and reduced its spending during boom periods. This strategy worked relatively well until the 1970s.

Monetary Policy and Demand-Pull Inflation

Monetarists such as Milton Friedman believe that inflation results from an increase in demand relative to supply. Simply put, inflation is always and everywhere a monetary phenomenon resulting from too many dollars chasing too few goods.

Monetarists argue that money supply can be controlled, within limits, by the Federal Reserve Board (Fed). The Fed controls the money supply by selling bonds in open-market operations, changing reserve requirements, or changing the interest rate it charges to banks. The Fed can take such actions on its own, but often it does so in conjunction with government spending. If the government runs a deficit, the Fed will often increase the money supply to finance the deficit. Thus many monetarists argue that fiscal policies, if effective, have been so because they have changed the money supply.

Even though the government can control the money supply, monetarists argue that money is neutral. In other words, money has no affect on real economic variables, such as real interest rates, employment levels, or real output. On the other hand, increasing the money supply raises the demand for goods and services by putting more money into people's pockets. This basic separation between demand and supply is given by

$$P \times Q = M \times V \qquad (2.3)$$

where P is the general price level, Q is real output, M is money supply, and V is the velocity of money (how often it is turned over).

Rearranging Equation (2.3), monetarists solve for the price level by assuming that velocity is relatively constant.

$$P = \frac{M}{Q} \times V \qquad (2.4)$$

According to Equation (2.4), increasing the money supply causes price levels to rise because it increases demand while leaving supply unchanged—thus the term demand-pull inflation. Using Equation (2.4), Milton Friedman (1974) was able to explain 93% of the changes in price level according to changes in money supply. Fisher used the same equation but focused his attention on velocity instead of money supply with much less empirical success (Makinen, 1977).

Inflation from an Energy Perspective

An energy-based explanation of inflation follows on the monetarist position, uniting the importance of the money market with the physical underpinnings of economic production. As we saw in the last section, changes in the amount of fuel used has been highly correlated with real output in the United States [Equation (2.2)]. Substituting Equation (2.2) for Q in Equation (2.4) and introducing efficiency, we derive the following equation:

$$P = \frac{M}{E} \times \frac{V}{n} \qquad (2.5)$$

where P is the general price level, M is money supply, E is total energy use, V is velocity, and n is apparent energy efficiency. According to this equation, inflation should be related closely to the ratio of money supply to energy use since V/n is relatively unimportant due to the relative constancy of velocity and energy efficiency.

Examination of the historical record indicates that the ratio of money supply to energy consumption in Equation (2.5) explains 99% of the level of the Consumer Price Index in the United States from 1890 to 1982, the entire period for which data are available (Figure 2.18). Periods with a relatively constant ratio of money supply to energy consumption (1890–1915, 1920–1940, and 1948–1960) also were periods of relatively stable price levels. Conversely, when money supply increased at a much more rapid rate than energy consumption, price levels rose, as evidenced by the post World War I era (1919–1922), World War II (1941–1946), and (1969–1981).

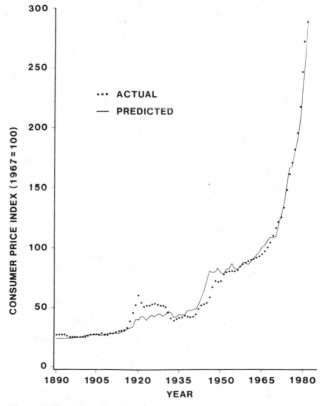

Figure 2.18. Actual values for the consumer price index (dotted line) compared to the results of a regression analysis based on Equation (2.5) of the ratio of money supply (*M*2) to fuel use (circles).

From an energy perspective inflation can be defined as "too many dollars chasing too few products of economic work" (e.g., goods and services)—or simply "too many dollars chasing too little economic energy." As consumers we perceive the increase in dollar supply relative to energy use as an erosion of the buying power of a dollar. As the money supply expanded more rapidly than energy use, the energy used per nominal dollar of GNP (fuel/nominal GNP ratio) dropped. As the fuel/nominal GNP ratio dropped, each dollar represented less and less economic energy and therefore less economic work. Indeed, the fuel/nominal GNP ratio tracks the CPI very closely (Figure 2.19). The declining amount of economic work represented by each dollar, and hence the declining amount of goods and services represented per dollar, is the essence of inflation.

In the early and middle parts of this century expansionary monetary and fiscal policy were largely correlated with, and perhaps responsible for, increases in both economic output and energy consumption (Andersen and Jordan, 1968). Earlier periods of insufficient demand created by internal economic conditions could be solved by fiscal and monetary policies because the energy and other natural resources needed to satisfy increased demand were drawn easily from nature. But recent stimulatory policies have not had the same effect because much of our economic problems of the post-1973 period stem from our declining supply of high quality fuels and other natural resources. As a result monetary and fiscal policies that stimulate demand cause inflation because the natural resources needed to satisfy this extra demand no longer are drawn as readily from the environment. In an economy constrained by natural resources, government spending competes with and may crowd out the private sector, causing interest rates to rise as they have in the late 1970s and early 1980s. Monetary and fiscal policies that attempt to stimulate demand without regard to the supply and quality of fuel and other natural resources may simply fan the flames of inflation. As Wilson (1961) stated, inflation results from "a failure to constrain competing monetary demands with limits set by real resources" and that inflation is a symptom of "a nation trying vainly to enjoy more goods and services than it has at its disposal." Although the 1980s brought lower inflation rates as the energy supplies stabilized and the growth of the money supply was sharply curtailed, the very large debts produced may be very difficult to repay since it is not clear that we will have future energy resources to make good our debts. We are living beyond our energy means.

SUMMARY

In 1952 the Paley Commission conducted what was until then the most thorough investigation of the nation's natural resources. The commission concluded that the "drama of the industrial revolution and a century of remarkable progress in the United States' living standards can be written in terms of constantly improving technology and ever increasing use of energy, mineral fuels, and water power in our factories, farms, and homes." Therein lies the crux of the material covered in this chapter, for it is from these natural resources that we obtain our material necessities and luxuries. It is ironic that modern economics, a discipline by its own admission concerned with how scarce goods and services are produced and allocated so as to meet our material needs, has ignored in large part the physical and biological underpinnings of the economic process. The physical laws that set limits on the formation, distribution, and transformation of energy and matter necessarily limit a society that depends on the

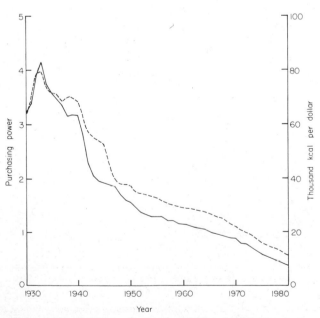

Figure 2.19. Fuel use per nominal dollar of GNP versus purchasing power of the dollar. (Data from U.S. National Accounts, *U.S. Historical Statistics,* and *Monthly Energy Review.*)

by-products of energy and matter for its survival and enjoyment of life. Despite Frederick Soddy's (1922) reminder that the "principles and ethics of human convention must not run counter to those of thermodynamics," economic and political policies today are based on an economic theory that does not consider natural resources to be qualitatively different than capital or labor, when they consider resources at all.

The purpose of Chapter 2 has been to emphasize the unique and important role that natural resources, and fuel in particular, have in the production of goods and services. The main points of this section can be summarized as follows:

1. The supply of natural resources, especially fuels, is determined by the laws of energy and matter, and these laws set upper bounds on economic production.

2. Technological advance in the United States historically has taken the form of using more fuel, both directly and as embodied in capital and industrial equipment, per laborer in the production process. Technological change therefore is not a self-generating phenomenon but is limited by the quality and availability of natural resources just like any other factor of production.

3. The factors of production are not independent. Capital and labor require fuel and other natural resources for their production and maintenance. When indirect labor, government, and capital energy costs are included in the total energy cost of producing a good or service, the differences in energy intensities for most goods and services are markedly reduced. Improvements in fuel efficiency due to a shift toward a more capital- or service-oriented economy will therefore be less than is commonly believed.

4. There is a strong correlation, both temporally and cross-sectionally, between fuel use and economic output. This is true not only for the United States for the period 1890 to 1982, but also for other industrialized nations for which data exist.

5. Improvements in the efficiency with which fuel is used to produce economic output have been smaller than is commonly believed. The bulk of the historical and the cross-sectional differences in fuel efficiency can be explained by changes in the types of fuel used, a shift of fuel use between intermediate and final sectors, fuel prices, changes in the product mix, and book-keeping vagaries of national accounts and fuel use.

6. Because of the strong connection between fuel use and economic activity, inflation is influenced strongly by the quality and availability of our natural resource base, and fuel resources in particular.

7. Any use of energy has an opportunity cost, which restrains total economic possibilities.

APPENDIX 2.1: MAXIMUM POWER: THE TRADE-OFF BETWEEN RATE AND EFFICIENCY

Howard Odum has hypothesized that *maximum power* relationships are very general in the physical and biological world (Odum and Pinkerton, 1955; Odum 1971; Odum 1983). The basic idea, which was derived from an early paper by biologist Alfred Lotka (1922), is that for a particular physical or biotic system faster rates of processes result in a greater proportion of the input energy being lost to heat or other nonuseful forms of energy. The assumption is that any organism, or system, that invests energy very rapidly but inefficiently, or very efficiently but not at a high rate, will be less competitive in natural selection than that which works at some intermediate, but optimal, efficiency, so that the useful power output is maximum at an intermediate process rate. Although it is easy to show many examples where there is a trade-off between rate and efficiency, it is not as easy to assess that there is in fact selection for the middle of the curve or, where that is possible, that such relations are general.

The Parable of the Three Industrialists

Although the maximum power concept has been applied most often to biological systems, it has been most clearly defined, and is most easily understood, by examining physical systems. Figure 2.A1 is a simple mechanical device, sometimes appearing in physics textbooks, known as Atwood's machine. If we assume that such a machine is analogous to a generic production process within an economic system, we can examine how the principle might be

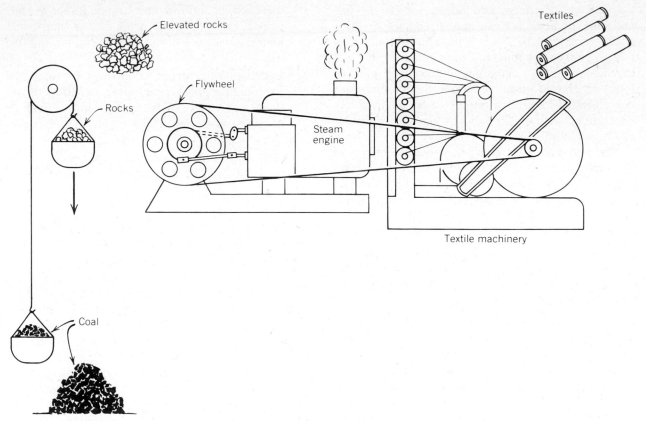

Figure 2.A1. Atwood's machine, by which potential energy is invested in exploiting other resources to produce economic gain.

important in understanding how real economic systems operate. Let us assume a group of industrialists in England in 1840, each with a source of energy (in this simple case rocks elevated with respect to some lower position) that they wish to invest in bringing to market their commodity, for our example, textiles made with coal-powered textile machinery. (The same argument could be made using the energy in the elevated rocks directly.) The coal comes from an underground mine. Each industrialist would like to corner the market, or to at least not lose their market share to the competition. This requires that he or she produce textiles more rapidly than his or her competition.

In this parable industrialist number one decided that it is important to bring the coal to his factory at the surface as quickly as possible, so he invests his existing energy resources (the elevated rocks) into rapid returns by loading his uphill basket with a load of rocks much heavier than the coal load in the downhill basket (Figure 2.A1). Sure enough, this rate of investment of energy brings the coal to the surface very quickly—but not much coal arrives.

He throws this coal into his boiler to make steam to make textiles and repeats the process. Although the coal is moved rapidly, most of the energy invested in Atwood's machine does not do useful work (moving the coal) but is instead dissipated as heat when the rock-laden basket forcibly strikes the ground.

A second industrialist, observing the first, decides to load the uphill basket with just barely more weight in rocks than the downhill basket of coal so that when the basket does arrive (finally), it brings with it lots of coal which he then uses to try to outcompete merchant number one, who is still furiously moving small loads of coal with very large loads of rocks. His system is very efficient because little energy is lost to heat.

A third industrialist has been carefully watching the first two and, having heard of Lotka's principle, loads her top basket with twice the weight of the load of coal in the bottom one. Over time this system—operating at an intermediate rate and intermediate efficiency—is the one that delivers the most coal, and she wins the competition to produce the most textiles per unit time. We say that this system delivers the maximum power (defined here as *useful* work done per unit time, or specifically, coal delivered per hour). The second industrialist's system is the most *efficient,* for most of the energy invested in the system (i.e., the energy in falling rocks) goes into doing useful work (i.e., lifting the coal). The first industrialist's system has the highest *rate* of operation, but most of the energy invested goes into the production of waste heat when the rock-loaded basket strikes the ground. The efficiency of the first industrialist's operation is therefore low. Figure 2.A2 plots the useful power output of an Atwood's machine loaded with different relations of driving energy (elevated rocks put into the top basket) to load (coal in the lower basket).

Some Examples of the Maximum Power Principle

An interesting experimental generation of a power-efficiency relationship was undertaken in seminatural experimental streams (Brockson et al., 1968; see Warren, 1971, for synthesis). The streams were stocked with different levels of predatory cutthroat trout. When predator density was low, there was considerable invertebrate food per predator and the fish used relatively little maintenance or food searching energy per unit of food obtained. With a

higher fish stocking rate food became less available per fish and each fish had to use more energy searching for it. Thus their efficiency of turning food into growth declined. At very high trout density the energy in all food obtained by the fish was approximately equal to the energy cost for the fish to live and to search for food. Thus no production occurred since production (a power term) was equal to growth times biomass. Maximum production occurred at intermediate fish stocking rates which means intermediate rates at which the fish utilized their food. Although these data clearly demonstrate the operation of the maximum power principle it is not known whether in nature there is a selection process for the fish to regulate their own biomass at the intermediate level that would result in maximum growth or *maximum power,* and the relationship could be complicated by other factors. Thus, although it is easy to show many examples where there is such a trade-off between rate and efficiency, it is not as easy to assess whether there is in fact selection for the middle of the curve or, if there is, whether such relations can be extrapolated wholesale to economic systems.

It is clear, however, that something resembling Lotka's principle has a great deal of applicability to certain energy-related problems in modern industrial society. The potential importance of Lotka's principle is that people who wish to survive and prosper in a competitive environment (such as our business world) must be interested in not only the efficiency but also the rate at which something is done, given the design constraints of our machines. We deal with examples every day. We must continually shift the gears on our automobiles, especially as we accelerate, in a trade-off of rate and efficiency of gasoline combustion. The same was true when faster jets replaced more fuel-efficient piston airplanes and railroads in the late 1950s and 1960s. It is not surprising that the very fast, but very inefficient SSTs have proved too expensive to compete in the airline industry without government subsidies. The business person using a more efficient but slower means of travel loses business to the person who can do more business in a day, but, on average, the increase of rate of travel in an SST is not worth the inefficiency and high fuel (and hence ticket) cost. Perhaps an argument could be made that the prevailing type of transportation is the one that is of intermediate efficiency, given existing technology, although of course there are many other factors of importance in this relation.

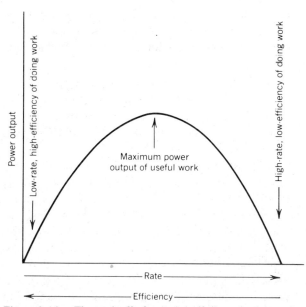

Figure 2.A2. The tradeoff of rate and efficiency in the production of power. See text for explanation.

Another application of the Lotka concept to modern industrial conditions is the relation between power and efficiency for a given energy-producing technology, or even for comparing one technology against another. According to Gilliland and Lavine (1980) the same concept applies for regular fossil fuel-powered electricity-generating plants. The upper limit of efficiency for any thermal machine such as a turbine is determined by the Carnot (or Rankin) efficiency (p. 6). For a steam turbine and associated generator to run at close to the nearly 80 percent or so maximum efficiency, it would need to operate at a nearly infinitely slow rate. Obviously humans are interested in processes that operate on a finite time scale. No one would invest a lot of money in a machine that generated revenues infinitely slowly no matter how efficiently. Actual operating efficiencies for modern steam-powered generators are closer to 35–40%, roughly half the Carnot efficiency. Yet, although power plants increased their efficiency considerably during this century, they appear to be asymptotic at a little less than 40%. It may be that very high efficiency cannot be selected for in a market economy as long as fuel sources are reasonably abundant, and it may be a reasonable explanation for the actual operating efficiencies encountered with modern power plants. Of course a power plant is not a simple machine but rather several complicated ones hooked together with losses in linkages, and so on. Such additional complications, however, appear small in at least this case relative to the basic relation between efficiency and power output. Additional trade-offs between rate and efficiency appear to have guided the selection of efficiency in other industrial power systems. Curzon and Ahlborn (1975) give examples for several engines that seem to fit the hypothesis very nicely.

Such considerations are important when viewing proposed new technologies such as solar power. Very high efficiencies of some devices are obtainable in the laboratory under certain conditions, but these may not be applicable where large quantities of output are required at a rapid rate. Thus it is unlikely that solar collectors that have higher efficiencies at lower light intensity would be placed in low light to increase efficiency when higher power output (accompanied by less efficiency) can be obtained by putting the unit in bright sunlight.

The maximum power principle is an extremely interesting and important hypothesis. It appears to operate for at least some straightforward systems but its generality and, especially, applicability to more complex systems have not yet been demonstrated fully. See Odum (1977, 1983) for a more comprehensive discussion.

APPENDIX 2.2: SOME ECONOMIC TERMS DEFINED FROM AN ENERGY PERSPECTIVE

Capital.   Economic energies used in the past to organize matter into goods such as machinery, buildings, factories, and other tools that empower human labor in the process of producing new economic goods or services.

Conservation.   A reduction in the total amount of economic energy used by a household, firm, or economy. Energy conservation may or may not be associated with a change in energy efficiency.

Curtailment.   A reduction in fuel use achieved by reducing economic activity.

Declining (or Diminishing) Returns.   As the rate at which a resource is exploited increases beyond a certain point, so does the energy cost of recovery, with other factors held constant. For nonrenewable or *stock* resources, the energy cost per unit tends to increase over time as the resource is depleted and at any one time as exploitation is increased. For renewable or *flow* resources, the energy cost per unit increases as a function of use at any given time.

Demand.   The amount of natural or economic energy a household or firm is willing and able to buy in the market in the form of goods and services.

Depreciation.   The spontaneous degradation of natural and economic energies embodied in capital toward a greater state of entropy in accordance with the second law of thermodynamics.

Direct Energy.   The free energy (i.e., solar, fossil, and atomic fuels) used irrevocably in the production process at the physical location of that production.

Economic Energy.   The form in which fuel energy is used by the economy to do economic work. Economic energies include the direct and embodied energy of capital, labor, and fuel.

Economic Good or Service.   A highly organized state of matter and/or information whose value was made possible by the previous upgrading of natural resources using economic energies.

Economic Growth. A rise in the production of economic goods and services made possible by either an increase in the quantity or quality of fuels used or in the efficiency with which fuel is used to do economic work in combination with capital, labor and technology.

Economic Production. The process by which economic energy is used to upgrade the organizational state of natural resources to economic goods or services.

Economic Work. The process of using economic energies to produce economic goods and services.

Embodied Energy. The direct plus indirect energy that has been used to produce an environmental or economic good or service.

Energy Discount Rate. The value of fuel corrected for people's preferences concerning the use of energy in the present relative to some future date. Fuel used now is often perceived as more valuable than fuel used in the future for it can be invested to produce more fuel or goods in the future. On the other hand, fuel actually may be more valuable in the future when an increase in energy efficiency could increase the amount of economic goods or services produced per unit of fuel or when high quality fuel may be scarcer.

Energy Efficiency. The amount of economic goods or services produced per unit of economic energy. Energy efficiency usually is measured by the amount of economic energy used per unit of real GNP produced.

Energy Transformation Process. The process by which society invests some of its already extracted surplus energy to make available additional amounts of new energy.

Externalities. The diminution (or enlargement) of natural energy flows caused by economic production but not included in the monetary cost or benefit of production.

Factors of Production. Economic energies of fuel, capital, labor, technologies, and natural resources used to produce economic goods or services.

Firms. The part of the economic system that combines economic energies to produce economic goods and services.

Fiscal Policy. The adjustment of government spending to influence the rate at which fuel and other natural resources are extracted from the environment and converted to economic goods and services.

Fuel. Thermal, chemical, atomic, potential, or kinetic energies recovered from the environment with an EROI greater than 1:1.

Households. Owners of economic energies.

Indirect Energy. Free energy used in the past to produce and maintain capital and labor used in the present to produce goods and services.

Inflation. An increase in the money supply relative to the amount of fuel used to produce economic goods and services. As a result demand for goods and services by dollar holders exceeds the supply, forcing prices to rise.

Investment. The use of economic energy in the present to produce capital equipment, labor, and fuels that will be used in the future to produce economic goods or services.

Infrastructure. The sum total of (the embodied energy of) society's physical public and private capital equipment including buildings, bridges, trucks, factories, machines, and so forth.

Marginal Productivity. The additional output produced by using an extra unit of embodied fuel, capital, labor, or natural resource in production.

Monetary Policy. Adjusting the money supply to control the rate at which fuel and other natural resources are extracted from the environment and converted to economic goods and services.

Money. A medium of exchange that entitles its possessor to a wide array of economic goods and services. It's value is in large part determined by the economic energies available to produce those goods and services.

Natural Resources. The products of nature, such as fuels, minerals, crops, and fibers, that are available to people. Their availability to people is determined in part by their energy opportunity cost—the total amount of energy needed to find, extract, and refine them, which in turn defines their quality.

Nominal Dollar. The amount of embodied economic energy represented by one of today's dollars.

Opportunity Costs. The laws of thermodynamics state that energy used to produce a good or service no longer is available to produce an alternative good or service. The opportunity cost associated with good *A* is the value of the alternative goods that could have been produced by the economic energy used to produce good *A*. In other words, an energy opportunity cost is a foregone energy benefit. In a Pareto-efficient economy the value of a good or service produced must be equal to or greater than the value of the alternative goods or services which could have been produced with the same amount of economic energy.

Productivity.   The amount of output produced per unit of fuel, capital, labor, or natural resource used in production. Labor productivity depends in large part on the amount of nonhuman economic energies used to subsidize a worker's effort.

Real Dollar.   The amount of embodied economic energy represented by a dollar after correcting for the effects of inflation.

Resource Quality.   Resource quality is determined by its energy opportunity cost—the amount of economic energy that must be invested by people to locate, extract, and refine one unit of natural resources to a raw material that is then ready for use by our economy.

Static Law of Diminishing Returns.   Successive additions of fuel, capital, or labor, with the amount of the other factors of production held constant, will eventually result in a reduction in the output produced per additional unit of fuel, capital, or labor used.

Substitution.   Replacing one form of economic or embodied energy with another form, often one that has a lower energy opportunity cost.

Supply.   The amount of economic energy a household or firm is willing and able to trade in the market.

Technology.   The method or recipe by which economic energy is used to upgrade and transform natural resources to economic goods and services.

Value.   Energy flows, information, goods, or services that contribute to the survival of individuals, societies, and natural systems. Beyond this, human economic value also is influenced by that which contributes to the enjoyment of life.

Value Added.   Value is added to preexisting materials by increasing its organizational state. The embodied energy in a material increases as value is added.

3

VALUE

Throughout this book the term *value* is used to compare goods, services, processes, or energy flows with one item occasionally described as more valuable than another. In a general sense an entity, whether it be a material good, an energy flow, or a piece of information, has value if it can be exchanged for another good (or goods) and/or if it is in some way deemed useful or important by its possessor. Useful or having worth can mean either that the entity contributes to the survival of the owner or simply that it enhances the quality or enjoyment of his or her life. Yet this definition still does not answer the question why one person or society values a particular commodity more highly than another. For although you and I make such judgments everyday, we seldom think about the criteria on which we base our decisions.

As Aristotle first noted, and as Adam Smith later made famous, the word value has two meanings for economists:

> The word VALUE, it is to be observed, has two different meanings, and sometimes expresses the utility of some particular object, and sometimes the power of purchasing other goods which the possession of that object conveys. The one may be called value in use; the other value in exchange. (Smith, *Wealth of Nations*, Bk. 1, Chap. 4, p. 131.)

The differences between *use value* and *exchange value* are important because they form the base of two opposing economic paradigms. The *classical* school of economics sees economic value originating in the *production* process. According to this view the value of an object is determined by the quantity of labor used to produce it (the embodied labor). Classical economists like David Ricardo and Karl Marx emphasized exchange value because they believed the amount of labor embodied in goods and services determined the ratio at which they exchange. The other paradigm, the *neoclassical* school, emphasizes use value because they view value as originating when goods are exchanged *and used,* as opposed to when they are produced. According to this view the ratio at which goods and services exchange is based on the *utility* objects convey to the traders. In other words, a good's utility is a function of the benefits it confers to its owner. Because goods are exchanged based on the utility they provide their owners, the neoclassical paradigm emphasizes use value. Differences in the definition of value are important, as the economic historian David Meek states:

> The particular definition of value with which an economist begins is almost invariably a set of shorthand expressions of the basic attitude which he is going to adopt towards the phenomena he seeks to analyze and the problems he seeks to solve. (Meek, 1979.)

Our analysis of economic events described in Chapters 1–6 is very different from either one of the two established economic paradigms because we do not believe unequivocally in either the use or exchange value as the best measure of value. No single factor, be it labor, utility, or energy, is both a necessary and sufficient condition for economic value. Using labor as a measure of value, as do classical economists, is not appropriate for an economic system in which the amount of economic work done by human labor is very small relative to other prime movers and energy sources. In an industrialized society, fossil fuels used to increase the productivity of labor have made each unit of human labor increasingly powerful in the production process. Labor is now a trigger action, controlling and directing larger but variable quantities of fossil fuels so that labor is no longer a relatively constant measure of producibility. A standard of value based on subjective human preferences also is incomplete, for this approach fails to ask the important questions, What forces determine or influence people's preferences? And are society's preferences consistent with the physical and biological guidelines of our natural environment that must be adhered to if our culture is to survive and maintain a minimum quality of life?

Among the early energy analysts Frederick Soddy, Leslie White, and Fred Cottrell believed that the energy resources available to society influenced the development of cultural values. Cottrell (1955) stated that "the idea that energy influences what man will do implies that within the limits so set, the supply of energy is also a factor at work influencing human choice." Cottrell went on to note that "the preservation of a system of values requires a continuous supply of energy equal to the demands imposed by that system of values." In a similar vein Soddy (1912) stated: ". . . whether [human aspirations] can achieve realization . . . is in the end a question of the physical resources rather than the psychical attitudes of men."

The relationship between energy and human values, however, is not a strictly deterministic one, a condition recognized by both Soddy and Cottrell. The important point is that the type, quality, and quantity of natural resources, and fuel in particular, set general but definite limits on the development of human values and the physical implementation of human ideas. Numerous examples of this relation were cited in Chapter 2. Neoclassical theory tends to ignore these constraints while it assumes that the ultimate end of the economic process is the satisfaction of our material wants and that the ultimate means for achieving this end are stocks of goods and services. On the other hand, a physical theory of value based on energy also is incomplete because many characteristics of human tastes and human moral and ethical behavior are not determined or influenced directly by resource availability. We believe, however, that much can be gained by analyzing human values from a physical perspective. In order to understand where an energy perspective fits in the intellectual economic landscape and how it can contribute to a more complete understanding of economic processes, the basic tenets of each paradigm must be understood. This chapter describes the most important aspects of the two fully developed economic paradigms as well as the emerging body of literature of a new paradigm based on the importance of fuel and other natural resources.

CLASSICAL ECONOMICS

Value

The fundamental tenet of classical economics, that labor is the source of all value, was first formalized by Adam Smith in 1776 in his book *The Wealth of Nations*. Smith states:

> The value of any commodity . . . is equal to the quantity of labor which it enables him to purchase or command. Labor, therefore, is the real measure of exchangeable value of all commodities. (Smith, p. 133.)

To this formulation he adds:

> Among a nation of hunters, for example, it usually costs twice the labor to kill a beaver which it does to kill a deer, one beaver should exchange for or be worth two deer. (Smith, p. 150.)

This strict equivalence between the labor required to produce a good or service (embodied labor) and the rate at which they exchange did not always hold, however, for several reasons. Smith recognized that the value of all labor was not alike:

> . . . (the) amount of another's labor that an object could command also depended on the quality of the work involved. If the one species of labor should be more severe than the other, some allowance will naturally be made for this superior hardship; and the produce of one hour's labor may frequently exchange for that of two hour's labor in the other. (Smith, p. 150.)

Thus a pair of shoes that required 4 hours to produce may have been traded for coal that required 2 hours to produce because of the risk and exertion associated with mining coal.

David Ricardo followed Adam Smith and refined Smith's labor theory of value. Like Smith, Ricardo stated:

> . . . it is the comparative quantity of commodities which labor will produce that determines their present or past relative value. (Ricardo, *The Principles of Political Economy and Taxation,* Chap. 1, p. 9.)

Ricardo's contribution to classical economics lay in his explanation of economic rent consistent with a labor theory of value.

The cornerstone of Ricardo's theory of rent was based on his explanation of prices. Ricardo wrote that the quantity of labor used by the least efficient producer of a good (the producer that used the most labor to produce a good) set that good's market price:

> The exchangeable value of all commodities . . . is always regulated . . . by those who continue to produce them under the most unfavorable circumstances. (Ricardo, p. 37.)

This pricing scheme is especially important in setting the price of natural resources because the quality of the deposit influences greatly the amount of labor required to extract it from the environment and fashion it into a useful good or service.

In agriculture, for example, the soil with the lowest native fertility set the price of crops: "corn which is produced by the greatest quantity of labor is the regulator of the price of corn" (Ricardo, p. 49). Landlords who owned relatively fertile land could charge rent because a farmer on less fertile land would be willing to pay to use the more fertile land where more corn could be grown per hour of the farmer's labor. The rent a landlord could charge depended on the fertility of his or her land relative to the least fertile land under cultivation. Rent on land only slightly more productive than the least fertile land under cultivation was less than rent on the most productive land because only a small amount of extra crop could be grown per labor hour on the marginally superior land. Similarly, rent could be charged by all owners of natural resources whose quality was greater than the lowest quality deposit in use. For example, owners of gold mines with 7% ores could charge rents if gold mines with 5% ore also were in use.

In using such a pricing scheme, Ricardo was one of the first economists to describe the economic interaction between demand and resource quality. As demand for a resource grew, either cumulatively for nonrenewable resources or at a given point in time for renewable resources, lower quality deposits had to be developed. Lower-quality deposits, whether they are a ton of copper or a bushel of wheat, required more labor per unit extracted. Since the labor time needed to extract resources from the lowest-quality deposit set their price, using lower-quality deposits raised the price of natural resources. Conversely, when demand subsided, the lowest-quality deposits no longer were used, less labor was required to produce resources from the lowest-quality deposit, and the price of natural resources declined.

Ricardo also attempted to explain the origin of profit as a return to capital based on a labor theory of value. Expanding on Smith's example of the value of a hunter's catch relative to the time used to catch it, Ricardo wrote that the value of a good also included the labor embodied in capital used by the laborer:

> The value of these animals would be regulated, not solely by the time necessary for their destruction, but also by the time and labor necessary for providing the hunter's capital. (Ricardo, p. 13.)

Thus Ricardo recognized the importance of indirect labor costs. Ricardo's attempt to explain the origin of profit based on the labor embodied in capital failed to be accepted because of internal inconsistencies not considered here.

Like his intellectual forebears, Karl Marx also believed that labor was the source of all value:

> Commodities . . . in which equal quantities of labor are embodied, or which have been produced in the same time, have the same value. (Marx, *Capital*, Vol. I, Chap. 1, p. 34.)

Marx's explanation for profit consistent with a labor theory of value was one of his most notable contributions to classical economic theory. Marx stated that the value of labor was determined by the same rules that applied to all commodities—labor's value was set by the quantity of labor required to produce it. The quantity of labor required to "produce" labor was the time a laborer had to work to produce or earn wages sufficient to purchase all the commodities needed by a laborer and his or her family to feed, clothe, and reproduce. This wage rate was the subsistence wage rate. The subsistence wage set a lower limit on wages because the worker's earnings had to be at least equal to "the value of the commodities without the daily supply of which, the laborer cannot renew his vital energy consequently, by the value of the means of subsistence that are physically indispensable" (Marx, p. 173).

Marx pointed out that the time a laborer worked per day for a factory owner always was greater than the time a laborer needed to produce goods and services to sustain his or her self. This advantage was achieved through the capitalist's superior bargaining position. Laborers needed to work every day to survive and had only their labor to sell, whereas capitalists do not need to hire labor constantly. Capitalists therefore could pay laborers a wage equal to (or sometimes below) the subsistence wage rate, and this wage was less than the value of the goods that a laborer produced. Laborers worked longer than the time needed to produce subsistence commodities but got paid only the equivalent of subsistence commodities. Marx used the term *surplus value* to refer to the commodities produced above and beyond subsistence. Surplus value belonged to the factory owners and showed up in their ledgers as profit.

Unit of Analysis

Because capitalists could extract surplus value from workers by virtue of their superior bargaining position, social classes defined by the process of production became the unit of analysis for Marx and most of the classical economists that followed him. According to Marx's theory, workers and capitalists struggle over the disposition of surplus value created by workers. Capitalists try to increase their share, so they can reinvest and reproduce their capital. On the other hand, workers want a greater share of the surplus value to raise their "socially defined level of subsistence" (standard of living). Classical economists use the idea of class struggle to explain many economic events, such as the length of the working day, the wage, profit rate, and the type of technology used in production.

Objective

As implied by the full title of Adam Smith's book, *An Inquiry into the Nature and Causes of the Wealth of Nations,* the objective of classical economists was to identify the source of economic wealth and point out ways by which it could be increased. Thus classical economists focused on the production process and the importance of labor. Marx recognized the importance of machinery for increasing surplus value and pointed out that machinery and nonhuman energy increased the rate at which surplus value could be extracted from workers:

> The difference between tool and machine is that in the case of a tool, man is the motive force, while the motive power of a machine is something different from man, as is, for instance, an animal, water, wind . . . . (Marx, p. 372.)

Marx also recognized the critical role that fuel played in the types of machines and industrial organizations possible:

> Wind was too inconsistent and uncontrollable . . . the use of water could not be increased at will, it failed in certain seasons of the year, and above all it was essentially local. Not till the invention of the steam engine was a prime mover found that begat its own force by the consumption of coal and water, whose power was entirely under hand control . . . that permitted production to be concentrated. (Marx, p. 377.)

Although Marx and other classical economists recognized the importance of fuels, important differences remain between classical economists and an emerging body of thought organized about an energy theory of value.

NEOCLASSICAL ECONOMICS

Value

To neoclassical economists the notion that value was "put into" goods and services by labor during production was unacceptable. Instead of assuming that all goods and services produced had some *use value,* as did classical economists, neoclassical economists made *subjective human wants* the overriding factor in the valuation of goods and services. For them, the value of an object lay in its desirability to people, so its value cannot be determined until it was exchanged in the market. Uniting all neoclassical economists is the belief that market forces fix the value of goods and services based on the supply of that good, which depends on the willingness of people to work, and the demand for that good, which depends on the utility that good provides to people.

Before the writings of early neoclassical economists, such as Jean Baptiste Say (1767–1832) and Nassau Senior (1790–1864), could be considered on the same footing as the classical school, economists in the emerging neoclassical paradigm had to overcome a theoretical roadblock proposed by Adam Smith. Smith dismissed a theory of value based on utility on these grounds:

> Things which have the greatest value in use have frequently little or no value in exchange; and on the contrary, those which have the greatest value in exchange frequently have little or no value in use. (Smith, p. 131.)

Specifically, Smith referred to a seeming contradiction in the value of diamonds relative to water:

> Nothing is more useful than water; but it will purchase scarce anything; scarce anything can be had in exchange for it. A diamond, on the contrary, has scarce any value in use, but a very great quantity of other goods may frequently be had in exchange for it. (Smith, p. 132.)

It was not until the 1870s, when Stanley Jevons, Carl Menger, Böhm Bawerk, and Leon Walras independently developed the concept of marginal utility could the water–diamond paradox be resolved and the neoclassical school develop a consistent model for explaining value and other economic events. According to neoclassical theory *utility* is the benefit derived from consuming a good or service, whereas *marginal utility* refers only to the utility obtained from the last unit consumed. Because neoclassical theory assumes that the utility derived per unit of good declines for each additional unit consumed, marginal utility diminishes as the quantity consumed increases. The price consumers are willing to pay declines as the quantity purchased increases because of the diminishing returns obtained from additional purchases.

The concept of marginal utility was used to resolve the diamond–water paradox. The marginal utility of the first units of water consumed are great, but as our thirst is satiated and our bellies bloat, the marginal utility of water becomes very small, even negative. The same is true for diamonds. The first few (diamonds) we own are very precious, but as our jewelry boxes swell, the utility obtained from another diamond diminishes. The difference in the value of diamonds and water arises because most of us have plenty of water so additional units have relatively little value. On the other hand, most of us have very few diamonds, so additional diamonds are very dear.

Armed with the analytical tool of marginal utility, the neoclassical paradigm set out to explain the value of all goods and services traded in the market. Neoclassical economists believe that the value of a good or service is set by the balance between the desire to consume and the unwillingness to work. In other words, because we do not like to work, we must be paid more to work extra hours. Our decisions as to the time we are *willing* to work depends on how much we dislike our last hour of work (versus leisure) compared to how much we enjoy the good or service purchased with the income earned by the last hour worked. According to neoclassical theory, individuals balance these goals so the disutility (pain) of the last hour worked equals the utility (pleasure) derived by spending the last dollar earned and the utility derived from an additional hour of leisure. The point at which the two are equal is the equilibrium point. At equilibrium the amount of good or services produced equals the amount of productive services offered, and this equality sets the price at which they exchange.

Unit of Analysis

The individual is the unit of analysis for the neoclassical economist because neoclassical economists take the existing distribution of wealth as a given, and different individuals are assumed to have unique tastes and preferences that are determined outside the market where bargaining and exchanges take place. According to this view individuals make rational decisions as to how much of their labor, capital, or land they offer on the market and how they rank their purchases based on personal tastes and preferences. Neoclassical theory treats macroeconomic phenomena as the sum of individual decisions based on tastes and preferences exogenous to the economic system rather than the result of class struggle. For example, employment levels and total output represent the sum choice of millions of individuals who decide how much of their productive factors they offer in the market and current levels of productivity and capital stock.

Objective

The objective for neoclassical economists is to ensure that the market operates efficiently. Neoclassical theory maintains that the market integrates individual supply and demand curves, setting the value of goods and services, although such ideal values can be biased by oligopolistic industries, externalities, and hidden subsidies, such as taxing policies, production incentives, price floors, and price ceilings. Once such disruptions are identified, neoclassical economists attempt to evaluate the magnitude of the disruption and suggest policies for correcting or ameliorating their effect.

Summary

The differences between classical and neoclassical economic paradigms are important because they lead classical and neoclassical economists to ask very different questions about the economy and approach economic problems from a different point of view. In the following section we describe the beginnings of an economic paradigm based on the importance of energy. Although we do not subscribe to all the theoretical aspects of this emerging field of thought, our own analysis does spring from its theoretical principles concerning value, units of analysis, and objectives.

ENERGY THEORY OF VALUE

The laws of thermodynamics were formalized by physicists in the middle of the 19th century. Since then the principles derived from these laws have been used by many analysts in the social, biological, and physical sciences. Classical and neoclassical theories recognize the importance of fuel and other natural resources as necessary factors of production but believe(d) that these factors are not important enough to warrant the incorporation of the laws of energy and matter into economic theory. Two notable exceptions in the economics profession are Georgescu-Roegen (1971) and Daly (1977) who both describe the lack of a biophysical foundation as a major deficiency of conventional economic theory. But in general, the people who give energy a major role in economic systems are outsiders, such as biologists, physicists, sociologists, and engineers.

As described in Chapter 2, much of our work can be traced back to ideas first expressed by noneconomist pioneers, the attitude of whom can be best expressed by Frederick Soddy (1926):

> If we have available energy, we may maintain life and produce every material requisite necessary. That is why the flow of energy should be the primary concern of economics.

The intent of Soddy and other analysts with a biophysical perspective of economics has been to synthesize a new paradigm, making the supply and quality of fuel and other natural resources the focus of analysis.

Two ecologists, Howard Odum and Robert Costanza, have been in the forefront of those attempting to forge an alternative economic paradigm based on an energy theory of value. These ecologists and others are currently attempting to fashion these ideas into a complete paradigm for explaining economic systems. In this section we describe the attempts at such a synthesis.

Value

As set forth by Odum (1971) and Costanza (1981), an energy theory of value assumes that individual and societal tastes, preferences, and economic decisions are influenced and often directed by environmental factors like natural resource quality and availability, and in particular free energy. In this line of reasoning energy is the organizing principle from which all values—economic, social, and political—ultimately are derived. Odum (1977) states that the formation of human tastes and preferences,

a process explicitly beyond the realm of neoclassical economics and insignificant to the classical economist's view of class struggle, are critical to a thorough understanding of economics. Odum's energy theory of value draws on Lotka's (1924) hypothesis that natural selection is driven by differential rates and efficiencies of energy use relative to competitors. Odum proposed that moral, ethical, and all psychological phenomena are derived from incorporating surviving patterns as ultimate values, simply because patterns selected against eventually disappear. Because of the importance of free energy he proposed that changes in the rate or efficiency of energy use lay at the heart of many human behaviors. According to Odum's hypothesis, changes in energy availability and the rate and efficiency of energy transformations are the principal mechanism in natural and cultural selection and economic development.

According to Odum *maximum power* (see page 63) is the criterion by which systems, whether they are ecological or economic, select behavior patterns. Surviving patterns are those that enable an organism, culture, or any unit of a system to transform energy from its environment into useful power at a rate and efficiency that enables an organism to compete successfully with those around it. The maximum power principal is based on the observation that for most energy conversions the efficiency of the conversion decreases as the rate of energy conversion increases (Odum and Pinkerton, 1956). At very slow rates energy is used very efficiently, but because of this slow rate, not much power is generated. Furthermore, at low rates of use much of the energy resource remains available for potential competitors. At very rapid rates of use energy is used relatively inefficiently, and the power generated is small because not much of the energy consumed is converted to useful power. Thus at some intermediate rate there is a trade-off between efficiency and speed that maximizes power. Odum (1971, 1983) hypothesized that maximum power is the optimum that all living systems strive for.

Costanza (1980, 1981) has extended Odum's original analysis and offered empirical evidence that supports one necessary condition for an energy theory of value: the relative prices of goods can be explained by their relative embodied energy cost. Costanza hypothesized that solar energy is the only net input into the biosphere; therefore a perfectly functioning market would, through a complex evolutionary selection process, arrive at prices that were proportional to their embodied energy content. In support of this hypothesis Costanza showed that the fossil fuel energy embodied in goods and services is closely correlated with market-determined dollar values when the analysis is sufficiently comprehensive (see Figure 2.12).

According to Odum, systems that maximize power outcompete those that do not. As Cottrell (1955) proposed, some cultures switched from one energy source to another because such a substitution offers the potential to generate a greater amount of useful economic and social power. Thus some energy analysts hypothesize that economic systems (through the actions of their individual members) attempt to maximize power just as neoclassical economists hypothesize that individuals attempt to maximize utility and just as classical economists hypothesize that economic classes struggle for a larger share of surplus value.

Dollar Profits in Energy Terms

The nature and origin of economic profits can be explained with the theoretical construct of an energy theory of value. An energy theory of value suggests that profit in human economic systems is the unpaid work of nature measured in dollar terms. Profit accrues to those who control the products derived from the flow of energy. Such a view is analogous to Marx's theory of surplus value where profits were equated with the unpaid services of labor.

Natural resources are highly ordered, thermodynamically improbable arrangements of matter. Biogeochemical cycles, which are driven by solar energy and energy derived from radioactive decay in the Earth's interior, occasionally have organized the various elements into concentrations far greater than would be found in a system at maximum entropy. Had this work not been done by nature, or had it been done to a lesser degree, humans would have to invest far more of their economic energies simply to upgrade natural resources to a state where they could be used as a factor of production. Humans reap the benefits of nature's work but do not pay nature a dollar amount for its services rendered. According to an energy theory of value, the uncompensated effort of nature is what makes profit possible in human economic systems.

The work done by nature in creating and upgrading fuel deposits is especially important. Surplus energy available from fuel is measured by its EROI and represents the energy available to produce nonenergy goods and services. As the energy surplus grows, it is possible to increase the per capita material standard of living. The distribution of the products of surplus energy depends on individual levels of productivity and the relative bargaining position of economic classes. It is not possible to predict the distribution of the products of surplus energy or to judge whether a given distribution of surplus energy is just or not. An energy perspective can, however, identify the source of societal surpluses and how changes in the natural resource base affect this surplus.

Unit of Analysis

Inherent in an energy theory of value is the principle that a society, or individual, maintains its existence by securing energy and natural resources from its environment for the purpose of producing those goods and services that facilitate survival. Individuals choose items freely in the market, but they may be penalized for choices that may jeopardize their survival or the survival of their culture in the long run. Because of this ongoing selection process, whole cultures may institutionalize choices that have contributed greatly to their survival, in a manner analogous to natural selection in biotic systems.

Survival of a system requires the survival of its support system (e.g., a human economy and its environment). Each unit of a system must contribute certain amounts and qualities of work to its support system, or else that unit will drain resources away from the system causing that unit and the entire system to be selected against. The requirement for survival of any part within the system therefore is some type of service fed back to some other unit of the system. Since low entropy energy is the primary requisite for the survival of any living system, the useful work contributed by an individual unit is a measure of its value to the system as a whole.

Objective

The objective of energy analysts is to understand how energy availability and use influence the rate and direction of economic and cultural development. The interactions between components of economic systems also interest energy analysts because it is important to understand how the actions taken by one part of the system affect the functioning of the other parts and how this interaction affects the type and amount of economic work done. For example, extracting fuels from the environment may disrupt other natural energy flows which the economy also depends on. Thus the net amount of useful power generated by extracting fuel from the environment depends in part on the amount of fuel extracted relative to the magnitude of the natural energies lost due to the extraction process. Because such impacts of fuel use occur during the production process, energy analysts tend to focus on production just like classical economists.

An energy theory of value emphasizes basic physical laws rather than the psychical attitude of humans to explain economic phenomena. Energy, however, is not a necessary and sufficient cause of economic value because many economic decisions are not influenced by energy quality and availability. Criteria other than biophysical laws also must be considered when developing models to explain and predict human economic behavior. Soddy (1926) observed:

> . . . the interest of the average man will be and must continue to be in a just appreciation of the relations of several worlds, the spiritual and the mechanical, to his own life . . . they do not meet in common ground in him. His is the unfortunate body from which, during life, neither the aspiring soul can altogether soar, nor the wheels of scientific materialism can be unmeshed. He has to make his peace with both, as he is the sufferer if his soul gets caught in the gear.

Like Soddy we emphasize the physical aspect of life at the expense of the humanistic aspect but not because we deny or belittle the existence of the latter, rather because the former has been greatly ignored in the history of economics. Although a biophysical perspective is very different from either one of the two economic paradigms that hold political power in different nations, we feel that it is indispensable to solving many economic problems faced by all societies.

4

RESOURCE QUALITY

The issue of natural resource quality and availability is vital to human existence and economic well-being, particularly in industrial societies that rely heavily on many nonrenewable resources. We are told often, however, that modern technology has *freed* humans from their dependence on natural resources, and most modern economic theory lacks a sophisticated and systematic treatment of the role of natural resources in human economic affairs. As one extreme example of this view Solow (1974) suggests that "the world can, in effect, get along without natural resources." Many other resource analysts argue that technological advances and human ingenuity make the issue of resource quality essentially irrelevant.

Nothing could be further from the truth. Modern technology has made the link between natural resources and human existence less apparent to most of us by separating the consumer from the resource extraction and refining process, but it has not changed the importance of the relationship. As consumers, we are often unaware of the direct link between our demand for material goods and services and the need to extract and process additional natural resources from the environment. We do not perceive readily that such demands not only deplete existing stores of many types of resources but also impose additional demands on the environment to process the wastes and by-products which result from the processing and use of large quantities of energy and matter. We often ignore the undeniable fact that most environmental problems can be traced directly to our own consumption, and that our consumption is someone else's pollution.

In agrarian societies people were much more in tune with changes in the quality of their natural environment because those changes directly affected their welfare. If a fisheries failed, or there was a decline in soil quality due to erosion of fertile topsoil that lowered crop yields, there was less food for one's family. But today we can use fuel to find new fish farther away, and we can subsidize declining soil quality by applying increasing amounts of petroleum-derived fertilizers and by building larger and more sophisticated farm implements to till the soil and harvest the crops. The decline in quality of many other resources has been offset to a large degree in a similar manner—we use increasing amounts of high EROI fossil fuels to exploit increasingly lower-grade resources, both fuel and nonfuel. In many cases the decline in quality of these resources goes largely unnoticed since the potentially adverse economic effects of using lower-grade resources has been offset by the use of more fuel, directly or indirectly as fuel-derived capital equipment and industrial machines. For example, improvements in pumping technology, refining, and transportation in the oil industry, as well as the large physical extent of previously-found reserves, have until quite recently offset the cost-increasing effects of the depletion of high-quality deposits. The results were declining real prices to the consumer for petroleum products, such as gasoline, throughout much of this century despite the fact that the cost of finding a *new* barrel of oil has risen steadily. This is possible as long as the quality of the *extracted* fuel resources does not decline substantially.

But when fuel itself requires more and more energy to recover and process, our ability to subsidize the exploitation of other resources is diminished, and increasing amounts of whatever fuel supplies are available must be diverted to the extractive sectors simply to supply the same amount of resources to the economy. When this occurs, we are less able to avoid the negative social impacts of declining resource quality. Thus modern technology has in a sense *increased* the sensitivity of economic activity to resource quality issues, for industrialization has shifted our dependence from direct solar energy and renewable fuel sources, whose rate of use is controlled by the solar flux but whose total supply is essentially infinite, to nonrenewable fossil fuels and minerals. Although the rate of exploitation of nonrenewable resources can be controlled to a degree by humans, the total supply has very definite physical limits which may eventually pull the economic rug out from under those who have become dependent on them. Because of humanities' dependence on many resources whose stocks will eventually run down, it is important for us to be able to identify, catalog, and assess empirically the types, quantities, and qualities of natural resources that potentially are available for human economic purposes and to examine which are and which are not subject to serious depletion.

It is important to realize that how one perceives the scarcity of natural resources dictates the importance attributed to resources in models of the economic process. In general, if resources are believed to be available in copious amounts at reasonable prices and will continue to be so in the future, then resources do not play a major role in models of production. Instead, labor and capital inputs are the focus since their availability is assumed to be more constrained and costly than resources. This approach assumes that our material standard of living is relatively insensitive to changes in natural resource quality. Conversely, if one assigns greater importance to the fact that the physical characteristics of resources ultimately determine their availability for human economic purposes, the modeling approach is quite different. In this approach, natural resources are the cornerstone of our material standard of living because the availability of capital and labor themselves are a function of resource quality and availability. The first approach characterizes standard neoclassical economic theory which assumes that technological advances and the market mechanism will overcome any physical constraints imposed by nature. The second approach is adopted by many physical and biological scientists who, while acknowledging the impressive achievements of technology, realize that in the final analysis it is nature that dictates the terms of our material standard of living. Physical limits are absolute limits. Economic problems are relative problems—we choose between alternatives that are physically and technologically feasible. As will be discussed below, both approaches have theoretical and practical advantages and disadvantages which limit their usefulness. When assessing resource quality issues, it is important to keep both approaches in mind and pay particular attention to the assumptions underlying each.

Among those who analyze natural resources, there is no consensus as to whether resources are increasing or decreasing in availability. In fact there is a wide range of opinions by experts from many fields who argue vehemently one side or the other on the issue of whether we even need to consider resource quality. In an extreme version of the standard economic view Simon (1981) would have us believe:

> . . . natural resources are not finite in any economic sense . . . resources will progressively become less scarce, . . . and will constitute a smaller proportion of our expenses (GNP) . . .

The view of Simon and other *resource optimists* is based, to varying degrees, on the pioneering work of Barnett and Morse (1963; see also Barnett, 1979), whom Simon quotes selectively. Their work, *Scarcity and Growth: The Economics of Natural Resource Availability,* has become the bible for resource and technological optimists and was the pioneering empirical work done on economic measures of natural resource quality. Barnett and Morse analyzed temporal trends in the quantity of capital and labor inputs per unit output across the entire extractive sector of the U.S. economy (forestry, fishing, agriculture, and mining). The authors found that labor and capital inputs per unit output remained stable or declined in all these industries (except forestry) over the period 1857 to 1957. The authors concluded that ''nature imposes particular scarcities, not an inescapable general scarcity.'' They believed:

> Advances in fundamental science have made it possible . . . [and] feasible, without preassignable limit, to escape the quantitative constraints imposed by the character of the earth's crust. A limit may exist, but it can be neither defined nor specified in economic terms.

According to Barnett and Morse and other resource analysts, resource depletion creates its own remedy. Society avoids potentially adverse effects of declining resource quality by developing resource-augmenting technologies and by substituting more abundant resources for scarcer ones. Price increases of the scarce material provide the stimulus for both these changes. Extreme advocates of this position envision the mining of ordinary country rock to meet future mineral needs at a cost not much greater than those incurred today (e.g., Brown, 1967; Brooks and Andrews, 1974; Goeller, 1979; Simon, 1981).

Taking the other side of the issue, some physical scientists including Brobst (1979) and Cook (1976) argue that geological and energetic constraints, not human ingenuity, ultimately determine the availability of nonrenewable resources. Brobst and Pratt (1973) state:

> . . . as the mining industry turns to lower and lower grades of many ores, the cost and availability of the required energy are probably the single most important factor that will ultimately determine whether or not a particular mineral deposit can be worked economically.

Similarly Cook (1976) observed:

> . . . [most ore and fossil fuel] deposits do not show a compensating increase in tonnage of reserves . . . as grade decreases . . . and that the energy or work cost of recovery increases exponentially with decreasing grade of the ore or with increasing cumulative recovery in the case of crude oil.

Resource management and economic policies based on one or the other of these two conflicting views of natural resource availability would clearly be vastly different. If resources continue to be available in large quantities at relatively low cost, as Barnett and Morse and Simon believe, increases in our material standard of living in the future would be limited only by advances in human technical capabilities. If, however, they are definite physical limits to our exploitation of nonrenewable resources, as Cook and Brobst and Pratt suggest, economic and political policies geared toward forcing economic growth under conditions of declining resource quality must eventually lead to further economic stagnation, unemployment, and inflation. These issues are implicit in many of our modern economic debates but are rarely considered explicitly. Natural resource quality is therefore an important, if not critical, issue for our society, for it in large part will determine the future direction of economic development in industrialized society.

MEASURING NATURAL RESOURCE QUALITY

The essence of resource depletion in any reasonably well-explored region is the movement from relatively cheap, high-quality deposits to higher cost, lower-quality deposits—which may or may not be more abundant. The problem of accurately tracing this movement is essentially a problem of defining and measuring what is meant by resource quality. Fisher (1979) suggests that any indicator of natural resource quality should reflect all the sacrifices (opportunity costs), both direct and indirect, made by society in order to obtain an additional unit of that resource. Most resource analysts would certainly agree with this basic statement and that, in general, higher-quality resources *cost* society less than lower-quality resources. Yet this general definition still begs a more analytically precise definition of resource quality and also a measure by which those sacrifices can be measured.

The energy and environmental events of the 1970s spurred a flurry of research by a variety of analysts in the area of natural resources and their role in economic production. Much of the recent theoretical and empirical work in this area has been focused on (1) deriving an unambiguous index of resource quality and (2) applying that index to particular resources to calculate past trends in their availability and also to estimate their future availability. As the preceding discussion suggested, there are two basic approaches to analyzing natural resource quality: economic and physical models. Resource economists employ measures including price, extraction cost, and resource rental rates to model resource availability. Physical scientists emphasize changes in the physical characteristics of resources and how those changes affect the cost and type of technology used to develop the resource. When applied over time to a particular resource, different indexes of resource quality generated by the two approaches often give conflicting signals about the resource's availability. In order to understand and evaluate such differences, it is necessary to study the models and assumptions of the economic and physical approaches to measuring resource quality.

Economic Measures of Quality

With few exceptions, neoclassical economists are very optimistic about the future supply of natural resources, seeing no foreseeable resource-related constraints to continued economic expansion. This optimism is based in large part on four factors which are hypothesized to mitigate any resource constraints. As summarized by Smith (1980) these factors are (1) as the quality of a resource declines, its price increases which in turn stimulates substitution of other resources; (2) such price increases are incentives for increased exploration for and recycling of the resource; (3) the physical occurrence of most resources is such that the greatest quantities exist in the lower-grade deposits; and (4) technological change continuously increases our ability to develop lower-grade deposits at reasonable cost, and also leads to the substitution of more abundant resources for less abundant ones. Whether they are explicitly stated or not, these assumptions underlie most of the resource quality models in the economic literature.

It should be noted that the first two factors are market mechanisms which occur in response to changes in resource price, much the same way the supply or demand of a regular good or service changes when its price changes. The third factor is purely physical in nature and can be analyzed only by examining the actual mode of occurrence and distribution of elements in the Earth's crust. The fourth factor is a technological issue, not an economic one, an area in which economists have no particular expertise advantage over other resource analysts. In many economic models, a causal link is assumed to exist between resource price increases and technological inventions which circumvent the decline in resource quality (Barnett and Morse, 1963; Solow, 1978). In retrospect, it is true that many technical advances appear to have occurred in response to economic problems, resource or otherwise. As Aage (1984) observes, however, the mere existence of an economic problem (i.e., resource depletion) is not a sufficient condition for its solution, let alone its most humane and effective solution. Price increases do not ensure technological feasibility. The latter is a function of the physical characteristics of the resource, and also the type and quantity of energy available to make that technology a reality. Technological feasibility is therefore not solely a function of market forces. It cannot be determined with price and supply and demand information. The existence of an economic problem is a necessary condition for its solution, while technological feasibility is a sufficient condition.

It is obvious that to consider technological change as a self-generating phenomena of the market mechanism is inappropriate, and that assertion of the price mechanism as the sole driving force behind all solutions to resource depletion problems is a gross oversimplification. Physical properties of the resource and technological issues are also important to consider. These two factors are treated in greater detail in the energy-based models of resource quality presented below and in Chapters 7–14. As we begin a more detailed analysis of economic and physical measures of scarcity, it is important to keep in mind that neither approach by itself yields complete and unequivocal results. Indeed, we will see that even when all the available information is considered, the issue of resource scarcity is not entirely resolved.

In economic theory an important distinction is made between the production of consumer goods and the production or extraction of nonrenewable resources. The market price of an ordinary good or service is determined by its marginal cost of production. Included in that cost to the producer is the foregone benefits he or she could have earned by selling their factory, machines, and so on, and investing that money in the market. The extraction of a unit of nonrenewable resource, however, involves an additional opportunity cost: the foregone services the resource could have provided at some future date. Economic theory of exhaustible resource use holds that resource owners are compensated for the fact that their product is finite and that extraction of a unit today leaves less to be extracted in the future. The price of a nonrenewable resource is therefore determined by its production cost plus the foregone future profits of extraction:

$$\text{price} = \text{marginal extraction cost} + \text{rent}$$

The foregone future profits of extraction are referred to as economic rent. It is not the same rent that a tenant pays a landlord as we commonly use the term, but rather a measure of the value of an untapped resource *in situ*.

Based on this relation there are three possible economic measures of resource quality: price, marginal extraction cost, and rent. Although in theory all three measures are calculatable, the first two have been used most frequently, owing to the difficulty of observing actual rental rates of natural resources. The most commonly used economic index of resource quality is the *unit extraction cost*. Devised initially by Barnett and Morse (1963), the unit cost index measures the quantity of capital and labor used to extract a unit of natural resource. Unit cost of extractive output was defined as $(\alpha L + \beta K)/Q$, where L is labor, K is reproducible capital, Q is extractive output, and α and β are weighting factors. As mentioned previously, Barnett and Morse applied this index to all the extractive sectors of the U.S. economy (mining, forestry, fishing, agriculture) from 1870 to 1957 and found little evidence for declining resource quality, except for forestry. Barnett (1979) and Johnson et al. (1980) extended and updated the original analyses to 1970 and concluded that the overall trend was the same: natural resources in general were not becoming scarcer (Figure 4.1).

The price of most natural resources over time indicates a similar trend (Figure 4.2). In this case what is of interest is not the market price of the resource itself but rather its price *relative* to some benchmark index such as the wage rate or the consumer price index. Relative resource price is more appropriate, for even if a resource's absolute price were increasing, it would be of little importance if the prices of all other goods were increasing at a more rapid rate.

The analyses of Barnett and Morse have received considerable attention from resource analysts and policymakers, and rightfully so, for it placed the issue of natural resource quality on firm empirical ground. Their analyses have been used, to varying degrees, as an intellectual springboard by many subsequent resource analysts who believe that resource limitations will in no way pose a threat to future economic growth (e.g., Kahn et al., 1976; Stiglitz, 1979; Simon, 1981). It is important to realize, however, that the conclusions of Barnett and Morse are not cast in stone. Both their methodologies and conclusions have been challenged by various researchers, including other resource economists. The current research on natural resource availability, having evolved in large part from Barnett and Morse, suggests that the issue is far more complex than Barnett and Morse originally envisioned.

A major problem with all measures of resource quality, both economic and physical, is that no single measure of resource quality is everywhere unambiguous. In the case of unit extraction cost, technological change that lowers extraction cost may

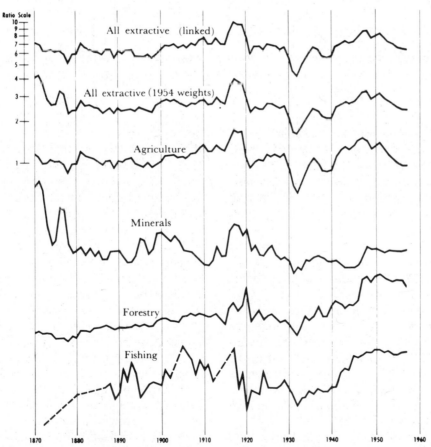

Figure 4.1. Trends in unit prices of extractive output relative to nonextractive output in the United States, 1870–1957. (From Barnett, 1979.)

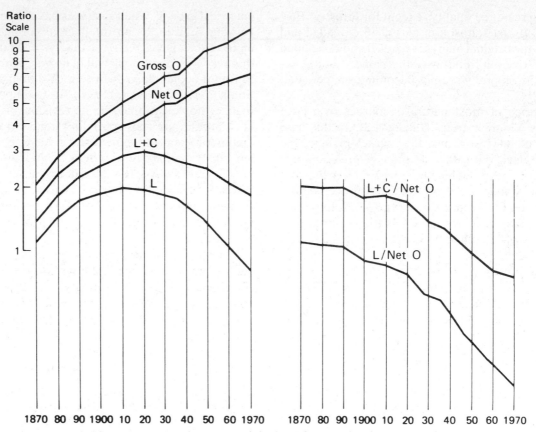

Figure 4.2. Labor and capital inputs, output, and cost per unit in the U.S. extractive sectors, 1870–1970: O = output, L = labor input, and C = capital input. (From Barnett, 1979.)

not warn us of impending physical exhaustion because it accounts for extraction costs only, not exploration and discovery costs. In other words, extraction costs reflect only the difficulty of getting a resource out of the ground, not the difficulty of finding a *new* replacement unit of the resource. Another problem with the unit cost measure is that it is not forward-looking; it provides no information on future extraction costs. For these and other reasons Brown and Field (1978) concluded that unit cost is an "ambiguous indicator of scarcity."

Relative price is generally preferred over cost as an index of scarcity because it is forward-looking to the degree that expected future costs of exploration, discovery, and extraction are included in the price. The principal drawback of price, however, is that the choice of the benchmark index is crucial because different indexes produce different indications of scarcity (Brown and Field, 1978).

Another problem with using cost and price data is that even when applied to the same resource, analysts using different weighting schemes or methodologies may arrive at conclusions different from the original Barnett and Morse analyses. Smith (1978, 1979) found no stable trends one way or the other using relative resource price data. Crosson (1979) found similar instability in agricultural prices from 1910 to 1978 using average cost data. In an update of Barnett and Morse's original work, Hall et al. (1981b) found increasing scarcity in the 1970s. Fisher (1981) concluded that we have already reached a "historical turning point . . . at which the resource base, after having effectively expanded for several decades, will begin to shrink." Smith (1980), after an extensive review of the subject, concluded that the "verdict on the adequacy of U.S. resources is not in" due to the inability of any single index to provide unambiguous results.

A problem related to the choice of a scarcity measure is the issue of *where* in the resource transformation process an index should be applied. Do we want an indication of the actual physical scarcity of the resource in nature, or rather the price consumers have to pay after the resource has been transformed into a good or service? The stage of processing one chooses affects not only the scarcity index most appropriate but also the trend in scarcity reflected by that index. Rent is some reflection of the quality of the resource *in situ*. On the other hand, production cost applies to the resource product after discovery, extraction, and some degree of processing. Applied to the same resource, these two indexes could give conflicting and ambiguous signals. For example, technological progress could decrease costs in the processing phase but could mask increasing scarcity in the extraction phase that would be reflected in rental rates. The supply of U.S. petroleum is an example of this problem. Since at least the 1940s petroleum deposits have becoming increasingly expensive to discover—whether measured in economic or physical terms (Chapter 7). At the same time the cost of petroleum products to the consumer declined, as measured by both the real price of motor fuel and the wellhead price of oil. This suggests that technical improvements in sectors downstream from the discovery phase (i.e., pumping, refining, and transportation) more than made up for increased oil discovery costs.

As measured by relative prices (Figure 4.2), most resources appear to have become increasingly more abundant, at least until the 1970s. Trends in mineral exploration and discovery costs, however, tell a different story. Cranstone and Martin (1973) found that marginal discovery costs per pound of aggregate metal in Canada increased by 3%/yr between 1946 and 1971 (Figure 4.3). The authors attributed the increase to the inability of technical advances in exploration to keep pace with the decline in prospect quality. Harris and Skinner (1982) stated that although such data are not readily available for the United States, the trend is probably toward even higher discovery costs because our mineral industry is in a more advanced stage of development. Nevertheless, real prices of metal after production declined during this same period of increasing discovery costs. Cranstone and Martin suggested that increases in the productivity of capital and labor downstream from the discovery process served to more than offset the deleterious effects of rising discovery costs, enabling prices to decline or remain steady. After reviewing these and other data Harris and Skinner (1982) concluded that the balance between increasing productivity and declining discoveries was probably only a temporary one and that metal prices must eventually rise. Another factor

influencing prices is aggregate demand, so that during economic recessions, as was the case for much of the decade after 1973, demand is relatively less, causing prices to soften.

The question of where to measure resource quality has no definitive answer, suggesting that information on all phases of resource transformation is desirable. Simon (1981) argues that what is really important is what consumers must pay for resources in the form of final demand goods and services. Simon further argues that increasing physical scarcity is unimportant if the final cost to consumers is prevented from increasing by technological change. Such an argument is not without merit. On the other hand, it could be argued that what is important is the cost of society of finding a *new* unit of a resource to replace the unit just consumed by society. As Davis (1980) observed, we have not solved the problem of declining oil reserves by increasing the efficiency of our pumping technology. What is of concern is the cost of finding a replacement unit or a comparable alternative energy source. Moreover, we should not be so quick to revel in our impressive technological advances. As Aage (1984) observed, the fact that depletion of high-quality deposits makes exploitation of less valuable deposits profitable does not constitute a solution to the scarcity problem, but only adaptation to a new, poorer situation.

It is appropriate then to gather as much data as possible when evaluating the quality of natural resources. Having discussed briefly various economic measures of scarcity, we next turn our attention to the evaluation of resource quality from a physical perspective.

Resource Quality from a Physical Perspective

The ultimate economic limit to the exploitation of nonrenewable resources can be defined in physical terms. For fuel resources (petroleum, coal, uranium, shale oil, etc.) the ultimate limit is the energy break-even point (EROI = 1:1; see page 28). When the energy required directly and indirectly to recover and process a fuel equals the energy released by the combustion of the fuel, then that resource effectively becomes a nonresource, at least as a fuel (but see p. 103 for a consideration of fuel quality). The limit for society as a whole is probably considerably higher than 1:1, for energy must be used to produce the machines that use energy, support workers, and so on.

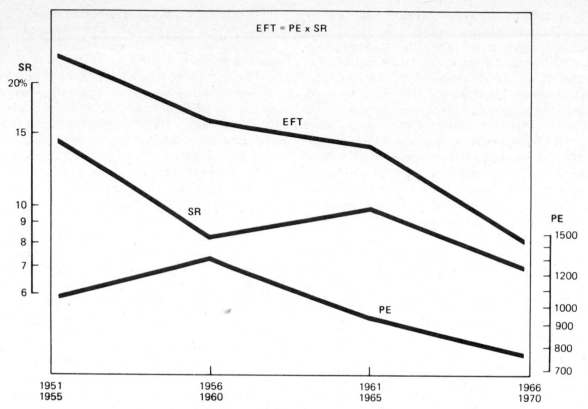

Figure 4.3. Trends in exploration and discovery costs for metals in Canada:

$$EFT = \frac{\text{value of discoveries}}{\text{expenditures for all exploration projects}} = \text{exploration effectiveness}$$

$$PE = \frac{\text{value of discoveries}}{\text{expenditure for project}} = \text{project effectiveness}$$

$$SR = \frac{\text{expenditure for successful projects}}{\text{expenditure for all projects}} = \text{success ratio}$$

(From Harris and Skinner, 1982; based on data from Cranstone and Martin, 1973.)

For nonfuel resources (copper, aluminum, gravel, etc.) the limit of exploitation is described by the energy opportunity cost of their extraction and beneficiation. As the energy costs of extraction rise, so does the amount of other goods and services that must be foregone elsewhere in the economy, for that energy can be used for only one task. At some point society may decide that the diversion of energy to the extraction of a resource is too large a sacrifice to make: the foregone benefits of other goods and services have become too large a price to pay for extracting another unit of the resource. At this point the resource effectively becomes a non-resource.

As Cook (1976) observed, before energy limits are reached, one of two other limits can and usually

do come into play. The first is resource substitution. When the EROI of petroleum declines to a level below that of alternative fuels such as shale oil, the less dollar and energy-costly resource will be substituted for petroleum, provided it is available in sufficient quantities. Copper, for example, has long been used in the communications industry due to its high conductivity and resistance to corrosion. The energy costs of producing copper, however, have risen sharply due in large part to the depletion of high-grade copper ore deposits. The energy opportunity cost of copper in this application has increased to the point where previously uneconomic, higher dollar and energy cost materials like aluminum are now cost competitive with copper for some electrical uses.

8

Cook (1976) cites examples where, for various cultural reasons, certain societies were unwilling to pay the energy costs of extraction, even when the EROI was greater than unity. An example of this type of limit might be an unwillingness of society to pay the environmental costs of disturbing vast amounts of landscape in order to mine a low grade mineral.

Therefore resource quality can be analyzed in energy terms. The sacrifices referred to by Fisher that society must make are economic sacrifices: economic work must be done to locate, extract, and process natural resources. Economic sacrifices can be represented by energy opportunity costs. As high-grade ores are depleted, more economic work must be done per unit of resource, producing a necessary diversion of economic work from the production of other goods and services. Because economic energy must be expended to perform economic work, it is the diversion of energy from other tasks that must be sacrificed. This increased energy cost may or may not be offset to some degree by technological improvements in efficiency.

We define *natural resource quality* in terms of the amount of economic energies required to locate, extract, concentrate, refine, and otherwise upgrade a unit of resource to an economically useful state. A decline in resource quality implies an increase in the amount of economic energies required to make available a unit of the resource. For nonfuel resources the index used is the mass (tons) of resource extracted per unit of energy invested in the extraction process—such as tons of refined copper metal per million kilocalories used. For fuels the appropriate measure is energy return on investment (EROI), the ratio of fuel made available by a fuel supply process to the total energy used to extract and make available that resource—for example, kilocalories of petroleum per kilocalorie invested. Higher-quality resources require less economic energy per unit extracted and processed than lower-quality resources. Thus a given resource in nature is valuable because of both its quantity and quality.

It should be emphasized that physical considerations such as the grade of a particular ore deposit or the energy required to locate and extract a barrel of oil do not in and of themselves determine when the use of a particular resource will be abandoned in favor of a substitute. Rather, society collectively decides just how high a price it is willing to pay (i.e., what dollar and energy opportunity costs it is willing to make) for the use of a given resource. Physical considerations in large part determine what that price to society will be by dictating the type and cost of the technology required to develop the resource. They also influence the form and direction of change in the technology used to develop and utilize the resource. Information on the physical characteristics of resources and the economic energies required to develop resources are therefore useful in evaluating the real cost to society of changes in resource quality.

Economic measures of labor and capital inputs per unit output of resource—such as price, cost, and rent—do not always assess the role that energy plays in the extractive sectors and may hide some real changes in scarcity. Thus, although Barnett and Morse (1963) and Barnett (1979) showed clearly that the dollar value of capital and labor inputs per unit output decreased in agriculture and mining (Figure 4.1), due primarily to technical advances in those industries, the question must be asked as to *how* those technical advances worked to reduce the amount of capital and labor inputs in these and other natural resource sectors. Were they due simply to the ingenuity of human minds, manifested in technological advances as Barnett and Morse and their colleagues suggested? Georgescu-Roegen (1975) provides us with the answer:

> . . . a great stride in technological progress cannot materialize unless the corresponding innovation is followed by a great mineralogical expansion . . . Still more important, all mineralogical discoveries have included a substantial portion of *easily* accessible resources . . . Energy thus becoming cheaper, substitution-innovations (technological change) have caused the ratio of labor to net output to decline. Capital also must have evolved toward forms which cost less but use more energy to achieve the same result.

The trend in the use of human labor in extractive sectors is similar to that in the entire economy presented earlier (see Figure 2.6). The proportion of total economic energy used in the extractive sectors accounted for by fossil fuels and electricity has increased dramatically, whereas the relative contribution of the physical work of labor has declined. With more fuel available to direct toward his or her task, each laborer could extract more material per unit time, and the productivity of labor increased dramatically (Figure 4.4). Thus much of the "breathtaking" decline in the labor cost per unit of natural resource which Simon (1981) and others present as an indicator of increasing resource quality has been due rather to the substitution of cheap, abundant fuel for labor in the extraction process. One cannot

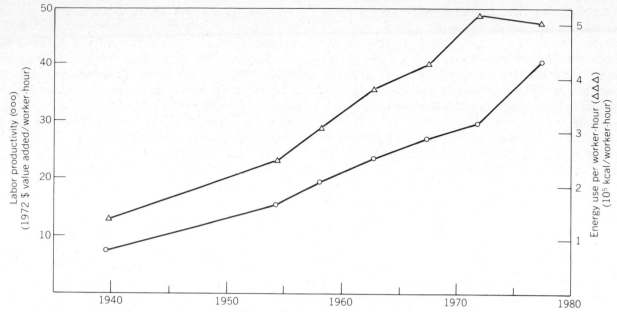

Figure 4.4.   The link between direct fuel use per laborer and the productivity of labor in the U.S. extractive sectors, 1939–1977. Fuel use includes fossil fuels and quality-corrected electricity. (Data from Census of Mineral Industries.)

deny that, in a metaphoric sense, technological change has created resources by developing new methods for developing lower-grade deposits that were previously too expensive. It was, however, human ingenuity as manifested in technologies designed to use increasing amounts of fuel that enabled total extractive output and capital and labor productivity to increase while the average grade of many resources steadily declined. This was a perfectly logical and inevitable trend while the price of fuel declined relative to labor. Our ability to mitigate the potentially adverse effects of declining deposit grade, however, has been reduced substantially now that fuel is no longer as abundant or cheap. By measuring only the contribution of capital and labor in dollar terms, as did Barnett and Morse, one cannot substantiate the claim that technological progress will continually counter declining resource quality by making even lower grades of ores economically feasible to exploit. The contribution of energy must be considered as well.

PHYSICAL PATTERNS OF RESOURCE AVAILABILITY AND ENERGY REQUIREMENTS FOR MINING

From an energy perspective there are several definite patterns that occur in the development of natural resources for human economic purposes. These trends are outlined here and discussed in greater detail in sections to follow.

1. Geological factors ultimately determine the amount of economic energy that must be used to develop a resource deposit. Human technical capabilities or other economic factors may limit the development of a resource at a particular time or location, but they too are ultimately determined by the physical characteristics of the resource. As was mentioned, short of these limits, other economic or social factors can and often do determine the limits of resource development.

2. In any reasonably well-explored region, as the United States is for many nonrenewable resources, the highest-grade deposits tend to be developed first because they are the most profitable. This is followed by a gradual decline in the average grade of deposit mined. Over time the energy costs per kilocalorie of fuel or per ton of mineral generally increase except when frontier regions are first opened to development, or where the resources occur abundantly in high concentrations (e.g., sand, stone, gravel) (Figures 4.5 and 4.6). For many metals the relationship between ore grade and fuel requirements is an inverse exponential one,

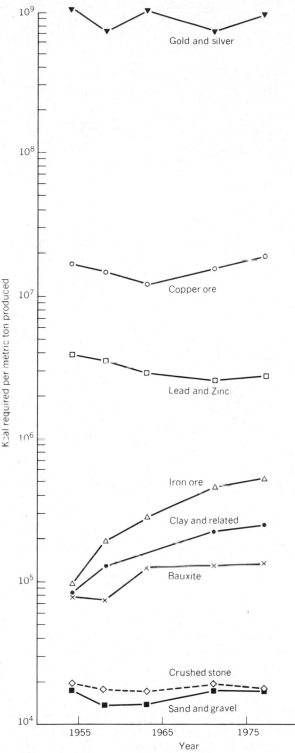

Figure 4.5. Trends over time in direct fuel and electricity use per ton of output for some metal and nonmetal minerals. Direct electricity use was multiplied by 3. (Data on mineral production from Bureau of Mines, *Minerals Yearbook*. Data on fuel use from Bureau of Census, Census of Mineral Industries, 1954, 1958, 1963, 1967, 1973, 1977.)

meaning that as ore grade goes down linearly the energy cost increases exponentially. Metals tend to be more fuel intensive than nonmetals, and scarcer elements (as measured by their crustal abundance) tend to be more fuel intensive than more abundant elements.

3. Technical improvements in the mining sectors, with few exceptions, have evolved toward processes that replace and subsidize human labor with fuel in the extraction process.

Types of Resources

Nonfuel resources exist in both renewable and nonrenewable forms (Figure 4.7). Nonrenewables include mineral resources, both metal (copper, bauxite, titanium, etc.) and nonmetals (sand and gravel, clay, chemical fertilizers, etc.). Renewable resources include forestry, fishery, and agricultural products, as well as a wide variety of nonmarketed environmental services such as clean air, water, and wind. In this section we examine more closely how trends in the quality and availability of these resources are evaluated in terms of the energy required to extract and upgrade them to an economically useful state. Nonrenewable resources are considered first, followed by renewables. The energy costs of fuel resources will be discussed individually at length in Chapters 7–14.

ENERGY COSTS OF NONRENEWABLE RESOURCES

Considerations of Geologic Availability

Geologic availability is the ultimate determinant of mineral potential because there is no economic availability if there is no geologic availability first (Brobst and Pratt, 1973). Advances in technology can create economic reserves from the resource base, but we cannot physically alter the amount or characteristics of our mineral endowment. An assessment of the energy costs of extracting minerals from the environment must begin with a discussion of the geologic factors that control the formation and distribution of minerals in the Earth's crust. Too often appraisals of mineral availability are limited to considerations of economic availability

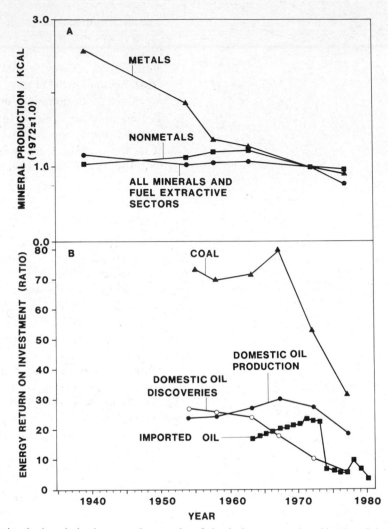

Figure 4.6. Trends over time in the relation between the quantity of physical output produced by the mining sectors of the economy to fuel and electricity use by those sectors. Values are conservative in that indirect energy, costs of labor and environmental disruption are not included. Data for domestic oil for 1982 shows a continuation of the downward trend while imported oil has recovered slightly. (From Cleveland et al., 1984.)

alone. The supply of a particular mineral at a particular place and time does depend on economic factors: the demand for the mineral, the cost of fuel, capital and labor to the supplier, the current state of extraction technology, and so on. Considering only the economic factors that influence mineral availability, however, often leads to the erroneous generalization that technological advances and mineral price increases will provide the means and stimulus for developing near-infinite quantities of increasingly lower-grade deposits. Geologic factors ultimately determine the economic potential of a mineral by determining the very existence, grade, depth of burial, and mode of occurrence of mineral

deposits. Geologic factors therefore affect directly the economic energies we must invest to develop a particular mineral deposit, and this is an important component of price and hence economic availability.

A discussion of geologic availability must begin with a distinction between the terms resources and reserves. The U.S. Geological Survey (USGS) defines a *resource* as a concentration of naturally occurring solid, liquid, or gaseous material in the Earth's crust in such a form and amount that economic extraction of a commodity from the concentration is currently or potentially feasible. The *reserve base* is that part of an identified resource that

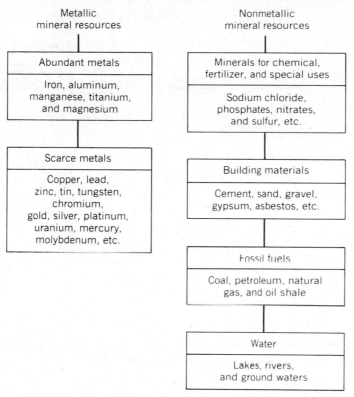

Figure 4.7. Kinds of nonfuel and mineral resources. (From Skinner, 1976b.)

meets minimum physical and chemical criteria related to current mining and production practices. Actual *reserves* are the part of the reserve base that could be economically, technically, and legally extracted at the time of determination. The relationship between categories of resources is shown schematically in Figure 4.8. This resource classification scheme was first proposed by McKelvey (1972) and, with modifications (USGS, 1980), has been adopted by the USGS as its official method of cataloging the nation's resources. One important feature of Figure 4.8 is that a given mineral deposit can move from resource to reserve status, or vice versa, depending on both economic and geologic conditions. For example, as the grade of a particular mineral declines, the energy costs of extracting and processing it may become so large that the deposit no longer produces a net profit in energy and/or dollar terms. Then the deposit may move from reserve to resource status, or even drop from resource status altogether. Conversely, price increases or technological advances can make previously uneconomic deposits profitable or feasible to develop. The sudden oil price increases in 1973 and again in 1979 moved many already discovered, but small, oil fields from resource to reserve status, or from indicated to measured reserves.

Brobst (1979) has expanded the standard resource classification scheme to include the relation of resources to noneconomic deposits (Figure 4.9). The inclusion of several new boundaries should be noted. Most important is the mineralogical threshold, where for many metals the energy costs of extraction increase sharply. Mineralogical threshold refers to the minimum natural conditions that permit formation of separate particles of specific minerals of the type that could be mined and processed by current methods (Brobst, 1979). As the concentration of chemical elements decreases to levels below that which allows the formation of separate particles of ore minerals, the elements are dispersed in the crystalline structure of other kinds of minerals that comprise the ordinary rocks that make up most of the Earth's crust. At this level of concentration elements are said to occur at their *crustal abundance,* or *Clarke concentration.* Simply put, crustal abundance refers to the concentration of a particular element in an average rock of the Earth's crust. Table 4.1 lists the crustal abundances for some important minerals. To recover minerals occurring at

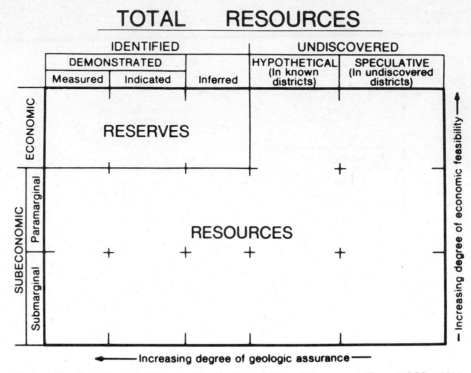

Figure 4.8. U.S. Geological Survey classification of Natural Resources. (From USGS, 1976.)

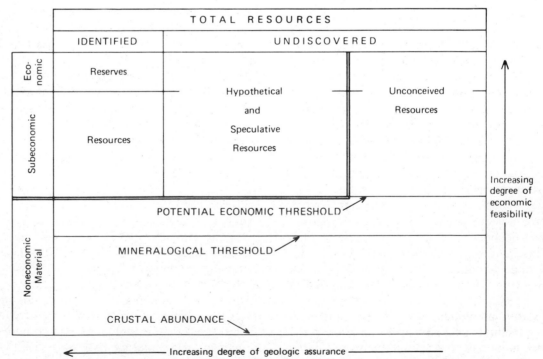

Figure 4.9. The relation of resources to noneconomic mineral materials. (From Brobst, 1979.)

Table 4.1.   Abundance of Some Mineral Resources in the Earth's Crust

Element	Percent by Weight of Earth's Crust
Abundant Elements	
Oxygen	46.60
Silicon	27.72
Aluminum	8.13
Iron	5.00
Calcium	3.63
Sodium	2.83
Potassium	2.59
Magnesium	2.09
Titanium	0.44
Hydrogen	0.14
Total	99.17
Scarce Elements	
Copper	0.007
Zinc	0.008
Lead	0.0016
Silver	0.00001
Gold	0.0000005

Source: B. Mason, 1958. *Principles of Geochemistry,* Wiley, New York.

their crustal abundance, very large investments of energy are required to break down the crystalline structure of the host rock.

Many, but by no means all, minerals exist in vast quantities at their crustal abundances (Table 4.1). The large amounts of minerals occurring at low levels of concentration suggest to some that human mineral needs of the future will be met by mining vast tonnages of average crustal rock or seawater (Brown, 1967; Goeller, 1979). This possibility must be analyzed in terms of the energy requirements to perform such operations.

Figure 4.10 is our resource classification scheme based on the direct and indirect energy costs of finding and extracting resources. Those deposits in the *reserve* base can be extracted with relatively minor inputs of energy. Moving to *marginal* resources, energy costs begin to rise until the mineralogical threshold is reached, at which point energy costs rise sharply. This suggests that meeting society's increasing needs by mining minerals at concentrations approaching their crustal abun-

dances will not be possible without the development of very large and relatively cheap additional fuel sources.

Technology plays an important role of course in the relation between resource quality and the energy costs of recovery. As a natural resource becomes economically important, a period of rapid technical change ensues as more is learned about exploiting that resource. But mature industries often have mature technologies, and the large capital investments in place associated with that technology make it difficult to produce large changes in efficiency rapidly, especially when existing capacity is not fully utilized. Although technology can on occasion keep up with or stay ahead of the rising energy costs of extracting lower-quality deposits (e.g., as apparently happened for lead and zinc from 1952 to 1972, Figure 4.5), it must nevertheless deal with the increased quantity of work required to recover the resource because more material must be processed per unit of final product. In addition, more wastes (air pollution, mine tailings, etc.) are produced per unit of production, which generally increase both direct and indirect energy and environmental costs.

The energy costs of developing lower-grade mineral deposits is determined in large part by their concentration in the Earth's crust. The *geochemically abundant* elements, which comprise over 99% by weight of the Earth's crust (Table 4.1), are found as small quantities in virtually every rock. A glance at Table 4.1 indicates, however, that some elements essential to our industrial economy are present in only minute amounts relative to the abundant elements. Elements with crustal abundances less than 0.01% are called *geochemically scarce* elements and include important industrial commodities such as copper, lead, zinc, nickel, gold, silver, cobalt, tin, and platinum. Scarce metals such as copper and zinc would rarely exceed concentrations of a few grams per ton of rock if they were evenly distributed throughout the Earth's crust (Parker, 1967).

Scarce elements rarely form separate minerals. Rather, they normally exist in the structures of common rock-forming minerals, usually the silicates, with an atom of scarce metal substituting for an atom of an abundant element (Skinner, 1976). To recover a scarce element occurring in this form, the entire structure of the host rock must be physically and chemically broken down, an extremely energy-intensive process. On very rare occasions, however, the right geological and geochemical condi-

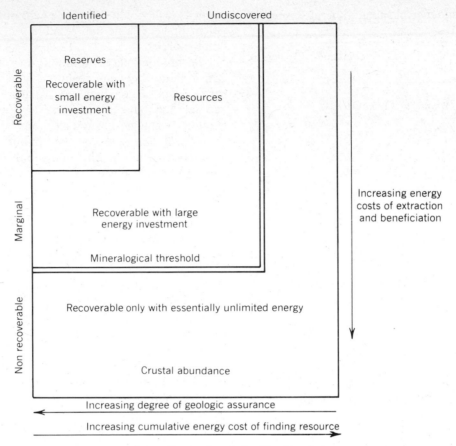

Figure 4.10. Energy-based classification of resources.

tions exist for scarce elements to form separate minerals, which can then be mined using conventional mining and beneficiation procedures. Less than 0.1% of the scarce metals in the Earth's crust occur other than as atomic substitutes in common minerals (Skinner, 1976).

The distinction between scarce and abundant minerals is important because their different modes of occurrence affect directly their energy costs of recovery. Figure 4.11 represents Skinner's (1976) idea of the distribution of geochemically abundant metals like iron, aluminum, and titanium. Because abundant metals are relatively widespread in common rock, many of their present ore deposits decrease gradually in grade into average rock. This general relationship for abundant elements is termed a grade-tonnage ratio, and was formalized initially by Lasky (1950) who concluded

> It may be stated that as a general principle that in many mineral deposits in which there is a gradation from relatively rich to relatively lean material, the tonnage increases at a constant geometric rate as the grade decreases.

Lasky's grade-tonnage formula was based on his analysis of porphyry copper deposits. Despite his warning that he was not ready to "claim it applies to every example of a particular type," Lasky's grade-tonnage ratio has experienced wholesale use by many to argue that there exists huge amounts of undiscovered, low-grade deposits. Recall, for example, that one of the four scarcity-mitigating factors cited by economists was the alleged great abundance of most minerals at very low grades. This is an assumption, not a universal truism, based in large part on a selective interpretation of Lasky's original work.

In fact, analyses by Singer et al. (1975) and Whitney (1975) suggest that the occurrence of large tonnage, very-low-grade deposits may not be typical of all elements as is commonly assumed. In their analyses of the massive sulfide-type copper deposits, Singer et al. included not only past data on ore

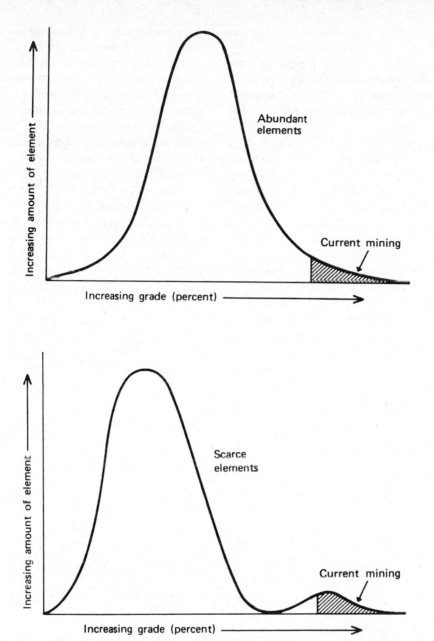

Figure 4.11. Skinner's (1976a) conception of the possible geochemical distribution of scarce and abundant elements. (Brian J. Skinner, *Earth Resources,* 2nd Ed., © 1976, p. 10. Reprinted by permission of Prentice-Hall, Inc., Englewood Cliffs, N.J.)

grade and production as Lasky did but also estimates of future availability of copper ore at expected grades and tonnages for the three most common forms of copper ore. The authors found no significant correlation between tonnage and grade for porphyry or strata-bound copper deposits, but a significant negative correlation between grade and tonnage for massive sulfide copper deposits. Singer et al. concluded that the often hidden assumption of

large amounts of low-grade ore in many mineral supply models could have a "devastating effect on the orderly supply of minerals if large tonnage–low grade deposits do not exist."

Similarly, Whitney (1975) found that 70% of the copper metal in North America porphyry deposits exists in grades above 0.7% copper. This indicates that lower-grade deposits do not contain increasingly larger amounts of copper, but just the opposite. Lovering (1969) found that in the case of mercury, total mercury metal in the numerous low-grade, but large tonnage deposits would be only one-half that of the smaller high-grade deposits. These and other analyses suggest that questions of mineral supply are best addressed on a case-by-case basis.

The mode of occurrence for geochemically scarce elements is much different than for abundant elements. Figure 4.11 is Skinner's (1976) conception of the possible bimodel distribution of scarce elements. As was noted before, scarce elements rarely form separate minerals. Instead, they exist as randomly distributed atoms trapped by isomorphous substitution in minerals of geochemically abundant elements. The area beneath the small hump in Figure 4.11 represents the rare amounts of a scarce metal that actually occur as separate minerals in rock. The large hump represents the amount of metal trapped within the crystal cages of abundant silicate minerals. The region between the two humps in Figure 4.11 is a mineralogical barrier be-

yond which any further reduction in the grade of ore mined entails significantly increased energy costs.

The difference between cost occurrences of a scarce element and that amount which rarely forms separate minerals is an example of a physical discontinuity that radically alters the cost of developing the resource. Copper, for example, forms separate minerals at a concentration of about 0.1%, below which the energy costs of mining and processing copper ore jump by an order of magnitude (Figure 4.12, Table 4.2). The recovery of copper from common silicate rocks has vastly greater energy requirements due to the larger volume of material that needs to be processed, and also because the copper must be separated from the atomic structure of the silicate. Normal smelting techniques cannot be used with these very-low-grade deposits. Harris and Skinner (1982) stated that almost every scarce element has the same type of mineralogical barrier shown in Figure 4.12, where energy costs can be expected to increase by one to two orders of magnitude. The authors concluded that considering the

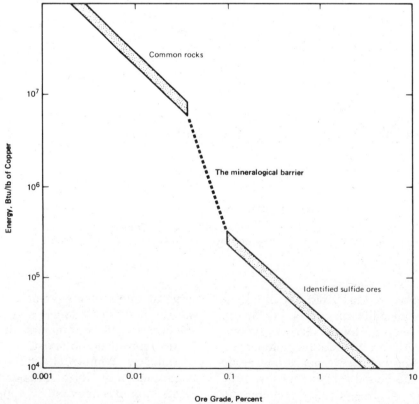

Figure 4.12.   The mineralogical barrier that technology must overcome to produce copper from ordinary rocks. All scarce metals have the same type of barrier. (From Harris and Skinner, 1982.)

Table 4.2.   Energy Used to Mine and Process Copper Ore in which Copper Is Present as the Mineral Chalcopyrite (CuFeS$_2$)

	Energy Used (Btu/lb of copper)		
Grade, % Cu	0.70	0.10	0.01
Mining plus concentration	33,040	231,280	2,312,800
Smelting and refining	20,000	20,000	20,000
Total	53,040	251,280	2,332,800
Equivalent thermal energy in bituminous coal (lb of coal)	4.1	19.3	180

Source: D. P. Harris and B. J. Skinner, 1982. The assessment of long-term supplies of minerals, pp. 247–326, in U. K. Smith and J. V. Krutilla (Ed.), *Explorations in Natural Resource Economics,* Johns Hopkins University Press, Baltimore.

Note: An overburden stripping of 2.5 tons of waste rock per ton of ore is assumed, and concentration is 80% efficient.

"entire crust or any portion of it as a future resource simply cannot be countenanced with any hope that it will come to pass."

The existence of a mineralized threshold appears to have been confirmed for copper metal by the National Academy of Sciences Committee on Mineral Resources and the Environment (COMRATE, 1975). The report found that only when copper metal has been concentrated to at least 0.1% do individual copper minerals begin to form as separate mineral constituents. It also estimates that no more than 0.01% of the total copper in the continental crust will be found in ore bodies of 0.1% copper or less. Thus a grade of 0.1% appears to be the mineralogical threshold for copper metal. At present the mean grade of copper mined in the United States is about 0.5%.

Goeller (1979) states that even when society is forced to use very-low-grade resources, "energy and other costs will not increase by a factor of more than about two because near infinite resources are not too different in either quality or concentration from presently used ores." Although this may be the case for abundant metals like iron ore and aluminum, it is clearly not the case for scarce metals like copper as the preceding discussion indicates. The mineralogical threshold is an example of a physical discontinuity that precludes the existence of a smooth and continuous gradation between high- and low-quality deposits, a gradation that would allow for gradual and predictable cost increases and the development of new technologies to offset those cost increases. The existence of steep geochemical gradients at the margins of many ore and fossil fuel deposits implies a steplike increase in costs at the point of the physical discontinuity. This suggests that mining some minerals from country rock is unlikely to occur without the development of cheap and virtually unlimited fuel supplies, itself an unlikely occurrence.

Energy Cost of Nonfuel Minerals

The ease with which a mineral deposit is extracted and purified is related directly to the amount of natural energies used in the past to form the deposit. The higher the degree of concentration by natural energies, the less economic energies humans must use to extract and process the mineral matter contained in the deposit. The theoretical minimum amount of energy or work required to extract a mineral from its ore is represented by its Gibbs free-energy change. For example, the Gibbs free-energy change associated with the extraction of copper metal from sulfide ore is the minimum amount of energy required to convert 1 mole of the reactants to products in the reaction

$$Cu_2S + O_2 \rightarrow 2Cu + SO_2$$

Table 4.3 lists the Gibbs free energy associated with various grades of three important industrial metals. Because the free-energy changes for the metals listed in Table 4.3 are negative, in theory we should be able to extract metal *and* energy from the ores. In practice, however, these ores are not sources of fuel as well as metal because of the large amounts of economic energy that must be used in extracting the

Table 4.3.   Gibbs Free-Energy Changes for Three Important Industrial Metals

Ore Grade	(kcal/g ore)	G (kcal/g mineral)[a]	(kcal/g metal)[b]
Iron (%Fe)			
5	0	0	0
10	−0.41	−2.87	−4.1
15	−0.95	−4.43	−6.37
17.5	−1.29	−5.15	−7.37
20	−1.66	−5.80	−8.28
25	−2.47	−6.91	−9.88
30	−3.44	−8.02	−11.47
Aluminum (%Al)			
8.13	0	0	0
10	−0.53	−1.83	−5.3
15	−2.49	−5.75	−16.6
20	−5.16	−8.93	−25.8
25	−8.54	−11.83	−34.16
30	−13.61	−15.71	−45.38
32.52	−15.57	−16.57	−47.88
Copper (%Cu)			
0.007	0	0	0
0.010	−0.0005	−1.75	−5.0
0.025	−0.0043	−6.02	−17.2
0.05	−0.0091	−6.37	−18.2
0.075	−0.0153	−7.14	−20.4
0.10	−0.0244	−8.54	−24.4
0.25	−0.0793	−11.10	−31.72
0.50	−0.1701	−11.91	−34.02
0.75	−0.2647	−12.35	−35.29
0.98	−0.3952	−14.11	−40.33

Source: M. W. Gilliland, 1982. Embodied energy studies of metal and fuel minerals, in M. J. Lavine and T. J. Butler (Eds.), *Use of Embodied Energy Values to Price Environmental Factors: Examining the Embodied Energy/ Dollar Relationship*. Report to the National Science Foundation. PEA-8003845.

[a] Minerals are Fe_2O_3, $Al(OH)_3$, and $CuFeS_2$ for the three cases.
[b] Metals are Fe, Al, and Cu for the three cases.

mineral particles from the ore. In this case the theoretical energy requirements are of little use in predicting the much larger empirical energy requirements.*

Economic energy is used by the extractive sectors to extract and purify minerals from ores in two general steps. In the *mining* phase the ore is extracted from the ground and transported to a processing center. In the *beneficiation* phase the mineral is purified by mechanical and/or chemical processing to the concentration necessary for the particular use it is intended. Within each general stage there are a number of intermediate stages, some of which are common to all minerals and others of which are mineral specific. Most minerals, for example, must be mined via underground or surface mining techniques. Other techniques, like the calcination of aluminum oxide in the preparation of aluminum metal from bauxite, tend to be mineral specific.

*The free energy of fuel minerals is, in effect, used when we extract and process those minerals, but it is a small amount of energy compared to the free energy of fossil fuels. Joyce (1979) calculated that amount of energy for the U.S. economy in 1967, and found that the quanity of free energy derived from nonfuel minerals was less than 1% of fossil fuel use. Such energies operate essentially to reduce somewhat the fossil fuel cost of beneficiation.

The energy costs of producing a mineral in a concentrated form therefore depend on the mineral being considered and are best understood by analyzing each mineral individually and empirically. Table 4.4 lists the direct fuel requirements associated with the extraction and beneficiation of some essential minerals. The production of copper metal will be used as an example of how energy is used to extract and upgrade a mineral resource.

Copper Ores Industry

Copper is one of the most important nonferrous metals in industrialized economies. Its high electrical conductivity, corrosion resistance, good malleability, and nonmagnetic properties make copper especially valuable in electrical applications. About one-half of all copper used is for electrical purposes (Bravard et al., 1975). Copper's availability for economic purposes, however, is severely limited by its physical scarcity. On a weight basis, copper represents only 0.007% of the Earth's crust, a condition that places distinct energetic limits to the economic exploitation of copper.

Most domestic copper is mined from sulfide ores. Sulfide ores also contain small but recoverable quantities of silver, gold, platinum, selenium, and tellurium. The extraction and purification of copper metal from sulfide ores can be divided into three steps: mining and milling, smelting, and refining. The mining and milling phase uses the most fuel of three general stages of copper production. Over 85% of domestically mined copper ore is extracted by surface mining techniques. After the ore body has been drilled, blasted, and excavated, it is transported to the beneficiation plant which concentrates the ore to between 20 and 35% copper metal. In this stage the copper minerals are separated from the gangue (waste rock) by several methods. The ore is first crushed and ground, and then agitated with water in what is called the froth flotation operation. The copper and other minerals rise to the surface of the liquid mixture and are removed.

The copper concentrate is then sent to a reverberatory furnace in the beginning of the smelting phase. In the furnace most of the remaining iron oxide in the ore is removed in the form of slag, producing a matte of 30–45% copper metal. The matte is then heated in air (smelted), which further oxidizes any remaining sulfur and iron. The product is a blister copper which is 98–99% pure.

The blister copper produced by the smelting operation is still too impure for most electrical applications of copper and is oxidized further in an anode refining furnace. In the final steps of the refining stage, copper is electrolytically refined to cathode copper, which is 99.9% + copper. The cathodes are then sold and/or melted and reshaped into other desirable forms (wirebar, wirerod, etc.). Battelle Laboratories (1975) estimated that in the United States in the early 1970s, it took about 28 million kcal to extract and purify copper ore into one net ton of 99.9% refined copper (Figure 4.13). This estimate represents only direct fuel and electricity use and is therefore a minimum estimate of the energy costs of producing copper.

The energy costs of extracting and purifying copper as outlined in Figure 4.13 are not constant. Changes in the energy costs of copper metal are due in large part to changes in the average grade of copper ore mined (Kellogg, 1974; Page and Creasy, 1975). Most of the increased energy costs associated with lower-grade ores occur in the mining and milling stage of production. In particular, the energy costs of grinding ore, which constitutes about one-half of all energy used in the mining and milling stage, is very sensitive to the grade of ore. Lower-grade ores also have higher waste-to-ore ratios, which necessitate large amounts of energy to haul nonmetal ore waste. The 30% decline in copper ore grade between 1958 and 1972 had associated with it a 43% increase in the amount of material hauled per ton of ore (Bureau of Mines, *Minerals Yearbook*, various years). This translates into a greater degree of environmental disruption associated with mining larger quantities of low-quality ore.

The Relationship between Ore Grade and Energy Costs

The relationship between ore grade and energy costs is an inverse exponential one for some, possibly most, metals (Page and Creasy, 1975). The average grade of copper ore mined in the United States has declined substantially throughout most of this century (Figure 4.14). As a result the energy costs of extracting copper ore and upgrading it to pure copper metal have increased (Figure 4.15). The energy cost per ton of refined copper will probably continue to increase in the future, due to the need to extract more ore per ton of copper metal and also due to continued increases in the waste-to-ore ratio.

The sharp increases in energy requirements per ton of metal associated with relatively small decreases in ore grade for copper and other important

Table 4.4.   Energy Costs of Extraction and Beneficiation for Some Important Minerals

Metal	Main Source	Direct Fuel Cost[a] (10^6 kcal/ton)
Copper	1% sulfide ore	13.0
	0.3% sulfide ore	25.6
	98% scrap recycle	0.7
	Impure Cu scrap recycle	1.3
Aluminum	50% Bauxite	54.9
	30% Bauxite	62.7
	Clays	67.3
	Anorthosite	74.3
	Aluminum scrap recycle	1.1
Iron	High-grade hematite	36.9
	Magnetic taconite	40.7
	Specular hematite	44.6
	Nonmagnetic taconite	46.3
	Iron laterites	54.4
	Iron and steel scrap recycle	14.3
Magnesium	Seawater	89.2
	Mg scrap recycle	1.6
Titanium	High-grade rutile ore	133.1
	Ilmenite rocks	160.0
	High Ti soils	224.9
	Ti scrap metal recycle	45.1

Source: J. C. Bravard, H. B. Flora, and C. Portal, 1972. Energy Expenditure Associated with the Production and Recycling of Metals. Oak Ridge National Laboratory. ORNL-NSF-EP-24.

[a] Includes only fuel and electricity requirements. Indirect energy costs of capital and labor are excluded.

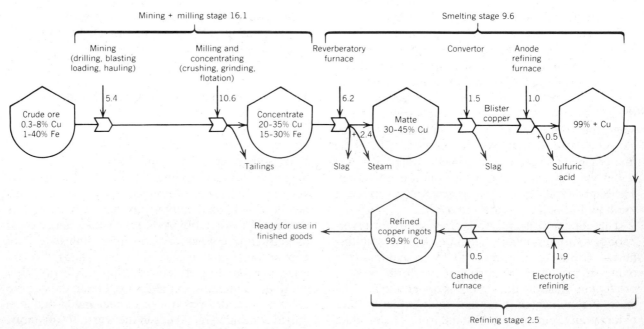

Figure 4.13.   Diagram of the energy costs of producing refined copper metal from sulfide ore. Numbers represent energy used in each step in millions of kilocalories. (Data from Battelle, 1975.)

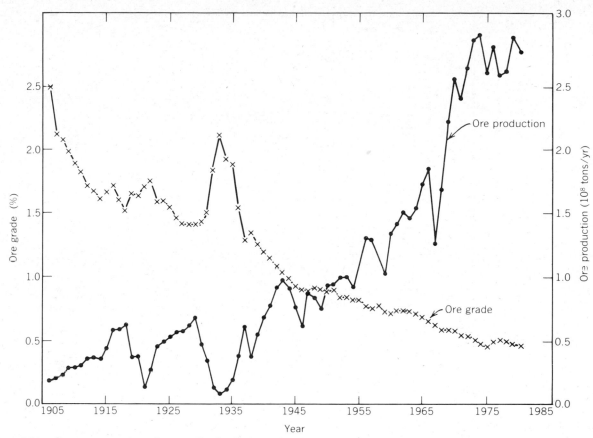

Figure 4.14.   Copper ore grade and ore production in the United States. (Data from U.S. Bureau of Mines, *Minerals Yearbook.*)

metals may well determine the minimum ore grade that can be mined economically. The relation between ore grade and energy use for the processing of metals has been formalized by Page and Creasy (1975):

$$E_t = \frac{E_m}{g} + E_s \qquad (4.1)$$

where E_t is the total fuel required to process 1 ton of metal from its ore, E_m is the fuel required to mine and mill 1 ton of ore, E_s is the fuel required to smelt and refine the concentrate to produce 1 ton of metal, and g is the grade of the ore. The fuel required for smelting and refining (E_s) is relatively constant and independent of fluctuations in ore grade because mill concentrates used in most smelting and refining operations have a fixed or narrow range of mineral compositions. Conversely, the energy required for mining and milling (E_m) is strongly influenced by ore grade. As the ore grade for a particular metal declines, fuel requirements increase slowly at first until mining and milling fuel require-

ments approach those for smelting and refining, after which total fuel use increases rapidly due to larger mining and milling fuel costs (Page and Creasy, 1975). Figure 4.15 shows the evolution of the energy requirements for deriving three metals using Equation (4.1).

The direct energy costs of some important minerals other than copper have also increased over time due to the depletion of high-grade deposits (Figure 4.5). For U.S. metals as a whole physical output of metal per unit of fuel declined 60% between 1939 and 1977 (Figure 4.6; Cleveland et al., 1984). For the entire mining sector of the economy (metals, nonmetals, coal, and petroleum) output per energy input declined by about one-third during this period. There are of course some important exceptions to the trend of increasing unit energy costs of extraction. Kakela (1978) found that for Lake Superior iron ores technological advances in the industry more than offset the decline in ore grade, leading to a reduction in total energy costs per ton of iron ore extracted. Improvements in the pelleti-

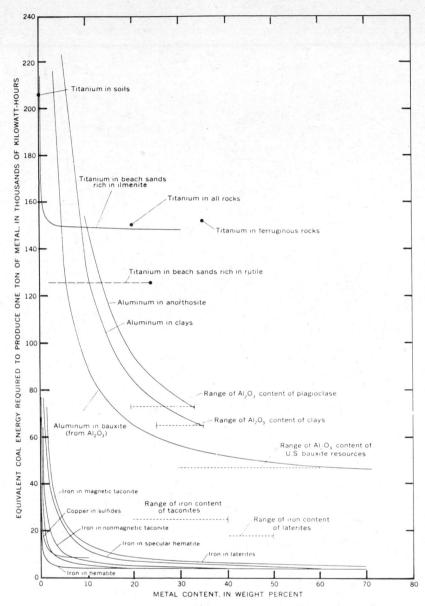

Figure 4.15. Direct energy requirements for recovery of iron, aluminum, and titanium at different grades. (From Page and Creasy, 1975.)

zation of iron ore from taconite decreased direct fuel requirements compared to the use of naturally occurring, more concentrated hematite ores despite the fact that taconite is only about 25% iron compared to 50% in hematite and geotite ores. The gains occurred because of the energy content of the taconite ore itself and because pelletization increased the efficiency of heating.

The taconite example illustrates how technical improvements can decrease unit energy costs in some instances. The overall trends for the nation,

however, are quite clear (Figures 4.5 and 4.6). Improvements like the one described by Kakela have been more than offset by the decline in quality of other minerals. Over the mining sector as a whole, technological advances have dampened but not in any sense countered the rate of increase in the unit energy costs of extraction. The net impact of declining grade is that higher-grade foreign resources are often cheaper, even given transportation costs and the continued abundance of low-grade resources in the United States. This of course moderates the impact of declining domestic grade on price, but also increases our dependence on foreign countries for our own economic well being.

CHANGES IN QUALITY WITH RATES OF EXPLOITATION

As we have seen the energy cost for recovering a unit of nonrenewable resource tends to increase over time—starting from near the time of first major industrial exploitation and continuing until physical or economic exhaustion—as the resource becomes depleted. This phenomenon is caused by the physical properties of the resource deposit and because humans have traditionally exploited the highest grades first.

Another mechanism that operates concurrently with the secular decline in resource quality relates to the rate of exploitation at a given point in time. In general, the higher the rate of exploitation at any given time, the lower the return to that effort. This effect is observable for both nonrenewable and renewable resources. Note in Figure 4.14 that when demand for copper decreased during the Depression, the average grade of copper ore rose as only the mines with the highest grade, and hence most profitable, ores were kept open. The converse occurred during the war when increased demand reopened mines with lower-quality ores, decreasing the average grade of ore. We show later that this inverse relation between the effort expended to develop a resource and the returns to that effort is also true for petroleum and uranium (pp. 179 and 274). Thus increasing demand at a point in time above some threshold level often forces exploitation of poorer-quality reserves.

Exploitation Rates and Renewable Resources

Although changes in the rate of exploitation at a given point in time affect extraction quality of nonrenewable resources, these changes have their greatest impact on renewable resources, such as agricultural goods or fish (see Chapters 18 and 20). For the purpose of this discussion, renewable resources are defined as materials (i.e., groundwater) or organisms (i.e., fish) that replenish or reproduce on time scales of the same order as the rate at which people use them. Since their rate of reformation is influenced by the rate at which they are harvested,

it is important to examine mechanisms by which changes in the rate of exploitation affect their production quality in the future.

If a resource is used more rapidly than it is reformed, its abundance declines, and more energy must be expended to recover the resource from a shrinking or lower-grade pool. For practical purposes, then, a renewable resource exploited at a rate faster than it can replenish itself is in effect a nonrenewable resource. As will be described in detail in Chapter 18, the production history of U.S. fisheries illustrates the relation between the exploitation rate at any point in time and future production quality. From 1950 to about 1978 the total catch of U.S. commercial fisheries remained about constant while the energy used to catch them increased by two- to threefold. As a result the kilocalories of fish caught per kilocalorie of fossil fuel used to catch them decreased from a ratio of about 0.13 kcal of fish per kilocalorie of fuel to 0.07 kcal per kilocalorie. Much of this decline in quality was caused by overfishing, which forced fishermen to travel farther and trawl longer for the same catch, although it is possible that some of the decline was due to the large increase in foreign fishing effort.

Another reason for the frequently observed decline in the average quality of renewable resources extracted is related to the expansion of the resource base necessitated by the increased effort needed when demand increases. In other words, when production is high, a larger proportion of low-quality resources must be used. This phenomenon often occurs for agriculture, forestry, rangeland, water, and other renewable resources, and is examined in Chapters 19–22. For example, if agricultural production in a settled nation is relatively small, then only the best grade, most economic lands will be used. If food production is increased, then more and more marginal land will be brought into production. Since, in general, this steeper, shallower-soiled, nutrient-poorer, and/or drier soil is less productive, more and steeper hectares must be plowed and more fertilizers and more water must be used for irrigation to produce a bushel of crop. Increasing food demand thereby increases the per bushel energy cost of most crops, at least relative to the trend that would have occurred if demand had been constant. On the other hand, the general increase of production that has occurred on all lands due to increased fertilizer use has meant that we can grow the food we need on less land and hence need to use less of our marginal land than would have been the case without the increasing industrialization of agriculture. This also has been an important factor in encouraging forest and wildlife conservation on marginal lands.

Another reason that increased effort often does not elicit a corresponding increase in yield is the asymptotic or saturation relation that is commonly observed in biological systems. This relation is formulated empirically by the Michelis–Menten relation, which describes the response of biological systems to many inputs. For example, the response of physiological processes to enzyme concentrations or plant growth to phosphorus concentrations is asymptotic, and as the saturation point is reached, output per unit of input decreases—an example of diminishing return on investment.

This relation is complemented by Liebig's law of the minimum, which states that many biological processes are limited by the substance that is in least supply compared with other materials and relative to the requirements of the organism. As a result increasing other substances without increasing the limiting substance will have very little effect on the process. For example, Redfield (1958) summarized extensive data that showed that oceanic plants require carbon, nitrogen, and phosphorus in the ratio of 106:16:1. Nitrogen is often the limiting nutrient, although it need not be the one present in the smallest absolute quantity. For example, in an environment of 10 atoms of nitrogen to one of phosphorus, nitrogen is likely to be limiting.

The complementarity of the Michelis–Menten relation and the Liebigian law of the limiting nutrient is especially important in agriculture, where crops need a very specific mix of materials and conditions, such as water, nutrients, soil texture, and light, to achieve specific yields. By changing the relative supply of each of these inputs, different materials may become limiting at different times. The relation of the yield of rice relative to the use of nitrogen fertilizer is an example of diminishing returns on investment in agriculture (Pimentel et al., 1974). At first, adding 15 kg of nitrogen per hectare increases the yield of rice by 350 kg, but the next additional 15 kg per hectare increases the yield by a smaller amount (Figure 4.16). Finally, the last addition of nitrogen fertilizer has very little effect on yield—some other factor has become limiting. Nitrogenous fertilizers, which are responsible for much of the increased productivity associated with the Green Revolution, can be very effective at relatively low levels of application but ineffective at high rates of application. As we have noted, however, energy is flexible, so that as when nitrogen becomes no longer limiting, energy resources can be switched to supply phosphorus, water, or other inputs. But there are ultimate limits to even this as we show in Chapter 6.

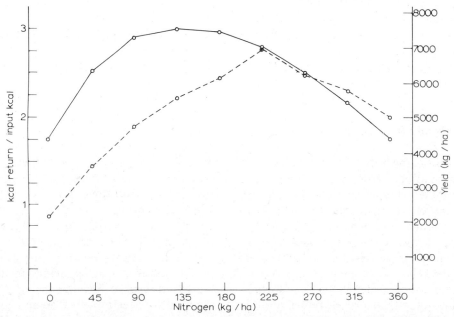

Figure 4.16.   The relation between rice yields (dashed line) and nitrogen fertilization. (From Pimentel et al., 1974.)

SUMMARY

Resource quality can be examined in both economic and physical terms. In the past we have relied mainly on economic measures of resource quality like cost, price, and rent. Technology and physical factors, however, are equally important because they interact with economic factors to determine the rate and pattern of resource utilization. Physical factors, which ultimately determine the type of technology adopted and the cost to society of developing a resource, are at least as important to consider in a comprehensive assessment of resource quality. The technologies we adopted that enabled us to maintain constant or declining dollar costs for resources often required ever-increasing amounts of direct and indirect fuel. Now that fuels themselves are becoming scarcer, this luxury becomes a costly necessity, requiring that increasing proportions of our national income be diverted to the resource-processing sectors in order to supply the same quantity of resource (Figure 4.17).

APPENDIX 4.1: FOSSIL FUEL QUALITY: PETROLEUM IS SPECIAL!

There are clearly many sources of energy available, and in the United States we have progressively developed higher and higher quality resources for our major industrial energy supply, from wood and feed for horses to coal to petroleum. A number of other energy sources have been proposed as substitutes for oil as it is increasingly depleted. Oil, however, has a number of characteristics that would make a shift to other energy sources difficult, and in some cases impossible, for many specific uses. First, oil is a liquid, and has a high energy content per unit volume. This makes oil especially useful for transportation purposes (in trucks, tractors, ships, etc.) and allows oil to be shipped relatively easily to points distant from oil fields or refineries. Second, oil is a versatile fuel: it can be used for a wide variety of purposes (for transportation, electricity generation, manufacturing, heating, etc., as well as for chemical feedstocks) with only moderate capital investment. The cost of adapting the world's machines from oil to coal burning is impossible to estimate, but it certainly would be enormous. Third, oil is special in that mining, shipping, and burning oil is relatively clean, especially when compared to coal. Finally, and perhaps most important, oil still can be

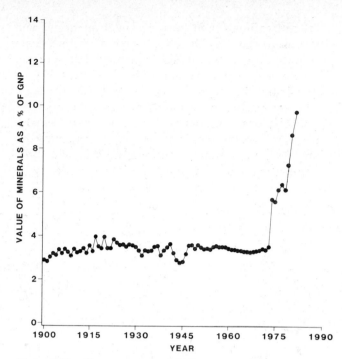

Figure 4.17. The proportion of GNP accounted for by fuel and nonfuel mineral purchases from 1900 to 1981. (From Cleveland et al., 1984.)

mined in much of the world with a relatively small energy investment compared to the energy produced and compared to almost any alternative fuel. As a result of all three factors a great deal of our existing capital equipment is designed for oil use, and one of the reasons that oil is special is that we have developed a society dependent on inexpensive oil, and until recently there has been no particular incentive to develop alternatives.

Gaseous petroleum is also special for it is very clean and inexpensive to produce and burn, although it is not especially suitable for transportation because it is difficult to store and transfer small quantities cheaply and safely. Although the shipping of gas is not physically difficult, it is very capital expensive, requiring the construction of massive pipelines. Yet once the investments in pipelines are made, the shipping of gas is not energy intensive. Natural gas is especially useful for fertilizer production.

Coal has few of the advantages of oil and gas aside from its high energy content per unit mass. It is especially dirty, due to its high sulfur content and to particles resulting from combustion. Coal is not useful for most transportation systems because it is difficult to ignite and burn in small quantities and because it requires elaborate machinery for its use. In addition it is bulky and hence difficult to store, handle, and load into combustion chambers. On the other hand, once coal is converted to electricity, it is a very useful and valuable energy source.

Other forms of energy are not normally as flexible as petroleum and frequently have only one principal use. For example, there is no practical application for nuclear energy other than in electricity generation, and solar energy can be used almost exclusively only for space heating and, at a very high price, production of electricity. That is not to say that such uses are not extremely important or that energy produced from other forms cannot free oil or gas for alternate uses, but it does point out the very special nature of oil. It was abundant. Cheap oil was in large part responsible for making the United States a wealthy nation. Finding substitutes will not be easy.

5

ENERGY AND MANUFACTURING

INTRODUCTION

Chapter 2 showed how fuel, capital, labor, and technology are different forms of economic energies and how they are combined in the production process to yield finished goods and services. The purpose of this chapter is twofold: (1) to examine in greater detail how energy is used to produce and maintain fuel, capital, and labor and (2) to present several methods for calculating the embodied energy cost of goods and services.

THE STAGES OF PRODUCTION

In order to calculate the energy cost of a good or service, the energy used in each step of production must be identified and measured. The economic energies of fuel, capital, labor, and other natural resources are used to produce economic goods and services from raw materials in three general steps: *extraction,* in which natural resources are recovered from the environment and refined to raw materials, *processing,* in which raw materials are converted into intermediate goods, and *fabrication,* in which the intermediate goods are manufactured into final goods (see Figure 2.1). In each of these steps economic energies are used to upgrade the organizational state of the preexisting materials. The energies used to produce goods and services are referred to as being *embodied* in the object.

Extraction, processing, and fabrication are done by a myriad of industries, each of which is classified according to the type of product it produces (Standard Industrial Classification, SIC).* Extracting natural resources and refining them to raw materials accounts for a significant percentage of the eco-

*The interested reader is referred to the *Standard Industrial Classification Manual* (1972), which describes in detail how the various industries are classified. Most federal government economic statistics are gathered and reported by SIC classifications.

mineral extraction sectors of the U.S. economy—comprised of coal, petroleum, metal, and nonmetal mining—used 19% of all fuel consumed by the manufacturing sector.

Once extracted and processed, most raw materials still are not in a form amenable for final consumption by the consumer. For example, bauxite must be refined and rolled, drawn, or extruded into intermediate forms, such as sheets, fuel, or plates, by aluminum foundries. In 1977 intermediate processing of aluminum used 4.5 times more energy per dollar of output than the mining of bauxite itself.

In the third and final step, intermediate goods are combined to produce final demand goods and services. For example, refined aluminum can be processed further into metal cans, parts for airplanes, buildings or cars, or aluminum foil. When calculating the energy cost of goods and services, it is often hard to make the distinction between industries that produce intermediate products versus final products without input–output tables, which trace the flow of materials from one sector to another throughout the economy.

Finished goods are transported to wholesalers and retailers who market these goods using fuel directly for lighting, heat, and transportation and indirectly to produce the buildings and other capital structures and equipment associated with the storage and sale of their goods. Interstate trucks, for example, used an average of 5110 kcal to move 1 ton of freight 1 mile in 1974 (Aerospace Corp., 1977). Energy used for such purposes normally is classified as commercial or transportation energies.

ENERGY REQUIREMENTS FOR FUEL, CAPITAL, AND LABOR

To calculate the total quantity of fuel energy used in the extraction, processing, and fabrication of an economic good or service, we must be able to quantify in energy terms the fuel, capital, and labor used by each industrial sector. The energy equivalents of the factors of production can be calculated in one of several rather straightforward methods to be discussed shortly, using readily available data on annual flows of fuels and other materials through the economy.

Fuel

The energy costs associated with a fuel have two components: the quantity of energy released on combustion (enthalpy) and the energy required to extract, process, and deliver it to the customer. The first component is called the *chemical energy* of the fuel and usually is measured by the heat given off by combusting the fuel. For example, a 42-gal barrel of refined crude oil releases approximately 1.5 million kcal of heat when combusted.

The second component of energy consists of *processing* energy, the energy used by an energy transformation process to secure additional energy supplies. The sum of all such processing energies is the embodied energy cost of a fuel. The embodied energy in a barrel of oil includes the energies required to produce and operate drilling rigs, pipelines, and refineries, as well as the energies used to house and support the geophysicists and other individuals who work in the oil business. For example, in 1977 the United States used about 4.8 quadrillion kcal of fuel to recover, process, and make available in an economically useful form about 94 quadrillion kcal of fuel. Based on this relation, 0.05 units of fuel had to be invested in 1977 to make available 1.0 additional unit of fuel to the rest of society.

It is important to note that it is a fuel's enthalpy, or heat of combustion, that makes it economically useful. A fuel's embodied energy is *not* a thermodynamic property of that fuel. Rather, it is an accounting procedure that measures the direct plus indirect energy content of fuel. When a fuel is combusted, only the potential energy stored in its chemical bonds is able to do economic work. Nevertheless, both the embodied energy of a fuel and its heat of combustion are energy costs of commodities whose production requires the combustion of fuel.

Capital

Capital can be viewed as economic energy invested in machines and tools that replace or empower hu-

man labor, thereby making workers more productive and, ultimately, allowing for higher wages. Capital can be used to do economic work that is difficult and/or dangerous for human labor to do alone. Capital also can be used to reduce the quantity of time, fuel, or labor used to produce a good or to provide a service. The energy equivalent of a unit of capital equipment normally is calculated as the quantity of fuel energy used to produce and maintain it. Alternatively, the energy cost of capital can be thought of as the fuel energy or labor energy saved by capital equipment at the margin (Baumol and Wolf, 1981).

The energy required to produce a dollar's worth of capital equipment for a given industry is calculated by summing all the fuel used to produce each year's new capital according to the appropriate standard industrial classifications. For example, in 1979 SIC sector 3553 used 3.1 trillion kcal of fuel (at both the site of manufacture and embodied in the materials it purchased from other sectors) to produce $727 million worth of woodworking machinery. There were therefore 4264 kcal of fuel embodied in each dollar's worth of woodworking machinery produced in 1979. If a particular industry purchased $1 million of this type of machinery, the embodied energy of that equipment would be about 4.3 billion kcal.

The economic usefulness of capital tends to diminish as it wears out, requiring increasing amounts of energy to maintain and repair it. Eventually, capital equipment becomes so worn out that it must be discarded altogether. The rate at which capital depreciates or becomes less useful is not constant but depends in part on the rate at which the capital is used to do economic work, the type of work it performs, and the rate at which it is economically efficient to replace. Although a piece of capital continues to degrade even when idle, the rate at which it wears out increases as its rate of doing economic work increases.

Energy analysts often assume that capital degrades at a constant rate and that this rate is caused by, and spread evenly over, each unit produced by a machine. Based on this assumption, one method for estimating the embodied energy in capital used to produce a good or service is to divide the total energy cost of capital used in production by the number or value of goods and services that the capital produces over its lifetime. This ratio is the quantity of embodied energy used to produce one unit of a good or service. Unfortunately this procedure would be a difficult and tedious task for every piece of capital equipment used to produce a good or service.

A second method used in most energy analyses assumes that industries replace their capital at a more or less constant rate. The monetary value of the capital bought by an industry per year can be converted to its energy equivalent as an estimate for the rate at which capital degrades or is replaced within the industry. This quantity of embodied energy is prorated or spread evenly across the number or monetary value of units produced per year. If an industry is expanding or contracting its capital infrastructure, this method can lead to errors by charging the energy cost of capital expansion or contraction on current production, which introduces an unknown but probably, on average, small error into the analysis.

Human Labor

Labor normally is not viewed as being *produced* like capital. Instead, it is generally assumed that labor is an exogenous input to production—that is, neoclassical economists assume that labor inputs are available for use in economic production without considering how they got there. An alternative explanation is that like any other factor of production, labor has direct and indirect energy costs of production. Households produce and support human labor just as firms produce capital. In doing so, households invest energy and other resources to produce and maintain labor in its economic role. Therefore, labor also has an energy cost associated with its use. These energy costs can be separated into three components: (1) the caloric value of the food the worker consumes, (2) the embodied energy of that food (i.e., the direct plus indirect fuel used to produce food), and (3) the fuel purchased with the wages and salaries of labor. Obviously, there are important differences between human labor and other factors, but this does not alter the fact that labor requires a continuous input of energy to sustain itself.

Biological Energy Equivalent of Labor

Human labor does mechanical work, and like any other work process the amount of work done can be measured by the quantity of fuel consumed. This quantity of fuel can be measured directly by a respirometer, a device that measures the rate at which oxygen is combined with food to produce carbon dioxide and work. The amount of fuel used depends in part on the type of work being done. For example, a person doing desk work uses about 71 kcal/hr whereas a manual laborer uses about twice that quantity (Lehninger, 1975).

Energy Embodied in Food

A comprehensive analysis of the energy equivalent of labor would include the economic energy required to grow the food metabolized by the worker. Chapter 6 will detail how energy is used in modern agricultural systems. In 1970 the U.S. agricultural and food-processing system used about 9.5 kcal of economic energy to produce 1 kcal of edible energy (Steinhart and Steinhart, 1974). Based on this use of fuel, the embodied energy cost of food used by human labor could be estimated as the quantity of fuel used to grow and distribute the amount of food consumed by each person: 1354 kcal/hr based on the average American's diet of 3400 kcal/day in 1980. Some methods for energy cost accounting have used this biological measure as the energy equivalent of human labor (see Odum, 1971).

Embodied Energy of Income

A still more comprehensive assessment of the energy cost of labor would account for all the direct and indirect uses of fuel that are used to produce the goods and services a worker buys with his or her paycheck. This is a measure of the energy cost of producing and maintaining labor in the household sector. This method overestimates the energy cost of labor because it assumes that all the energy used to support a laborer's paycheck is necessary to produce and maintain the laborer in his or her economic role. Certainly there is a portion of almost everyone's income that, if taken away, would not reduce his or her ability to perform on the job (although it would certainly make the laborer less well off). Equally as certain, however, a portion of wages and salaries exists beyond that required for basic needs (food, shelter, clothing) that, if removed, would diminish a laborer's economic performance. For example, a business woman must receive a salary sufficient to cover the costs of food, shelter, and clothing for her and her family to survive. She also needs enough income to purchase transportation to and from her job. She may also subscribe to several journals in order to stay abreast of advances in her particular field. The reader can see that the list could be expanded substantially.

Suffice it to say that a portion, and probably a substantial portion, of the energy represented by a laborer's paycheck is a necessary and unavoidable energy cost of producing labor. And, as most businesses recognize, if skilled labor is to be secured for one's own business, competitive wages must be paid, requiring a substantial portion of the nation's total energy resource toward giving meaning to those paychecks.

When calculating the energy costs of goods and services, the energy cost of labor should not include all the fuel and nonfuel goods and services bought by a worker's paycheck because hiring a new worker does not always create new demands for goods and services. It may instead only increase existing demand—that is, people buy fuel and nonfuel goods and services whether they are employed or not. The impact of a new worker on the demand for fuel, which is implicit in the demand for goods, can be estimated by examining the energy used to support an employed laborer versus an unemployed laborer. For example, in 1979 the medium income of families in the United States with one employed wage earner was $15,607, whereas the medium income of $7648 for households with no wage earner was 49% less. Although the percentage of income spent on direct versus indirect purchases of fuel varies with income (see Landsberg and Dukert, 1981), a rough approximation is that the employed earner used about 51% more fuel and bought 51% more goods and services than did an unemployed laborer. Based on this assumption the energy cost to society of having an average new laborer can be estimated as 51% of the fuel represented by all dollar purchases of that employed laborer. Since some new jobs are filled not by unemployed workers but by individuals who change jobs, the estimate for the energy equivalent of labor must be corrected for employment levels.

If, on the other hand, the total amount of energy used by society is approximately constant, as it was in the U.S. from 1973 to 1982, each new worker divides up existing supplies of economic energy into smaller parts. Thus the exact amount of energy that should be attributed to labor is a complex issue without an unequivocal solution. Because of these problems we do not include an energy cost of labor for most of the calculations used in this book. This produces conservative estimates of the energy costs that we give. Future research will probably give us more precise ranges, although high and low estimates (i.e., with and without labor) can presently be estimated in a relatively straightforward manner (e.g., see Hall et al., 1979 and Table 14.3).

METHODS OF ENERGY COST ACCOUNTING

There are two general methods of calculating the embodied energy cost of goods and services: process analysis and input–output analysis. Most of the economic data used in these approaches are based on information collected and distributed by the U.S. Department of Commerce and published in the *Annual Survey of Manufacturers* and *Census of Manufacturers*. This information includes estimates of the value of products shipped, value added, purchases of raw materials, investment in capital equipment, labor hours worked, payroll, and energy used by each industry for over 450 industrial sectors (depending on the year) of U.S. manufacturers.

Both process and input–output analyses condense the use of energy in production into two general types: fuel burned at the site of manufacture (*direct* fuel use), and fuel burned in other sectors to produce the materials purchased and used as inputs at the site of manufacture (*indirect* fuel use). Process analysis and energy input–output analysis, and variants of each, differ in the way the flow of material is traced through the manufacturing process, the types of energy costs included in the analysis (e.g., just fuel, or fuel, capital, and labor), and the energy equivalents assigned to the three factors of production. As a result the methods give somewhat different values for embodied energy even when applied to the same set of data.

Process Analysis

In theory, process analysis provides the most detailed information on the energy cost of goods and services. There are several practical problems, however, such as data limitations, that effectively limit its applicability. Process analysis assesses the energy used directly in each successive step of the production of a good or service. Depending on the data available, energy requirements may be calculated per unit mass or per dollar value of the input. Consider Figure 5.1, which depicts schematically how the energy cost of producing a power plant might be calculated by the process analysis approach. The same general format would apply to any good or service. Direct energy costs of a power plant refer to the fuel burned at the site of production—diesel fuel burned by cement trucks and cranes, electricity and gas used by welders, and so on. Indirect energy costs are those incurred in the production of the steel, cement, and other raw materials. Going one step further, indirect energy costs also include the energy used to mine the iron ore used to make the structural steel of the power plant. In general, roughly half of the embodied energy of a good or service is used at the site of production (level *I* of Figure 5.1). The more indirect the energy cost becomes, the smaller its contribution to the total energy cost of the product. This leads to one of the problems with process energy analysis, namely,

where should the system boundaries be drawn? This is referred to as the *truncation problem* because there is no standard procedure for determining the level at which indirect energy costs become small enough to neglect.

Energy Input–Output Analysis

The energy input–output approach is more comprehensive than process analysis and is analogous to and derived from the input–output matrix used in standard economic analyses. The economist Wassily Leontief developed economic input–output analysis in the United States during World War II

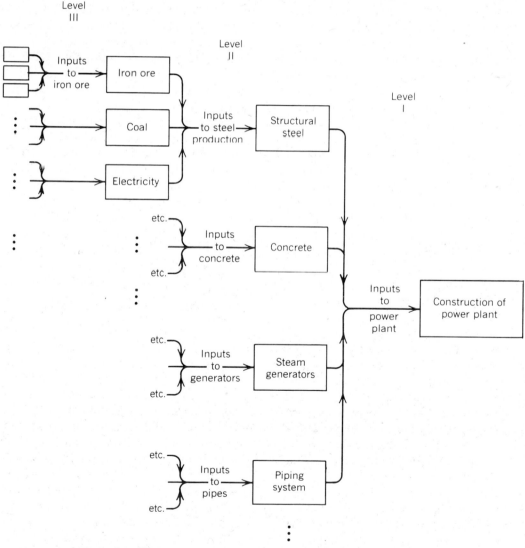

Figure 5.1. Process energy analysis of the construction of a power plant.

when the government was concerned with preventing shortages of materials critical to the war effort. Because they provided a very detailed breakdown of interindustry transactions, input–output tables were useful in determining how much consumption of critical materials in the household sector would need to be reduced in order for there to be sufficient raw materials to build tanks, airplanes, and other military equipment. Planned economies such as the Soviet Union use input–output tables as the basis for their economic planning.

Herendeen and Bullard (1975), Hannon et al. (1981), and others at the Energy Research Group at the University of Illinois modified the economic input–output table published by the Bureau of Commerce—which are based on dollar flows—to energy input–output tables based on embodied energy flows between industries. Energy input–output tables have been developed for the years 1963, 1967, and 1972,* the only years in which the requisite economic input–output tables have been published. The input–output table breaks the economy into about 400 different sectors. The numbers in the table represent the quantity of direct plus indirect energy that each industry purchases from all the other sectors in order to manufacture its product. Combined with the dollar flows between industries, the energy intensity factor (kcal/$) of each good or service can be calculated (Figure 5.2). These results give a comprehensive and reasonably accurate representation of both the direct and indirect energies used to manufacture a product. The input–output approach does not suffer from the truncation problem of process analysis. By the way the data are collected, and the energy intensities calculated, in the input–output methodology, all levels of energy costs are included: direct fuel use, and the indirect fuel embodied in intermediate goods purchased by each sector. The energy input–output table developed by the Energy Research Group provides the most comprehensive empirical assessment of the direct and indirect energy costs of goods and services and, as such, stands as a monument to the assessment of energy use in the U.S. economy.

Hall et al. (1979a, b) modified the energy input–output analysis so that they did not need to specify the specific upstream sector(s) from which a sector producing final goods purchased materials. Instead, their method is based on a very aggregated estimate of the energy embodied in the intermediate goods

purchased at the site of final manufacture. As a result their estimate of energy embodied in a dollar's worth of purchased materials is the same regardless of what intermediate goods were purchased. They used three different methods that reflect different assumptions about that source(s) of intermediate goods. One method assumes that all materials purchased by sectors producing final demand come from extractive sectors (e.g., metal or mineral industries; SIC sectors 10–14), so that the energy used per dollar value added by extractive industries is used as an estimate of the energy embodied in a dollar's worth of purchased materials. Since the input–output tables show that all manufacturing sectors receive some materials from nearly all other manufacturing sectors, a second method sets the energy embodied in materials as the energy used per dollar value added by all manufacturing sectors. The last method developed by Hall et al. (1979a, b) estimates the energy embodied in materials as the ratio of U.S. fuel use to nominal GNP. These three methods are thought to give a high, low, and medium estimate for upstream energy costs, respectively.

Although the method developed by Hall et al. (1979) is not as comprehensive as the Herendeen and Bullard (1975) analysis, it can be calculated relatively easily from data published annually. As a result estimates made using this method are much more up to date and include many more years than the Herendeen and Bullard method. The estimates for embodied energy calculated by the two methods differ by less than 10 percent for most final demand goods (Hall et al., 1979b; Kaufmann and Hall, 1981).

Costanza (1980) modified the boundaries of the traditional input–output tables so that fuel energy (including solar energy) would be the only net input to the economy, with labor and government classified as internal transactions within the economic system and gross capital formation as the principal output. Costanza estimated the quantity of fuel energy used to support labor and government services—as well as that used directly for production—and divided this amount of fuel use by employee compensation and indirect business taxes to estimate the energy used to support a dollar's worth of labor and government service, respectively. Double counting was avoided by constraining the analysis to use only actual total energy use. In other words the energy used directly by labor was included, but not the energy embodied in goods and services purchased by labor.

*1977 will be available soon.

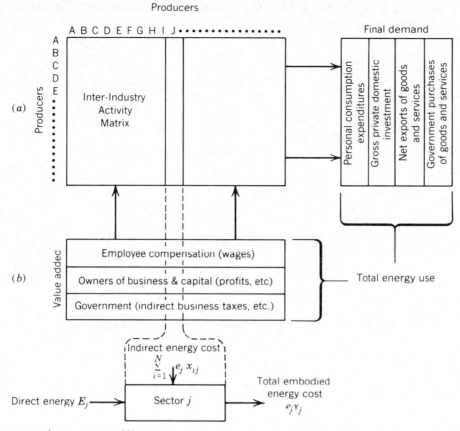

Figure 5.2. (a) An energy input–output table as developed by Herendeen and Bullard (1975) and others at the University of Illinois Energy Research Group. Entries in the table represent the embodied energy purchased by the industries listed across the top of the table from the industries listed down the left side of the table. Units are kcal/$. (b) For any sector j, the total embodied energy cost per dollar of its output is $e_j x_j$, where e_j is the embodied energy intensity (kcal/$) of sector j's output, and x_j is the total output of sector j in dollars. e_j is the sum of two components: direct fuel burned by sector j (E_j) plus all the fuel burned in the other sectors from which sector j purchases intermediate goods in order to produce its output. x_{ij} is the transaction from sector i to sector j.

Table 5.1 gives values for embodied energy for various sectors of the U.S. economy calculated by the methods developed by Herendeen and Bullard (1975), Hall et al. (1979), and Costanza (1980) for 1967, the year of overlap for all analyses. The values calculated by Herendeen and Bullard (1975) and Hall et al. (1979) are similar because they include only the direct use of fuel and because the specifics of how energy is used in upstream sectors is swamped by the interdependency of most of our economy. The values for embodied energy calculated by Costanza (1980) are considerably greater than these analyses because they include an estimate of the energy used to support labor and government services.

THE ENERGY COST OF SERVICES

The quantity of economic energy used to produce goods like copper tubing, refrigerators, or a bushel of corn is relatively easy to measure. The amount of economic energy used to render services is not as easily measured because the fuel component of medical, business, or educational services is less obvious than is the fuel used to produce, for example, a steel ingot. Despite such difficulties research by a number of investigators (e.g., Bullard et al., 1978) has shown that rendering services also requires large quantities of fossil fuels. In several cases the quantity of energy used to provide a dollar's worth of services is greater than the quantity

Table 5.1.   The Energy Costs (kcal/1967 $) for Various Industrial Goods in 1967 Calculated by Various Methods of Energy Cost Accounting

Product	Sector BEA	Sector SIC	Herendeen and Bullard (kcal/$)	Hall et al. Method 1 (high) (kcal/$)	Hall et al. Method 2 (medium) (kcal/$)	Hall et al. Method 3 (low) (kcal/$)	Costanza (kcal/$)
Textiles	17	22	16,982	18,907	15,826	11,905	195,413
Apparel	18	23	9789	13,931	11,160	7634	245,599
Lumber and wood	20	24	13,648	19,720	16,871	13,246	876,670
Shaped wood	22	25	10,715	13,138	10,709	7619	282,202
Paper	24	26	22,246	23,806	21,027	17,492	255,465
Printing	26	27	9085	9569	7817	5553	193,208
Chemicals	27	28	55,044	22,315	20,036	17,137	225,049
Rubber	32	30	23,976	15,364	12,903	9771	205,141
Skin and leather	33	31	22,667	14,684	12,187	9009	158,243
Nonmetal	36	32	24,602	26,683	24,504	21,732	198,904
Primary metals	37	33	48,300	28,465	25,450	21,613	220,790
Fabricated metal	42	34	17,335	14,251	11,782	8641	190,626
Machinery	49	35	13,735	12,542	10,282	7406	187,099
Electrical equipment	53	36	11,775	12,456	10,193	7312	186,582
Transport equipment	59	37	13,726	15,752	12,696	8808	197,977
Instruments	62	38	9419	9909	8098	5794	186,662

Sources: C. W. Bullard and R. A. Herendeen, 1975. Energy costs of goods and services, *Energy Policy* **3**:263–278; C. A. S. Hall, E. Kaufman, S. Walker, and D. Yen, 1979. Efficiency of energy delivery systems: II. Estimating energy costs of capital equipment. *Env. Management* **3**:505–510. R. Costanza, 1980. Embodied energy and economic valuation, *Science* **210**:1219–1224. The original data of Hall et al. were corrected for inflation.

used to provide a dollar's worth of goods. For example, in 1975, 11,287 kcal of energy (fossil fuels and electricity) were used to produce a dollar's worth of fabricated metal products, whereas 15,905 kcal were used to provide a dollar's worth of state and local government services (Hall et al., 1979). The latter energy cost includes the fuel used to build and maintain roads, schools, and sewer systems.

Establishing consistent and distinct system boundaries for defining the energy costs of services is difficult. Some of the more obvious costs clearly should be included, such as the fuel used to build and heat buildings or to manufacture books and medicines. Other energy costs of a service are more difficult to evaluate. For example, should indirect costs such as transportation of the patients and the staff to and from the hospital, the energy costs of research to develop a new medicine, and/or the energy cost of the researcher's salary also be included in the energy cost of medical services?

The definition of consistent system boundaries for evaluating the indirect energy costs of the transportation component of a service is especially difficult. If included, the energy costs of transportation for social services, especially in rural areas,

Table 5.2.   The Energy Costs of Various Services in 1967

Service	Energy Cost of Services (kcal/$)
Railroad and related services	15,175
Radio and TV broadcasting	5222
Water and sanitary services	17,161
Finance and insurance	4403
Real estate and rental	6643
Hotels and lodging	10,543
Business services	5833
Automobile repair	10,384
Medical and educational	8128
Wholesale and retail trade	7384

Source: C. W. Bullard and R. A. Herendeen, 1975. Energy costs of goods and services, *Energy Policy* **3**:263–278.

would be a very large component of the energy cost of the service because transportation is fuel intensive and many services require clients, who tend to be spread out over a large geographical area, to come to a centralized facility, or require the service to come to the people. Table 5.2 estimates the quantity of energy used to render services.

In this chapter we have tried to show how energy is used to produce goods and services and how these energy costs can be estimated. Clearly these energy costs are estimates and are subject to error. But so too are dollar costs. The following chapter describes how energy costs can be used to analyze the U.S. agriculture and food system.

6

FOOD, ENERGY, AND AGRICULTURE

OVERVIEW

People use food energy for maintenance metabolism and to provide fuel for human activity. Today four to five billion people live on Earth, a vastly larger number than in the past (Figure 6.1). The average person requires at least 2000 kcal of food energy per day to survive, so that each day a minimum of 8–10 trillion kcal of food are consumed. An approximately equal amount of food energy is either grown, caught, or slaughtered—not counting food grown for domestic animals. Without that food energy the world's people would starve, but on the other hand, without the people most of the food would not be grown. This chicken-or-egg relation characterizes any inquiry into the past of human food production, for increased food energy is required for population growth while the increased population can grow more food—at least as a generality. Therefore human populations are affected greatly by food-producing technology, food-exploiting social activity, the cultural activity and cultural change made possible, or necessitated, by changes in food availability, and the relation of all of these to energy.

Only a limited portion of the Earth's surface is especially suited for agriculture: alluvial (river) valleys, rich prairie soils, the edges of many volcanoes, and many glaciated regions. These regions often are characterized by young soils, for old soils tend to have their nutrients leached out by millennia of rains. By definition the highest-quality soils tend to require less human and fossil energy inputs per unit of food produced than poorer soils. Food production on both good and poor soils can be improved markedly by investing increasing amounts of energy as is the case, for example, with recent corn yields in the United States, which are about five times per hectare what they were in colonial times. Most of the increases in global food production since 1950 have come from increased use of energy and energy-intensive inputs on good soils rather than increased use of poorer grade soils (Brown, 1981). A general problem facing farmers is that agricultural activities on many soils tend to degrade the properties of the soil that made them desirable in the first place—for example, clearing vegetation increases the loss of soil by erosion as well as the leaching of nutrients. Humans invest their energy and fossil energy in attempting to mitigate these problems, but the good soils often are degraded further by energy-intensive farming. The net result is that agriculture uses increasing amounts of fossil fuel, and in many cases yields cannot be sustained without the continued use of energy-intensive inputs.

HISTORICAL AND ECOLOGICAL PERSPECTIVE

The relation between human population growth and food production is obscured by the relatively poor records left by most cultures. Ancient China and Egypt are exceptions, for they had sophisticated cultures that left records that have allowed the more or less successful reconstruction of their past. Two such reconstructions are presented here along with modern-day suggestions as to the most important factors responsible for the population changes (Figure 6.2). Although human populations respond to a complex interaction of biological, cultural, environmental and political factors, the availability of food energy is one of the most important. Historically, increases in agricultural technology have allowed (or at least are correlated with) increased population growth, whereas wars, disease, and famine are associated with decreases in populations. Presumably energy, especially food energy, was important in the growth and decline of earlier human populations, but the specific forces that caused these human populations to grow and decline are lost to

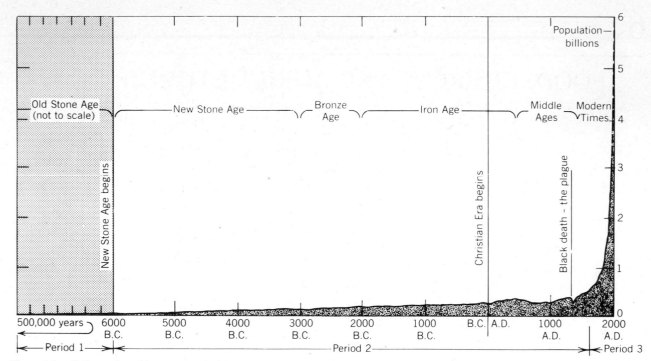

Figure 6.1. Estimated total human population numbers for the world for different cultural stages. (From *A Geography of Populations: World Patterns* by Treworth, 1969. Reprinted by permission of John Wiley & Sons Inc.).

antiquity. Changes in the technology and sociology of resource acquisition and utilization were important (e.g., Harris, 1979) as were the military aspirations and resources of one's neighbors (e.g., see Colinvaux, 1975; McFarland et al., 1966), and many other factors. Clearly, however, when human populations increased, it was necessary to increase the flow of economic energy to feed those people.

Exponential Growth

Human beings have the physiological capacity to double their population numbers in less than 30 years. We know that because many modern populations do just that. When this occurs for extended periods, the actual population increases in a compound interest fashion called exponential growth (i.e., China since 1900, Egypt since 1800). A simple hypothetical example serves to illustrate the potential for human populations to expand under unlimited growth conditions. Future populations of exponentially growing populations, such as the present human population, may be predicted by the equation

$$N_{t+\Delta t} = N_t e^{rt}$$

where $N_{t+\Delta t}$ = population at Δt years in future
  N_t = population at present time t
  e = the natural log (2.718)
  r = the annual rate of increase, which is about 0.02 for the human population in recent years (see Figure 6.1)

If there were but one man and one woman only 1100 years ago, and if they and their descendants doubled every 30 years, they could have produced as many descendants as the total present population of the Earth by now. Clearly throughout human history people have increased much more slowly than their potential, and for long periods the effective r was near zero or even negative. Why? There is no simple answer of course. Angel (1975) argues from the evidence of ancient skeletons that it was relatively rare for women thousands of years ago to live long enough to have more than two or three children, a condition that effectively prevented rapid population growth. Diseases, wars, and conquerings took their toll as suggested by Figure 6.2. In other cultures, social or cultural pressures that discouraged large families may also have been important limiting factors.

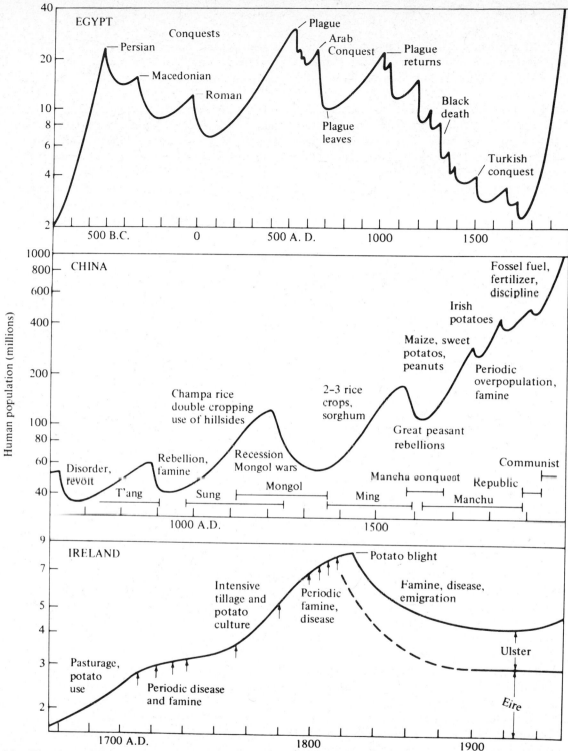

Figure 6.2. Historic populations of several civilized societies. Estimates of population numbers, together with some events that are thought to have affected them, are plotted for three nations; the population numbers are on logarithmic scales. *Top:* Egypt, 800 B.C. to the present, as interpreted from various historic records and estimates. (From Hollingsworth, 1969.) *Middle:* China, 600 A.D. to the present, based on censuses in some periods and estimates from other historic records. (From Cook, 1972; see also Durand, 1960; Ho, 1959; Clark, 1967.) *Bottom:* Ireland, 1650 to present, based on censuses from 1821 to present and interpretation of less reliable household counts before 1800. (Connell, 1941; Reinhard et al., 1968.) (Summary and references from Whittaker and Likens, 1975.)

117

We wish to emphasize a third constraint on human population growth, one that also is often a determinant of the first two: the limits imposed by the energy investments required to produce human food supplies. It is not possible to say that such limits were *the* most critical ones limiting human growth, but clearly they have been extremely important. Later we show how the industrialization of agriculture has allowed, at least temporarily, many populations of people to escape these limits.

Carrying Capacity

We develop this argument again from a biologist's perspective, beginning with the ecological concept of *carrying capacity,* which population biologists often use to examine populations of plants and animals in nature. Population biologists find that over long time periods the size of many populations appears to oscillate about some average level. Game managers often speak of the carrying capacity of a unit of landscape as the number of deer, pheasants, or some other species that a unit of landscape can support. This level, which was once thought to be more or less constant, is now known to vary in response to many environmental variables, such as season, weather, and the abundance of other species. One way that environmental variables affect population size is by influencing the quantity of available food, which in turn influences the number of organisms a unit of landscape can support. Thus the quantity of usable food produced by a surrounding ecosystem sets an upper limit on potential population size, particularly during extreme periods (e.g., see Pulliam and Enders, 1971). Although human population size often is influenced by the effects of social and political decisions, environmental factors affecting the availability of food have had direct impact on the population growth of many societies throughout human history.

In a more general sense animal populations are considered controlled in part by *density-dependent* factors. During periods of low population density, excess food is available, births will be favored over deaths, and population levels can rise. During periods of relatively high population density, less food is available per capita, diseases can spread more easily, and so on, so that deaths will be favored over births and population levels will decrease. A considerable body of research has found that many plants and animals in nature have developed very sophisticated means of population self-regulation that help to maintain the populations at a relatively constant number, evidence that seems consistent with the idea of carrying capacity. For example, territoriality in birds is a mechanism that keeps many bird species from overexploiting their food resources (e.g., Lack, 1954). For many decades much of theoretical and applied ecology was based on the concept of self-regulation and also on the idea that a fixed carrying capacity (called K) existed for all populations in nature. The concept has been expressed as the logistic equation

$$N_{t + \Delta t} = N_t + N_t \times r \left(1 - \frac{N_t}{K}\right)$$

where r is the intrinsic rate of population increase per Δt units of time, N_t is the present population level, K is the population level at the carrying capacity, and t is the time interval. This equation assumes that a population grows exponentially when it is at very low levels, linearly when at intermediate levels, and then stops and (depending on intervals between reproduction) fluctuates about the carrying capacity when it approaches that value. This results in an *S-shaped* growth pattern over time for a population starting at low density (see Chapter 18, Figure 18.8 for a further example and a critique).

In practice, the logistic equation is not an accurate descriptor of the growth of most real field populations, for both the potential and the actual number of organisms present in an area changes constantly in response to factors not explicit in that equation: the resources of that landscape, demographic factors internal to that species, competition from other species, and physical changes in the environment. Even though the logistic equation is not especially useful for describing real populations, it is important to examine the potential utility of the carrying capacity idea for understanding human populations. Clearly an upper limit exists to the number of organisms (including people) that can be fed from a given unit of land, and even though this changes over time, we may wish to call this limit carrying capacity. A critical determinant of human carrying capacity is the development of food and other resources by investing energy in social infrastructure, including modern agriculture and much of modern industry. Such investments have increased human carrying capacity by greatly modifying the environment so that the production is increasingly of substances of direct use by humans. This ability to modify the

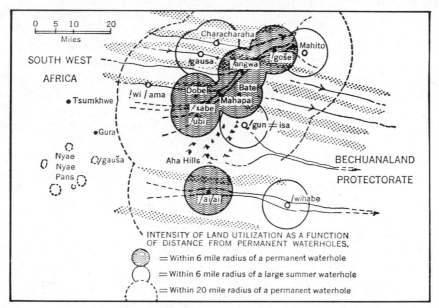

Figure 6.3. Intensity of land utilization by !Kung as a function of distance from permanent water holes. Within six miles of a water hole the food resources can be considered of higher quality and are utilized more intensively, because the energy investment required for exploitation is less. (Map from *Environment and Cultural Behavior* by Andrew P. Vayda. Copyright © 1969 by Andrew P. Vayda. Reprinted by permission of Doubleday & Company Inc.

environment on a large scale and in rapidly evolving ways makes the consideration of carrying capacity for humans very different from that of other species.

Preagricultural Societies

In contrast to the complex economies of modern societies, the economies of preagricultural, hunter–gatherer human societies were relatively simple. The relatively small number of tools used were made by human labor, generally from local materials, and virtually all energy resources within such societies were derived from direct solar energy. The major economic activity was the collection and distribution of food, and the environmental and cultural knowledge necessary to exploit diverse natural resources was often very complex. Since the total food resource available was only the rather small proportion of the energy flows of local ecosystems that was edible (e.g., fruits, nuts, game), the population densities of preagricultural societies were (food) energy limited, at least in comparison to today's population numbers. Howard Odum (1971) estimated that the density of human populations rarely exceeded one person per few square kilometers in societies that did not practice agriculture,

and it is not unreasonable to use this value as a rough estimate for the carrying capacity of natural environments that support hunter–gatherer-type societies.

The energy investment to gather the edible productivity of natural ecosystems was relatively small, such as a day's hike in search of nuts or game (Figures 6.3 and 6.4; Lee, 1969). Regardless of the magnitude of the energy investment, the upper bound for the quantity of food that could be collected was determined by the productivity of the ecosystem and the small proportion of that which was edible for people. Several comprehensive studies of the energetics of contemporary hunter–gatherers have been made, and the values derived probably are similar to those for our preagricultural ancestors (Lee, 1969; Rappaport, 1967). These studies generally have shown that although the energy flow per hectare that went to human food production is low, the energy returns on investment (EROI) for hunter–gatherer societies is large (often 10 food kcal obtained per 1 human kcal invested). The resultant energy surplus often was spent in leisure (as in the !Kung studied by Lee), in production of luxury protein (as in the pigs in the Tsembaga tribe studied by Rappaport), or in ritualistic and relatively nonlethal war. Human populations had the potential to increase during periods of food surplus,

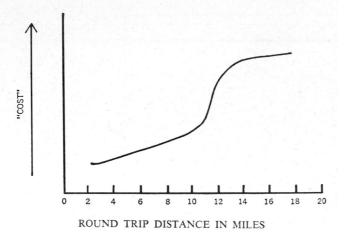

ROUND TRIP DISTANCE IN MILES

Figure 6.4. Energy cost of foraging for !Kung bushmen. The large increase in cost at about 12 miles occurs when an overnight trip is required. (Chart from *Environment and Cultural Behavior* by Andrew P. Vayda. Copyright © 1969 by Andrew P. Vayda. Reprinted by permission of Doubleday & Company Inc.)

but periodic failures of the food production system, real wars, or pestilence normally kept populations low relative to modern times. Thus the carrying capacity for some populations was established not during times of average food abundance but rather during low-food bottlenecks. Direct starvation because of food depletion is only the extreme example of this type of population control. For example, severely malnourished people are more susceptible to disease than are well-fed people, and Thomas (1973) found that Indian children of the Peruvian highland are more likely to freeze during cold winter nights when the family food supply is insufficient for all to get their daily caloric requirements.

Agricultural Societies

By contrast, early agriculturists modified their environment through the application of technology so that it produced more food per unit area. The general term used for the work humans do to control and redirect environmental energies to desired species is *management*. Probably the first major modification in human food technology was the corralling and breeding of animals solely for the purpose of food production. Since most domestic animals can eat a wider variety of vegetable matter than can people, the amount of natural plant production available to human societies increased. A second major innovation was, of course, planting seeds which, in effect, meant foregoing current food con-

sumption as an investment for more production in the future. Deevey (1960) suggests that there have been three large jumps in the human population in the last 10,000 years, the first as a result of the development of herding, the second due to the development of agricultural technology, and the third resulting from modern industrialization. The development of herding and agricultural technology were milestones in human history because they enabled our species to escape from the carrying capacity imposed on it by the limited food energy supplied by natural ecosystems. Table 6.1 gives a time and energy budget for a group of people in Peru who make their living by a combination of gathering and farming.

Agriculture also marks one of the first relatively long-term energy investments conducted by human society as a group. In order to convert a natural ecosystem to an agroecosystem, energy must be invested to remove the old system (i.e., chop down trees), to create the new system (i.e., sow the seeds), to defend against other competitors (i.e., pest control), and to maintain the system against the natural forces that initiate regrowth of the natural system (i.e., weed control). Although agriculture required a larger energy investment than a hunter–gatherer lifestyle and had a longer waiting period before returns on that investment began to be realized, agricultural systems in general produced much more edible food per unit area. The increased return on investment increased the carrying capacity per unit area of land, making possible population densities many times in excess of the carrying capacity of unmanaged systems (Table 6.2). The increased number of people in turn could clear more land and produce more crops. This kind of human–environment interaction allowed human populations to expand to much higher levels. Whereas 1 km^2 of most forest ecosystems are needed to support 1 person, less than 1% of that area of an agroecosystem produces the same quantity of edible energy (MacElroy and Averner, 1978).

SOME PROBLEMS WITH AGROECOSYSTEMS

Natural ecosystems tend to be characterized by a series of evolved mechanisms that operate to stabilize, protect, and maintain that system through time. Of particular importance to our present interests is that most natural ecosystems are characterized by both a relatively tight cycling of essential

Table 6.1. Daily Time Allocations and Energy Expenditures for Adults by Season

Major Activity Category	Dry Season				Wet Season			
	Men		Women		Men		Women	
	min	kcal	min	kcal	min	kcal	min	kcal
Eating	73	139	44	57	69	131	63	82
Food preparation	11	21	141	268	12	23	140	266
Child rearing	3	7	72	101	0	0	66	92
Manufacture	67	248	112	190	95	352	137	233
Wild foods	117	620	48	211	128	678	55	242
Garden labor	110	550	43	112	182	910	61	157
Idle	180	234	157	157	98	127	141	141
Hygiene	22	53	50	125	18	43	20	50
Visiting	87	113	41	41	35	46	48	48
Other	112	358	70	126	142	454	48	86
Daylight hours[a]	782	2343	778	1388	779	2764	779	1397
Night hours	660	660	660	528	660	660	660	528
Totals[a]	1442	3003	1438	1916	1439	3424	1439	1925

Source: E. Montgomery, J. V. G. A. Durnin, and J. E. Ellis, 1976. *Energy Consumption*. Workshop volume manuscript. The Institute of Ecology, Madison, Wis.

[a]Totals vary slightly from expected due to rounding.

nutrients and a relatively high degree of protection from insect outbreaks.

Nutrients

Recent studies of undisturbed natural ecosystems have found that in general the quantities of nutrient elements (e.g., nitrogen, phosphorus, potassium) leaving these ecosystems, principally as dissolved forms in rivers, are about the same or less than the quantities entering the ecosystem from precipitation and rock weathering. Experiments that have removed natural vegetation (e.g., Likens and Bormann, 1970) have shown that the vegetation is an important determinant in maintaining the nutrient balance of these systems. When the vegetation is experimentally removed, the nutrients tend to be lost much more rapidly.

An important mechanism for holding on to the

Table 6.2. Density of World Population Over Time

Years before Present	Cultural Stage	Persons per 100 km²
Present	Industrial and farming	1700
80	Farming and industrial	1100
180	Farming and industrial	620
230	Farming and industrial	490
330	Farming and industrial	370
2,000	Agriculture-based urban; subsistence village	100
6,000	Subsistence village; early urban	75
10,000	Mesolithic	4
25,000	Upper Paleolithic	4
300,000	Middle Paleolithic	1.2
1,000,000	Lower Paleolithic	0.4

Source: Adapted, with permission, from E. Deevey, 1960. The human population. *Sci. Amer.* **203**:195–197.

nutrients in undisturbed systems is the uptake of nutrients by the roots of the vegetation. The deep roots of forests, in particular, can capture nutrients that enter with rain water and/or that are released from decaying vegetation as they percolate downward in the soil horizon over time (Figure 6.5). When ecologists speak of the tight nutrient cycle of many natural systems, they are referring to the continuous recycling of nutrients through leaf fall, or death and decay, that is balanced by root uptake.

Agroecosystems, however, are normally composed of short-lived herbaceous plants that have much shallower root systems (Figure 6.5). Additionally the bare ground exposed during agriculture increases the loss of soil and nutrients with surface runoff and erosion. Finally, nutrients are further depleted during the harvesting of the crops. Maintaining soil fertility on agricultural land is therefore an extremely difficult problem on many of the world's soils. Frisell (1977) presents a fairly comprehensive

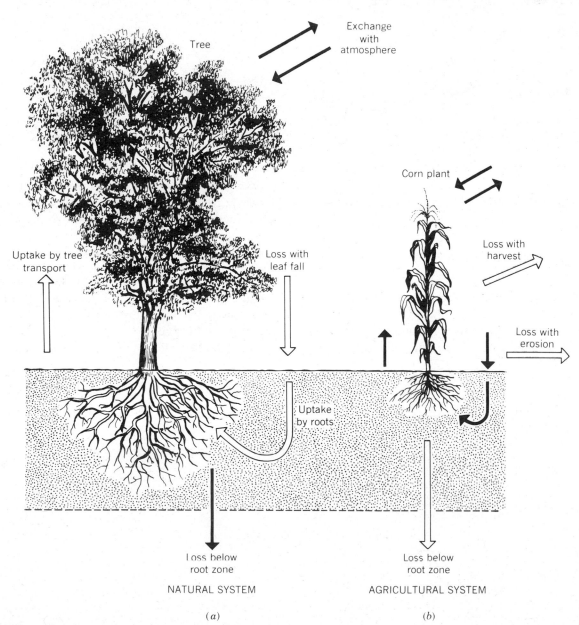

Figure 6.5.   Comparison of nutrient inputs, storages, and outputs in a natural forest and an agricultural ecosystem. As a rule nutrients tend to be recycled in natural systems and added and lost in most agricultural systems.

review of nitrogen, phosphorus, and potassium budgets for agricultural ecosystems.

The loss of nutrients is often especially severe in the tropics and tropical soils are often nutrient poor for several reasons. Most tropical soils were never glaciated, a process that renews the fertility of soils by replacing leached out topsoils with new material. Tropical soils also tend to be very old (because they have not been glaciated), and they generally receive large amounts of rainfall, which has leached nutrients out of the soil for a very long time. Yet tropical forests are often among the most productive and spectacular forests of the world despite poor soil conditions. The reasons for this has been elucidated in recent research (e.g., Richards, 1966; Stark and Jordan, 1978; Jordan and Herrara, 1981; Nadkarni, 1981) which has shown that (1) the very complex association of epiphytes growing on trees captures nutrients from the rain, which then are taken up by the trees (which even may have roots growing into the soil formed by epiphytes up in the trees), and (2) the surface root complexes and the fungus–root associations, known as mycorrhizoids, are especially efficient at capturing any available nutrients and at breaking down organic matter at the soil surface and making it available to the tree. Thus in tropical forests, in particular, symbiosis of different species appears to be very important to the tightness of nutrient cycles. All of this is lost when the natural systems are replaced with agricultural systems.

Historically, the most common human solution to the problem of declining soil fertility of cultivated lands was simply to move to new areas. One important variant of that solution, still so prevalent in many areas of the tropics that it produces food for 5–10% of the Earth's population, is called *shifting cultivation, slash and burn, swidden,* or *milpa* (see Nye and Greenland, 1959; Watters, 1971). In this cultivation scheme the virgin or second-growth forest is cut, the vegetation allowed to dry, and then burned. The nutrients once held in the vegetation are made available to crop plants as ash residual on the ground. Farmers cultivate a diverse and complex series of crops for about two years, and then as insect pests and weeds increase and as the soil becomes less fertile, they clear a new plot and allow the original plot to go fallow for (ideally) one to two decades. As the forest grows back, a new inventory of nutrients is established on the site as the natural vegetation and soils sequester nutrients brought in with rain and produced from underlying rocks. Then the cycle is repeated. Unfortunately as populations grow and demand for food increases, the fallow cycle tends to be shortened, fewer nutrients are replenished, and soil fertility declines.

Another solution to the problem of decreasing fertility is various forms of hydraulic farming, where crop cultivation is closely associated with water management. On a global scale, paddy rice cultivation, or *sawah,* is the most common example of this type of agricultural practice, and there are fields in China that produce very high yields of rice (5–10 tons/ha) even after having been farmed for many thousands of years. Hydraulic farming methods emphasize recycling and resupply of nutrients on the same site via green manure, nitrogen-fixing algae and bacteria, animal manure, night soil, and so on. Plants such as water hyacinths are planted in the effluent canals to recapture nutrients. These plants are composted and later spread on the crops. Geertz in Vayda (1969) presents an interesting comparison of paddy rice versus swidden agricultural system emphasizing how the former is much more flexible and adaptive to increases in population. It is also interesting to examine these two different agricultural solutions to the nutrient problem from our energy perspective. In swidden the economic work of concentrating the dilute nutrients in rain is done by the natural ecosystem using solar power. The human energy investment takes place when the natural system is chopped down so that the nutrients can be utilized for human ends. In the paddy rice system human energy must be constantly invested to replace nutrients lost from the agroecosystem.

Pests

We saw in Chapter 1 that one adaptation by which plants avoid insect pests is by filling their tissues with chemicals that are poisonous or distasteful to insects. Unfortunately these *secondary compounds* are also often distasteful to humans. The strong chemicals that distinguish garlic, tea, mustard, and *Cannabis* make interesting dietary supplements to us, but they would hardly do as primary sources of food energy. As a result over the past 5 to 10 thousand years of agricultural history, people have tended to select crops for taste—meaning in part a diminished concentration of unpleasant secondary chemicals. This probably also had the undesirable secondary effect of making the crops tastier to— and more susceptible to—insects. Additionally the planting of monocultures makes it easier for pests to find the crops upon which they feed. Preindustrial people attempted to mitigate these problems in several ways. Insects were picked by hand, crops were grown in complex spatial arrangements rather than monocultures, fowl were placed in fields to eat the insects, and so on. Nevertheless, crops were lost continuously to insects, reducing the efficiency of agriculture. Modern agriculture uses pesticides, but it has not always decreased the proportion of crops lost to insects. There was not, and probably cannot be, an entirely satisfactory solution to the problem of loss to insects.

INDUSTRIAL AGRICULTURE

Since the first agriculturists, farmers have been able to increase the productivity of their agroecosystems by overcoming environmental limitations and by dampening environmental fluctuations. These improvements in agriculture are perceived as technological advances, but we can view them also as energy investments. The sun drives the hydrological cycle and the metabolism of many different species drives the nutrient cycle. Humans invest their energies to increase the productivity and reliability of the natural energy flows and to channel solar energy toward species that are useful as food.

The increasing ability of humans to manage nature for their own purposes has come from the increasing ability of humans to amplify their own physical labor with nonhuman energy sources. Originally, domesticated animals and wind and water power were used. More recent technology has allowed the use of the much more concentrated energy of fossil fuels, which in turn has enabled people to increase greatly their energy investment in agriculture. Howard Odum (1967) was among the first to point out the relation between fossil fuel input and the large modern agricultural yields (Figures 6.6 and 6.7). About 3–5% of our national energy is used to grow food, which is a surprisingly small amount of energy considering how important food is to our citizens and our economy. The total industrial energy used in our total food system, however, including energy used for growth, processing, and distribution, is from 13 (Steinhart and Steinhart, 1974) to 17% (NAS, 1980) of the total energy use in the United States.

The use of fossil fuel energy to produce fertilizers, herbicides, pesticides, tractors, irrigation systems and other farm products has made each farm worker more efficient by subsidizing his or her effort in agricultural production. From 1940 to 1970 the worker-hours of farm work per year decreased from 20 billion hours to about 6 billion hours, whereas the total annual fossil energy used on the farm in the United States increased fourfold, from 125×10^{12} kcal to 526×10^{12} kcal (Steinhart and Steinhart, 1974). Fossil fuels now are used to subsidize human labor in all processes of the modern agricultural system, resulting in much higher productivity per farmer (Figure 6.7). Thus the quantity of food produced by one farmer increased from enough to feed 11 people in 1940 to 78 people in 1980 (USDA, 1982). Table 6.3 shows the energy investment in different components of the agricultural system between the years 1940 and 1970.

The use of herbicides is a specific example of a labor-saving energy subsidy. Farmers can weed fields by hand, but in many industrialized nations (and industrial goods importing nations) herbicides, which require considerable energy to manufacture from petroleum-based products, have replaced human labor. Other important inputs, as can be seen from Table 6.3, are the energy resources used to produce and transport fertilizers and power tractors, and to produce fuels and machinery. Nitrogen fertilizer is an especially petroleum-intensive input, and to produce one ton of nitrogenous fertilizer requires a quantity of natural gas equivalent to six barrels of oil. Figure 6.8 shows how energy is used both directly and indirectly in agriculture.

The EROI for food—in this case the quantity of edible energy produced per unit of energy used to produce that unit—must have been greater than one for one with early agricultural methods, for if early agricultural methods required more human energy than the quantity of edible energy produced, agricultural society would not have survived. But the EROI for modern agriculture, where humans increased greatly the total energy investment, is no longer greater than 1 edible kcal per fossil fuel kcal input (Figure 6.12). From 1947 to 1979 the fossil energy input to the U.S. agricultural system increased by almost a factor of 10, whereas food production increased by only a factor of 2, so that the quantity of food energy return per unit of input has decreased by nearly a factor of 5. In recent years the U.S. agricultural system has used about 9.5 kcal of nonedible fossil energy (about one-third on the farm, one-third in processing and transportation, and one-third in preparation) to produce 1 average kcal of edible food. To put this energy cost in more

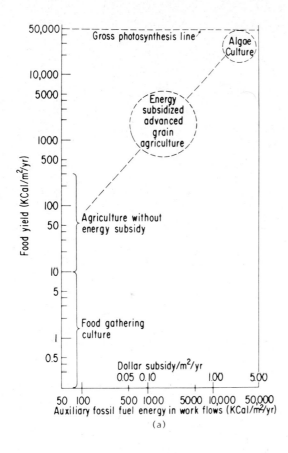

(a)

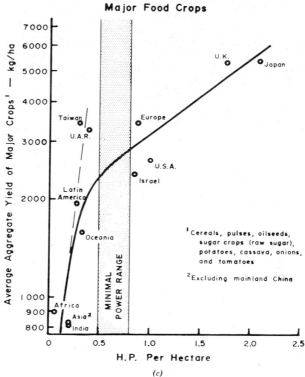

(c)

FERTILIZER/YIELD REGRESSION
Japan, India, & United States — 1952, 1958, 1963

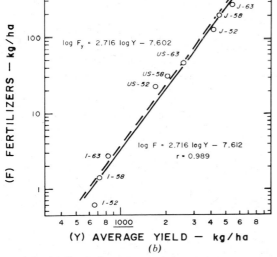

(b)

concrete terms, the energy equivalent of nearly a gallon of crude oil is used to grow, distribute, and prepare a diet of 3400 kcal/day for a U.S. consumer. Some nations, such as Japan, use even more energy per hectare than the United States, and they often have higher yields (Figure 6.6a and 6.6b). Thus the general pattern has been to increase the yield per hectare by trading large quantities of fossil fuels, which have relatively little intrinsic worth, for smaller quantities of high-quality food energy. This has decreased the EROI (expressed as food kilocalorie yield per invested human plus fossil fuel kilocalories) from the 10:1 or so of early hunter–gatherers and agriculturists to the ratio of about 1:10 that characterizes the modern United States agricultural system. The total yield, the total yield per hectare, and the carrying capacity for people, however, have been increased enormously.

Figure 6.6. (a) Generalized diagram of net food yields to people as a function of the subsidy of fossil fuel industry. (From *Environment, Power, and Society* by H. T. Odum. Copyright © 1971. Reprinted by permission by John Wiley & Sons, Inc.) The relation between (b) agricultural yield and fertilizer. (From Ennis et al., 1967.) (c) Yield and horsepower (From Giles, 1967.)

Limiting Factors

Increasing inputs of a given factor of production (e.g., nitrogen fertilizer) often do not elicit a cor-

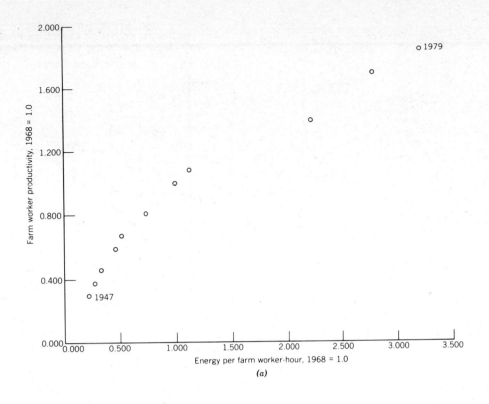

(a)

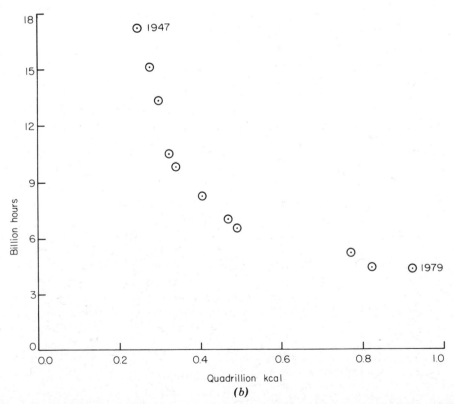

(b)

Figure 6.7. *(a)* Productivity per U.S. farm worker versus direct energy use on farms per farm worker (1968 = 1.0). *(b)* Trend in time (1947 = upper left to 1979, lower right) of annual energy and labor use on U.S. farms. (Data from USDA, various years.)

Table 6.3a. Energy Use in U.S. Food Industry

Component	1940	1947	1950	1954	1958	1960	1964	1968	1970
On Farm									
Fuel (direct use)	70.0	136.0	158.0	172.8	179.0	188.0	213.9	226.0	232.0
Electricity	0.7	32.0	32.9	40.0	44.0	46.1	50.0	57.3	63.8
Fertilizer	12.4	19.5	24.0	30.6	32.2	41.0	60.0	87.0	94.0
Agricultural steel	1.6	2.0	2.7	2.5	2.0	1.7	2.5	2.4	2.0
Farm machinery	9.0	34.7	30.0	29.5	50.2	52.0	60.0	75.0	80.0
Tractors	12.8	25.0	30.8	23.6	16.4	11.8	20.0	20.5	19.3
Irrigation	18.0	22.8	25.0	29.6	32.5	33.3	34.1	34.8	35.0
Subtotal	124.5	272.0	303.4	328.6	356.3	373.9	440.5	503.0	526.1
Processing Industry									
Food processing industry	147.0	177.5	192.0	211.5	212.6	224.0	249.0	295.0	308.0
Food processing machinery	0.7	5.7	5.0	4.9	4.9	5.0	6.0	6.0	6.0
Paper packaging	8.5	14.8	17.0	20.0	26.0	28.0	31.0	35.7	38.0
Glass containers	14.0	25.7	26.0	27.0	30.2	31.0	34.0	41.9	47.0
Steel cans and aluminum	38.0	55.8	62.0	73.7	85.4	86.0	91.0	112.2	122.0
Transport (fuel)	49.6	86.1	102.0	122.3	140.2	153.3	184.0	226.6	246.9
Trucks and trailers (manufacture)	28.0	42.0	49.5	47.0	43.0	44.2	61.0	70.2	74.0
Subtotal	285.8	407.6	453.5	506.4	542.3	571.5	656.0	787.6	841.9
Commercial and Home									
Commercial refrigeration and cooking	121.0	141.0	150.0	161.0	176.0	186.2	209.0	241.0	263.0
Refrigeration machinery (home and commercial)	10.0	24.0	25.0	27.5	29.4	32.0	40.0	56.0	61.0
Home refrigeration and cooking	144.2	184.0	202.3	228.0	257.0	276.6	345.0	433.9	480.0
Subtotal	275.2	349.0	377.3	416.5	462.4	494.8	594.0	730.9	804.0
Grand total	685.5	1028.6	1134.2	1251.5	1361.0	1440.2	1690.5	2021.5	2172.0

Source: J. S. Steinhart and C. E. Steinhart, 1974. *Energy Sources: Uses and Role in Human Affairs.* Duxbury Press, Duxbury, Mass.

Note: All values are multiplied by 10^{12} kcal.

responding increase in yield because biological laws limit the growth of most renewable resources. One commonly observed biological relation is the *saturation relation,* described empirically by the Michaelis–Menten equation. The response of biological systems to a given variable, for example, enzyme concentration or phosphorus concentration, is asymptotic, and as the saturation point is reached output per unit of input decreases—an example of diminishing return on investment.

This relation is complemented by Liebig's *law of the minimum,* which states that many biological processes are limited by the substance that is in least supply compared with other materials and rel-

ative to the requirements of the organism. As a result increasing other substances without increasing the limiting substance will have very little effect on the process (see pp. 8 and 102).

The complementarity of the Michaelis–Menten relation and the Liebigian law of the limiting nutrient is especially important in agriculture, where crops need a very specific mix of materials and conditions, including water, various nutrients, soil texture, and light to achieve high yields. Each factor individually often shows a diminishing return on investment, and the relation of the yield of rice relative to the use of nitrogen fertilizer is a good example. At first, adding 15 kg of nitrogen per hectare

Table 6.3b.   Energy Costs of Fertilizer Production (kcal per kg of N, P, K or CaCO₃)

Type	Production	Transportation, storage, transfer, and distribution	Total
Nitrogen			
Anhydrous ammonia	11,700	300	12,000
Urea (prilled or granular)	13,600	700	14,300
Ammonium nitrate (prilled or granular)	13,900	800	14,700
Phosphorus			
Phosphate rock	400	900	1300
Normal superphosphate (0–20–0)	600	1700	2300
Triple superphosphate (0–46–0)	2200	800	3000
Potassium			
Potash	1100	700	1600
Limestone			
Crushed and ground	15	300	315
Burned limestone	1908	500	2408
Hydrated lime	1908	500	2408

Source: D. Pimentel, 1984. Energy Flow in the Food System. In *Food and Energy Resources,* D. Pimentel and C. W. Hall, Eds. Academic, New York.

increases the yield of rice 350 kg, but the next additional unit increases the yield by a smaller amount (Figure 6.10). This general type of response is common for many crops and many fertilizers. Finally, the last addition of nitrogen fertilizer has very little effect on yield—some other factor is limiting. Thus nitrogen fertilizers, which are responsible for much of the increased productivity associated with the Green Revolution can be very effective at relatively low levels of application, but their effectiveness diminishes at high rates of application. Energy itself, however, is flexible, so that as nitrogen becomes no longer limiting, energy resources can be diverted to supply phosphorus, water, or whatever. Thus the ultimate limiting factor for agricultural production is in most cases the amount of energy that can be diverted to whatever is limiting.

A relatively new problem is that increasing certain types of energy-intensive inputs may actually backfire and produce smaller yields over the long run. The buildup of pesticide resistance in insects is the classic example. The process operates by straightforward Darwinian selection where susceptible insects, initially the great majority, are eliminated. Those few with resistant genes remain to produce the next generation, which will have a larger proportion of resistant genes. Figure 6.11 shows how the number of insects and other pests resistant to specific pesticides has increased over

time. Since the pesticides also kill the natural predators of the pests, the net effect often is to make the pest problem worse (Bottrell, 1979). Presently considerable research is under way to apply more sophisticated biotic methods to the problems. For example, much smaller quantities of pesticides can be used if enough is known about the pests to spray them when they are most susceptible. Sophisticated computer models can help in this process, especially in determining economic trade-offs for treating versus not treating pests (e.g., Shoemaker, 1977). Another basic approach is to learn enough about the biology of the predators and the pests, so that natural predation is encouraged and crops can be cultivated in such a way as to decrease susceptibility to pests (van de Bosch et al., 1982). Such approaches should be able to decrease the amount of petroleum and petroleum products used to manage pests, but the problem has not been studied explicitly from that perspective. Meanwhile there is considerable effort by the agrochemical industries to maintain high pesticide usage through the support of research on chemical-intensive pest management.

Steinhart and Steinhart (1974a) suggested that the yield of edible energy per unit of fossil fuel used in the entire U.S. agricultural and food system shows signs of diminishing returns. Input of fossil fuel subsidies to the U.S. agricultural system increased

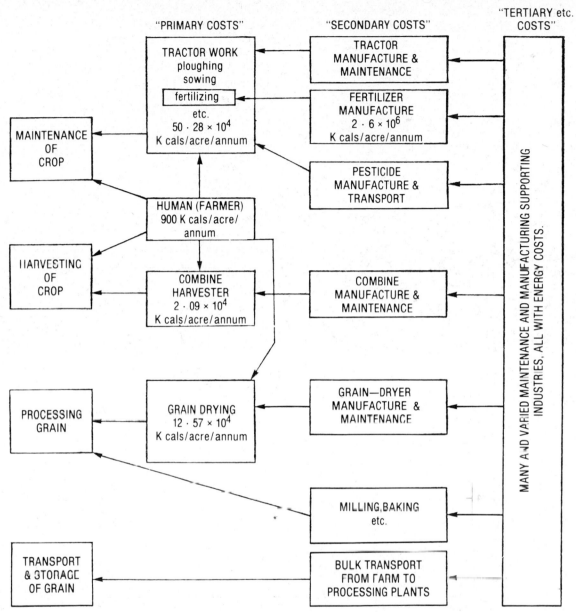

Figure 6.8. Diagram showing how energy is used both directly and indirectly in enhancing agricultural production. (From Lawton, 1973.)

steadily from 1920 to 1970 (Figure 6.9). The output of edible food energy matched this increase in fuel use until 1959, after which the output of edible energy did not increase as rapidly as the increased use of fuel and finally became nearly asymptotic. We do not know the precise reasons for the decline in food production, but soil depletion, increased use of poor quality farmland due to expanded production, and bad weather are possible candidates.

Quality of Output and Energy Requirements

Different food crops and agricultural methods have different energy requirements for producing an edible calorie (Figures 6.12 and 6.13). As a rule animal products (beef, eggs, etc.) require more fossil energy per edible calorie than plant food, due principally to the low efficiency (5–40% or so) with which an animal converts its food into biomass. The re-

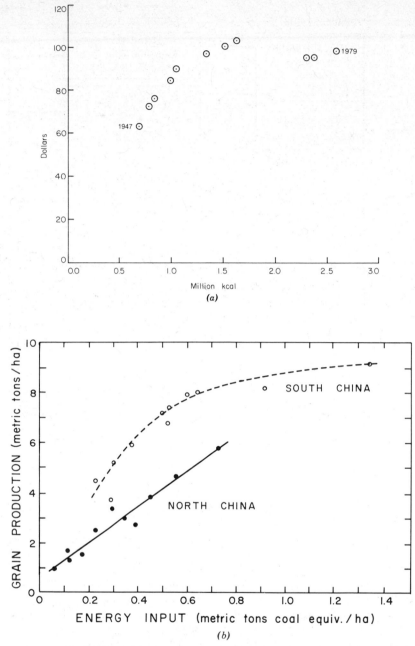

Figure 6.9. (*a*) Farm output (per hectare) as a function of energy input (per hectare) to the U.S. food system, 1920 through 1970. (From USDA, various years.) (*b*) Farm output vs. energy input for different regions in North and South China, about 1980. (From Shen Hong-Li, personal communication.)

maining 60% + is used for maintenance metabolism (Figure 1.2).

In general, expensive foods (lobster, shrimp, beef, cauliflower, winter tomatoes in the North) are energy intensive. The quantity of fossil energy required to produce a unit of certain edible plant foods can be as great as for animal food energy (Figure 6.13). Notice that grains such as corn and wheat have a relatively high EROI (as food energy per unit of fossil fuel used), whereas many greens and luxury foods such as broccoli, lettuce, melons, and tomatoes require more fossil energy to produce than the quantity of edible energy they yield. This difference is especially important in the production of animal proteins, which use the relatively non energy-intensive grains as a feedstock. If grains had

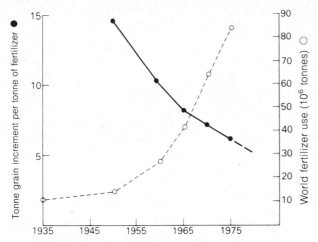

Figure 6.10.   Global fertilizer use and increment in yield per unit of fertilizer input. (From "Food, Energy and Society" by D. and M. Pimentel, in *Resource and Environmental Science Series*. Published by Edward Arnold, Ltd. and Brown, L. R. World-watch Institute, Washington, D.C.)

the same energy intensity as greens, animal protein would be prohibitively expensive, at least as produced by current industrialized farming methods. Changing technology can change the efficiency of food production. According to Heichel (1976) the conversion efficiency of hogs, cattle, and sheep declined between 1950 and 1974, whereas the conversion efficiency of broiler chickens increased (Figure 6.13a).

Just as with fossil fuels the qualities of different kinds of food energy are not identical. Humans are unable to synthesize all the amino acids needed for protein synthesis and therefore must have an external source of these essential amino acids in their diet. Without these amino acids humans develop diseases known as marasmus or kwashiorkor, which are common in many poorer nations where protein-deficient diets are common or where total caloric input is so low that any protein ingested is

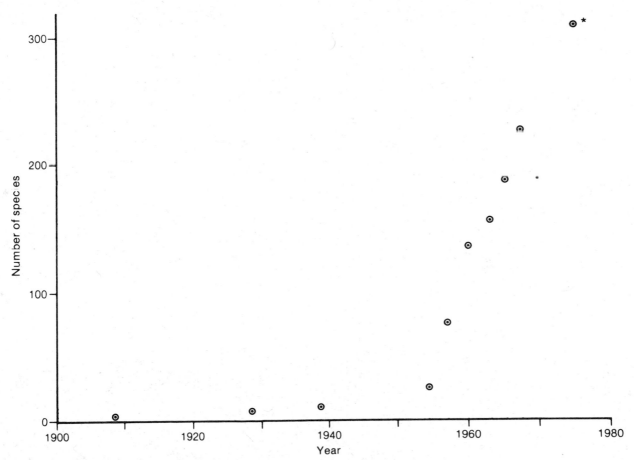

*Resistance confirmed in laboratory studies; resistance of an additional 59 species not examined in laboratory studies.

Figure 6.11.   Numbers of arthropod species that have developed resistance to one or more insecticides worldwide. (From Bottrell. 1979.)

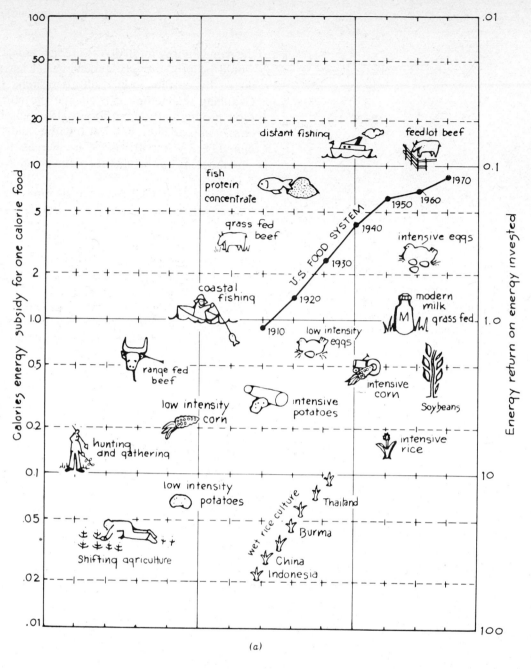

(a)

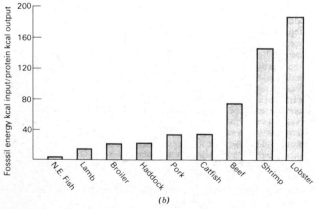

(b)

Figure 6.12. Energy subsidies for various food sources. The energy history of the U.S. food system is shown in *a* for comparison. (Part *a* from "Energy Use in the U.S. Food System," by J. S. and C. E. Steinhart. *Science,* **184:**307–316. © 1974 by the AAAS. Part *b* from "Food, Energy and Society" by D. and M. Pimentel, in *Resource and Environmental Science Series.* Published by Edward Arnold, Ltd.)

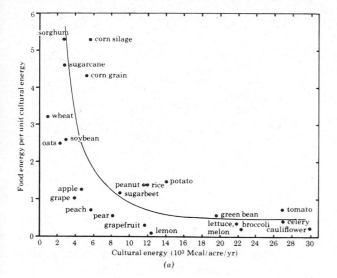

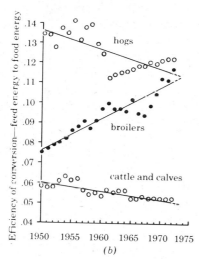

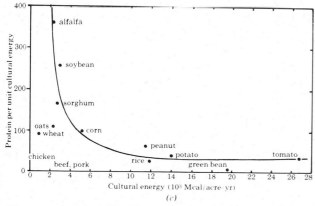

Figure 6.13. (a) Food energy output per cultural (fossil) energy input plotted versus total cultural energy used per acre: 1 acre = 0.405 ha. (b) Efficiency, and trends in efficiency, for several major protein sources. (c) Energy cost of protein production. (From G. H. Heichel, 1976, "Agricultural Production and Energy Resources," *American Scientist* Vol. 64, pp. 64–72, Fig. 2. Reprinted by permission, *American Scientist,* Journal of Sigma Xi.)

used for energy rather than an amino acid source. Either animal proteins, which have a good balance of amino acids, or the right balance of plant proteins can eliminate this disease if eaten in sufficient quantities. Animal proteins are generally a very energy-intensive source of protein compared to beans and soybeans (Figure 6.13b). Because of the preferences of most Western societies for animal proteins, large quantities of energy are used in agricultural systems to provide what we consider a balanced diet. In the United States we eat directly only about 10 percent of the grains we grow—the rest goes for exports or feeding domestic animals (Table 6.4). Thus we have enormous food reserves in the United States should we choose to decrease exports and/or change our diets.

Agricultural methods used by nonindustrialized societies commonly depend on much more human labor in relation to fossil energy and have large EROI (food calories produced per human and/or fossil energy input). Most crops and agricultural methods of the industrialized world use large quantities of fossil fuel relative to human labor and have smaller EROIs. At the extreme some of the methods used by industrial societies to produce animal protein (e.g., feedlot beef, deep-sea fishing) use large quantities of fossil fuel to produce 1 kcal of food (Figure 6.12). The diet of a nation, especially the proportion of protein that comes from animals, therefore affects and is affected by the energy that it uses in its food production system. The small return of food energy per unit of fossil energy invested is not an inherently bad characteristic of industrial agricultural systems. It is yet another example of using petroleum to produce something we do want, food. But it is important to recognize that our agricultural system is energy intensive and that, in particular, it requires large quantities of energy to sustain the Western world's protein-intensive diet.

Although fossil fuels are used increasingly to replace human labor on the farm, more laborers are employed elsewhere in the economy to produce the industrial materials used in modern agriculture. Only 2% of the U.S. population works on the farm, but another 6 or 8% work in industries associated with food production (e.g., the food industry, manufacturing farm machinery) and about 20% work in the whole food industry (D. Pimentel, 1984, personal communication). If and when the supply of fossil energy subsidies decrease, the United States may have to revise the energy-intensive agricultural methods that have allowed 98% of its population to

Table 6.4.   Comparison of Percentages of Food Items Consumed by Humans and Their Use as Animal Feed Stock

	North America	Latin America	Western Europe	Africa	South Asia
Wheat	59/13	64/2	49/16	71/0	79/1
Corn	1/94	33/47	3/87	77/5	75/4
Rice	53/13	59/0	54/1	64/0	59/1
Starchy roots	70/9	43/17	50/29	67/0	67/8
Vegetables	90/0	81/7	84/1	91/0	90/0
Pulses	87/1	90/0	45/37	84/0	78/8

Source: Reprinted with permission of Macmillan Publishing Company from *Human Ecosystems* by W. B. Clapham, Jr. Copyright © 1981 by W. B. Clapham, Jr.

work away from farms and have enabled a square kilometer of farmland to support 302 people. Some day we may have a great deal to relearn about low-energy farming from the farmers of less industrialized nations. Will we ever plow up our well-fertilized and irrigated suburban lawns to grow food?

THE FUTURE OF U.S. AGRICULTURAL PRODUCTION

Julian Simon (1981), an economist with a perspective many would call extremely optimistic or cornucopian, has predicted a very rosy future for U.S. agricultural production (and about everything else) based on the extrapolation of past trends. For example, the price of food relative to other goods generally declined over the past century, and over the past 45 years total agricultural output has increased dramatically while the total amount of land in cultivation has remained steady or declined. Both productivity (output per worker-hour) and crop yields per acre have increased dramatically. Simon extrapolated these trends into the future and concluded that all of these positive trends will continue indefinitely.

In an industry where overproduction is a chronic problem, the U.S. population is in no danger of starving. The relevant question is, however, are Simon's optimistic predictions justified based on *all* the pertinent data available? In particular, given the expected long range increases in demand for U.S. agricultural products for export, and the possible use of agricultural land for energy production (Chapter 13), can food production costs and food prices remain at their historically low levels relative to our purchasing power? Simon says yes, unequivocally. We believe that a more careful review of the data indicates that agricultural costs can be expected to rise in the future, although not at a rate fast enough to threaten the viability of our agricultural system or our supply of food. Rising fuel prices are a major factor contributing to the expected cost increases.

Simon chose the data he used rather carefully, and two omissions in the data he used stand out in particular. The first is the series compiled by the Soil Conservation Service (SCS) of the U.S. Department of Agriculture for the years 1958, 1967, 1975, and 1977, in which a wide variety of data were collected on total output, land use, and agricultural productivity. The last two studies in this series, the Potential Cropland Study (USDA, 1977) and the Natural Resources Inventory (USDA, 1980) emphasized the potential for new cropland formation and the problems associated with bringing such land into cultivation. The second major omission from Simon's analysis is the National Agricultural Lands Study (USDA, 1981), from which preliminary reports were available in 1979 and 1980. Many of the agricultural economists cited in this report believe that higher costs in the future are inevitable due to the effects of higher energy costs, increased soil erosion, declining rates of increase in crop yields, and the increased costs of irrigation.

Recent Patterns of U.S. Agricultural Production

Until the 1970s the history of modern agriculture in the United States was one of steadily increasing yields for most types of crops. Between 1950 and 1972 the annual rate of increase in total crop yields was 2.5%. Between 1973 and 1981, however, the rate declined to 0.3%/yr (Crosson and Brubaker,

1982). The analyses of Horsfall (1975), Brown (1978), Thompson (1979), and Crosson (1979) also indicate a decline in the rate of increase in crop yields. This historic break with past trends following the energy events of the 1970s is overlooked by Simon who states that "yields per acre have been going up sharply."

The continuing declines in the rate of yield in the 1970s indicate to some that diminishing returns have set in to investments in yield-increasing inputs like fertilizer and irrigation technologies. Others believe that the leveling in yields is only a response to higher fuel prices and scarcer and more expensive water in the West. Finally, the use of more low quality lands as crop exports increased may have had an impact. Nevertheless, the energy events of the 1970s clearly have altered the economics of U.S. agriculture, making extrapolations of historical trends questionable. As Crosson and Brubaker (1982) stated,

> . . . modern agriculture, with its massive use of fertilizers and pesticides, was developed in large part to take advantage of cheap fossil fuel energy. Not only is the technology itself keyed to energy from fossil fuels, but the research establishment that developed the technology also is oriented to exploitation of this resource. As a consequence, the scientific knowledge and technical skills needed to develop low cost energy substitutes for fossil fuels . . . are in short supply.

Energy Prices and Factor Productivity

Agricultural Technologies

Between 1860 and 1910 new agricultural technologies tended to be land using (Heady, 1982). Production was increased simply by increasing the total acreage under cultivation. Between 1910 and 1940 emphasis began to go toward the use of larger and more sophisticated implements. This was made possible by cheap, abundant petroleum supplies and the rapid development and dissemination of the internal combustion engine. Between 1940 and 1970 the trend was toward even more intensive use of land-saving technologies—fertilizers, herbicides, irrigation, improved seed varieties, and so on. Again, the development of these technologies was encouraged and made possible by declining real energy prices which made the production of nitrogen-based fertilizers and the transportation of irrigation water extremely cheap. In addition the nitrogen-

fixing facilities developed in World War II by the U.S. government for munitions were readily adapted to produce nitrogen fertilizer. These technologies literally substituted for land as taxpayers, through government support programs, paid farmers to remove land from active cultivation. This was supposed to reduce surpluses but did not address the real cause of the large surpluses.

The observed trend of technological change in U.S. agriculture is consistent with the standard economic theory of production, and it is also consistent with an energy-based perspective. In response to increased demand farmers will either (1) convert nonfarmland into farmland or (2) increase yields from existing farmlands by using more fertilizers and other high-energy inputs. The farmer's choice between land-saving and land-using technologies is determined by the relative price of land and nonland inputs and by the productivity of each at the margin. Between 1945 and 1972 the price of nonland inputs (i.e., fossil fuel-derived inputs) declined relative to the price of land. Therefore the use of nonland inputs was favored. Between 1950 and 1972 the amount of cropland harvested declined from 338 million acres to 289 million acres (USDA, 1980). At the same time inputs, total output and yields per hectare increased substantially. For example, the use of machinery in corn production nearly doubled from 1950 to 1970 while the use of nitrogen fertilizer, irrigation, and pesticides increased by factors of from 6 to 10 (Table 6.5).

Since 1972 the price of land-saving inputs like fuel and nitrogen fertilizer increased rapidly relative to the price of land, thereby discouraging further increases in their use. For example, although fertilizer use increased at 7%/yr between 1951 and 1972, the increase was only 1.4%/yr between 1972 and 1980 (Crosson and Brubaker, 1982), and Crosson (1979) estimated that the cost of direct and indirect fuel use accounted for 33% of the value of corn at mid-1975 prices. The effect of higher relative prices for nonland inputs has been a shift back to more land-using technologies. The amount of harvested cropland increased from 289 million acres in 1972 to 346 million acres in 1980 (USDA, 1980). Because the average hectare is now subsidized by less fuel, fertilizer, and so forth, the rate of increase in crop yields per hectare has declined as noted previously.

Some analysts argue that we are seeing a slowing in the growth of productivity realized per unit input. Pimentel and Pimentel (1979) found diminishing re-

Table 6.5. Energy Used in U.S. Corn Production, 1700–1983 (1000 kcal/ha)

Component	1700	1910	1920	1945	1950	1954	1959	1964	1970	1975	1980	1983
Labor	653	65	65	31	24	23	19	15	12	10	7	6
Machinery	19	278	278	407	555	648	777	907	907	925	1018	1018
Draft animal	0	886	886	0	0	0	0	0	0	0	0	0
Fuel												
Gasoline	0	0	0	1200	1350	1500	1550	1250	1200	600	500	400
Diesel	0	0	0	228	275	342	399	741	912	912	878	855
Manure	0	0	0									
N	0	0	0	168	357	630	966	1365	2625	2478	3066	3192
P	0	0	0	50	69	82	113	126	221	410	466	473
K	0	0	0	15	28	50	85	80	168	188	240	240
Lime	0	3	3	46	61	39	50	64	69	69	134	134
Seeds	44	44	44	161	322	421	470	520	520	520	520	520
Insecticides	0	0	0	0	7	20	54	74	110	200	300	300
Herbicides	0	0	0	0	3	7	20	40	100	400	700	800
Irrigation	0	*	*	125	125	250	375	625	1125	2000	2125	2250
Drying	0	0	0	9	10	15	54	145	376	458	640	660
Electricity	0	*	1	8	16	24	36	60	80	90	100	100
Transport	*	25	25	44	58	67	79	89	84	82	90	89
Total	716	1301	1302	2492	3260	4118	5047	6101	8509	9342	10,784	11,037
Yield	7520	7520	7520	8528	9532	10,288	13,548	17,060	20,320	20,575	26,000	26,000
Ratio	10.5	5.8	5.8	3.4	2.9	2.5	2.7	2.8	2.4	2.2	2.4	2.4

Source: D. Pimentel, personal communication.
Note: * means somewhat greater than 0.

turns in worldwide yield increases in response to increases in energy inputs like fertilizer (Figure 6.10) as did a comprehensive study on agricultural production efficiency by the National Academy of Sciences (1975). The reason for that diminishing return may be inherent biological limitations or the failure to supply other, more complex, factors that may have become limiting. Thus it is interesting that the NAS study also found evidence for diminishing returns to all direct plus indirect energy use based on cross-sectional data, and concluded that "progressively larger expenditures of cultural energy produce less crop output." Corn was the only notable exception, probably due to the type of technological transformation of corn production made possible by the hybridization of corn. But even the energy efficiency of corn production has declined. Pimentel et al. (1973) found that the energy efficiency of corn production (kilocalories of corn per kilocalorie of fossil fuel) declined 24% between 1945 and 1970, although others have argued that 1970 was a somewhat below average year for the early 1970s.

Irrigation

Julian Simon sees no limit to continued expansion of irrigation because ". . . . in the future, cheap water transportation . . will transform what are now deserts into arable lands." We believe that the evidence suggests just the opposite. In the United States the development of irrigation technology, an extremely important component in the development of increased crop yields, has been stimulated by cheap energy and easy access to cheap, abundant water. This water which has commonly been treated as a free good—with no internal or external costs—available to anyone willing to pay to transport it to the place of use (Frederick, 1982). Water development costs to farmers were reduced further by enormous subsidies from the federal government. Farmers have paid only 3% of the almost $4 billion spent by the Bureau of Reclamation on federal water and irrigation projects (LeVeen, 1978). Low-energy prices made it inexpensive to move that water, but in most western states the days of cheap water are over. Surface water withdrawal for irrigation peaked at 88 million acre-ft in 1955. Frederick (1982) found that throughout most of the agricultural regions in the West, including 66% of Western irrigated land, total water use was greater than total surface runoff. As a result western agriculture has become increasingly dependent on groundwater, with much of the expansion of western irrigation in the past 20 years based mainly on the mining of groundwater. Most aquifers in the region are effectively nonrenewable because recharge rates are negligible relative to withdrawal rates. Although most of these aquifers are not in any immediate danger of depletion, their levels are being lowered significantly which increases the energy costs of pumping water (see Figure 22.7).

An example cited by Frederick (1982) illustrates the impact of energy costs and water availability on food prices. For a typical irrigated corn and dryland wheat farm in western Kansas, the returns to risk for corn in 1977 was $14.50/acre, assuming a yield of 104.2 bu/acre, a price of $2.47/bu, pumping 2 acre-ft of water to height of 215 ft, a surface distribution system, and natural gas at $1.20/ft^3. The return for dryland wheat was $6.88/acre. If pumping depth increased by 2 ft/yr, and as energy prices increased, then the returns to risk in 1980 would have been reduced to $1.35 for corn and $3.82 for dryland wheat. By the year 2000 both crops would yield substantial losses.

This example overstates the problem because farmers can take measures to avoid rising water prices. The point is, however, that the economic feasibility of irrigation is influenced strongly by the physical availability of water and by energy prices and that both these factors have taken turns for the worse. Frederick concluded that despite all the adjustments available to farmers, they "are not likely to alter the basic facts that rising water costs are reducing the profitability of irrigated farming relative to dryland techniques and that the combination of higher energy costs and increasing water scarcity is limiting the opportunities for profitably expanding irrigated acreage in the West." Heady (1982) reached similar conclusions.

Potential Supply of Cropland

There is no physical shortage of potentially arable land in the United States (USDA, 1977, 1980). Of concern, however, is that the quality of potential farmland is of lower native fertility and of higher susceptibility to erosion than land already under cultivation (Heady, 1982; Crosson, 1982a) and the fact this land is generally being used for other activities such as grazing or watershed protection. The Natural Resources Inventory (USDA, 1980) estimated that 125 million acres exist with high or medium potential for conversion to cropland. The Potential Cropland Study (USDA, 1977) estimate for the same figure was 111 million acres (Table 9).

But Larson et al. (1983) noted that most of the land defined by the Natural Resources Inventory as potential cropland is lower quality land that is highly susceptible to erosion. Soil erosion is undoubtedly the largest noninternalized cost of modern agriculture. Pimentel et al. (1976) estimated that one-third of the nation's topsoil has been lost over the past 200 years, and Larson et al. (1983) estimated that on 30% of all cropland, erosion rates exceeded the level that permits a high level of productivity to be maintained economically and indefinitely. Pimentel et al. also estimated that 50 million tons of nitrogen, phosphorus, and potassium are lost each year due to erosion and that corn yields are reduced by 4 bu/acre for each inch of topsoil lost from a base of 12 in. or less. Pimentel et al. (1981) estimated that over a 30-year period a 37% increase in direct plus indirect energy inputs to corn production would be needed simply to maintain yields at their current levels if erosion continued at present rates.

The preceding discussion suggests an uncertain future for crop yields. One point can be stated definitely: due in large part to rising energy prices, the historic trend of 2–3% annual increase in crop yields has been broken. This alone is sufficient to question Simon's conclusions on future food prices, for they are based on extrapolations of yield trends that no longer exist, due in large part to constraints in energy supplies and perhaps also to declining soil quality.

A related issue pertaining to the future of United States agriculture seems especially important as of this writing in the spring of 1985. The issue has to do with the mounting debt, and resultant foreclosures and farm losses that face a large number of American farmers. The common response of most farmers to this problem seems to be an increase in their use of inputs, especially nitrogen fertilizer. One way of looking at what this means is that we are, as a nation, depleting our petroleum reserves at a more rapid rate in order to increase both the productivity of our farmers and their debts while we produce additional amounts of surplus foods. Is this free market solution to agricultural problems equitable and in our long term national interests? Would not it be better to regulate farm production by limiting the quantities of energy intensive inputs that we use so that each farmer was less productive, but would have lower costs and would need to borrow less? While the problem is certainly more complex than that it does seem to be an idea that might help the farmers while reducing our rate of depletion of our limited national petroleum reserves.

ENERGY, AGRICULTURE, AND WORLD POPULATION GROWTH

The publication in 1968 of Paul Ehrlich's *The Population Bomb* focused national and international attention on the enormous problems associated with human population growth and associated environmental problems. Although the problem had been clearly defined earlier (e.g., Malthus, 1778; Vogt, 1948; Borgstrom, 1969; Brown, 1978), the rapid advances in agricultural production, the pervasiveness of public faith in new technology, and the economic incentives for new workers to develop undeveloped farmland defused public and governmental interest in the problems associated with rapid population growth until Ehrlich's book. As of this writing, however, famine is pervasive in at least some parts of the world.

There are at least three dimensions to the population problem as it relates to and interacts with energy. First, in the industrialized world, where many raw materials are to some degree depleted, increased numbers of humans may mean less resources and less potential wealth per capita (e.g., see Ehrlich and Holdren, 1972). Second, in less developed, less industrialized parts of the world the same problem exists except that the deficient resources are not luxury items, as is the case for the developed world, but basic necessities of life such as food. Third, even if additional energy resources can be developed to meet global needs for food and for increased material standards of living, the additional energy use needed to feed the additional people would deplete premium energy stocks more rapidly and cause increased pollution.

A different view of population growth is offered by Simon (1981), who suggests that more people provide more workers, more ideas, and more wealth for everyone. Clearly this view is not supported by the analyses in this book, for we believe that more people and more per capita wealth deplete fundamental resources more rapidly than they add to it. Cook (1982) and Miernyk (1982) also take particularly sharp objection to Simon and others of that view, the ideas of whom Cook refers to as "economic creationism."

It is interesting to reconsider the Malthusian premise—that the human population increases geometrically (see p. 116) while food increases arithmetically—from an energy perspective. Although there is clear support for Malthus' assumption of geometrical population growth, it is not clear that food production increased at only an arithmetic rate. Since the time of Malthus the world human population has increased geometrically (Figure 6.1), but the mean level of nutrition has not clearly declined. Although the land area in cultivation has expanded as new continents have been colonized by new groups of people, it has not increased as rapidly as human population growth.

Often arguments are made that it is technology that has produced the large increases in food pro-

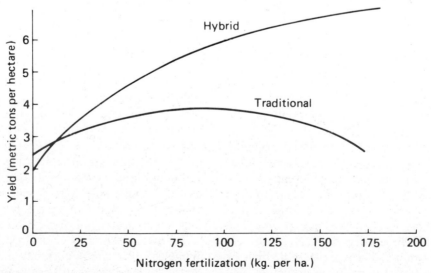

Figure 6.14. Yield of a typical hybrid wheat compared to a traditional variety as a function of nitrogen fertilization under controlled conditions. (From Chapham, 1981.)

duction that we observe, and that all we need to do is increase technology. For example, Green Revolution strains of wheat and other crops have been genetically engineered to produce much higher yield per hectare. In contrast, we believe, as stated in Chapter 2, that most modern technology is essentially a subsidy of human or natural energies with more industrial energy. The new genetic strains mentioned earlier often yield *less* than traditional varieties without large quantities of fertilizer (Figure 6.14; Brown, 1981). In other words, the technology selects those strains most responsive to external inputs. Thus most agricultural technology to date requires additional energy inputs for its realization, and although we have increased yields per hectare enormously, we are not increasing yield per unit of fossil energy used. Possibly new technologies might be developed that are not as energy intensive. For example, some experimental genetic changes select for strains that use nitrogen more efficiently. Other approaches attempt to combine nitrogen-fixing bacteria with higher-yield cereal grains, although the fixers themselves use a substantial portion of the grain's own energy, which decreases yields compared to what can be obtained with fertilizers. Thus far we have no evidence that there is any general decrease in the energy we use to produce 1 kg of food (e.g., Figure 6.9). Our own travels and research in the less-developed world convince us that, in general, these nations are industrializing their agriculture as rapidly as possible (Schlichter et al., in preparation). For example, most Central American countries use from 50 to 150% as much fertilizer per year per hectare as does the United States. The importation of these energy-intensive products has, since their large increase in price following the 1973 energy price increases, caused enormous economic problems for these countries.

SUMMARY

It appears that the widespread development and application of fossil-fuel-subsidized agricultural technologies has allowed a roughly geometric increase in food supply to match the world population growth. How far into the future this can occur is a subject for debate. If human populations continue to increase geometrically at current rates, the global population will be about 8 billion by the year 2000, double the number of 1970.

If agricultural production is increased chiefly through intensification of production on existing agricultural land, the energy required simply to feed the human population could be greater than the availability of fossil fuels. Hall (1975) developed a simple model to estimate the energy that would be required by a continually growing world population, using different assumptions about fossil energy required per kilocalorie of human food. The results of this simple analysis (Figure 6.15) indicate that if human population growth continues as it has in the past, and agriculture continues to be industrialized, the energy required for human food production will eventually exceed available fossil energy resources.

Another implication of this relation is that although the Malthusian crunch has been avoided for the time being, it is only because of the ready availability of cheap petroleum and petroleum-derived

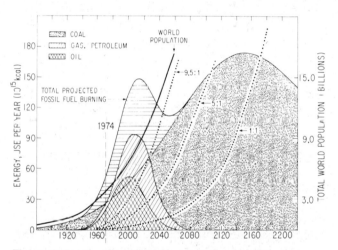

Figure 6.15. If present world population growth continues (solid line), how much industrial energy will be needed for food by people? We calculated this energy requirement (dotted lines) under two assumptions: Three billion people is about the maximum number that can be supported with nonindustrial agriculture, and all people above three billion will require industrial energy to raise the productivity of existing farmland or to grow food on soils not presently cultivated. Three estimates are used of the energy required for feeding these people: 9.5 industrial kcal per food kcal (U.S. average), 1 industrial kcal per food kcal (minimum estimate for a food producing, processing, and distribution system with no animal protein), and an intermediate, perhaps most likely, value of 5 to 1. These are plotted along with very rough estimates of potential world use of fossil fuels (Hubbert, 1968). The intersection of the energy required for food production and the available industrial energy might be considered as some sort of indication as to the maximum ability of the Earth to support people. (From C. A. S. Hall. © 1975 Educational Foundation for Nuclear Science, *Bulletin of the Atomic Scientists*.)

products such as fertilizers. As prices of these products have risen, most of the poorer nations that had become for a brief time food exporters, following the introduction of *Green Revolution* strains, have again become net importers (*Time,* October 17, 1981). This is one example of a common theme of this book—that energy as applied to agricultural production, and not technology per se, has allowed us to avoid the consequences of the mismanagement of agricultural land, the depletion of high-quality soil resources, and the steady increase in human numbers.

Finally, one school of thought argues, convincingly we think, that patterns of land distribution are more important in determining the amount of food people get to eat than population numbers or energy resource availability (see Lappe and Collins, 1977; Murdoch, 1980). Much of the land in hungry parts of the world (e.g., Guatemala) is devoted to large holdings that produce cash crops for export and for monetary gain to relatively few individuals. Food production could be increased there without more fossil fuels if subsistence crops were planted in smaller land holdings worked more intensively by many farmers. On the other hand export crops are not necessarily bad from the perspective of local food production because they are essential to get foreign exchange to trade for fertilizers. The People's Republic of China has essentially solved its problem of starvation, once perennially chronic, by rapidly increasing its use of fossil fuels and by intensive and sophisticated use of human labor.

We cannot predict now with certainty whether increasing agricultural productivity can (or can possibly) keep up with the large increases in the world's population that are expected. Recent trends are somewhat encouraging, but petroleum supplies must eventually become severely limited. Most of the world's most favorable farmlands are already developed for agriculture and the remaining farmlands of lower quality will—as we have pointed out—be more energy intensive to develop because they will require large inputs of fertilizers, water, or mechanical power to overcome their inherent deficiencies. An additional problem is that many of the fuel-intensive technologies that seem to work well in temperate regions (at least so far) may prove unproductive in other regions or over longer time spans. For example, the implementation of expensive fossil fuel technology to grow cotton in the Sudan *decreased* cotton production compared to traditional procedures (*Wall Street Journal,* No-

vember 25, 1981). Another well-known problem is that chemical pesticides, which often are very effective initially, may lose their effectiveness as natural predators of pests are killed and as pests develop resistance. Some cotton fields in Guanacoste province, Costa Rica, are now sprayed dozens of times a year with DDT when previously once or twice a year was sufficient. Finally, soil erosion can reduce the efficiency of turning fossil or human energy into food, and most fossil-fuel-intensive agriculture is damaging to soil (see Chapter 20; Brown, 1981). In many mountainous regions of Central America subsistence crop agriculture increasingly is replaced by cattle production for export to U.S. hamburger chains, producing severe soil erosion. Joseph Tosi of the Institute for Tropical Studies, San Jose, Costa Rica (personal communication), estimates that for every kilogram of beef produced on certain Costa Rican soils 100 kg of topsoil go down the rivers, degrading the future potential for food production from that land unless large fertilizer inputs are used.

Feeding the world's people in the future would be much easier if population growth diminishes or if more equitable distribution patterns can be arranged. Unfortunately recent history does not leave us optimistic. We have in general bailed ourselves out of our basic food (and other resource) problems with petroleum. In the future such solutions might not be available.

The United States has immense total energy reserves, especially as coal, compared to most of the rest of the world and, in addition, large agricultural production, both realized and potential. Many third world nations are becoming increasingly dependent on energy-intensive agriculture to keep up with population growth. As world oil production peaks and declines over the next several decades, moral and political pressure may build for the United States to supply coal, coal-derived fertilizers, and/or coal-grown food to many countries in the third world that have little to trade in return. Will we in the United States be called upon to deplete our own coal resources, strip-mine our lands, and acidify our skies to feed the world's hungry? To what extent will we be willing to do that? To what extent should we?

High input agriculture is increasing throughout the world because fossil-energy-intensive systems tend to be very productive, because there is a great deal of corporate and social inertia associated with our input-intensive agricultural systems, and because most of our research establishment is geared toward finding ways to increase production using petroleum and petroleum-derived chemicals. A superior alternative exists: develop less fossil-energy-intensive agricultural processes through the interactions of modern technologies, such as computer models for managing insect pests (e.g., see Shoemaker, 1977), with indigenous knowledge of local agroecosystems and cultures. Such schemes need not—indeed, cannot—turn their backs on the importance of energy. For example, an emphasis on renewable energy in development schemes may make long-term sense even when standard economic analysis would suggest that it is cheaper simply to purchase fertilizer. In many mountainous tropical regions hydroelectric power could be developed to produce fertilizers locally rather than continuing the importation of fossil-fuel-intensive fertilizers. Finally, a new generation of young scientists are combining ecological and agricultural sciences to develop systems of knowledge for understanding and managing agricultural ecosystems with an emphasis on lower energy and chemical inputs and on long-term protection of the soil. Such approaches could, if successful, replace the present bludgeon of fossil-fuel- and chemical-intensive agriculture with the rapier of agroecological and sociological knowledge. This is the world's great hope for food production in the long run, but it will be very difficult to implement.

EDITORIALS

ENERGY VERSUS ECONOMIC ANALYSES: WHO IS RIGHT?

The preceding chapters have developed a general method for analyzing human economic activity from a biophysical perspective. Among the most important concepts introduced were (1) energy investment, (2) the enhancement of human labor, and hence productivity, in the production process by the increasing use of energies, including especially fossil fuels, and (3) the interaction of energy and resource quality. Energy investment refers to the simple physical condition that free energy is required to upgrade matter into a more organized state, regardless of whether that matter is associated with natural or industrial ecosystems. The subsidization of human labor with other energies has made labor many times more productive than would otherwise be the case and therefore is largely responsible for our material intensive standard of living. Resource quality refers principally to the fact that individual deposits of natural resources differ in the amount of energy required to upgrade them to an economically useful state. Changes in resource quality, especially energy quality as measured by its EROI, play a significant role in determining what is and what is not possible in the human economy.

While developing an energy framework for economic analysis, we have sometimes criticized traditional economic analyses for ignoring and occasionally contradicting the physical laws that govern all economic processes. To adherents of traditional economic analysis, our criticism may appear too severe and often unjustified. In turn economists from the entire spectrum of economic thought—from strict neoclassical growth paradigm economists to steady-state economists—have criticized energy analysts for ignoring the role of human free will in developing technologies that have enabled us to overcome many of the scarcities imposed by na-

ture. The tone and objectivity of the debate between energy analysts and economists have ranged from coherent, rational review and criticism to irrational and empassioned rhetoric. Much of the acrimonious exchange has no doubt been due to a clash of paradigms, for it is difficult to assess objectively another person's approach and suggested solutions to economic problems when one views the economic process (and possibly the world in general) from a completely different perspective.

Nevertheless, analyzing the issues in this debate point by point is important because energy analysts and economists sometimes advocate different solutions to problems that affect us all. Yet there is no clear answer to the question, Who is right? Clearly energy analysis in the last decade has become established as a discipline that at a minimum has developed the tools to identify and assess empirically the flow of energy through the economy, and so it can be useful in the evaluation of some economic and energy policies. It is also true that traditional economic theories and policies have served us well since at least the 1930s. Equally clear, however, is the fact that during the past decade some of these theories and policies have been substantially less effective relative to previous years. It is no small coincidence that the beginning of this period of reduced effectiveness coincided with two important energy events: first, the peak in the early 1970s of U.S. domestic oil and gas production and, second, the quadrupling of real energy prices between 1973 and 1981 after a century of decline.

We present the debate between energy analysts and economists in a criticism-rebuttal format derived mostly from published accounts (e.g., Gilliland, 1975; Huettner, 1976; Huettner, 1982; Costanza, 1982; and others quoted later). Each section begins with a criticism of energy analyses levied by economists, followed by our evaluation of the validity of those criticisms. The reader should be aware of our own bias. We do our best, however, to present both sides as accurately as possible while trying to reduce some of the unnecessary polemic.

Statement. Energy analysis treats free energy as the only scarce resource (Webb and Pearce, 1975). In doing so, energy analysis ignores capital, labor, technology, matter, and other inputs, as well as substitution possibilities between them. This argument has been made by a number of analysts. Berndt (1983) states that energy analysis assumes that "there are no substitutes for energy." Alessio (1981) states that an energy theory of value fails to "recognize that production requires inputs other than energy." Huettner (1982) asserts that "the assumption that energy is the ultimate limiting factor is not only unacceptable but arbitrary," and that "one can just as easily . . . view technological change as the ultimate limiting factor."

Reply. An energy approach to economic analysis does not assume that "all commodities are nothing more than embodied energy," as Alessio asserts. Instead, an energy approach views economic production as a physical work process that requires free energy for sustainment. The quantity of energy needed depends on the quality of the energy used, the quality of the resources transformed, the prevailing technology, and the fuels used by laborers in their personal lives. We emphasize the importance of energy because, if the boundaries are expanded to portray the economic system as but a subsystem of a larger global system (see Figure 2.4), free energy is the only factor of production that must be supplied from *outside* the economic system and that is nonrecyclable and nonproducible (in a physical sense) by humans. Free energy, and natural resources in general, is the only factor of production that some combination of the other factors of production cannot physically create. An additional, more practical, reason for its importance is its new scarcity as manifested by the declining EROI of all major U.S. energy resources. The physical interdependence between free energy and the other factors of production has existed since humans first began to modify their natural environment for survival and other economic ends. The relation took on greater significance when civilization shifted its dependence from renewable to nonrenewable fuel sources. The dwindling stores of high quality fossil fuels on which we have become so dependent means that energy may now be the limiting factor in many economic processes. Such a case cannot be made for labor, at least at this time, because there is large unemployment and populations are growing, or capital equipment, because there is presently an underutilization of existing capital equipment. Using energy as an index does not in any way ignore capital, labor, and technology but instead recognizes that all other factors require free energy for their production and maintenance. If energy becomes scarcer, it will be difficult to develop new capital because of the energy opportunity cost.

That other factors or other natural resources have in the past (and will in the future) become limiting before energy is not an effective argument against the importance of energy analysis as Huettner (1981) and Alessio (1981) have argued. Previously shortages of other resources often have been mitigated by new technologies that nearly always required more energy directly and indirectly. We overcame (temporarily at least) water shortages in the West and Southwest by investing huge amounts of energy in wells, pumps, and pipeline networks to pump groundwater and to divert water from Rocky Mountain streams. As Chapter 4 documented, we mitigate the adverse effects of declining mineral quality by investing ever more energy to extract and beneficiate lower grade ores. As Chapter 6 illustrated, we upgrade the productivity of poor quality soil or soil depleted by erosion by using more energy in the form of direct fuel, pesticides, herbicides, irrigation, fertilizers, and sophisticated farm implements. Most important, we have offset to some degree the declining quality of fossil fuel by investing increasing amounts of energy in producing capital and technologies for the energy industries themselves. Meanwhile the United States is still running on fuels with high EROI found long ago. For example we still get more than half of our oil from fields discovered before 1940 (Nehring, 1981).

Georgescu-Roegen (1979), a leading contributor to a biophysical economic perspective, nevertheless criticized energy analysts for ignoring the importance of matter in the production process, stating that "matter matters, too" and that the "energetic dogma" assumes all matter is completely recyclable (i.e., that the dissipation of matter is completely reversible if enough free energy is available). The importance and supposedly neglected role of matter led to the formulation of Georgescu-Roegen's (1977) so-called *fourth law of thermodynamics*—no reversal of material entropy can be 100% complete.

Georgescu-Roegen's point is well taken—to a

degree. Clearly we cannot use energy without a "material lever, a material receptor, or a material transmitter." But it is *ordered* matter that makes it useful as a lever, receptor, or transmitter, and ordering matter is a work process that requires a minimum quantity of free energy—that is, the Gibbs' free-energy requirement. Moreover the qualitative difference between energy and matter seems to escape Georgescu-Roegen's analyses. Although friction, wear, and tear, and so on, degrades matter, making it less available in a manner analogous to the entropic degradation of free energy, matter can be recycled, often to a substantial degree. The second law of thermodynamics instructs that the recycling of free energy is not possible *at all*. Furthermore no energy analyst that we know has suggested that matter is 100% recyclable. The recycling of matter clearly is commonplace in the U.S. economy.

Statement. "When energy analysts undertake to evaluate projects using the criterion of net energy, various energy types are aggregated using (heat equivalent) conversion rates, i.e., it is assumed that all [heat equivalents] are identical in value to society" (Berndt, 1983). Other critics have made similar statements. Hyman (1980), for example, states that an "energy theory of value ignores variations in the quality of energy."

Reply. Berndt's (1983) assertion is correct to a degree, for many EROI calculations do not explicitly include quality factors for intercalibrating different fuel types. In contrast, price reflects many aspects of energy quality, or at least those reflected in the market. Energy analysts sometimes ignore the fuel quality value due to a lack of agreement on what the actual quality factors are rather than making the assumption that all kilocalories of all fuels are of equal quality. But contrary to what the critics maintain, energy analysis may be the best way by which differences in fuel quality can be identified. In Chapter 2 we described how equal quantities of different fuels can do different amounts of economic work and how those energy quality differences affected the fuel/real GNP ratio. In Chapter 2 we also presented various estimates that have been calculated for fuel quality factors. Regardless of what the exact values of the quality factors are, it is clear that wood is a lower quality fuel than coal, which is lower quality than petroleum, which in turn is lower quality than electricity for many appli-

cations. Also liquid fuels may have a high quality for, for example, transportation.

Energy analysis is a useful method for distinguishing between different qualities of fuels used in the production process, whether for the manufacturing of consumer goods or the extraction of additional fuel from the environment. In a report to Congress, the Government Accounting Office (1982) stated that energy analysis allows policymakers to identify processes that "result in the highest quantity or quality of fuels." For example, one of the often-stated goals of national energy policy is to reduce our dependence on foreign sources of energy, and high-quality liquid petroleum in general. Toward this end energy analysis can identify those domestic energy technologies that minimize the use of invested fuels or maximize the output of high-quality liquid fuel. For example, one analysis of the H-coal liquefaction process calculated an EROI of 4.3 when energy inputs and outputs of *all* qualities are included (GAO, 1982). However, the *high-quality fuel* EROI—the ratio of the liquid fuel produced to the liquid fuel used—was 9.4, more than twice the EROI for the overall process. Thus, given a particular energy policy objective, energy analysis can help identify those technologies that will achieve the stated objective. Contrary to the critics' claims, energy analysis is well designed to account for fuel quality differences, although an exact quantitative assessment is difficult. Of course price also has various biases, such as subsidies and a failure to reflect externalities, that make it an imperfect index of value.

Statement. Energy analysis is not better (in fact it's worse) than market prices in tracking natural resource scarcity. Webb and Pearce (1977) state unequivocally that energy analysis cannot identify future limits to exploitation of a resource.

Reply. Chapter 4 was devoted to discussing how energy can, in both theory and in practice, be used as a measure of natural resource quality. It also showed how relying solely on economic models of scarcity might be dangerous and misleading. Part Three will go into much greater detail on how the quality of many natural resources has declined in energy terms.

Contrary to Webb and Pearce's claim, energy analysis unambiguously identifies the ultimate economic limit to the exploitation of fuel resources and even gives predictions of when that will occur for a particular fuel. An EROI of 1:1 is an absolute physical and economic limit, since no economic work could be performed with a fuel that did not yield an energy profit, although a lower-quality fuel might be used to extract fewer calories of a higher quality fuel. The energy break-even point is an absolute limit that cannot be overcome by simply raising the price of the resource. Of course the actual EROI at which society as a whole cannot run is greater than 1 because fuel must be available both to recover new quantities of fuel and to make the machines that use the fuel.

A measure analogous to the EROI exists for measuring the quality of nonfuel mineral resources. The ultimate limit to the exploitation of these resources is in large part set by their energy opportunity cost, a concept developed in Chapters 2 and 4. As the quality of a resource declines, more energy must be diverted from other economic tasks to extract, beneficiate, and otherwise upgrade it to an economically useful state. The real cost of declining resource quality is the sacrifice of the other goods and services that could have been produced using the extra energy required to extract that resource from lower-quality deposits. The foregone use of energy to produce goods and services due to a decline in resource quality is the energy opportunity cost of those resources. Eventually society will decide that the energy opportunity costs exceed the utility derived from the actual use of the resource, at which point the resource effectively becomes a nonresource. As Cook (1976) notes, other factors, such as substitution of a less-energy-costly material or inclusion of environmental costs, may intervene before society reaches the limits imposed by energy opportunity costs. Energy costs, however, constitute an economic threshold that is becoming increasingly important in the use and degree of importation of many natural resources.

Statement.   Energy analysis ignores the issue of intertemporal allocations of resources (Webb and Pearce, 1977; Hyman, 1980). When economists evaluate a particular economic activity or project whose costs and benefits occur at different points in time, they use a *social discount rate* to calculate summed costs and benefits relative to the present. The discount rate reflects the assumption that consumers value present consumption more than future consumption, due to a variety of economic and social factors. That is to say, consumers discount the value of future consumption. For example, the promise of a car 20 years from now is not perceived

as valuable as a car received this year because conditions 20 years from now are less certain—we may have died by then, gasoline may be very scarce, and so on. Economists account for this phenomenon by correcting the dollar values of future costs and benefits with a discount factor, so that the same benefits in the future are made less valuable than those realized in earlier periods. The U.S. government even publishes official discount rates each year. Nevertheless, decisions based on cost-benefit ratios calculated with a discount are extremely sensitive to the discount rate chosen.*

Based on human time preferences, economists argue that the costs (energy inputs) and benefits (energy outputs) of energy supply technologies are spread over time, but most EROI calculations assumed that a kilocalorie supplied 10 years from now is equally as valuable as a kilocalorie supplied today. For example, an electric power plant takes several years to build (during which energy is used for its construction), but no electrical energy is generated until it is completed. This is followed by a 15- to 25-year period during which the facility generates power. During this period energy must also be used to operate and maintain the physical plant. The question is, Even if the price of electricity remains constant, do consumers of electricity value a kilocalorie of electrical energy today more, less, or the same as a kilocalorie 15 years from today? The application of a discount rate will make an investment in an energy-generating facility appear less favorable than a nondiscounted analysis.

Reply.   When and if society does value present use of energy more highly than future use, then a discount factor should be included when calculating the EROI of an energy supply process whose energy inputs and outputs are displaced in time. Since there is no *a priori* reason for assuming that consumers do not discount their consumption of energy like they discount their consumption of other commodities, then energy analysts may be justly criticized for not discounting their EROI calculations.

---

*Despite its widespread use there is by no means a consensus among economists concerning, on the one hand, what the appropriate rate of discount should be or, on the other, whether discounting helps to maximize the rate of resource use in the present at the expense of future generations. See Page (1977) for a discussion of the discount rate and its effect on the rate of resource use. Hall et al. (1979a) even have argued that oil, for example, should have a *negative* discount rate, as it is likely to be scarcer in the future.

This issue, however, has not gone undetected by all energy analysts. In general, society would rather consume energy now than invest it in a technology to produce energy in the future. Hannon (1981) recalculated the EROI for a variety of technologies using an energy discount rate. He describes the energy discount rate as the mechanism by which society implicitly expresses its desire to convert a present energy surplus into an energy supply process so that a greater surplus of energy can be created in the future, rather than consuming the energy now for transportation, home heating, and so on. Hannon's work breaks new theoretical ground in the field of energy analysis and marks the beginning of trying to deal with the important issue of social discounting from an energy perspective. Regardless a discount rate can be applied to energy analysis essentially as easily as to monetary analyses.

Statement. Energy analysis and an energy theory of economic value, like any other single factor theory of value, ignores the importance of subjective consumer preferences (Alessio, 1981). This criticism has been levied by a number of economists in a variety of forms. Hyman (1980) states that the "most serious flaw in the [energy analysis] approach is that it maximizes net energy while we are interested in maximizing social welfare" and that net energy "has no necessary connection to social welfare whatever." Georgescu-Roegen (1971) states that the "true 'product' of the economic process . . . [is] a psychic flux—the enjoyment of life by every member of the population." Similar arguments have been raised by a variety of economists representing a wide range of economic thought (see Huettner, 1981; Arrow, 1981; Daly, 1981). The argument boils down to this: energy analysts are said to emphasize a physical entity (free energy or embodied energy) at the expense of a psychic flux (human preferences), and economists are not comfortable with this method of economic analysis.

Reply. As we concluded in Chapter 3 on value theory, there is no unambiguous definition of economic value. Each definition has much to offer, and the proponents of each theory weaken their respective cases by overstating their virtues and ignoring the validity of particular aspects of the opposing theory. The authors of this book do not themselves embrace every aspect of the energy or embodied energy theory of economic value as proposed by, for example, Odum (1971) and Costanza (1980, 1981). We find even more unacceptable, however,

theories based solely on subjective human preferences when these theories implicitly ignore the biophysical constraints on human economic behavior which we develop in detail in this book. We agree with the extreme energy analysts that many human subjective wants may have little to do with those factors that ultimately lead to survival. We also agree with those many economists who believe that most day to day decisions on value are in fact purely subjective.

Consider the neoclassical statement that the economic process aims to maximize social welfare. What does the maximization of social welfare really mean and how can it be measured? In economic terms, it means that every individual maximizes his or her utility.* Put another way, the objective is to maximize the well-being of society. According to consumer theory people increase their well-being by increasing their consumption of goods, services, and leisure time. The psychic flux of consumer satisfaction is therefore related directly to physical entities: the utility derived from consuming goods and services and the utility of the goods and services foresaken by enjoying leisure time. Goods and services are the output of the production process, which is the process by which free energy is used to reorder natural resources to meet human needs and desires. As Chapter 2 documented, the production process is grounded in the biophysical world and therefore is limited by physical realities such as the entropy law. The laws of energy and matter impose fundamental rules on any process that depends on the use of energy to alter the organizational state of matter. It follows directly that the satisfaction of human wants and needs, derived from consuming the outputs of the production process, also must be influenced by basic physical laws. This is not to say that the laws of energy *determine* all human behaviors, for certainly they do not. It is equally arrogant, however, to ignore these physical principles when deriving models of the economic process. It is this glaring defect of standard economic theory with which we take exception.

Arguments about the absolute virtues of one approach versus the (an)other, however, miss the greater power that can be brought to bear on analyzing any problem when several different perspec-

---

*In consumer theory maximization of social welfare is achieved by maximizing $f(U_1, U_2, U_3, \ldots, U_n)/(1 + i)^t$, where t = time, U_1, U_2, etc., = the utility functions of individual members, and i = social discount rate.

tives are used. The increasing availability of computerized assessment programs should soon allow routine assessments of projects and policies using both economic and energy criteria. (See Hall et al., 1979a, for an example.) If the results of such analyses agree, the course of action is clear. If they disagree, then the reasons for the disagreement, if understood, provide a good point for further discussion and analysis.

POLITICAL PHILOSOPHIES AND THEIR RELATION TO ECONOMIC THEORY AND ENERGETIC CONSTRAINTS

Two opposing philosophies have dominated the American political scene during the past century; that of the Democratic party and that of the Republican party, representing the American peculiarities of the so-called liberal and conservative approaches to political issues. In the past decade the distinction has seemed less apparent to at least some observers. Nevertheless it is useful to examine the relation of these dominant philosophies to the energy situation over the last hundred years of U.S. energy and economic growth and to the recent decade of relative economic stagnation.

Traditionally the basic difference between Republicans and Democrats on domestic issues has been that the former group advocated a philosophy of less government intervention in domestic economic activity and the latter favored (in a relative sense) such interventions. Before about 1880 government intervention in the economy by either party was almost nonexistent—there appeared relatively little need for it (with some important exceptions including regulations of the railroad and the Homestead Act). The development of monopoly control and the resultant shift in income distribution under this hands-off policy, however, led to the beginning of federal government interventions in business operations and of course indirectly affected the way in which people interacted with resources. Theodore Roosevelt (a Republican) has at least the reputation of being the first president to regulate business and resources on a large scale through various trust-busting activities and by setting up a national system of parks and forest management reserves. One of Roosevelt's most renowned acts was the breakup of the Rockefeller's Standard Oil Trust in 1911. Widespread government regulation remained the exception, however, throughout the pe-

riod of economic, social, and industrial expansion that occurred throughout the early decades of this century, even though there was common acknowledgment of the problems of boom–bust economies.

Cyclical fluctuations in output and employment were recognized and considered undesirable by economists, but the rapid overall expansion of the economy in the 19th and early 20th centuries made such problems seem relatively unimportant. Per capita income in the United States, England, and other Western industrialized nations was at a record level and increasing at an unprecedented rate. Economic theory of this period was characterized by the classical economic doctrine as espoused by David Ricardo (1817), John Stuart Mill (1848), Alfred Marshall (1890), and others. A dominant theme in this doctrine was that government influence in the market should be kept to a bare minimum. Another important tenet of the classical doctrine was that chronic market gluts, characterized by deficit consumer spending and high unemployment, were essentially impossible. This theory was based on the hypothesis known as Say's law, named after the French economist Jean Baptiste Say, which held that new economic supply creates its own demand. As it was interpreted by many economists of the period, Say's law denied the possibility of general overproduction and economic depression. A notable exception to such analysis was the theory of Thomas Malthus (1820), who expressed concern over the possibility of economic gluts. Known more for his famous *Essay on the Principle of Population*, Malthus' macroeconomics were largely overlooked for 100 years because he was unsuccessful in debating the issue with Ricardo and other economists of his time. Keynes, however, acknowledged Malthus as an intellectual predecessor in his *General Theory*.

The collapse of the stock market in 1929 and the ensuing decade-long Depression clearly indicated that Say's law was no longer an appropriate model of existing economic realities. Economic thought was revolutionized in the 1930s by John Maynard Keynes (1936), who flatly rejected the classical economic doctrine and called for increased government involvement in the economy to stabilize business cycles and prevent severe depressions. The primary economic goals became full employment and the highest level of national income (GNP) possible given the level of employment. A principal means of increasing national income was increased government expenditures to compensate for perceived investment deficiencies in the private sector.

Franklin Roosevelt, who originally ran on a platform of reducing government intervention, turned increasingly to Keynesian tools in his New Deal policies. The Tennessee Valley Authority (TVA) and the Public Works Administration (PWA) are two examples of massive government expenditures aimed at creating jobs and stimulating economic growth. A very fundamental change had taken place in American politics based in large part on the Keynesian revolution in economics. People now *expected* the federal government to take a large and active role in managing the economy through the use of monetary and fiscal policies. Since the time of the New Deal, that view has been increasingly interpreted as maintaining conditions favorable for economic growth. Republican and Democratic administrations alike have used massive government spending as a means of maintaining high employment rates via a rapidly growing economy.

Economic policies based on Keynesian principles appear to have been effective in dampening oscillations in output and maintaining high rates of growth through the 1960s (Figure 2.11). Some recent economic history, however, is either unexplainable by traditional economic models or directly contradicts those models. These events have caused many analysts, both economists and noneconomists, to take a hard look at the relevancy of traditional Keynesian economic doctrines to rapidly changing economic conditions. There has always been a debate over the proper role of government in the economy, but the debate has taken on renewed vigor after the economic events of the 1970s and early 1980s. The supply-side economics of the Reagan administration is a clear shift toward reversing the historic trend of increasing government involvement in the economy. Supply-side economics, however, has not (at least yet) stimulated the sustained economic growth it once promised, and problems such as high interest rates, high unemployment and especially large deficits have worsened compared to preceding administrations.

We need to examine the Keynesian hypothesis (and its alternatives) from our view of energy and resources. The Keynesian concept in recessionary times was to *prime the pump* by government spending—if the government poured money into the economy, then domestic spending and production would be encouraged, providing more paychecks and further encouragement of production, jobs, and

so on. Assuming this process was successful, it translated into a mechanism for, in essence, pumping more petroleum from the ground to fuel the machines of production. As the dollars in the economy increased, so would the energy to support that economic activity, although probably few if any people at that time considered that relation. The system worked more or less, for more energy easily could be brought into the economy (although it took World War II with its very large government spending to really end the Depression). On the other hand, reducing aggregate demand through monetary policies and by inducing recession reduced the amount of general economic activity, which was clearly associated with a drop in the energy used to support economic activity (Figure 2.11). Since World War II, however, increased government spending became institutionalized long after the particular need for increased aggregate demand had been forgotten. But relatively few complained, for America and most Americans indeed became richer, year after year. One can argue from Figure 2.11a that the application of Keynesian economics worked, for the fluctuations of the economy empirically can be seen to have been reduced.

Increasingly, and most characteristically in the Lyndon Johnson presidency, there has been a tendency for federal spending to be aimed at certain social improvements, with the effect on aggregate demand a secondary, but related, issue. Thus a larger proportion of our society's energy use has been in the public sector, which is good, bad, or neutral depending on one's personal and social philosophy.

A strictly conservative philosophy of government, such as is more or less represented by the campaign rhetoric of recent Republican presidential candidates, would tend toward minimizing government control of aggregate demand and lower government spending. According to these principles a relatively unregulated market economy is the best way to allocate resources to the production of consumer products: the consumers, through their preferences in the marketplace, will create demand for products most needed or desired, and that will provide incentives through profit opportunities for allocating production capacity. To the degree that aggregate demand is regulated, it would be regulated through controlling money supply (a monetarist view), not government spending. Actions sometimes speak louder than words, however, and the massive debt spending of the conservative Reagan administration certainly can be viewed as a Keynesian approach to attempt to end stagnation.

One liberal view in opposition to this system of allocating resources is that historically such mechanisms for allocating production (again in our parlance, allocating where energy is invested) is manipulated by narrow mercantile interests. This results too often in an emphasis on the production of trivial consumer goods that give immediate satisfaction, often at the whims of manipulative advertising and often at the expense of perceived greater goods such as public health and environmental quality (which are not especially available through conventional markets) and education.

Whichever one's philosophy, it is important to recognize that completely unregulated markets within this century are basically a myth and that even the greatest advocates of unregulated market activity (characteristically big business and independent businesses) live in a world in which the corporate goal is to regulate raw material availabilities and markets—often through intensive government regulation of both aggregate demand and specific markets, a view that has been espoused and documented especially by the economist John Kenneth Galbraith (1968). Additionally the nations that we compete with economically most directly (Germany and Japan) are characterized by even stronger interconnections between government and industry. Increasingly, despite popular rhetoric to the contrary, the question is not whether government should or should not regulate but whether a specific regulation, or regulations in general, is favorable or not to a given industry or corporation. The presence of legions of defense and other industry lobbyists in Washington attests to that fact. Conceivably recent political trends might reverse the drift toward increasing governmental regulation, as has happened recently with the trucking and airline industries. The net economic benefits of this deregulation is uncertain. Following deregulation, air fares fell, then rose again.

What does all this have to do with our central theme, that of energy and its relation to economic processes? Obviously the allocation of productive capacity—through free markets, central planning, or the complex mix that characterize contemporary Western societies—is a means of allocating energy resources. Consumer preferences for vacations versus larger automobiles, or education versus larger homes, or government encouragement of medical care versus weapons result in energy resources being diverted to supply more of one and less of another. Likewise, governmental policies aimed at increasing defense or education spending, or reducing taxes, or protecting American automobile or textile industries from foreign competition change the allocation of available energy resources from one sector of the economy to another, or even from one nation to another. Furthermore, governmental programs designed to increase aggregate economic activity through policies of low interest rates, high spending, and so on, will, if successful, mean that more oil will be pumped out of the ground, coal mined, or imported oil purchased, than had they not been successful or had such an agenda not occurred at all. Since both major political parties have had as a goal the increased economic activity of the United States, and since that activity has generally occurred, a result has been the increased depletion rate of oil reserves and high quality ores. This is not to say that such economic policies are not desirable but only that they have certain costs that are borne increasingly by future generations, each of which presumably will continue to attempt to maintain and improve its own level of economic prosperity with yet lower grades of resources. A countervailing opinion is that as time goes on, we become increasingly clever in our uses of energy and other resources so that we can maintain a higher standard of living per unit of resources. This is our greatest hope for economic prosperity in the future, a concept whose desirability is very difficult to argue with and one that we support wholeheartedly. But as we showed in Chapters 2 and 4, it is not clear that a reduction of energy use per unit of real economic activity can occur on a scale sufficient to make up for the depletion of high quality resources.

Thus the energy approach to economic analysis has important information for both liberal and conservative approaches to the management of our economy. It is clear to us, at least, why the basic ideas of both liberals and conservatives could work before 1970 and have not worked too well since then. The reason is simply that high EROI fuels were available before 1970, and they could be drawn from nature readily to support whatever policy was in political vogue. But such resources are no longer readily available, and they extract an ever higher energy, environmental, and economic price, as more and more economic activity must be diverted to processing the energy and other raw materials that are our basis for wealth production. Thus future politicians may find increasing constraints on their campaign promises.

PART THREE

CHARACTERISTICS AND MAGNITUDE OF U.S. ENERGY RESOURCE SYSTEMS

Earlier chapters examined some basic laws of energy, the way these laws constrain biotic and economic systems, and how an energy perspective can be used to understand these systems. We emphasized that all ecological and economic systems have sunlight as a major energy input but that many natural ecosystems (e.g., oyster reefs) and our own industrial systems are subsidized by other energy resources—in these examples tidal currents and fossil fuels, respectively. In Part Three we examine the fossil and other energy resources available to the United States from the perspective of their characteristics and, where appropriate, their methods of formation, their modes of occurrence, the technology and history of their use, our methods for estimating their magnitude, and, where possible, the energy required for their production and the trends in their EROI over time.

THE FORMATION OF FOSSIL FUELS

The largest energy source used on earth by both biotic and industrial systems is the sun. Direct solar energy provides light that enables plants to grow and that drives weather systems. A second major source of energy on the earth is fossil (meaning old) fuels, which are the solar-derived residues of ancient ecosystems that have been saved from complete oxidation by what we might consider accidents of geology. Tidal motions and radioactive decay deep within the earth are also important sources of energy, but their direct effects in most ecological and economic systems are small except for the obvious case of tidally affected marine ecosystems.

THE ENERGY BALANCE OF ANCIENT ECOSYSTEMS

Living systems of the earth are in a delicate balance between the processes of construction (photosynthesis) and destruction (respiration, predation, and/or decay) of organic matter. Living plants generally have a positive balance between photosynthesis and respiration—that is, they fix more carbon and energy in the process of photosynthesis than they use in their own respiration. This is balanced elsewhere in the ecosystem by the respiration of animals (or consumers), bacteria, and decomposers and also by the periodic death of trees. On average there is no large change in the quantity of living biomass on earth from one year to the next.

Occasionally some energy-rich organic material is trapped in sediments with no access to atmospheric oxygen, so that decay organisms cannot entirely decompose the organic material. It is thought that these anoxic conditions are most often present in swamps, marshes, or shallow coastal basins, where anaerobic standing water retards oxidation and where river-born or coastal sediments can cover material and begin the process of fossil fuel formation. The large quantities of fossil fuels that are found inland (Appalachian coal, Colorado oil-rich shales) are in areas that were once part of vast shallow inland seas or marshes, and oil-rich desert areas such as Saudia Arabia have moved through continental drift from climates that were once biotically productive.

The global rate of both photosynthesis and respiration is about 10^{18} kcal/yr, some 30 times the rate at which global industry uses fossil fuels. The entire resource of fossil fuels represents the plant production of about 50 years. Since there have been extensive forests on the face of the earth for at least 500 million years, it is obvious that only a tiny fraction of plant production has been transformed into fossil fuels. Coal, gaseous and liquid petroleum, and oil shale are all derived from partially decayed organic matter, but the geologic conditions that lead to the formation of each of these fuels are different (Figure III.1).

Coal formation is favored in certain coastal and inland swamps and bogs where plant materials accumulate after a drop in sea level to form large, thick layers of organic matter, rich in cellulose and lignin, materials characteristic of terrestrial trees and other plants. In these sediments the amount of mineral matter deposited is far less than the amount of organic matter laid down. Conversely, the formation of petroleum and oil shales is favored in shallow marine environments, particularly in geologic times when a rise in sea level results in the flooding of shallow continental depressions. Under these conditions the organic matter is of marine origin and high in lipids, proteins, and carbohydrates. In these types of marine environments, the amount of mineral matter deposited is far greater than the amount of organic material. The differences in the relative amounts of organic matter and mineral matter are responsible for the different types of geologic formations the two fuels are found in. Coal is generally found in thick continuous seams, while petroleum is found dispersed in the pore spaces of sedimentary rock.

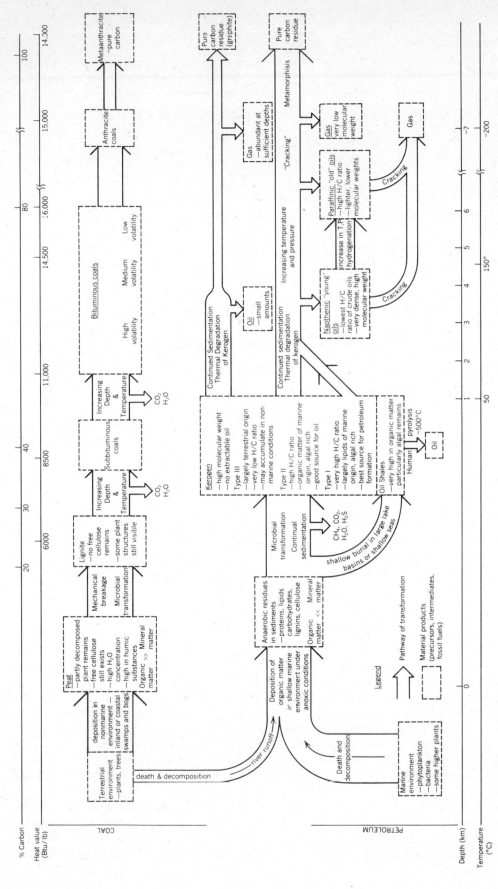

Figure III.1. The formation of fossil fuels. In general plant residues are transformed to higher quality fuel through the influence of heat and pressure over geological time, represented from left to right in the figure.

Special geologic conditions would sometimes concentrate these organic materials into deposits that are of high enough reduced-carbon concentration to be considered *economic fossil fuels*—fossil carbon that is economic to mine and use. Coal beds are common relative to oil or gas, extensive, and represent the remnants of organic material that have probably not moved very much relative to adjacent sediments. Oil and gas, being fluid, are much more mobile and are found only where peculiar conditions cause materials to be trapped—such as permeable sandstone overlain by impermeable shale. Since the special geological conditions that produce minable fossil fuel resources are relatively rare, it is important for society to assess their extent. About a third of the oil and gas industry's annual field expenditures of about $50 billion are soley for exploration procedures, through which geologists, geophysicists, and petroleum engineers gather and interpret geologic data in order to determine where economic deposits exist. Much less effort is needed to find coal because it tends to occur in much more continuous deposits.

Petroleum Formation

Petroleum, from the Greek meaning rock oil, refers to fluid fossil hydrocarbons that are available as fuel, either as oil (liquid petroleum) or natural gas (gaseous petroleum), or their derivatives. Petroleum formation begins with the death and subsequent deposition of organic matter in shallow marine environments or landlocked aquatic basins, and many of the major oil-bearing regions today are located in coastal environments. Since estuaries and continental margin regions are typically areas of high primary productivity, sources of organic material are abundant in these environments. Typical sources of organic matter in marine sediments are phytoplankton (especially diatoms), bacteria, and, to a lesser extent, higher aquatic plants and zooplankton. In addition, organic matter from terrestrial plants is transported and deposited via rivers in marine sediments. As will be discussed later, the source of the organic matter in marine sediments (terrestrial vs. marine) is a major factor in the process of crude oil and gas formation.

Retention of organic material is enhanced in anoxic environments, and such environments are necessary for the preservation of organic matter in freshly deposited sediments. The presence of oxygen would allow the consumption of organic matter to occur by both animals and aerobic bacteria as it filtered down through the water column and benthic scavengers as it was deposited on the bottom (Demaison et al., 1980). Upon deposition in the oxygen-limited environment, the organic matter begins to be transformed through anaerobic microbial activity. Anaerobic bacteria reduce sulfates and nitrates to gain their necessary electron acceptors and produce methane gas as one of their metabolic by-products; methane (CH_4) is the only hydrocarbon produced in these young sediments. The residues are progressively combined to form *kerogen*, an insoluble organic compound that is the main precursor to petroleum in marine sediments. Sedimentation continues to occur throughout these early stages of transformation so that by the time kerogen is formed thousands of years have passed and the sediments may be buried up to a kilometer deep. Kerogen and oil shales are similar substances. Oil shale is simply the name given to rock that contains large amounts of kerogen derived from deposits formed at shallow depths in lake basins, bogs and lagoons, or shallow seas, where the organic matter was not subject to temperatures and pressures sufficient to transform it completely into liquid oil or gas. The distinction between kerogen-rich source beds and the more restrictive term *oil shales* is determined by the relative degree of energy and dollar investment required to extract, process, and deliver each to society in an economically suitable form and content. Oil shales are kerogen-rich beds thought to be economically recoverable as a fuel.

After millions of years increasing temperature and pressure produced by sedimentation replace microbial activity as the main agent of transformation. In this stage the organic matter is still in the form of kerogen. With increasing depth the sediments containing the kerogen, termed *source rocks,* become increasingly compacted as water is expelled and porosity and permeability decrease. The carbon:oxygen ratio of the kerogen steadily increases as O_2 is driven off through the removal of oxygen-rich carboxyl (carbon–oxygen) groups. Extensive thermal breakdown of kerogen cannot occur until this decarboxylation process occurs (Tissot and Welte, 1978). With increasing temperature more and more chemical bonds are broken, molecules are rearranged, and purer and purer hydrocarbons are produced as other compounds containing sulfur, oxygen, and nitrogen are driven off. At some variable and as of yet ill-defined depth kerogen begins to be thermally degraded until the first liquid petroleum is formed. Most oil is formed at temperatures between 60 and 150°C, corresponding to burial depths of 1500–4500 m (Saxby, et al., 1974). As depth, temperature, and pressure continue to increase, many carbon–carbon bonds begin to be broken, molecular weight decreases, and the first gaseous petroleum is formed, either from the kerogen itself or from the further *cracking* of liquid petroleum. (Cracking refers to the temperature and pressure-induced breaking of carbon–carbon bonds to produce hydrocarbons of lower molecular weight. It is the same process used extensively by the petroleum-refining industry to transform heavier crude oil into more socially desirable products such as gasoline.) Even today it is relatively rare to find liquid petroleum below a 2500-m depth, although gaseous petroleum is common at that depth. Perhaps the heat and temperatures at that depth have cracked all of the liquids to gas. At even greater depths organic matter exists only in the form of metamorphized solid carbon, or graphite, as it is more generally known.

Not all of the hydrocarbons produced by the petroleum formation process are in forms or quantities suitable to meet the diverse requirements of modern industry. Society therefore invests large quantities of dollars and energy in the petroleum-refining industry to put the finishing touches on what the geologic energies of temperature and pressure did not or had not yet the time to accomplish. High molecular weight, low-quality fuel ores such as heavy oils or oil shale must undergo extensive and energy-intensive thermal and chemical processing before they have an energy quality and content high enough for them to be useful to modern industrial society.

Since liquid petroleum is the most important primary fuel input into the United States and world economies, it is important to understand the conditions that favor the formation of oil instead of gas. The source and nature of the original organic matter and the temperature under which it was transformed are the two most important factors governing this process. Petroleum (oil and gas) formation is dependent on the nature and abundance of kerogen, which in turn is controlled by the composition of the original organic input and by the type of microbial transformation that took place in the newly deposited sediment (Stephens and Spencer, 1957). Specifically, because kerogen is not a homogeneous substance, but rather exists in a variety of forms, each form leads to varying amounts and types of oil

and gas being formed. Kerogen can be subdivided into three types (Tissot and Welte, 1978). Types 1 and 2, derived from *marine* organic matter, have a high original hydrogen:carbon ratio and low oxygen:carbon ratio, and they usually result in the production of more oil than gas. Type 3 is derived mainly from *terrestrial* organic matter, has a low hydrogen:carbon ratio and high oxygen:carbon ratio and therefore has a low potential for oil formation, though it can yield substantial amounts of gas at greater depths. In turn the chemical transformations of kerogen are temperature controlled, so that the combination of these two factors—nature of organic matter and temperature—makes the entire

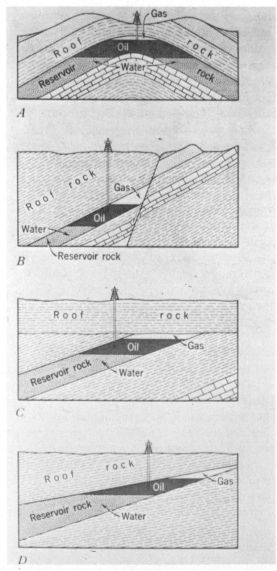

Figure III.2. Formation of oil and gas accumulations. (From R. F. Flint and B. J. Skinner, *Physical Geology*, 1974. Copyright © 1974. Reprinted by permission of John Wiley & Sons, Inc.)

petroleum formation process very specific to site and condition.

The thermal evolution of kerogen results in oil and gas, though normally at depths and in sediments that are not economically or technologically feasible to exploit. The pools and reservoirs of petroleum that we currently recover are the end result of the upward movement or migration of petroleum from the *source beds* where it is generated, through more porous and permeable *carrier beds,* to *reservoir rocks* where the oil and gas become trapped beneath a layer of impermeable sediment (see Figure III.2). It is only from these *traps* that oil or gas can be economically extracted. In the tiny pore spaces of the fine-grained sedimentary rocks of the source beds, petroleum exists in the form of minute droplets. As the sediments continue to become compacted, the droplets of oil are squeezed out of the fine-grained sediments into the larger pore spaces of the rocks above. On entering these larger pore spaces, the small oil droplets accumulate into large globules that tend to cling to each other as they move through the rock. Oil will continue to migrate upward, or laterally, in this fashion, always moving from regions of higher pressure to regions of lower pressure until it reaches an impermeable layer of sediments, where the oil accumulates in formations called traps. Traps are hydrocarbon pools that can be mined commercially. In very large oil- and gas-producing regions such as the U.S. Gulf Coast and the Middle East, this type of migration may occur over distances of hundreds of kilometers (Tissot and Welte, 1978). If the oil is not trapped, it reaches the surface and evaporates, often leaving a tarry residual. Regions where petroleum has been found are located throughout the United States, but

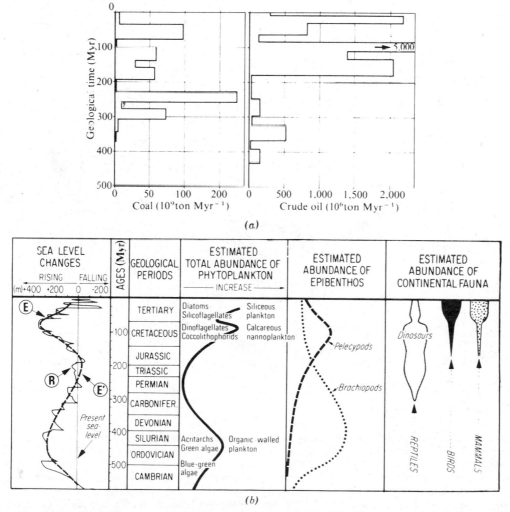

Figure III.3. (*a*) Formation of fossil fuel as a function of geological time. (*b*) Fossil fuel formation as a function of the abundance of phytoplankton, epibenthos, and continental fauna. (From Tissot, 1979).

most of the petroleum has come from Texas, Oklahoma, Louisiana, California, and Alaska (Figure III.4).

Hydrocarbon pools can be found at depths ranging from surface seeps to over 7000 m deep. The estimated average depth of all oil fields is about 1500 m (Tissot and Welte, 1978). It is interesting to note that the average depth of all petroleum wells drilled in the United States has steadily increased over time, at least until recently, as the industry has selectively found, developed, and exhausted those fields closest to the surface. One implication is that it is becoming more dollar and energy intensive for the industry to deliver the same amount of petroleum to society from these deeper and hence lower-quality fields. The recent reversal in the trend of deeper wells is due to the dramatic price increases of the 1970s which provided economic incentive to

exploit many shallow but small deposits. (For a more comprehensive discussion of the energy costs involved in producing oil and gas in the United States see p. 178.)

Coal Formation

The formation of coal is similar to that of petroleum in that it also is the result of the incomplete decay of plant organic matter that was subjected to burial through sedimentation and subsequent chemical and physical alteration induced by increasing temperature and pressure. For reasons that are poorly understood, the generation of coal has occurred more frequently than the generation of petroleum throughout geologic time, so that on a kilocalorie basis coal represents 82% of the U.S. known and

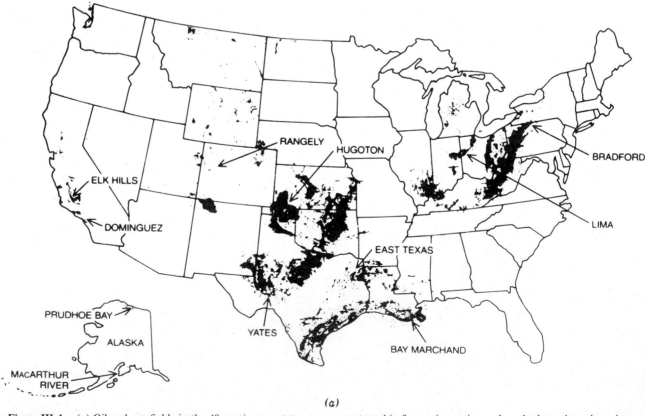

(a)

Figure III.4. (a) Oil and gas fields in the 48 contiguous states are concentrated in five major regions where hydrocarbons have been trapped in sedimentary rocks. The biggest oil discoveries in the 48 states have been in California and in east Texas. The map identifies 11 giant fields, most of which contained more than 500 million bbl of recoverable oil (or gas equivalent of oil). The Prudhoe Bay field on Alaska's North Slope, the largest on the continent, holds some 9.6 billion bbl. Of the total land area in the 48 contiguous states, 60%, or 4.7 million km², consists of sedimentary basins with oil-bearing potential. (From "Toward a Rational Strategy for Oil Exploration" by H. William Menard, *Scientific American,* January 1981. Reprinted with permission, W. H. Freeman and Company.) (b) Coal fields in the United States. (From DOE, 1980.)

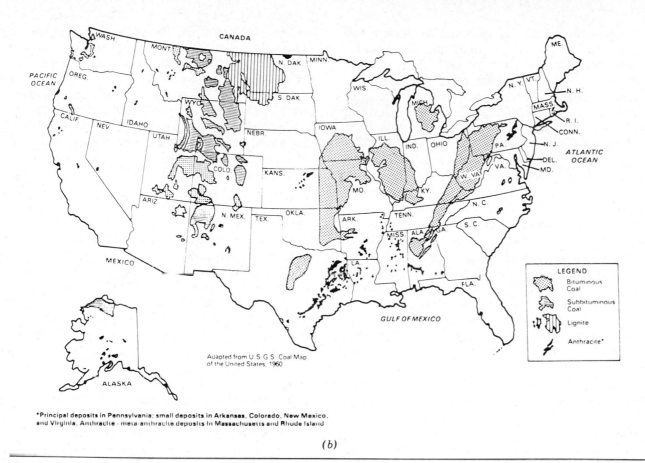

Adapted from U.S.G.S. Coal Map
of the United States, 1960

LEGEND

Bituminous Coal

Subbituminous Coal

Lignite

Anthracite*

*Principal deposits in Pennsylvania; small deposits in Arkansas, Colorado, New Mexico, and Virginia. Anthracite - meta-anthracite deposits in Massachusetts and Rhode Island

(b)

presently recoverable fossil energy reserves (Office of Technology Assessment, 1979).

Whereas petroleum was produced from the accumulation of lower aquatic plants and bacteria in marine sediments, coal was produced from the accumulation of terrestrial plants at the bottom of freshwater swamps and bogs. Unlike petroleum deposits, where the compressed fatty remains of marine organisms are dispersed between the granules of the sediment, coal forms from thick layers of plant material that contain far more organic matter than mineral matter. These original layers of plant material were laid down hundreds of millions of years ago and had initial thicknesses of up to 100 m.

The chemical composition of coal and petroleum also differs, due in part to the differences in their respective source materials. Petroleum has a carbon:hydrogen ratio of roughly 1:2 by atoms, similar to that of the original plant material from whence it came, although the specific number of carbon atoms linked together varies considerably. Large numbers of carbon atoms in chains or cyclic structures yield heavy crude oils and tars, and increasingly shorter chains yield lighter crude oils and eventually CH_4, which is the simplest of the fossil fuel hydrocarbons. Except for carbon, hydrogen, and a little sulfur, most other elements are virtually absent from oil and virtually all others including sulfur, are absent in gas (Teller, 1979). Conversely, the carbon:hydrogen ratio in coal is very different than that of the original plant material. Higher *ranks* (quality) of coal have higher carbon:hydrogen ratios and higher heating value per unit mass. In addition to carbon and hydrogen coal contains many other elements, such as sulfur, nitrogen, and oxygen, as well as many trace elements. The observed differences between coal and petroleum have been related to the role that oxygen played in the transformation of the original organic matter (Teller, 1979). In marine sediments, where petroleum is formed, bacterial transformation occurs under predominantly anaerobic conditions. In swamps and bogs anaerobic conditions predominate, but small amounts of oxygen also may be present during the bacterial decay of the plants. Since oxygen reacts more readily with hydrogen than with carbon, some of the hydrogen may have been consumed through its reaction with oxygen, the result being an increase in the carbon:hydrogen ratio in coal.

A final difference between petroleum and coal is the geologic formation that each is found in. Petroleum migrates from the site of formation to pools where the hydrocarbons exist as small globules that collect in the pore spaces of permeable sedimentary rock. Coal does not migrate from its source bed but instead is found in seams which are sometimes over 35 m thick at the site of original deposition. Cycles of alternately rising and falling sea levels resulted in successive layers of coal buried between layers of sedimentary rock in specific geographic regions.

As sedimentation continued, the resultant increases in pressure and temperature drove off water and organic gases (volatiles) and caused the fixed carbon content to increase relative to depth and time of burial. Coals are classified according to their carbon content, since the fixed carbon (or rank), and to a lesser extent volatile matter, determines the amount of heat released by coal on combustion. The amount of fixed carbon in coal varies from about 20% in some lignite coals to almost 100% in some anthracite coals. Generally, coal that has undergone the greatest compression for the longest period has the highest carbon content, lowest water content, and therefore, the greatest heating value. Figure III.1 illustrates the transformation of coal from peat to anthracite as a function of increasing depth, time and temperature.

Tissot (1979) illustrates the different geologic conditions necessary for the formation of petroleum rather than coal, showing that both are unevenly distributed through geologic time. Periods of rapid petroleum formation also were periods of low coal formation, and vice versa, due to the different settings under which each is favored (see Figure III.1). Tissot related this phenomena to the alternating periods of transgressions and regressions of the ocean that have occurred since the Precambrian Period, which ended some 580 million years ago. Periods of rapid petroleum source rock accumulation occurred during periods of global marine transgressions. During these times shallow epicontinental seas were formed which produced nutrient-rich waters that supported large phytoplankton blooms, the main source of organic matter in petroleum source rocks. Conversely, rapid coal formation was favored within coastal lowlands in swamps that developed after a drop in sea level. Large accumulations of the higher terrestrial plants necessary for coal formation flourished in these environments which prevented their dilution by inorganic materials (mineral matter). The ages of the major coal deposits of the world correspond to global regressions of the sea whereas the major petroleum source rocks of the world correspond to periods of global sea transgressions.

7

PETROLEUM

INTRODUCTION

Petroleum resources, including crude oil, natural gas, and natural gas liquids, are industrialized society's most important fuel. Since its discovery in the United States in 1859, the use of petroleum has increased rapidly in both relative and absolute terms to the point where it accounted for about 75% of total fuel use in the 1970s. We have already seen how the transition to a petroleum-based economy greatly increased the rate and amount of economic work done as opposed to coal-, wood-, or agrarian-based economies. Despite the fact that petroleum has been a significant source of fuel for less than a century, high rates of petroleum use and rapid economic expansion have become the expected way of life for many people. From 1900 to 1973 petroleum use grew by about 3%/yr (Figure 7.1) while real GNP grew at about 4%/yr.

The idea that future economic growth might be constrained by fuel availability has not been considered a real possibility by most analysts until very recently. Even now many resource analysts, economists, and political leaders appear quite unconcerned about the trends of U.S. petroleum discoveries and production. Until the world economy and its political and economic leaders were shocked by the economic and social effects of the oil embargo in 1973 and the Iranian situation in 1979, little importance was attached to the dependence of our economic well-being on the availability of petroleum resources. This delusion was fostered by government reports of substantial amounts of undiscovered petroleum, analyses unsupportable by the physical realities of the oil discovery process. A 1972 report by the U.S. Geological Survey (USGS) reassured us that undiscovered liquid hydrocarbons amounted to 450 billion barrels (billion bbl), a rather comfortable amount compared to cumulative domestic production of crude oil through 1972 of

about 103 billion bbl, proven reserves of crude oil of 36 billion bbl (Figure 7.2), and annual domestic use of about 6 billion bbl. In addition crude oil prices in 1972 were \$3.40/bbl in constant dollars, about what they were in 1910 (Figure 7.3). The illusion of abundant domestic oil supplies was exposed in 1973, when the United States and other oil-dependent nations were thrust into an economic crisis due to the actions of a few seemingly unimportant Middle Eastern nations. But lessons of the 1973 embargo were lost on political and economic leaders and the general public despite the severe economic repercussions caused by the embargo. After modest declines in 1975 and 1976 oil imports increased rapidly in the late 1970s, reaching 3.2 billion bbl (48% of U.S. petroleum use) in 1977 (Figure 7.1). Domestic crude oil prices again declined in real terms between 1976 and 1979. Lulled into a false sense of resource security, oil-importing nations were again shocked by the effects of the Iranian situation in 1979.

Clearly, domestic petroleum availability did not change radically overnight in the autumn of 1973. The energy crisis was not due to a sudden domestic shortage of petroleum. Instead, economic and political policies did not account for our steadily increasing dependence on foreign sources of oil as domestic demand outstripped domestic supply. This trend increased the economic leverage of foreign suppliers of oil, allowing them to increase the price they could charge for a barrel of oil. The energy crisis and ensuing economic and social disruption was due not to a conspiracy of the monopoly power of the multinational OPEC oil cartel, misguided government regulatory policies, or lack of proper economic incentives in the domestic exploration and extraction industries, although those factors exacerbated the situation. It was the failure of economic and political models to account for and accommodate the sensitivity of the economy to

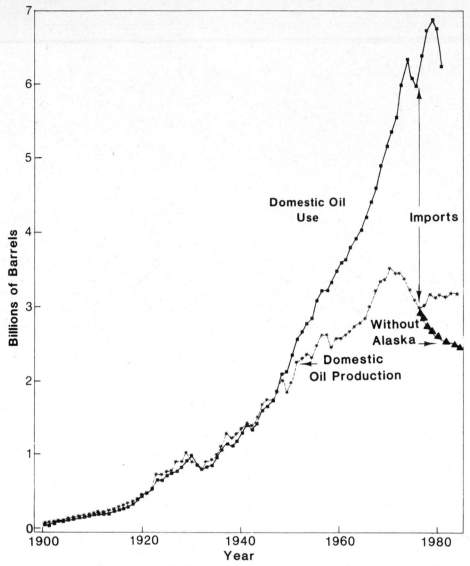

Figure 7.1. Crude oil production and use in the United States, 1900–1983.

changes in the physical availability of its resource base that made us so vulnerable to interruptions in foreign petroleum supply. The price shocks resulting from those events finally awoke decision makers to the resource availability problem that had been slowly developing since at least the mid-1940s, when domestic demand for oil first exceeded domestic supply.

By the time of the oil embargo in the fall of 1973, the following landmarks had already passed in the U.S. petroleum industry:

1. Oil discoveries peaked in the decade of the 1930s and have declined steadily since.

2. Domestic oil production peaked in 1970.

3. Proved reserves of oil in the lower 48 states peaked in 1961. Including Alaska, reserves peaked in 1970.

4. Domestic oil demand exceeded domestic production in the years immediately following World War II. The gap widened steadily through the mid-1970s to the point where imports comprised almost half of domestic oil use by 1977 (Figure 7.1), although they have decreased since then.

The economic significance of these events was largely unnoticed except by a few resource analysts

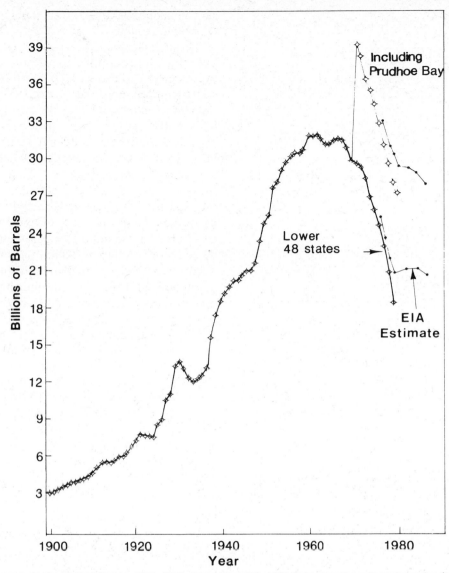

Figure 7.2. Oil reserves in the United States, 1900–1982. (Data for 1900–1979 from American Petroleum Institute, for 1976–1982, Energy Information Administration, USDOE.)

but will undoubtedly prove to be a historic landmark in the economic and social development of the United States. We begin this chapter on petroleum by examining some of the reasons why the four events cited above occurred when they did, based on the pattern of physical distribution of petroleum resources and how those resources are located and extracted over time. Different views as to how much petroleum remains to be discovered in the United States will be presented and discussed. Because accurate assessments of oil and gas availability are vital to sound economic planning, we examine the different methods available to analyze past

events in the oil and gas industry and then methods used for predicting future behavior. Finally, we end with a discussion and comparison of various estimates of remaining oil and gas resources, with special attention to the different methodologies employed and the role that the federal government has had in shaping public perception of petroleum availability.

Although the domestic petroleum industry has been one of the most intensively analyzed sectors of the economy, there exists a wide range of opinion on how much longer petroleum can be a major fuel source for the nation. Nehring (1981) states that

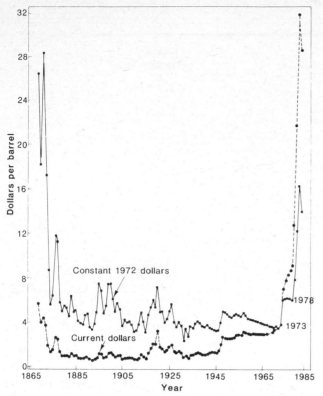

Figure 7.3. Domestic wellhead prices for crude oil in both current and constant dollars.

set the scene for the 1973 oil crisis and the ensuing economic troubles.''

These disagreements would clearly lead to vastly different energy and economic policies. If the situation is closer to the way Hubbert and Nehring describe it, then there must be relatively rapid development of alternative fuel systems with reasonably high EROI ratios if historic rates of economic growth are to be maintained in the future. Conversely, if domestic supplies of petroleum are as abundant as McKelvey and Commoner purport them to be, it is unlikely that economic activity in the near future will be constrained by the availability of fuel. To resolve such disagreements, the relation between two important parameters must be analyzed: the manner in which oil and gas deposits occur in the Earth's crust and how that distribution, combined with the way the industry searches for and extracts petroleum, influences the rate at which we discover and extract petroleum over time. These factors can be combined to determine how the energy costs and gains from petroleum drilling and extraction change over time, as reflected by the EROI for domestic oil and gas. The EROI for fuel resources will eventually be the most important determinant of the economic usefulness of that fuel because petroleum and other fuel resources will cease to be net sources of energy to industrialized economies long before they are physically exhausted.

''most of the conventional petroleum that will ultimately be produced in the United States has already been discovered and made available.'' M. King Hubbert (1974), one of the foremost analysts of the industry, instructs us that ''it may well be, therefore, that the year 1970 will prove to have been the most fateful year in the complete cycle of petroleum production in the U.S.—that dreaded year which marks the divide between an era during which crude oil production could be depended upon to increase . . . and a following era during which the rate of production may be expected to decline.''

On the other hand, in the same year as Hubbert's statement, V. McKelvey, former director of the USGS stated that ''our identified reserves of oil and gas and our potential reserves of other types of hydrocarbons in presently subeconomic resources are large enough to meet projected consumption through and beyond the end of this century.'' In the following year Commoner (1976) asserted that ''we now know that there is no *physical* reason for the sharp decline in the rate of domestic oil production that made the country dependent on foreign oil and

THE AVAILABILITY OF OIL AND GAS: SOME BASIC CONCEPTS

The introduction to Part Three outlined the unique geochemical processes that led to the formation of hydrocarbons and the special geologic conditions required to trap the hydrocarbons in formations that can be located and tapped by humans. The largest accumulations of petroleum in the United States cover immense areas. For example, the largest natural gas deposit in the Western Hemisphere is the Hugoton-Panhandle field in Oklahoma which covers an area of over 18,000 km[2] (Nehring, 1981). The majority of the deposits, however, encompass much smaller areas, often only a few square kilometers, an extremely small figure compared to the 4,700,000 km[2] of outcrops of sedimentary rocks that have oil-bearing potential in the 48 contiguous United States (Cram, 1971).

Because petroleum deposits are dispersed widely throughout the total potential petroleum-bearing sediment, the oil industry has developed progressively more sophisticated exploration techniques. In early stages of petroleum exploration before the turn of the century, oil and gas were found through relatively nonscientific methods such as drilling near surface oil seeps or by drilling a random well. These techniques were followed by the development of surface geology mapping techniques that relied on structural mapping of surface features, the presence of oil or gas seepages at the surface, and also by geologic interpretation of aerial photographs and other remote sensing data. Later developments enabled geologists to estimate contours of subsurface rock formations by monitoring the echoes of shocks produced by explosives detonated below the surface or from special vibrators. Geophysical techniques also involve the analysis of gravity or magnetic surveys, often by remote sensing.

Despite the great advances that have occurred in oil exploration technology, the only sure way to determine the presence or absence of petroleum is to actually drill into the formation of interest. A well drilled in a currently nonpetroleum-bearing region in search of petroleum is called a wildcat exploratory well. The process is called *exploratory drilling* or wildcatting. Wildcatting is at best a very risky venture. In the early 1980s only one in six to seven wildcat exploratory wells yielded producible quantities of oil or gas. Additions to reserves made by exploratory drilling are called *new field* discoveries or *new pools in old fields*. Oil or gas in a known field added to reserves after the initial year of discovery of the field are called *revisions* or *extensions*. Large deposits tend to be underestimated initially, and small fields tend to be overestimated, although the industry has become increasingly more accurate in its initial assessments of the ultimate size of newly discovered deposits. Once a deposit has been located, *development drilling* is done to define more precisely its size, and also to begin extracting the petroleum. In many cases development drilling subsequent to a discovery indicates that either more or less petroleum actually exists than was initially estimated, leading to a revision and/or extension of reserve estimates.

Only a fraction of the total amount of oil or gas in a deposit (termed oil-in-place) can actually be extracted, usually between 30 and 35% and occasionally as high as 60%. In some deposits the hydrocarbons flow freely from within the pore spaces of the reservoir rock into the well hole and to the surface without pumping due to the pressure exerted by natural gas which collects above the liquid hydrocarbons. This process is known as gas drive. Figure 7.4 shows recovery efficiencies for U.S. oil fields as calculated by Steinhart and McKellar (1982). Recovery efficiency has generally declined since the 1930s because many fields discovered now are much deeper and smaller. Deep formations tend to be less permeable than shallower ones, which results in a reduction in the ability of fluids to flow and therefore lower recovery rates. *Enhanced recovery* techniques like water flooding or steam or CO_2 injection can increase the percentage of original oil in place that is extracted but are very expensive in both dollar and energy terms. Such tertiary oil recovery techniques often require at least 2.5 times as much energy to recover a barrel of oil than conventionally produced crude oil, and with present techniques recovery rates are rarely increased by more than 2–10%.

Physical Factors Affecting the Supply of Oil and Gas

There are at least five factors that limit the supply of oil and gas that can be recovered at a net energy profit:

1.   The most obvious is that petroleum is a nonrenewable or *stock* resource. Because it is a stock resource, its rate of use by humans must eventually fall to zero after a certain period of use. Increases in human technical capabilities can increase only the rate, or sometimes the total proportion of that stock eventually recovered with a net energy profit via improved exploration and recovery techniques. Technical advances cannot, of course, increase the amount of resource that actually exists in the crust of the Earth.

2.   The majority of petroleum that can be recovered at a net energy profit is found in very large, relatively rare deposits. Due to their sheer size, most of the larger deposits (and hence most of the oil and gas) tend to be found in the early stages of exploration and development. This is true worldwide for any geologic region or nation.

3.   As large deposits within a region are discovered, the rate and amount of annual discoveries of new oil and gas deposits reach a peak, followed by

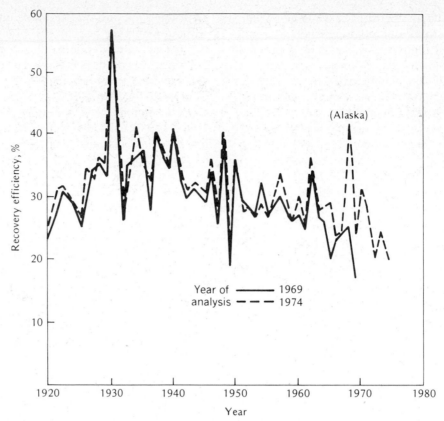

Figure 7.4.   Oil recovery efficiency in the United States (all oil is credited to year of discovery). (From Steinhart and McKellar, 1982.)

peaks in the amount of proven reserves and the rate of oil and gas extraction.

4.   As exploration and development continue to increase, the average size of new discoveries declines, and the larger, higher quality deposits are pumped harder in an attempt to compensate for declining discoveries. The net result is that there is a decline in the amount of petroleum found per unit of effort expended in exploration and extraction. This decline is commonly calculated in terms of the number of barrels of oil and gas discovered per unit of drilling effort, or simply yield per effort (YPE).

5.   Finally, and most important, the combination of increasing energy costs associated with high levels of drilling and other types of exploration effort required to discover and extract less accessible deposits causes a decline in the EROI for petroleum extraction as the industry matures.

The Finitude of Petroleum Resources

It may seem hardly noteworthy to observe that the exploitation of a nonrenewable resource like petro-

leum must follow a pattern of beginning at zero, rising slowly to one or more maxima, and then declining gradually to zero. Yet a comprehensive analysis of nonrenewable resources based on this principle can yield important information and even predictions regarding the precise timing and magnitude of the cycle. The first to apply the concept of the *production growth cycle* to petroleum resources was Hubbert (1949, 1956) who modified Hewett's (1929) model of cycles in metal production and applied it to U.S. resources of fossil fuels and uranium. Hubbert proposed that the production cycle for a nonrenewable resource follows the pattern illustrated by the bell-shaped curve in Figure 7.5. Production rates begin slowly as exploration and extraction technologies start to develop, and because markets for the new resource are relatively limited. The initial phase is followed by a very rapid increase in the rate of production, culminating in one or more maxima, after which production again slowly declines to zero as the resource is depleted. The area beneath the curve in Figure 7.5 is equal to the total amount of the resource produced. Hubbert found that those fuel extraction industries like an-

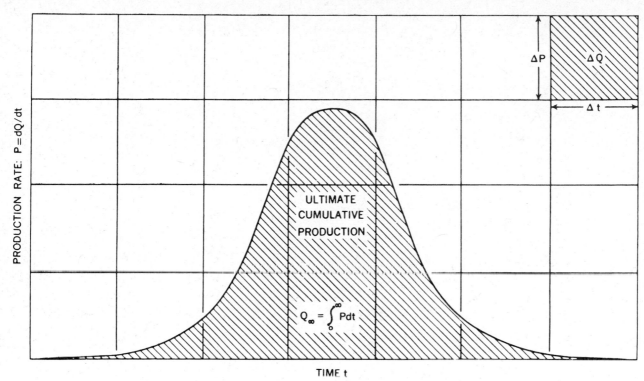

Figure 7.5. The production growth cycle curve for the extraction of nonrenewable resources. The area beneath the curve represents the total amount of the resource to be extracted and made available to society. (From Hubbert, 1956.)

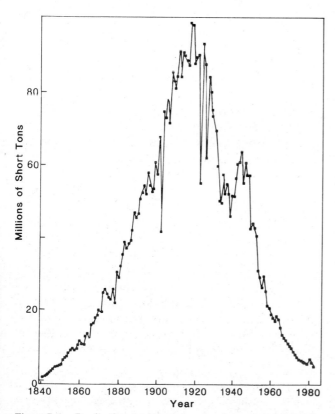

Figure 7.6. Production growth cycle curve for anthracite coal in the United States, 1840–1982.

thracite coal mining that were in a stage of development mature enough to test this hypothesis did in fact show a production cycle similar to the one predicted (Figure 7.6).

The production growth cycle curve can be used to estimate the period(s) of maximum production rates, after which production must begin to decline. To make such calculation(s), however, first requires an estimate of the total quantity of the resource that will ever be found and produced, a quantity referred to as Q_∞. In the early 1950s Hubbert and other oil analysts, among them Wallace Pratt, estimated that ultimate production (Q_∞) for the contiguous lower 48 states would be between 150 and 200 billion bbl of crude oil. Hubbert and his colleagues' estimate was based on their subjective evaluation of the petroleum-bearing potential of unexplored sedimentary basins in the United States. Using this estimate Hubbert (1956) fit the model shown in Figure 7.5 to production data through 1956 and estimated that domestic oil production would peak around 1966 for Q_∞ of 150 billion bbl or 1971 for Q_∞ of 200 billion bbl, respectively. History has shown this early estimate to be quite prophetic, for oil production actually peaked in 1970, and subsequent analyses performed independently of Hubbert's original technique confirm the general accuracy of his original estimate

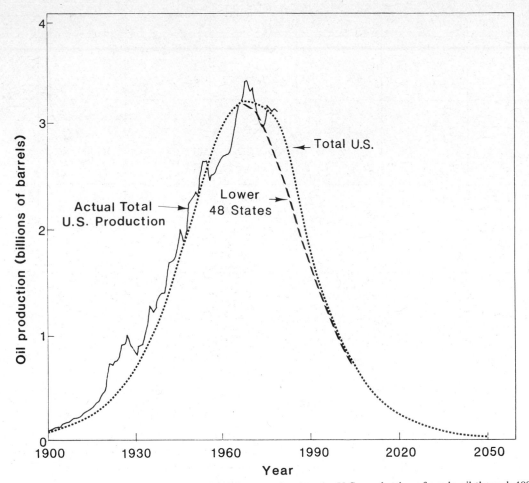

Figure 7.7.   The production growth cycle curve fitted to the U.S. production of crude oil through 1980.

of Q_∞. Figure 7.7 shows the actual oil production history for the United States and the Hubbert growth cycle curve fitted to that data.

The Significance of Large Oil and Gas Deposits

The dominant role that relatively few large petroleum deposits play in the total supply of petroleum has been documented by Menard and Sharman (1975), Meyerhoff (1976), Nehring (1978, 1981), Root and Drew (1979), Menard (1981), Drew et al. (1980), and Wiorkowski (1981). The trend is the same regardless of whether one considers the whole world, a single nation, or a specific geologic province within a nation: the few very large fields are discovered first, even with relatively unsophisticated technologies and small amounts of drilling effort. Discovery rates, as a function of both time and the level of drilling, tend to be very high in this

initial period. As we shall see, so is the EROI. As development of the resource continues, the average size of new fields decreases while exploration and development technologies become more sophisticated and drilling rates increase. Thus the yield per effort begins to drop. Finally, in the waning phase of the industry, large amounts of drilling effort and the most advanced technical capabilities are required to locate increasingly smaller fields and to pump harder the old high-quality fields.

The American Association of Petroleum Geologists (AAPG) devised a classification scheme for aggregating fields of similar size for analytical purposes (Table 7.1). All fields that contain at least 10 million bbl of liquid or liquid equivalent hydrocarbons are termed significant oil or gas fields.* Fields

---

*In order to express all hydrocarbons in the common unit of barrels, natural gas is often converted to its thermal equivalent in barrels based on conversion rates of between 5800 and 6000 ft^3 of gas per barrel.

Table 7.1.   Nehring's (1981) Modification of the American Association for Petroleum Geologists' Field Size Classification Scheme

Field Size Category	Field Size[a]
Class AAAA	500 million bbl or more
Class AAA	200–500 million bbl
Class AA	100–200 million bbl
Class A	50–100 million bbl
Class B	25–50 million bbl
Class C	10–25 million bbl
Class D	1–10 million bbl

Source: From The Discovery of Significant Oil and Gas Fields in the U.S. by R. Nehring, 1981. Reprinted with permission of the Rand Corp.

[a] Expressed in million barrels (mbl) of petroleum liquids and natural gas expressed in liquid equivalents.

with more than 50 million bbl are referred to as *large* fields, and those with more than 500 million bbl are called *giant* fields.

Nehring (1981), in the most comprehensive empirical analysis of U.S. oil resources, found that about 14,000 fields of all sizes in about 20 major petroleum provinces had been discovered through 1976. Of these only 2471 are classified as significant (i.e., more than 10 million bbl). Only 81 out of the 2471 significant fields are giants, but these contain 40% of all known recoverable petroleum. Less than 10% of known recoverable petroleum is found in the many thousands of smaller fields. Some 63% of recoverable oil is found in class AAAA fields (500 million bbl of oil equivalent), but these fields represent less than 5% of all the size C or greater fields discovered through 1975 (Figure 7.8). The two largest fields discovered to date, Hugoton-Panhandle and Prudhoe Bay, contain 45% more liquid petroleum than all the Class D and E fields combined. The four largest petroleum provinces (Gulf Coast, Permian, Andarko-Amarillo, and East Texas) contain 61% of known recoverable petroleum resources. Simply put, we rely on a very small percentage of known fields to supply a very large percentage of our oil and gas. Most of these fields were discovered before 1940 with the notable exception of Prudhoe Bay, discovered in 1968.

Nehring also found that total oil discoveries peaked in the 1930s corresponding to the peak in discovery of giant fields, while discoveries of significant oil fields peaked between 1940 and 1950 (Figure 7.9). Significant discoveries lagged behind

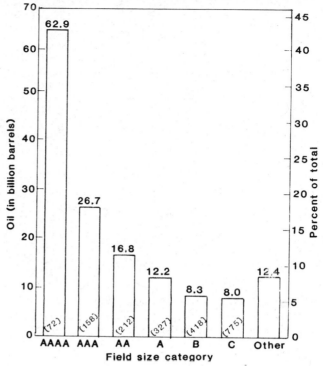

Figure 7.8.   Distribution of crude oil discovered before 1976 in the United States as a function of field size. The numbers in parentheses refer to the number of fields in that particular field size. (From Nehring, 1981.)

total discoveries because the large land areas covered by giant fields make them easier to find than the smaller classes (B–D) of significant fields. Nehring found the situation to be similar for natural gas, although the trends for gas tend to lag behind those for oil by one or two decades. Gas discoveries also are highly concentrated in relatively few, large fields. For example, the Hugoton-Panhandle field in the mid-continent region contains 10% of all known domestic recoverable gas resources. Total gas discoveries peaked around 1940, and significant gas discoveries peaked in the late 1950s. Gas discoveries generally lag behind oil discoveries because there were few markets for natural gas until after World War II and because technical capabilities of drilling very deep wells (>15,000 ft), where many gas deposits are located, took many years to develop. There was a slight reversal in the decline of gas discovery rates in the early 1970s when the offshore waters of Texas and Louisiana were opened for extensive exploration, but this reversal proved to be a temporary one.

Another moderate reversal in gas discovery rates may occur after 1985 when the wellhead price of gas

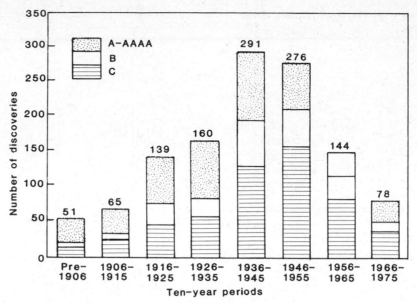

Figure 7.9. The number of significant oil discoveries in the United States by the field size category through 1975. The number of significant discoveries peaked in the decade around 1940. Total oil discoveries, however, peaked a decade earlier, illustrating the overriding importance of the very large fields. (From Nehring, 1981.)

will no longer be held below its market value by federal government regulation. Exploration for natural gas has undoubtedly been discouraged somewhat by artificially low prices, so an increase in exploratory effort for gas seems likely after 1985. The extent to which such drilling will add to our gas reserves is unknown. Deregulation of oil prices in 1980 helped stimulate the search for new oil, but did not result in significant new oil discoveries. On the other hand, natural gas has not been developed as long as crude oil, so a moderate surge in discoveries after deregulation is possible.

The moderate reversal in gas-finding rates when certain offshore regions were made available for development illustrates an important point: if there is to be a significant reversal in petroleum discovery rates, it will come only from *frontier* areas—areas where the oil industry has yet to explore fully. In the United States these areas include sediments deeper than 15,000 ft, the outer continental shelf region, most of Alaska (onshore and offshore), and the overthrust belt in the Western United States. Because it is very likely that the vast majority of the petroleum we will ever use from fields in the lower 48 states has been found and made available (Nehring, 1981; Hubbert, 1974), it is only from frontier areas that new major discoveries are likely to occur. The recent discovery of the Point Arguello field during 1982 in the offshore waters of California is an

example of the importance of exploring frontier areas, although even if Point Arguello turns out to be a billion barrel field, it would be equivalent to only a 2 to 3 month supply of oil at current rates of use. Figure 7.10 illustrates how the discovery of large oil and gas deposits in frontier regions (especially offshore) has helped mitigate the consequences of declining discovery rates for large fields in mature areas in the 1970s. Nehring (1981), however, notes that even large and giant offshore discoveries declined by over 60% during the 1970s.

Despite the need for continued exploration of new frontier regions, two important considerations should be noted. First, rapid oil price increases following the oil crises of the 1970s and deregulation of domestic oil prices in 1980 have encouraged and made financially possible unprecedented levels of drilling effort (Figure 7.11). Many frontier provinces originally thought to be quite promising have instead proved to be very expensive nonproducers. Examples are Cape Sable, Georges Bank, the Baltimore Canyon, and the Gulf of Alaska. Second, and more important, exploration and development in most frontier regions is much more energy and material intensive than in more hospitable environments in the lower 48 states. It requires five to seven times as much energy and materials to drill a foot of hole offshore in the Gulf of Mexico compared to onshore (Cleveland and Costanza, 1983)

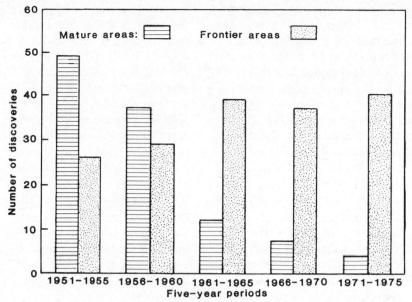

Figure 7.10. The decline in the discovery rate of large and giant oil fields in mature petroleum-producing provinces during the 1950s was partially offset by an increase in the amount of large discoveries in frontier areas. Offshore areas by far were the most important frontier region, particularly the Gulf Coast region. Since the early 1970s, however, the number of large and giant fields discovered in frontier areas also has declined precipitously. (From Nehring, 1981.)

despite dramatic technical improvements that have enabled us to tap previously unaccessible deposits in sediments lying under several hundred feet of water. As a result, the EROI for new discoveries in frontier regions is likely to be significantly lower than comparable sized discoveries in mature regions.

Time Variation of Cumulative Discoveries, Cumulative Production, and Proven Reserves

Due to the petroleum discovery patterns just described, the amount and rate of discoveries, production, and proven reserves follow rather predictable patterns that can be quantified and used as the

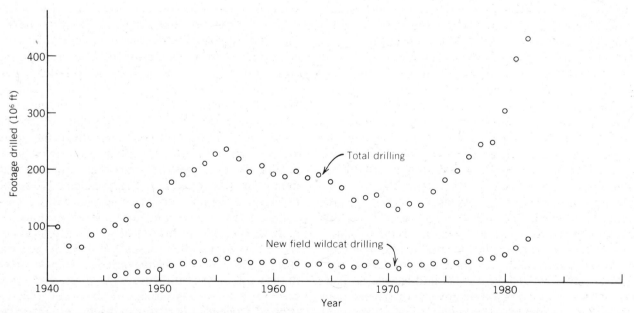

Figure 7.11. Total drilling effort (new field wildcat exploratory and development) in the United States. (From American Association of Petroleum Geologists and World Oil, *Annual Review*, various years.)

basis for another method of estimating ultimate petroleum production (Q_∞). Hubbert (1962) was the first to employ this method for estimating both Q_∞ and the peak year of production for the U.S. petroleum industry. In part the development of this technique was motivated by the response to Hubbert's 1956 analysis in which he first estimated Q_∞ and then back-calculated the production curve. Due to its economic and political implications, the date of peak oil production was "looked upon by oil companies and government alike as an evil which must be exorcised if possible; or if not, at least driven into the remote future" (Hubbert, 1974). Thus oil company and government analysts have good reason for maximizing their estimate of Q_∞. Hubbert's estimate in 1956 of a peak in crude oil production between 1965 and 1971 was criticized by some because it was based on a subjective estimate of Q_∞.

It was necessary therefore to devise a means of estimating independently Q_∞ based on the available statistical data for production, proven reserves, and discoveries. Hubbert (1962) began with the observation that in any given year the three variables were related in the following manner:

$$Q_D = Q_P + Q_R$$

where Q_D is cumulative discoveries, Q_P is cumulative production, and Q_R is proven reserves in any given year. Hubbert also found that the time variation of these variables was described accurately by variations of the logistic growth equation. The relations among Q_D, Q_P, and Q_R are shown graphically in Figure 7.12a. The yearly rates of change for these quantities follow a predictable path and are shown in Figure 7.12b. These rates are simply the first derivative of the cumulative curves in Figure 7.12a. The future behavior of the production curve, which is usually the parameter of most interest, can be estimated by examining the current behavior of the cumulative discoveries curve. Hubbert found that discovery precedes production by 10 to 12 years in the United States. Because of this time lag (Δt) and the fact that the curves for Q_D and Q_P are similarly shaped, the discovery curve gives an approximate preview of the behavior of the production curve Δt years in the future. In 1962 Hubbert analyzed the annual proven discoveries curve for the United States (Figure 7.13) and calculated that the annual rate of new oil discoveries had peaked in the mid-1950s. Hubbert calculated that oil production would peak 10 to 12 years after the peak in

annual discoveries, or in the late 1960s. By fitting the models shown in Figure 7.12 to the existing data, Hubbert also estimated that Q_∞ would be about 170 billion bbl of oil. Thus via a totally different and independent method Hubbert supported his original 1956 estimates. A 10-year review of this analysis (Hubbert, 1974) confirmed the accuracy of his 1962 estimates: oil discovery rates peaked in the early 1950s (Figure 7.13), production peaked in 1970 (Figure 7.1) and Q_∞ for the contiguous 48 states was still estimated to be 170 billion bbl, even with 10 years of additional data. In his most recent update Hubbert (1980) found that the annual rate of change in proven reserves and cumulative discoveries had fallen below the trend predicted by the data used in his 1974 analysis. Accordingly Hubbert revised his estimate of Q_∞ for the lower 48 states downward from 170 to 163 billion bbl.

Discoveries as a Function of Drilling Effort

Because petroleum is found only through drilling, it is useful to examine the historic success rate of drilling by the oil industry. This rate is usually measured in terms of yield per effort (YPE), the number of barrels found per unit of drilling effort. Yield per effort calculations are an especially useful tool for analyzing historic costs and returns and also for estimating future behavior because they do not depend directly on time as are the approaches described earlier. This is important since discoveries or production per unit time are likely to be influenced by political or economic policies. Discoveries per unit of drilling effort depend more on the amount of oil remaining to be found and its distribution among fields of various sizes and on the current state of exploration technology in the oil industry.

In general, total drilling tends to lag behind discoveries. As Nehring (1981) notes, this is the opposite relation normally attributed to the two, but it accurately describes the importance of the tendency for the few giant and large fields to be found first. Since the few very large fields are discovered early on in the development of a given geologic region with relatively little drilling effort, the incentive exists for rapid increases in drilling in the area. But by the time a major drilling campaign gets underway, most of the very large fields, and hence most of the oil, has already been discovered. As drilling proceeds, increases in drilling effort are required simply to locate increasingly smaller,

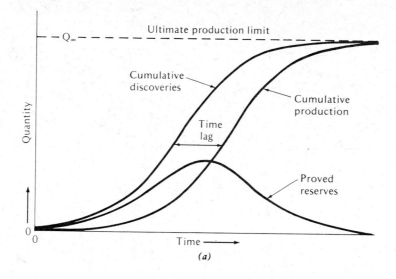

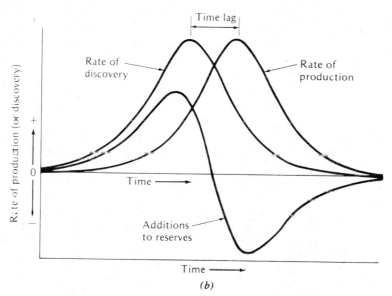

Figure 7.12.   Logistic growth curves as developed by Hubbert (1962) for the analysis of petroleum production, reserves, and discoveries. (From Steinhart and McKellar, 1982.)

deeper, and more dispersed fields. As a result the return to drilling effort, as measured by barrels found per foot drilled, tends to decrease as a function of cumulative drilling effort. As we will see in a later section, this trend also decreases the EROI for drilling as exploration and development continue.

The application of the YPE methodology to the U.S. oil and gas industry is generally attributed to Hubbert (1967) who most fully developed the idea and publicly emphasized the importance of its results. The first published analysis using discovery rates per unit of drilling effort, however, was by

Davis (1958), a petroleum engineer for Gulf Oil. In retrospect, Davis was amazingly ahead of his time not only in his use of the methodology but also the conclusions he reached concerning what the results of his analysis meant for the future productive capacity of the U.S. oil industry.

Davis plotted annual additions to reserves (new fields, new pools in old fields, revisions, and extensions) per foot of total drilling effort (exploratory and development) as a function of cumulative reserves developed (Figure 7.14). The data indicated that returns to drilling effort had declined steadily

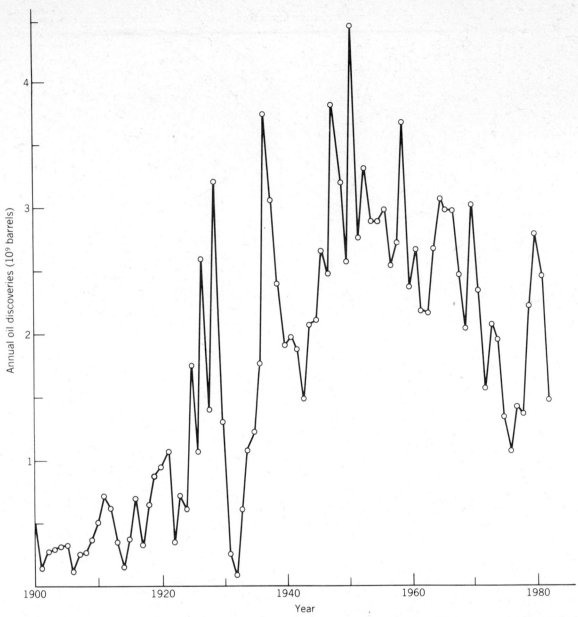

Figure 7.13. Total annual oil discoveries in the lower 48 states from 1900 to 1982. Annual additions to reserves include new field discoveries, new pools in old fields, revisions, and extensions. In general, over three-quarters of the oil added to reserves each year comes from revisions and extensions rather than from oil discovered through true wildcat drilling. (Data for 1900–1979 from American Petroleum Institute; for 1980–1982, Energy Information Administration, USDOE.)

from the mid-1930s, a trend Davis attributed largely to the net result of two opposing forces. First, less oil remained to be found, and what remained was in smaller and otherwise more difficult to locate deposits. Second, technical improvements in exploration procedures tended to offset the decline in field size and availability. The trend, which is shown in Figure 7.14, suggests that since at least the 1930s, most technical improvements in the oil exploration

industry have been more than offset by a decline in size and availability of undiscovered petroleum resources. Had the technical improvements not been made, the rate of decline in YPE would probably have been much more rapid.

Davis also was the first to notice an inverse relation between the level of drilling effort at any point in time and the return to drilling effort as measured by YPE, a relation later confirmed by Hall and

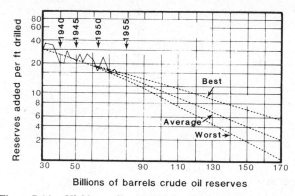

Figure 7.14. Yield per effort (YPE) for the domestic crude oil as a function of cumulative oil reserves; YPE is total barrels added to reserves divided by total drilling effort (exploratory plus development drilling). (From W. Davis, *Oil and Gas Journal,* Vol. 56, 24 February 1958.)

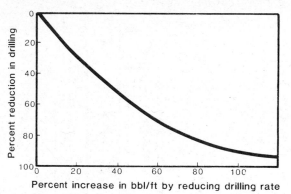

Figure 7.15. Inverse relation between the level of total drilling effort and the returns to drilling, as measured by yield per effort. (From W. Davis, *Oil and Gas Journal,* Vol. 56, 24 February 1958.)

Cleveland (1981). Davis found that in a given year the higher the rate of drilling effort, the lower the quantity of oil found per foot of drilling (Figure 7.15). This occurs because in years of high drilling effort, many more areas are drilled that have a low probability of yielding oil compared to years of reduced effort. Therefore, contrary to what many believe, high levels of drilling may not offset the rate of decline of new oil and gas discoveries. It is interesting to note that Davis ended his graph (Q_∞ in Figure 7.14) at 170 billion bbl, the same figure cited by Hubbert in his 1962 and 1967 analyses.

The most comprehensive analysis of the oil and gas industry based on extrapolations of finding rates was done by Hubbert (1967), which has proved to be the most controversial analysis among those engaged in estimating the potential for future oil and gas production in the United States. Although Hubbert's analyses have proved to be the most accurate and reliable to date, his work has probably not received the attention and credit it deserves, particularly from the economic and political leaders to whom this information would seem of primary importance. Reasons for the general lack of acceptance of the implications of Hubbert's analyses will be discussed in an editorial at the end of Part Three.

Hubbert (1967) began his analysis of YPE data as another means of estimating independently Q_∞ for the lower 48 states. The principal concept is similar to that employed by Davis, except that Hubbert compared the amount of oil discovered per foot of exploratory drilling to *cumulative* exploratory drilling effort expended in the search for oil over time (Figure 7.16). Hubbert found that, as one might guess by now, the rate of discovery dropped rather

rapidly as the few very large fields were discovered first with relatively little exploratory drilling effort. In the 1930s the U.S. found about 300 bbl of new oil per exploratory foot drilled. By the late 1960s the rate had declined to about 25 bbl/ft, although there appeared to be a leveling off of the rate of decline in YPE.

The data in Figure 7.16 can be approximated mathematically by a negative exponential equation, which indicates that discoveries through 1972 declined at the rate of about 10% per 10^8 ft of exploratory hole drilled.* Integrating the area between the curve and the x axis, Hubbert calculated Q_∞ to be 168 billion bbl of oil. By yet a different method of analysis Hubbert's estimates of Q_∞ in his 1967 and 1974 analyses exhibit remarkable fidelity with his original work in 1956 and 1962, and also with Davis' 1958 analysis. Hubbert (1974) refined his 1967 analysis by including an estimate of 43 billion bbl for the ultimate recovery from Alaska. In a recent update Hubbert (1980) and Root (1980) confirmed the original accuracy of Hubbert's YPE analysis. After the slight stabilization in the late 1960s mentioned above, the YPE for oil declined to about 9 bbl per exploratory foot in 1979. But Q_∞ was still estimated to be about 163 billion bbl for the lower 48 states.

Alaska

The case of Alaska is an important one for it illustrates the promise and frustration of expanding the

---

*The quantity of 10^8 ft of exploratory footage has come to be known as a *Hubbert unit.*

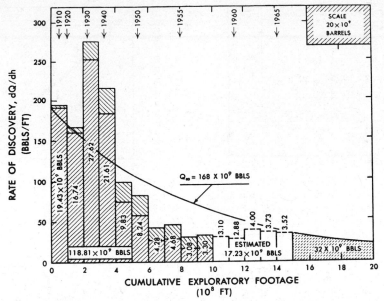

Figure 7.16. Average discoveries of crude oil (new fields, new pools in old fields, revisions, and extensions) per foot for each 10^8 ft of exploratory drilling (a Hubbert unit) in the lower 48 states from 1860 to 1967. (From Hubbert, 1980.)

search for oil and gas into frontier regions. Due to the common occurrence of surface seeps, it was long believed that Alaska harbored potentially large oil reserves. Commerical exploratory drilling began in earnest in Alaska in the mid-1960s, and the Prudhoe Bay find was announced in 1968. With estimated reserves of 9.6 billion bbl and 26 trillion ft³ of natural gas, Prudhoe Bay is the largest oil field in North America. The controversial Alaska pipeline, designed to transport oil from the North Slope region to shipping ports in southern Alaska, was delayed for a number of years due to environmental and construction difficulties, but the pipeline began delivering about a million barrels a day to the Valdez shipping terminal in 1978, almost all of which was from the Prudhoe Bay field. This rate was increased to about 1.7 million bbl a day by 1983—a figure equal to about 18% of U.S. daily production of crude oil and also the maximum attainable production rate from Prudhoe Bay. The stabilization of total U.S. crude oil production in the early 1980s since its peak in 1970 can be attributed entirely to the sharp increase in Alaskan oil production. Oil production from the lower 48 states has continued to decline since 1970 (Figure 7.1).

Although the Prudhoe Bay reserves should last for many decades to come, the total oil reserve at Prudhoe is the equivalent of less than a 2 year supply for the United States based on its current use of about 5.5 billion bbl of crude oil per year. Prudhoe

Bay has already begun to show signs of aging. Oil that once flowed freely into the wells is now being coaxed out via the injection of gas bubbles and flooding techniques, an expensive necessity considering the field encompasses some 650 km². After the mid-1980s production of crude oil from the Prudhoe Bay field is expected to decrease by about 10–15%/yr, although it is possible that other reserves might be added from adjacent offshore regions (Weeks and Weller, 1984).

It is not clear whether there are other large discoveries waiting to be made in Alaska, although it is significant that there have been no large ones reported in the 17 years since the Prudhoe Bay discovery—during a time when there have been many economic incentives to look for petroleum in this and other frontier regions. Two hundred wells drilled in Alaska following the Prudhoe discovery in the 1970s failed to find commercially significant quantities of oil, and since the drilling is very expensive, less drilling is now being done (*Wall Street Journal*, 26 Nov. 1979). The Endicott field located near Prudhoe Bay is estimated to contain 300 million bbl, which would make it a giant field in the lower 48 states. But on the North Slope of Alaska development costs dwarf those in other regions of the country to the degree that the Endicott field may not be profitable to develop (Lowenstein, 1983). Drilling operations in the once-promising Gulf of Alaska have been curtailed and have been largely an expensive disappointment, especially following the expensive failure in 1983 of the Mukluk field. It is possible that large reserves remain to be found under the adjacent Bering or Chukchi Sea, but dollar, energy, and environmental costs associated with any resource development in this region will be very large.

Random Drilling Models

A recent improvement in our ability to analyze and estimate oil and gas resources was made by Menard and Sharman (1975; see also Menard 1981) who combined the YPE based methodology of Davis and Hubbert with data on the relative distribution of oil deposits by field size. The authors developed a computer model of the surface area of the contiguous United States, dividing it into sedimentary (potentially oil-bearing) and nonsedimentary rock. With knowledge of the location, area, and amount of oil in each field discovered through the early 1970s, the authors reconstructed in the computer model the actual distribution patterns of oil and gas deposits. The computer "drilled" according to a *random* pattern rather than the directed spatial distribution that the oil industry actually had used. The discovery rates of the random drilling program could be compared to the actual success rate demonstrated by the industry itself.

The results were rather surprising. On the whole, there was no significant difference between the discovery rate of the random search and the historical behavior of the industry. Interestingly, the random search of the computer was four to five times *more* successful than industry in the very early stages of resource development. The reason was that the probability of finding giant fields via a random search was quite high due to their large area and because most actual drilling did not recognize the monotonous and often unique terrain of the giant fields as oil-bearing. By 1938 the discovery rate of the industry had caught up with that of the random drilling pattern. Menard and Sharman attributed this pattern to changes in the mode of discovery used by the industry. In the earlier years surface geology and surface seeps were the primary exploratory methods used by the industry. During this period the rate of discovery of new oil was below that of the computer's random search. The introduction of subsurface geology and geophysics in the 1920s and 1930s, however, brought the industry's discovery rate up to that achieved by random sampling.

The effects of technological changes in the method of discovery on the historic finding rate for oil deserves further elaboration. The four general methods of discovery are based on surface seeps, surface geology, subsurface geology, and geophysics. Menard and Sharman found that upon its initial introduction, each method quickly discovered most of the remaining giant oil fields it was capable of locating (Figure 7.17) until around 1950, when there was no more "cream to skin." This was the point at which the total discovery rate curve began to decline slowly after half a century of steady increase (Figure 7.13). For example, between the early 1920s and early 1940s Menard and Sharman found that the rate of discovery of giant oil fields per exploratory effort increased. The same increase is seen in Hubbert's analysis of oil discovered per foot of all exploratory drilling (Figure 7.16). Both authors attribute this increase to the use of new geophysical techniques that were able to locate deposits not amenable to discovery by surface geology. Since the 1940s, however, discovery rates per drilling effort have declined precipitously despite further technical advances in the petroleum industry.

This trend illustrates clearly the opposing forces of declining natural resource quality and increasing human technical capabilities. In this instance the

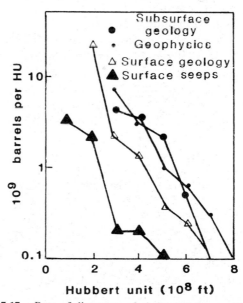

Figure 7.17. Rate of discovery of giant oil fields per Hubbert unit of drilling effort (10^8 ft of exploratory footage) by method of discovery. Soon after its introduction, each new mode of discovery finds all of the giant fields it is capable of locating. (From H. W. Menard and G. Sharman, *Science*, Vol. 190, Fig. 4, p. 339, 24 October 1975. Copyright 1975 by AAAS.)

rapid technical advance in the oil exploration industry, particularly from 1945 to 1965, has been effective only in decreasing the rate of decline in oil and gas discoveries. This has important implications for the impact of declining prospect quality on the cost of finding new oil. Norgaard (1975) found that technical advance in the oil industry between 1939 and 1968 was able to offset only partially the cost of finding a new deposit. Norgaard found that without any technical improvements, the real cost of successful wells would have increased 233%. Even with the actual improvements made during this period, the impacts of the decline in prospect quality outweighed the impacts of the new technology, and the real costs of successful wells increased 64%.

Net Energy Analysis of the U.S. Oil and Gas Exploration Industry

It was noted previously that production and proved reserves of crude oil and natural gas peaked in the early 1970s. Dramatic increases in both exploratory and development drilling effort made possible by oil price increases during the 1970s have been able only to moderate the downward trend in both production and proved reserves. High levels of drilling effort are likely to continue in the future because imports carry a heavy economic and political price and because oil companies will continue to have substantial amounts of working capital from high oil prices, although the recession of 1981–1982 and resulting declining world oil prices have somewhat dampened drilling effort. It is important therefore to analyze the potential of the recent increases in drilling effort to add to our reserves of petroleum.

Although Davis (1958) and Hubbert (1967) found a decline in the rate at which petroleum was found per foot drilled, during the 1960s this trend stabilized and even increased, contrary to Hubbert's predictions of a continued decline. Because of this apparent contradiction of Hubbert's predictions, some analysts thought that his analysis was no longer applicable. It gave encouragement to those who thought that large new quantities of petroleum would be found in the contiguous United States, for the increase in drilling effectiveness could be attributed to the many improvements in geophysical theory and exploration technology. During the 1970s, however, the ratio of petroleum found per unit drilling effort fell to levels at and below those of the 1950s (Root, 1980; Hall and Cleveland, 1981).

Hall and Cleveland (1981) offered a revision and extension of the classic Hubbert analysis, based on the returns of petroleum per total drilling effort as a function of both time and drilling effort at any one point in time. As Davis (1958) recognized, and as developed in this section, the variable success rate of drilling for any given year can be explained simply as a downward trend in YPE as the resource is depleted over time (i.e., as a function of cumulative effort), coupled with an inverse linear relation to drilling effort at any point in time. Periods of high drilling effort are associated with low YPE, and vice versa. The inverse relation between yield and effort is similar to the one for some minerals like copper described in Chapter 4 and has been a basic tool in fisheries analysis for many years (Schaeffer, 1957).

The data used in Hall and Cleveland's analysis are essentially the same used in the previous analyses described in this chapter. Estimates of annual additions to proved reserves are available from the American Petroleum Institute (1980), subdivided according to *new fields* (NF), *new pools in old fields* (NPOF), *revisions* (REV), and *extensions* (EXT). The analysis is complicated somewhat by the fact that about 80% of the oil added to reserves each year is through revisions and extensions, some of which are confirmed through exploratory drilling (D_e), and others via development drilling (D_d) and production history. Since what is of interest is total gains (additions to reserves) related to total drilling effort, the YPE ratio used by Hall and Cleveland was:

$$YPE = \frac{NF + NPOF + REV + EXT}{D_d + D_e}$$

where YPE is measured in barrels per foot. This is the identical ratio used in Davis' (1958) original YPE analysis. In a sense this ratio overestimates the effort used to find petroleum because some of that effort was solely for production and not for discovery, but it does give a running average of the total effort used to bring new petroleum to society.

Figure 7.18*b* shows the relation of yield per effort for oil, and oil plus gas, for the years 1946–1978. Periods when the finding rates were high relative to the historical trend (the mid-1960s) were also periods of relatively low drilling effort for both oil alone and oil plus gas. Conversely, less oil was found per drilling effort during periods of high drilling (Figure 7.18*a, b*). The inverse relation between yield and effort is shown more clearly by connect-

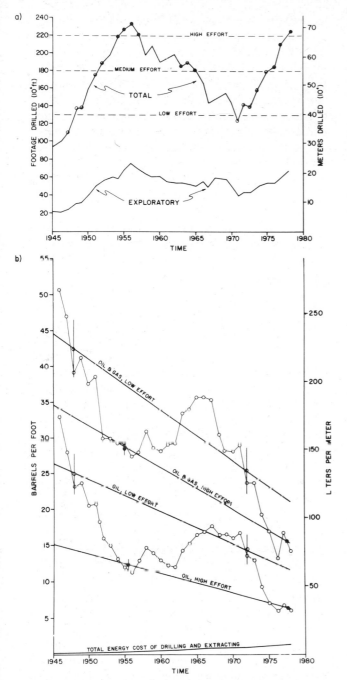

Figure 7.18. (*a*) Development and exploratory drilling effort for the U.S. oil and gas industry. (*b*) Energy gains and costs for oil and gas exploration and development in the United States. The topmost irregular line is annual yield per effort for oil plus gas, the middle irregular line is yield per effort for oil alone, and the bottommost line is an estimate of the mean energy cost of exploration and development per foot drilled. Energy cost of drilling includes fuel use plus an estimate of some of the capital structures and equipment used by the industry, as reported by the U.S. Bureau of Census, Census of Mineral Industries. (From C. A. S. Hall and C. J. Cleveland, *Science,* Vol. 211, p. 578, Fig. 2, 6 February 1981. Fig. 1a, p. 577. Copyright 1981 by the AAAS.)

ing points in Figure 7.18*b* for all years with high (about 220 million ft/yr), all years with medium (about 180 million ft/yr—not plotted for clarity), and all years with low (about 130 million ft/yr) drilling effort as they have occurred at various times since 1946. These results indicate an approximately parallel decline in YPE for high, medium, and low rates of effort. The actual yield per effort for any one year is an inverse function of both the year (and hence the amount of reserves left to be found) and the drilling effort for that year. The important trends of the year-to-year yield *per effort* for both oil and oil plus gas can be explained as a secular decrease of about 2%/yr in the rate of petroleum added to reserves per foot of drilling effort and about 5% increase or decrease for each 10 million ft of effort, i.e., effort had little effect on *total* yield.

Why should yield per effort be related to effort? This makes sense for fish, for the fish can recover through reproduction and growth when not fished. Petroleum obviously cannot, at least on time scales of interest to our species. One possible explanation is that when drilling rates are low, the petroleum industry drills at locations where present information suggests that success is most likely. During years of high drilling rates, drilling is done there plus at other, less promising locations. Presumably the development of exploration theory, as well as seismic charting and interpretation, occurs at a more constant rate than drilling effort, so that when drilling effort (i.e., economic incentive) is low, it is concentrated in areas where success appears more likely. When drilling effort (and economic incentive) is high, much of that effort is directed at targets less likely to produce a large find. In a sense it is promising but untested geologic information that is depleted as wells are drilled and that accumulates in the absence of drilling.

The decrease in drilling rates after 1956 is associated with government taxation and regulatory policies that decreased profitability of finding new domestic oil and encouraged the importation of foreign oil. Had that not occurred, our present finding rates probably would be considerably lower than they are now. Since 1974 much of the increased effort has been concentrated in mature, well-known fields where chances of some success are great but where the chances of large new discoveries (and hence large additions to proved reserves) are very low. Another factor is that less efficient drilling companies contribute a higher percentage of all drilling when economic incentives are high.

There are two ways in which Hall and Cleveland's analysis fails. First, when the previously excluded Prudhoe Bay find (the largest field ever found in the United States and a very atypical find) was included in the analysis, the yield per effort for oil and oil plus gas jumped to, respectively, 80 and 120 bbl/ft for 1968. Although exploratory drilling in Alaska in the 17 years since the discovery of Prudhoe Bay has been one expensive disappointment after another, Alaska overall still has an average yield per effort since 1965 of some 1580 bbl of oil per foot due to the importance of the Prudhoe Bay field.

The second way in which their analysis fails is that estimates of YPE for 1979, based on the 1946–1978 data and the linear time and effort components used for the initial analysis, were low, predicting less oil than was actually found. There are three possible reasons beyond a statistical quirk. First, we may be becoming suddenly more clever at finding oil. Second, we may be exploring new provinces more rapidly than in the past. Third, because of the increased value of oil relative to the cost of finding it, there are economic incentives for upgrading previously known, but previously uneconomic, fields to the status of reserves. A decline over time in developed reserves due to the depletion of higher quality deposits stimulates the development of safe regions already known to contain some petroleum. This in fact happened in 1979 with a portion of the Kern River field (discovered in 1888) whose revision in 1979 was a large contributor to new reserves attributed to 1979.

Trends in Energy Return on Investment

Since the principal use of petroleum is as a fuel, the point at which domestic petroleum will no longer, on average, be a fuel for the nation is not when the wells run dry but rather when the average energy cost of drilling a foot of petroleum well and delivering that petroleum to society equals the energy value of the petroleum found by that drilling. Aggregate statistics on the energy use and the economic activity of the petroleum exploration and development industry are available (U.S. Census of Mineral Industries and *Annual Survey of Manufacturers*) from which estimates of the direct and indirect energy cost of drilling and extraction can be made. A quantity of energy equivalent to about 1½ bbl of petroleum was used per foot of drilling by the petroleum exploration and development industry in 1977, a bit more than half directly as fuel and a bit less than half as fuel to produce the equipment and services used. This quantity has been increasing in recent years (Figure 7.18*b*) as the petroleum industry has increasingly drilled deeper, offshore, and in hostile environments such as Alaska and as a larger percentage of petroleum is produced using energy-intensive secondary and tertiary recovery. An additional 0.6 bbl equivalent per mean foot was used in 1977 for refining petroleum. The energy investments, yields, and their ratio (EROI) are given in Figure 7.19.

Hall and Cleveland used a linear extrapolation of the trends in energy cost and energy gained to project the energy break-even point for domestic petroleum exploration and extraction (Figure 7.20). Based on this extrapolation, if we were to decrease drilling rates to a low level of 130 million ft/yr, the lines would intersect in 2004. If we continue to drill at 1978 levels of about 220 million ft/yr, the linear extrapolations intersect in 2000. Oil alone could reach the break-even point within about a decade. These estimates, however, are probably overly pessimistic because some of the energy included in the denominator of the EROI ratio is used solely for oil production and not exploration, and the numerator represents only the energy in annual additions to reserves and does not include the energy in oil produced in that year. Including the energy content of oil production from old fields would increase the EROI ratio as well as postpone the intersection of the energy cost and energy return curves in Figure 7.20 by about 10–15 years (see Figure 4.5).

Hall and Cleveland's analysis assumes that the rate of technical improvements in the petroleum industry's exploratory and development methods remains constant. There also could be significant deviations from the projections of Figure 7.20 if, for example, new provinces (such as the Bering or Chukchi Sea, or overthrust belts) were explored more rapidly or more effectively than in the past. Most remaining new frontier provinces will be very energy intensive to develop, so intensive new province drilling could work either to increase or decrease the time until intersection of the energy costs and gains. But it is only from such unexplored areas that we reasonably can expect to find the very large oil fields which are necessary if a change in the sharply downward trend of the yield per effort in Figure 7.20 is to occur. One possible conclusion of this analysis is that it might be advisable relatively

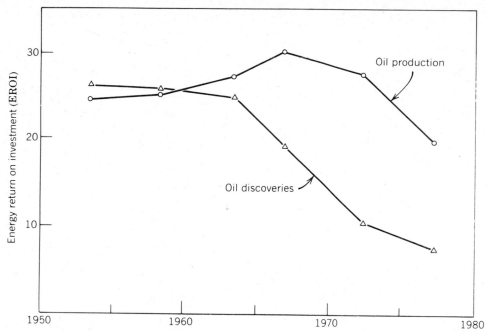

Figure 7.19. Energy return on investment (EROI) over time for the U.S. oil and gas exploration and development industry. The top curve is the ratio of the energy in annual discoveries and domestic production to energy use by SIC sector 13, the oil and gas exploration and development industries. The bottom curve is the energy in discoveries only to energy use by the oil and gas sector. Recently acquired data for 1982 show a continuation of these downward trends. These values are conservative as only direct energy use is included. (From Cleveland et al., 1984.)

soon to do *most* exploration only in new provinces. A somewhat similar conclusion was reached by Menard and Sharman.

Most oil (and presumably gas) that is now produced in the United States comes from fields discovered before 1940 as the petroleum industry tries to compensate for declining discovery rates by developing mature regions more intensively. The preceding analysis and those of Davis, Hubbert, Nehring, and Menard and Sharman give little hope for changing this picture significantly through increased conventional drilling effort. In fact large increases in such effort could decrease the total energy delivered to society by the petroleum industry by lowering the efficiency of that energy-intensive industry. Integrating the extrapolated regions of Figure 7.20 for the period 1980 to the intersection of the energy cost and gain lines gives a projected ultimate additional net yield of 29 billion bbl equivalent for a low drilling effort and 27 billion bbl equivalent for a high drilling rate. Thus, developing our remaining reserves slowly could increase somewhat our projected ultimate net yield. Concentrating new drilling effort in new provinces might change the trend. On the other hand, after the energy gained from petroleum drilling decreases below the energy cost, pe-

troleum could still be pumped at a monetary profit for feedstocks, or lower quality fuels such as coal could be used to pump the more valuable liquid fuels even at a net energy loss.

The results of this analysis indicate that increasing conventional exploration effort by the oil industry may not be in the best interest of the nation as a whole due to the lower efficiency with which the industry will deliver petroleum to society at these higher rates of drilling and also because such efforts appear to offer a "solution" to the decline in domestic conventional production. In fact it appears that no genuine long-term solution exists unless there is a dramatic change in the way that we go about finding petroleum.

Ultimate Recovery of Hydrocarbons in Louisiana: A Net Energy Approach

The state of Louisiana, including its offshore waters, has been the leading state in terms of petroleum production, supplying about 17% of all the oil and gas discovered through 1979 (Nehring, 1981). Louisiana has been the most important state as a source of natural gas and the fourth leading pro-

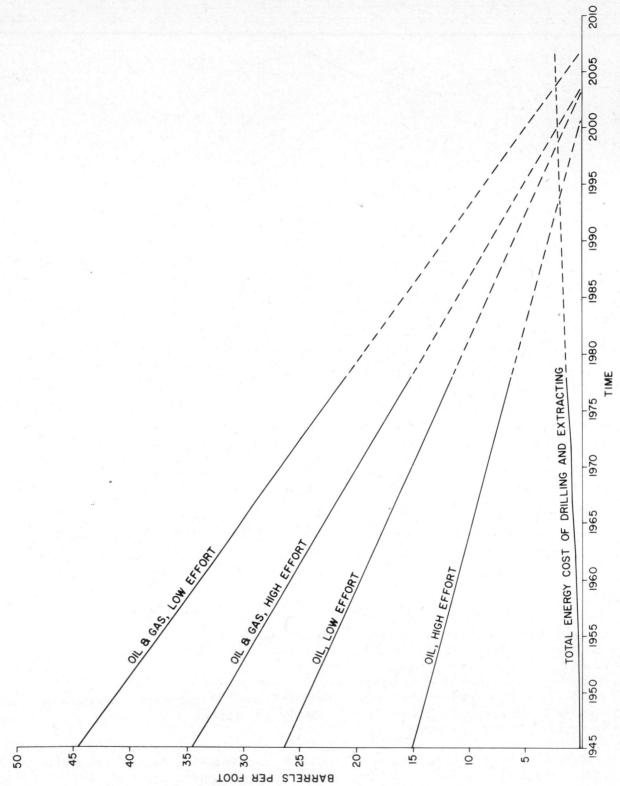

Figure 7.20. Linear extrapolations (dashed lines) of energy costs and gains of Figure 7.18*b* for high and low drilling effort. The inclusion of Prudhoe Bay into the extrapolation would extend the time of intersection by about 6 years. Since 1980 both energy gains and energy costs per foot drilled have increased somewhat due to an acceleration of secondary recovery compared to the data base used to derive this graph. This appears to have little effect on the timing of the intersection. (From C. A. S. Hall and C. J. Cleveland, *Science*, Vol. 211, p. 578, Fig. 2, 6 February 1981. Copyright 1981 by AAAS.)

182

ducer of crude oil. Since the state's petroleum resources were developed relatively early, it's production history might be considered as a model for the U.S. petroleum industry. When plotted as a function of either time or cumulative extraction, petroleum resources in Louisiana follow very closely the production growth cycle trend described by Hubbert (1956) (Figure 7.5). Crude oil and natural gas extraction both peaked in 1970 but have declined precipitously since then (Figure 7.21a, b).

Cleveland and Costanza (1983) developed a model to assess and compare historic trends in total energy use and total hydrocarbon energy extracted in Louisiana from 1953 to 1981. The authors' goal was to project the time at which petroleum extraction reached the energy break-even point, in a manner similar to that of Hall and Cleveland (1981). Note that Cleveland and Costanza were concerned with the *extraction* phase of the industry, not the exploration phase. The EROI in this case was the ratio of oil or gas produced to the direct and indirect fuel used in drilling, pumping, and so forth. To calculate the indirect energy cost of drilling, the authors used data on the dollar costs of all wells drilled in Louisiana published by the Joint Association Survey. Dollar costs of drilling were converted to energy costs using energy-intensity factors ($/kcal) for goods and services calculated by Hannon et al. (1981).

Figure 7.22 shows the EROI for total hydrocarbon (oil plus gas) extraction in Louisiana as a function of cumulative hydrocarbon extraction (and time), as calculated by Cleveland and Costanza. Note that the trend in the EROI has the same general shape of the production growth cycle curve itself (Figure 7.5). The EROI for total hydrocarbon extraction in Louisiana peaked at about 42:1 in 1970 and declined rapidly to about 8:1 in 1981. Extrapolation of the model used by Cleveland and Costanza to analyze the historic behavior of the EROI predicts that the energy break-even point will be reached when about 210 quads of energy have been produced. To estimate the year in which this would occur, cumulative extraction (plotted along the x-axis in Figure 7.22) was plotted as a function of time (Figure 7.23). A logistic model similar to the one used by Hubbert (1962) which was described earlier (Figure 7.5) was applied to the existing data on cumulative extraction of hydrocarbons. Extrapolation of this model predicts that 210 quads of energy will be produced by the mid-1990s, when the energy break-even point as predicted by the EROI model

will be reached. Based on these models, Cleveland and Costanza concluded that total hydrocarbon extraction in Louisiana could cease to be a net source of fuel within the next 12–15 years.

In a regional breakdown the authors found that natural gas extraction in southern Louisiana had an EROI of over 100:1 in the late 1960s, although it too has declined precipitously since then. Petroleum extracted in offshore waters had the lowest EROI because it required five to seven times as much energy to produce a barrel offshore compared to onshore. Constructing and operating an offshore platform is extremely energy and material intensive relative to onshore oil and gas operations, and its EROI never exceeded 13:1.

The predictions of the model used by Cleveland and Costanza are probably overly pessimistic for at least one important reason. The rapid increases in drilling effort from 1979 to 1981 created shortages of drilling equipment in many regions of the country, thus driving its price up. If the use in nominal dollar costs of drilling due to short-term supply and demand interactions was greater than the general inflation rate, then Cleveland and Costanza's model would overestimate energy costs and underestimate the EROI for those years. Regardless of the magnitude of this error, if in fact it does exist, their analyses clearly document a rapidly declining EROI for oil and gas extraction in Louisiana, a principal supplier of the nation's petroleum.

SUMMARY

The purpose of this chapter was to describe the factors that affect the supply of petroleum for human economic purposes. The important points are summarized as follows:

1. The vast majority of petroleum is contained in a relatively few large deposits. These deposits are discovered early in the exploration history of the industry.

2. As exploration and development continues, more drilling effort is required to find many small fields. The result is a decline in returns to drilling effort, commonly measured as yield per effort.

3. The net effect of items 1 and 2 is a decline in the EROI for petroleum as the large, high-quality fields are discovered.

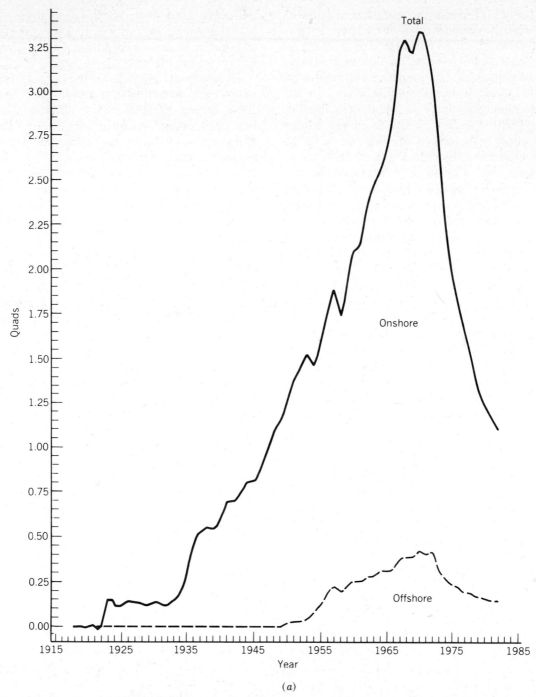

Figure 7.21. (*a*) Production of oil, and (*b*) production of gas in Louisiana, excluding federally owned offshore leases. One Quad equals $0.252 \cdot 10^{15}$ kcal or about 168 million bbl of oil. (From C. Cleveland and R. Costanza, 1983).

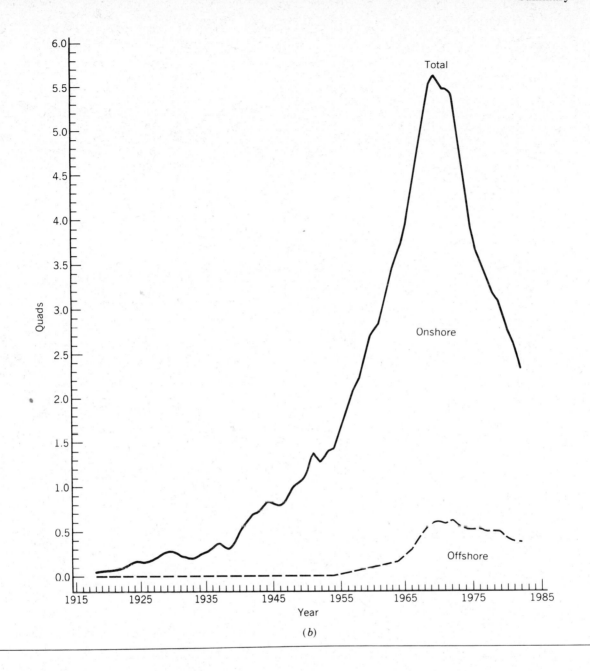

(b)

4. In the United States it appears that intensive exploration of frontier regions is the only means by which the decline in YPE and EROI could be stabilized or reversed.

The widespread economic problems resulting from petroleum supply shortfalls in the 1970s awoke the oil-importing nations to their precarious dependence on petroleum resources for economic and social well-being. In the United States, Project Independence and other programs were initiated to stimulate development of domestic fuel resources (both fossil and alternative technologies) as a means of decreasing dependence on foreign sources. Such policies have had modest success. By 1982 the United States was still importing 30% of its liquid hydrocarbon requirements. By March 1984, however, our dependence had increased again to about 36% of our liquid fuel budget. As the nation pulled itself out of the 1980–81 recession, rising economic activity spurred demand for fuel, a demand that our depleted domestic supplies could not meet.

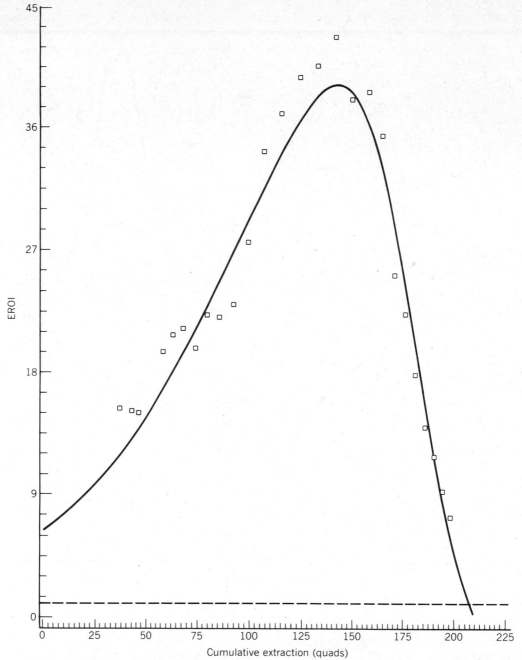

Figure 7.22. EROI for oil and gas extraction in Louisiana. The points are computed values for EROI. The curve represents a best-fit curve to the data, extrapolated to the energy break even points (----). (From C. Cleveland and R. Costanza, 1983).

Many of the policies aimed at reducing our import dependence have relied principally on economic incentives to the domestic petroleum industry to overcome the physical limitations of petroleum supply described in this chapter. These policies have so far proved to be unrealistic in their expectations. Singer (1974), for example, stated that the goal of total energy self-sufficiency by 1985 was entirely plausible given higher real oil prices and less government intervention in the petroleum industry. Simon (1981) states that domestic energy supply is very sensitive to price—the higher the price for energy, the more of it will be delivered to consumers. These scenarios and many others place complete faith in free-market mechanisms to overcome any physical constraints. The popular

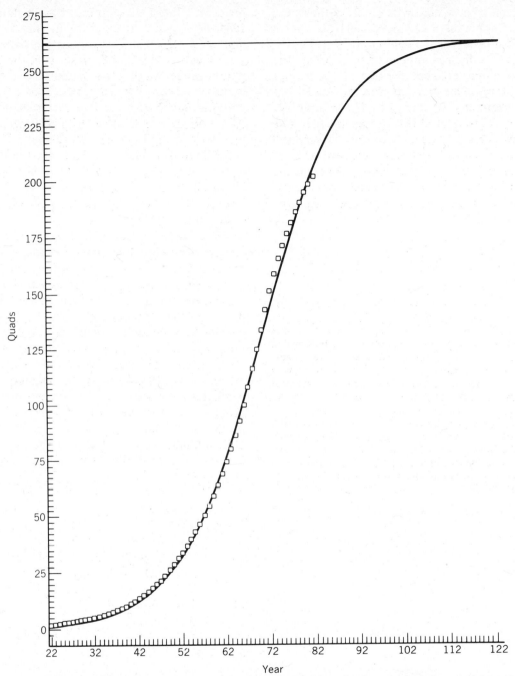

Figure 7.23. Time-series data of cumulative oil and gas extraction in Louisiana fitted to the logistic growth equation. (From R. Costanza and C. Cleveland, 1983. Final Report to Center for Energy Studies, Louisiana State University, CES 85-01-20, under U.S. Department of Energy Grant No. DEFG 058 OER 10197.)

scenario is as follows: higher oil prices due in part to deregulation will increase oil company revenues. Increased revenues will provide both the incentive and means to invest in more exploration and development, resulting in more petroleum discoveries.

A review of industry behavior since 1973 provides a means of evaluating this hypothesis. Be-

tween 1973 and 1980 the real wellhead price of domestically produced oil quadrupled. Expenditures on exploration and development increased by at least 500%. Total drilling effort increased by 280% between 1972 and 1981. So far, so good. The largest increase ever in domestic drilling even produced an increased success rate in finding *new* deposits, as the percentage of new field wildcat wells that were productive of some oil or gas increased from 10% in 1970 to 20% in the early 1980s (AAPG, 1981). At first glance these figures appear to substantiate the ability of free market incentives to overcome our energy problems. The bottom line, however, has remained essentially unchanged. The *amount* of new petroleum discovered by the increased drilling and increased success rate has continued its overall downward trend because the new fields are smaller. Proven reserves continue to decline, although the rate of decline in gas reserves has apparently slowed somewhat.* Domestic production from the

lower 48 states has continued to decline through 1984. The overall decline for the nation has been mitigated somewhat by production from Prudhoe Bay which is now being pumped at the maximum rate possible, and by the increased use of energy-intensive secondary recovery (Figure 7.1). More important, YPE for oil has continued to decline. After a modest reversal in 1979, YPE declined again in 1980 and 1981, years that saw the two largest percentage increases in drilling effort in the history of the industry. By 1982 we were finding only 3.5 bbl of oil per foot drilled, or 10.8 if gas is included. Energy costs per foot are not available yet but clearly are very high.

These simple facts demonstrate the inappropriateness of economic theories that do not include the physical and energetic limits to the development of natural resources. Sharp increases in working capital and drilling have not produced the results predicted by these theories because those predictions considered economic availability only. Certainly those who were listening to the analyses of Hubbert (1956) and Davis (1958) were not shocked by the events of the 1970s, nor by the inability of market mechanisms to alleviate domestic petroleum supply problems. Unfortunately for the users of petroleum products the economic and political leaders in the United States were not among those heeding the economic implications of the analyses conducted by Hubbert and others.

---

*Comparison of post-1979 data on discoveries and reserves to the historic trend is made difficult by the fact that prior to 1979, the American Petroleum Institute published the only such data. API discontinued its series in 1979, and the Energy Information Administration of the USDOE replaced it with their own series. Comparisons of reserve-discovery data for the 4 years in which the two series overlap (1976–1979) showed the EIA estimates to be about 10% higher than API. Nehring (1984) attempted to reconcile the differences between the two series. His analysis suggests that for oil the higher discovery rates reported by the EIA are due primarily to revisions of previous estimates rather than to a significant increase in the actual discovery of new oil deposits. Thus part of the slowing of the decline in finding rates of oil and of U.S. oil reserves in the late 1970's was definitional.

8

IMPORTED PETROLEUM

The United States relies on two sources for the petroleum that provides about 70% of its energy supplies: domestic resources and imports. Domestic consumption of petroleum has increased more rapidly than production in recent decades, and this difference has been made up by increasing imports. Since 1948, the first year in which the United States was a net importer of oil, the percentage of total oil consumption obtained from foreign sources increased to nearly 50 percent in 1977 (Figure 8.1). Since then it has dropped as the demand for oil has diminished.

As described in Chapter 7, annual domestic production of both oil and gas generally has diminished since the early 1970s, and very large additions to domestic reserves are unlikely. Furthermore, if trends of the past 30 years continue, the energy discovered by exploratory drilling will approach the energy used to find and extract the petroleum found within the next several decades, so that even if domestic petroleum continues to be found, it may not serve as a net fuel for the nation. Consequently, it appears unlikely that domestic production of oil and gas can be increased to replace imported petroleum, especially if demand for oil again increases.

A number of politicians and economists have advised the federal government to reduce dependence on imports by intensifying efforts to find new domestic petroleum supplies and by developing domestic alternatives, including nuclear power, coal-derived synfuels, oil shale, alcohol fuels, and solar-powered satellites (see Chapters 10–13). Yet, despite the large economic and political costs of relying on foreign oil, federal programs aimed at reducing that dependence, and a decreased demand for petroleum since 1979, imported oil still accounted for about 35% of the liquid petroleum consumed by the United States in 1983.

In addition to limitations imposed by its origins in politically "unstable" regions and its high dollar and energy cost, imported petroleum is a finite nonrenewable resource. Its supply can be analyzed using the same distributional traits used by Hubbert and other petroleum analysts to estimate the size of domestic petroleum resources and the rate at which they are discovered and produced. Since the rigorous and dangerous conditions involved in transporting natural gas severely limit the quantity the United States can import we restrict our analysis to oil alone. Most of the information in the following section was taken from Nehring (1982), and the interested reader should consult that paper for further details.

WORLD OIL SUPPLIES

The most important feature relating to estimating the size of world oil deposits is their concentration in a few, relatively small geographical areas called provinces. As described in Chapter 7, oil forms only under special geological conditions, so oil fields are found only in provinces that contain particular sedimentary formations. Approximately 600 such provinces exist, of which about 420 have been explored. Two hundred and forty of these 420 regions contain oil, and most of the remaining 180 also indicate such possibilities.

But the large number of provinces with oil fields does not automatically indicate significant quantities of oil contained there. Such a conclusion probably is false because most of the world's oil is found in a few provinces with extremely large reserves. Of the 420 provinces that contain oil, only seven contain more than 25 billion bbl, which is about one year's world oil consumption at its peak in 1980 (Table 8.1). Together, these seven provinces contain over two-thirds of known world oil supplies. The largest of these provinces, the Arabian–Iranian, contains nearly half of world oil supplies. Summing the known oil deposits in the 18 major provinces listed in Table 8.1 indicates the special-

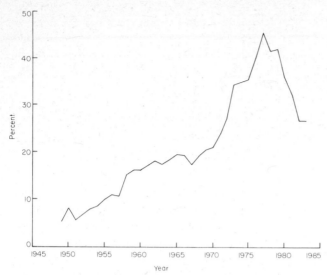

Figure 8.1. Percent of U.S. oil consumption supplied by imports between 1949 and 1982. Prior to 1949 the United States was a net exporter of oil.

ized conditions needed for significant oil formation. Nearly 90% of the world's oil supply is located in 6% of the 420 explored oil provinces.

The concentrated deposition of oil also is reflected by the distribution of oil between fields within each of these provinces. Most of the oil in a province is concentrated in giant and super giant fields (Table 8.2). All together, giant oil fields contain more than 75% of known world supplies, of which nearly half are contained in only 35 super giant oil fields. In nearly 70% of the 25 major provinces, the 10 largest fields contain over 60% of that province's supply. Conversely, in all but two U.S. provinces, the amount of oil in small but numerous fields is less than 20% of the province's total.

World oil discovery rates have peaked, are declining, and probably will continue to decline. Between 1935 and 1970 (except during World War II), 25–30 billion bbl were found per year, averaged over 5-year intervals. Since 1970, however, the rate has dropped to 15–18 billion bbl/yr. Oil experts base their prediction for a continuing decline in the rate of world oil discoveries on two important characteristics of oil exploration: most of the world's oil is found in a few regions in a few giant and super giant oil fields and these giant and super giant oil fields usually are among the first fields discovered following the onset of significant exploration (Table 8.3). In addition to their large size, which greatly increases the probability of their discovery, most of the giant and super giant oil fields are found in asso-

ciation with obvious geological clues, such as anticlines, salt domes, and reefs. As a result, in explored provinces most of the significant fields already have been discovered, reducing the probability for significant new discoveries in these areas. Just as with exploration in the United States, unexplored regions offer the best hope for finds that could yield significant additions to world oil supplies.

Estimates for ultimate world oil supplies have varied over a wide range (Table 8.4). In a pattern similar to the history of estimates for domestic oil reserves, early estimates tended to be low when scientists' understanding of the geological conditions necessary for oil formation was small and exploratory techniques were crude. On the other hand, some of the higher estimates for world oil supplies were issued following the oil embargo in 1974 when it was politically expedient to project large supplies of oil. Based on the distribution of oil between provinces and within their respective fields, recent estimates for the recoverable supply of world oil fall into a more narrow range, about 2000 billion bbl, plus or minus 400 billion bbl.

Estimates near 2000 billion bbl are not etched in stone, and slight upward revisions are possible. If oil prices continue to rise, extraction efficiency may rise, making an ultimate recoverable yield of 2400 billion bbl more likely. Even this increase, however, is limited by the EROI for pumping oil from the ground. Nor will increased drilling in response to high prices raise ultimate finds significantly, since small and very small fields are very unlikely to provide more than 10% of total oil reserves. Finally, drilling novel types of sedimentary functions probably will not increase recoverable supplies significantly, since most giant oil fields are found in obvious formations. For example, only 8% of known U.S. giant oil fields occur in subtle formations, such as stratigraphic traps, and the rest are in seismically obvious formations. Despite these possibilities for slight upward revision, estimates greater than 3000 billion bbl are extremely unlikely to be correct (Nehring, 1982).

Although the statistics on proven reserves are not sufficiently complete to allow us to estimate the complete production cycle by means of the curve for cumulative production, Figure 8.2 shows a Hubbert curve for world oil production based on yearly production rates since 1900 and an ultimate recoverable yield of 2000 billion bbl. The curve peaks about the year 2000. This differs slightly from some more recent estimates for the peak of world oil pro-

Table 8.1. The 25 Largest Oil Provinces Ranked According to Their Cumulative Production Through 1980

Province	Location	Known Recovery as of 1/1/81 (billion bbl)	Age of Major Source Rock(s)
Megaprovinces (100 billion bbl plus)			
1. Arabian–Iranian	Arabian–Persian Gulf	626.3	Cretaceous, Jurassic
Superprovinces (25–100 billion bbl)			
2. Maracaibo	Venezuela–Colombia	49.0	Cretaceous
3. West Siberian	Soviet Union	45.0	Jurassic, Cretaceous
4. Reforma–Campeche	Mexico	42.2	Jurassic, Cretaceous
5. Volga–Ural	Soviet Union	41.0	Devonian
6. Permian	United States	32.6	Permian, Pennsylvanian
7. Sirte	Libya	28.0	Cretaceous, Paleocene
Superprovinces subtotal		237.8	
Major Provinces (7.5–25 billion bbl)			
8. Mississippi Delta	United States	22.4	Miocene–Oligocene
9. Northern North Sea	U.K.–Norway–Denmark	22.4	Jurassic
10. Niger Delta	Nigeria–Cameroon	20.8	Oligocene–Miocene
11. Eastern Venezuela	Venezuela–Trinidad	19.5	Cretaceous
12. Texas Gulf Coast–Burgos	United States–Mexico	18.7	Oligocene–Miocene, Eocene
13. Alberta	Canada	17.0	Cretaceous, Devonian
14. East Texas-Arkla	United States	15.2	Cretaceous
15. Triassic	Algeria–Tunisia	13.5	Silurian
16. San Joaquin	United States	13.0	Miocene
17. North Caucasus–Mangyshlak	Soviet Union	12.0	Oligocene–Miocene, Jurassic
18. South Caspian	Soviet Union	12.0	Miocene
19. Anadarko–Amarillo–Ardmore	United States	10.8	Pennsylvanian
20. Tampico–Misantla	Mexico	10.7	Jurassic, Eocene
21. Arctic Slope	United States	10.3	Cretaceous
22. Central Sumatra	Indonesia	10.0	Miocene
23. Los Angeles	United States	8.9	Miocene
24. Chautauqua	United States	8.5	Pennsylvanian
25. Sung-liao	China	8.5	Cretaceous
Other major provinces subtotal		254.2	
All major provinces subtotal		1118.3	
All other provinces subtotal		146.7	
World total		1265.0	

Source: From ''Prospects for conventional world oil resources'' by R. Nehring. Reproduced, with permission, from *Annual Review of Energy*, Volume 7, p. 181. Copyright © 1982 by Annual Reviews, Inc.

duction that predict world oil production will peak within the first two decades of the 21st century because recent growth in oil use has been less than was once projected.

Even though the Hubbert curve gives us an idea of the rate at which world oil may be pumped in the future, determining how much of this oil is available to the United States is more difficult. Much depends on the rate at which the third world develops, for as underdeveloped countries industrialize, they will certainly demand more equitable access to world oil supplies. In addition the threat of oil as a political weapon or a radical change in a petroleum-exporting country's government or government policies cloud the future. For past years we can assess the quality of imported oil as fuel source by looking at its EROI, which ultimately will determine its value to the United States in the future.

Table 8.2. Distribution of Oil in the 25 Largest Oil Provinces

Province	Province Type	Proportion of Oil — Largest Field	Proportion of Oil — Ten Largest Fields	Supergiant and Near-Supergiant Oil Accumulations
1. Arabian–Iranian	Downwarp	13%	56%	Ghawar (82.0), Burgan (79.0), Safaniyah-Khafji (45.0), Rumaila (24.2), Zakum (22.5), Raudhatain-Sabriya (17.5), Kirkuk (17.0), Marun (16.5), Fereidoon-Marjan (15.6), Berri (15.5), Manifa (15.5), Abqaiq (14.5), Ahwaz (14.4), Qatif (13.7), Agha-Jari (13.0), Gach Saran (12.8), Khurais (12.8), Zuluf (11.8), Minagish (11.1), Shaybah (8.2), Bu Hasa (8.0), Abu Sa'fah (6.6), Asab (5.0), Jawb (5.0)
2. Maracaibo	Median	76%	97%	Bolivar Coastal Field (37.0)
3. West Siberian	Craton margin	40%	69%	Samotlor (18.0)
4. Reforma–Campeche	Downwarp	36%	78%	Cantarell (15.0), A. J. Bermudez (6.5), Abcatun-Kanaab (5.0)
5. Volga–Ural	Craton margin	34%	62%	Romashkino (14.0)
6. Permian	Craton margin	7%	42%	NSG: Eastern Central Basin Platform Group (5.0), Western Central Basin Platform Group (3.4)
7. Sirte	Rift	18%	76%	Sarir (5.0)
8. Mississippi Delta	Delta	3%	20%	None
9. Northern North Sea	Rift	11%	63%	NSG: Statfjord-Brent (4.7)
10. Niger Delta	Delta	4%	29%	None
11. Eastern Venezuela	Downwarp	8%	45%	NSG: Orinoco heavy oil belt
12. Texas Gulf Coast	Downwarp	5%	30%	None
13. Alberta	Craton margin	11%	43%	NSG: Pembina (8.0 OOIP), Massive oil sand deposits
14. East Texas-Arkla	Downwarp	40%	64%	East Texas (6.0)
15. Triassic	Craton margin	63%	92%	Hassi Messaoud (8.5)
16. San Joaquin	Subduction	17%	75%	NSG: Westside Group (4.4)
17. North Caucasus–Mangyshlak	Downwarp	21%	66%	NSG: Uzen-Zhetybay Group (3.0), N. Caucasus Group
18. South Caspian	Subduction	20%	78%	NSG: Apsheron Peninsula Group
19. Anadarko–Amarillo–Ardmore	Craton margin	39%	74%	NSG: Hugoton-Panhandle (4.2 + 80 trillion ft^3)
20. Tampico–Misantla	Downwarp	38%	92%	NSG: Chicontepec Area (c. 100.0 OOIP)
21. Arctic Slope	Downwarp	95%	100%	Prudhoe Bay (9.8)
22. Central Sumatra	Subduction	40%	72%	NSG: Minas (7.0 OOIP)
23. Los Angeles	Subduction	31%	84%	NSG: Wilmington Trend (11.0 OOIP), Wilmington–Long Beach–Huntington Group (5.2)
24. Chautauqua	Craton margin	10%	41%	None
25. Sung-liao	Rift	82%	98%+	Ta'ching (7.0)

Source: From ''Prospects for conventional world oil resources'' by R. Nehring. Reproduced, with permission, from the *Annual Review of Energy*, Volume 7, pp. 186–187. Copyright © 1982 by Annual Reviews, Inc.

[a]Near supergiant (NSG) accumulations include nonconventional (e.g., oil sands) accumulations with more than 10 billion bbl of original oil-in-place (OOIP), conventional fields with more than 5 billion bbl OOIP, groups of neighboring fields with at least 2.5 billion bbl of petroleum liquids of total recovery, and individual fields with total recovery of at least 2.5 billion bbl. Sizes of supergiant or near-supergiant accumulations are given in billions of barrels.

Table 8.3. Peak Decades, in Order of Declining Importance, of Oil Discovery and Giant Field Discoveries in the 25 Largest Oil Provinces

Major Province	Peak Discovery Decade(s)	Discovery Year	
		First Giant	Latest Giant
1. Arabian–Iranian	1950s, 1960s	1908	1979
2. Maracaibo	1910s, 1950s	1914	1958
3. West Siberian	1960s	1961	1973
4. Reforma–Campeche	1970s	1972	1980
5. Volga-Ural	1940s, 1950s	1973	1958
6. Permian	1920s, 1930s, 1940s	1926	1949
7. Sirte	1960s, 1950s	1959	1971
8. Mississippi Delta	1930s, 1950s, 1940s	1930	1950
9. Northern North Sea	1970s	1969	1980
10. Niger Delta	1960s	1958	1973
11. Eastern Venezuela	1940s, 1950s	1913	1958
12. Texas Gulf Coast	1930s	1931	1938
13. Alberta	1950s	1947	1960
14. East Texas-Arkla	1920s, 1930s	1922	1940
15. Triassic	1950s, 1960s	1956	1964
16. San Joaquin	1910s, 1900s	1899	1928
17. North Caucasus–Mangyshlak	1960s, 1910s	1915	1961
18. South Caspian	1890s	1848	1963
19. Anadarko–Amarillo–Ardmore	1900s	1904	1944
20. Tampico–Misantla	1920s, 1900s	1901	1956
21. Arctic Slope	1960s	1968	1969(?)
22. Central Sumatra	1940s	1941	1970
23. Los Angeles	1920s, 1910s	1919	1922
24. Chautauqua	1920s, 1910s	1912	1928
25. Sung-liao	1950s	1959	1973

Source: From "Prospects for conventional world oil resources" by R. Nehring. Reproduced, with permission, from the *Annual Review of Energy,* Volume 7, p. 188. Copyright © 1982 by Annual Reviews, Inc.

Note: Decades listed are those in which 20% or more of the province's known recovery was discovered. Question mark indicates a possible but as yet unconfirmed discovery.

EROI FOR IMPORTED OIL

Obtaining petroleum from domestic or foreign sources requires an investment of dollars. Less obviously, but equally important, obtaining petroleum from either source also requires an investment of energy. Domestic fuels are recovered by an expenditure of energy to search for, drill, extract, and refine fuels. Imported fuels are obtained by using energy to produce goods and services that are exchanged directly or indirectly on the international market for the chemical energy of imported petroleum. Although foreign trade satisfies many objectives, in this section we examine its role as an energy resource for the United States in the past, present, and future. To do so, we calculate the EROI for imported petroleum between 1963 and 1981.

Energy return on investment (EROI) for petroleum is calculated in a similar manner for both domestic and foreign sources: it is the ratio of the energy delivered to society from a particular source divided by the quantity of energy required to make it available for use in that economy. The EROI for a domestic source of fuel is calculated by dividing the quantity of energy delivered to the nonfuel producing sectors of the economy by the quantity of energy it takes to find, extract, process, and deliver

Table 8.4.   Major Estimates of World Supply of Recoverable Oil

Year of Estimate	Estimator	Affiliation	Estimate (billion bbl)
1946	Duce	Aramco	500
1946	Pogue	—	615
1948	Weeks	Exxon	617
1949	Levorsen	Stanford	1635
1949	Weeks	Exxon	1015
1958	Weeks	Exxon	1500–3000
1959	Weeks	—	2000–3500
1965	Hendricks	USGS	1984–2480
1968	Weeks	Weeks	2200–3350
1969	Hubbert	USGS	1350–2000
1970	Moody	Mobil	1800
1971	Warman	BP	1200–2000
1972	Jodry	Sun	1952
1973	Odell	Erasmus	4000
1974	Kirkby, Adams	BP	1600–2000
1975	Moody, Esser	Mobil	1312–2000–3237
1975	Moody	Moody	1705–2030–2505
1976	Grossling	USGS	1960–5600 (method 1) 2200–3000 (method 2)
1976	Klemme	Weeks	1600
1977	Parent, Linden	IGT	2130–2480
1977	Delphi	IFP	1240 (low group mean) 1799 (middle group mean) 2117 (poll mean) 3050 (high group mean)
1978	Nehring	Rand	1700–2300
1979	Halbouty, Moody	—	1421–2128–3556
1979	Nehring	Rand	1600–2000

Source: From "Prospects for conventional world oil resources" by R. Nehring. Reproduced, with permission, from the *Annual Review of Energy,* Volume 7, p. 177. Copyright © 1982 by Annual Reviews, Inc.

that fuel. For imported oil or gas Kaufmann and Hall (1981) used the following equation:

$$EROI = \frac{CE_i}{EE_e}$$

$$= \frac{(\text{kcal imported fuel/\$ imported fuel})}{(\text{kcal embodied in exports/\$ exports})} \quad (8.1)$$

$$= \frac{\text{kcal imported}}{\text{kcal exported}}$$

in which CE_i is the chemical energy in an average dollar's worth of imported fuel and EE_e is the embodied energy in an average dollar's worth of export, assuming that the mix of commodities exchanged for fuel, or for the foreign exchange used to purchase fuel, is similar to the overall mix of exported commodities.

Petroleum imports include liquid fuels (both crude and refined oil products) and small amounts of natural gas. Data on the physical quantities and price of imported petroleum were used to calculate the numerator in equation 8.1, which is the quantity of energy (in kilocalories) purchased by each dollar spent on imported liquid fuels and natural gas between 1963 and 1981. The energy used elsewhere to extract, transport, and/or refine the fuel imported is not included because the United States does not pay this energy cost directly.

The United States exports a wide variety of goods that are, in a net sense, used to purchase imported fuel, and each dollar's worth of export represents a somewhat different quantity of em-

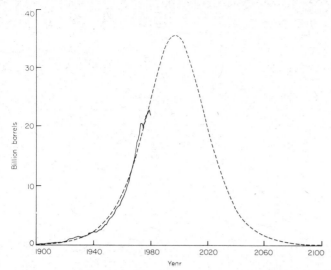

Figure 8.2. Hubbert curve for world oil production (dashed line) and observed values (solid line). According to Hubbert's analysis production peaks in 1998 but the world economic slowdown starting in the 1970's may delay the peak. (Data from *Petroleum Facts and Figures* and *Basic Petroleum Handbook*. Published by American Petroleum Institute.)

bodied energy (Table 8.5). Ideally, physical units of goods exchanged and their energy equivalents would be used, but trade statistics are compiled by dollar volume. Without such data, Kaufmann and Hall used one method developed by Bullard and Herendeen and three methods developed by Hall et al. (1979) (described in Chapter 5) to calculate the energy embodied in each type of exported goods, which is the denominator in Equation 8.1. We use an average of all four methods for Equation 8.1 because there is no unambiguous rationale for the use of any one method, but the choice of method used for this assessment is not important to its conclusions.*

Industrial products are not the only goods that the United States exports. Separate calculations were made for agricultural products, which ranged between 12.8 and 17% of total dollar value of exports between 1963 and 1981. The embodied energy in four crops—corn, wheat, rice, and soybeans—

*The average standard deviation, as a percentage of the mean yearly EROI for imported petroleum calculated by the four methods—average of (standard deviation/mean EROI)—is 6.41%. This indicates that the uncertainties associated with specifying the sectors from which intermediate goods originate do not have a large effect on the values of EROI calculated for imported petroleum.

were included, and these four crops account for approximately 60% of the total monetary value of agricultural exports. The physical quantities traded (i.e., bushels) and their monetary values were converted to their embodied energy content using coefficients for the embodied energy in each type of crop. All these energy costs are conservative because they do not include the solar and other environmental energy used to produce goods, nor do they account for the environmental disruption, such as soil erosion, occasioned by their production (see Part Four).

Figure 8.3 shows that the EROI for imported liquid fuels peaked at approximately 23.1:1 in 1971. Most of the early increase was caused by a 28% decline in the energy embodied in an average dollar's worth of exported goods between 1963 and 1971 due to an increase in the price of U.S. exports versus the price of imported oil. Following the OPEC oil embargo in 1973–1974, the EROI declined because the quantity of chemical energy in a dollar's worth of imported oil decreased 70% while the embodied energy in an average dollar's worth of export decreased only 22%. A period of relative price stability occurred between 1974 and 1978, as the embodied energy in a dollar's worth of exported goods decreased (i.e., the price inflated) at a rate similar to the chemical energy in a dollar's worth of imported petroleum. The EROI dropped to about 4.2:1 in 1980 as oil price levels rose following the Iranian crisis and the resultant world shortfall in 1979. The EROI rose slightly to 4.6 in 1981, due to an easing of world oil prices brought on by a worldwide recession (leading to a relative surplus of oil) and a decline in the embodied energy in a dollar's worth of exports. The EROI continued to rise through 1984 as the price of oil dropped and inflation continued to reduce the embodied energy in an average dollar's worth of export. It is important to note, however, that although we have observed declining oil prices from 1980 to 1985 in the United States this is almost entirely attributable to the relative strength of the dollar. During that period the price of imported oil had been *increasing* in most other countries. The general trend since 1973 in Figure 8.3 is not too different from the EROI for a running average of finding and producing new domestic petroleum given in Figure 7.19.

The EROI for natural gas peaks twice, at 36:1 in 1968 and 38:1 in 1973, after which it declined rapidly (Figure 8.4). Unlike the EROI for liquid fuels, which decreased gradually following the 1973–1974

Table 8.5. Dollar Value Per Dollar Energy Intensity of Exported Goods and Imported Petroleum

	1965		1970		1975		1979	
	Exports (million $)	Energy Equivalent of Exports (kcal/$)	Exports (million $)	Energy Equivalent of Exports (kcal/$)	Exports (million $)	Energy Equivalent of Exports (kcal/$)	Exports (million $)	Energy Equivalent of Exports (kcal/$)
Exported Goods (SIC)								
Textile mill products (22)	528	15,982	603	15,012	1624	11,250	3189	7340
Apparel, clothing (23)	143	9892	198	8636	403	5893	931	3736
Lumber and wood products (24)	124	17,775	366	16,365	751	10,121	1782	6011
Furniture and fixtures (25)	119	11,128	193	10,402	413	6639	1015	4716
Paper and allied products (26)	389	21,802	622	20,373	1447	16,228	1967	11,098
Printing and publishing (27)	226	7669	327	7026	548	5048	956	3,303
Chemical and allied products (28)	2403	21,538	3826	18,507	8691	16,164	17,306	10,909
Petroleum and coal products (29)[a]	895	538,573	1450	463,492	4164	149,080	5242	141,109
Rubber and miscellaneous plastic (30)	166	13,318	186	11,985	544	9389	353	6343
Leather and leather products (31)	2533	15,391	2154	9770	4332	6890	7581	4735
Stone, clay, glass products (32)	302	25,889	475	23,021	964	17,683	1949	11,993

	1965		1970		1975		1979	
	Imports (million $)	CE[b] Imports (kcal/$)	Imports (million $)	CE Imports (kcal/$)	Imports (million $)	CE Imports (kcal/$)	Imports (million $)	CE Imports (kcal/$)
Primary metal industry (33)	1234	26,877	2234	25,657	3770	12,164	4495	12,443
Fabricated metal industry (34)	553	12,652	744	11,993	1891	7483	3431	5659
Machinery except electricity (35)	6002	10,355	10,295	9817	25,573	5915	39,151	4421
Electric, electronics (36)	993	9956	1390	9367	3642	6129	6763	4248
Transportation equipment (37)	3204	11,723	6197	10,941	16,452	7171	24,577	4883
Instruments (38)	610	7874	1007	6900	2398	4643	5516	3306
Corn	833	22,509	818	19,156	4422	12,892	6059	14,156
Soybeans	650	29,492	1228	26,995	2865	14,796	5444	11,504
Wheat	1184	36,756	1112	34,038	5293	11,321	4862	13,920
Rice	244	21,320	313	17,850	858	7911	884	11,281
Total	23,335	—	35,838	—	91,045	—	143,453	—
Mean	—	35,867	—	31,824	—	15,220	—	11,865
Imported Petroleum								
Imported liquid petroleum	2092	644,335	2764	682,334	24,814	131,504	56,048	79,294
Imported natural gas	105	1,113,040	258	826,326	1081	226,823	2765	114,197

[a] Energy per dollar exported goods includes both chemical and embodied energy.
[b] CE = Chemical Energy of Imports.

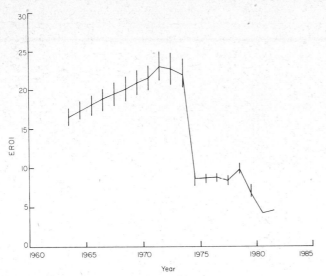

Figure 8.3. Energy return on investment for imported liquid petroleum fuels. Lines connect the mean EROI calculated by the four methods. Error bars indicate the range of values calculated by the four methods. No error bars are possible for 1980 and 1981 because data necessary for all four methods were unavailable.

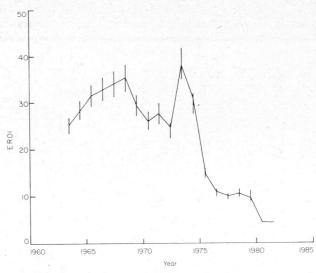

Figure 8.4. Energy return on investment for imported natural gas. Lines connect the mean EROI calculated by the four methods. Error bars indicate the range of values calculated by the four methods. No error bars are possible for 1980 and 1981 because the data necessary for all four methods were unavailable.

embargo, the EROI for imported natural gas continued to decline as the price of a kilocalorie of gas was pegged closer to the price of a kilocalorie of liquid fuel.

CAN ALTERNATIVE DOMESTIC FUELS REPLACE IMPORTED OIL?

Monetary values are the most common, and most appropriate, measure for quantifying economic transactions and the price of imported fuels probably will be the chief determinant of how much oil the United States imports. The EROI approach presented in the last section does not have any startling new policy implications—everyone knows that the price of imported oil has risen a great deal since 1973 even with the drop in price between 1981 and 1984. But further information is obtained by assessing energy costs and gains that are not captured by economic analyses. For example, unless coal replaces petroleum as the predominant fuel source for producing exports, the real energy cost of imported oil cannot rise by a factor of five as it did during the 1970s because, assuming its current mix of exported commodities, the United States would not gain energy in that exchange.

As we have pointed out in previous chapters,

numerous government subsidies, tax structures, and regulations, as well as market preferences, have disrupted the link between the dollar cost of fuel and the energy cost of fuel. Despite such complications a number of analyses have attempted to predict the prices for imports that would trigger the substitution of domestic alternatives (e.g., oil shale) using economic criteria alone. One such study, Adelman et al. (1974), predicted oil shale would be a viable substitute for imported petroleum when the cost of imports exceeded $10.00/bbl (1973 dollars). A more recent study by the Rand Corporation (1979) predicted a trigger price of $29.00/bbl (1979 dollars). As prices for imported liquid fuels have exceeded these predictions, however, the anticipated substitutions have not taken place.

We hypothesize that for domestic alternatives to substitute for imports in a competitive market, the EROI for imports must fall to levels well below those of alternative fuels. The EROI for imported liquid petroleum fuels has not yet declined to levels equal to—let alone below—the range for most domestic alternatives (see Table 2.1). Estimates of the EROI for alcohol production range between 0.7 and 2.6. The EROI estimates for oil shale in Kentucky and Colorado range between 0.7 and 13.3. Domestic alternatives are not yet viable substitutes for imported petroleum because most are unable to produce the same return of energy per unit or energy invested and, we believe, such relations must be reflected, ultimately, in market price.

Although neither energy nor economic analyses alone can evaluate all the political implications of continued U.S. dependence on imported petroleum, the effects of another oil embargo, or the impact of depleting our domestic petroleum reserves more or less rapidly, energy analyses can indicate physical limitations that are constraining attempts to displace imports with alternative domestic fuel sources. Because the EROI for domestic alternative fuels is less than the EROI for imported fuels at this time, greater quantities of energy would have to be invested to replace a barrel of imported petroleum with a barrel from a domestic alternative, such as alcohol. Since the additional quantities of energy needed for alternative domestic fuel production are not likely to be forthcoming from U.S. petroleum reserves, fuels would have to be diverted from other sectors of the economy, reducing nonfuel economic production. Thus any significant attempt to displace imports with domestic alternatives that have a lower EROI would decrease the quantity of net energy available to the U.S. economy and further constrain useful economic activity.

9

NATURAL GAS

INTRODUCTION

In many respects natural gas is our most valuable energy source because of its versatility, abundance, ease of transport, and relatively low price, and because its extraction and combustion products are less polluting relative to oil or coal. Additionally natural gas normally requires no local storage facilities like other fossil fuels and therefore is readily available to consumers. Finally, the overall conversion efficiency of natural gas is higher than other fossil fuels and electricity for some uses (see Table 9.1).

The United States consumed about 17.5 trillion ft^3 of natural gas (or about 4.4×10^{15} kcal) in 1984, which is equivalent to about 25% of total domestic energy consumption in that year. Proven reserves were about 200 trillion ft^3, but potential future reserves are more difficult to estimate than for oil because gas can be found at much greater, hence less explored, depths than oil. Some of the most important information about past U.S. natural gas use is not its physical availability but rather how federal regulations have affected its use.

Despite its attractiveness as a fuel source, natural gas has not always been perceived by industry or the federal government as a high-quality fuel source. The reasons were purely economic: markets and transportation networks for natural gas developed more slowly than they did for oil, and natural gas was viewed as an unwanted, undesirable by-product of crude oil production (and still is in some parts of the world). Crude oil was more valuable and more easily marketed in the early part of this century due in part to the rapid growth of the transportation sector and also because after World War II there was considerable development of chemical feedstock industries. As a result in the early days of the petroleum industry large quantities of natural gas were simply released into the atmosphere via venting or flaring. From an economic standpoint this behavior was understandable, even justifiable. From the perspective of efficient management and use of our natural resources, however, large quantities of our highest-quality gas were wasted in the early days of the industry. Large quantities of natural gas in the Middle Eastern countries are still wasted in this fashion for the same reasons.

Natural gas production increased 4–8%/yr during the late 1950s and early 1960s (Figure 9.1) while natural gas discoveries increased an average of about 3%/yr (Figure 9.2a). The abundance of gas combined with very generous estimates of future natural gas availability led most planners and policymakers in and out of government to dismiss casually any mention of possible future natural gas scarcities as antigrowth doomsaying. Federal government regulatory policies rendered during this period of natural gas (and overall fuel) abundance led to the establishment of mandatory price ceilings that eventually fell below the true market value of the resource. The notion of indefinite domestic energy abundance also was fostered by extremely optimistic federal government estimates of future oil and gas availability that were not reconcilable with the history and data of the industry itself (p. 343).

In 1954 when federal regulation of the wellhead price of natural gas began, the real price of gas was 16.9 cents/10^3 ft^3* (Figure 9.3). Fifteen years later it had increased to only 19.3 cents/10^3 ft^3—a price increase much less than the general inflation rate and less than the decline in the rate at which new gas deposits were being discovered. The net effect that government regulation of natural gas prices has had on present-day supplies is difficult to discern precisely. On the one hand, natural gas was probably consumed at a more rapid rate than had it not been regulated, since it cost less per kilocalorie than alternate fuel resources. On the other hand, low

---

*1000 ft^3 = 28.3 m^3; thousand cubic feet is the unit normally used for natural gas.

Table 9.1. Thermal Efficiencies of Domestic Fuels

	Production (1)	Transmission (Distribution) (2)	Delivered at Consumer's Door (3) (1 × 2)	Usage (e.g., furnace efficiency) (4)	Final Efficiency (3 × 4)
Coal	96	97	93	50	47
Oil	84	97	81	60	49
Natural gas	96	97	93	75	69
Electricity (coal-based)	30	85	25	100	25

Source: E. N. Tiratsoo, 1979. *Natural Gas.* Reprinted with permission of Plenum Press, New York.

prices paid to producers were a disincentive to explore for, and develop, new natural gas deposits, so today we have more gas left in the ground to look for. The former practice contributed to the depletion of our gas resources, and the latter probably conserved them.

WHAT PETROLEUM GAS IS

As we described at the beginning of Part Three, petroleum forms when large amounts of organic material are deposited in an anoxic environment and subsequently transformed by microbial and geologic activity. When the organic matter is buried at a shallow depth (1–2 m) substantial amounts of methane (CH_4 or *marsh gas*) and other decomposition gases may be produced. As the depth of burial increases up to several kilometers, the oxygen content of the gases decreases while the methane and nitrogen proportions increase. These deeply buried or *fossil* gases are usually termed natural gas, and they contain varying amounts of hydrocarbon and nonhydrocarbon constituents.

Natural gas is often, but not always, found in the same structural formations as crude oil. To this extent it usually is considered to be the gaseous phase of crude oil. Natural gas that overlies and is in contact with crude oil in a reservoir is termed *associated* or gas cap gas. Natural gas not in physical contact with crude oil in a reservoir is called *nonassociated* gas. *Dissolved* gas is natural gas that exists in solution with crude oil at prevailing reservoir conditions (Tiratsoo, 1979). In the United States only 23% of natural gas is associated with oil, although the percentage varies from region to region. In California, for example, 60% of the gas is associated with oil. The largest dissolved gas deposits are

in the Gulf Coast region of Texas and Louisiana, where 50% of all known U.S. gas reserves are located (McGraw-Hill, 1980).

The chemical composition of natural gases is dominated by members of the paraffin or alkane series (C_nH_{2n+2}) of hydrocarbon, with methane being the simplest and most common constituent of natural gas. Although methane generally accounts for 70–90% by volume of most natural gas, ethane, propane, butane, nitrogen, and carbon dioxide also may be present in varying amounts. Nonassociated natural gases tend to have higher proportions of methane than do associated or dissolved gases. Nonassociated gases also tend to flow to the surface at higher pressures, which makes their transportation in pipelines easier than associated gases, which often need artificial boosting just to get them to the surface. This condition, plus the fact that associated gas also must be separated from crude oil, makes nonassociated gas the most commercially attractive type of gas deposit to develop.

When the higher molecular weight components of natural gas such as butane (C_4H_{10}), pentane (C_5H_{12}), and hexane (C_6H_{14}) are present in significant quantities, they are normally extracted from the gas in liquid form during the production process. These *natural gas liquids* (NGL) (isobutane, pentane, etc.) and *liquefied petroleum gas* (LPG), especially propane and *n*-butane, often are commercially valuable as separate by-products of the natural gas production process. When the amount of extractable liquid hydrocarbons is less than 0.1 gal/10^3 ft^3 of gas, the natural gas is called *dry*. Nonassociated gas is usually of this type. *Lean* gases are those with 0.1–0.3 gal/10^3 ft^3 of extractable liquids, whereas *wet* gases are those with a ratio of 0.3 gal/10^3 ft^3 or greater (Tiratsoo, 1979).

Liquefied natural gas refers to natural gas (con-

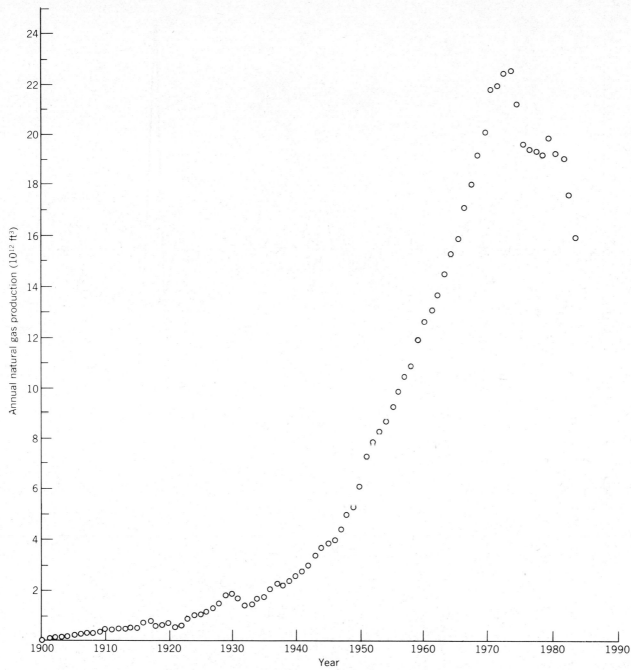

Figure 9.1. U.S. natural gas production, 1900–1983.

sisting primarily of methane) that has been liquefied under specific temperature and pressure conditions. It should not be confused with the LPG that consists of principally propane and butane and has a higher critical temperature. Because of its low critical temperature ($-75°$ C), liquefaction of natural gas cannot occur simply by increasing pressure. Cryogenic temperature must be reached and main-

tained in a well-insulated storage system in order for the natural gas to remain in a liquid state.

The most common use of LNG is in a peak-shaving system, where the LNG is used to supplement pipeline supplies of natural gas during times of peak demand when supplies are short. In this type of system gas is collected and liquefied during the summer months when demand is low, and held in

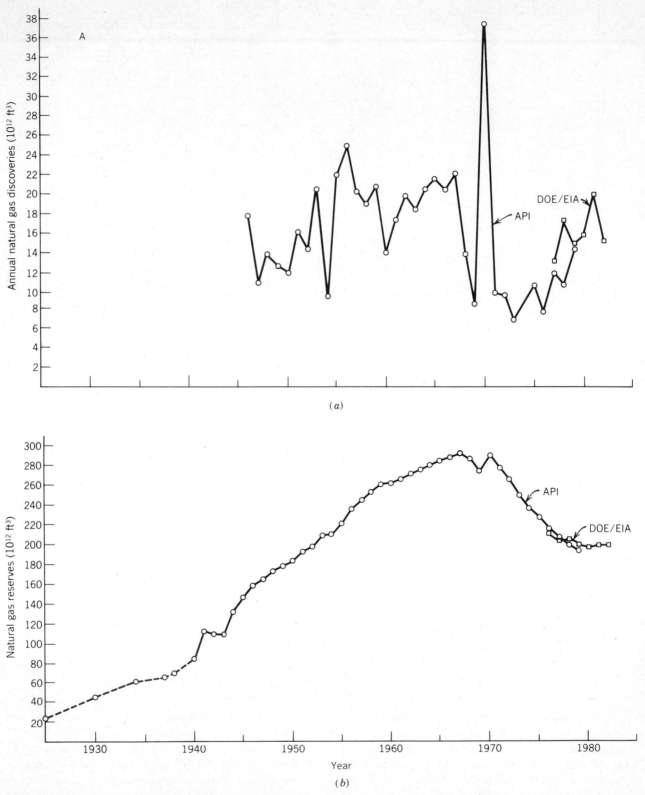

Figure 9.2. (*a*) U.S. annual gas discoveries, 1945–1982. (*b*) U.S. natural gas reserves. Data exclude Prudhoe Bay, Alaska, reserves which would add about 26 trillion ft³ of gas in 1970. (Data for 1930–1979 from American Petroleum Institute; for 1977–1982, Energy Information Administration, USDOE. Note that the latter agency assigns a higher yield to a given initial discovery.)

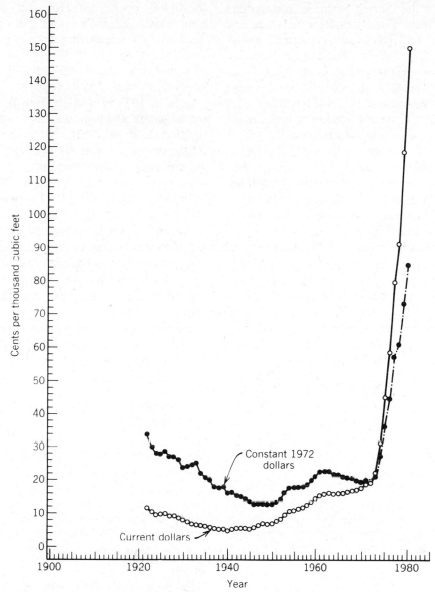

Figure 9.3. Wellhead price of natural gas in United States. (From USDOE, 1984.)

storage tanks until winter time when it is vaporized and sent back into the pipeline system. This is a convenient way to store natural gas since 600 ft^3 of natural gas condenses into less than 1 ft^3 of liquid (McGraw-Hill, 1980).

The first methane liquefaction plant and peak-shaving system was built in Cleveland, Ohio, in 1941 where it operated successfully for several years until one of the storage tanks blew up. The accident killed many people and destroyed much of the facility. Several decades passed before the LNG approach to peak shaving was used again, but storage techniques were improved greatly. In 1978, 65

of the world's approximately 100 LNG storage facilities were located in the United States, with a combined storage capacity of about 330 × 10^6 ft^3/day (Tiratsoo, 1979). No LNG facility has had a serious accident since the 1941 Cleveland disaster.

HOW GAS IS FOUND AND EXTRACTED

The Extraction Process

Discovery and extraction of crude oil and natural gas often are related very closely and about 18% of the gas produced in the United States is from crude oil wells (NAS, 1980). The distinction between an oil well and a gas well is not always a clear one, especially before the well is drilled. In situations where gas is the only hydrocarbon being extracted, the production process is simplified because gas normally flows freely out of a drilled hole to the surface and therefore does not need to be artificially lifted to the surface as crude oil often does.

In many instances where oil and gas are both present, gas that reaches the surface through the well casing or tubing is reinjected into the formation to maintain the flow of crude oil flow. Gas injection is used most often to increase oil flow rates from wells that have natural flow rates that are considered to be too slow. In the early 1980s about 6% of the natural gas produced or held in surface storage was used for repressuring. In oil wells that do have high natural flow rates, natural gas or water pressure are usually the forces responsible for crude oil flow. Natural gas facilitates the natural flow of crude oil in two ways. The first is gas cap drive where the pocket of gas that lies above the crude oil in the reservoir forces the oil into a well hole that is drilled through the gas and into the oil itself. The Prudhoe Bay Field, the largest field in North America, is an example of this type of drive. The second type of oil flow is dissolved gas expansion, where the gas is actually in solution with the oil and forces the oil to move from a region of higher concentration within the formation to a region of lower concentration in the well itself, such as the point of extraction. In most reservoirs both of these natural drive mechanisms, as well as water and gravity drive, operate simultaneously.

Flaring

In the early years of the petroleum industry when natural gas was viewed and treated as a nuisance—as an unwanted by-product of crude oil extraction—much more gas was discovered than could be sold, since there was little or no market for it. Additionally there was no pipeline technology to transport gas to markets that were geographically far from the centers of oil and gas production. The associated gas that came to the surface with the crude oil was therefore commonly burned off or *flared*. Today the United States flares less than 1 percent of its natural gas. Ninety-four percent of its domestic production is sold directly and another 5% is used

for oil field operations. However, as late as 1949 the United States flared over 11% of its annual gas production. The rapid increase in gas pipeline construction during the 1930s and 1940s helped create new markets for gas and led to the decrease in the wasting of this valuable fossil fuel.

Yet in 1977 about 7 trillion ft^3 or 12% of total world gas production was flared (down from an estimated 23% in 1962). An estimated 125 trillion ft^3 (or about a 7-year supply for the United States at current consumption rates) has been flared worldwide since 1935 (Tiratsoo, 1979). As Table 9.2 indicates, in the 1970s OPEC countries flared up to three quarters of their natural gas, for their domestic market is small relative to the huge volumes of gas they produce annually. Natural gas produced in the Middle East is not exported to the degree that crude oil is because gas must first be liquefied and placed in special containers to maintain liquidation during transport, an extremely expensive and hazardous process that makes it uneconomical. One possible use of this gas would be to make nitrogen fertilizer which is easily transported.

HISTORICAL TRENDS IN GAS USE AND REGULATIONS

Several aspects of natural gas discovery and extraction in the United States make an analysis of this industry more complex than that for crude oil. First, as we mentioned earlier, natural gas was originally produced as an undesired by-product of crude oil production, and much of it was flared. Two political decisions in the early 1950s put a stop to these practices and led to the construction of large pipeline systems that made natural gas available to the major industrial and commercial energy users in the east. The first decision, made by the Texas Railroad Commission, regulator of the nation's largest oil- and gas-producing state, forbade flaring of natural gas. The second, made by the Supreme Court in 1954, mandated the Federal Power Commission (FPC) to regulate the wellhead price of natural gas to be sold in interstate markets.

The FPC's decisions were influenced heavily by two factors. Testimony of experts from the USGS estimated over 2000×10^{12} ft^3 of remaining U.S. gas resources. Compounding this error was the commission's belief that, in setting and holding gas price levels, its primary duty was to protect consumer interests, which was interpreted as setting gas

Table 9.2.   Natural Gas Flared in 1976

	Gas Flared (billion ft³)	Percentage of Gross Production
United States	131.9	0.6
Canada	59.7	1.7
Mexico	185.2	24.0
South America and Trinidad	442.8	17.6
Western Europe	302.1	3.6
Soviet Union	614.1	5.1
Africa	1394.1[a]	57.2
Middle East	3844.6[b-e]	73.2
East Asia	261.7	8.4
Australasia	1.1	0.5
World total	7237.3	12.2

Source: E. N. Tiratsoo, 1979. *Natural Gas.* Reprinted with permission of Plenum Press, New York.

[a] Includes Nigeria, 745.5 billion ft³.
[b] Includes Iran, 982.5 billion ft³.
[c] Includes Iraq, 310.5 billion ft³.
[d] Includes Abu Dhabi, 463.0 billion ft³.
[e] Includes Saudi Arabia, 1321.1 billion ft³.

prices at their lowest possible levels. Prices were initially set at 16 cents/10^3 ft³ and were allowed to increase only very slowly over the next quarter century (see Figure 9.3). That the price of natural gas was held artificially well below its market value can be seen by the price differential between gas sold in the interstate market and gas sold in the intrastate market, which was not subject to price regulation by the FPC. Almost immediately after regulation began, the price of unregulated intrastate gas rose relative to regulated gas prices, to roughly the price of alternate unregulated fuels. This point is further substantiated by comparing the cost of natural gas versus the cost of alternate fuels on a per kilocalorie basis (see Figure 9.4). Natural gas has been consistently the cheapest of all fossil fuels.

The most predictable consequence of the price regulation by the FPC has been the subsidization of natural gas consumption, presumably resulting in more gas being consumed than had there been no price regulation. Another consequence of price regulation has been a decreased incentive to explore for natural gas since about 1960 due to the low prices paid to natural gas producers. This claim seems to be substantiated by the recent flurry of drilling for natural gas in the deep basins of Louisiana, Texas, and Oklahoma after the price of *deep* gas (15,000 ft or deeper) was deregulated as part of President Jimmy Carter's National Energy

Plan in 1977. Aspects of the natural gas industry cited in the foregoing discussion are crucial to a full understanding of any statistical analysis of the natural gas industry, since drilling, discovery, and production rates were all affected by the regulatory circumstances under which the industry was forced to operate.

U.S. RESOURCES OF NATURAL GAS

As was the case for crude oil, the United States was endowed with relatively large reserves of natural gas, and until the late 1960s the United States had the largest proven reserves of natural gas in the world. Since that time the United States has fallen to third, behind the Soviet Union and Iran, due mainly to the fact that the United States had been more thoroughly explored for oil and gas in the past (see Table 9.3) so that we are no longer finding the very large, high-quality fields that we did even 15 years ago. Proven reserves of natural gas peaked in 1967 at 293 trillion ft³ (Figure 9.2*b*) and have declined steadily since that time to 200 trillion ft³ in 1980, despite a 280% increase in annual drilling between 1972 and 1982. Annual gas consumption in the United States averaged about 20 trillion ft³ during the 1970s, but declined to about 17 trillion ft³ by the mid 1980s. Natural gas discoveries have gener-

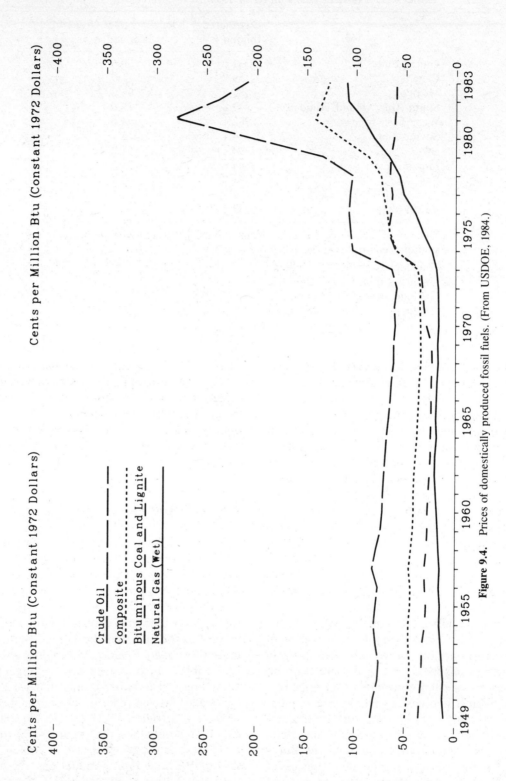

Figure 9.4. Prices of domestically produced fossil fuels. (From USDOE, 1984.)

208

ally declined since the late 1960s, although the decline is more erratic and less pronounced than is the case for oil (see Figure 9.2a). Six states—Texas, Louisiana, New Mexico, Kansas, Oklahoma, and California—account for almost 75% of all domestic proven reserves. Texas and Louisiana alone account for about 52% of total proven reserves of natural gas.

As with crude oil the distribution of and discovery rates for natural gas are strongly influenced by large and giant gas deposits, which tend to be discovered early in the exploration process. The four largest petroleum provinces alone have over 60% of recoverable gas resources, and within these regions large and giant fields account for most of the gas (Nehring, 1981) (Figure 9.5). The 867 giant and large fields with natural gas account for 551.5 trillion ft^3, over half of all estimated recoverable gas resources (Nehring, 1981). Nehring also estimates that the 2471 *significant* (>10 mbl of oil or its equivalent) fields with natural gas contain 663.5 trillion ft^3 of gas, or almost 90% of all natural gas discovered and made recoverable through 1979. As we mentioned in Chapter 7, the Hugoton-Panhandle field alone has 79.9 trillion ft^3 of gas, or almost 11% of the national total.

The historical pattern of annual gas discoveries is similar to that for crude oil, although the trend for gas lags behind oil by one to two decades. The lag is due to the slower development of markets for natural gas described earlier and the gradual development of deep drilling technologies. Total gas discoveries peaked in the 1940s, although significant gas discoveries didn't peak until the late 1950s (Figure 9.2a), except for a temporary reversal when the offshore areas of Texas and Louisiana were opened for exploration (Nehring, 1981).

Since almost every oil field also contains natural gas, large amounts of associated gas are found in giant oil fields. Gas:oil ratios range from several to over 8000 ft^3 of gas per barrel of oil. Table 9.4 lists the major giant oil fields of the world that also contain at least 3.5×10^{12} ft^3 of natural gas. The United States has only two such fields, Prudhoe Bay, Alaska, and the East Texas field, the two largest domestic petroleum accumulations found to date. Table 9.5 lists the 96 giant nonassociated gas fields discovered as of 1977. It is interesting to note that the largest domestic field, Prudhoe Bay, which has reserves of about 9.6 billion bbl of oil and over 26 trillion ft^3 of gas, is nowhere near as large as the immense fields of the Soviet Union and Middle East.

Finding Rate versus Drilling Effort for Gas

Analyzing natural gas resources, and their availability, can be performed in a manner similar to that for

Table 9.3. Number of Petroleum Wells Drilled Through Mid-1970s

	Wells Drilled	Percentage of Total Wells Drilled
United States	482,000	74.7
Soviet Union	100,000	15.5
Canada	20,000	3.1
Latin America	14,000	2.2
Western Europe	12,500	1.9
		97.4
Africa	6,500	
South and Southeast Asia	5,000	
Middle East	2,000	2.6
China	2,000	
Japan	1,000	
Australasia	500	
	645,500	100

Source: E. N. Tiratsoo, 1979. *Natural Gas*. Reprinted with permission of Plenum Press, New York.

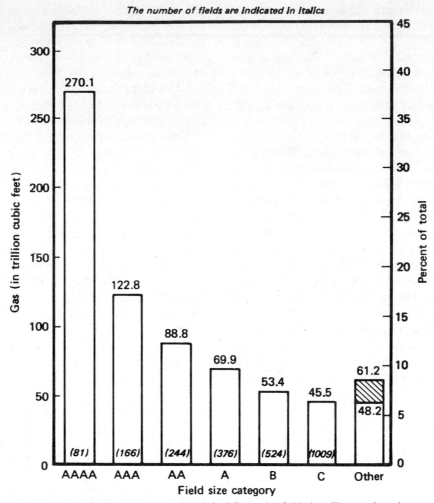

Figure 9.5. The distribution of natural gas resources in the United States by field size. The numbers in parentheses represent the number of fields in each size category. (From Nehring, 1981.)

oil, using methods developed by Davis (1958) and Hubbert (1962, 1967), and later modified by Hall and Cleveland (1981). Both methods document quantitatively the trend of increasing scarcity in the availability of new high-quality natural gas deposits. The results of these two methods, which extrapolate from past experience rather than attempt to estimate future resources directly, can be compared to more optimistic estimates made by the USGS and by Osborne (1984) who claimed that we have hundreds, and possibly even thousands, of years' supply of natural gas remaining to be discovered and utilized. Because the use of different methodologies leads to widely different policy choices, we as a nation must choose carefully among these different methods when formulating energy policy.

Since data on proven reserves for natural gas were not collected prior to 1945, Hubbert (1962)

modified his crude oil analysis slightly for his analyses of natural gas. Using data for oil production and estimates of gas:oil ratios that ranged from 6250 to 7500 ft^3 of gas found per barrel of oil, Hubbert estimated that the ultimate recovery of natural gas in the lower 48 states (Q_∞) would be about 1000 × 10^{12} ft^3. Hubbert's calculated curves for rates of discovery, production, and increases in proven reserves were performed in a manner similar to his analysis of crude oil (Figures 9.6 and 9.7). From these curves the peak in the rate of proven discoveries was at about 1961, the peak in proven reserves at about 1969, and the peak in the rate of production was predicted to occur about 1977.

Hubbert's predictions can be compared with the actual behavior of the natural gas industry to evaluate the accuracy of his methodology. In 1962 Hubbert predicted that proven reserves of natural gas

Table 9.4. World Giant Oilfields with at Least 3.5 Trillion ft³ of Ultimately Recoverable Associated Gas

Field	Country	Year Discovered	Age of Reservoir	Ultimate Oil Recovery (billion bbl)[a]
1. Ghawar	Saudi Arabia	1948	Jurassic	83
2. Burgan	Kuwait	1938	Cretaceous	72
3. Bolivar Coastal	Venezuela	1917	Miocene-Eocene	32
4. Safaniya-Khafji	Saudi Arabia–Neutral Zone	1951	Cretaceous	30
5. Rumaila	Iraq	1953	Cretaceous	20
6. Ahwaz	Iran	1958	Oligocene-Miocene, Cretaceous	17.5
7. Kirkuk	Iraq	1927	Oligocene-Eocene, Cretaceous	16
8. Marun	Iran	1964	Oligocene-Miocene	16
9. Gach Saran	Iran	1928	Oligocene-Miocene, Cretaceous	15.5
10. Agha Jari	Iran	1938	Oligocene-Miocene, Cretaceous	14
11. Samotlor	Soviet Union	1966	Cretaceous	13
12. Abqaiq	Saudi Arabia	1940	Jurassic	12.5
13. Romashkino	Soviet Union	1948	Cretaceous	12.4
14. Berri	Saudi Arabia	1964	Jurassic	12
15. Zakum	Abu Dhabi	1964	Cretaceous	12
16. Manifa	Saudi Arabia	1957	Cretaceous-Jurassic	11
17. Fereidoon-Marjan	Iran–Saudi Arabia	1966	Cretaceous-Jurassic	10
18. Prudhoe Bay	United States	1968	Cretaceous, Triassic, Mississippian	9.6
19. Du Hasa	Abu Dhabi	1962	Cretaceous	9
20. Qatif	Saudi Arabia	1945	Jurassic	9
21. Khurais	Saudi Arabia	1957	Jurassic	8.5
22. Zuluf	Saudi Arabia	1965	Cretaceous	8.5
23. Raudhatain	Kuwait	1955	Cretaceous	7.7
24. Sarir	Libya	1961	Cretaceous	7.2
25. Hassi Messaoud	Algeria	1956	Cambrian-Ordovician	2
26. Shaybah	Saudi Arabia	1968	Cretaceous	7
27. Abu Sa'fah	Saudi Arabia–Bahrain	1963	Jurassic	6.6
28. Asab	Abu Dhabi	1965	Cretaceous	6
29. Bab	Abu Dhabi	1954	Cretaceous	6
30. Ta-ch'ing	China	1959	Cretaceous	6
31. East Texas	United States	1930	Cretaceous	5.6
32. Umm Shaif	Abu Dhabi	1958	Jurassic	5
33. Wafra	Neutral Zone	1953	Eocene-Cretaceous	5

Source: From The Discovery of Significant Oil and Gas Fields in the U.S. by R. Nehring, 1981. Reprinted with permission of the Rand Corp.

[a] Primary recovery only.

would peak in 1969. Actually they peaked in 1967. Production was predicted by Hubbert to peak in 1977. The production rate peaked in 1973—an error of only four years which can be compared to official USGS estimates of virtually unlimited supplies until at least the year 2000. Like his estimates for crude oil, Hubbert's predictions for the natural gas industry have, to date, been remarkably accurate.

Hubbert also analyzed natural gas discoveries per foot of exploratory drilling just as he did for crude oil (Figure 9.8). As was the case for oil, gas discoveries per foot drilled declined exponentially,

Table 9.5. World Giant Nonassociated Gas Fields

Field	Country	Age of Reservoir	Date Discovered	Ultimate Gas Reserves (trillion ft³)
1. Urengoy	USSR (W. Siberia)	Cretaceous	1966	176.5
2. Kangan	Iran	Permian	1973	170.0
3. Yamburgskoye	USSR (W. Siberia)	Cretaceous	1969	155.3
4. N-W Dome	Qatar	Permian	1976	100.6+
5. Zapolyarnoye	USSR (W. Siberia)	Cretaceous	1965	94.0
6. Krasniy Kholm	USSR (Orenburg)	Permian	1966	74.0
7. Hassi R'Mel	Algeria	Triassic	1956	70.0
8. Hugoton-Panhandle	USA (Kansas-Texas)	Permian	1926	70.0
9. Kangan	Iran	Permian	1973	70.0
10. Groningen	Netherlands	Permian	1959	60.9
11. Medvezhye	USSR (W. Siberia)	Cretaceous	1967	55.0
12. Bovanenko	USSR (W. Siberia)	Cretaceous	1971	53.0
13. Pars	Iran	Permian	1973	50.0
14. Pazanan	Iran	Oligocene-Miocene	1938	50.0
15. Kharsavey	USSR (W. Siberia)	Cretaceous	1974	42.4
16. Taz	USSR (W. Siberia)	Cretaceous	1962	40.4
17. Dorra	Neutral Zone	Cretaceous	1974	35.0
18. Bahrain	Bahrain Island	Jurassic/Permian	1931	20.0
19. Kangiran	Iran	Permian	1968	20.0
20. Semakovskoye	USSR (W. Siberia)	Cretaceous	1971	19.0
21. Vyuktyl	USSR (E. Siberia)	Permian	1964	17.7
22. Layavozh	USSR (Pechora)	Carboniferous	1965	17.5
23. E J Bermudez	Mexico	Cretaceous	1976	17.5
24. Gazli	USSR (C. Asia)	Cretaceous	1956	17.0
25. Shebelinka	USSR (Ukraine)	Permian	1956	16.4
26. Severo-Urengoy	USSR (W. Siberia)	Cretaceous	1970	16.0
27. Komsomolskoye	USSR (W. Siberia)	Cretaceous	1966	16.0
28. Sredne-Vilyuy	USSR (E. Siberia)	Triassic-Jurassic	1963	15.9
29. Yansovey	USSR (W. Siberia)	Cretaceous	1970	15.0
30. Nar	Iran	Permian	1974	14.0
31. Messoyakha	USSR (W. Siberia)	Cretaceous-Jurassic	1967	14.0
32. Arun	Indonesia	Miocene	1971	13.0
33. Kirpichli	USSR (C. Asia)	Cretaceous	1972	12.5
34. Gubkin	USSR (W. Siberia)	Cretaceous	1965	12.3
35. Leman Bank	UK (North Sea)	Permian	1965	12.3
36. Frigg group	Norway (North Sea)	Eocene	1971	12.1
37. Hateiba	Libya	Cretaceous, Cambrian-Ordovician	1963	12.0
38. Blanco Mesaverde–Basin Dakota	USA (N. Mexico)	Cretaceous	1927	11.0
39. Rhourde Nouss group	Algeria	Triassic-Ordovician-Cambrian	1962	11.0
40. Vyngapurovskoye	USSR (W. Siberia)	Cretaceous	1968	10.6
41. Russkoye	USSR (W. Siberia)	Cretaceous	1968	10.6
42. Gomez	USA (Texas)	Ordovician-Cambrian	1963	10.0
43. Bagazhda	USSR (C. Asia)	Cretaceous	1971	9.6
44. Sui	Pakistan	Eocene	1953	8.6
45. Noviy Port	USSR (W. Siberia)	Cretaceous-Jurassic	1964	8.5
46. Krestichenskoye	USSR (Ukraine)	Permian-Pennsylvanian	1968	8.2
47. Jalmat-Monument-Eunice	USA (California-N. Mexico)	Permian	1929	8.1
48. Naip	USSR (C. Asia)	Cretaceous	1972	8.0
49. Indefatigable	UK (North Sea)	Permian	1966	8.0
50. N. Rankin	Australia (NW Shelf)	Triassic	1972	7.9

Table 9.5.   (Continued)

Field	Country	Age of Reservoir	Date Discovered	Ultimate Gas Reserves (trillion ft³)
51. Shih-you-kou group	China	Triassic-Jurassic	1955	7.8
52. Severo-Stavropol-Pelagiada	USSR (Caucasus)	Oligocene-Miocene	1950	7.3
53. Monroe	USA (Louisiana)	Paleocene-Cretaceous	1916	7.0
54. Lacq	France	Cretaceous-Jurassic	1951	7.0
55. Severo-Komsomol-skoye	USSR (W. Siberia)	Cretaceous	1969	7.0
56. Pelyatinskoye	USSR (W. Siberia)	Cretaceous	1969	6.6
57. Puckett	USA (Texas)	Ordovician	1952	6.5
58. Maastakh	USSR (E. Siberia)	Jurassic	1967	6.4
59. Badak	Indonesia	Miocene	1975	6.2
60. Drake Point	Canada (Arctic Islands)	Triassic-Jurassic	1975	6.1
61. Carthage	USA (Texas)	Cretaceous	1936	6.0
62. Bintulu	East Malaysia (Sarawak)	Miocene	1975	6.0
63. Katy	USA (Texas)	Eocene	1964	6.0
64. Rabbit Island group	USA (Louisiana)	Miocene	1940	6.0
65. Soleninskoye	USSR (W. Siberia)	Cretaceous	1969	5.7
66. Maui	New Zealand	Eocene	1969	5.6
67. Urtabulak	USSR (C. Asia)	Jurassic	1963	5.4
68. L-10	Netherlands	Permian	1975	5.3
69. Achak	USSR (C. Asia)	Cretaceous	1966	5.0
70. Kenai	USA (Alaska)	Tertiary	1959	5.0
71. Old Ocean	USA (Texas)	Oligocene	1934	5.0
72. Süd-Oldenburg	W. Germany	Permian	1968	5.0
73. Yefremovka	USSR (W. Siberia)	Permian	1965	4.6
74. Dhodak	Pakistan	Eocene	1976	4.5
75. Hecla	Canada (Arctic Islands)	Triassic-Jurassic	1975	4.2
76. Seredny-Yamal	USSR (W. Siberia)	Cretaceous	1970	4.1
77. Hewett group	UK (North Sea)	Triassic-Permian	1966	4.0
78. Gidgealpa	Australia	Permian	1964	4.0
79. Moomba	Australia	Permian	1964	4.0
80. Mari	Pakistan	Eocene	1957	4.0
81. Mocane-Laverne	USA (Oklahoma)	Pennsylvanian-Mississipian	1952	3.8
82. Samantepe	USSR (C. Asia)	Jurassic	1964	3.7
83. Marlin	Australia	Eocene-Paleocene	1966	3.6
84. Rio Vista	USA (California)	Eocene-Paleocene	1936	3.5
85. Bierum	W. Germany-Netherlands	Permian	1963	3.5
86. Meillon	France	Cretaceous-Jurassic	1965	3.5
87. Khadzhiy-Kandym	USSR (C. Asia)	Jurassic	1967	3.5
88. Bayou Sale	USA (Louisiana)	Miocene	1940	3.5
89. Gugurtli	USSR (C. Asia)	Cretaceous-Jurassic	1965	3.5
90. Tang-e-Bijar	Iran	Oligocene-Miocene	1975	3.5
91. Kettleman Hills North	USA (California)	Miocene-Eocene	1928	3.5
92. Gassi Touil	Algeria	Triassic	1971	3.5
93. Yetypur	USSR (W. Siberia)	Cretaceous	1971	3.5
94. Zapadno-Tarkoso-linskoye	USSR (W. Siberia)	Cretaceous	1972	3.5
95. Bakhrabad	Bangladesh	Miocene	1963	3.5
96. Barqan	Saudi Arabia (Red Sea)	Miocene	1969	3.5

Source: From The Discovery of Significant Oil and Gas Fields in the U.S. by R. Nehring, 1981. Reprinted with permission of the Rand Corp.

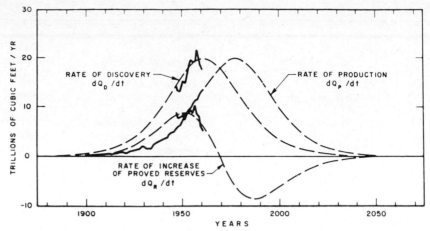

Figure 9.6. Rates of natural gas production, discovery, and increase in proven reserves by U.S., 1900–1962, superimposed upon predictions of Hubbert (1962). (From Hubbert, 1980.)

although the rate stabilized during the mid to late 1960s. Using this method to estimate Q_∞ for natural gas, Hubbert (1974) estimated that about 1100 trillion ft[3] of gas would be produced in the lower 48 states. Adding an estimate for natural gas recovery for Alaska, Hubbert calculated a Q_∞ of 1184 trillion ft[3] for the entire United States.

A reasonable explanation for the stabilization in the trend of decreasing yield per effort (YPE) over time for natural gas in the 1960s may be the decrease in drilling effort, itself an artifact of natural gas price regulation by the Federal Power Commission as discussed earlier. If, as Hubbert (1974) suggested, there was decreased incentive to explore for natural gas after about 1960 due to the low price

paid to producers, then the increase in YPE during this time is consistent with the inverse relation between YPE and effort described by Davis (1958) and Hall and Cleveland (1981).

In partial summary, the use of natural gas, a premium and widely used fuel, has been encouraged by economic regulations based in part on the assumption that very large reserves of gas remain to be discovered. Data of the last several decades, however, indicate that we have been using gas at nearly twice the rate we have been finding it. Conventional drilling in mature regions has not yet led to significant reversals in the trends documented by Hubbert (1974) and Hall and Cleveland (1981) despite a more than doubling of drilling effort after the

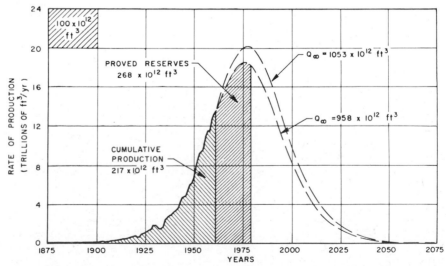

Figure 9.7. Two complete cycles of U.S. natural gas production based on high and low estimates of Hubbert (1962). (From Hubbert, 1980.)

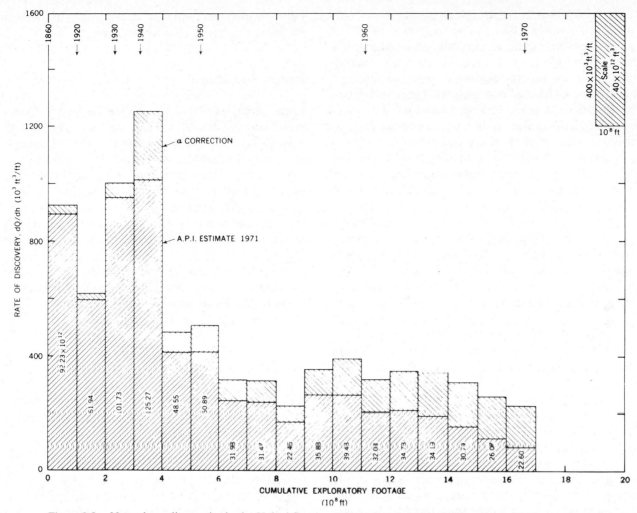

Figure 9.8. Natural gas discoveries in the United States per foot of exploratory drilling. (From Hubbert, 1974.)

energy price shocks of the 1970s. Thus very large additional reserves from frontier regions must be found if there are to be significant reversals in these trends.

UNCONVENTIONAL SOURCES OF GAS

Several additional sources of natural gas that are currently feasible to exploit are being evaluated by industry for future use. *Unconventional sources* of gas already provide about 1 trillion ft^3 (5% of current use), with what would initially appear to be a potential for much greater output in the years to come. The feasibility of these so-called unconventional sources are constrained by two basic criteria. The first and most commonly used criterion is economic incentive. Estimates for recoverable gas

are usually made in relation to projections for gas prices. Most analyses of unconventional gas sources find that the price of natural gas will have to be several times higher than its 1983 level of about $2.60 per 10^3 ft^3 before enough incentive will exist for gas producers to exploit lower quality deposits. One problem with the use of economic analyses, however, is that they require predictions of future economic conditions like inflation, interest rates, and relative price levels, a tenuous task at best even under promising and stable economic conditions.

Some geologists assume that vast amounts of unconventional natural gas exist dwarfing current proven reserves of gas and that these resources can be exploited if the incentive is sufficient. Paul Jones, geologist and gas industry analyst in Louisiana, has given a conservative estimate of gas in the geopressured aquifers alone as equivalent to about 2500 times current consumption rates. Commoner (1979) cites a figure of up to 9800×10^{12} ft^3 of gas in unconventional formations, about nine times the amount likely to be recovered from conventional sources. But the portion of this vast resource that can be extracted economically should dampen the optimism of some geologists and politicians toward predictions more in line with the more conservative estimates of Hubbert and others.

A second criterion for assessing future availability, less commonly used but equally important for resource estimates, is a consideration of the EROI for each of these proposed sources of natural gas. Because energy analysis is based on physical rather than monetary flows, the results of a net energy analysis do not become obsolete in five to ten years regardless of inflation or the discount rate. Only the changes in the energy costs of the discovery and extraction technology can change significantly EROI. Crude oil and natural gas at the wellhead currently have an energy return on investment of between 8 and 24 for one (Figure 7.19), though both declined rather steadily from a level between 27 and 45 to one in the mid-1950s. Presumably unconventional sources of gas have EROI ratios less than that of conventionally produced gas, otherwise the industry would have already exploited it. In a market with unregulated wellhead gas prices, we do not expect these unconventional sources to be tapped until the EROI (and hence, approximately, unsubsidized economic cost) for conventional gas drops to a level equal to or lower than the EROI for unconventional sources. The situation is complicated because the wellhead prices for natural gas were maintained until 1985 at an artificially low level for over 25 years, thereby depressing the incentive to explore for natural gas. Thus the lack of data on the dollar or energy cost of their development makes it difficult to assess their feasibility. The following sections investigate some of the possibilities.

Coalbed Methane

Controlling the release of methane from coal mines has always been looked at from the perspective of miner safety, with little thought given toward tapping this resource to augment local supplies of natural gas. With an average gas content of about 200 ft^3 per ton of coal, gas production may be feasible from existing mines. Methane production from coal mines would be concentrated in deep Appalachian

mines which account for nearly 90% of current methane emissions (Kuuskraa et al., 1978).

Eastern Gas Shales

Large amounts of natural gas are trapped in Devonian Shale accumulations and can be extracted with present technology by fracturing—by "shooting" the formation with either explosives or hydraulic foam—and water injection. Although this is a proven technology that could be expanded, the total quantities produced so far are very small. Wells in the Appalachian region have produced gas in this manner since before 1900 and have yielded about 2.7 trillion ft^3 of gas to date. There are currently over 9600 Devonian shale wells producing gas wells in Kentucky, West Virginia, and Ohio that combined produced an estimated 0.1 trillion ft^3 of natural gas in 1976 (McGraw-Hill, 1980).

Tight Gas Formations

So-called tight reservoir rocks are the most promising of the proposed unconventional gas sources because of the large recoverable quantities of gas involved. Large quantities of natural gas are trapped in sandstone or limestone reservoirs. Hydraulic fracturing (MHF) of the reservoirs by pumping large amounts of fluid into the formation until it cracks appears to be the best method for extracting the gas. Propping agents such as sand are also pumped into the fractures so that, when the fracturing fluid drains away, the induced fracture can be held open to allow for the flow of gas into the well.

Areas such as the Green River Basin in Colorado and the Piceance Basin in Wyoming contain the most promising tight formations of natural gas. In 1977 tight gas formations produced about 0.8 trillion ft^3/yr, and there has been about 15 trillion ft^3 of cumulative production from these sources.

Geopressurized Reservoirs

A fourth type of unconventional natural gas occurs in what are called *geopressured reservoirs*. The term geopressured refers to fluids trapped in sedimentary strata (usually sandstone or shale) under pressure gradients greater than the normal 0.465 psi/ft. The high temperature and pressure in

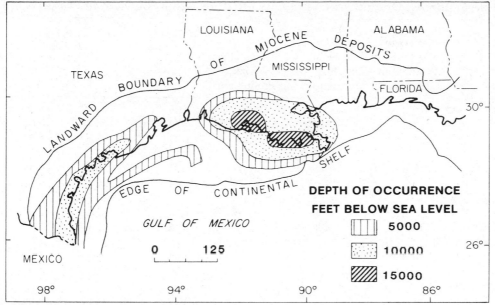

Figure 9.9. The geopressured gas zone in the U.S. Gulf Coast Region. (From Cleveland and Costanza, 1984.)

these reservoirs cause large quantities of hot salty water, called brine solutions, to form—in which are varying amounts of dissolved methane gas. It has been proposed that the methane gas can be extracted by drilling into geopressured aquifers using technologies very similar to conventional oil and gas operations, pumping brine solutions to the surface, and separating and collecting the methane from the fluids. Estimates for the magnitude of recoverable geopressured gas exhibit a wide range from 3 to 1300 trillion kcal (Cleveland and Costanza, 1984), but none of these analyses consider the energy costs of exploiting this vast, but diffuse, fuel resource.

CASE STUDY: EROI OF GEOPRESSURIZED GAS RESOURCES

The Department of Energy, in conjunction with Louisiana State University and the University of Texas at Austin, sponsored a test drilling program in order to gather specific information on the actual range of physical reservoir parameters for geopressured aquifers in the Gulf Coast region (Figure 9.9). Using the preliminary information obtained through this program, Cleveland and Costanza (1984) analyzed the potential net energy yield from the extraction of methane from geopressured fluids. A closer look at their analysis shows how energy analysis is a useful tool for analyzing not only existing fuel

supply technologies, as we saw for conventional petroleum, but also how to evaluate and rank proposed alternate energy technologies.

Energy Inputs and Outputs

Energy used to recover geopressured gas falls into five main categories: drilling, separation and disposal, plant facilities, operation and maintenance costs, and power costs for reinjection. Because geopressurized gas systems extract thousands of barrels per day of brine fluids, disposing of these fluids in an environmentally benign way is an important component of the operation. Reinjecting fluid back into the ground through what are called reinjection wells is the most practical disposal method. Two to three reinjection wells are required for each production well in the system currently being considered. Separation and disposal plant facilities consist of a system of pipes leading from the source well to methane separators that extract the methane from the water and collect it and pipes that carry the fluid back to the reinjection wells for disposal.

Cleveland and Costanza found that the largest energy cost is the drilling portion of the project, which involves the drilling of one source well to between 10,000 and 20,000 ft, and 2 to 3 reinjection wells drilled to shallower depths, generally 3000 to 7000 ft (1000 ft = 305 m). Methane production for individual geopressured wells was calculated based

Table 9.6.   Energy Inputs, Outputs, Net Yield, and EROI for Geopressured Well Prospects in Louisiana

Prospect	Average Fluid Production (bbl/day)	Average Gas Production (10^3 ft^3 day)	Total Gas Production[a]	Total Energy Cost[a] A	B	Net Energy Yield[a] A	B	EROI (ratio) A	B
Bayou Hebert									
10 years	27,478	386	1.44	0.74	0.77	0.69	0.67	1.94	1.87
20 years	24,567	344	2.56	1.03	1.47	1.53	1.09	2.48	1.74
Pleasant Bayou									
10 years	18,631	388	1.44	0.73	1.00	0.71	0.44	1.97	1.44
20 years	9832	206	1.54	0.84	1.20	0.70	0.34	1.83	1.28
Gladys McCall									
10 years	36,627	960	3.57	1.05	1.48	2.52	2.09	3.40	2.41
20 years	18,772	490	3.65	1.16	1.68	2.49	1.97	3.14	2.17
Lafourche Crossing									
10 years	40,000	1050	4.56	0.95	1.30	3.61	3.26	4.80	3.50
20 years	36,960	984	7.23	1.34	1.79	5.89	5.44	5.39	4.03
Base Case									
10 years	15,452	297	1.02	0.79	1.08	0.23	−0.06	1.29	0.94
20 years	9468	181	1.19	0.91	1.30	0.28	−0.11	1.30	0.91

Source: Reprinted with permission from *Energy,* 9, C. Cleveland and R. Costanza, "Net energy analysis of geopressure gas resources." Copyright © 1984, Pergamon Press, Ltd.
[a]Values are in 10^{12} Btu.

on the regional physical parameters—salinity, temperature porosity, and permeability—that determine the methane content of the brine.

Net Energy from Geopressured Gas

Cleveland and Costanza evaluated five cases to determine the potential for net methane production from geopressured aquifers. The first four cases used data from sites in Louisiana and Texas that had been or were being drilled as part of the U.S. Department of Energy's test program. These sites had been selected because they had the most promising physical characteristics of all prospects evaluated in the Gulf Coast. Therefore estimates of their potential EROI should be viewed as an upper bound or most optimistic case. The fifth case evaluated was a base case scenario involving estimates of average or most likely reservoir conditions, which probably is a more realistic estimate of what the typical geopressured aquifer is like.

Values for EROI ranged from about 5.5:1 down to approximately the energy break-even point for the base scenario (Table 9.6; Figure 9.10). These can be compared to returns of between 8 to 23 to one for conventionally produced natural gas calculated by

Cleveland et al (1984), and gas produced from coal gasification with a 5 to 6:1 return on energy investment (Gilliland et al., 1981). Total net methane energy supplied ranged from about 6×10^{12} ft^3 over 20 years to about 1×10^{12} ft^3 for the base scenario. The main reason for the low EROI for the base scenario was its relatively small area compared to the other sites, which caused fluid production to decline rather rapidly after production began. Cleveland and Costanza found that net methane energy supply from geopressured aquifers is sensitive to assumptions made about reservoir parameters, but the best current information suggests that average reservoir conditions lead to EROI ratios in the range of the energy break-even point. Figure 9.11 shows the EROI for the base scenario as a function of both reservoir area and salinity.

The results of Cleveland and Costanza's analysis may change somewhat as more accurate and complete information on the geology of geopressured aquifers is obtained. Based on their results, however, it appears that geopressured aquifers do not offer a significant potential to supplement dwindling supplies of conventional natural gas. This would be possible only if large numbers of prospects with the most optimistic characteristics analyzed became available. Current information suggests that this

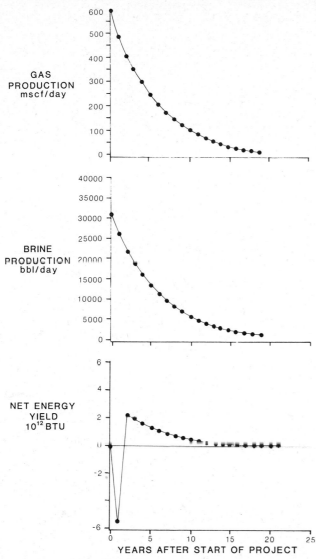

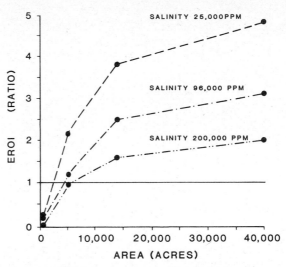

Figure 9.11. EROI for geopressured wells in Louisiana as a function of reservoir area and salinity. (From Cleveland and Costanza, 1984).

Figure 9.10. Brine and methane production and net energy yield for a geopressured well prospect in Louisiana. (From Cleveland and Costanza, 1984.)

possibility is unlikely and that reservoir parameters are most likely to be in the range of the base scenario. Based on the results of Cleveland and Costanza's preliminary analysis, it is not surprising that geopressured gas is not currently being exploited because its EROI is still well below that of conventionally produced natural gas (see Figure 4.6).

10

SHALE OIL

Chris Neill and Charles A. S. Hall

INTRODUCTION

Large quantities of energy are stored in carbon-rich sedimentary rocks known as oil shales and tar sands. The organic matter in oil shales and tar sands contains an insoluble organic polymer called kerogen, which yields a viscous type of crude oil (syncrude), gas, and water when it is heated to 480°C (pyrolized). The oil and gas produced can be refined and/or used directly in a fashion similar to crude oil and gas from conventional sources.

The quality of oil shale is determined chiefly by the quantity of recoverable oil in gallons per ton (GPT) of shale. Although most oil shales contain carbon in such dilute amounts that they can never be considered a net source of fuel, some high quality deposits yield large quantities of oil and gas when pyrolyzed. Oil shale convertors can be viewed as expensive investments of fuel and materials by humans to complete the transformation of kerogen in shale oils to crude oil, a process not completed by natural energies due to insufficient geologic conditions or time. In this chapter we assess the potential of oil shale to serve as a fuel source by examining the quantity and quality of oil shale in place and the methods used for recovering them. Further limits on the production of oil shale may be imposed by environmental constraints and are discussed in Chapter 15.

RESOURCE BASE

The United States possesses about two-thirds of the world's identified oil shale resources. However, reserves on other continents are poorly documented, and lower grade deposits of oil shale are probably unrecorded (Table 10.1). Estimates of domestic oil shale resources in place are as high as 28 trillion bbl

of oil, an amount three orders of magnitude greater than domestic reserves of crude oil. Estimates expand to 170 trillion bbl if low quality shales, containing 5–10 gal of oil per ton, are included. Known reserves of oil shale have been most reliably estimated between 2.0 and 2.2 trillion bbl, but even these estimates have been criticized as imprecise: "It should be apparent . . . that shale oil estimates commonly quoted in the literature represent only the crudest first approximations" (Marland, 1979). But estimates of total oil shale in place are misleading for evaluating oil shale as a fuel resource unless corrections are made for the quality of the shales present and the quantity of overburden. These qualities of the oil shale in part help determine the type of technology used for recovery, which in turn ultimately determines the EROI for an oil shale operation (Table 10.2).

Most of the oil shale located in the Eastern United States is low grade (less than 25 GPT), and known resources in Alaska are small, even though extensions could eventually bring the total quantity of Alaskan reserves to 0.5 trillion bbl. About 80% of the high grade (greater than 25 GPT) oil shale in the United States is located in the 1.1 million acre Green River formation of western Colorado, eastern Utah, and southern Wyoming (Figure 10.1). Of these high grade resources, 80–90% of the oil is located in the Piceance Creek Basin of western Colorado. Notice that the total quantity of oil shale resource in the Green River formation and most other regions increases as the quality of the resource declines (Table 10.3).

The high grade resources of the Green River formation are part of the Mahogany geologic zone. Deposits in this zone were formed by sedimentation of Lake Uinta, a freshwater lake that covered the region during the tertiary period. The thickness of the Mahogany zone reaches a maximum of 210 m in the

Table 10.1. Comparison of U.S. and World Shale Oil Resources (in billions bbl)

	Identified			Hypothetical			Speculative		
U.S. Deposits									
Range in grade (oil yield, in gallons per ton of shale)	25–100	10–25	5–10	25–100	10–25	5–10	25–100	10–25	5–10
Green River Formation Colorado, Utah, and Wyoming	418	1400		50	600		—	—	
Devonian and Mississippian shale, Central and Eastern United States	None	200		None	800		—	—	
Marine shale, Alaska	Small	Small		250	200		—	—	
Other shale deposits	Small	Small		ne	ne		600	23,000	
Total, United States	418+	1,600+		300	1600		600	23,000	
World Resources (by continent)									
Africa	100	Small		ne	ne		4000	80,000	
Asia	90	14		2	3700		5400	110,000	
Australia and New Zealand	Small	1		ne	ne		1000	20,000	
Europe	70	6		100	200		1200	26,000	
North America (excludes USA)	Small	Small		50	100		1000	23,000	
South America	Small	800		ne	3200		2000	36,000	
Total, World	678+	2420		450	8800		15,200	318,000	

Source: G. Marland, 1979. Shale Oil: U.S. and World Resource. Oak Ridge National Laboratory. ORAU/1 EA-79-8(e).

Note: ne = no estimate. Estimates and totals rounded.

Table 10.2. Recovery Methods for Shales of Varying Quality and Projected Yield

Technology	Oil Shale Quality (gal/ton)	Recoverable Reserves (billion bbl)
Underground room-and-pillar mining plus surface retorting	35 or more	20
	30 or more	54
Open pit mining plus surface retorting	25 or more	380
	20 or more	760
Conventional in situ retorting (after mining plus explosives)	20 or more	300
Nuclear in situ	20 or more	200

Source: From "Oil shale: the prospects and problems of an emerging energy industry" by S. Rattien and D. Eaton. Reproduced, with permission, from the *Annual Review of Energy*. Volume 1, p. 185. © 1976 by Annual Reviews Inc.

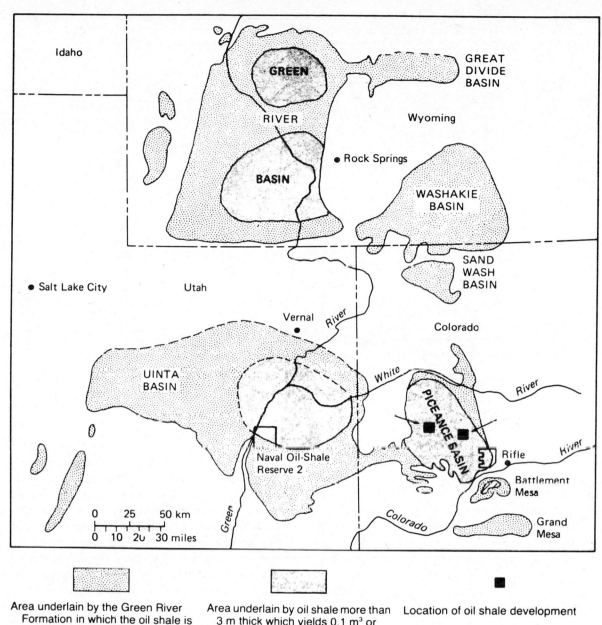

Figure 10.1. The major U.S. oil shale deposits. Most of the high-grade shale is found in the Piceance Creek Basin in western Colorado. Black areas indicate tracts of federal land leased for private development.

eastern Uinta Basin and 520 m in the Piceance Creek Basin. The thickness of overlying soil and rock varies from thousands of meters to almost zero, but most of the deposits in the Piceance Creek Basin are overlaid by at least 300 m of overburden.

Like oil shales, tar sands are low-grade petroleum deposits and are, as the name implies, kerogen-type materials mixed with sand. Domestic re-

sources of tar sands are smaller than resources of oil shale. Major deposits of tar sands lie in eastern Utah and contain approximately 19–24 billion bbl of oil, which is equivalent to about 90% of conventional U.S reserves. Most known reserves of tar sands in North America are found in Canada, where already small quantities are being recovered from the large deposits at Athabasca.

Table 10.3. Estimates of Oil Shale Resources in Place in the Green River Formation (billion bbl)

Shale with yields greater than X gal/ton	Colorado	Utah	Wyoming	Total
$X \geq 30$	355	50	13	418[a]
	311		4	315
				468[b]
$X \geq 25$	480	90	30	600
	607	69	60	731
				490
				1200[b]
$X \geq 15$	1521		260	1781
				860
$X \geq 10$	1280	320	430	2030[a]
				1818
				2468[b]
				4000[b]
$X \geq 5$				4000[a]
				8000[b]

Source: From "Oil shale: the prospects and problems of an emerging energy industry" by S. Rattien and D. Eaton. Reproduced, with permission, from the *Annual Review of Energy.* Volume 1, p. 185. © 1976 by Annual Reviews Inc.

[a] Known (only).
[b] All (including speculative).

RECOVERY

The recovery and processing of oil shale involve two general steps: mining, during which the oil shale is extracted from the environment, and retorting, in which the oil shale is pyrolyzed to separate the crude oil, gas, and water from the shale. In this section we examine the different methods used to mine and retort oil shale. Four general methods have been proposed for recovering oil shale resources: underground room and pillar mining plus surface retorting, open pit mining plus surface retorting, conventional *in situ* retorting and nuclear *in situ* retorting.

The first two methods—underground mining plus surface retorting and open pit mining plus surface retorting—use conventional mining techniques to bring oil shale to the surface where it is retorted. Room and pillar mining is very similar to underground techniques for coal mining (see Chapter 11). Underground rooms are dug, and oil shale is brought to the surface via some type of rail transportation system. Open pit mining is very similar to the strip mining of coal, which we describe in Chapter 11.

The second set of methods—conventional *in situ* retorting and nuclear *in situ* retorting—do not bring the oil shale to the surface. Instead, the oil shale is retorted in place underground. As a result the energy and money invested in mining are reduced. Using *in situ* techniques, voids are created within the oil shale deposit by removing a fraction of the shale via conventional mining techniques. The remaining shale is caved into these voids (rubblized) to create a permeable region so that air can be pumped through large underground regions to sustain burning. Conventional *in situ* retorting uses conventional explosives to rubblize the shale, whereas nuclear *in situ* techniques would use small nuclear explosions. The area of underground retorts would be about 60 m square and 70 m high, and they would be burned for about 200 days before the oil produced could be pumped to the surface. Using microwaves to heat the shale underground also has been suggested. Other *in situ* approaches, including true *in situ* retorting which eliminates all conventional mining, are being researched. None of these other *in situ* processes are ready for commercial production and probably will not be available before 2000 at the earliest.

Substantial U.S. oil shale production before 2000 is unlikely: "Optimistic estimates of the maximum likely rate of production is about two million barrels per day—ten percent of present oil demand. Some very optimistic figures go as high as four million barrels per day. Such production cannot possibly be reached before late in this century at the earliest" (Wishart, 1978). The first demonstration projects for oil shale recovery are located within the Green River formation, and this region will undoubtedly be the focus of the first U.S. attempts at commercial production of oil from shale. Three companies were involved in producing experimental quantities of shale oil in the late 1970s: the Paraho Development Corporation, which is developing an experimental retorting facility for evaluation by the Department of Energy and the Department of Defense, the Occidental Oil Company, which is making field tests of its modified *in situ* technology, and Geokinetics, which is field testing its modified *in situ* process. Morton Winston, president of The Oil Shale Corporation (TOSCO), said in January 1980 that he expected to announce TOSCO's plans to undertake the first commercial oil shale project and to build a 47,000 bbl/day plant (about 0.2% of U.S. oil use) in northwest Colorado at a cost of $1.2 billion (*Forbes*, January 21, 1980). Later events proved him wrong, however, and in 1982 all U.S. oil shale operations closed down for at least the forseeable future due to the declining prices and demand for oil and the higher than expected costs of constructing and operating an oil shale operation. This was just the last in a 50-year series of disappointments for the oil shale industry, for the costs of production keep rising even as the costs of conventional oil rise. Thus the trigger price that would enable oil shale to be developed has not yet arrived despite many predictions that it would. This situation is captured in an interesting review of the history of oil shale "Shale Oil, Always a Bridesmaid" (Fearon and Wolman, 1982; see Figure 10.2). We believe that the trigger price must rise as the price of conventional fuels rises, since more money will be required for the direct and indirect energy used for production.

EROI FOR OIL SHALE

The most efficient method of oil shale extraction cannot be used on all deposits because the quality of an oil shale deposit determines the methods that can be used (Table 10.2). Because different mining techniques have different extraction efficiencies (i.e., recover different percentages of the oil shale in place), this efficiency determines the ultimate quantity of oil shale in place that can be recovered. Room and pillar mining has an extraction efficiency of about 50–70%. Strip mining extracts 80–90%, which is similar to estimates for the extraction efficiency of strip mining coal (Averitt, 1975). Despite its relative inefficiency, room and pillar mining is the most likely candidate for much commercial production because the depths of most oil shale deposits (210 to 520 m) make surface mining too expensive. Probstein and Gold (1979) estimate that no more than 10 to 15 percent of oil shale reserves are recoverable by strip mining.

Modified *in situ* mining is estimated to have a recovery rate of about 21% if the shale carved out from the underground regions is discarded, but the percentage recovery increases to about 37% if the shale mined and brought to the surface prior to underground burning is eventually retorted. Estimates of the fraction of total reserves that can be retorted at the surface by modified *in situ* mining range between 20 and 35% (Metz, 1974). The efficiency for recovering oil shale and coal, at least from surface mining, is therefore considerably greater than the mean of 33% extraction efficiency for the recovery of crude oil via primary and secondary methods.

Some recent analyses suggest that surface-mined shales containing less than 8 GPT require more energy just to retort than they yield. Shales containing 3 GPT is the lowest grade of oil shale that can be retorted by modified *in situ* techniques and produce net quantities of energy, even under the most favorable conditions. Our review of six proposed or projected oil shale operations and one actual operation indicates that the energy return on investment ranges between 0.68 and 19.7 (Figure 10.3). One thorough study gave a value of 8.6:1, but this estimate did not include the energy cost of refining (Marland et al., 1978). If some of the measurable environmental costs associated with oil shale production (in Kentucky) are included, the EROI drops another 3–9% (Lind and Mitsch, 1981). Finally, if real dollar cost estimates continue to rise, estimates for the EROI will decline.

The only analysis of an actual operation, the Paraho retort in Rifle, Colorado (Applied System Corp, 1975, in Gardiner, 1977), yielded the lowest EROI (0.68). Much of the energy costs of the operation, however, were primarily for research. Data on the known energy flows of the Bureau of Mines

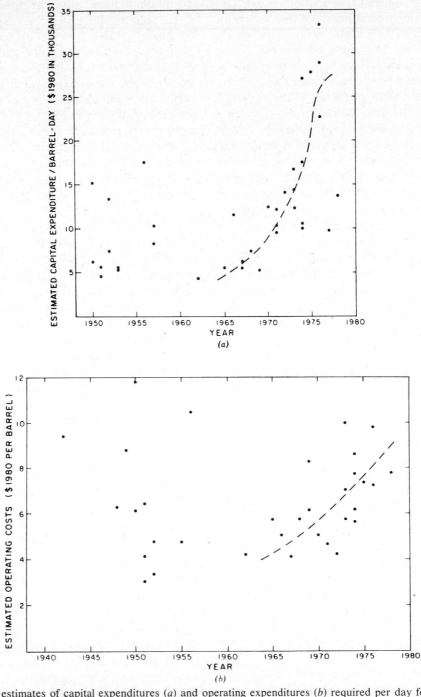

Figure 10.2. Various estimates of capital expenditures (*a*) and operating expenditures (*b*) required per day for projected oil shale operations (in thousands of 1980 dollars). (From J. G. Fearon and M. G. Wolman, 1982. *Energy Systems and Policy*, Volume 6, Number 1, page 83/page 85, Crane, Russak & Company, Inc., New York 10017.)

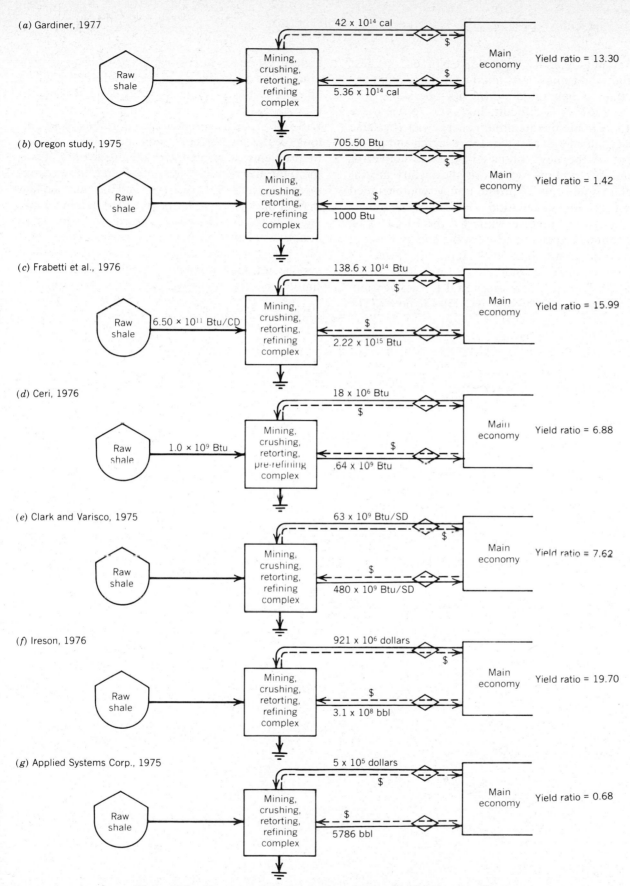

Figure 10.3. Energy return on investment of proposed oil shale operations.

Anvil Points Facility at Rifle, Colorado, suggest that the facility has consumed 82.4 times more energy than it generated during its operation from 1944 to 1967. It is difficult, however, to use these data to estimate the maximum energy yield because much of the energy cost was for research and development of the new, untested oil shale convertor system. In addition records at the facility are incomplete, and some oil production was not included in the preceding calculation (Gardner, 1977).

Although the precise value for the EROI of oil shale cannot be calculated from the energy flows of an operational production system, the results of most energy analyses indicate that its EROI will be below 10:1, well below the EROI for conventionally produced crude oil discussed in Chapter 7. The highest grade resources are likely to become economic when EROI for conventional and imported oil approaches 5:1 for extended periods. The energy capital costs, however, for supplying 10% of U.S. oil production would be enormous. The outlook for the development of oil shale also is clouded by problems associated with the availability of water in the arid regions where it is found, the environmental impacts of its recovery and use, and the fact that all operations so far have ended in a mothballing of facilities. Therefore, regardless of the large quantities of oil shale in place, it seems unlikely that oil shale can substantially supplement dwindling supplies of conventionally produced crude oil anytime within the next several decades, if at all.

11

COAL

COAL FORMATION AND CLASSIFICATION

As outlined in the introduction to this part, coal is formed from large accumulations of trees and other plants in inland swamps and bogs that were overlaid and compressed by inorganic sedimentation after their deposition. The precise geologic conditions necessary for the formation of coal occurred far more frequently than those for petroleum formation, and as a result coal represents on a kilocalorie basis more than ten times as much recoverable energy as conventional petroleum in the United States. Like petroleum the specific chemical and physical characteristics of coal are determined largely by the nature of the original plant material and by the length and magnitude of the pressure it is subjected to by overlying strata. Coal is classified by *type* and by *rank,* the former being determined largely by the nature of the original plant material and the latter by the coal's stage of chemical alteration, which is directly related to the duration and depth of burial. In general, the *quality* of a given coal, as measured by its heat content per kilogram, and hence its usefulness in the economic process, is more related to the geological pressure that has been exerted upon it than its chemical content. The coals that have undergone the greatest degree of compression have the highest percentage of carbon, the lowest amount of water and volatile hydrocarbons, and therefore possess the highest heating values.

The rank of a particular coal is determined largely by its percent of elemental carbon, and to a lesser extent on the amount of volatile hydrocarbons (gases), since these two components are the primary determinants of the heat energy released on combustion. The terms lignite, subbituminous, bituminous, and anthracite are names given to various ranks of coal based on their respective heating values (see Figure 11.1).

Anthracite is the highest ranking coal with a fixed carbon content of over 90% and a heating value between about 13,000 and 15,000 Btu/lb (7200–8330 kcal/kg). Due to its high carbon and low sulfur content (97% of the resource base contains less than 1% sulfur), anthracite was an attractive source for residential and industrial space heating throughout the early part of the 20th century before the rapid increase in the use of petroleum products. Anthracite is produced under unique geologic conditions that subject the coal deposits to unusually high pressures and temperatures, usually by means of extensive folding and faulting. As a result most anthracite deposits were concentrated in a small geographic area in Pennsylvania where 96% of all U.S. anthracite reserves were found. Total anthracite reserves in the United States represent only 2% of the total coal reserve base. The anthracite coal industry peaked in about 1918 and currently represents less than 1% of total U.S. coal production (Figure 11.2). Note that the pattern of anthracite coal use very closely resembles the production growth cycle curve for nonrenewable resources developed by Hubbert (1956). Although substantial anthracite resources remain, they are in deep, narrow, and economically unattractive seams.

Bituminous coals are the most physically abundant and geographically widespread of the ranks of coal and range in heat value from about 12,400 to 15,600 Btu/lb (6900–8700 kcal/kg). Bituminous coals are also the most extensively used coal in the economic process because of their abundance and suitability for a wide variety of economic tasks. Applications range from industrial process heat to the production of coke in the steel industry to electricity generation in power plants. Bituminous coals currently generate the steam that produces over 50% of all the electricity in the United States. Low sulfur (1% or less) bituminous coals are most common in Central Appalachia, Colorado, Utah, Oklahoma, and Alabama, and high sulfur bituminous coals are found in Northern Appalachia and the

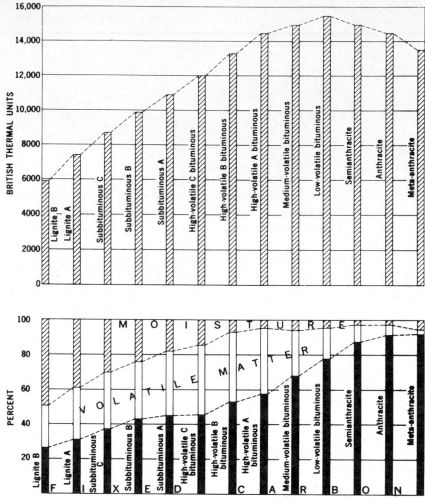

Figure 11.1. Coal ranks and characteristics. (From Averitt, 1975.)

Midwest. Since bituminous coals are much older (300 million years) than the lower ranks of coal, most of the water and other impurities have been driven off leaving its high fixed carbon content.

Bituminous coals also are classified according to their coking properties which describe a coal's usefulness as a raw material in steel production (see Figure 11.3). Coking coals, termed *metallurgical* grade coals, produce a hard carbon residue (coke) when heated to a minimum of 850°C. About 60% of the coal that the United States currently exports (about 9% of total U.S. production in 1984) is high-quality, metallurgical grade bituminous coal.

Subbituminous coals are the next lower rank of coals with heating values ranging from about 8800 to 12,100 Btu/lb (4900–6700 kcal/kg). Subbituminous coals are found most frequently in the Northern Great Plains regions and in the New Mexico–

Arizona region. Subbituminous coals have low fixed carbon and high moisture levels, and they rapidly deteriorate on exposure to air. These coals also have high levels of ash and volatile matter which yield large amounts of smoke on combustion. These qualities generally limit the use of subbituminous coals to steam generation in mine mouth power plants.

Lignite is the lowest quality coal with an average heating value of about 7600 Btu/lb (4200 kcal/kg). The most expansive deposits of lignite occur in the Northern Great Plains and Gulf Coast regions. Being of its relatively recent geologic origin (about 150 million years old), lignite still retains many of the characteristics of peat, the main precursor of coal. Lignite often contains recognizable woody plant remains, indicating its early stage of chemical and physical alteration. A typical lignite coal contains

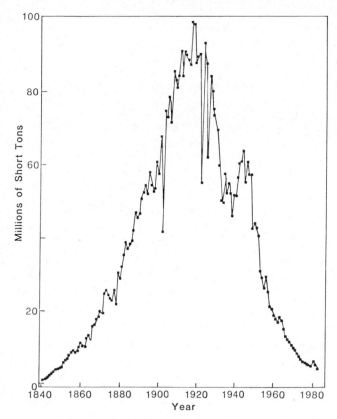

Figure 11.2. Production growth cycle of U.S. anthracite coal.

(not including the ash content) 33% fixed carbon, 26% volatile hydrocarbons, and 41% moisture. Like the subbituminous coals, lignite is most often used in mine mouth power plants because of its low heating value and high transportation cost. Due to its high percentage of hydrogen in relation to other ranks of coal, lignite is an attractive source for coal gasification, an energy transformation technology discussed later in this chapter.

Although carbon is the most important chemical constituent of coal, since it largely determines a particular coal's heating value, there are other constituents that affect its use and contribute to the quality of the resource. Most coals contain varying degrees of ash, hydrogen, oxygen, nitrogen, and sulfur and a variety of minor and trace elements present in minute quantities. Of these, sulfur is by far the most important. Sulfur combines with oxygen on combustion to form sulfur dioxide, one of the most pervasive and dangerous air pollutants identifiable today. Pyritic sulfur that is exposed to air and water vapor also contributes to acid rain, a severe pollutant in several regions of the country (see p. 398). Much of the air pollution control legisla-

tion in this country, particularly the Clean Air Act, is aimed at controlling the emission of sulfur dioxide, so a large part of coal's future in the U.S. economy will be determined by the path that air pollution control legislation follows. A more thorough discussion of the impact of coal use on the environment and the effect of the Clean Air Act and other environmental legislation on the coal industry can be found in Chapters 15 and 16.

Sulfur in coal normally exists in two forms: *combined* with the organic portion of the coal or *attached* to the mineral or inorganic matter in the coal. This distinction is an important one because most of the attached inorganic sulfur can be removed before combustion through washing and other coal preparation processes. The organic form of sulfur, however, cannot be removed through simple washing techniques and therefore requires expensive postcombustion removal, usually through flue-gas desulfurization devices (NAS, 1980). The sulfur content of U.S. coals varies from 0.2 to 10% by weight (Table 11.1) and, unlike the fixed carbon content, cannot be correlated directly with the coal's rank. Of the total U.S. reserve base 46% is low sulfur coal (less than 1% sulfur by weight), 21% is medium sulfur (1–3%), 21% is high sulfur coal (3%), and another 12% is undetermined. Most of the low sulfur coal in the United States, some 75% by weight, lies in beds west of the Mississippi River. Accounting for sulfur on a percent by weight basis, as the U.S. Bureau of Mines (USBM) does, may be an inappropriate method for classifying the sulfur content of coals because the combustion of coal of various heat contents may change dramatically the sulfur released per unit of energy produced. For example, although most Western coals are low in sulfur content, they are low in heat content as well (see Table 11.1). A lower-quality lignite coal with 3300 kcal/kg and 1% sulfur contains as much sulfur per kilocalorie as a bituminous coal with 6700 kcal/kg and 2% sulfur. Although Western coal, in general, has lower sulfur content than Eastern coal, its lower heat content means more of it must be burned to produce the same amount of heat and useful work as Eastern coals, which negates some of its apparent advantage over high sulfur Eastern coals.

DOMESTIC RESOURCES OF COAL

Much optimism currently exists concerning coal's potential for replacing decreasing supplies of do-

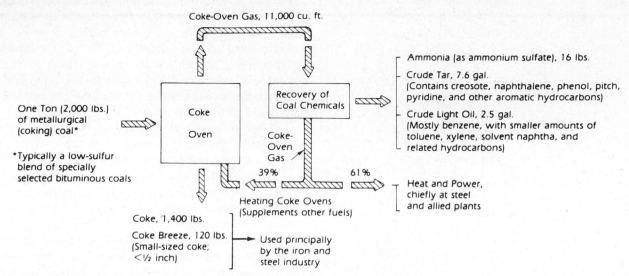

Metallurgical (coking) coal is converted in special high-temperature ovens into coke, which is used in smelting iron ore in a blast furnace. The coking process also yields useful coal chemicals as by-products.

Figure 11.3.   Production of coke and coke chemicals (approximate yields are per ton of raw coal. (From DOE, 1982.)

mestic oil and gas and diminishing current high levels of politically undesirable imports. This optimism is based on the apparent abundance of U.S. coal resources and our currently low consumption levels of coal. We consume much less than 1% per year of our recoverable coal reserve base, while we consume almost 10% per year of our known oil and gas reserves. The U.S. Geological Survey (USGS) estimates that the United States has 3.9 trillion tons of coal resources, an apparently enormous amount of coal when compared to current production rates of about 800 million tons per year. On an area basis

Table 11.1.   Comparison of Coal Reserves and Quality in Different Coal Regions of the United States (from OTA, 1979)

	Coal Quality	
Region	Heat Content (kcal/kg)	Sulfur Content (%)
Northern Appalachia	6670–7270	1.3–4.9
Southern Appalachia	7215–7440	0.5–4.6
Midwest	5550–7492	0.8–7.8
Northern Plains	3600–4660	0.3–3.5
Rocky Mountains	5710–6660	0.4–3.5
Gulf Coast	2940–4600	0.5–4.2
Pacific Coast	4830–5940	0.2–1.3

coal deposits are more widespread than oil and gas deposits (see Figure III.4). The existence of large amounts of coal, however, does not mean it will all be economically or energetically feasible to extract, or even that coal will necessarily once again assume a dominant role in the U.S. energy market. Economic, technological, social, and environmental constraints in the future will all affect the extent to which coal use is expanded in the United States.

Unlike reserve estimates for other fuels like oil and gas that are based on *proven* reserves, which are estimates of the contents of already discovered and operating properties, coal reserve data are based on inferences gained from surface geological exploration. The USGS bases its coal resource estimates on three criteria: (1) the thickness, rank, and quality of a coal bed, (2) the depth of the bed, and (3) the proximity of actual coal resource data to the area on which the estimates are based. The depth and thickness of a seam usually determine the economic, technologic, and energetic feasibility of its recovery. Smaller and deeper seams have a lower return on dollar or energy investment than larger seams that are close to the surface.

The USGS total resource estimate of 3.9 trillion tons of coal includes all coal beds that are at least 14–30 in. thick, depending on the coal's rank, and less than 6000 ft deep. This estimate also includes all thinner and deeper beds of coal that are currently being mined or that may be feasibly mined in the future. Based on this information, the USGS estimate is an approximation of the total physical availability of coal in the United States. The USBM has taken the USGS total resource base estimate and identified the more economically attractive portion of that base—by seam thickness and depth—to come up with an estimate of the coal *reserve* base. The USBM (1976) estimates that 483 billion tons, or about 12% of the total resource base, of coal are recoverable under current economic and technological conditions (see Table 11.2). Of the 483 billion tons identified as the reserve base, estimates vary on the net amount of coal that can actually be extracted and delivered to consumers. Of this amount, about 45% is surface minable, while the rest must be extracted with underground mining methods. The Office of Technology Assessment (1979) estimated that only 283 out of the 483 billion tons can actually be recovered, whereas the National Coal Association (1979) places the recoverable reserve base at 218 billion tons based on an average extraction efficiency of about 50% for the entire industry. A considerably more conservative estimate of 150 billion recoverable tons was made by the National Petroleum Council, who applied more stringent criteria, especially those concerning the minimum minable seam thickness. Using their standards, only 100 billion tons are recoverable by underground methods and about 50 billion tons via surface mines, indicating that available low-cost coals may be less plentiful than the USBM purports it to be.

The Bureau of Mines (USBM) method of reserve estimates can be misleading, since there is such a large difference in heat content between ranks of coal and even within a given rank. Up until 1972 the USBM used an average heating value of 26.2 million Btu per ton of coal, an assumption that ignores the fact that the average heating value of all bituminous and lignite coals has declined steadily over at least the past 25 years (see Figure 11.4). Based on the average range of heat content historically used by the USBM, 45% of the originally mapped U.S. resource base by tonnage is bituminous coal, 27% subbituminous, and 28% lignite. Anthracite is much less than 1%. Based on a specific average heating value, however, the corresponding fractions are 57, 25, and 18% percent, respectively.

Coal-bearing strata underlie 458,600 mi^2 (1.2·10^6 km^2) of the United States, or 13% of the total surface land area (Department of Energy, 1980). As was the case for petroleum, the majority of recoverable coal is located in relatively small geographical regions and in a few very large beds. In 1975, for example, over 70% of production came from less than 10% of the mines (Batelle, 1978). Recoverable quantities of coal exist in 32 states and are mined in 26, but 90% of recoverable reserves exist in only 10 states—West Virginia, Pennsylvania, Ohio, Kentucky, Illinois, Indiana, Montana, Wyoming, Colorado, and North Dakota (Table 11.2). Over 300 coal beds currently are being mined, but less than 50 of them produce the majority of coal. A single bed, the Pittsburgh bed, which covers some 6000 mi^2 (16,000 km^2) in four states, has produced almost 20% of the nation's cumulative output through 1973 (NAS, 1980). The Wyodak bed, located mainly in Wyoming, which penetrates the surface continually for over 190 km and has a thickness of over 50 m, is the largest coal bed in the country, with strippable reserves of 150 billion tons, almost 180 times total U.S. production in 1984.

Broad distinctions between coals that lie east of the Mississippi and those that lie to the west can be drawn on the basis of mining method, heating value, and to a lesser extent sulfur content (see Tables 11.1, 11.2). Although the bulk of U.S. coal reserves are west of the Mississippi River, the Appalachian region has produced over 90% of the coal mined to date. Of the 5600 active mines operating in 1983, 37% are underground mines, and 63% are surface mines. Although Eastern surface mines currently produce 44% of current output, only 17% of remaining reserves in the East are strippable. The percent of strip mining in the East in the future can therefore be expected to decline.

HOW COAL IS FOUND AND EXTRACTED

A bed of coal can be extracted from the ground in one of two general mining methods depending on its distance from the surface and the type of geologic formation within which it is found. The underground method, where miners work beneath the surface of the earth in a series of shafts and tunnels constructed to gain access to the coal face, is the oldest mining method and is widely used in most Eastern coal fields. Surface mining in the United States, on the other hand, is a relatively new practice used to extract coal from beds that lie close to the surface, generally less than 60 m. Surface min-

Table 11.2. **Demonstrated Reserve Base of Coal in the United States by Area, Rank, and Potential Mining Method, in Short Tons (from DOE, 1984)**

Region and State	Anthracite Underground and Surface[b]	Bituminous Coal[a] Underground	Surface	Lignite Surface[c]	Underground	Total Surface	Tot.
Appalachian							
Alabama	7.0	1.7	2.4	1.1	1.7	3.5	5
Kentucky, Eastern	0	8.5	4.1	0	8.5	4.1	12
Ohio	0	13.0	5.9	0	13.0	5.9	18
Pennsylvania	7.1	21.9	1.1	0	28.8	1.2	30
Virginia	0.1	2.4	0.8	0	2.5	0.8	3
West Virginia	0	34.2	5.1	0	34.2	5.1	39
Other[d]	0	1.4	0.4	0	1.4	0.4	1
Total	7.2	83.1	19.8	1.1	90.1	21.0	111
Interior							
Illinois	0	63.1	15.7	0	63.1	15.7	78
Indiana	0	8.9	1.6	0	8.9	1.6	10
Iowa	0	1.7	0.5	0	1.7	0.5	2
Kentucky, Western	0	16.9	4.0	0	16.9	4.0	20
Missouri	0	1.5	4.6	0	1.5	4.6	6
Oklahoma	0	1.2	0.4	0	1.2	0.4	1
Texas	0	0	0	13.5	0	13.5	13
Other[e]	0.1	0.3	1.1	f	0.4	1.1	1
Total	0.1	93.7	27.8	13.5	93.8	41.3	135
Western							
Alaska	0	5.4	0.7	f	5.4	0.7	6
Colorado	f	12.3	0.8	4.2	12.3	5.0	17
Montana	0	71.0	33.6	15.8	71.0	49.4	120
New Mexico	f	2.1	2.6	0	2.1	2.6	4
North Dakota	0	0	0	9.9	0	9.9	9
Utah	0	6.2	0.3	0	6.2	0.3	6
Washington	0	1.3	0.1	f	1.3	0.1	
Wyoming	0	42.6	27.1	0	42.6	27.1	69
Other[g]	0	0.1	0.3	0.4	0.1	0.6	
Total	f	141.0	65.5	30.3	141.0	95.7	236
U.S. Total	7.3	317.7	113.1	44.8	324.9	158.0	483
States East of the Mississippi River	7.2	172.2	41.1	1.1	179.2	42.3	22
States West of the Mississippi River	0.1	145.6	72.0	43.8	145.7	115.8	26

Source: Energy Information Administration, *Coal Production–1982,* September 1983.

Note: Sum of components may not equal total due to independent rounding. Includes measured and indicated resource categor.. representing 100 percent of the coal in place. Recoverability varies between 40 and 90 percent for individual deposits. About one-half the demonstrated reserve base of coal in the United States is estimated to be recoverable.

[a] Includes subbituminous coal.

[b] Includes 133.9 million short tons of surface mine reserves, of which 118.3 million tons are in Pennsylvania and 15.5 million tons are Arkansas.

[c] There are no underground demonstrated coal reserves of lignite.

[d] Includes Georgia, Maryland, North Carolina, and Tennessee.

[e] Includes Arkansas, Kansas, and Michigan.

[f] Less than 0.05 billion short tons.

[g] Includes Arizona, Idaho, Oregon, and South Dakota.

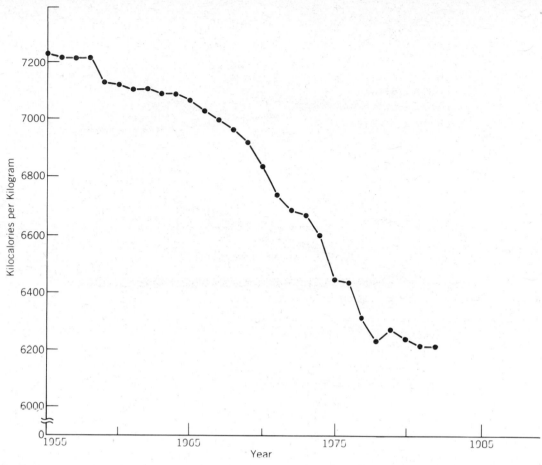

Figure 11.4. Average annual heating values of mined U.S. bituminous coal. (From *DOE Monthly Energy Review*, various issues.) Note that mean resource quality declines most rapidly as production expands (see Figure 11.11).

ing techniques are extensively exployed in the Western United States where the beds are often nearly horizontal and close to the surface.

A type of underground mining new to the United States but long utilized in Great Britain is the longwall method of mining which recovers over 90% of the coal in place. Longwall mining currently accounts for about 7% of total U.S. coal production although the technique is expected to become more widespread.

The extraction efficiency of each mining method (the percentage of coal in place that is actually recovered) depends in part on the mining method used, the geologic characteristics of the coal bed itself, and any legal or regulatory restraints applicable to mining practices in a given region. It is estimated that only about 50% of the 483 billion tons of coal that make up the U.S. proved reserve base

are recoverable by current mining practices. Underground mining methods, which currently produce about 40% of total U.S. coal (53% of the coal in the Eastern United States, and 10% of the coal west of the Mississippi), generally recover only 50–60% of the coal because a portion of the coal seam is usually left in place to support the roof of the mine.

Surface mining has extraction efficiencies of 80–90% because no coal need be left behind to ensure mine safety. As late as 1940 surface mines accounted for less than 10% of total U.S. production, but the recent development of surface-mined coal reserves in the western United States increased the proportion of all coal mined by the surface method to over 60% by the mid 1980s. Surface mining dominates in the western United States, accounting for 90% of production there, and about 47% of the coal currently produced east of the Mississippi.

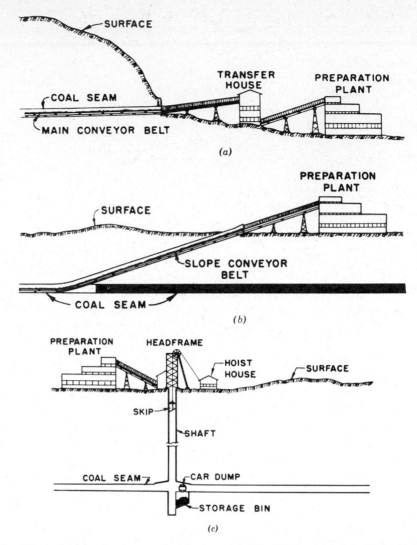

Figure 11.5.   Principal types of underground coal mines: (*a*) drift; (*b*) slope; (*c*) shaft. (From Schroeder, 1973.)

Underground Mining

Types of Underground Mines

Underground mining methods are employed where coal beds lie deeper than 60 m or where they are tilted at sharp angles from the surface due to folding and faulting, as is the case with most of the anthracite and bituminous coal beds in the eastern United States.

Depending on the geologic conditions and the type of coal to be mined, one of several basic types of mines may be constructed (see Figure 11.5). A *shaft* mine is utilized where the terrain is fairly level and the coal seam lies very deep. In this approach two or more vertical shafts are dug down to the seam, one to transport the miners, equipment, and coal back and forth from the seam and another with a large fan on top to provide ventilation in the mine. The average depth of U.S. shaft mines is about 100 m, although in Great Britain, where the coal mining industry is much older, shaft mines are commonly over 250 m deep (Lindberg and Provorse, 1977).

Slope mines are used in areas where the coal lies close to the surface but is too deep to be mined by surface methods. In this type of mine, the access shaft is dug toward the seam on a gradual incline in a way that enables the machinery to move up and down from the surface to the coal face itself. The miners often travel by rail car while the extracted coal travels to the surface either by rail or by a conveyor system.

Drift mines are constructed when coal seams occur in the side of a hill, necessitating that the mine be drilled horizontally into the seam. Electric locomotive-driven rail car systems are used to transport the miners and equipment up to the mine, while the coal is removed via rail cars, conveyors, or even trucks. Drift mines have an advantage over other types of underground mines in that they do not require the construction of shafts through layers of rock overlying the coal seam, thereby requiring less money and energy to construct.

Extraction Techniques

Once access to the coal seam itself has been gained in an underground mine, the actual extraction of the coal can be accomplished with one of four methods: conventional, continuous, longwall, or shortwall (Figure 11.6). Over 200 years ago conventional picks and shovels were used to excavate the mine and to extract the coal, and wooden push carts were used to transport the coal to the surface. Increasing mechanization, particularly as it occurred in the early 1900s, rapidly subsidized human labor in underground mines with cutting machinery, blasting, high speed electric and oil powered coal haulers, and high speed conveyor systems which extract and transport the coal at much faster rates than labor-intensive methods.

Conventional mining is the oldest and most complex method of extracting coal underground and also one of the most dangerous (Figure 11.6a). The steps of conventional mining method include: the undercutting of the seam itself, the drilling and blasting of the seam, the loading of the dislodged coal, the transport of it to the surface, and, finally, the stabilization of the roof in the mined out area for safety purposes. Despite the latter safety precaution most conventional mines still employ the *room and pillar* method of mining where large sections or pillars of a seam are left in place in order to support the mine. This limits the extraction efficiency to about 50%. Conventional mining historically produced over 90% of all coal produced underground, but it has been replaced gradually by continuous mining methods to the point where it now accounts for only 30% of annual underground production (National Coal Association, 1980).

During the 1940s and 1950s advances in coal mining technology led to the development of *continuous mining* machines capable of combining many of the operations of the conventional mining process into one continuous operation (Figure 11.6a). Continuous mining machines are now used to varying degrees in most underground mines and produce almost 65% of all coal from these mines (National Coal Association, 1980). In this type of operation the coal is ripped from the face by large carbide-tipped rotating heads that break the coal up, gather it, and then transfer it directly back over the machine to shuttle cars or a conveyor system. These machines, capable of carving 12 tons of coal per minute out of a seam, increase the mechanical efficiency of extraction because they reduce the transfer of several pieces of equipment to and from the face as is necessary in conventional mining. In many cases the supporting pillars of coal left for support can be mined as the machine retreats backward, allowing the roof to cave behind the retreating machine. The major constraint to the continuous mining method is the hauling of coal away from the face, a condition that limits the use of continuous mining to only 30% of the total time it could be used in underground mines (OTA, 1979).

The most recent and efficient means of underground coal extraction is the *longwall* mining method which removes 90% of the original coal in place. Although utilized in Great Britain for some time now, the longwall method was not introduced into the U.S. mining industry until the mid 1960s, and still produces only about 7% of total U.S. underground production. The longwall method is quite different from the conventional room and pillar method in that the coal is ripped from the face by a shearing machine that cuts up to a meter of coal from across an entire 100–300-m long face (see Figure 11.6b). The feature of this method that makes it so efficient is the self-advancing, hydraulically controlled roof support system which extends heavy steel beams in front of the cutting head as it moves across the face, creating a protected space for the equipment and its operators to work. As the shearer passes by, the supports and conveyor system automatically move forward along the face, allowing the roof behind the supports to cave in. The elimination of the cutting, drilling, and blasting and of the need for support pillars used in conventional and continuous mining creates a much higher extraction efficiency and has tripled miner productivity in mines where the longwall method is employed. The *shortwall* method is similar to the longwall method but employs shuttle cars to transport the coal from the machine to a fixed point conveyor system. The advantage of the shortwall method is that it is very

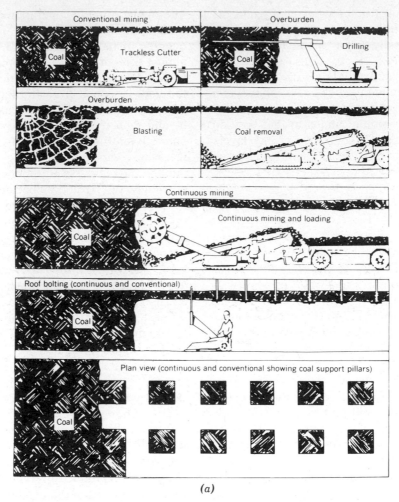

(a)

Figure 11.6. Underground coal extraction technologies: (*a*) conventional and continuous methods; (*b*) longwall mining method. (From National Coal Association, 1980.)

flexible and can be used in smaller areas where ease of movement to and from the face is restricted.

The underground coal mining industry has been surrounded by a cloud of public controversy and internal labor unrest since its beginnings in the mid-1700s, which is centered around the health and environmental impacts of coal mining and utilization. All underground mining methods are very labor intensive relative to surface mining—at current annual productivity rates underground mining requires 550 miners to produce a million tons of coal compared to surface mining which needs 160 miners to produce the same amount (OTA, 1979). Even with increased mechanization in underground mines, miners are still subject to working conditions that involve large health risks ranging from loss of fingers or limbs to black lung disease resulting from inhalation of coal dust to large-scale fatalities from

occasional mine cave-ins. In 1978 the rate of non-fatal disabling injuries in the underground mines was 2.5 times the rate in all surface mines (MSHA, 1979). Although significant reductions in both fatal and nonfatal injuries have been achieved since the passage of the U.S. Mine Safety and Health Act in 1969, the health impacts of underground mining still remain a serious consideration in any plan for expanded coal production and are also important in determining efficiency and EROI for coal (see Chapter 17).

Surface Mining

Although surface or strip mining in the United States is the oldest and most efficient means of removing coal from the ground, it has only recently

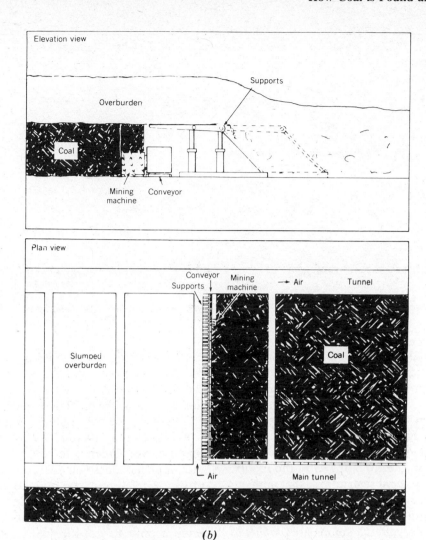

(b)

become responsible for a significant portion of total U.S. coal production (Figure 11.7). Since it is not restricted by safety or space limitations like underground mining, labor productivity tends to be much higher in surface mines. In 1984 the average laborer produced about 1.6 tons/worker-hour in underground mines compared to almost 4 tons/worker-day in surface mines.

Surface-mined coal's share of total production grew from near zero in 1950 to 64% in 1979 (Figure 11.7). Several factors contributed to this rapid increase in surface-mined coal. First, the Coal Mine Health and Safety Act of 1969 placed mandatory safety guidelines on the underground mining industry which resulted in an increase in the cost of coal produced from underground mines. Second, the Clean Air Act, passed in 1970, empowered individual states to establish emission limitations for sulfur

dioxides, the main pollutant produced by coal combustion. This made low sulfur Western coal, most of which is strip minable, more attractive for electric utilities, the primary consumer of all coal. Technological changes in the methods of production have been more effective in keeping costs down in surface mines compared to underground mines.

Due to technological advances in surface mining techniques it has become economically feasible to mine coal as deep as 60 m from the surface. In 1977 overburden thickness (the amount of soil lying above mined coal) ranged from 7 to 25 m in the East and from 15 to 40 m in the West. The national average overburden to seam thickness ratio was 8.9 m^3 of soil removed for every ton of coal extracted (DOE, 1979). In 1970 the overburden to coal thickness ratio was 11:1, a 15% increase from the 1950s, indicating that more energy is now being expended

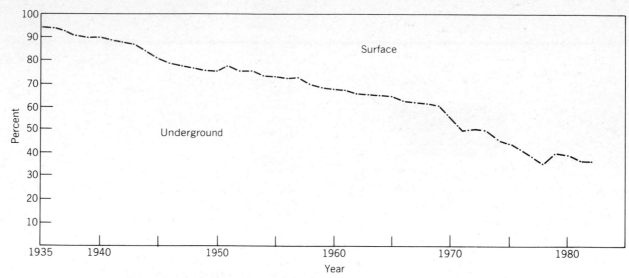

Figure 11.7.   Coal production in the United States by method of mining. (From National Coal Association, 1980, and DOE, 1984.)

to remove additional amounts of soil in order to extract the same amount of coal, which results in a lower amount of net energy being produced at the mine mouth.

Once it has been determined that a given coal deposit is minable by surface techniques, one of four different methods can be used depending on the depth of the coal, the topography of the region, and any regulatory or legal restrictions that apply to that region. *Area* strip mining is the major method used for western and midwestern coals where the terrain is generally flat and the seams are parallel to the surface (Figure 11.8). Large draglines remove the overburden and then extract the coal, which is then removed by truck to nearby stockpiles or treatment plants. Reclamation begins as soon as an area is mined. The dragline immediately moves on to the next area to be mined and the overburden in the spoil banks is put back into the mined-out trench and reshaped by bulldozers back to (more or less) their original land forms. The Surface Mine Control and Reclamation Act (SMCRA) of 1977 mandates that impacted land must be restored to at least as good a use as that prior to mining, and that the land must be returned to its original contour and stabilized against erosion through revegetation practices.

Contour surface mining is used in hilly or mountainous regions where coal seams commonly outcrop from the side of hills, as is often the case in the Appalachian region (see Figure 11.9). A wedge of the hill is actually cut away by removing the over-

burden starting at the outcrop and then proceeding along the contour of the bed in the hillside. Until the SMCRA was passed in 1977, overburden was dumped down the hillside, creating both an eyesore and severe erosion problems in many areas.

When contour mining has cut so far into the hill that the overburden is too deep to remove, or when the slope is too steep for contour mining, *auger* drills are used to extract the coal up to 200 ft into the overburden. The drills used in auger mining churn holes into the coal up to 2 m wide but can remove only 40% of the coal in place.

Open pit mining is very similar to area mining, except that it works over a larger area, often several thousand feet wide. This method is used only for very thick and extensive Western coal seams.

Coal Preparation and Treatment

After coal is extracted from the ground by either surface or underground methods, it is usually transported to a nearby processing plant where the coal is modified to suit the particular needs of the customer in terms of size and moisture, mineral, and heat content. Some types of coal preparation serve as a precombustion form of pollution control since they remove some of the ash and inorganic sulfur in the coal. The fate of a shipment of coal through a typical preparation plant is illustrated in Figure 11.10.

About 40% of total bituminous coal production is

Figure 11.8. Area strip mining. (From OTA, 1979.)

currently cleaned in some manner, and 90% of all bituminous coal is crushed to provide uniform size, to increase the surface area for combustion and to remove impurities (OTA, 1979). The refuse fraction of raw coal produced from the cleaning process has increased steadily since 1960, primarily due to the fact that lower-quality, dirtier coal is being mined today. The refuse presents a large solid waste disposal problem at the site of the preparation plant.

HISTORY OF COAL USE IN THE UNITED STATES

The history of coal use in the U.S. economy is one of fluctuating demand and uncertain markets, a history that has led to fluctuating production levels (Figure 11.11) and, until the late 1970s, a steadily decreasing proportional role in total U.S. energy supply. Prosperity in coal fields has never been predictable—it has fluctuated according to the availability and price of alternate fuels, the development of new energy technologies, labor problems, and more recently to the enactment of environmental legislation.

After its discovery in the early 1700s in what is now Pennsylvania, West Virginia, and Kentucky, the only use of bituminous coal prior to 1830 was in a few areas in Pennsylvania, where it was utilized by blacksmiths, while anthracite was used in a few homes for space heating purposes. Even by 1850 wood still supplied 90% of all the energy consumed in the country, while coal accounted for only 10% (Schurr, 1969). Coal began to become a more attractive energy source as increased deforestation moved the remaining forests farther and farther away from population and industrial centers. Between 1850 and the Civil War the use of coal tripled, due primarily to the rapid expansion of the railroad system after its introduction here in the 1830s.

The increasing availability of coal to the young U.S economy in the 1850s, initiated by its use as a fuel for steam locomotives, fostered the growth of several large energy-consuming industries that later served as the base for the rapid expansion of the industrial sector of the economy. This was especially true in the iron and steel industry, which by the 1860s had abandoned the use of charcoal in favor of coke (a derivative of coal) in the production of pig iron. The substitution of coke for charcoal as

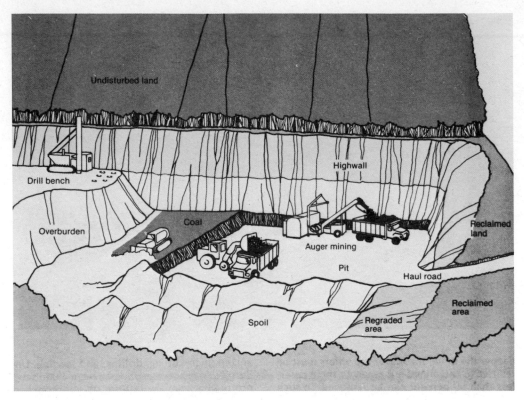

Figure 11.9. Typical contour strip mining and auger mining methods. (From OTA, 1979.)

a furnace fuel helped increase the output of the iron mills, thereby stimulating coal demand and the movement of the iron industry to Pennsylvania, Ohio, and West Virginia where it would be nearer to the coal fields. Concurrent with the growth of the iron industry was the further development and expansion of the network of railroads across much of the nation. Not only were the railroads the largest consumer of coal, but they were able to move the coal from the distant fields to the rapidly growing population centers in the East and Midwest. This made a cheap and high quality (relative to wood) energy resource readily available and helped encourage the rapid expansion of all industry. Throughout the Civil War and immediate postwar years, continued expansion of the railroads created a coal boom, with production reaching 200 million tons by 1890. During this time anthracite and bituminous coal enjoyed almost equal shares of the coal market, with anthracite used primarily for home heating and bituminous used extensively by the railroads and the steel industry for the production of coke.

The final use of coal that contributed to its rapid growth occurred when the steam turbine was ap-

plied to the generation of electrical power in the 1880s. By the year 1910 coal accounted for 90% of total energy consumption, compared to fuel wood's 10% share, the opposite of the ratio in 1850. Three newcomers to the energy market also appeared at this time. Oil, gas, and hydropower came into use but accounted for only a very small percentage of the market. Although coal production had reached 580 million tons by 1918, these new energy sources had begun to make headway into coal's share of the market, which by then had slipped to 70%.

Between 1920 and 1940 declining consumption by railroads and coke ovens along with the absence of new types of machines that used coal had progressed to the point that, by the time of the nation's entry into World War II in 1941, oil and gas consumption was almost equal to that of coal. The technological and industrial changes following World War II, especially the continued development of the internal combustion engine, further enhanced petroleum's role as the economy's primary fuel. Petroleum burns cleaner than coal, was readily available due to the discovery of many giant fields in Texas, Louisiana, and California in the 1930s and 1940s, and was easily transported. Petroleum rap-

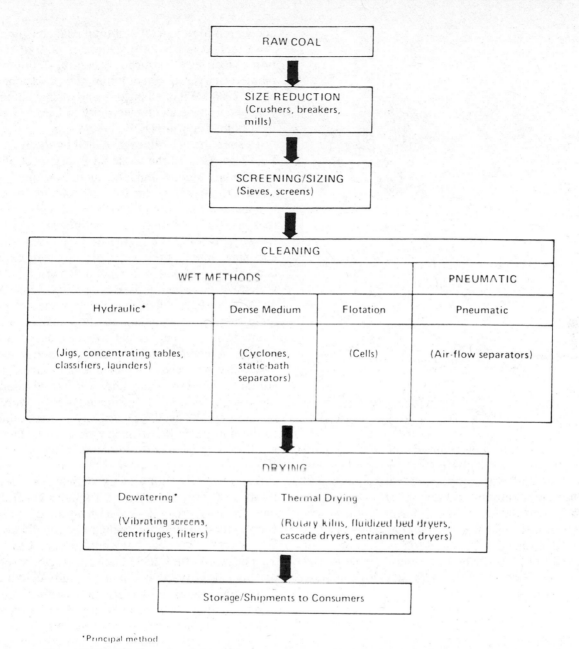

RAW COAL

SIZE REDUCTION
(Crushers, breakers, mills)

SCREENING/SIZING
(Sieves, screens)

CLEANING

WET METHODS			PNEUMATIC
Hydraulic*	Dense Medium	Flotation	Pneumatic
(Jigs, concentrating tables, classifiers, launders)	(Cyclones, static-bath separators)	(Cells)	(Air-flow separators)

DRYING

Dewatering*	Thermal Drying
(Vibrating screens, centrifuges, filters)	(Rotary kilns, fluidized bed dryers, cascade dryers, entrainment dryers)

Storage/Shipments to Consumers

*Principal method

Figure 11.10. The flow of coal through a coal preparation and treatment facility. (From DOE, 1982.)

idly began to displace coal as the preferred fuel source for locomotives, industry, utilities, and space heating—functions that had been established earlier and whose initial growth had been powered by coal. Coal use plummetted during the 1950s and 1960s, reaching a low of about 17% of total energy consumption in 1972, despite the fact that from 1950 to 1974 the price of coal per kilocalorie was consistently about one-half the price of domestically produced crude oil (DOE, 1980).

The steady decline in coal use and its replacement by oil and gas is illustrated by the home and commercial space heating sector. As late as 1948 coal still provided over 50% of the energy to this sector. By the mid-1970s, however, it had dropped to less than 2%. The use of oil and gas also was encouraged because they did not require coal's daily stoking, dirty delivery and storage procedures or smoke generation on combustion.

Thus the production history of the U.S. coal in-

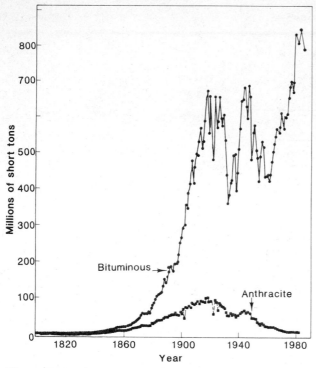

Figure 11.11. Annual coal production in the United States, 1800–1983.

dustry is an extremely varied one (see Figure 11.11). The Arab oil embargo and the ensuing realization that domestic supplies of petroleum would not continue to support current levels of economic growth for more than another quarter century stimulated a rebirth, of sorts, in the coal industry. The 1947 production of 630 million tons of coal was not surpassed until 1975, when production reached 648 million tons. Production in the mid 1980s was over 800 million tons.

Present Patterns of Use of Coal

Combined with the decline in absolute coal consumption there has been a relative shift in the end uses of coal (Figure 11.12). As outlined in the previous section, coal use in the transportation and residential/commerical end-use sectors has declined to the degree that they now represent about 1% of total coal consumption in 1984. Government energy policy, particularly the Industrial Fuel Use Act of 1978, which discouraged the use of oil and gas in utility and industrial boilers, has virtually tied the future viability of the coal industry to electric utility demand for coal. In the early 1980s electric utilities

accounted for about 80% of total coal consumption. As a fuel, coal currently generates 60% of the nation's electricity produced by central station power plants. An average ton of bituminous coal generates about 2000 kWh of electricity and a single pound of coal generates enough electricity to light ten 100 W light bulbs for 1 hr (DOE, 1980).

Despite the incentives provided by the Industrial Fuel Use Act and the National Energy Act of 1977 to stimulate utilities and large industrial energy consumers to switch to coal from petroleum, the move toward coal by utilities has been slow for several reasons. First, it is difficult for utilities to respond in the short term to the increased incentives for coal use, since much of the installed generating capacity is comprised of boilers designed to burn oil or gas, which was installed, indeed encouraged, during the 1950s and 1960s when domestic sources of petroleum were more abundant and less costly. Government subsidy of natural gas to consumers via wellhead price ceilings and of nuclear power-generated electricity, via research, development, accident insurance, and other subsidies further discouraged the use of coal. Other disincentives for increased coal use are the need for vast storage areas adjacent to the utility and the stringent air quality standards relating to coal combustion under the Clean Air Act.

The uncertain future of the demand for electricity as an energy source is also a more general hinderance to increased coal utilization by utilities. Electricity consumption, which historically grew at a rate of 7% per year, suddenly dropped to about 3% per year in the mid and late 1970s. Many regions of the country that planned and constructed generating plants based on the historical growth rate now found themselves with generating capacity in excess of current demand. With the future market for electricity somewhat clouded, utilities are more reluctant to make any large-scale, long-term investment, and the conversion of boilers designed to burn petroleum to ones able to burn coal is a costly undertaking for utilities.

The second largest consumer of coal is the steel industry in the production of coke for use in the production of iron ore. Coke production is the only end use where coal enjoys special advantages over all other fuel types. In 1983 about 37 million tons of bituminous coal, or about 5% of national consumption, was used in coking plants, well below the 1950 levels of 114 million tons and a 1% share of the coal market. Technological progress in the production of

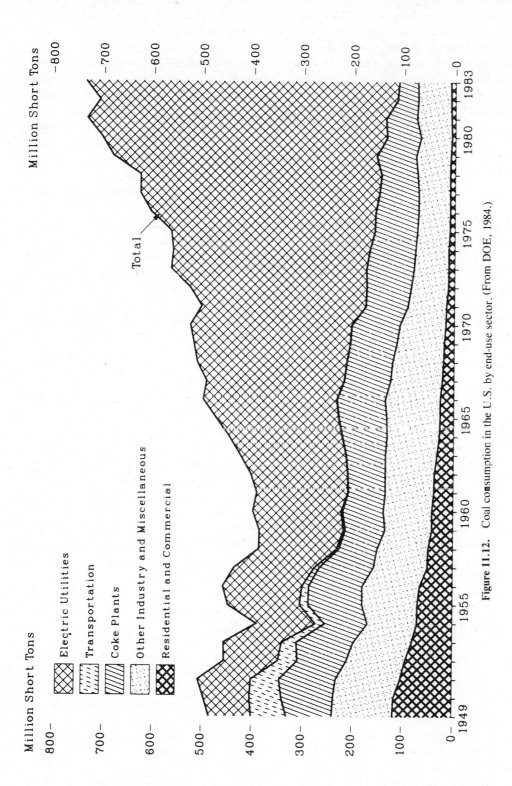

Million Short Tons

Electric Utilities

Transportation

Coke Plants

Other Industry and Miscellaneous

Residential and Commercial

Million Short Tons

Total

Figure 11.12.   Coal consumption in the U.S. by end-use sector. (From DOE, 1984.)

iron ore has reduced the coal input per ton of iron output, a trend that has more than offset the impact of increasing iron ore output. Coke can be produced only from certain low volatile bituminous coals with specific ash and sulfur contents and very high heat contents. Recoverable reserves of coking coals amount to 12–16 billion tons (about 4% of total U.S. reserves), found mainly in West Virginia and eastern Kentucky.

Industrial heat applications are the third largest end use of coal and account for less than 10% of total coal consumption in 1984. In terms of total energy consumption coal contributes less than 10% of the total required by industry. This ratio has declined steadily for the past 30 years, again because cheaper oil and gas were available but also because small industrial users found it cheaper to burn oil or gas which, although sometimes more expensive, precluded the high capital and labor costs involved with handling and storing the coal, treating wastes, and meeting emission limitations.

Labor Productivity

Related to the production history of the coal industry is the trend of labor productivity in coal mines, usually measured in tons of coal produced per worker-day. This ratio rose steadily from 5 tons/worker-day in the early 1940s to about 20 tons/worker-day in 1969 but then dropped precipitously to about 15 tons/worker-day in 1980 (see Figure 11.13). Productivity since 1980 has increased slightly to about 18 tons/worker-day. Labor productivity in coal mining is an important barometer of the overall economic health of the industry because labor inputs still represent a significant portion of the final cost of a ton of coal. An examination of the causes underlying the observed changes in labor productivity can help explain past and present relationships between coal prices, work stoppages due to strikes, depletion of high quality coal deposits, and government regulation of the industry.

As Baker (1981) notes, labor productivity in coal mining is important for several reasons. First, a decrease in labor productivity means that more labor is required per ton of output. This would increase the possibility of labor-related constraints limiting future coal industry growth. Second, other parameters remaining constant, increased labor requirements per ton would lead to a higher price per ton of coal. Third, increased labor inputs per ton of output

could affect injury and fatality rates by increasing the number of miners required to produce a ton of coal. Since all of these factors represent potential constraints to the expanded use of coal, the problem of decreasing productivity in underground bituminous coal mines is an important one to consider when discussing coal's potential to replace diminishing supplies of petroleum.

Most analyses of the decrease in labor productivity has been centered around the impact of the 1969 Coal Mine Health and Safety Act (CMHSA) on deep underground mining (Baker, 1981). Although the CMHSA has undoubtedly contributed to the decline in productivity by increasing the number of nonproduction workers in the mines, other conditions such as contract strikes and price changes also were important factors. From 1970 to 1973 most of the changes in productivity probably were attributed to the effects of CMHSA, the passage of which was prompted by the Farmington, West Virginia, disaster in 1968 where 78 miners lost their lives (Baker, 1981). Coal mine health and safety inspectors in underground mines reached an all time high of over 72,000 in 1973 (CMSHA, 1980). It is interesting to note that the results of two studies indicated that continuous mining methods were more affected by CMSHA regulations than were other types of underground mining methods (David, 1972; Baker, 1979). Continuous mining methods may be more dangerous than other extraction techniques due to high dust levels and the rapid advance along the coal face. Continuous mining grew from 2% of total underground mine output in 1950 to 64% by 1975.

As indicated by the precipitous drop in fatalities and disabling injuries after the passage of the CMHSA (Figure 11.14), some of the productivity gains of the 1960s came at the expense of the health and safety of the miners themselves. Before the passage of the act the costs of injuries and black lung disabilities were external to the coal-producing companies and the price of coal. Once these costs were internalized, miner safety was enhanced, production costs increased, and productivity rates fell. Baker (1981) found a strong negative relationship between real coal prices and productivity—not only during the 1970s when productivity rates were falling but also during the 1960s when rates were increasing. The sharp increase in real coal prices in 1974 and 1975 led to increases in coal production, total employment, and the total number of mines operating. Many of the new mines that opened as a

Average Short Tons per Man-Day

Average Short Tons per Man-Day

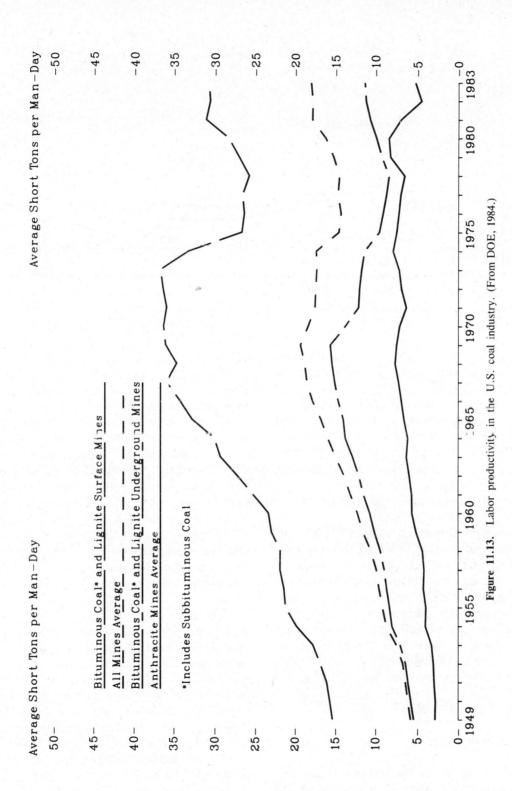

Bituminous Coal* and Lignite Surface Mines

All Mines Average

Bituminous Coal* and Lignite Underground Mines

Anthracite Mines Average

*Includes Subbituminous Coal

Figure 11.13. Labor productivity in the U.S. coal industry. (From DOE, 1984.)

247

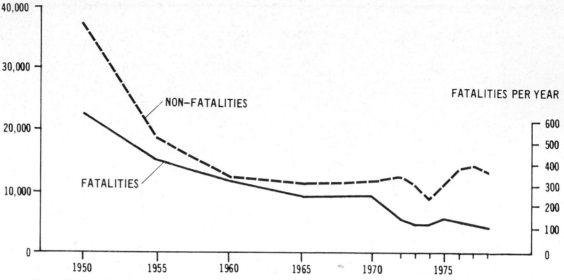

Figure 11.14.   Trends in coal miner injuries and fatalities. (From President's Commission on Coal, 1980.)

direct response to price increases were small, relatively inefficient, labor-intensive mines that had low productivity rates. Economic prosperity in the coal industry also tends to increase confrontations between labor and management. High prices and profits lead to more work stoppages due to strikes, since greater profits provide incentive for both labor and management to seek a larger slice of the economic pie for themselves.

During the 1960s, when real coal prices were stable and even decreasing, the opposite situation existed. Depressed profit levels drove the small labor-intensive companies out of business, leaving only the most productive and efficient mines in operation. Labor-management relations also were less strained during the 1960s, since it was in the best interest of both groups to maintain productivity during a period of low profit levels. Increased mechanization in the form of continuous mining machines that both replaced and subsidized human labor also helped to increase productivity levels during the 1960s.

COAL TRANSPORTATION

On leaving the mine or the preparation plant, coal can be transported to its point of final consumption by one of several modes. Most coal travels in unit trains along railroads, but significant quantities are also transported via waterways, highways, or slurry pipelines. Coal also can be burned by mine mouth generating plants and the electricity transmitted to population centers. Unlike petroleum products, which are relatively easy to transport long distances through pipelines, coal is a very bulky commodity not easily transported. Transportation costs may be over 20% of the delivered cost of coal hauled by rail to consumers. It costs three times as much to transport a shipment of coal from Montana to Indiana by unit train as it did originally to mine the coal (OTA, 1979). Also, due to its high density and bulkiness, coal is energy intensive to transport, although some modes are much more energy efficient than others. The average shipment of coal travels 480 km (OTA, 1979).

Railroads are still the primary mode of coal transport, currently hauling about 70% by weight of all coal produced. In turn coal is the railroads' primary commodity, accounting for 25% of total rail tonnage and 18–20% of total gross freight revenues in the late 1970s. Three-quarters of all coal moved by rail is delivered to electric utilities. Trains that haul only coal, called unit trains, are economically feasible only when very large quantities of a commodity are moved between two fixed points on a continuous basis. A typical unit train consists of up to 110 hopper cars, each weighing 100 tons when loaded, being pulled by up to 6 locomotives of 3000 horsepower each. On average, two unit trains per day are needed to haul 1 million tons of coal per year. Water transport by barge along inland river and lake systems currently moves over 10% of the coal produced, with the average trip by barge being 770 km (OTA, 1979). Barges up to 60 m long and 11 m wide with capacities of 1000–1500 tons are usually linked together and pushed by tugboats capable of moving as many as 36 barges each (Lindbergh and Provorse, 1977). On the Great Lakes system, bulk cargo carriers with an average capacity of 20,000 tons are used to move Western-mined coal to Eastern markets. Most of the 110 million tons of coal exported in 1981 were moved by large ocean-going freighters.

Movement of coal in trucks along the highway is very energy and dollar intensive and therefore is used only when the distance from the mine mouth to consumer is very short. In 1979 trucks transported 13% of all coal produced, with the average trip covering between 80 and 120 km.

Slurry pipelines currently transport less than 1% of the coal in the country. Only one slurry pipeline system is presently in operation in the United States. The Black Mesa pipeline, which runs 440 km from Arizona to southern Nevada, has the capacity to move 4.8 million tons of coal per year. At the terminal end, the slurry must be dewatered through centrifuge processing and the coal dried in large ovens (see Figure 11.15). Slurry pipelines are a very expensive way to transport coal and are used only when a large volume of coal needs to be transported over a long distance from the mine area to a single user, such as a utility, in a region where no barge or rail system exists. Another major problem with slurry pipelines is their impact on freshwater systems. About 1 ton of water is required for every ton of coal to be moved (President's Commission on Coal, 1980), all of which must be reclaimed to a usable condition upon dewatering at the terminal end. Two-way pipeline slurries that return the water back to the mine have been proposed as well as a scheme utilizing eastward flowing oil as the fluid for transport.

An alternative to transporting coal long distances is to burn it in a mine mouth power plant and then transmit the electricity directly to the consumer. In 1979, 12% of all coal produced was burned in this manner. For most low-quality coals such as lignite,

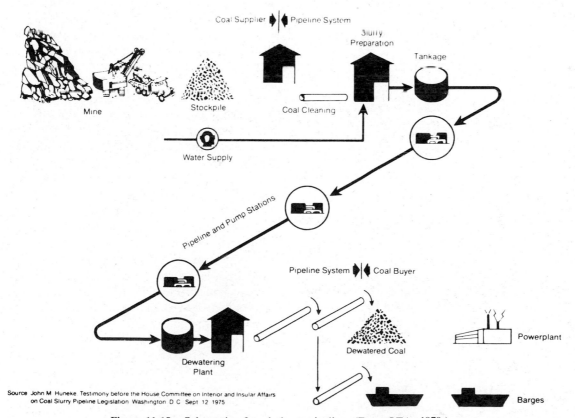

Source John M Huneke Testimony before the House Committee on Interior and Insular Affairs on Coal Slurry Pipeline Legislation Washington D C Sept 12 1975

Figure 11.15. Schematic of coal slurry pipeline. (From OTA, 1979.)

this is the only energetically feasible form of utilization since their low heat content make long distance transport via railroads or barges prohibitive.

There are, however, deterrents to the building of mine mouth plants west of the Mississippi where most of the lignite and subbituminous deposits of coal exist. Many of these deposits occur in pristine or wilderness areas that are given special protection against air pollution under the Clean Air Act of 1970. Line losses during long-distance electricity transmission are also a limiting factor. A 1000-mi, 600-kV dc power transmission line suffers the equivalent of a 3.5×10^{11} kcal/yr line loss or about 6% of the energy in the coal burned to generate electricity (Bayley and Zucchetto, 1977).

In addition to their different applicabilities based on the location of the coal mine in relation to its market, coal transportation modes also differ in their energy requirements to transport an average ton of coal (see Table 11.3). In general, either the unit train or barge mode of transportation is the most energy efficient means of coal transportation, although the results depend on specific localities, distances, and operating conditions. Slurry pipelines, trucks, and electrical transmission are the least efficient means of delivering coal. The equiva-

lent of between 2 and 4% of the energy contained in the coal is consumed for a 1000-mi coal haul by rail depending on the rank of the coal; 12.5% would be used if the coal were hauled by truck. The only advantage that a slurry pipeline has over barge or rail is that it consumes electricity that could be generated with a nonpetroleum energy source, whereas railroads and barges are powered by diesel fuel. Barge transport may be the most efficient mode of transportation under certain conditions, but its application is limited to routes along existing waterways. Conversely, railroads are very efficient along routes where tracks already exist but extremely inefficient where energy must be invested to build new tracks or upgrade existing ones.

Coal mining in the East has the advantage of being near the major coal consuming areas. The same cannot be said for Western coal, a problem compounded by the fact that most Western coals have a considerably lower heat content per ton than Eastern coals to begin with. A ton of 4.3 million kcal/ton, low sulfur coal, hauled 1000 mi (1600 km) at 150 kcal/ton-mile requires at least 3.5% of its original energy content for transportation (OTA, 1979). In this case a utility that decides to import low sulfur coal from the West to comply with SO_2 emission limitations uses more energy for the coal's transport than it would if high sulfur Eastern coal were burned and treated with an energy expensive flue-gas desulfurization process.

Expansion of any type of coal transportation system to meet projected increases in coal production will encounter difficulties, including the aforementioned geographic problem facing Western coal in relation to the fact that major consuming sectors are located primarily in the East. There also exists a potential shortfall in railroad capacity due to the long lead times required for building or upgrading rail systems. Slurry pipelines face problems in gaining access to the large amounts of water necessary for their operation in the West and of obtaining rights-of-way from railroads who view slurry pipelines as a threat to their coal revenues. In short, transportation considerations are one component of myriad problems facing a rapid expansion of the coal industry in the United States.

Table 11.3. Energy Costs of Various Modes of Coal Transportation (from Bayley et al., 1977)

Mode	Energy Cost (kcal/ton-mile)[a]		
	Direct	Indirect	Total
Barge transport			
Average conditions	171	88	259
Dedicated tow[b]	63	22	85
Railroad			
Unit train	100	30	130
Unit train w/new track[c]	100	686	786
Coal slurry pipeline[d]	—	—	190
Electrical transmission[e]	1512	58	1570
($\pm$ 600 KV DC)			

[a] Direct energy is fuel used in operation. Indirect energy costs are those associated with capital investment, goods, labor, and natural system destruction. Based on transporting coal with heating value of 5550 kcal/kg a distance of 1609 km.
[b] Dedicated tow refers to barges reserved for coal transport only.
[c] Based on constructing 1609 km of new track.
[d] Extrapolated from 439 km pipeline carrying 5 million tons coal per year.
[e] Direct energy costs are line losses.

COAL COMBUSTION, CONVERSION, AND CLEANING TECHNOLOGIES

The oldest and still most commonly used form of coal combustion is the pulverized coal-fired boiler furnace, used by most utilities in the generation of electricity. In this process crushed coal is blown into the burners of the furnace where it burns at about 1500°C. The hot furnace walls are in contact with a system of tubes conducting water. Only about 50% of the thermochemical heat stored in the coal and released on combustion is actually transferred to the furnace wall tubes (OTA, 1979). The rest of the heat energy is dissipated elsewhere in the system or into the environment and therefore is not available to perform any useful work. On contact with the walls of the furnace the water is converted to steam, which then passes through a turbine that in turn powers an electrical generator. A 1000-MW$_e$ coal-fired plant consumes roughly 2.5×10^6 tons of coal annually to produce 6.5×10^9 kWh of electricity.

A large steam electric plant also requires large amounts of land and water in its 35- to 40-year life. For every ton of coal consumed, about a ton of fresh water is required for cooling. In addition between 600 and 700 acres (240–280 ha) of land are required for solid waste disposal by a large steam electric plant operating with flue-gas desulfurization equipment (NAS, 1980).

Flue-gas desulfurization (FGD) is a technology frequently used in conventional coal combustion furnaces used by utilities. A FGD system is designed to remove the organic sulfur portion of the coal, which cannot be removed by precombustion washing techniques, by passing the hot exhaust gas from the furnace through a series of *scrubbers* as it passes through the stack enroute to the atmosphere. In this process the sulfur-laden exhaust gases are brought into contact with a substance (usually containing limestone) with which they react and are thereby removed from the gas leaving the stack.

The use of FGD systems is implicitly required by the Clean Air Act and its amendments in 1977. The actual efficiency and reliability of these units, however, is currently a topic for debate between the Environmental Protection Agency, charged with enforcing the provisions of the Clean Air Act, and industry who must install, operate, and maintain the expensive FGD systems. Several requirements of the system, however, are known. A 500-MW$_e$ coal-fired unit burning Eastern coal with 2% sulfur will produce 200 lb of sludge for every ton of coal burned, requiring over the facilities' lifetime a 220-ha sludge pond 12 m deep to store the waste. So in essence the environmental damage caused by coal combustion is only altered in form, from a gaseous state (air pollution), the mechanics of which are poorly understood and therefore difficult to manage, to a solid state (solid waste disposal) which may not be any less harmful to the environment or to humans, at least on a local level, but which is certainly more easily managed. In addition, FGD systems reduce the efficiency of the chemical to electrical energy conversion from about 40% to 35–38%. FGD systems combined with other pollution control devices required by law may consume 5–7% of the facility's total electrical output.

Coal Gasification and Liquefaction

Due to the abundance of coal and because energy in the form of liquid or gaseous hydrocarbons is more economically useful, various technologies have been proposed to convert coal into synthetic oil or gas (see Figure 11.16). Although technologies exist for coal gasification and liquefaction, large quantities of energy are lost during conversion processes. These losses, combined with the large amounts of energy embodied in the construction and maintenance of the large conversion facilities, greatly reduce the initial attractiveness of coal gasification and liquefaction. Other new technologies, such as fluidized bed combustion and magnetohydrodynamic generation are in the process of being implemented in research projects for the generation of electricity directly from coal.

Techniques for converting coal to liquid or gaseous fuels, originally developed in the early 1930s, are now being more thoroughly explored as a means of supplementing diminishing supplies of domestic oil and gas (Figure 11.16). In both coal gasification and liquefaction the basic objective is the same—to decrease the C:H ratio of the original solid coal through a hydrogeneration process in which hydrogen from an outside source is chemically bound to the coal under very high temperatures. Coal has a C:H weight ratio ranging from 12 for lignite to 20 for most bituminous ranks. By the addition of hydrogen the ratio can be lowered to 10 to produce the heavier synthetic crude liquids and to 6.5 for lighter liquids. Decreasing the C:H ratio yields progressively lighter molecular weight fractions until methane gas, with a C:H of 3, is formed:

Fuel:	Coal		Liquid		Gas
Formula:	$(CH_{0.8} + 0.4H_2)$	$\rightarrow$	$(CH_{1.6} + 1.2H_2)$	$\rightarrow$	(CH_4)
C:H ratio:	15:1		10 to 6.5:1		3:1

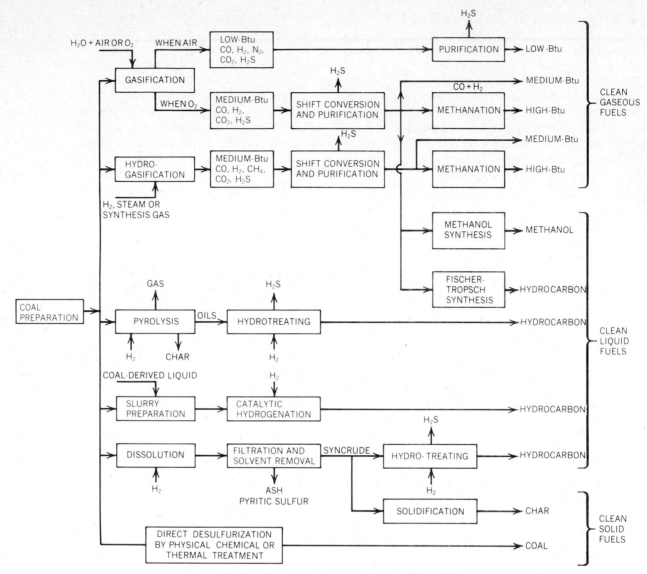

Figure 11.16.   Clean fuels from coal. (From DOE, 1980.)

Hydrogen to feed this reaction is often produced by reacting char, a carbon residue, with steam in the following endothermic reaction:

$$C + H_2O \rightarrow CO + H_2.$$

Coal Gasification

The process of coal gasification can yield gases with varying degrees of heating values, few of which, however, have as high a heating value as natural gas (9100 kcal/m³). The gasification process, represented schematically in Figure 11.16, involves the burning of the coal at very high temperatures in the presence of hydrogen to produce a gas. From a thermodynamic standpoint, lignite coals, which are considered to be of low quality due to their low C:H ratio, are a good source for the production of synthetic gas because less hydrogen is required to affect the desired transformation. A typical synthetic gas process, outlined here, has a energy conversion efficiency of 60–70% (McGraw-Hill, 1980):

$$Coal \xrightarrow{\text{Heat}} \text{gases (CO, CO}_2\text{, CH}_4\text{, H}_2\text{)}$$
$$+ \text{ liquids + char} \quad (1)$$

$$Coal + H_2 \xrightarrow{\text{Catalyst}} \text{liquids + (char)} \quad (2)$$

$$Coal + H_2 \text{ (from a hydrogen donor)} \rightarrow$$
$$\text{liquids + (char)} \quad (3)$$

$$Coal + H_2 \xrightarrow[\text{destruction}]{\text{Noncatalytic}} CH_4 + char \qquad (4)$$

Char reactions

$$C(char) + 2H_2 \rightarrow CH_4 \quad \text{exothermic} \qquad (5)$$
$$C(char) + H_2O \rightarrow CO + H_2 \quad \text{endothermic} \qquad (6)$$
$$C(char) + CO_2 \rightarrow 2CO \quad \text{endothermic} \qquad (7)$$
$$C(char) + O_2 \rightarrow CO_2 \quad \text{exothermic} \qquad (8)$$

Gaseous reactions

$$CO + H_2O \xrightarrow{\text{Catalyst}} H_2 + CO_2 \quad \text{exothermic} \qquad (9)$$
$$CO + 3H_2 \xrightarrow{\text{Ni}} CH_4 + H_2O \quad \text{exothermic} \qquad (10)$$
$$CO_2 + 4H_2 \xrightarrow{\text{Ni}} CH_4 + 2H_2O \quad \text{exothermic} \qquad (11)$$
$$xCO + yH_2 \xrightarrow{\text{Fe}} \text{hydrocarbon gases and/or}$$
$$\text{liquids} + zCO_2 \quad \text{exothermic} \qquad (12)$$

The hydrogeneration stage of this process (reaction 6), where the requisite hydrogen is produced from the reaction of carbon with steam, producing carbon monoxide and hydrogen, is endothermic, meaning that an input of heat is supplied from the burning of the coal or char, as shown in reaction 8. The first gaseous product yielded is a low Btu gas ($1330-3100$ kcal/m^3) composed mainly of CO and H compounds (reaction 9). The heating value of the gas can be increased through the process of methanation, where hydrogen gas is reacted directly with the CO (reactions 10, 11). This procedure produces either medium (4450 kcal/m^3) or high (up to 9100 kcal/m^3) grade synthetic gas.

Another approach to coal gasification is the underground or *in situ* gasification method, where the coal is converted to gas without bringing the coal to the surface. Underground coal gasification is generally considered most appropriate where low-rank coals lie in steeply tilted beds, so the utilization of the coal via conventional mining methods is impossible. In this process oxygen and H$_2$O are pumped down to the bottom of the coal seam through wells drilled from the surface. A fire is ignited, and the burning coal provides the necessary heat to transfer hydrogen from the water, producing gas just as in the process described in Figure 11.16. The fire burns its way up the seam producing gas along the way which is captured and collected at the surface. A shortcoming of this method is that the hot burning gas is very buoyant and therefore tends to burn a path along only the top of the seam and then stops, leaving large amounts of noncombusted coal behind. Also, the low heating value of the gas produced ($450-2600$ kcal/m^3) restricts the subsequent use of the gas to nearby facilities.

Coal Liquefaction

The conversion of coal to produce liquid, low sulfur fuels is in many ways more promising than gasification because liquefaction requires less chemical transformation than gasification and has an energy-conversion efficiency approaching 80%. A wide range of liquid products can be produced including heavy fuel oils for use by electric utilities, distillate fuel oil, and even gasoline. Several coal liquefaction processes also yield gaseous and solid by-products as well.

Direct liquefaction of coal was developed in Germany in the late 1920s and was commercially used in that country to produce over 100,000 bbl/day during World War II. During direct liquefaction or direct hydrogenation, pulverized coal is slurried and then mixed with hydrogen in the presence of a catalyst in a reactor under moderate temperature ($455°C$) and pressure (1.4 to 2.8×10^6 kg/m^2). The vapor and liquid phases produced are then cooled to separate the products and refined to remove any by-products.

The solvent extraction method of liquefaction involves the partial dissolvement of coal in a hydrogen-rich solvent, with the undissolved solid being filtered out. The carbonaceous solids are reacted with steam to produce hydrogen, which then reacts with the excess solvent to produce liquid fuels.

A third method of producing liquid fuels involves the pyrolysis of coal in the absence of oxygen, a condition that allows for the recovery of products through the application of heat without the direct addition of hydrogen. Most of the carbon is rejected as a solid (char) in this process. This process produces significant quantities of gas and char as by-products that must be utilized or disposed of. Another method of liquefaction is through indirect liquefaction, whereby a hydrogen-carbon monoxide mixture (medium Btu synthetic gas) is produced, followed by a catalytic liquefaction process that produces a wide variety of liquid fuels. The yield of products such as gas, LPG gasoline, kerosene, diesel fuel, fuel oil, wax oil, methanol, and acetone is dependent on the catalyst and operating conditions (ERDA, 1975).

A variety of technical, economic and energetic problems currently exist that constrain rapid development of a massive synfuels program in the United States. The availability of adequate water supplies is a major problem, particularly in the arid western regions. A single synfuel plant may require $32.5 \times$

10^9 gal of water per year, little or none of which is directly returned to its source. Seventy-two percent of this amount is lost to the atmosphere as steam, and the remaining 28% is used either in the mining process itself or the chemical reactions producing synthetic oil or gas (Lindbergh and Provorse, 1977). Capital costs also are extremely high. In the United States estimates for a plant capable of producing 7 $\times 10^6$ m^3 per day of high Btu gas are on the order of $1.5 billion at 1979 prices (McGraw-Hill, 1980). Although many of the technological details of these energy conversion processes can no doubt be overcome, the fact remains that all synfuels suffer from low energy return on investment ratios (relative to conventionally extracted fossil fuels), a thermodynamic constraint that may not be substantially improved by future technological advances. The fact that massive government subsidies have been necessary to generate private sector activity in the area of synfuels development is also a clue to the potential feasibility of a large-scale synthetic fuels industry, at least given present alternatives. Declining real oil prices in 1983 and 1984 and the enormous capital investment required by these operations have led to the abandonment of several prototype liquefaction plants. A more thorough net energy analysis, including all environmental costs involved, is needed to determine the actual energy yield of these processes so that they may be ranked with and compared to other existing and proposed energy production systems.

Fluidized Bed Combustion

One of the most promising coal conversion technologies is through the use of a fluidized bed reactor (see Figure 11.17), in which a bed of inert ash or limestone is fluidized (held in suspension) by the uniform injection of hot air through the bottom air distribution grid. When fluidization occurs, the bed of material expands and exhibits the properties of a liquid. As air velocity increases, the particles mix more violently, and the surface of the bed takes on the appearance of a boiling liquid (McGraw-Hill, 1980). Once the bed is fluidized and sufficiently hot, pulverized coal is blown into the bed and combusted.

The main advantage of fluidized bed combustion of coal, other than its high energy conversion efficiency, is that most of the sulfur oxides produced when the coal is burned react with the limestone bed material and fall to the bottom of the reactor in the form of a dry calcium sulfite solid.

Fluidized bed combustion is therefore a relatively efficient pollution control technology that greatly reduces SO_x emission into the atmosphere and thereby replaces expensive flue-gas desulfurization devices, although the sludge produced does present a solid waste disposal problem.

YIELD PER EFFORT AND NET ENERGY ANALYSIS OF THE U.S. BITUMINOUS COAL MINING INDUSTRY

The EROI for coal at the mine mouth is higher than it was for oil and gas at the wellhead in the late 1970s. For example, in 1977 the EROI for coal, taking into account only those direct and indirect energy costs necessary to bring the coal to the surface (i.e., not including labor and transportation), was 42 kcal returned per kilocalorie invested. The same ratio for oil and gas in 1977 was about 10:1 (Figure 4.5). When transportation and an energy assessment for labor was included the ratio was 20:1 for coal in 1977. The apparent EROI advantage of coal is due to several factors. First, petroleum resources are far more depleted than coal resources. Second, the EROI for coal does not include an estimate of the severe environmental and human health impacts of mining and burning coal. Including these costs would substantially reduce the EROI for coal.

The relationship between effort and EROI for the domestic coal industry is not a consistent one as has been the case for the oil and gas industry, where the two are consistently and inversely related to each other. Technological advances in coal extraction techniques may explain some of the observed fluctuations in the relationship between effort and EROI. Based on the trends of effort and EROI over time, it is convenient to divide the history of the coal industry into three phases: 1929 to 1954, 1954 to 1969, and 1969 to 1977. The numbers given below for EROI include not only direct and indirect energy costs but labor and transportation costs as well.

From 1929 to 1954, EROI remained relatively constant at about 30 kcal returned per kilocalorie invested (Figure 11.18b) while total energy costs declined slightly (Figure 11.18c). During the same period effort, as measured by production, fluctuated greatly and peaked at over 600 million tons in the mid-1940s, due mainly to increased wartime energy needs (see Figure 11.18a). Periods of low effort occurred during the depression and after World

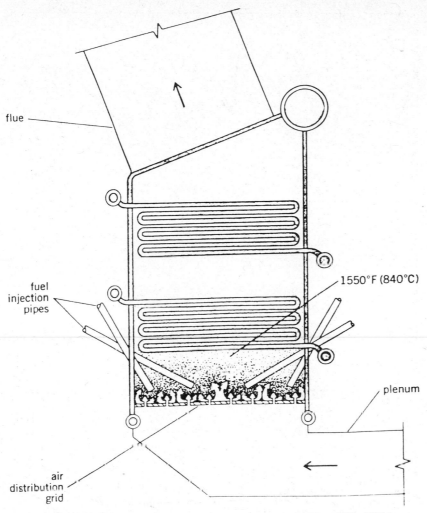

Figure 11.17.   Fluidized bed steam generator. (From McGraw-Hill, 1980.)

War II due to the expansion of oil and gas use. Effort in the coal industry also is influenced by factors other than demand, such as labor-management relations, which historically have had great impacts on the behavior of the industry as a whole. During this time period, for example, the number of worker-days lost due to strikes averaged over 5 million per year (Bureau of Census, 1978).

The consistency of the EROI from 1929 to 1954 may be due to the static nature of coal production technology during this period relative to later years. Conventional underground mining methods produced the vast majority of coal from the early 1900s until well into the 1950s. Unchanging extraction techniques could lead to the observed constant EROI for two reasons. First, in the early stages of the life cycle of a large nonrenewable resource, the energy-intensifying effects of declining resource

quality and availability do not have a significant impact. Second, in the coal industry, unlike the oil industry, only a small percentage of the total effort is expended in exploration activities. Coal deposits of roughly equal quality and availability, whose location is already known and which are extracted with similar technologies, could produce a relatively constant EROI.

During the second time period of interest, 1954 to 1969, EROI steadily increased to 36 kcal returned per kilocalorie invested while production increased from 390 to 560 million tons. The reason for the simultaneous increase of both EROI and production probably is due to technological advances in coal extraction techniques that greatly increased the efficiency of coal production. For example, continuous mining machines by 1954 accounted for only 6 percent of total production but by 1969 they

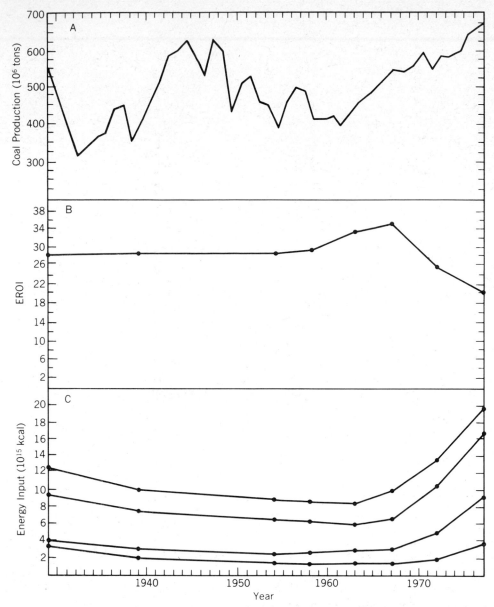

Figure 11.18. (*a*) Coal production and (*b*) energy return on investment for U.S. coal production, including estimates of the energy costs of capital structures, labor, and coal transportation costs; (*c*) total energy use in coal mining. Bottom line is direct fuel use, and the second line includes the energy embodied in purchased capital structures. The third line includes an estimate of the energy costs of transportation, and the top line includes an estimate of the energy costs of labor. (From Hall et al., 1981.)

produced 50% of total output (National Coal Association, 1979). Employment in the coal industry decreased 55% during this period, indicating that increased mechanization was able to replace human labor while also increasing production levels by over 40%. This displacement is also reflected in the changes in the energy costs of production (Figure 11.18*a*). From 1954 to 1969 indirect energy costs increased slightly, reflecting the increase in mech-

anization, while labor costs declined due to a sharp drop in employment. These data, and especially the increased EROI, indicate that the increased total energy costs incurred due to increased mechanization were more than offset by increases in extraction efficiency, resulting in an increase in the EROI. Another reason for increased production during this period was the relative stability of labor-management relations. Worker-days lost due to strikes averaged about 220 thousand per year from 1954 to 1969, a sharp decrease from the 5 million worker-days per year from 1929 to 1954.

During the final period of interest, 1969 to 1977, EROI (Figure 11.18c, line 3) and effort exhibit the same type of inverse relationship previously shown for oil and gas and also for uranium. EROI decreased from about 36 kcal per kilocalorie in 1969 to about 20 in 1977, while at the same time production increased steadily to about 690 million tons in 1977, the highest output up to that time in the history of the coal industry. It is interesting to note that production levels in the 1940s were about equal to those in the mid-1970s (about 600 million tons), but that the EROI for the latter time period was only two-thirds the level it was during the mid-1940s (Figure 11.18a).

There are several reasons behind the rapid increase in energy costs during this time. Continued increases in mechanization of both surface and underground mining methods resulted in a sharp increase in direct energy costs. Surface mining in particular, which increased from 38% of total production in 1969 to over 60% in 1977, experienced rapid increases in energy costs due to the development of sophisticated mechanical equipment such as immense draglines and other earth- and coal-moving machines. Labor costs also increased due to a growth in employment and real wages. But the most important reason for the rapid drop in EROI is probably the declining quality of the resource itself. Especially in the underground mines of Appalachia, lower quality coal seams were being exploited. Decreasing seam thickness, deeper mines, and exploitation of seams at more severe angles all led to increased energy costs per ton. At the same time there was a steady decrease in the heat content of the average ton of bituminous coal produced in the United States. From 1969 to 1977 the heating value of a ton of bituminous coal decreased about 10% from about 6.3 million kcal/ton to 5.7 million kcal/ton (DOE, 1979; Figure 11.4). Unlike the period from 1954 to 1969, increases in direct and indirect energy costs in the form of increased mechanization resulted in decreases in the efficiency of energy delivery by the industry, since these advances did not result in large enough increases in output to counter their increased energy costs. It is not known for sure that the trends of recent years will continue into the future, although it is certain that the coal industry will be faced with increasingly smaller seam thicknesses and more remote deposits.

12

NUCLEAR ENERGY

Paul T. Jacobson and Charles A. S. Hall

Nuclear power in all of its forms has the potential to supply vast amounts of energy for the United States. If technically, socially, and energetically feasible, fusion and breeder reactors could yield far more energy than do current fission reactors, which now supply about 2% (uncorrected for quality) of the U.S. fuel budget, perhaps even more energy than that presently derived from fossil fuels. At one time it was thought that nuclear energy would be so cheap that its metering would be unwarranted, and several ambitious government plans in the 1960s and early 1970s called for nuclear energy to supply 30–40% of the total electrical generating capacity of the United States by the end of the 1980s (Bupp, 1979).

Such optimism no longer appears warranted. The cost of electricity produced by nuclear power plants is increasing rapidly and was already approximately equal to the cost of electricity supplied by coal-fired plants by the late 1970s (NAS, 1979). At least one analysis indicates that it could become much higher (Komonoff, 1981). Additionally, the industry has been plagued by a series of accidents and blunders that have cost public confidence and brought out into the open many of the less apparent social and capital costs of generating electricity by nuclear fission. At present, the capital costs, environmental safeguards, and the long lead times associated with the construction of nuclear power plants precludes such plants from generating a large percentage of the total electrical generation in the United States by 1990, even if a decision were made today to build them. No new reactor units have been ordered since 1978, one year prior to the accident at Three Mile Island. Fifty eight reactors that had been ordered were canceled (Weinberg et al., 1985).

Many advocates of nuclear power lament the long lead times for constructing new nuclear power plants and feel that the period could be shortened by streamlining the permit and licensing procedures. Furthermore, many of them blame cumbersome government regulations and the opponents of nuclear power, claiming they overemphasize the risks associated with nuclear power and make the plants much more expensive than they would otherwise be. Opponents, on the other hand, counter that the cost of the plants and the fuel to run them does not include many costs associated with nuclear power systems. They argue, for example, that long-term costs such as waste disposal and plant decommissioning should be paid *today* by those who are benefiting from the use of the electricity from nuclear facilities. Others suggest that the catastrophic consequences of a core meltdown (however improbable and incalculable) would be so severe that nuclear fission of any type should be considered a high-risk option. No other form of energy is subject to so much controversy or such strongly held opinions, both pro and con.

NUCLEAR CAPACITY

In early 1983, 80 nuclear power plants operating in the United States had a total generating capacity of approximately 64 GW_e (DOE, 1983). These plants generated about 13% of the electricity and approximately 1.5% (unadjusted for quality) of the total energy consumed in the United States in 1982 (DOE, 1983a). At that time 51 reactors were in various stages of construction. Construction of nine had been indefinitely deferred, three were undergoing construction permit reviews, and two were on order. It is important to keep in mind that nuclear power is useful for large-scale electricity production only, except for rather limited applications such as naval propulsion systems. For this reason nuclear power conserves petroleum only to the extent that nuclear plants displace existing or future oil-fired generating plants.

Why hasn't nuclear power fulfilled the expectations of its proponents? The answer lies partly in the fact that nuclear power plants have become very expensive to build and maintain. The high energy and dollar cost of a reactor reflects exacting specifications required for design, materials, construction, and also the added cost of various safety equipment and environmental safeguards. Furthermore producing nuclear fuel is a complex, energy-intensive process. The original cost estimates of nuclear-supplied electricity were low mostly because they were based on inaccurate and optimistic estimates of plant and fuel costs, costs that changed dramatically when the price of fossil fuel used to run and maintain the nuclear fuel cycle increased and when safety problems and concerns mounted. The magnitude of the original underestimates were apparent only after many years because of the long lead time associated with nuclear power stations. Additionally, several years of actual plant operation and maintenance were needed to determine the true cost of running these plants. Finally, the as yet unresolved issues of waste disposal, plant decommissioning, and many aspects of reactor safety may add a great deal to the eventual cost of the energy systems. For example, 6 years after the accident at Three Mile Island the cost of the cleanup has not been established, although it may exceed $1 billion, and no one knows who will pay for it.

REACTOR TYPES

An analysis of nuclear power must consider the reactor type involved and the degree of development of the technology. There are two fundamentally different ways to produce electrical energy from nuclear energy. Nuclear *fission* is the process of splitting radioactive isotopes of heavy elements such as uranium and plutonium. Nuclear *fusion* is the process of fusing heavy isotopes of the hydrogen atom to form helium. Fusion has many very desirable characteristics, at least compared to fission, but is technically complex and at least several decades away from commercial reality.

Fission

Nuclear fission is the method now used to produce electricity. In the United States virtually all nuclear power is generated in light water reactors (LWRs). In these reactors ordinary (*light*) water is used to slow down (i.e., moderate) the velocity of neutrons released by the nuclear reaction and to carry heat away from the reactor core. Since the reaction will not occur with fast neutrons, the water is required for operation. The rate of the reaction is determined using neutron-absorbing control rods generally made of cadmium. The water is either boiled off directly in the core of a so-called *boiling water reactor* or, more commonly, kept under pressure and used to produce steam in a separate unpressurized water supply in a *pressurized water reactor* (Figure 12.1). The second water loop is used in the pressurized water reactor to prevent radioactive contamination of the generating equipment, thereby facilitating maintenance and repair. In both reactor types the steam is then used to drive electrical generating turbines similar to those used in a fossil fuel power plant.

Fuel for LWRs is made from naturally occurring uranium ore (Figure 12.2a and b). The ore is mined using conventional mining equipment and techniques. Uranium constitutes about 0.1% by weight of the ore and is concentrated during the *milling* process, in which the ore is crushed and chemically leached using sulfuric acid or sodium carbonate. The uranium is chemically precipitated and recovered. The concentrated uranium product (U_3O_8), sometimes called yellowcake or natural uranium, is 85 percent pure uranium in an isotopic mixture of 0.7% fissionable ^{235}U and 99.3% nonfissionable ^{238}U. The isotopic ratio (ratio of ^{235}U to ^{238}U) as well as the molecular structure of the purified uranium is the same as it was in the original ore—thus the name natural uranium.

The natural uranium for LWR fuel must be *enriched* in the fissionable or *fissile* ^{235}U isotope. To separate the two isotopes, natural uranium (U_3O_8) is converted to gaseous uranium hexafluoride (UF_6) and run through a long series of diffusion barriers. Since the less dense ^{235}U diffuses more rapidly than ^{238}U, but barely so, the UF_6 on the far side of each barrier is enriched in the ^{235}U isotope by a factor of 1.0043 (Day and Johnson, 1974; Figure 12.3a). The diffusion process is repeated until the uranium gas contains about 3% fissile ^{235}U (Figure 12.3b). The

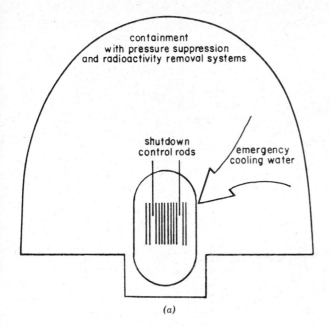

(a)

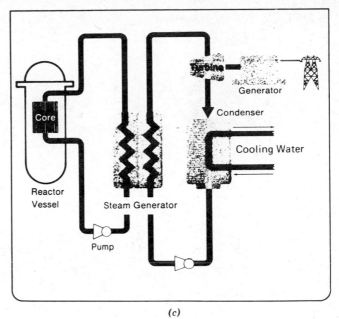

(c)

Figure 12.1. General schematic diagrams of various light water reactors: (*a*) a general schematic of the relation of the reactor to the containment dome; (*b*) boiling water reactor, in which steam generated by reactor drives the turbines directly; (*c*) pressurized water reactor, in which steam from the primary loop heats water under pressure which in turn operates the turbine. All designs ultimately use cooling water drawn from the environment to get rid of waste heat and condense the water behind the turbine. (From Nero, 1979.)

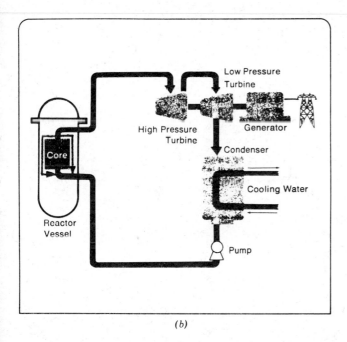

(b)

resulting *enriched uranium* is converted to uranium dioxide (UO_2), fabricated into ceramic pellets, and placed in hollow fuel rods called cladding.

Uranium-238, which constitutes about 97% of the fuel, is called a *fertile isotope* because it can be made fissile by absorbing a neutron, which occurs during operation of a LWR. As uranium-235 undergoes fission, it releases neutrons that, if captured by a ^{238}U atom, produce a third isotope of

uranium, ^{239}U. This quickly decays to fissile plutonium-239 (^{239}Pu). The ratio of fissile ^{239}Pu produced to ^{235}U consumed is called the conversion ratio, which is about 0.5–0.65 in a light water reactor. Plutonium produced in this manner in the reactor core supplies about one-third of the fission energy used in an ordinary LWR.

Breeder Reactors

In the *breeder reactor* production of plutonium is enhanced by eliminating the moderating water and surrounding the core with a blanket of pure fertile material (^{238}U). Neutrons released from the core by the fissioning fuel are caught in the blanket and plutonium-239 is *bred* from the ^{238}U. On the average slightly more than two neutrons are released per atom of fuel fissioned. One neutron continues the chain reaction in the reactor core while additional neutrons breed plutonium in the blanket. In this

NUCLEAR FUEL CYCLE

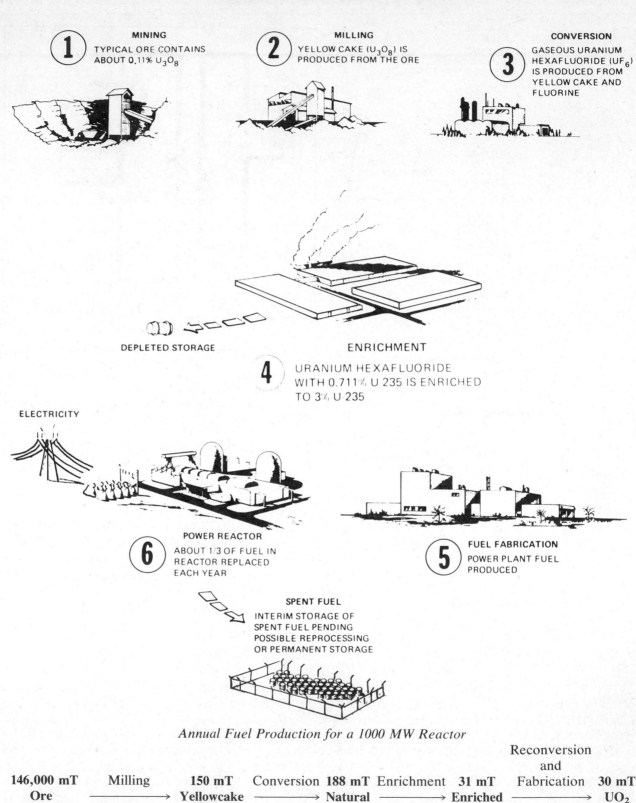

1 MINING
TYPICAL ORE CONTAINS ABOUT 0.11% U_3O_8

2 MILLING
YELLOW CAKE (U_3O_8) IS PRODUCED FROM THE ORE

3 CONVERSION
GASEOUS URANIUM HEXAFLUORIDE (UF_6) IS PRODUCED FROM YELLOW CAKE AND FLUORINE

DEPLETED STORAGE

ENRICHMENT

4 URANIUM HEXAFLUORIDE WITH 0.711% U 235 IS ENRICHED TO 3% U 235

ELECTRICITY

6 POWER REACTOR
ABOUT 1/3 OF FUEL IN REACTOR REPLACED EACH YEAR

5 FUEL FABRICATION
POWER PLANT FUEL PRODUCED

SPENT FUEL
INTERIM STORAGE OF SPENT FUEL PENDING POSSIBLE REPROCESSING OR PERMANENT STORAGE

Annual Fuel Production for a 1000 MW Reactor

146,000 mT Ore (Average grade 0.11% U_3O_8) $\xrightarrow[\text{93\% recovery}]{\text{Milling}}$ 150 mT Yellowcake (U_3O_8) $\xrightarrow{\text{Conversion}}$ 188 mT Natural UF_6 $\xrightarrow{\text{Enrichment}}$ 31 mT Enriched UF_6 $\xrightarrow[\text{Fabrication}]{\text{Reconversion and}}$ 30 mT UO_2 Fuel

Figure 12.2. The nuclear fuel cycle, including the quantity of ore and ore products required for an average 1000 MW reactor for one year. (From DOE, 1976.)

262

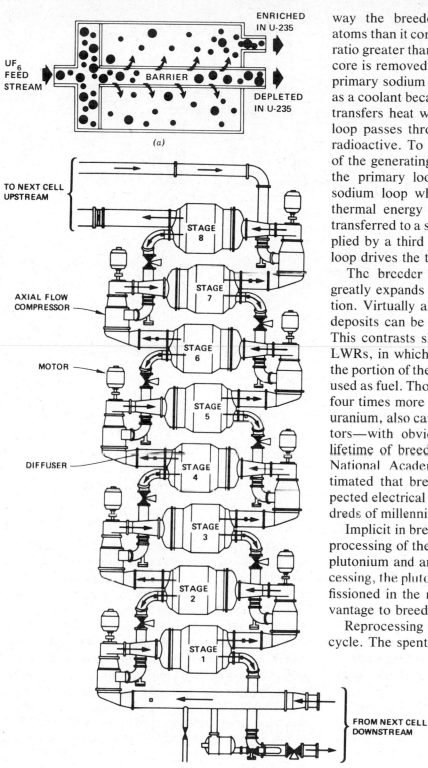

way the breeder reactor produces more fissile atoms than it consumes—that is, it has a conversion ratio greater than one. Heat produced in the reactor core is removed by liquid sodium circulating in the primary sodium loop (Figure 12.4). Sodium is used as a coolant because it absorbs neutrons poorly and transfers heat well. As the sodium in the primary loop passes through the reactor core, it becomes radioactive. To prevent radioactive contamination of the generating equipment, the heat absorbed by the primary loop is transferred to a secondary sodium loop which remains nonradioactive. The thermal energy of the secondary sodium loop is transferred to a steam generator where water is supplied by a third loop. Steam produced in the third loop drives the turbines that generate electricity.

The breeder reactor is of interest because it greatly expands the resource base for fuel production. Virtually all the uranium recovered from ore deposits can be used as direct fuel or fertile feed. This contrasts sharply with the use of uranium in LWRs, in which the ^{235}U component (0.7%) limits the portion of the total uranium resource that can be used as fuel. Thorium, a fertile element that is about four times more abundant in the Earth's crust than uranium, also can be used in modified breeder reactors—with obvious implications for the possible lifetime of breeder use as compared to LWRs. A National Academy of Sciences study (1979) estimated that breeder reactors could meet the expected electrical needs of the United States for hundreds of millennia from domestic uranium reserves.

Implicit in breeder operation, however, is the reprocessing of the spent fuel and blanket to recover plutonium and any remaining ^{235}U. Without reprocessing, the plutonium bred in the blanket cannot be fissioned in the reactor core. Thus there is no advantage to breeder reactors without reprocessing.

Reprocessing is also an option for the LWR fuel cycle. The spent fuel contains a certain amount of

Figure 12.3. (*a*) The gaseous diffusion process. Gaseous uranium containing both isotopes is fed in from the left. The lighter ^{235}UF$_6$ passes more easily through the barrier forming a stream that is slightly enriched in ^{235}UF$_6$, leaving behind a stream slightly depleted in the ^{235}U isotope. (*b*) Gaseous diffusion units connected in series. (From DOE, 1976.)

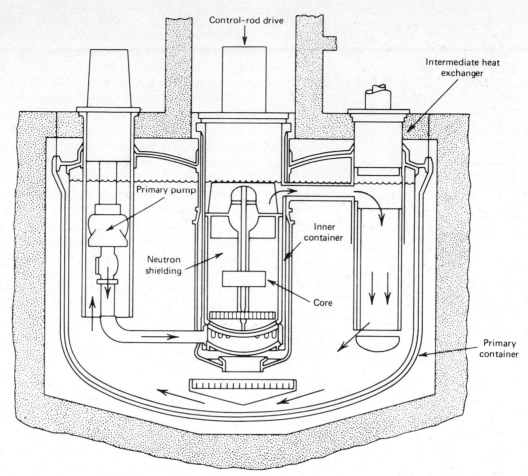

Figure 12.4. Schematic diagram of a breeder reactor. (From "Superphenix: A Full-Scale Breeder Reactor" by Georges A. Vendryes. Copyright © March 1977 by Scientific American, Inc. All rights reserved.)

plutonium produced in the core that wasn't fissioned. Spent fuel also contains some unused ^{235}U, the important fissile isotope of the starting fuel. Reprocessing isolates these isotopes for reuse in reactor fuel by grinding up the fuel elements to expose the fuel and then dissolving the fuel in nitric acid to separate the spent fuel from its metal cladding. A solvent-extraction process is used to separate the uranium and plutonium from the other components of the spent fuel.

Plutonium deserves special mention, because several of its characteristics make it potentially quite dangerous. Plutonium in very small quantities is both directly toxic and carcinogenic. More important, it is a highly explosive material used in the production of nuclear weapons. Since plutonium is fissionable by fast neutrons, the inherent safety of the negative alpha-t reaction (see below) is absent, and consequently there is even a (remote) possibil-

ity that a breeder reactor could explode, something not possible with nonbreeder power reactors. Although plutonium is virtually nonexistent in nature, it is now relatively abundant because it is produced in the cores of normally operating nuclear power plants. Reprocessing spent fuel concentrates this element. The large amounts of plutonium that would be concentrated during reprocessing are a potential source of plutonium; furthermore, only a small proportion need be diverted to produce a bomb. Procurement of plutonium, or some other weapons grade material, such as highly enriched uranium, is considered by some people to be the only major impediment to the construction of nuclear weapons by many less technologically advanced countries or by terrorist groups (Willrich and Taylor, 1979). This point was illuminated when the magazine *The Progressive* published plans to build a nuclear bomb written by an undergraduate physics student at Princeton. Because the United States routinely loses small quantities of plutonium even under the strictest safeguards of the nuclear weapons program, the Carter administration initiated a moratorium on reprocessing of spent fuel citing the danger of nuclear proliferation associated with a plutonium economy. Construction of the Clinch River breeder reactor, a demonstration project, was stalled as well, in large part because of the reprocessing of large amounts of plutonium inherent in its fuel cycle. In the fall of 1983 the project was cancelled for a variety of technical and economic reasons.

A consequence of the no-reprocessing policy is that spent fuel from nuclear reactors is presently treated as waste, and only 0.7% of the fission energy potentially available is used. If the spent fuel was reprocessed, and the recovered plutonium and ^{235}U were used as reactor fuel, the efficiency could be raised to 0.9 or 1%. The use of breeder reactors and recycling could recover up to 70% of the energy in the uranium ore, essentially multiplying by a factor of 100 the energy available in our uranium resources usable by LWRs and a once through fuel cycle (NAS, 1979). Furthermore, breeder reactors would permit economic recovery of lower grade ores and the use of thorium as a fuel source. Thus we are left with a difficult dilemma. Commercialization of breeder technology offers the possibility of enormous quantities of energy with little or no CO_2 or acid rain production, but it also opens the possibility of plutonium contamination of the environment, the theft of plutonium by terrorists, nuclear blackmail or a catastrophic accident. The stakes involved with breeder reactors and the inherent reprocessing of fuel are enormous and deserve careful consideration.

Nuclear advocate Alvin Weinberg considers the option a "Faustian bargain" whose enormous potential energy gains are worth the significant social risk—even if a considerable tightening of security is required. On the other hand many people consider the implications of such a nuclear-powered society abhorrent. Yet one must also consider the non-nuclear alternatives in an equally critical manner. For example, the risk of war presented by a greatly increased dependence on imported oil may be as great as the risks of catastrophic accidents and nuclear proliferation associated with commercial nuclear power. Furthermore most experts agree that renewable energy sources, despite their desirable characteristics, cannot meet our energy needs in the

coming decades, and coal, though in great supply, has environmental impacts that many will find unacceptable on a large scale. At present the social and economic environment that has virtually eliminated construction of new light water reactors probably also precludes construction of a significant number of the breeder reactors that would have all the benefits and dangers of LWRs—plus additional ones. Even the ambitious breeder programs of France and West Germany are being scaled back.

Fusion

Fusion energy (Figure 12.5) offers the possibility of long-term electrical production with probably much smaller, but still substantial, risks from radiation, toxicity, and nuclear weapons proliferation. Commercial reality for even a few reactors, however, is at least 35–40 years into the future. The problems with development arise from the conditions required for the fusion reaction to occur. In the method that is presently considered the best possibility for development, deuterium and tritium are fused to form helium. The reaction requires temperatures about 10 times that of the sun's core. At this temperature matter exists as plasma—atoms stripped of their electrons. In order for the reaction to occur the plasma must be isolated from materials that could contaminate or cool it. These two conditions, confinement and high temperature, are very difficult to create simultaneously. In addition structural materials must be capable of withstanding the intense heat and neutron flux, and be able to do so with little maintenance once the reactor is turned on, for the internal parts soon become too radioactive for a worker to repair. Herein lie the scientific and engineering challenges to fusion development. Because of the complexity involved in controlled fusion, the capital costs are expected to be very high. But the rewards could be enormous.

The principal advantages of fusion energy are the virtual inexhaustibility of fuel, the absence of the dangerous bomb materials associated with fission reactors, and the greatly reduced radiation compared with fission reactors. The resource base for fusion fuel is very large, because although deuterium makes up only 1/7000 of the hydrogen in water and tritium, it can be bred from abundant lithium (NAS, 1979). Tritium is made by bombarding lithium-6 with a neutron, but, unfortunately, most lithium is lithium-7. As with uranium, the desirable

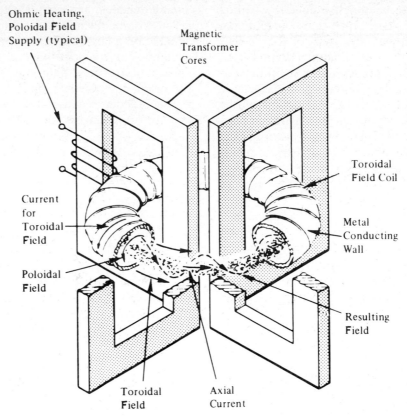

Figure 12.5. Tokamak fusion reactor. Strong magnetic fields are used to keep the fusion reaction from expanding. (From *Energy in Transition 1985–2010* by the Committee on Nuclear and Alternative Energy Systems. W. H. Freeman and Company. Copyright © 1980.)

isotope is only 7% of the total, and the D–T reaction will probably be limited by the supply of lithium-6. Any long-term use of fusion will have to be based on the D–D reaction which requires temperatures of about 1×10^9 °C. The D–T reaction is easier to develop because it requires temperatures of only about 1×10^8 °C or perhaps less. For many people the development of fusion power offers a great, perhaps the only, hope for humankind. Others consider it a scientific and engineering rat's nest, with little chance of attainment.

HISTORY OF NUCLEAR ENERGY IN THE UNITED STATES

Early efforts to tap the energy of the atom were initiated by the U.S. government with the intent of developing an atomic bomb. The development program, begun during World War II and known as the Manhattan Project, led to the construction of the first experimental nuclear reactor which consisted of a complex array of graphite blocks and natural uranium. Rods could be inserted into the reactor to control or stop the nuclear reaction. Early reactors such as this were dubbed *piles*, and the first was called CP-1 (chain-reacting pile-1). The CP-1 was built in the doubles squash courts under the football stands of the University of Chicago. There, on December 2, 1942, the first self-sustaining chain reaction was achieved, thereby initiating the controlled release of nuclear energy.

Emilio Segre, in his biography of Enrico Fermi, a Nobel laureate who was instrumental in the Manhattan Project, describes CP-1:

The first pile had no device built in to remove the heat produced by the reaction, and it was not provided with any shield to absorb the radiations produced by the fission products. For these reasons it could be operated only at a nominal power which never exceeded two hundred watts. It proved, however, two points: that the chain reaction with graphite and natural uranium was possible, and that it was easily controllable. (Segre, 1970)

The first reactor that had a substantial power output (approximately 1 MW$_{th}$) began operation on November 4, 1943, at the government laboratory in Oak Ridge, Tennessee.* This was a graphite pile as well and was built to produce plutonium for examining ways to chemically separate plutonium from other elements in the spent fuel. On September 27, 1944, large-scale production of plutonium for use in nuclear weapons began at Hanford, Washington. This reactor differed from the pile at Oak Ridge in that water was circulated through the reactor to remove the heat produced by the reaction.

About 10 months after the Hanford reactor began operation, enough plutonium had been produced to construct the first atomic bomb, exploded at Alamogordo, New Mexico, on July 16, 1945. On August 6, 1945, a bomb made of uranium-235 was dropped on Hiroshima, Japan, destroying that city completely. Three days later a bomb made of plutonium destroyed Nagasaki. The destruction of Hiroshima and Nagasaki introduced the world to the awesome power of the atom and firmly established in public consciousness a link between the controlled release of nuclear energy for electricity generation and nuclear weapons.

Until 1952 the development of nuclear technology focused on weapons applications. Subsequently this emphasis shifted to include the development of reactors for nuclear-powered submarines:

> The military advantages of using atomic energy for submarine propulsion are very great: the submarine is given a tremendous range of action and can develop full power for long periods when submerged. To secure such advantages, it is necessary to use expensive equipment and expensive fuel which could not be justified in purely commercial applications. For the submarine it is, however, essential to obtain low weight per (horsepower), compactness of plant and flexibility of operation, at the expense of all else. (Hinton and Moore, 1959)

To meet these criteria, a new type of reactor was needed. The light water reactor (LWR) was designed specifically for the first nuclear submarine, the USS *Nautilus*. It used ordinary water circulating through the core as both moderator (replacing graphite) and coolant. The light water design fulfilled the navy's requirements for compactness and low weight at the expense of fuel cost and fuel efficiency, and, possibly, safety. Although LWRs

*The following section is adapted from Dawson (1976) and Bupp (1979).

are inherently safer than graphite reactors other options might have been safer yet. In a LWR, or pressurized water reactor, the fission also is controlled by the temperature of the water or moderator. Since hot water is less dense than cold water, hot water slows or "moderates" fewer neutrons, more neutrons leak out of the core, and the neutron flux, hence fission, decreases. If cold water enters the core as it returns from the steam generator more neutrons are slowed, and fission (power) increases, and the converse. This negative feedback (called negative alpha-t) is an automatic, inherent regulator of the rate of fission and contributes to reactor safety.

Commerical Nuclear Power

The first *full-scale* land-based nuclear power plant project was approved by the Atomic Energy Commission (AEC) in July 1953. The plant, known as the Shippingport reactor, was to have an electrical capacity of 60 MW$_e$ and be located at Shippingport, Pennsylvania. Although it was not expected to produce electricity economically, its construction represented the first step toward civilian nuclear power generation. The light water reactor system (LWR) was chosen for this power station because the design had already been well developed by the naval propulsion program (Dawson, 1976). In fact the Shippingport reactor project was conducted by the AEC's naval reactors group. An unfortunate consequence of this decision to use the light water system is that today virtually every power reactor in the United States utilizes the fuel-inefficient, fuel-costly, and relatively dangerous light water design. Had a reactor been specifically developed for civilian electricity generation, it is quite possible that a safer, more fuel-efficient design that did not require enrichment, such as, at least according to some accounts, the Canadian CANDU heavy water reactor, would be prevalent today. But the discovery of large deposits of uranium in the United States, Canada, and Australia in the 1950s and the initial enthusiasm of nuclear power's proponents appeared to promise a bright future for nuclear power and reduced any concern over the fuel efficiency and safety of LWRs.

In December 1953 President Dwight D. Eisenhower presented his historic Atoms for Peace proposal, and the Atomic Energy Act which implemented many of Eisenhower's proposals was signed into law August 30, 1954. In the following 10 years several policy changes and legislative actions reduced the federal government's direct involvement in nuclear power development and encouraged the entry of private industry into the nuclear energy market. The Price-Anderson Act of 1957 limited the liability of the nuclear industry in the event of a catastrophic accident (Dawson, 1976). Without such legislation, which in effect is a multimillion dollar subsidy to the industry, it is doubtful that private companies would insure the nuclear industry, no matter how small the probability of a severe accident is thought to be.

The first plant to be built without any direct federal support was purchased by the Jersey Center Power and Light Company in 1963. The utility claimed nuclear power could generate electricity more cheaply than other methods. About that time the AEC began to phase out development and assessment of the LWR and shifted its emphasis to breeders and other advanced reactors (Bupp, 1979), and by 1970 it had completely phased out its direct support of the LWR industry (Dawson, 1976).

"The commitment to the breeder reactor diverted the government's managerial attention, and fiscal resources away from other huge unsolved problems that remained in the development of light water systems" (Bupp, 1979). The critical point made by Bupp and others is that nuclear power technology involves much more than burning uranium in a power plant. The nuclear power reactor is a subsystem of a much larger, complicated, highly interdependent fuel cycle that extends far beyond the nuclear generating plant. The production of reactor fuel is itself a complex matter involving several different processes in facilities owned by the federal government and the private sector (Figure 12.6). The power plant must be designed, constructed, and operated, spent fuel must be treated and properly disposed of, and plants must be safely decommissioned when they cease operation. Until 1964 the government stressed the development of operating nuclear power stations, at which time it considered development of the technology complete. Actually only the technology to construct and begin operation of light water reactors was complete. Important components of the system were still missing and had not been given serious attention. Today there is still no permanent waste storage system established for radioactive spent reactor fuel. While the debates go on, spent fuel accumulates in temporary storage at the reactor sites. In

addition we have yet had to deal with the decommissioning of a large-scale nuclear plant.

Utilities ordered nearly 50 nuclear reactors between 1965 and 1967, induced by the belief that nuclear power was the cheapest source of electricity. But "by the end of the 1960s there was already considerable evidence that the 1964–1965 cost estimates by government and industry for electricity from light water nuclear power plants had been low" (Bupp, 1979). These findings were widely dismissed on the grounds that additional experience and larger plants would reduce the cost of nuclear-generated electricity. Nuclear plant orders continued to increase during the early 1970s, although concern over reactor safety already was growing and increasing the cost of installing nuclear plants. By contrast, no new reactors have been ordered since 1978, and many others have been cancelled (DOE, 1983). Growing safety concerns, growing expense, and the unexpected reduction in the growth rate of electricity demand contributed to this reversal. A more comprehensive treatment of environmental and safety concerns associated with LWRs is found in Chapters 15 and 17 and the appendices of this chapter.

U.S. URANIUM RESOURCES

Breeder reactors would be of less interest if it were not for the limited supplies of domestic high-quality uranium resources that LWRs depend on. Large untapped deposits of uranium exist in other parts of the world (Table 12.1), but uranium embargoes and exporting cartels analogous to oil embargoes by OPEC would be possible. For this reason our domestic resource base of uranium deserves careful attention.

Assessing domestic uranium resources is a complex task. Like many minerals uranium exists in virtually unlimited supply, but most of this is at very low concentrations. Seawater, for example, contains about 5 billion tons of uranium at an average concentration of 3–4 parts ber billion (ppb). Although state-of-the-art technology can extract uranium from these low quality deposits, energy and dollar costs are prohibitive. Very low-grade uranium sources such as these do not constitute an economic resource, at least presently. Geologic and economic parameters and the availability of other energy sources limit domestic uranium resources to relatively rare, highly concentrated deposits. But

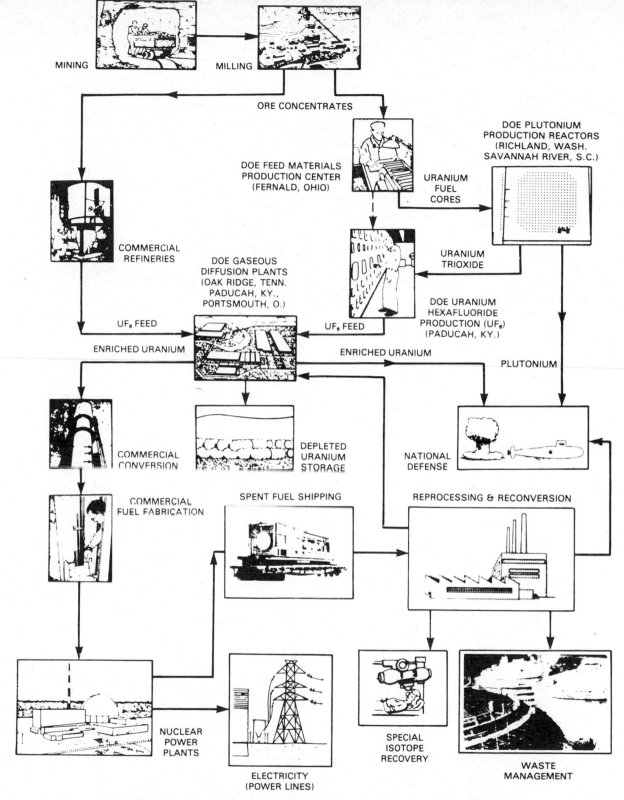

MINING MILLING

ORE CONCENTRATES

DOE FEED MATERIALS
PRODUCTION CENTER
(FERNALD, OHIO)

DOE PLUTONIUM
PRODUCTION REACTORS
(RICHLAND, WASH.
SAVANNAH RIVER, S.C.)

URANIUM FUEL
CORES

COMMERCIAL
REFINERIES

DOE GASEOUS
DIFFUSION PLANTS
(OAK RIDGE, TENN.
PADUCAH, KY.,
PORTSMOUTH, O.)

URANIUM
TRIOXIDE

UF$_6$ FEED UF$_6$ FEED

DOE URANIUM
HEXAFLUORIDE
PRODUCTION (UF$_6$)
(PADUCAH, KY.)

ENRICHED URANIUM ENRICHED URANIUM PLUTONIUM

COMMERCIAL
CONVERSION

DEPLETED
URANIUM
STORAGE

NATIONAL
DEFENSE

COMMERCIAL
FUEL FABRICATION

SPENT FUEL SHIPPING

REPROCESSING & RECONVERSION

NUCLEAR
POWER
PLANTS

ELECTRICITY
(POWER LINES)

SPECIAL
ISOTOPE
RECOVERY

WASTE
MANAGEMENT

PRODUCTION OF FISSIONABLE MATERIALS
FOR ELECTRICAL POWER PRODUCTION AND NATIONAL DEFENSE

Figure 12.6. The production, use, and disposal of fissionable materials. Wastes associated with each process and facility are not included. (From DOE, 1976.)

269

Table 12.1. Foreign Uranium Resources by Continent (Thousands Tons U_3O_8)

Continent/Country	Reasonably Assured		Estimated Additional	
	$30/lb U_3O_8	$50/lb $U_3O_8^a$	$30/lb U_3O_8	$50/lb $U_3O_8^a$
North America	303	374	470	1017
Canada	299	335	465	988
Mexico	4	4	5	8
Greenland	0	35	0	21
Africa	770	946	219	394
South Africa	321	465	109	227
Niger[b]	210	210	69	69
Namibia	155	176	39	69
Algeria	34	34	0	0
Gabon	25	26	0	13
C.A.E.	23	23	0	0
Zaire	2	2	2	2
Somalia	0	9	0	4
Egypt	0	0	0	7
Madagascar[b]	0	0	0	3
Botswana[b]	0	0.5	0	0
Australia	380	410	340	370
Europe	111	208	64	193
France	77	97	37	60
Spain	16	21	11	11
Portugal	9	11	3	3
Yugoslavia[b]	6	8	7	27
United Kingdom	0	0	0	10
Germany	1	7	2	11
Italy	0	3	0	3
Austria	0.4	1	1	2
Sweden	0	49	0	57
Finland	0	4	0	0
Greece	2	7	3	9
Asia	56	72	1	33
India	42	42	1	33
Japan	10	10	0	0
Turkey	3	6	0	0
Korea	0.05	14	0	0
Philippines[b]	0.4	0.4	0	0
South America	188	194	111	133
Brazil	155	155	106	106
Argentina	33	39	5	17
Chile	0.03	0	0	9
Bolivia	0	0	0	0.9
Totals (rounded)	1800	2200	1200	2140

Source: Based on *Uranium: Resources, Production, and Demand: A Joint Report*, OECD Nuclear Energy Agency and the International Atomic Energy Agency, February 1982.

[a] Includes resources at $30/lb U_3O_8.
[b] Reported in 1979 but not in 1982.

the relationship is not simple. For example, the cost of mining relatively high-grade ore having a very thick, hard overburden may prove greater than the dollar value of the ore. Conversely, a large deposit of relatively low-grade ore with little overburden may prove highly profitable to mine and may be termed a high-grade resource.

Another problem in determining the extent of our uranium resource is that assessments are made from incomplete information and are therefore only estimates, the accuracy of which depends on the amount of geologic information used in making those assessments. Just as with oil, new deposits are confirmed by exploratory drilling after preliminary assessments, using geological and radiometric information. The size and uranium content of new deposits are estimated by developmental drilling. These assessments, however are ongoing processes, and the total amount and grade of uranium ore in a given deposit is not known precisely until the deposit is mined entirely.

The classification scheme used by DOE divides uranium resources into reserves and potential resources. Potential resources are divided further into probable, possible, and speculative classes (see Figure 4.8 and Table 12.2).

Uranium *reserves* include known uranium deposits recoverable at less than a specified cost with state-of-the-art technology. Reserve estimates are calculated using data provided voluntarily by the uranium companies and are reported by DOE "in a manner which avoids disclosure of proprietary information" (DOE, 1983). More specifically:

Probable potential resources are those estimated to occur in known productive uranium areas based on: 1) extensions of known deposits, or 2) undiscovered deposits within known geological trends or mineralized areas.

Possible potential resources are those estimated to occur in undiscovered or partly defined deposits in formations or geologic settings productive elsewhere within the same geological province or subprovince.

Speculative potential resources are those estimated to occur in undiscovered or partly defined deposits: 1) in formations or geologic settings not previously productive within a productive geologic province or subprovince, or 2) within a geologic province or subprovince not previously productive (DOE, 1983).

In DOE publications uranium resources are broken down into cost categories that represent relative economic availability (Table 12.2). These figures are based on costs of power, labor, materials, royalties, and so forth, but do not include income taxes or expenditures made prior to the time of the estimate. As of January 1, 1983, reserves with the assigned costs of $30, $50, and $100 per pound U_3O_8 correspond to an average grade (percent U_3O_8) of 0.21, 0.10, and 0.06, respectively (DOE, 1983). An additional estimated 110,000 tons of U_3O_8 potentially will be available through the year 2010 as by-product uranium, that is, uranium recovered as a by-product of other industries. It is unlikely, however, that most of this uranium will reach the market (DOE, 1983). In 1980 five phosphate plants produced uranium as a by-product, and several others were in the development pipeline. Copper plants in Utah and Arizona are also producing U_3O_8 by treating copper-leaching solutions. The construction of similar facilities at other copper mines and plants could increase the uranium recovery capacity to about 500–1000 tons U_3O_8 annually by the mid-1980s (DOE, 1980). Most high grade uranium resources are in the West (Figure 12.7).

In the East, Chattanooga shale is a potential

Table 12.2. Uranium Resources in the United States, January 1, 1983 (thousand tons U_3O_8) (from DOE, 1984)

	$30/lb U_3O_8[a]	$50/lb U_3O_8	$100/lb U_3O_8	Total
Reserves				
	180	576	889	1645
Potential Resources				
Probable	654	513	720	1887
Possible	257	251	334	842
Speculative	216	175	261	652

[a] All forward costs are in 1983 dollars.

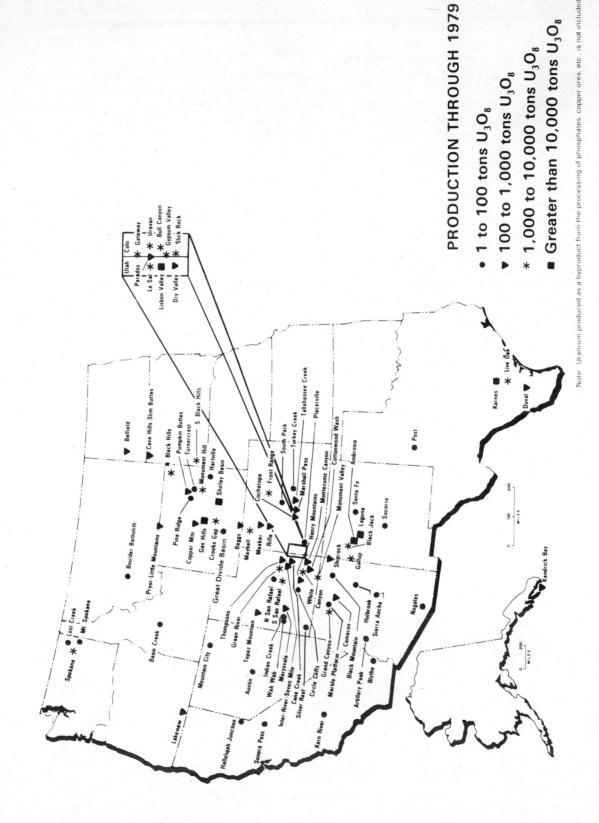

AREAS OF URANIUM PRODUCTION IN THE UNITED STATES

Figure 12.7. Areas of uranium production in the United States. (From DOE, 1983.)

source of large amounts of low grade uranium. Relatively little is known about this resource because assessments have concentrated on high and intermediate grade (greater than 0.01 percent U_3O_8) resources. Chattanooga shale contains more than 5 million tons of U_3O_8 at an average grade of 0.006%, and an additional 10 million tons at an average grade of 0.0045% U_3O_8. This resource is of relatively uniform grade and thickness covering hundreds of square kilometers of the East Central United States (DOE, 1978).

A single 1000 MW_e plant utilizing Chattanooga shale as a uranium source would require the mining of 12,545 metric tons of ore per day assuming 60% recovery at the uranium mill and unchanged reactor fuel requirements. A single 1000 MW_e coal plant on the other hand would require 7909 metric tons of coal per day, assuming 22,000 Btu/kg and 9500 Btu/kWh (DOE, 1978). Thus approximately 60% more Chattanooga shale would have to be mined compared to coal to produce an equal amount of electricity. Uranium production in recent years extracted ore containing approximately 0.11% U_3O_8, and at this concentration a 1000 MW_e nuclear plant requires the equivalent of about 400 tons of ore per day, or 146,000 tons/yr (Figure 12.2b). This amount is about 3% of the amount of Chattanooga shale that would be required for an equal period, and only 5% of the amount of coal (not including overburden) mined per day for a present-day coal plant of similar size.

Uranium Production and Geology

Formations appropriate for open pit mining contain 33% of the reserves in the form of 635,000,000 tons of ore with an average grade of 0.05% U_3O_8. Approximately 66% of the ore and 29% of the total

uranium produced in the United States in 1982 came from open pit mines (Table 12.3 and Figure 12.7). The average grade of ore extracted from those mines was 0.07% U_3O_8 (DOE, 1983). In 1980, 120 m was the maximum overburden being removed in open pit mines (DOE, 1980a).

Most (54%) domestic reserves exist in formations appropriate for underground mining. These deposits contain 515,600,000 tons of ore with an average grade of 0.09% U_3O_8. Forty-four percent of the reserves are at a depth of 122 m or less. Thirty percent of the reserves are found at depths exceeding 460 m. Approximately 34% of the ore and 46% of the uranium produced in 1982 came from underground mines. The average grade of this ore was 0.22% U_3O_8 (DOE, 1983). The deepest underground operation in the United States is owned by Gulf Mineral Resources, and extends to 1000 m (DOE, 1980a).

Eleven percent of uranium reserves are found in formations appropriate for solution mining (DOE, 1983), and solution mining provided 11% of the total uranium production in 1982. Uranium produced as a by-product of other industries provided 12% of the 1982 uranium production. Two percent of production came from other minor sources such as mine water, heap leach, and low-grade stockpiles (DOE, 1983).

The average grade of ore delivered to uranium mills in 1982 was 0.119% U_3O_8. Mill recovery rates averaged 96.2%. From 1966 to 1979 uranium processing rates increased more than threefold. During that same period the average grade of ore processed fell by 50% from greater than 0.23% U_3O_8 to 0.113% U_3O_8 in 1979 (Figure 12.8; DOE, 1983). Thus more than 40% of the increase in uranium ore processing went to offset the decline in ore quality. It is interesting to note that domestic uranium mills processed 17% more ore in 1979 than in 1978 while the

Table 12.3   Uranium Production in the United States in 1982 (from DOE, 1984)

Source	Thousands Tons Ore	Percent of Total Ore	Percent U_3O_8	Thousands Tons U_3O_8	Percent of Total U_3O_8
Underground mines	2809	33.8	0.22	6.3	46
Open pit mines	5504	66.2	0.07	3.8	29
Solution mining	—	—	—	1.5	11
By-product	—	—	—	1.6	12
Others: heap leach, mine water, and low grade stockpiles	—	—	—	0.2	2
Totals	8313	100.0	0.12	13.4	100

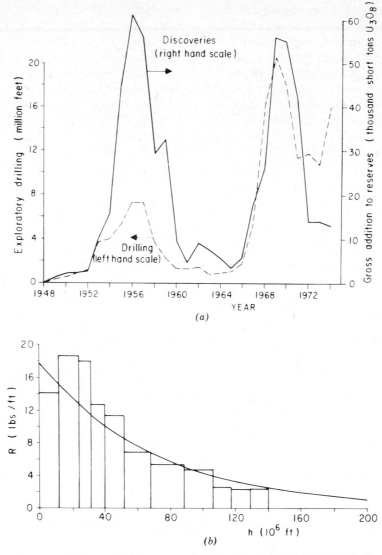

Figure 12.8. (*a*) Uranium exploratory drilling and finding rates in the United States, 1948–1974. (*b*) Finding rate (*R*) versus cumulative drilling effort (compare with Figure 7.16). (From Lieberman, 1976. Reproduced with permission from "United States Uranium Resources—An Analysis of Historical Data" by M. A. Lieberman, *Science*, Volume 192, pp. 431–436, Figs. 1 and 3, 30 April 1976.)

average grade of ore processed was 16% lower. Thus the total quantity of uranium recovered from milled ore was roughly the same (DOE, 1980b) because the entire increase in ore processing was needed to offset the decline in ore quality. Presumably as the quality of ore decreased the energy cost of mining and milling per ton of final product increases but this data is not provided by the Department of Energy. The discovery of uranium as a function of exploratory effort has also declined in recent years (Figure 12.8) in a fashion similar to that experienced by the petroleum industry (see Figures 7.16 and 7.18).

Both uranium ore processing and total uranium concentrate (yellowcake) production peaked in 1980 at 46,000 tons/day and 22,000 tons/yr, respectively. By 1982 the quantity of ore milled had fallen 48% to 24,000 tons/day, and uranium concentrate production had fallen 39% to 13,400 tons U_3O_8 (Figure 12.9). The disproportionate decline in milling rate was due to a partial substitution by unconventional concentrate production methods such as solution mining and by-product uranium recovery and to a lesser degree by improved recovery of uranium at the mill. The very modest increase in the grade of conventional millfeed, from 0.113 to 0.119% U_3O_8,

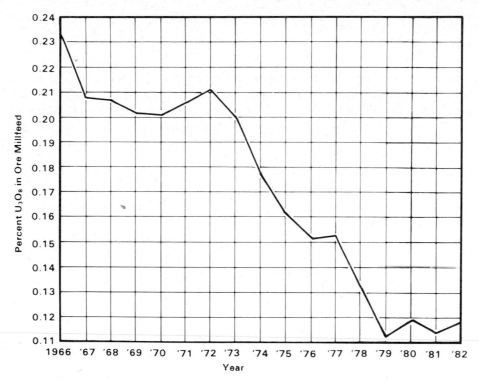

Figure 12.9.   Percent U_3O_8 in ore mill feed for all U.S. uranium production. Uranium production increased from 12,000 tons/day in 1966 to 46,000 tons/day in 1980, and then declined sharply to 24,000 tons/day in 1982. The effect on ore grade is analogous to that shown in Figure 4.13.

between 1979 and 1982 appeared to be another manifestation of the general inverse relationship between exploitation rate and average ore grade for extractive industries (see also Hall et al., 1981). This relationship was also found in historical data on the copper industry (Figure 4.14), and the petroleum industry (Figure 7.18).

Total Production

Preliminary calculations based on these estimates of economically recoverable uranium (i.e., 2.4 million metric tons of uranium and 5.8 million kcal electricity produced per ton) indicate that the total ultimate yield for electricity from U.S. uranium used in LWRs is on the order of a few hundred 1000 MW_e reactor lifetimes (NAS, 1979), equivalent to 5–10% of our 1980 energy use for 50 years. Thus fission reactors alone will not make a major contribution to our energy needs unless we import large quantities of uranium, at unpredictable prices, or institute a breeder program. If we choose to pursue either (or both) of those options, a great deal of energy could be produced from nuclear power. On

the other hand, if we make large energy investments in nuclear power and if subsequent public opposition, perhaps in response to accidents, mismanagement, or escalating costs causes the program to be curtailed or abandoned, the large fossil energy investments would have little return.

ENERGY RETURN ON INVESTMENT FOR NUCLEAR POWER

The entire nuclear fuel cycle is highly subsidized by fossil fuels, and it is difficult to get a truly comprehensive analysis of the energy balance of commercial nuclear power. Seven earlier studies compared the electric output generated by a nuclear power plant with the energy required as inputs in an attempt to determine the EROI for nuclear power. These analyses were prompted in part by the fact that the nuclear power research and fabrication industries are both very dollar and energy intensive, and apocryphal reports in the late 1960s that though nuclear power generated 4% of the nation's energy, it required some 4% of our electrical output just to run the gaseous diffusion building "K-9," at Oak

Ridge, Tennessee. Clearly, if nuclear power required as much energy as it produced there would be no particular reason to trade coal-generated electricity for an equal quantity of nuclear-generated electricity. On the other hand, if nuclear power's EROI is substantially greater than 1:1 uranium might make an effective "coal extender" by increasing the quantity of electricity that can be obtained from a ton of coal.

The results of these earlier studies normally were expressed as the ratio of energy produced to energy invested (E_p/E_i) during the lifetime of the plant. But these studies differed greatly in the types of energy inputs included. Consequently their conclusions produced confusion over what should be a more or less unambiguous analysis. Some studies stated that nuclear power will yield a large return on investment (e.g., Chapman and Mortimer, 1974: EROI = 15.6), others found that nuclear power yielded a much smaller return on investment (e.g., Lem et al., 1974: EROI = 3.23). The remaining estimates ranged over the interval (Table 12.4).

Berger (1981), Hall et al. (1981), and Howarth and Hall (1985) converted the energy input values

Table 12.4.   Results of Eleven Studies on the EROI for Nuclear Power Plants (summary analysis from Berger, 1981, and Hall et al. 1981)

Study	Energy Return on Investment	
	Original Numbers	Standardized[a]
Rotty et al.[b]	9.63	4.98
Rotty et al.[b]	8.93	4.49
Rotty et al.[b]	12.11	6.52
Rotty et al.[c]	3.40	2.31
Rotty et al.[c]	4.89	3.31
Rombough and Koen	6.31	5.39
Lem, Odum, and Balch	3.23	2.74
Chapman and Mortimer	10.9 ± 2	3.54
Chapman and Mortimer	15.6 ± 3	6.08
Chapman and Mortimer	12.9 ± 2	5.13
Chapman and Mortimer	16.5 ± 3	6.75

[a] Standardized by assuming the same quality factors (for electrical vs. fossil inputs), fuel types, and load factors (0.75). Numbers would be three times higher if electricity were given a quality factor.
[b] Using 0.176% uranium ore; several reactor types.
[c] Using 0.006% uranium ore (i.e., Chattanooga shale), several reactor types.

given in these studies from their original units into units of kilocalories of heat for each process used to produce nuclear power (mining, milling, conversion to uranium hexafluoride, enrichment, fuel rod fabrication, etc.). Ideally these costs included the energy used or purchased for carrying out a process and the energy required to produce the materials and capital equipment required for the process, including the energy requirements for production of raw materials. In practice, not all the studies included energy estimates of capital equipment, and this accounts for a small part of the variation among the results of the different studies. Furthermore none of the studies considered uranium as an energy input but rather assumed that investment energy of interest is fossil fuel because large amounts of fossil fuels are invested to run the nuclear fuel cycle. If the total energy content of the fresh uranium fuel was counted, the energy return on energy invested would be much lower. On the other hand, the EROI would be much higher if the energy content of the spent fuel was credited to the system.

When standardized to similar assumptions, operating parameters, fuel qualities, and boundary assumptions, the seven studies all gave much more similar results (EROI = 5 ± 1.5 kcal of electricity returned per kilocalorie invested (Table 12.4). In other words, the results of these EROI analyses depend on the assumptions used by the different investigators much more than on properties inherent in different power systems or in the data base. In particular, the most important assumption was whether or not a quality factor was given to the electrical output when compared to fossil energy inputs. Weighting the output by a factor of 3 to account for the quality of electricity gives an EROI of approximately 15:1. Other important considerations included the load factors assumed, boundaries selected for analysis (e.g., whether an energy cost was given for waste storage or potential accidents), and variations in the estimate of energy cost for some critical components such as the fuel rods. Details are given in Hall et al. (1981) and Berger (1981). It should also be noted that other energy costs such as decommissioning and government subsidies were excluded from these analyses.

No estimates for costs of transporting electricity to consumers were included in the original papers, but this presumably is about 3% transmission losses for medium distances (Hall et al., 1979a; Mau et al., 1978). The energy used by labor was not included in any analysis studied.

Projections of Future EROI for Fission Energy

If expansion of nuclear power generation conforms to recent government predictions (DOE, 1983), exhaustion of high-grade reserves will occur by 2020 unless significant additions are made to domestic reserves. It is unlikely, however, that such additions can be made (Lieberman, 1976). Although Chattanooga shale could be exploited to fuel LWRs, the vast scale of the requisite mining and milling operations would make such an endeavor very energy intensive, environmentally disruptive, and uneconomical.

Technical innovations in uranium enrichment may make significant reductions in the energy intensiveness of the fuel cycle. Uranium enrichment by gaseous diffusion is currently the largest energy input to nuclear-based generation of electricity, accounting for about half of the total lifetime energy inputs to a light water reactor built today. An alternative to gaseous diffusion is gas centrifuge enrichment, which requires 95% less direct energy. A $10 billion plant employing this technology is under construction at the DOE site at Portsmouth, Ohio, and scheduled for completion by 1994 (*Science,* August 19, 1983). Gas centrifuge enrichment could lead to a significant short-term improvement in the EROI of light water reactors. Over the longer term, however, the savings in enrichment energy are likely to be countered by the energy inputs to mining and milling of low grade deposits (Howarth and Hall, 1985).

Although $2 billion has been spent on the gas centrifuge plant through 1983, the wisdom of continuing construction has been called into question. Existing enrichment capacity presently exceeds demand and is expected to do so for decades. Additionally the numerous plant cancellations have led to a glut of reactor fuel, further reducing prices. A third factor is ongoing technological advance. An enrichment process currently under development involving lasers and gas centrifuge technology may make the Portsmouth plant obsolete before it comes on line.

Either of these advanced enrichment technologies could extend our uranium reserves by making more efficient use of natural uranium. Furthermore, if these technologies could reduce significantly the cost of fueling reactors, they could expand the resource base by permitting the economical use of a lower-grade resource base. Such technologies might include reprocessing and recycling of spent fuel and the use of advanced reactor designs that have higher conversion ratios (advanced converters). Although these measures could prolong the use of LWRs, all burner reactor systems are restricted to a relatively short-term contribution to our energy needs by the limited nature of the uranium resource base and, eventually, by the high energy costs of processing huge amounts of low-grade ore.

Other factors that have already stalled the expansion of the LWR industry, include the reduction in the growth rate of electricity use, the increasing safety-related capital costs of plants, the difficulties in producing an effective waste storage system, and public fear of real or potential accidents. These factors may be more important in restricting the growth of the industry and in determining the EROI of LWR systems. The future may bring the development and implementation of breeder and fusion technology, but the technical difficulties and high capital costs suggest that it will be many decades, if ever, before these systems make a major contribution to our energy budget.

APPENDIX 12.1: UNANTICIPATED PROBLEMS WITH NUCLEAR POWER*

The technologists claim that if everything works according to their blueprints, fission energy will be a safe and very attractive solution to the energy needs of the world. This may be correct. Hence, they consider all the objections to be due to "ignorance," "viciousness," or "hysteria." This is not correct. The real issue is whether their blueprints will work in the real world and not only in a "technological paradise."
— Nobel laureate Hannes Alfven, 1972

The analysis presented in this chapter points out that nuclear power yields the lowest energy return on investment (EROI) of all major fuel sources when expressed as thermal units, although this ratio is higher if the energy production is multiplied by a quality factor (Table 12.5). Some recent technical findings as well as some unresolved, long-term problems indicate that the actual net energy return for nuclear power may be considerably lower. This is especially true if the net energy analysis includes the costs of plant decommissioning and the long-

*This section is written by Scott G. Leibowitz.

Table 12.5.   Estimated Effects of Decommissioning and Several Problem Scenarios on the EROI of a Nuclear Power Plant. The Decrease in EROI Is Probably Very Conservative as Noted

Scenario	1980 Dollar Costs (millions)	Energy Costs[a]	EROI[b]	Note
Decommission	50	0.40	4.94	c
Decommission + steam tube corrosion	65	0.52	4.92	d
Decommission + reactor embrittlement	70	0.56	4.91	e
Decommission + TMI-type accident	1050	8.40	3.95	f

[a] Very approximate energy costs, in trillion kilocalories, calculated by using an energy per dollar ratio of 8000 kcal per dollar (this is roughly the average value for the U.S. economy in 1980.

[b] An EROI of 5:1 is assumed for a 1000 MW plant without quality correction and without decommissioning or any technical problems. For a plant with a 30-year lifetime and a 70 percent capacity factor the energy produced would be

$$1000 \text{ MW}_e \times \frac{860 \times 10^3 \text{ kcal}}{\text{MWh}} \times \frac{24 \text{ hr}}{\text{day}} \times \frac{365 \text{ day}}{\text{year}} \times 30 \text{ year} \times 0.70$$

This is equal to 158×10^{12} kcal. The energy cost of investment is assumed to be one-fifth of this based on the 5:1 EROI, or 31.6×10^{12} kcal. Energy returns on investment for decommissioning and the different problem scenarios are calculated by adding the energy cost of the scenario to the energy cost of investment and then dividing energy production by this sum.

[c] The cost of decommissioning a 1200 MW_e plant is estimated at $50 million to $100 million (Norman, 1982). The lower value of $50 million is used as a conservative estimate.

[d] Costs of dealing with steam tube problems have ranged from $25 million for the TMI-1 reactor (Anonymous, 1982) to $112 million for the two Virginia Electric Power Co. generators that were replaced (Marshall, 1981b). A conservative estimate of $15 million is used here.

[e] It is estimated that utilities with embrittlement problems may have to spend $20 million to take actions to avoid vessel rupture (Marshall, 1982d). This does not include costs of reduced plant efficiency and is therefore probably an underestimate.

[f] The cost of cleaning up the damaged Three Mile Island reactor is estimated as $1 billion (Marshall, 1981a). It is assumed that the accident occurs at the end of the plant's normal operating life.

term disposal of high-level radioactive wastes. Some recent problems, along with estimates of their potential impact on the EROI of nuclear power, are discussed in this section.

Technical Problems with Aging Reactors

Commercial nuclear reactors were originally designed to generate electricity for 30–40 years. The first commercial nuclear reactor at Shippingport, Pennsylvania, came on line only 28 years ago, in December 1957. Thus the nuclear industry has little direct experience with aging reactors. Already there have arisen two major age-related problems, costing many millions of dollars in repairs, increasing shutdown time, and possibly decreasing the life expectancy of reactors. These two problems are steam tube disintegration and embrittlement of reactor vessels.

Steam Tube Corrosion

In pressurized water reactors, steam tubes are used to transfer heat from the reactor core to a secondary water system that circulates around the steam tubes. The secondary water system supplies the steam to the turbines that generate electricity (Figure 12.1). Corrosion and/or mechanical problems weaken these steam tubes and can lead to leakage and rupturing. Leaking steam tubes can allow contaminated water to enter the secondary water system, where it can then be released to the environment (GAO, 1982). At least 40 of the nation's 48 pressurized water reactors have reported steam tube problems (Cheng, 1982). According to a study by the Nuclear Regulatory Commission (NRC), 23% of all nuclear plant shutdowns not due to refueling were caused by tube problems (*Science News*, **121**:283, 1982). For example, problems with steam tubes were partially responsible for the January 1982 "site area emergency" at the Ginna nuclear plant in Ontario, New York (Peterson, 1982a, 1982b, 1982c). Before the TMI-1 reactor could be put back in service (it was shut down for refueling at the time of the TMI-2 accident), 8000–10,000 steam tubes had to be replaced at an estimated cost of $25 million (*Science News,* **121**:283, 1982). Steam tube problems were so severe at two nuclear power plants run by Virginia Electric Power Company that the steam generators had to be replaced at a cost of $112 million (Cheng, 1982; Marshall, 1981b). Interestingly these plants were not even a decade old at the time the shutdowns occurred. Florida Power and Light Company is also in the process of replacing the steam generators at its Turkey Point Units 3 and 4 (Cheng, 1982). Besides leading to higher operating costs, the NRC concluded in a 1982 report that steam pipe problems are responsible for 10–60% of the annual occupational radiation exposures at plants experiencing such problems (Cheng, 1982).

Vessel Embrittlement

Another problem that may plague at least 18 of the older nuclear power plants is neutron bombardment of the steel vessel of pressurized water reactors, a process that causes the vessel to weaken and become brittle with time (Marshall, 1981b, 1982b, 1982d). Brittleness increases the probability of a vessel rupture, especially when the vessel is stressed by rapid cooling and high pressure. Events that could cause such a vessel rupture are not unknown: studies show that a 1978 "transient" at the Rancho Seco power plant in California could have caused the vessel to burst if the reactor had been only 7-years older (Marshall, 1981b). The NRC believes that some reactors are currently vulnerable to such a thermal shock accident. Ways of dealing with this problem are expensive, could cause long shutdowns, and may not be totally effective. Some utilities may end up spending $20 million to make changes such as rearranging the fuel. In addition these plants will probably experience a permanent loss in operating efficiency, causing further economic and energy losses (Marshall, 1982d). Ultimately, some reactors may have to be taken out of service sooner than planned because of the hazard posed by reactor embrittlement.

In addition to increased costs, aging nuclear power plants have caused increases in occupational exposures to radiation. A report by the Congressional General Accounting Office found that the average collective occupational dose of radiation received per reactor increased fourfold from 178 human-rems in 1969 to 791 in 1980 (GAO, 1982). This was due to several factors, one being that radiation levels in reactors increased over time. Although the average exposure received per individual worker has not increased, a greater number of individuals are being exposed. If the supply of workers does not keep up with the increased demand caused by these higher radiation levels, GAO predicts that individual exposures will increase.

Decommissioning of Aged Reactors

In order to compare the cost effectiveness of different energy techologies, it is necessary to be familiar with the complete fuel cycle of each technology. This includes the mining of the fuel, the building of the power plant, the operation and maintenance costs, and the costs of disposing of any noxious or hazardous waste products. The nuclear power industry is unique in that two important costs of the fuel cycle are still unknown: decommissioning of retired plants and storing spent fuels. Because the technologies for both of these activities have not been satisfactorily developed yet, their precise cost in energy and dollars has yet to be included in cost-benefit studies comparing nuclear power with other power-generating systems.

Nuclear reactors become radioactive from two main processes. First, radioactive deposits form on internal surfaces of portions of the reactor due to leakage and corrosion. Second, radioactive isotopes are formed when certain materials in the reactor vessel are bombarded with neutrons (Norman, 1982). Until recently it was believed that these radiation sources were short-lived and that the reactor could be decommissioned by entombing it in concrete until radiation levels declined. New research, however, suggests that the radioactive isotopes nickel-59 and niobium-94 may be present in quantities great enough to pose an unacceptable health risk (Norman, 1982). Because these materials remain at dangerous levels for thousands of years, they require a more permanent and more costly form of disposal. Two disposal options are currently being considered. The first is dismantlement, in which the entire plant is cut up and transferred to a burial site. The second is safe storage, in which the plant is placed under constant guard for

30–100 years, a period sufficient to allow the short-lived isotopes to decay, and then dismantled (Norman, 1982). Costs for decommissioning a 1200 MW reactor are estimated at $50–100 million, or nearly 10% of the (1970s) cost of building a plant (Norman, 1982). Yet because the nuclear industry has no experience in dismantling a large plant, this estimate may be conservative. Merrow (1979) found that past studies have routinely underestimated costs of nuclear waste disposal. Two small reactors, the Elk River plant in Minnesota and the Sodium Reactor Experiment in California, have been dismantled successfully at costs of $6 million and $10 million, respectively. The first large reactor to be dismantled will be the Shippingport power plant in Pennsylvania. Decommissioning of this plant will be a model for other plants and will provide valuable information on the economics of decommissioning.

Waste Disposal

One of the most serious charges against the nuclear industry is that it still has not implemented a safe procedure for long-term disposal of radioactive wastes. Waste disposal is a difficult problem because high-level wastes must be isolated from the biosphere for thousands of years. In addition the disposal method must be able to withstand constant stress caused by radiation bombardment and heat dissipation. The nuclear power industry has generated more than 7000 metric tons of spent fuel rods, all of which are temporarily stored in over-crowded surface sites (Garmon, 1981; Skinner and Walker, 1982). Because 95% of spent fuel is recyclable uranium, one disposal option is reprocessing (Garmon, 1981). Although reprocessing does not remove the need for permanent storage totally, it does produce radiation products that reach safe levels in a time period of about 1000 years. Fuel rods that are not reprocessed contain plutonium and uranium and are highly radioactive for hundreds of thousands of years (Garmon, 1981). In addition wastes from reprocessed fuel rods require less storage space.

Reprocessing of spent fuel is a politically sensitive issue. Because reprocessed plutonium can be stolen and used to make nuclear weapons, many people are opposed to commercial fuel reprocessing. Presidents Ford and Carter banned reprocessing for this reason. President Reagan has since lifted the ban and the Department of Energy (DOE) is trying to encourage commercial reprocessing (Garmon, 1981). The nuclear industry itself is reluctant to develop commercial reprocessing because of the many risks and uncertainties and also because of a desire to remain apart from military applications of nuclear energy.

If spent fuel is not reprocessed, other methods of treatment and long-term disposal are required. Treatment possibilities range from disposing of the cut-up reactor in an unconverted form to dissolving the pieces and converting them to glass or some other material (Garmon, 1981). A *multiple-barrier* approach will probably be used to isolate these materials, with the final barrier being some stable geologic formation. Formations being considered as candidates include crystalline rock such as granite (Bredehoeft and Maini, 1981), salt deposits (Gonzales, 1982), and subseabed sediments (Hinga et al., 1982; Hollister et al., 1981; Kelly, 1981). Although industry representatives and DOE officials are confident that these materials can be stored safely, any environmental contamination could have serious health effects. Even without health costs, the costs of developing, constructing, guarding, and monitoring these facilities will be large and will probably be paid for by the federal government.

Nuclear Accidents and Risk Analysis

Since the accident at Three Mile Island in 1979, public attention has focused on possibilities of other nuclear accidents. Until Three Mile Island many industry representatives and government officials considered the probability of such an accident to be so negligible that it did not warrant serious concern. The low probability rates were primarily based on a report referred to as WASH-1400, authored by Norman Rasmussen of the Massachusetts Institute of Technology (Marshall, 1982c; NRC, 1975). The Rasmussen report calculated risks by considering the theoretical probabilities of failure for different reactor components.

Actual operating experience, however, has shown that WASH-1400 substantially underestimated the probability of certain types of nuclear accidents. For example, WASH-1400 calculated the probability of an accident like the one experienced at the Ginna plant on Lake Ontario, N.Y. as one in every 40 years, given the number of reactors presently in the United States. Between 1975 and 1982, however, four Ginna-type accidents have in fact occurred (Marshall, 1982a). A more recent report for the NRC used the actual operating records of the nuclear industry to derive risk estimates. This report concluded that an accident producing core damage as severe as that at Three Mile Island would occur between 1.7 and 4.5 times per 1000 reactor-years (Marshall, 1982c). Since there are 74 commercial reactors in the United States, this means that there could be a severe accident as often as every 3–8 years.

There are many factors contributing to the higher than expected accident rates. Unforeseen problems related to plant aging such as steam tube failures and embrittlement can increase the chances of an accident. Faulty construction, poor quality control and maintenance, and sloppy management also can increase the chances of an accident. Examples of human-related errors are improper welds in reactor shield walls, air bubbles in concrete walls, a plant under construction in Michigan that has been sinking because of poorly compacted fill, and a reactor vessel installed backwards in a California plant that remained undetected for months (Mintz, 1983). The most infamous of these human-related problems is the case of Pacific Gas and Electric's Diablo Canyon plant in California. The NRC suspended the start-up license for this plant after several serious design problems were discovered, including the misplacement of structural supports because of a chart mix-up (Smith, 1981; Sun, 1981).

More recently the Salem-1 reactor in southern New Jersey experienced an "anticipated transient without scram" (ATWS), whereby the plant's safety system failed to shut down the reactor automatically when it should have (Marshall, 1983b). The control rods of a nuclear reactor are actively kept in the raised position by an electric current. When the circuit is broken, the rods are automatically lowered into the core by gravity. The circuit breakers that interrupt this current are supposed to do so during any major disturbance of the controls (Marshall, 1983b). An ATWS occurs if these breakers do not operate. The nuclear industry had argued for years that the probability of an ATWS was so negligible that it was not worthy of being considered as a design factor. As a result of the industry's opposition, the NRC never adopted any ATWS regulations (Marshall, 1983b). Yet a second ATWS occurred at Salem-1 3 days after the first, even though it was believed that such a failure should occur only once in a million reactor-years. The NRC has blamed these failures on poor maintenance of vital safety equipment and careless management (Marshall, 1983c).

Economic Costs

All of the foregoing problems associated with nuclear power plants have proved to be very costly, although they were not included in original cost–benefit calculations. Cleanup costs at Three Mile Island have already surpassed $200 million, and may reach as high as $1 billion (Marshall, 1981a). The shutdown of the two Three Mile Island plants has cost the utility millions of dollars in lost revenue and forced users to purchase more expensive electricity from other utilities. The owner of Salem-1 lost $330,000 a day during the 2 months that the plant was closed (Marshall, 1983d). In addition to these direct costs, the Three Mile Island accident caused a serious loss of confidence in nuclear power by both the public and investors. Also as a result of this accident an estimated 0.75% risk premium has been added to utility financial offerings (Marshall, 1981a). Any future large nuclear accidents will seriously cripple the already damaged nuclear power industry.

Developing a More Comprehensive EROI

As the introductory quote by Nobel laureate Alfven points out, building and operating a nuclear plant is considerably more difficult than planning one on paper. One result is higher dollar and energy costs than had been anticipated by the industry. The problems cited in this appendix could cause significant changes in the EROIs calculated for nuclear power (Table 12.5). The magnitude of these changes can be estimated very crudely by using the national mean energy:dollar ratio (p. 110) to convert dollar costs to energy costs. Such an analysis shows that if decommissioning costs are included for a plant with a 5:1 EROI, the EROI is lowered from 5 to 4.94 (Table 12.5). For a plant experiencing steam tube or embrittlement problems, the EROI would be lowered to 4.92 or 4.91, respectively. Finally, the EROI for cleanup and decommissioning of a reactor that had experienced a TMI-type accident would be lowered to 3.95, a 20% reduction in net energy yield. Even though this analysis is only approximate, it still neglects other important factors such as lowered plant efficiency, the possibility of shorter operating lives, the costs of storing high level wastes, and possible health effects. Therefore these EROIs are probably still too high.

Although all of the problems discussed in this section lower the EROI of nuclear energy, they are probably less important than the effect of declining fuel quality; and there are other potential changes that could counteract these problems to a degree. Reprocessing of fuel would probably increase the energy return of nuclear power, and incorporating knowledge gained through years of experience will probably lead to safer, less accident-prone reactors. For example, reactors being built in Japan incorporate the newest in nuclear technology and should therefore be safer and more efficient (Marshall, 1983a). Designs for *ultrasafe* reactors have been developed in Sweden and elsewhere, although the U.S. nuclear industry has shown little interest in switching to them. Experience to date indicates that nuclear energy has substantial energy costs, including some very large costs that are still unknown. The future of conventional nuclear power as a fuel source will depend largely on the ability of the industry to identify, assess, and rectify these problems to the degree that society demands.

APPENDIX 12.2: HOW DANGEROUS IS NUCLEAR POWER?

One of the reviewers of this chapter was Dr. Robert Bachman, who has had a long and distinguished career in the United States Navy, where he was, among other things, second-in-command of a nuclear-powered, nuclear-armed submarine. He now teaches a popular environmental science course at Penn State and he makes it very clear in that course that despite his nuclear experience he is strongly opposed to nuclear power. Dr. Bachman liked our chapter, but felt it was missing something, which, as is his manner, he let us know in no uncertain terms. With his permission we have reproduced the following from his review:

> Slow reactors, those that use U-235 as a fuel, and hence require a moderator, are marvels of engineering achievement, and are inherently safe because of the negative feedback of the moderator, water. But, fissioning of U-235 results in some 30 pairs of radioactive fission fragments, each of which has its own characteristic decay chains and radioisotopes. About 6% of the energy of a nuclear power plant comes from the radioactive decay of the fission fragments. Immediately after the plant is shut down, or scrammed, all fission stops, but the core continues to generate energy at 6% of the rate it was producing when on line. That means that a 1000 MW plant continues to generate *60 MW* of power with all of the control rods fully inserted. That's one hell of a lot of energy, and it must be continuously removed or the core will begin to melt within seconds. (If coolant is interrupted while fissioning, melting starts in a fraction of a second.) As long as cooling water (or sodium, if you are crazy enough to want to use sodium for a coolant) is supplied, and the decay heat is removed, the plant is undoubtedly the safest and the most environmentally acceptable way to make electricity. *BUT*, no acts of God are allowed (Alfven, 1974). The decay heat must be removed continuously. If not, the core *will* melt, and if it does, even for only a few minutes, restoring flow of coolant will *not* assure that the core will be properly cooled.

> That's what happened at Three Mile Island. Feedwater to the steam generators was interrupted, the steam generators boiled dry, a primary pressure relief valve stuck open, and the "incredible" happened. As a result of a series of operator errors, the reactor core was partially uncovered, and some of the core melted (Cantelon and Williams, 1982). *That the core failed to "melt down" in a China syndrome fashion seems to have been more a matter of luck than design.* Various accounts suggest that the core came within 20 to 30 minutes of a meltdown, and that the actions of the operators during the first two hours of the casualty *caused* rather than prevented the uncovering of the core (Ford, 1982; Martin, 1980). Indeed, recent studies reveal that part of the core reached 4800 degrees Fahrenheit, just 280 degrees short of a meltdown (*Centre Daily Times* (Pa.), November 7, 1984).

> Few people seem to know what is meant by a meltdown. The core of a typical pressurized water reactor such as that at Three Mile Island is composed of uranium dioxide pellets enclosed in zirconium alloy tubes. At the high temperatures encountered in the Three Mile Island accidents, water splits into hydrogen and oxygen, zirconium begins to oxidize, and the zirconium cladding ruptures releasing radioactive fission fragments into the primary loop. At about 5100 degrees, the uranium pellets start to melt, and various exergonic chemical reactions occur, causing the temperature to rise even higher (Ford, 1982). At such temperatures, water ceases to be a coolant, but rather becomes a *fuel*.

> A typical pressurized water reactor core contains close to a hundred *tons* of uranium dioxide. The amount of radioactivity in a mature core is mea-

sured in billions of curies, with radioisotopes having half-lives ranging from seconds to thousands of years. There is no physical or chemical means of turning off radioactivity, and no theoretical upper limit to the temperature or pressure that could be generated. How much of this murderously dangerous material would be released to the environment in a meltdown is unknown because no such test has been (or can be) conducted. The amount would vary with the core history (how long it had been operating at full power before the accident), the degree of confinement of the molten (or vaporized) core, the geological formations under the reactor site, and the proximity and amount of groundwater. It is almost certain that the most severe meltdown accidents would release vaporized radioactive materials into the atmosphere. Some studies have suggested that a major meltdown could create a disaster area the size of the state of Pennsylvania (Ford, 1982). No one can calculate the *probability* of such a meltdown but the *possibility* is certain.

A reactor need not melt down for a catastrophe to develop. At Three Mile Island, enormous quantities of radioactive material were released from the fuel rods and found their way onto the floor of the containment building. Approximately a quarter of a million gallons of highly radioactive water sat for months on an island in a large river feeding one of the most extensive estuarine systems of the eastern United States. The accident at Three Mile Island was a Class IX type accident, so unlikely that its occurrence was considered not to be "credible." One can only speculate on the ecological, economic, and social consequences the accidental spill of the water into the Susquehanna River would have had.

References

Alfen, Hannes (1974). Fission Energy and Other Sources of Energy. *Bulletin of the Atomic Scientists,* January, pp. 4–8.

Cantelon, Philip L. and Robert C. Willams (1982). Crisis contained: The Department of Energy at Three Mile Island. Southern Illinois University Press, Carbondale and Edwardsville.

Ford, Daniel F. (1982). *Three Mile Island: Thirty Minutes to Meltdown.* The Viking Press, New York.

Martin, Daniel (1980). *Three Mile Island: Prologue or Epilogue?* Ballinger Publishing Company, Cambridge, MA.

13

SOLAR ENERGY

Gary Wayne, with the assistance of Charles A. S. Hall and David Behler

Renewable, or income, energy sources are those that are continuously replenished by the sun or other energies at a rate similar to the rate at which people use them, and that are not depleted by their use as is the case for fossil fuels. Income energy includes direct solar energy and its corollaries (wind, elevated water, etc.) and also tidal power and geothermal power. The solar flux is enormous—some 1.8×10^{18} kcal reach the Earth's surface each day—but dilute. It is a critical and nonreplaceable energy flux for our economic system and performs economic services such as food and fiber production, water renewal, and the production of winds to disperse pollutants. For the majority of humankind solar energy is the principal energy input to economies. Our interest in this chapter, however, is the concentration of solar power so that it can be used to replace presently-used fossil and other fuels in industrial societies. The challenge for many solar advocates is to extend the degree to which we already use the sun. The following sections review the long history of solar technology, describe many present-day options, and consider, where possible, the EROI for these technologies and some of technical improvements that may be able to improve greatly the degree to which we can use the sun.

HISTORICAL PERSPECTIVE

As early as the fifth century B.C. the ancient Greeks were experiencing a fuel shortage. The forests surrounding heavily populated portions of Greece were depleted due to the increased and continued demand for wood for heating, cooking, shipbuilding, and house construction. By the fourth century B.C. legislation had been passed in Athens in response to the growing energy shortage regulating the use of firewood and charcoal and protecting the valuable olive trees. The Greeks, familiar with sundials and the seasonal path of the sun, began to build their homes and public buildings accordingly (Figure 13.1). Socrates explained:

> In houses that look toward the south, the sun penetrates the portico in the winter, while in the summer the path of the sun is overhead and above the roof so that there is shade. (Figure 13.2.)

Passive solar design was apparently the norm in much of ancient Greece. Modern excavations of planned Greek cities from that era revealed that the Greeks, sometimes overcoming very adverse topography, made every attempt to face their buildings toward the south. Their planned cities were so arranged that no building blocked another's view of the sun. The playwright Aeschylus considered this form of building a mark of civilization when he wrote:

> Though they had eyes to see, they saw to no avail; they had ears but understood not. But like shapes in dreams, throughout their time, without purpose they wrought all things in confusion. They lacked the knowledge of houses turned to face toward the sun, dwelling beneath the ground like swarming ants in sunless caves.

Solar architecture became possible in less temperate climates when the Romans introduced the concept of transparent window coverings of glass and mica. The Romans, who were especially fond of bathing, used the sun to heat their heliocami, or hot baths. The Justinian Code of Law specified:

> If any object is so placed as to take away the sunshine from a heliocamus, it must be affirmed that this object creates a shadow in the place where the

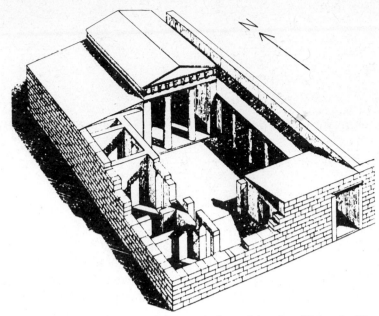

Figure 13.1. Reconstruction of a classical Greek home, from excavations of the city of Priene by Theodore Wiegand. The rooms behind the portico faced south into the courtyard. They receive sunshine in the winter when the sun is low in the sky but are in shade in summer (shown here) when the sun is higher. (From Buti and Perlin, 1980.)

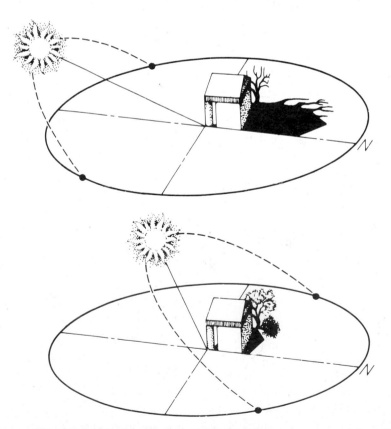

Figure 13.2. Path of the sun on December 21 (top) and June 21 (right), the winter and summer solstices. Solar position is shown for noon on both days. (From Buti and Perlin, 1980.)

sunshine is an absolute necessity. Thus it is in violation of the heliocamus' right to the sun.

It was known to the ancient geometers thousands of years ago that a mirrored parabolic surface would concentrate the rays of the sun into an intensely hot beam. Leonardo da Vinci is credited with formulating the earliest known plans for the industrial application of solar energy in the 16th century using this idea, and he proposed the use of large parabolic mirrors to create process heat for a dye factory. The idea of *burning mirrors* fascinated intellectuals for hundreds of years, but its refinement was limited by the inability of metal workers of the time to fashion single pieces of copper or brass much larger than half a meter in diameter. This limitation was finally overcome in the late 1700s when Peter Hoesen, Dresden's royal mechanic and carpenter, constructed a focusing mirror 3 m in diameter by fastening preshaped sections of polished brass to a hardwood frame (Figure 13.3). An experimentalist using a Hoesen mirror of less than 2 m in diameters discovered: ". . . copper ore melted in one second, lead melted in the blink of an eye, asbestos changed to a yellowish green glass after only three seconds,

and slate became a black glassy metal in twelve seconds.'' A writer of the time in London concluded there is hardly anything that is not destroyed by this fire (Butti, 1980). The practical application of this form of solar energy in preindustrial society was extremely limited.

By the middle of the 19th century coal-fired heat engines had become the prime mover of industrial society. The potential of direct solar power was forgotten by all but a few because coal had a much higher EROI than solar and could therefore do more economic work while producing the heat at all hours of the day and night year round. Coal of course had its limitations. Mining coal was very hazardous and unhealthy. Coal combustion was environmentally devastating, and coal is found only in certain geographical areas. In areas where coal had been mined for long periods of time, the effects of diminishing returns were experienced. In England, James Watt understood the implication of diminishing net energy return when he observed that more and more of the coal extracted from the ground had to be burned to power the pumps that were needed to empty water from increasingly deep coal mines. As abundant as coal appeared to be, the great

Figure 13.3.   Large concentrating solar mirror or burning mirror of the late 1700s. Like Peter Hoesen's prototype, this mirror was built in sections and made use of existing skills and local materials. A machine of this design and size could produce temperatures on the order of 3000°C and 3 kW of power. (From Buti and Perlin, 1980.)

Swedish-American engineer John Erickson wrote in 1868 that only the development of solar power could avert an eventual global fuel shortage. In 1872 Erickson, the designer of the ironclad battleship *Monitor,* applied direct heat from the sun to drive an external combustion air engine of his own design. He wrote:

> The world moves—I have this day seen a machine activated by solar heat applied directly to atmospheric air. In less than two minutes after turning the reflector toward the sun, the engine was in operation. As a working model I claim that it has never been equaled: while operating by direct application of the sun's rays it marks an era in the world's mechanized history. (Figure 13.4.)

By the early 20th century technology had advanced to the point that some of the limitations of solar power could be overcome, and solar energy commercialization had begun on several fronts. Frank Shuman, an American inventor, formed a corporation to market a solar-powered irrigation system to the sun-rich and coal-poor arid regions of the world, but the start of World War I ended his plan. The use of solar energy to heat domestic hot water was becoming widespread nearly 100 years ago, and by the turn of the century, 1600 units were in place in Southern California. For an initial investment of $25, the owner could save $9 of fuel per year. Storage and collector were combined in one unit so that hot water was available day or night

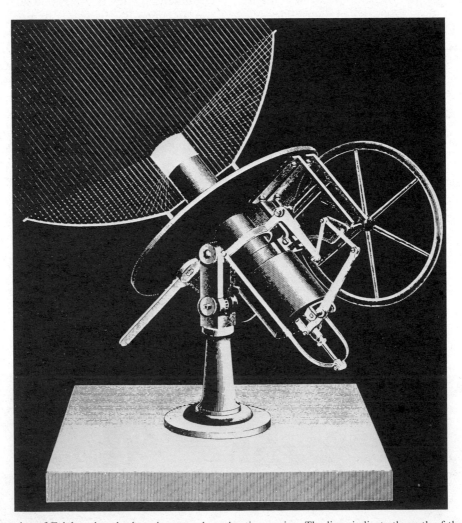

Figure 13.4. Engraving of Erickson's solar hot air external combustion engine. The lines indicate the path of the parallel incoming rays of sunlight being focused on an absorber surface. Because this engine is heated externally, any source of heat (solar, coal, natural gas, etc.) can power such a device. This means that for a small added cost the machine can run continuously as a conventional device might without any storage and still take full advantage of solar energy as it is available. External combustion engines, because they do not need valves to admit and expel fuel and exhaust, are very quiet. (From Buti and Perlin, 1980.)

Figure 13.5.   A Maryland gentleman of the 1890s enjoys a steaming hot bath provided by his solar water heater (shown on the roof above). (From Buti and Perlin, 1980.)

(Figure 13.5). The collectors were produced locally by many competitive manufacturers and by 1920, these units had become very common. But all this was changed by the discovery in Southern California of huge oil and natural gas fields which made cheap petroleum with very high EROI available. Gas companies trying to attract many new customers offered free installation, low rates, and financing. By the thirties the first solar boom in America had ended (Butti, 1980).

Thus the use of solar energy is not new, and at the present time the technical feasibility of many solar technologies has been well established. The question that remains is whether solar energy can be utilized to produce a significant quantity of economic energy for a modern industrial nation at an acceptable cost in dollars and energy. The next sections give a more comprehensive view of what constitutes the solar resource and discusses some of the principal solar devices most likely to make a significant contribution of net energy to society. It should be noted that solar energy use is not dominated by a single or small group of related ideas but is characterized by thousands of different possible approaches.

THE SOLAR RESOURCE

Solar energy is the product of thermonuclear fusion on the sun. It arrives at the Earth in the form of electromagnetic radiation distributed over a band of wavelengths principally between 300 and 2000 nanometers (nm) (Figure 13.6). At the upper reaches of the Earth's atmosphere the intensity is approximately 1.3 kW/m^2 (1120 kcal/m^2·hr), an amount of energy flux termed the solar constant. At sea level in a standard atmosphere this value is reduced by scattering and absorption of the sun's radiation to a maximum value on a clear day of about 1.0 kW/m^2. Instantaneous insolation is affected by local weather conditions and the changing incidence of the rays of the sun on its daily and seasonal paths across the sky (Figure 13.7). Light from the sun is composed of direct and diffuse light. Direct light travels in a coherent beam and can be focused or concentrated. Diffuse light is more random. It cannot be focused, and it does not leave a

distinct shadow. In cloudy environments a large fraction of solar energy is diffuse.

Although solar energy is an unconcentrated source, the total amount falling on the United States is 10,000 times greater than our fossil energy use. Therefore the use of solar energy will not be limited by scarcity but by the need for locally concentrated and instantly available energy and by the investment costs of solar energy systems. The seasonally available average level of total insolation for the United States is given in Figure 13.8.

Variation in solar radiation affects the instantaneous output of most individual solar devices on time scales of minutes to hours. For this reason energy storage is necessary for many solar energy systems. But storage problems are not limited to solar energy. The largest thermal power stations in this country are not available approximately 35% of the time due to maintenance and refueling (Lovins, 1977). Down times are on the order of weeks and months. The steady supply of conventional sources of electric power is possible only by integrating many sources to form a network. Similarly integrating different sources of solar and conventional energies can mitigate costs of storing solar energy. The availability of wind and hydropower in the United States shown in Figures 13.9 and 13.10 indicates how such an integration might occur. Areas of low insolation like the northeastern and northwestern United States tend to be areas of high wind potential, large realized and potential hydropower, and/or a large biomass resource. No area is without some large renewable resource base. Most areas have moderate amounts of all of these resources. Additionally, although high available sunshine is generally a summer phenomenon, high and prolonged wind speeds generally occur in the winter. In New York State, for example, average June direct normal insolation is 6 kWh/m^2·day, or two to three times the December value. Winter winds, however, are twice as powerful as average summer winds in this region. Finally, liquid and solid fuels derived from biomass are easily stored for year round use. Another factor that can help reduce the ill effects of intermittent availability is that almost all areas experience daily peak power demands in the middle of the day. Perhaps not coincidently, average solar and wind energies tend also to peak near the middle of the day. The minimum availability of these energies occurs at night when most human energy-using activity is also at a minimum. All of these characteristics enhance the possibility of load-source matching.

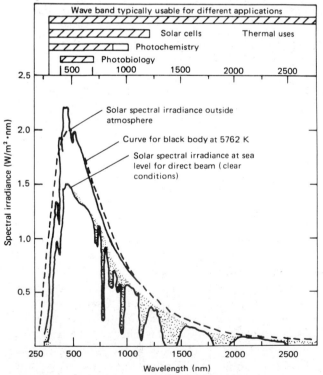

Figure 13.6. Spectral irradiance curves for direct sunlight just outside of the Earth's atmosphere (upper curve) and at sea level. The drop in the curve indicates the amount of energy lost due to absorption and reflection of sunlight traversing our atmosphere. The dashed line indicates the theoretical output of a sun which behaved liked a black body of perfect emitter–absorber.

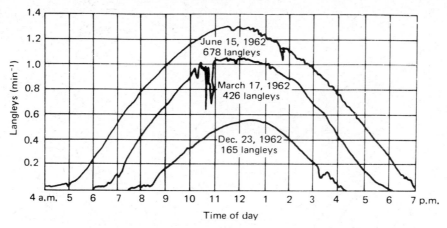

Figure 13.7. Typical solar radiation on clear days at 43° north latitude (between New York and Boston); 1.43 langleys/min equals 1 kW/m². The larger availability nicely compliments the summer availability of wind and water power (see Figure 13.18).

Solar energies will probably not provide the total quantity of economic energy anticipated by some projections. If, however, future energy requirements are not significantly higher than they are today, as many of the most recent projections indicate, then solar sources have the potential to provide a substantial fraction of our total energy budget (Lovins, 1980, Table 13.1).

SOLAR ENERGY COLLECTION

Many components of solar energy systems are similar to conventional energy systems. The one distinguishing feature of every solar power system is the collector, the device that intercepts incoming solar radiation. Solar collecting concepts number in the hundreds but generally can be categorized as either tracking or nontracking, concentrating or nonconcentrating. Tracking refers to rotating a collector to follow the sun's movement. The appropriate solar collector for a given application is determined to a large extent by the form of output desired, for example, low temperature heat, high pressure steam, or electricity, and by the local weather characteristics.

Flat Plate Collectors

The simplest and most common collector is a flat plate collector, which is simply an insulated box consisting of a dark absorber plate with attached pipes that circulate a fluid to remove useful heat.

Glazing the collector surface improves its performance in two ways: (1) by greatly reducing convective and conductive heat loss to the surroundings and (2) by selectively transmitting electromagnetic radiation. Light from the high temperature sun is of a wavelength predominantly within the transparent range of glass, which is about 400 nm. Approximately 90% of incident solar energy is transmitted through the glass, but much of the reradiated energy from the cooler absorber plate is longer than the 400 nm transmission cutoff. Thus the glazing appears transparent to incoming radiation but opaque to this reradiation, and as a result heat is not lost. Flat plate collectors can achieve working temperatures of about 70°C. Absorber plates can be made selective, absorbing more than they reradiate, with the application of a surface coating such as black chrome, and this improves their efficiency. Double glazed collectors with selective coatings can achieve operating temperatures near 150°C and can achieve high efficiency in cold climates where a large fraction of the radiation is diffuse. Individual flat plate collectors cost approximately $20–$100 (1980 dollars) per square meter. Many cost-reducing ideas exist, but, in general, the price of these devices depends largely on their level of production which presently is low.

Concentrating Collectors

Concentrating collectors focus incoming radiation on a small area. Normally this is done through the use of a parabolic reflector, which have achieved

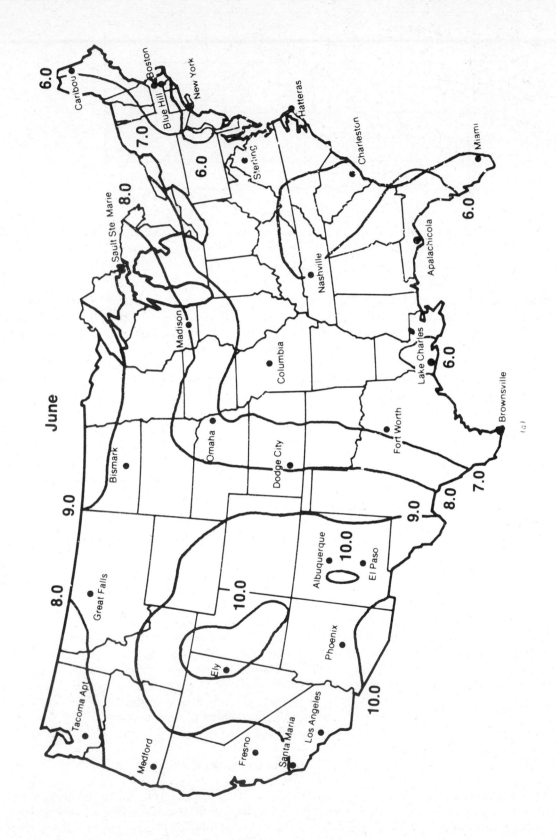

June

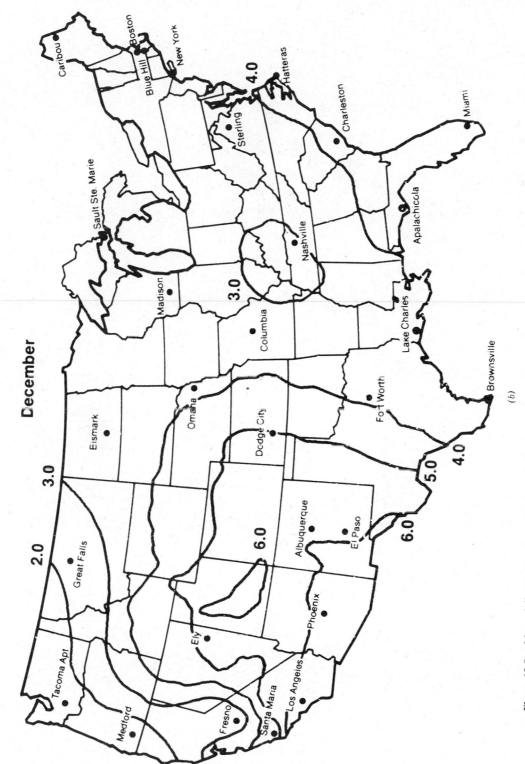

Figure 13.8. Mean daily total horizontal solar radiation in June (a) and December (b). Note that a tilted collector that is fixed at the optimum angle or a tracking collector will have larger fluxes than those indicated. (From OTA, 1978, and Boes, et al., 1976.)

(b)

293

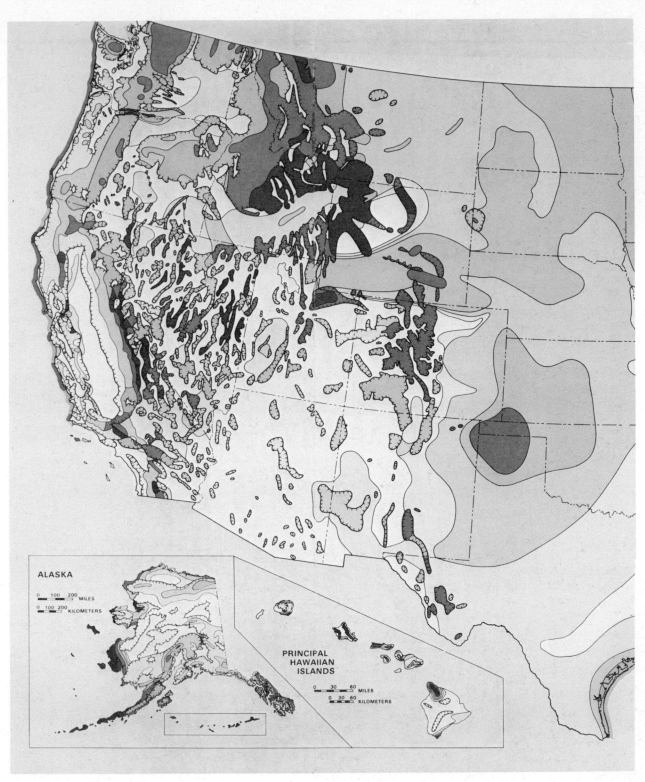

Figure 13.9. Annual average wind power in the United States. Winds are highest in coastal and mountain areas and on the great plains. Wind power is highest generally in winter when solar energy is lower and occurs in areas of low solar insolation. (From Battelle Memorial Institute.)

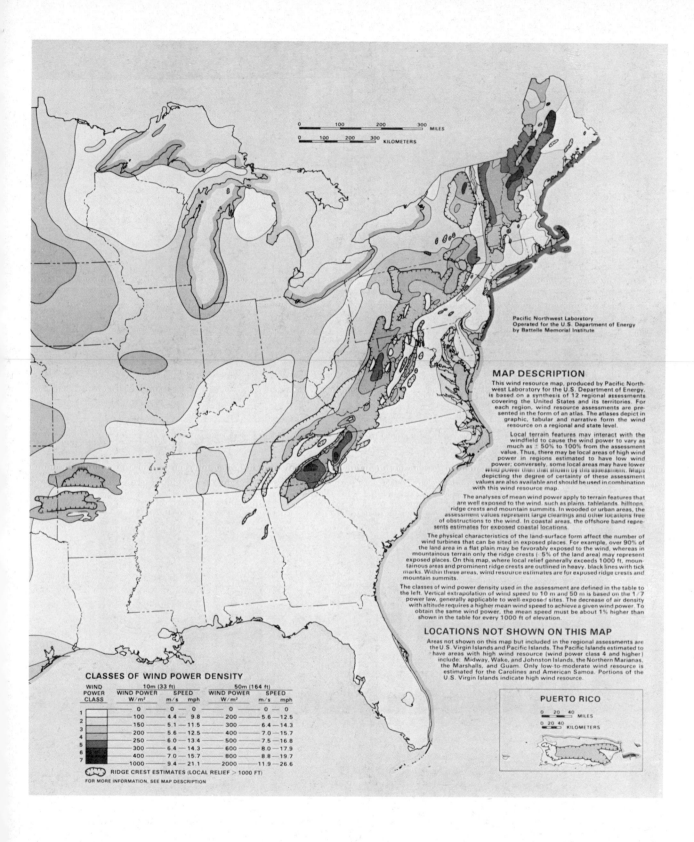

MAP DESCRIPTION

This wind resource map, produced by Pacific North-west Laboratory for the U.S. Department of Energy, is based on a synthesis of 12 regional assessments covering the United States and its territories. For each region, wind resource assessments are presented in the form of an atlas. The atlases depict in graphic, tabular and narrative form the wind resource on a regional and state level.

Local terrain features may interact with the windfield to cause the wind power to vary as much as ± 50% to 100% from the assessment value. Thus, there may be local areas of high wind power in regions estimated to have low wind power; conversely, some local areas may have lower wind power than that shown by this assessment. Maps depicting the degree of certainty of these assessment values are also available and should be used in combination with this wind resource map.

The analyses of mean wind power apply to terrain features that are well exposed to the wind, such as plains, tablelands, hilltops, ridge crests and mountain summits. In wooded or urban areas, the assessment values represent large clearings and other locations free of obstructions to the wind. In coastal areas, the offshore band represents estimates for exposed coastal locations.

The physical characteristics of the land-surface form affect the number of wind turbines that can be sited in exposed places. For example, over 90% of the land area in a flat plain may be favorably exposed to the wind, whereas in mountainous terrain only the ridge crests (5% of the land area) may represent exposed places. On this map, where local relief generally exceeds 1000 ft, mountainous areas and prominent ridge crests are outlined in heavy, black lines with tick marks. Within these areas, wind resource estimates are for exposed ridge crests and mountain summits.

The classes of wind power density used in the assessment are defined in the table to the left. Vertical extrapolation of wind speed to 10 m and 50 m is based on the 1/7 power law, generally applicable to well-exposed sites. The decrease of air density with altitude requires a higher mean wind speed to achieve a given wind power. To obtain the same wind power, the mean speed must be about 1% higher than shown in the table for every 1000 ft of elevation.

LOCATIONS NOT SHOWN ON THIS MAP

Areas not shown on this map but included in the regional assessments are the U.S. Virgin Islands and Pacific Islands. The Pacific Islands estimated to have areas with high wind resource (wind power class 4 and higher) include: Midway, Wake, and Johnston Islands, the Northern Marianas, the Marshalls, and Guam. Only low-to-moderate wind resource is estimated for the Carolines and American Samoa. Portions of the U.S. Virgin Islands indicate high wind resource.

CLASSES OF WIND POWER DENSITY

WIND POWER CLASS	10m (33 ft)			50m (164 ft)		
	WIND POWER W/m²	SPEED m/s	mph	WIND POWER W/m²	SPEED m/s	mph
	0	0	0	0	0	0
1	100	4.4	9.8	200	5.6	12.5
2	150	5.1	11.5	300	6.4	14.3
3	200	5.6	12.5	400	7.0	15.7
4	250	6.0	13.4	500	7.5	16.8
5	300	6.4	14.3	600	8.0	17.9
6	400	7.0	15.7	800	8.8	19.7
7	1000	9.4	21.1	2000	11.9	26.6

RIDGE CREST ESTIMATES (LOCAL RELIEF > 1000 FT)

FOR MORE INFORMATION, SEE MAP DESCRIPTION

PUERTO RICO

0 20 40 MILES

0 20 40 KILOMETERS

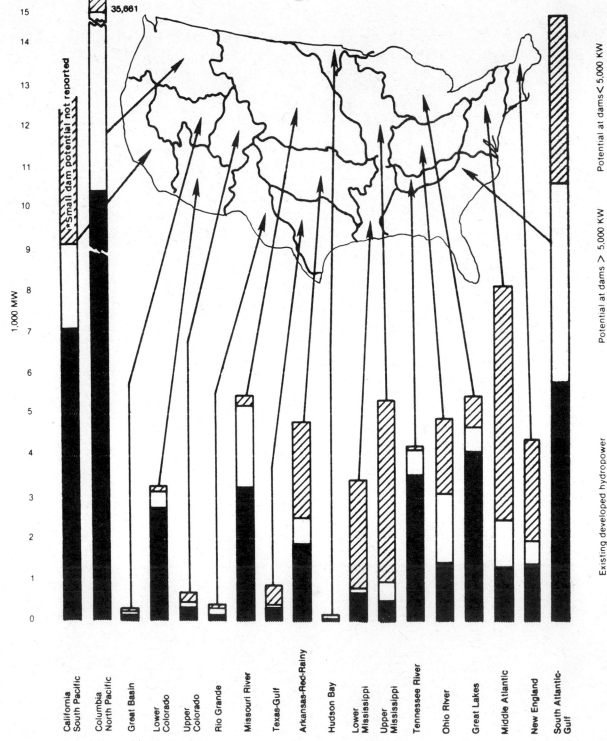

1,000 MW

36,418

35,661

*Small dam potential not reported

California
South Pacific

Columbia
North Pacific

Great Basin

Lower
Colorado

Upper
Colorado

Rio Grande

Missouri River

Texas-Gulf

Arkansas-Red-Rainy

Hudson Bay

Lower
Mississippi

Upper
Mississippi

Tennessee River

Ohio River

Great Lakes

Middle Atlantic

New England

South Atlantic-
Gulf

Potential at dams< 5,000 KW

Potential at dams > 5,000 KW

LEGEND

Existing developed hydropower

296

operating temperatures of 6000°C. This is close to the theoretical limit of concentration which occurs when a perfectly focused image of the sun is formed. The higher the desired temperature of thermal output, the more one must concentrate the reflected or focused beam and consequently the more precisely one has to track the sun's movement. Although high concentration can be more useful, it is more expensive to produce and cannot use diffuse light. Figure 13.11 illustrates a fully tracking concentrating collector. Although the sensors and motors required for tracking are expensive, many ingenious designs exist that simplify tracking collectors. Figure 13.12 shows a less expensive collector that tracks along only one axis and needs to move only a small amount each day. Figure 13.13 illustrates a collector capable of moderate concentration without any moving parts at all and using diffuse light. The cost of standard concentrating collectors ranged from \$100–\$500/m^2 in 1980.

APPLICATIONS OF SOLAR ENERGY

In the following section we will discuss a few of the most significant and promising applications of solar energy in order of increasing thermodynamic quality. Table 13.1 gives several projections of energy use and table 13.2 lists total end use of energy in the United States and four other industrial nations. This pattern is not qualitatively different from present energy end use. The major portion of energy end use in all of these countries is thermal. Countries less dependent on personal transportation than the United States have greater proportional needs for heat. Because solar energy is most efficiently converted to heat rather than either electricity or mechanical drive, and because the majority of energy needs of much of the world are thermal, the potential use of solar energy is very large. The United

States, because it is highly industrial and because the bulk of its industry is located in middle latitudes, forms a conservative basis to judge the potential contribution of solar energy to other nations' total energy needs. Table 13.3 lists the amount and form of energy used by different sectors of the U.S. economy. The total in 1973 (less energy used for feedstocks and thus having no direct substitution) is approximately 56.33 quadrillion BTUs, or quad. Low-temperature heat, below 100°C, constitutes about 19.49 quad, or 35% of the total. Half of the delivered energy in West Germany is in the form of heat less than 100°C. These temperatures can be achieved by the most simple and cost-effective nontracking, flat plate collectors. Low-temperature heat is stored easily and inexpensively, and in the case of domestic hot water it is already stored even when conventional fuel sources are used.

Solar Heating

Commercial and residential space heating accounts for nearly 20% of the total delivered energy in the United States. To maintain a house at constant temperature, all the sources of heat gain, including sunlight, internal gains such as cooking, lighting, and human metabolism, and auxiliary sources like the furnace must equal all the heat lost through the windows, walls, and so on. To minimize the consumption of delivered heating energy, we always have two options: we can either increase the gains or decrease the losses. It is not commonly appreciated that all buildings are solar heated to some extent. Thus any reduction of heat losses or increase in efficiency represents an intrinsic increase in the building's solar contribution. If even the most basic principles of better orientation and insulation were applied to new housing, a solar contribution of 40% is possible. If this were done for new housing in this country for 12 years, the cumulative gross saving of primary energy over an assumed 30-year life cycle of the houses would be approximately 54 quad, or about the same amount of gross recoverable energy from the Alaskan North Slope oil fields (Bliss, 1978). When similar but more comprehensive principles of solar design are applied for small additional cost and difficulty, the solar fraction can be increased from 40 to 80% even in cloudy winter climates. Solar building ordinances are beginning to appear throughout the country. Just as in Greek cities, in Davis, California, new construction can-

Figure 13.10. Conventional hydroelectric capacity potential at existing dams. The United States could potentially double its hydroelectric capacity at sites with existing dams by upgrading existing plants, restoring plants out of service, and adding capacity to dams that were built for other purposes and do not presently produce any power. Much of this capacity could be added at small facilities rated less than 5000 kW. It is estimated that another third of capacity could be reasonable developed at sites that do not presently have dams. (From OTA, 1978, based on McDonald, 1977.)

Figure 13.11. A complete solar power plant that could be located on the roof of a small manufacturing facility is built around this 6 m diameter tracking parabolic dish. Sunlight is concentrated 10,000 times. The system uses solar-heated steam to operate a 7.5 kW generator and also can cogenerate 8.8 kW of 82°C hot water. The large hot water storage tank below the collector might easily be insulated and placed below ground as are most septic tanks now. (From *Solar Energy in America*, eds., W. D. Metz and A. L. Hammond, 1978, Fig. 35. Copyright 1978 by AAAS.)

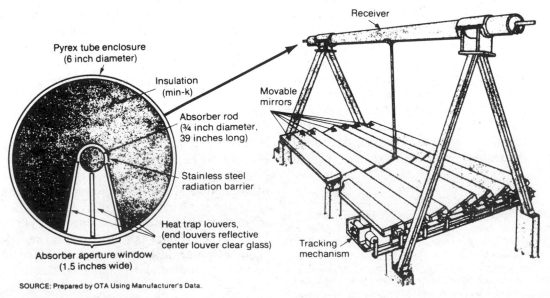

SOURCE: Prepared by OTA Using Manufacturer's Data.

Figure 13.12. The high performance receiver design. In this design the absorber is stationary while small louvered mirrors track and focus the sun to a line. The small nearly flat mirrors merely rock back and forth on one axis to track; only small amounts of power would be required to track the sun.

298

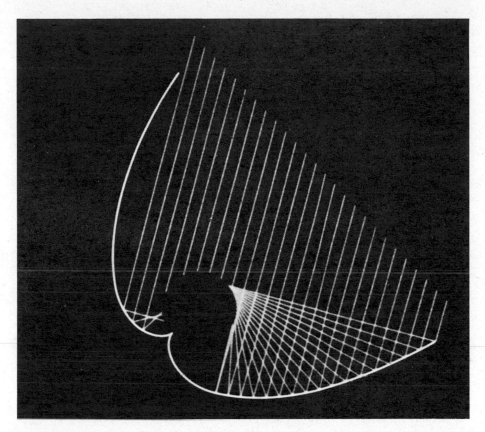

Figure 13.13. Schematic diagram showing how solar radiation is concentrated onto the collector tube using a compound parabolic concentrator. Because of its shape, the fixed collector can collect and concentrate light from a wide range of angles without any tracking mechanism. Systems like this might be manufactured in factory insulated, waterproof panels that could be used instead of a conventional roof material and collect solar energy at the same time.

not block solar gain to existing solar homes. In Boulder, Colorado, all new residential construction must include some passive solar heating.

Solar housing design, because it depends on naturally occurring and uncontrolled sources, is almost always site specific, although general principles exist. The two design points of most solar construction are to (1) achieve a comfortable average or equilibrium temperature by balancing heat gains and losses over the entire heating and cooling season and (2) minimize temperature fluctuations about this average temperature. In general, 75% of the heating requirements for a building occurs at night when the sun doesn't shine. Thus some heat received during the day should be stored for nighttime use. Solar gain is generally achieved with south-facing (in Northern Hemisphere) windows. Not only are winter gains maximum for south-facing collec-

tors when heat requirements are greatest in winter, but a south-facing vertical window with appropriate overhang receives little heat during the summer. The same principles of solar heating: good insolation, proper orientation, and increased thermal inertia (heat storage) also reduce building cooling loads and thereby reduce air conditioning cost as well. A well-designed solar building can also induce cooling breezes through the house in the summer.

Active versus Passive

Two distinct design strategies are commonly employed in solar construction. *Active* heating refers to the use of mechanical devices to achieve heat transfer. Externally mounted collectors are used to capture heat. Blowers and either ductwork or plumbing transfers this heat to the building. *Passive* heating uses natural mechanisms of radiation and

Table 13.1. Selected Estimates of Approximate U.S. Primary Energy Demand in 2000 (1973 Use = 70 Quads)
$(Q/y \equiv 10^{15} \text{ BTU/yr} \sim 10^{18} \text{ J/yr} \sim 33 \text{ GW}_t \sim .252 \cdot 10^{15} \text{ kcal/yr})$

Year in Which Analysis Was Published	Beyond the Pale	Sociological Category of Analyst		
		Heresy	Conventional Wisdom	Superstition
1972	125[a]	140[b]	160[c]	190[d]
1974	100[e]	124[f]	140[g]	160[h]
1976	75[a]	89[i]–95[j]	124[g]	140[h]
1978	33[k]	63[m]–77[m]	95[n]–96[m]–101[p]	123[q]–124[r]

Source: A. B. Lovins as cited by E. Marshall, *Science* **208**:1353–56 (20 June 1980), with typographical errors corrected and refs. *n* and *q* added.

Note: The agencies, institutes and corporations with direct involvement in the regulation and production of energy consistently project high energy end use. These same projections are revised downward at two year intervals as the economic advantage of conservation and efficiency improvements continues to be perceived.

[a] A. B. Lovins speeches (in 1972, J. P. Holdren was perhaps the only analyst estimating below 100 q/y).

[b] Sierra Club (a major national conservation group).

[c] U.S. Atomic Energy Commission.

[d] Federal Power Commission, Bureau of Mines, other Federal agencies (some were reportedly as high as 300 q/y, and Exxon was about 230 q/y).

[e] Ford Foundation Energy Policy Project, "Zero Energy Growth" scenario.

[f] Ford Foundation Energy Policy Project, "Technical Fix" scenario.

[g] U.S. Energy Research & Development Administration (forerunner of USDOE).

[h] Edison Electric Institute (and, generally, Electric Power Research Institute).

[i] F. von Hippel and R. H. Williams (Princeton University).

[j] A. B. Lovins, *Foreign Affairs*, October 1976.

[k] J. Steinhart (University of Wisconsin) for 2050 with lifestyle changes.

[m] CONAES Demand & Conservation Panel (*Science*, 14 April 1978), scenarios I–III (II and III are pure technical fixes) for 2010 assuming doubled real GNP.

[n] USDOE Domestic Policy Review of Solar Energy, assuming world oil price (1977$) of $32/bbl by the year 2000 (this is scarcely below the 1980 price).

[p] A. M. Weinberg (Institute for Energy Analysis, Oak Ridge), "low" case.

[q] As ref. *n*, but averaging the $18/bbl and $25/bbl cases.

[r] R. Lapp (prominent U.S. advocate of nuclear power and of fixed energy/GNP ratio).

free convection to transfer heat. Walls and floors are used in passive solar buildings to store excess heat, and windows are used in place of attached solar collectors with a great reduction in cost compared to active systems. In other words, sensible building design in passive buildings replaces expensive equipment. In passive homes south-facing windows, located adjacent to indoor spaces, provide a direct gain of sunlight and a large amount of natural lighting (Figure 13.14). Passive design has many advantages. Good integration of design with the other elements of a house, such as the use of walls as heat storage and windows as collectors, is less expensive than active devices. Passive systems have no moving parts to wear out or fail, so they are reliable and require no maintenance. Because there are no

Table 13.2. Approximate Deliveries of Enthalpy by End Use (percentage)

Category	U.S. 1973		Canada ~ 1973		F.R.G. 1975		France 1975		U.K. ~ 1975	
Heat	58		69		76		64		65	
<100°C		35		39		50		37		55
100–200°C		6[a]		19		6		}27		}10
>200°C		17		11		20				
Mechanical work	38		27		21		31		30	
Vehicles		31		}24		}18		}26		}27
Pipelines		3								
Industrial electric drive		4		3		3		5		3
Other electrical[b]	4		4		4		5		5	

Source: From Lovins (1977).

Note: The United States with our combined heavy reliance on personal transportation and our advanced level of industrialization uses by percent less heat energy than western European nations and substantially less than less industrial nations. Many of the data are rough and preliminary. See text for detailed reservations. The data for France and the U.K. are especially unsatisfactory, though probably among the best available.

[a] Includes appropriate share of process heat (see text), all clothes drying, and two-thirds of cooking.

[b] Includes all lighting, electronics, telecommunications, electrometallurgy, electrochemistry, arc welding, electric drive for public transport and for home appliances, and all other uses of electricity except low grade heating and cooling (under heat) and industrial electric drive (under work).

pumping losses and less reradiation losses due to the cooler, more massive design of the collector (the house is the collector in passive buildings), passive design is more efficient than active. Passive design is nearly always preferred in new building construction, and solar contributions of 80 percent of total heating requirements are obtainable even in unfavorable climates. Payback time for the small added cost of passive design is very short. Finally solar building can take many forms, so public aesthetic or cultural acceptance need not be an obstacle to widespread use. Since 1973 the number of solar homes has doubled every eight months on average, and this trend shows no sign of slowing (Metz, 1978).

Low-Temperature Heat

Solar hot water system retrofits are now competitive with electrically-produced domestic hot water and with deregulated natural gas in many areas. Hot water holding tanks for solar systems now are equipped with built-in electric or natural gas backup which combine needed components of both systems in one less expensive package. Larger commercial systems cost less per unit output due to certain economies of scale. Simple black tanks are common

on the roofs of houses in Israel where they serve to cool houses and warm waters. There is no technical or economic reason that such simple systems could not become widespread in the United States.

Intermediate Temperature Heat

Much of the intermediate temperature heat used in this country occurs in the food, chemical, and textile industries of the sun-rich Southeastern and Southwestern regions of the United States. Active solar technologies could have their greatest and most immediate impact in this area. A study by ITC Corporation concluded that intermediate temperature systems could displace 7.5 quad, or 36% of industrial process heat now supplied by oil or natural gas, by the year 2000 (Warrenton, 1977). There are several good reasons to believe that such a large and rapid conversion could take place. Single-axis tracking systems capable of producing temperatures of 300°C can cost as little as flat plate collectors because only a fraction of the entire collector is actually absorber surface. Tracking can be accomplished by moving only the smaller absorber surface, as is done with the distributed collector concept (Figure 13.15). It has been projected that these collectors could be mass produced for under $100/

Table 13.3.   1973 U.S. Deliveries of Enthalpy by End Use (10^{15} Btu/yr)

Sector	Heating and Cooling $\Delta T < 100°C$	Heating and Cooling $\Delta T \geqslant 100°C$	Portable Liquids	Miscellaneous Mechanical Work	Obligatorily Electrical[a]	Lost at Power Stations
Residential total	8.72	0.62			0.52	4.21
Space heat	6.48					0.71
Water heat	1.54					0.59
Air conditioning	0.32					0.68
Refrigeration	0.38					0.80
Cooking		0.46				0.17
Lighting					0.26	0.56
Clothes drying		0.16			0.01[b]	0.17
Other electrical					0.25	0.53
Commercial total	5.93	0.15			1.05	3.67
Space heat	4.28					
Water heat	0.61					
Air conditioning	0.76					0.87
Refrigeration	0.28					0.59
Cooking		0.15				
Lighting					1.05	2.21
Industrial total	4.84	12.46			2.78	5.58
Process heat	4.84	12.46				0.33
Electric drive					2.34	4.14
Electrolysis and other electrical					0.44	1.11[c]
Transportation total			17.10	1.81	0.02	0.03
Autos, trucks, buses, aircraft, ships			15.42			
Rail			0.58		0.02	0.03
Other			1.10	1.81		
GRAND TOTAL	19.49	13.23	17.10	1.81	4.37	13.49
(percent of 56.33 q)	(35)	(23)	(30)	(3)	(8)	—
(percent of 69.49 q)	(28)	(19)	(25)	(3)	(6)	(19)

Source: From Lovins (1977).

Note: Obligatory electricity, that which has not economic substitute, constitutes only 6% of our gross domestic energy use while heating and cooling at temperature differences less than 100°C constitute 28% of our gross primary energy use. Domestic and residential space and water heating constituted in 1973 nearly three times our obligatory electric end use. The "energy crisis" has been incorrectly perceived as a shortage of electricity by many. Present events indicate that if anything, we are suffering from the effects of financing an excess of electric power.

[a] All these applications are assumed to be obligatorily electrical even though some are likely in practice to be readily substitutable, *e.g.,* by gas mantle or solar lighting or compressed air drive.

[b] Assumed to be the drive (as opposed to heat) requirements.

[c] The 0.33 q of fossil fuel used to generate electricity at industrial sites is arbitrarily assumed to provide its output to "other electrical": the exact mix is unknown.

m² (Metz, 1978). Other factors that favor the expanded use of solar generated process heat are (1) industries already have in-house maintenance staffs that could operate such systems, (2) industry has been quicker to convert to less expensive heat sources than the domestic sector, and (3) the econo-mies of solar thermal systems increase with size up to several hundred kilowatts.

Battelle Labs and Northwestern Mutual developed an irrigation system that uses one-axis tracking parabolic collectors and a Rankine heat engine. Their own market survey indicated that there were

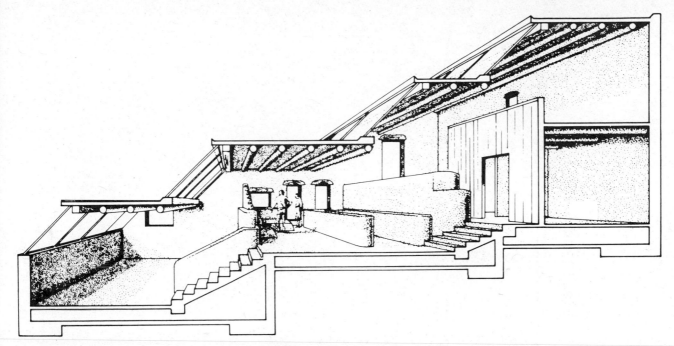

Figure 13.14. A section of the Karen Terry House, Santa Fe, New Mexico, designed by architect David Wright. All of the interior public spaces are naturally lit and heated by south-facing windows adjacent to each space. In this house rooms are divided with low walls made of water drums faced with adobe. Design and low cost heat storage are neatly integrated. The principles of good passive solar design are so simple and general that enormous variation and invention are possible. (Reprinted from *The Passive Solar Energy Book* © 1979 by Edward Mazria. Permission granted by Rodale Press, Inc., Emmaus, PA 18049.)

300,000 fossil-fuel-powered irrigation pumps of similar capacity in use in 17 Western states. Annual energy cost of pumping this water now runs around $700 million per year, and worldwide demand is estimated to be 10 times the U.S. level. Pumping irrigation water seems to be a particularly good solar application because average water demand coincides with peak solar availability and pumped water is very easily stored in ponds.

Absorption air conditioning systems could be the most significant aspect of intermediate temperature systems. This could be a very important application because these devices could help reduce summer electrical peaks which most utilities now experience. Peak cooling loads coincide roughly with maximum solar availability but are out of phase by several hours. These systems combined with small amounts of intermediate temperature storage could reduce the summer peaks substantially. Since many air conditioning systems presently are located on roofs, the necessary space and plumbing for roof-mounted solar air conditioners generally exists. This is a good example of how the direct use of solar energy on site could eliminate the need for the central electric generation that some coal and nuclear energy advocates believe is irreplaceable.

TOTAL ENERGY SYSTEMS

A number of intermediate temperature systems have been proposed to supply several forms of energy. These systems would provide mechanical drive or electricity and simultaneously thermal energy, or in some cases, cooling. When thermal input from the collectors is converted to mechanical drive by heat engines, low-temperature heat can be recovered for other useful tasks. Electricity generation from intermediate temperature collectors could take two forms. Either thermal output could be used to drive a heat engine that turns an electric generator, or direct solar energy could run photovoltaic cells along the focus of a concentrating collector. Cell efficiency is not reduced at high sunlight concentrations. Large electrical outputs can be obtained from a relatively few cells. In this way inexpensive concentrating collectors and a few photovoltaic cells take the place of many expensive

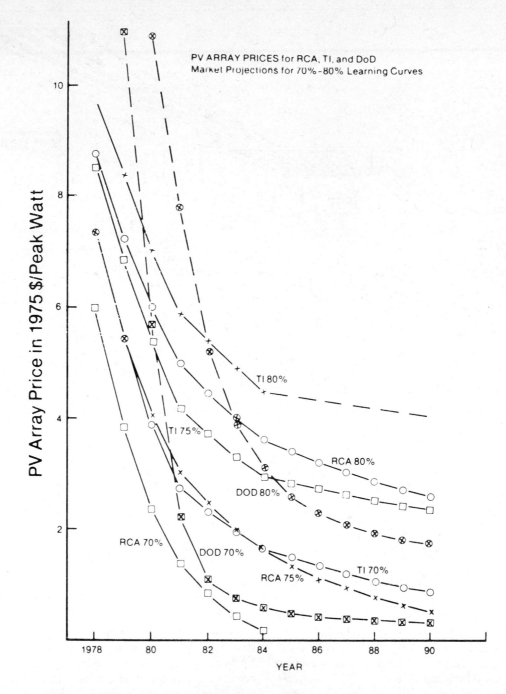

Figure 13.15. Several market projections of the cost of photovoltaic arrays in 1975 dollars per peak watt. (OTA, 1978.)

NOTE: With the exception of the "DOD sales" curves, it has been assumed that the Federal Government spends $129 million for photovoltaic systems between 1978 and 1981, of which 40 percent is actually used to purchase cell arrays (the remainder being spent for design, installation, storage, control systems, and other supporting devices). The forecasts assume that cumulative production of cells through 1975 was 500 kW and that average array prices at the end of 1975 were $15/Watt.

photovoltaic cells. The necessary cell cooling would produce useful cogenerated heat. Because these collectors can be located on site, waste heat from these cells can be used for other purposes. First law efficiency could be quite high, and such a device could be quite inexpensive in rated outputs of up to several 100 kW.

High Temperature and Advanced Design Heat Engines

The Sterling engine, invented and originally patented in 1816 by Robert Sterling, a Scottish minister, may be the best solar thermoelectric technology. This engine is a closed cycle, external combustion engine (combustion taking place outside the engine) similar to the Erickson air engine (Figure 13.4). Because the engine runs on external combustion, any source of high temperature heat, including concentrated sunlight, can supply the necessary operating temperature. The engine could be operated at times of inadequate sunlight either by including high temperature storage or by using a conventional fuel source. Sterling engines could operate from a number of sources of heat and may cost about the same when mass produced as a conventional power plant—but would operate without fuel when the sun was shining. The Sterling engine is extremely quiet because the Sterling cycle is closed and there are no intake or exhaust valve noises. Some air pollutants emitted by a normal combustion engine are not produced by a fossil fuel-driven Sterling engine because the fuel can be burned in an ideal atmosphere and have time to combust completely. Leaded gas fuels would not be necessary. In Sweden, Sterling engines are now used in some city buses. Cogenerated heat and work for a Sterling engine operating in the range of temperature that can be generated by a two-axis collector can approach 80%.

William Beale, an engineer and inventor from Ohio, has made substantial improvements on an advanced Sterling engine concept called the free piston engine. Beale combined the moving piston of an ordinary Sterling engine and the moving component of an electrical generator. The free piston bounces at an easily controlled frequency between a gas cushion at the end walls of the cylinder. The device is self-starting; a starter or starter switching is not needed. The heated gas inside the cylinder lubricates the piston without the need for mechanical lubrication. There are only two or three moving parts in the free piston design, so manufacturing and maintenance costs are reduced and expected lifetime of the engine is high. Because no mechanical linkage is needed to deliver power to the generator, a costly and bothersome high pressure seal is not necessary and causes no additional friction losses. Beale worked for many years yet neither the U.S. Government nor U.S. manufacturers recognized the importance of his design. At the present time Japanese and West German manufacturers have begun to fund his work. It is expected that the free piston engine will soon be commercially available for between \$100–\$200/kW$_e$, in units as small as 1 kW, similar to the one pictured in Figure 13.4. By the year 2000, at moderate levels of mass production, the price of this form of solar electricity is expected to be around, \$0.11/kWh (Caputo, 1977). If the waste thermal output is used as well, the cost of energy from a total energy system would be reduced further.

Photovoltaic Electricity

Photovoltaic cells have worked reliably in the space program for the past 25 years. Essentially photovoltaic cells are solid state devices that emit an electron when a photon, a particle of light energy, is absorbed by the cell. In order for useful current to flow, some type of electric field must exist within the device to keep the electron separate from the hole in the atomic structure it has just vacated. This barrier is normally created at the junction of two semiconductor materials with different electrical properties. There are many advantages to photovoltaic electricity. Sunlight is converted directly without any intermediate processes. There are no moving parts to wear out or break. Solar cells require very little maintenance and are simple to operate. They can operate at ambient temperatures and without auxiliary cooling. Cells could last indefinitely with proper encapsulation, but this has not yet been demonstrated. At present the high cost of these devices limits their widespread use. Solar cell electricity currently costs between \$0.50 and \$1.00 kilowatt-hour. While this cost is very high compared to, for example, the costs of coal or nuclear power at \$0.04–\$0.20/kWh, past cost reductions have been dramatic and may continue as production levels increase and new techniques emerge (Figure 13.15). The output from an individual cell is low amperage, low voltage direct current. Solar cells therefore must be carefully wired together in series and parallel during module fabrication to produce output at useful levels. Much of the cost of photovoltaics is associated with the large amount of handwork now involved which should be eliminated under advanced manufacturing techniques. Solar cell efficiency is generally low, in the range of 5–15% for prototype devices. To a large extent this is because the energy band of photons that can be absorbed by devices of a single compound is fairly narrow compared to the spectrum of solar energy. Therefore the theoretical efficiency of most cells is below 30%. Further losses occur when light is reflected off the front of the cell. Internal electrical resistance, electrons, and holes that recombine due to short circuits caused by imperfect cell crystals account for other losses. The latter effect means that cell crystals have to be of very high purity, or else they must be so thin that the path across the cell for an electron is short with respect to a typical distance to a crystal boundary or imperfection.

Most of the high performance cells available today are made from silicon because it is abundant and has high potential efficiency. Unfortunately silicon does not absorb photons very well. Silicon cells must be between 50 and 100 micrometers (μm) thick to be able to absorb an appreciable quantity of photons. This is quite thick by subatomic standards. Because of the large path length across the cell, silicon cells must be of either extremely high quality or have very large crystals. Such thick and high-quality crystals are very slow and costly to grow, which makes continuous mass production techniques difficult. Research is proceeding on mixing hydrogen with silicon to improve the cell's absorption. Amorphous silicon-plus-hydrogen cells can be produced substantially cheaper and require only one-hundredth the material necessary for the single-crystal design, but they have lower theoretical and measured efficiencies. For many years other compounds have been studied for use in photovoltaics, most notably gallium-arsenide and cadmium-sulfide. Because both materials absorb photons better than silicon, thin layers only 1 μm thick are required. Thin film cells have the potential for very inexpensive fabrication because the films need not be crystalline. The materials can be either chemically sprayed or deposited in vapor form on inexpensive substrates (Figure 13.16). Both processes are very rapid and can occur while electrical connections are applied continuously. Unfortu-

Figure 13.16. Thin-film cadmium sulfide cells. The advantage of thin films lies in their potential for inexpensive manufacturing. Much less material is used in thin-film cells, and this can be laid down in continuous vapor or chemical deposition techniques. Cells of this type have shown efficiencies above 10% and have a theoretical efficiencies of over 20%. Cadmium sulfide cells as well as others can maintain their high efficiencies even when exposed to highly concentrated sunlight. A small amount of these cells combined with an inexpensive concentrator system might prove to be the most cost-effective configuration.

nately cadmium and gallium are much less abundant than silicon, and this could limit their use. One interesting property of most solar cells, especially true for gallium-arsenide, is that they can operate at very high concentrations of sunlight without loss of efficiency. Gallium cells have demonstrated efficiencies of 19% at concentrations of sunlight greater than 1700 times the unconcentrated value. At this level of exposure, output of the cells is in the order of 0.25 MW/m^2 (OTA, 1978). A combination of concentrating collector and photovoltaic cells might provide a way to achieve low cost electricity in the near future in regions where a large fraction of direct sunlight is available. Another concept that increases the theoretical limit of efficiency and therefore could reduce the cost of photoelectricity is the multicolor cell. Here two or three different cell materials are deposited over each other, each with a different band width of operation, so that a larger proportion of incident light could be used. Multicolor cells have a theoretical efficiency of 40%. Photovoltaics are cost effective now for many special terrestrial applications. Current dollar volume is tens of millions of dollars per year. As production levels increase, prices should decline, and this will increase the available market. Progress toward the goal of making photovoltaics electricity generally competitive with conventional sources of electricity by 1990 continues.

Wind Energy

Wind energy is one of the most ancient forms of power used by people and has been used for irrigation, transportation, and the threshing of grain. Wind and hydropower were the prime movers in much of Europe until the development of the heat engine. As recently as World War II wind was used widely to pump water and produce electricity in many rural areas. It is estimated that over six million windmills have been produced and used in the United States (Metz, 1978). The Rural Electrification Act of the 1930s and 1940s caused the early retirement of many of these effective, reliable, and long-lived turbines. The first signs of energy shortage in the early 1970s sent many people in search of these abandoned machines. Since 1973 the federal government has sponsored a large wind research program, and the Canadians and most European nations are vigorously pursuing wind programs of their own. Government contractors project that the cost of wind electricity at good wind sites and at a reasonable level of production will approach $0.025–$0.04/kW$_e$.

The wind resource in the United States is substantial. Although good winds are locally specific, large areas of high and persistent winds have been surveyed (Figure 13.9). The total U.S. wind resource is estimated at about 7.5 GW$_e$, or 14% of present electric capacity. It would take 300,000 very large windmills of the type now being operated by NASA and DOE in the state of Washington (Figure 13.17) to extract this much power, but the capital cost of such a project should cost no more than investment in the same capacity of coal-fired plants. It has already been demonstrated in New England, Southern California, and Hawaii that wind farms with rated capacities around 100 MW can be built and put on line in shorter times than large, conventional fossil fuel plants, which take many years to complete. Because capacity can be added incrementally, so that load growth can be followed closely, wind power has two major economic advantages. Overcapacity caused by the planning uncertainty of 10-year time horizons and the stagger-

ing interest of money borrowed for these years is avoided. A kcal of electrical energy produced by a wind machine replaces three kcal of fossil energy burned at the thermal plant. Because the operation of wind generators do not produce the CO_2 and other emissions associated with fossil fuel plants, the large-scale development of wind-generated electricity could have a positive impact on environmental quality.

Wind farms are the most effective way for utilities to harness wind power. Siting turbines together in a cluster reduces many of the fixed and operating costs of wind energy. A single-site evaluation and improvement and a single electric substation can serve the entire farm. Special equipment needed to service the array would not have to be transported appreciable distances during scheduled maintenance of each machine. Finally, the electrical output of a single machine fluctuates widely as the wind gusts, and this variability is a problem for utilities. When many machines are placed in an array, the correlation of wind gust peaks at adjacent machines approaches zero, so that the combined output of the wind array is much steadier.

There are potentially millions of cost-effective applications for small wind machines. Remote stations, farm heating and pumping, and communities located at the far ends of the lines of energy distribution are examples. Wind energy could be very important for Hawaii, which imports 95 percent of its energy but has unusually steady winds. Numerous diesel-powered pumps on the 60 million acres in the wind-rich Great Plains could be wind driven. The USDA is now testing a small Darrieus wind machine in parallel with a diesel engine to pump water. The wind turbine, which cost $50,000 in limited production, saved $6000 per year in diesel fuel. Engines now used to pump stripper oil wells can be driven at considerable savings with a stand alone wind turbine. Many farm applications exist, for farms generally have the open fields necessary to site a wind turbine, and normal activities like grazing and field cropping can take place beneath a wind machine. In recent years hundreds of small wind turbine manufacturers have emerged in a way similar to the early auto industry. Not all of these machines or companies will be successful, although several excellent designs have occurred. More operational experience is necessary before the long-term impact of wind energy can be assessed. Mass production should lower prices relative to competing energy sources, and experience will allow lighter machines to be built more cheaply without sacrificing life expectancy. The cost of wind energy will not drop unless large enough markets exist to support the cost of the capital investments necessary to increase levels of production. This market, however, does not exist at the present time due to the high present cost of wind energy and, in part, because other energies are still subsidized or underpriced. The young industry is now threatened by possible repeal of the solar energy tax credit which was designed to overcome this dilemma.

Hydro

In 1900, 4% of our delivered energy supply (in thermal units) was in the form of hydroelectric power. Eighty years later it is still 4%, although much larger in absolute terms. At the present time the installed hydroelectric capacity in the United States exceeds 65,000 MW, but the potential for new hydropower development is still great. It has been estimated that another 55,000 MW of potential capacity exists at existing dams (OTA, 1978). Many dams developed in the past were not equipped to their maximum potential for lack of local demand and because of past economic disincentives for long distance electric power transmission. Therefore much added capacity could be derived by upgrading existing power plants. Second, many old facilities were retired during times of less expensive energy. With the increased dollar value of electricity and new automatic equipment, it is now economically feasible to restore or rebuild old power equipment at retired sites. Finally, a surprising number of dams, especially in the Northeast, were built for flood control purposes, recreation, and irrigation and have no power generation equipment of any kind. In many cases it is now economically attractive to develop much of this existing capacity. Figure 13.10 shows the distribution of this potential in the United States. Because these dams are already in place, the additional environmental impact of this development is small.

In addition to the capacity at existing dams, it has been estimated that there exists another 20,000 MW of potential capacity at sites where there is no dam now (OTA, 1978). Much of this new capacity is located at small sites, those with less than 5 MW potential. This large potential in the United States is not atypical worldwide. In 1965 the United Nation's assessment of European hydropower concluded

that only 32% of the European resource had been developed. In many parts of the world a vigorous hydropower development program is underway. New construction and power technologies have made the development of small hydro attractive. This small hydro is less environmentally disruptive and risky than large site development and requires far less planning and construction time.

The advantage of hydropower is that it can be turned on rapidly and can be efficiently used at outputs that are only a fraction of a site's rated capacity. Unlike any other electric generation technology it can follow widely fluctuating loads rapidly. This has value to the utilities in general and to the development of other solar technologies which are intermittent and would benefit from an integration with hydro. Other advantages of hydro are that well-planned dams can serve as flood control systems, for water management and irrigation, for fisheries enhancement, and recreation. Dams have long life spans with respect to other forms of electric power generation and require little maintenance. Dams of course also have negative impacts, but these normally are considered small compared to other forms of energy except for certain rivers with anadromous fish or with high recreation potential. The extent to which the potential of hydropower will be utilized is determined by economics and how well the damaging environmental impacts of each project can be reduced.

BIOMASS ENERGY

Biomass energy has been the traditional supplement to direct human energy for most of human history, and even now it remains the principal supplement for the majority of people on Earth. Presently it provides 15% of the energy used directly by people worldwide and about 2% of the energy used in the United States. Thus although there is considerable new interest in biomass energy in the United States, at this time it should be remembered that biomass energy is a well-developed technology that already, in some regions, is being fully utilized. Nevertheless the long-term maximum potential in the United States is unlikely to be much more than several times the present rate of utilization. Even the total energy fixed by green plants in the United States (about 13.6×10^{15} kcal) is less than the present total use of industrial energy (about 19.7×10^{15} kcal).

Biomass energy normally is defined as the energy associated with carbon and hydrogen bonds in plant, or occasionally animal, material—material that has been grown within the recent past, within years to at most decades. The most obvious sources are trees and herbaceous vegetative material such as agricultural products and by-products. Other sources that have been considered are domestic animal or human manure, and those municipal wastes that have a high percentage of paper and other biomass material. The principal attractiveness of biomass material is that it is, at least in theory, more or less indefinitely renewable, for it can be grown on the same site year after year. Other advantages include (for some applications) its relatively low pollutant output, its potential to assist in decreasing other environmental problems, and its labor intensiveness. There are also large potential disadvantages, however, including its inability to make up more than a relatively small percentage of present-day U.S. energy requirements without incurring significant environmental degradation, its frequent high cost, its relatively low energy density, and the fact that the largest proportion of the most readily available material is in a solid form, which can be converted to the much more desirable fluid forms only at a considerable loss of efficiency.

Until recently it was difficult to find comprehensive empirical assessments of biomass potential because there had been little research or interest in biomass fuels in the United States. Furthermore a number of the initial reports on the subject were too optimistic because they failed to consider the large energy subsidies from the general economy that would be required to produce a significant amount of biomass energy on the national scale, or the potential for severe environmental degradation. More recent and comprehensive studies (e.g., Energy Research Advisory Panel, 1981; Pimentel et al., 1981—together the source of most of the information in this section) indicate that biomass energy is best considered a supplemental resource that may have important local contributions if integrated into appropriate resource systems but that optimistically could contribute to no more than 1.3×10^{15} kcal (net)/yr to our national economy—an amount equivalent to about 13% (gross), or 5% (net), of the U.S. energy use in the late 1970s. Even this amount must be considered with respect to the trade-off between the production of fuel from biotic resources versus other uses of that material and land. A significant moral issue is raised if productive

farmlands are used to produce liquid fuel, principally for automobiles, in a world with half a billion or more severely malnourished people. Finally, most biomass energy production is normally at the low end of energy return on investments for the various energy technologies that we are aware of. It can be argued that trading corn for oil in international markets has a higher EROI. This argument does not apply of course to the production of fuel from otherwise unutilized organic material.

Sources of Biomass Energy

Biomass energy resources can be divided into two broad categories from two principal sources: (1) by-product sources and (2) deliberately grown materials from forests or agriculture. This next section describes briefly each of these resources and considers some of their desirable and undesirable characteristics. Table 13.4 summarizes the present-day and potential future contribution of crop and forestry residues to national energy supplies.

Forest Products

Wood is perhaps the oldest energy source for human economies except for sunlight itself and the use of the muscle power of people and grazing animals. In 1850 wood provided about 90% of the roughly 0.6×10^{15} kcal of fuel combusted in the United States, and interestingly, it provides a similar quantity, but much smaller proportion, of the energy that we use today. There are five principal ways in which wood can be used for fuel.

1. *Mill Residues.* Every step in the production of finished materials from wood creates wastes in the form of materials that cannot be utilized directly for products. For example, producing rectangular boards from round trees leaves residual materials that often have little use for the production of other wood products. Historically this material has been used extensively for the energy requirements of the forest products industry itself (see Chapter 19), and this has been especially true in recent years. But one problem for expanding the use of wastes as energy is that new technologies have increasingly used mill wastes to produce new products, like particle boards, of higher commercial value than fuel. It is estimated that all but 20 million tons of the 149 million tons of mill residues are currently being utilized, so the potential for additional

Table 13.4.   Potential Energy from Crop and Forest Residues

Source	Hectares Harvested ($\times 10^6$)	Crop Yield (ton/ha)	Residues (ton/ha)	Total Yield (tons $\times 10^6$)	Amount Readily Usable[a] (tons $\times 10^6$)	Potential Net Heat Energy[b] (kcal $\times 10^9$)	Potential Net Electrical Energy[b] (kcal $\times 10^9$)	Potential Net Ethanol Energy[c] (kcal $\times 10^9$)
Barley	3.8	2.4	3.5	13.3	3.6	6,940	2,652	758
Corn	28.3	5.6	5.6	158.5	39.6	71,874	19,800	9,308
Cotton	5.4	0.6	0.5	2.7	0	0	0	0
Oats	5.4	2.0	4.0	21.6	6.5	12,513	4,789	1,368
Rice	0.9	4.9	7.4	6.7	5.2	10,010	3,832	1,095
Rye	0.3	1.5	2.3	0.7	0	0	0	0
Sorghum	5.7	3.5	1.2	7.4	0	0	0	0
Soybeans	23.4	2.0	3.0	70.2	0	0	0	0
Wheat, winter	19.6	2.1	3.5	68.6	18.6	35,805	13,705	3,916
Wheat, spring	7.2	1.8	2.3	16.6	0	0	0	0
Other	56.0		1.1	62.0	0	0	0	0
Total crops	156.0			428.3[d]	73.5	137,132	44,778	16,445
Total forest	4.5		24.7[d]	111.2	44.0	71,032	20,307	8,372
Grand total				539.5	117.5	208,164	65,085	24,817

Source: From Pimentel et al. (1981).

[a] It is assumed that 40% of the land area in row crops is on a slope from 0 to 2% and that 50% of the land area in other crops is on a slope from 0 to 5%. An estimated 2000 kg of corn stover and 1600 kg of straw are left in the field to protect the soil from erosion and to maintain soil organic matter.

[b] Heat recovered from combustion of the biomass is calculated at 55%.

[c] Conservation tillage was assumed on all land and the ethanol or electrical energy yield from conversion of residues of all small grain crops was assumed to be similar to the yield from conversion of wheat residues.

[d] Calculated by multiplying yield data by dry weight ratio of residue to yields.

resources in this category is small. Some additional resources may become available if and as the wood products industry expands in future decades.

2. *Logging Residues.* Logging residues are comprised of tree tops, larger branches, dead trees, and other products of harvest that are not normally used by the logging industry. At present these materials are not being utilized primarily for economic reasons, which are in part related to the relatively low energy density of wood (Table 19.6). About 50 million metric tons, of a total annual production of about 180 million metric tons, could be utilized if a concentrated effort were made to bring this material into the energy market. This would provide perhaps 0.13×10^{15} kcal of energy to the United States.

3. *Forest Stand Improvements.* Intensively managed forests, especially monoculture forests in the Southeastern United States, normally are planted at densities that are greater than the densities at which the eventual greatest wood production takes place. Thinning these stands provides both better growing conditions for the remaining trees and a source of wood for fuel and fiber products.

Presently only about 5% of the thinning that does take place is for energy, but that percentage could be increased somewhat by deliberate planting for energy.

4. *Tree Mortality.* Only about 60% of the production of wood in the United States is harvested, the rest eventually falls to the ground and decomposes (USFS, 1978; Armantano and Hett, 1979). If a substantial portion of this energy could be harvested, it would represent a very large source of energy. This annual production is about 215 million metric tons, and the existing inventory is about 1 billion metric tons. The energy cost of extracting this wood, much of which is dispersed, often inaccessible, and often half rotted, would be very large.

5. *Short Rotation Energy Plantations.* Some trees grow very rapidly but have little commercial value due to the low density of fiber and the growth form. These trees have considerable commercial appeal for energy. Candidate species include Alder, *Eucalyptus, Platanus,* and poplar. Farming these species would be an energy-intensive procedure, and the amount of land that would be required

would be enormous even for very modest production. For example, nearly 4000 ha would be required to produce a very modest 250×10^9 kcal/yr, although a proper plantation could serve limited local needs with a reasonable EROI and little environmental degradation.

Residential Firewood

In certain parts of the United States, especially the Northeast, there has been a substantial increase in the use of wood for fuel through conventional harvest, often on a small scale using chain saws and small trucks. Presently about 30 million metric tons are harvested each year for this purpose. It is expected that this will increase to at least about 55 million metric tons by the year 2000. This is likely whether or not any governmental policies are enacted because wood is a proven source of residential heat and to many, an attractive technology, although not without its costs (see also Chapter 19).

There is a principal constraint to the use of any forest products for energy; that demand for forest products in general is expected to increase markedly in coming decades. This demand is likely to put severe pressures on using whatever wood resources are available for products. Even at present the United States imports 11% (net) of the forest-derived fiber it uses. Only an increased use of that part of the tree harvest that is not suitable for wood products (an increasingly small portion as new technologies use more and more of the tree) may be the only major source of biomass fuel from forests aside from small-scale harvest for local heating. Chapter 19 considers the energy cost of fuelwood and wood fiber harvest in considerable detail.

Agricultural Products

The use of agricultural products for energy is already familiar to many Americans through the promotion of *gasohol,* a blend of about 10% grain-derived alcohol and 90% gasoline. Although growing our fuel in our own rich farmlands appears to be a very desirable solution to declining domestic production of petroleum, the solution is not as simple as the sudden appearance of gasohol would make it seem. Studies available (e.g., Hopkinson and Day, 1979; Chambers et al., 1979) indicate that when all the agricultural and distillation energies are included, growing crops for alcohol in the United States does little better than break even energetically. On the other hand, Brazil, faced with the very large expense of importing oil for their expanding economy, but with some very favorable agricultural areas, has undertaken a large program of producing a large proportion of their fluid energy requirements from the growing of crops (mostly sugar cane) for the production of fuel alcohol. There are many automobiles in Brazil now that run completely (and with a considerable amount of backfiring) on pure alcohol. The program has a considerable subsidy from the government, and its success or lack thereof depends on who you ask. The future value of using agriculture to produce energy requirements for the United States also is not a clear-cut issue. It is likely that we can produce enough fuel from agriculture to help at least locally (e.g., on the farm itself; see Jewell, 1982). But it is unlikely that we could produce enough fuel to make a substantial contribution to our nation's energy requirements without simultaneously foregoing a substantial food production and producing considerable soil erosion and possibly using nearly as much (fossil) energy as is produced. Further investigations should clarify the potential future of this source of energy, and it is possible that technical innovations could make the resource much more favorable. In the meantime we include a brief summary of the possible resources.

Grains and Sugar Crops. The most readily available sources of fuel are routine crops, especially corn, wheat, sugar cane, and sorghum. The advantage of these materials is that we are well experienced with their cultivation and even their conversion into alcohol (although whiskey and rum would be a rather expensive fuel). One million metric tons of corn each year already are used to produce most of the 400 million liters of fuel alcohol produced in the United States each year (about one-tenth of 1% of the gasoline produced). It is possible that in the future President Carter's one time goal of 8 billion liters production per year will be obtained, but this would require about 10% of the nation's corn crop and would be equal to only about 1% (gross) of our present oil use. Therefore, although it is possible to produce a modest amount of fuel through the production of corn, a large-scale endeavor would require very large quantities of land, would preempt this land from other uses during a period when demand for food will undoubtedly be high, and would probably have rather severe environmental effects. Additionally there are large energy losses in conversion to liquid or gaseous fuels. A few crops can produce combustible oil directly, for example, *Euphorbia* and sunflowers. It is possible to harvest 1000 liters of such oil that can be used in a diesel engine from a hectare of sunflowers, but that is barely more than the energy it took to grow the sunflowers.

Crop- and Food-Processing Wastes. A rather desirable energy source is semiprocessed agricultural and food-processing wastes. The advantage of these fuels is that they have, in general, already been harvested and concentrated for various reasons. A good example is bagasse, or the residue from the production of sugar from sugarcane. Large piles of bagasse are burned in the open air in sugar-producing regions, causing some air pollution. If this material could be burned to produce useful heat near the sugar plants, then presumably there would be relatively little cost other than that of building the plant. Similarly alcohol can be produced from cheese wastes, and various other residues could be fermented to produce liquid or gaseous combustible products on a small scale for local consumption. Unfortunately none of these sources are thought to be especially significant even if fully developed. They do have considerable appeal to many people, however, by turning garbage into useful fuel, and they undoubtedly will be a common feature of many landscapes in the future.

Organic Wastes. Another appealing source of energy is the use of municipal or industrial wastes, much of which is composed of flammable material such as paper. The total resource at present is about 200 million metric tons produced per year, of which roughly half is combustible. The heat content of this fuel is somewhat less than low quality wood and only about a third of good quality coal. The advantages of this source is that the fuel is already collected by municipal refuse system and that it is low in many pollutants, such as sulfur. It also is possible that systems can be designed to dispose of refuse through energy production in a fashion that would be cheaper than conventional means of disposal. A principal disadvantage, at least in our experience, is that no one wants the plants in his or her own backyard, because of the heavy truck traffic and the pollutants produced, and because the total resource is, again, very small. Nevertheless, we look at this potential for energy production as one with a relatively high ratio of advantages to disadvantages and perhaps existing incinerator sites could be converted to energy production sites.

Energy Storage

The most persistent and bothersome question related to the use of solar energy in the minds of most people not working the field is that of energy storage. Because solar energy is intermittent, many reason that the cost of energy storage makes solar energy prohibitive. As a blanket statement this is incorrect for four reasons. First of all, enormous quantities of solar energy could be used in the economy at the present time with either very little or no added cost from the inclusion of storage. A well-designed passive solar house, for example, stores heat in well-placed floors and walls of the house. Heat engines powered by the sun but that also can burn other fuels can follow demand without storage. These engines in mass production will cost no more than conventional power plants for this added flexibility, and when the sun is shining, depletable fossil fuels are saved. Fuel-saver modes like these are available for electric utilities as well. In this case surplus solar generation is most cheaply stored by letting someone else on the grid consume the energy. When the solar device produces less than the local demand, the utility will provide backup power. Used this way, wind turbines connected to the grid, including those located in someone's backyard, become indistinguishable from any other generating plant on the grid. Utility engineers routinely adjust generating capacity to follow changing demand based on the time of the day, season, and temperature. As solar devices achieve greater penetration into the grid, generation requirement forecasts will be adjusted based on wind speeds and insolation as well. Until solar power penetration reaches 15–20% of a utility's capacity, grid stability and management should not be a problem (Metz, 1978).

This last observation underlies the second reason why energy storage will not limit the near-term development of solar power. Even the most optimistic rates of solar electric development do not predict a penetration of 15–20% before the first quarter of the 21st century. Indeed, the development of any new source of electric energy at this scale is so herculean that the same time scale exists for *any* alternative to fossil fuels. During this time, solar could contribute increasingly large quantities until this level of penetration is reached, saving vast amounts of valuable and irreplaceable fuels. By that time the world might look so different that the question of energy storage may not be important. Other forms of new energy may be economically available or energy storage technology might be at the point that increased solar electric contribution can economically proceed. Even if the continued development of solar electricity is halted 50 years from now by the limitations of energy storage, the cumulative and continuing fuel saving will be extremely valuable.

The third reason why energy storage is not a large problem is that energy storage is routinely used in conjunction with conventional forms of power. Almost all hot water systems have storage tanks. Storage buffers the system so that it need not follow the instantaneous demand for hot water which can be quite high for short periods of time. A small amount of inexpensive low-temperature storage replaces the much greater cost and complexity of the heater which would otherwise be necessary to match the peak load. Presently gas turbines are used in utility grids to meet the greatest anticipated or peak demand, even if this demand is very rare and of very short duration. This is because large base load plants in the network are not easily turned up and down or cycled to follow a dynamic load. To run the base load plants at less than full capacity often means a great sacrifice of efficiency and therefore excessive fuel consumption. Peak power is energy intensive. For these reasons storage serves a useful purpose for most utilities and is generally desirable. In many ways the well-planned addition of storage capacity to a grid coordinated with the development of solar electric generation could be beneficial for the entire network. The last point to be made concerning storage is that many inexpensive storage technologies already exist and that there are many possibilities in this conceptually rich area (Figure 13.17). Although many solar applications make use of inexpensive storage, any developments in the field will benefit both the solar industry and the general economy.

At the present time energy is stored in many different forms. Work can be stored as elevated water in reservoirs, batteries, super conducting magnets, or advanced-design flywheels. Most of these technologies are expensive but hydroelectric is not. Although hydropower is well developed in many areas, the potential for new hydro at both large and small sites exceeds existing hydro in many areas. In New York State, where 13% of the state's electricity capacity is from hydro, an inventory of untapped hydropower potential indicated that another 10% existed and that one-third of this amount could be

on line within 15 years (NYSERDA). Figure 13.18 shows the combined availability of river flow and the wind resource for the Salmon River area. These resources together are at a minimum in the summer when the solar resource is at a maximum. Integrated wisely renewable energy sources might reliably displace a great deal of conventional fuels.

The bulk of our present energy end use is in the form of thermal power. Thermal energy can be stored as sensible heat, latent heat, and in the chemical potential of molecular bonds. Low temperature heat is most easily stored as hot water. Inexpensive tanks typically serving a multifamily residence or a small business are quite small. The latent heat associated with the heat of fusion and heat of vaporation of many materials is now being investigated. Heat capacity of materials undergoing a phase change is very large. Water has a specific heat near 1 kcal/kg but has a heat of vaporization of nearly 860 kcal/kg. This is to say that for each kilocalorie stored in 1 kg of water below the boiling point, the temperature increases 1°C. This same kilogram of water can absorb 860 kcal at 100°C while going from liquid to vapor. Eutectic salts are used in some solar architecture to store large quantities of heat when they undergo a phase change from solid to liquid at 81°F. Many candidate substances exist for intermediate and high-temperature latent heat storage materials. Rocks can be used for very high sensible heat storage at a cost of about $1 per 5 kW of storage capacity. A good review of the many storage concepts being investigated is found in the Office of Technology Assessment report on solar technology (OTA, 1978). Storage research will benefit all energy sources and should continue. Successful development of new and inexpensive storage technologies will not limit the widespread and immediate development of solar energy but will effect the ultimate contribution of solar energy in any mixed-fuel economy of the middle of the 21st century.

ENERGY RETURN ON INVESTMENT FOR SOLAR

This section summarizes several studies of the renewable energy investments required to capture solar energy. Unfortunately deriving even reasonably accurate estimates of EROI for many solar converters at this time is difficult because most solar technologies are still in the planning or early

Figure 13.17. Two of the three 2.5 MW mod-2 wind turbines located in Washington state. The dam-controlled Columbia River is in the background. Integrating wind arrays with the existing transmission and switching gear of existing hydro plants not only reduces the amount of equipment necessary but increases the reliability of the wind systems. When the wind blows, water can be saved behind dams for later use when the wind is not blowing.

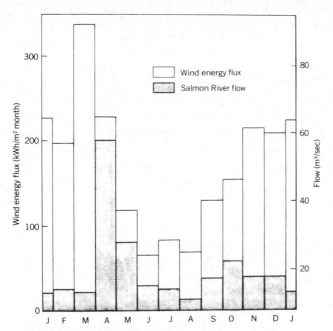

Figure 13.18. Comparison of the annual average wind and water supply of, and near, the Salmon River in New York. High combined water and wind availability takes place in the spring, winter, and fall. The summer low might be compensated for with the increased availability of summer sunshine (see Figure 13.7). The integration of different renewable energy sources potentially could comprise a substantial fraction of the optimum electric generation mix of the future. This seasonal availability of wind and sunshine is common in many parts of the country. (From *Solar Energy in America*, eds., W. D. Metz and A. L. Hammond, 1978, Fig. 83. Copyright 1978 by AAAS.)

development stage, technologies for many devices are changing rapidly, and most prototype devices have not operated for sufficient time to calculate their lifetime. Furthermore, any comprehensive assessment should include the energy costs of storage devices, such as pumped-storage reservoirs or batteries, or the cost of constructing and operating backup generating facilities such as gas turbines if these are required. In general, we cannot provide EROI assessments for small-scale solar energy generation. All of these factors make it nearly impossible to calculate EROI for solar energy in the future, when it is most likely to be important, and the results of the studies we review indicate a wide range in possible eventual assessments. Nevertheless, these studies can serve as a starting place for the assessments that eventually will be needed. Several studies appear thorough enough to suggest that some forms of solar energy almost certainly provide a good net return.

Hydroelectric Power

Assessments for EROI for hydroelectric plants indicate that they generally have a very favorable ratio. Where topographies are favorable, it does not take much energy to put a dam across a river valley and extract the energy contained in the elevated water. Existing studies suggest that characteristic EROIs are in the range of 10:1 to 35:1 (Table 2.1), but most of these studies are incomplete in that they do not include the important energy opportunity costs of foregone agricultural production of the regions inundated by the reservoir or the additional fertilizers that are needed downstream to replace the nutrients that were previously left on the rich alluvial soils that border most rivers. An important problem, again related to our general thesis that the first resources exploited tend to be the highest quality ones with the least energy cost for exploitation, is that the best dam sites in the United States were exploited long ago, so that most future large hydro developments are likely to be more energy intensive. On the other hand, one study (Gilliland et al., 1981) found that many small hydro plants had quite high EROIs.

Solar Water Heaters

We are aware of only two studies that have analyzed the EROI for solar water heaters, and although both conclude that in the proper climate solar water heaters are a favorable investment of fossil fuels, they come to different conclusions as to the magnitude of the advantage. Lencheck (1976) concluded that in an environment similar to Colorado the energy gained by a solar water heater would pay back the fossil energy invested in constructing it in 5–7 years. Substantial improvements could be made by using steel and wood, rather than aluminum, for construction, although it was not clear how long the devices were expected to last. Lenchek assumed that if these devices would last 30 years, the EROI would be at least 6:1.

A somewhat more explicit analysis gave a less favorable EROI for solar water heaters in Florida (Zucchetto and Brown, 1976). They calculated that solar water heaters would return from 1.3 to 2.1 kcal of hot water per kilocalorie of fossil energy invested, depending on initial costs, assumed interest rates, and whether the device lasted for 10 or 25 years. Although this does not appear especially favorable, it was considerably better than their 0.6 EROI calculated per unit of gas consumed, or the 0.17 EROI for electric water heaters, all calculated as thermal energy returned compared to fossil energy inputs to construct and run the devices. Thus, at least for Florida, solar water heaters appear to at least double the hot water returned per unit fossil energy invested, and they are much more favorable than electric water heaters. In conclusion it appears that solar water heaters in the proper environment are very favorable investments of fossil fuel.

Photovoltaic Cells

Assessments of the EROI of photovoltaic cells are subject to great uncertainty because the technology is changing so rapidly and because there has been little long-term experience with them. In theory the cells should last indefinitely, but in practice they can fail in many ways. Only experience will tell us how long they do in fact last. We can at this time, however, give a reasonable initial estimate of the energy costs of making solar cells. Our assessment is based principally on Iles (1974), Hunt (1976), Maycock and Stirewalt (1981), and Hay et al. (1981).

The leading manufacturing process—the Czochralski method—is divided into four steps:

1. The production of metallurgical grade silicon (MG-Si). This includes the mining of quartz (SiO_2) and refining it to 98–99% purity. This is carried out by placing the quartz in a large electric arc furnace with coal, coke, and wood chips and heating the mixture to 3000°C. Electricity used in the arc furnace accounts for 60 percent of the energy used in this subprocess.
2. The production of semiconductor grade silicon, or trichlorosilane (SeG-SiHCl$_3$) by placing MG-Si in a fluidized bed reactor with HCl.
3. The production of semiconductor grade polycrystalline silicon (SeG-poly-Si). By using the Siemens process, trichlorosilane is chemically vapor deposited (CVD) on a silicone substrate, and heated electrically to 1000–1200°C in a quartz vacuum bell jar to make silicon rods, which are subsequently cut into cylinders with a special saw. This subprocess is highly energy intensive.
4. The production of the cells themselves. Iles (1974) gives 15 steps, starting with the crystal to

environmental testing of the finished cells. The SeG-poly-Si is remelted, planted with a seed crystal and slowly pulled out to form a single crystal ingot about 20 cm in diameter and nearly a meter in length. The ingot is sawed into silicon wafers 20 mils (0.5 mm) thick, lapped or etched, and polished to remove surface damage and to reduce the wafer down to the desired size, about 0.3 mm thick. Interconnecting wires are applied in a grid, and the cells are encapsulated in glass so that the finished product is about 3 mm thick.

Arrays can be assembled using many different materials with a wide range of reliability, durability, and energy costs. Mounts can be made from wood, steel, aluminum, and the like, and may be mounted alone (*stand alone mount*), above a roof (*standoff mount*), by replacing shingles (*direct mount*), or as integrated units that include structural components of the roof (*integral mounts*). There are many trade-offs, for aluminum structures designed to withstand severe weather and to last for many years are more energy intensive to build. Less energy intensive cells tend to be less efficient and less durable; hence they require more mounting material per unit of energy delivered and the replacement and maintenance costs are higher.

Hunt (1976) calculates that the energy cost of assembling solar arrays is much less than the energy cost of the materials that they are made from. Hay et al. (1981) have estimated the amount of both the processing and the embodied energy required by each process to produce 1 kg of silicon. About three-quarters of the energy required to produce SeG-poly-Si is process energy (nearly all as electricity). The energy cost of step 4 is largely due to the high losses of SeG-poly-Si, which is converted to cells at an efficiency of only about 18 percent, so that a great deal of embodied energy is lost at this step, at least as it is presently done.

In this assessment electrical energy costs have been converted to kcals using a conversion factor of three, and there was no estimate included for the energy used to support labor, to produce the capital equipment used, to provide the R & D, to mount the collectors, or to transport the materials or the finished products. Hence assessments are minimum energy costs. On the other hand, improvements in manufacturing technology promise to continue to decrease a cell's energy cost so that a failure to include these energy costs may not be important.

The total energy cost of producing a kilogram of completed single-crystal silicon cells was estimated to be about 9.3 million kcal, equivalent to the energy in about six or seven barrels of oil. Silicon is very energy intensive per kilogram compared to other materials. The energy cost translates into about 6.6 million kcal per square meter of collector. Similarly, Herendeen (1979) estimated that it took 12 million kcal to produce 1 kg of silicon.

Hay et al. (1981) also examined the energy cost of producing new, promising technologies, including single-crystal silicon, ribbon silicon, amorphous silicon, and cadmium sulfide/copper sulfide cells, using data from Hunt (1976). Considerable energy savings in the production of solar cells are at least theoretically possible. In most processes considerable energy savings could be obtained by increasing sawing efficiencies and by recovering volatile gases from the furnace. For example, Wolf et al. (1978) suggest that by 1982 reducing losses by using lasers and increasing process efficiencies should reduce energy costs by 40%. An independent assessment by OTA (1976) suggests that at that time an upper bound for the energy requirements for solar collectors was about 32 million $kcal/m^2$ and that an obtainable bound with advanced manufacturing techniques was about 860 thousand $kcal/m^2$. These figures seem to be roughly consistent with those given previously.

The calculation of EROI for many solar convertors is difficult, as previously stated, because the life span of the collectors or of the ancillary equipment are not yet known. One common method that avoids such a problem calculates the *energy pay back time,* the time that the device must operate to produce the energy used to manufacture it. Hunt (1976) calculated an energy pay back time of 11.6 years for technologies available at that time, assuming 5 hr solar insolation per day at 1 kW/m^2, a level fairly typical of many areas in the Southeast United States. We believe that this estimate is too optimistic by perhaps 30%, for it assumes that the collector was always pointed directly at the sun, and it does not include an energy cost for tracking devices. The results indicate that EROI for solar collectors probably range between 1:1 to 10:1, assuming that the cells lasted for about 10–20 years, depending on cell type. Technologies now becoming available probably could double these estimates, but corrections for mounting and tracking devices, power conditioning systems, backup systems, and so on, will reduce new estimates by an

unknown degree. Based on current information and technology, we conclude that photovoltaic collectors can achieve a moderate EROI, one that is probably not yet competitive with coal but one that at least the upper estimate of may be near the energy return for fission-derived nuclear electricity. On the other hand, the EROI for fossil fuels is declining while that for solar technologies is increasing. At some point in the future, presumably, they will cross, and we might then find powerful financial incentives for developing large-scale solar energy.

Centralized, High Technology Solar Collector

Several schemes have been advanced for developing massive, high technology solar collectors, including solar collecting satellites, power towers, and floating power plants that use temperature and current speed differences between surface and deepsea waters—such as the Gulf stream off of Florida—to generate electricity. The most comprehensive analysis, in our opinion, is that of Herendeen et al. (1979), who found that a solar satellite would require about 1 kcal in rocket fuel, materials, and so on, for every two delivered to Earth.

Much of the U.S. federal solar energy R & D budget has been used, inappropriately perhaps, to fund highly technical, centralized systems that some have called solar energy in the image of nuclear power. This is because many of these endeavors ignore the best features of solar energy, namely free delivery and distribution, economics of small-scale, matched energy quality and end use, and the flexibility of these small systems that allows them to be physically and functionally integrated into existing buildings while making use of waste heat. The power tower requires 15 times more construction material per megawatt than a nuclear power plant and 35 times more material per megawatt than a coal plant. There are potentially hundreds of millions of applications worldwide of small- and intermediate-scale solar energy, but these are presently limited by high capital costs. This high cost reflects to a large degree the very low levels of present production.

Potential for Cost Reduction

During the 1930s when nearly 6 million windmills were in use in America, the real cost of a farm-sized machine was one-fifth of the present cost of a similar machine due to the well-known economics of mass production (Metz, 1978). Senior project engineers who estimate the cost of production of a future technology of the complexity of an automobile, tractor, and so on, at high levels of production use figures between $1.00 and $1.50/kg. Small solar heating equipment now produced at very low levels of production typically costs $3000–$4000 and weighs between 200 and 250 kg ($16/kg). A mass-produced solar collector at $1.50/kg should be achievable due to the comparative simplicity of solar collector manufacture. This represents a cost reduction potential of 10 times. These technologies could be mass produced simply by much of our presently idle industrial capacity. Many unions such as Sheet Metal and Glassworkers presently favor such an approach to the energy situation (OTA, 1978). At a time when most of the conventional technologies are suffering from declining EROIs, and many other alternatives seem distant, the increasing EROIs of small solar systems are encouraging. On the other hand, Frank Kreith, a director at the United States Solar Energy Research Institute, expressed to us in 1984 his disappointment that solar technologies were not decreasing in price as rapidly as was once predicted. He thought that the high energy cost of the materials required might be a principal reason.

Conclusions

The widespread use of solar technology is limited presently by its relatively low EROI which is reflected in its high cost. Although many more complex consumer goods such as tractors are sold for $1.00–$1.50/kg (1980 dollars), simple solar technologies produced in small quantities now cost approximately $16/kg. Hopefully the cost can be lowered through mass production. Once the investment is made, fuel and maintenance costs are low. But most consumers and even many commercial ventures choose alternatives on the basis of first costs and not life cycle costs. There is, however, much optimism in the field of solar energy because there is a great potential for cost reduction in many areas and because solar energy is so diverse a technology that it can be used in thousands of different ways. When compared to the increasingly high marginal cost of other energy sources, and not historic average costs, the direct use of solar energy in many cases can be the least expensive alternative. The high present cost of many solar systems will be reduced through technology refinement, but the greatest potential for cost reduction will occur only if the level of production of these simple technologies is increased.

At a time when our current industrial overcapacity is large and many factory jobs may have been permanently lost, large-scale solar development might simultaneously substantially lower the cost of solar energy and provide badly needed jobs. Such development is possible given existing public opinion. In Gallup, National Geographic, and other polls, solar energy was considered the second most preferred form of new energy, closely behind conservation. Many unions, such as the Sheet Metal and Glass Workers, have publicly supported the large-scale development of solar energy. Such a program might succeed because the manufacture of solar devices is well suited to our existing industrial base and would not require massive retraining and retooling or the high capital investment of a super high technology reindustrialization.

If the energy crisis is considered from the point of view of a supply crisis, solar energy is discounted as being too diffuse and intermittent a resource to tap. However, when the question is posed, What do we need, want, and what is the least expensive way of achieving this? the best alternative is often a combination of conservation and direct solar use. Our heating and cooling needs can be met better by weatherizing existing structures and incorporating passive polar design in new ones instead of simply building additional power plants. Recent economics suggests solar combined with conservation are a whole order of magnitude less expensive than centralized heat supply alternatives (see Chapter 14). Additionally such an approach makes consumers and our nation much less vulnerable to embargoes or strikes.

Because solar is a relatively diffuse and intermittent form of energy, large centralized solar facilities such as the government-sponsored power tower and ocean thermal projects may have poor economies and little chance for cost reduction. When one recognizes the unique characteristics of the solar resource, however, then real opportunities present themselves—namely solar energy is freely and globally distributed, and it is most readily and inexpensively converted directly to heat energy. The significance of this is that any centralized generation of power must produce an energy source of a quality equal to that of the most demanding user. Electricity is a high quality source that has no substitute in some important applications in our total energy budget. However, central production supply schemes impose the high cost of transmission on all end users. We in the United States mostly have need for the direct use of heat energy. Solar can provide this energy on site less expensively. Locating equipment on site allows for the possibility of cogeneration and the utilization of waste heat, leading to very high first law efficiencies. Low and intermediate heat is inexpensively stored. And although the solar resource is diffuse, so is the density of our end use. Most residential and small commercial and industrial facilities have over their parking lots and buildings the space to collect the necessary amounts of thermal energy which, outside of transportation, is the bulk of our nation's heating needs. Solar energy and biomass hold little promise to solve our eventual shortage of liquid fuels for our transportation sector, but then *no* alternative looks attractive compared with historic inexpensive high EROI oil.

When one considers the potential expansion of the hydro resource plus the possibilities of the integration of many diverse forms of renewable energy sources through the existing electric utility grid, most of our obligatory electricity requirements might be satisfied with a combination of solar and conventional resources without resorting to expensive storage technology.

The Office of Technology Assessment finds that solar technology has many positive impacts not found in other energy forms such as the high labor to output ratios and the lower environmental impacts per unit of production. When all of these factors are taken in the context that solar energy is even more appropriate in much of the rest of the world than in the United States, and that solar energy will never be the source of an international competition for a limited and depleting resource, one may begin to appreciate the potential significance of an expanded use of solar energy.

14

CONSERVATION

Conservation is not normally thought of as an energy source, but using a barrel of oil twice as efficiently can be as economically useful as producing another barrel from a new well—and it has the advantages of not depleting oil reserves as much or, in general, creating as much pollution. All studies of conservation as a source of energy have concluded that it represents by far the cheapest (and most reliable) method of producing new energy (Ford Foundation, 1974; Ross and Williams, 1977; Living, 1977; Gibbon et al., 1978; Stobaugh and Yergin, 1979; CONAES, 1979; Lovins, 1985). We also believe that conservation is one of our most potential energy sources, but it will not come as easily or cheaply as some people have suggested.

WHAT IS CONSERVATION?

Technically, and in most people's minds, conservation means saving or using less of something, and by that definition the United States has been very successful in conserving oil because in 1984 we used only about 5.5 billion barrels compared to 6.8 billion barrels in 1978. In a sense, though, no conservation took place at all. Oil was not saved—it was still pulled out of the ground, but at a slower rate. And most analysts would agree that the major reason that we are using less oil is the economic slowdown that has occurred since 1973. So by the definition given at the beginning of this paragraph the most conserving nations are third world nations with low GNP (Figure 2.12).

Energy *intensity* refers to the amount of energy required to produce a single unit of economic output (e.g., the kcals required to produce 1 ton of steel), and that is how we define conservation for our purposes. It is essentially identical to the concept of efficiency. The amount of *activity* is simply how much of a particular good or service is produced (e.g., 100 tons of steel produced or 100 bbl of oil used annually). Thus

total energy use = energy intensity of activity
× amount of activity

Since conservation also can be defined as using less energy to produce a comparable level of per capita goods and services, our focus will be on decreasing the intensity of a particular sector rather than decreasing the amount of activity. Nevertheless, it is important to remember that the United States could conserve a great deal of energy by reducing economic activity, which may in fact happen.

INTERNATIONAL COMPARISONS OF EFFICIENCY

International comparisons of the energy-GNP relations of other industrialized countries are the most important reason that many analysts have concluded that we can enjoy our present standard of living but consume far less energy per capita in doing so. President Carter, for example, stated: "Germany, Sweden, Japan and other countries have the same standard of living that we do, as far as the material things are concerned. They consume only one-half as much energy as we do. So I think we need to cut back on the consumption and waste of energy." The ecologist Paul Ehrlich maintains that countries like Sweden achieve this superiority of using one-half as much per capita energy because they are "cleverer than the U.S. in extracting more benefit from less energy." Past energy administrator John Sawhill claimed that 30% of our energy use is waste, which is why the United States has emerged as an "energy extravagant society," consuming over 30% of world energy annually while comprising only 6% of the world's population. Such statements leave one with the impression that a 30% difference in per capita energy use between the United States and certain other nations with as high a GNP, such as Sweden, *ipso facto* demonstrates that 30% of the energy consumption in the United States is pure waste.

Chauncey Starr, president of the Electric Power Research Society, believes that such a simplistic analysis of international GNP–energy relations is inaccurate because only one of the input variables (energy) is considered while ignoring international differences in the others (capital, labor, raw materials) (Starr 1979). Many other countries use energy *less* efficiently to produce wealth than does the United States (Figure 2.14). Each society had industries that combine their own diverse inputs in very diverse combinations for diverse ends. Starr points out that a more significant measure may be how well these inputs are integrated to establish national well-being, although this is more difficult to represent empirically. Starr and others have pointed out that differences in E/GNP ratios among countries are due to other variables as well, including resource endowment, geography, climate, population density, living style, cost of energy relative to capital and labor, structure and history of the economy, age of equipment, mix of agriculture and industry, range of imports and exports, and social values. Although there is a correlation between energy use and economic output, the importance of our abundant high-quality domestic energy resources and of falling energy prices in the United States from 1900 to 1970 has too often been ignored as a factor in determining our high level of energy consumption, for we have had little reason *not* to use large quantities of this useful stuff. Also at any given income level there exists a wide variation in energy use per capita, particularly as energy used directly, so that no firm rules can be set on the relation between energy use and GNP. We have examined empirically some of the factors suggested by Starr (and others) and found them to explain most of the differences in this ratio among nations (see Figure 2.17). Some of these factors are considered in more detail below.

The Importance of Product Mix

When trying to account for the observed variability in energy–output ratios among nations, it is of primary importance to distinguish between the energy intensity that characterizes particular economic activities and differences in structure, or product mix, between the economics as a whole. The energy intensity of a given economic activity refers to the amount of energy used to produce a single unit of a good or service. The structure (output or product

mix) of a country's economy refers to the way production is distributed across different economic sectors, that is, how much steel versus how many computers. The respective roles of these two factors, intensity and structure, are important parameters of any international comparison of energy use because aggregated comparisons of energy and output obscure the more relevant aspects of the relationship. For example, it is quite easy for a nation to have a high E/GNP ratio simply because its economy is dominated by energy-intensive industrial activities like steel making (e.g., Luxembourg) or petroleum refining (e.g., Saudi Arabia) compared to a nation that pursues largely banking activities. In such a comparison E/GNP ratios are not appropriate measures of energy efficiency due to the structural differences between the countries' economies and because some countries export most of what they produce. Thus as an index of efficiency it is less appropriate to compare two countries' aggregate energy consumption level, or even the level for a single consuming sector. Rather it is more informative to look at the relative intensities with which the two nations produce a similar unit of output or accomplish similar economic tasks. When such comparisons have been made, the United States appears to be essentially no less efficient than other countries (Dunkerley et al., 1980).

Nevertheless, some countries' aggregate economy do seem to be particularly efficient. The Sweden–United States example is the most familiar of such comparisons (see Table 2.14). Sweden uses 40% less energy per each dollar of GNP, uses less fuel to heat its homes, and gets better gas mileage from its cars (Schipper and Lichtenberg, 1976). The greatest differences in energy intensities—taking into account the structure of the economy and other features such as distances, amount of fuel extraction, climate, and energy embodied in foreign trade—appear in the areas of space heating, transportation and industrial process heating.

The greater efficiency with which the Swedes heat a square foot of house or apartment is not due to the fact that there are more apartments in Sweden, since the energy intensity of apartment heating in Sweden is roughly equal to that in single-family dwellings, nor can the difference be attributed to differences in living space per capita, since the figure is about the same for both countries. Sweden's higher efficiency stems from better-constructed building shells that maintain higher levels of thermal integrity and because fewer appliances are used per household. Transportation, dominated by the automobile in both countries, is in general much less energy intensive in Sweden. Swedes traveled 60% less than Americans and used 60% less energy per passenger mile in 1972. Both mass transit and intercity rail systems are more widely used in Sweden, whereas energy-intensive air travel is significantly greater in the United States. The only area of transportation in the United States that is more efficient is the trucking industry, which uses less fuel per mile traveled although the total use of energy by this industry is larger.

Industrial activity in Sweden consumes 40% of all primary energy and is geared toward the production of energy-intensive manufactured products. Industry by industry the energy intensities in the United States and Sweden are not too different. On a product by product basis, however, the United States is more energy intensive than Sweden (see Table 14.1). Process heat requirements are lower because Sweden uses newer equipment that employs heat recovery systems. Sweden also has relatively large amounts of cheap, relatively efficient hydroelectric power.

There are several conclusions drawn from the preceding comparisons:

1. Demographic and structural factors as well as energy efficiencies shape energy use in each country.
2. Energy use and energy efficiency are dependent in part on the price of energy relative to other resources, which has encouraged or discouraged efficiency in the past.
3. Institutional factors, shaped by government policies such as strict building codes, can encourage efficient energy use.
4. Financial policies can help industry to cope with increasing energy prices by providing the necessary capital to purchase more efficient equipment.

The Swedish example suggests that energy use for important tasks is quite flexible, given the right combination of time, technology, policies, and prices. It should not be assumed, however, that determining the right combination will be easy, and very large, energy-intensive investments would be required for a comprehensive conservation effort.

Table 14.1. Energy Intensities and Costs in Industry

Industry	T_j (kwh/capita) United States	Sweden	E (kwh/$) United States	Sweden	P (¢/kwh) United States	Sweden	P_e/P_f^a United States	Sweden
Five energy-intensive industries (excluding feedstocks)[b]								
Fuel	16,840	12,275	40	37.4	0.15	0.20	5.4	3.75
Electricity	1840	2930	4.4	8.9	0.81	0.75		
Other manufacturing[c]								
Fuel	3475	900	2.6	1.1	0.19	0.36	6.3	3.1
Electricity	1050	710	0.8	0.9	1.2	1.1		
Total manufacturing								
Fuel	20,315	13,175	11.7	11.6	0.16	0.22	6.25	3.7
Electricity	2890	3640	1.7	3.2	1.0	0.82		

Source: From "Efficient Energy Use and Well-Being: The Swedish Example" by L. Schipper and A. J. Lichtenberg, *Science* Vol. 194, pp. 1001–1013, Table 11, 3 December 1976. Copyright 1976 by the AAAS.

Note: The U.S. data are for 1971; the Swedish for 1970. Prices are for purchased fuels only. Electricity data are for purchased electricity, except for the Swedish paper industry. To change kwh/$ to kcal/$, multiply by 860.5.

[a] Subscripts e mean electricity and f, fuel. *P* is price.

[b] Energy-intensive industries include paper [Standard Industrial Classification (SIC) 26, Svensk Näringsgrenindelning (SNI) 341], chemicals (SIC 29, SNI 353 and 354), stone glass and clay (SIC 32, SNI 36), and primary metals (SIC 33, SNI 37).

[c] Other industries include SIC 20 to 25, 27, 30, 31, and 34 to 39, and SNI 31 to 33, 342, 355, 356, 38, and 39.

WILL THE SOFT PATH WORK?

The extreme view of the utility of conservation as an energy source is probably represented in the works of Amory (and Hunter) Lovins (1976, 1977, 1981), who argues strongly and persuasively for the virtues of achieving the energy requirements of the United States and other industrial nations through a combination of conservation, increased efficiency, small-scale systems built for specific jobs and the use of solar power—a so-called *soft path* versus the more conventional *hard path* of fossil and nuclear fuels. Lovin's arguments and predictions, once dismissed out of hand by many planners and politicians, have gained new respectability as certain of his predictions have come to be reality. Specifically, his predictions of electric power demand for 1980, made in the mid-1970s, were lower than anyone else's but were by far the most accurate when 1980 arrived. Clearly conservation, or something like it, has become a new and important component of the U.S. energy situation, and one might think that we are "on the soft path" (Table 14.2).

A closer examination of certain aspects of the position of the more enthusiastic supporters of the *soft path,* however, leaves us with a number of disquieting questions that appear not to have been answered in their writings to date. Specifically, although it is clear that energy can be saved in a gross sense by the *soft path,* it also is clear to us that many of these approaches have operating, investment and, environmental energy costs that have not been evaluated explicitly.

The following are some specific facts that demonstrate that conservation, at least so far, is not quite all it has been cracked up to be:

1. Although U.S. industries have used less energy directly since 1973 they have used more indirectly through a higher use of capital equipment.

2. The principal reason that electricity and other energy growth has decreased is that there was relatively little real economic growth in the U.S. economy between 1973 and 1983 (Hirst et al., 1983).

3. A recent study done by members of the U.S. Department of Energy found that much of what people had been attributing to conservation was instead *doing without,* that is, driving less and heating fewer rooms to a lower temperature (Marland, 1982).

4. Energy and other resources are required to build and maintain even energy efficient capital. We have seen in Chapter 13 how the energy requirements for constructing solar collectors are not trivial at all, and Chapter 15 indicates that the environmental impacts are not trivial either.

On the other hand, the sorts of things that Lovin's advocates—increased insulation, less building lighting, and the like (see Lovins, 1985)—certainly have contributed to the cessation of energy growth in the United States, and any investment in conservation or improved efficiency has the added advantage that it will continue to save energy far into the future. We are enthusiastic about the idea of the soft energy path and hope that it can be realized to some large degree, but we are realistic about its constraints. We are not of the opinion that the United States is about to deindustrialize or become simply a high-tech society (e.g., Hawken, 1983) because of the energy intensity of our most fundamental requirements. Atari games are no substitute for food, shelter, and transportation, and the energy cost of many basic resources is increasing.

CONSERVATION IN INDUSTRY

In 1984 industry used about 7.0×10^{15} kcal, equivalent to about 38% of the energy used in the United States. This is 15% less than peak industrial energy use in 1979. A prevailing view in the first three quarters of this century was that maintenance of a healthy industry necessarily meant continued high energy consumption, and U.S. energy policies focused on measures that would increase the supply of energy to society rather than encourage widespread conservation measures. Investment in increased energy supply is commonly believed to produce the benefits of expanded employment and increased output of goods and services, whereas investments in conservation were often viewed as costs that decrease economic well-being.

Energy consumption in the industrial sector of the United States and other industrialized countries is dominated by a few energy-intensive industries, and four general end uses account for almost 85% of the energy consumption in U.S. industry: process steam, direct heat, electric drive, and electrolysis. The generation of process steam alone consumes more than one-third of industry's total. The top eight energy-using industrial sectors in the nation are chemicals and allied products; primary metals; paper and allied products; petroleum and coal products; stone, clay, and glass products; food and kindred products; fabricated metals; and transportation equipment. These industries consume about 86% of the industrial energy. On average, these industries require 9700 kcal to produce one 1976 dollar of output (i.e., value added by manufacture) while producing 55% of industry's total economic activity. In contrast, the remaining industries consume 14% of the industry's total economic energy while requiring only 1900 kcal to produce one 1976 dollar of value-added output and accounting for 45% of industry's economic activity.

It has been suggested that secondary industries are more similar to the commercial sector than to the primary industries in that they use more energy for space heat and lighting than for fabrication of their product. It follows that conservation measures appropriate for the commercial sector could also succeed when applied to the secondary industries. Of course these secondary industries must purchase energy-intensive raw materials from the heavy industries, so that when indirect energy costs are included the various industries are not so different.

The Impact of Low Energy Prices

For years, up until the early 1970s, U.S. industry was faced with declining real energy prices while the prices of labor and capital were increasing. In particular, the use of natural gas and electricity were, in effect, subsidized through promotional pricing and government regulation. The availability of cheap energy in many cases led to an energy-inefficient capital base. In an attempt to minimize costs, cheap energy, and technology and process equipment designed to use it, were exploited liberally to produce cheaper goods and services. Large increases in unemployment did not occur due to the increase in economic activity of the nation as a whole—the loss of jobs caused by energy substitution was offset by an increase in demand for goods and services. An example of this substitution of energy for labor is the centralization of most production facilities such that more energy is required due to increased transportation costs. Centralization, however, allows capital, labor and on-site energy to be used more efficiently in the production process.

The situation has not been the same in all other industrialized countries for in 1972 prices for various fuels in other countries tended to be higher while energy intensities of steel, cement, aluminum, and pulp and paper were lower. It is apparent that Japan and the European countries are generally able to produce a unit of product with less energy than the North American countries. Numerous studies (e.g., Dunkerley, 1980) have suggested two major reasons for the difference in energy intensities: energy prices and age of capital stock. In Europe and Japan energy prices historically have been higher than in North America, primarily due to the lack of domestic supplies. The capital stock in these energy-efficient nations is newer for two reasons: (1) much of the original equipment in some of these nations was destroyed in World War II, and (2) increasing energy prices have made it economically efficient to replace old equipment. Therefore the international comparisons of energy intensities point to the im-

Table 14.2. Energy Savings and Procedures to Conserve Energy as Reported by Some U.S. Firms in the 1970s

Firm	Reduction	Time Frame	Comment
Burger King	17%	1974–77	50% by housekeeping
Lockheed (Los Angeles area factory complex)	59%	1972–77	Almost no investment
Tenneco	17%	1972–77	50% by housekeeping; rest by recycling waste heat
Colgate-Palmolive	18%	1973–76	Mostly housekeeping
Exxon (U.S. refineries)	21%	1972–77	80% with little or no capital investment (11.3 million bbl a year)
Western Electric (Kansas City plant)	38%	1972–77	Almost no investment

Source: From Stobaugh and Yergin (1979).

portance of energy prices in determining energy efficiencies. In short, energy was not substituted for labor and other resources in Europe and Japan to the degree it was in the United States and Canada where energy was cheap.

Nevertheless, there have always been strong incentives to reduce energy use in U.S. industry, and a great deal of engineering activity and engineering pride is based on designing efficient machines and processes. Probably the single most important conservation activity ever done was the invention of petroleum cracking processes that allowed the use of a much greater proportion of the energy in a barrel of crude oil. Another very important process was the electrification of mechanical power in manufacturing industries that allowed higher efficiencies than older processes that used huge building-long drive shafts, belts, countershafts, and so forth (Devine, 1982). These older transmission devices often consumed up to three quarters of the output of the steam engines that drove them. Costanza (1980) argued that industry has always had incentive to conserve its fundamental resource, energy, even well before the large price increase of energy in 1973, but that the incentive would be to conserve *both* embodied and direct fuel energy, which would sometimes make it appear fuel inefficient. Recent analyses by Bruce Hannon and his associates at the Energy Research Group at the University of Illinois, covered in more detail later, indicate that improving energy efficiency in the U.S. steel industry is not easy even though the Japanese are able to produce a ton of steel with only half the energy as the United States. For another example, an increase in cogeneration—whereby combined units produce various combinations of industrial process steam, hot water, and electricity—is extremely useful in certain industries but less so in others because the excess heat still must be wasted to the environment in summer and because cogeneration designs have a lower efficiency of production of electricity than do plants designed to produce only electricity.

CONSERVATION IN INDUSTRY: AN ENGINEER'S PERSPECTIVE*

Conservation has been defined in this chapter as the use of less resource to make the same amount of

*The authors would like to acknowledge Alan Rojer for contributing this section.

product. Often discussed in reference to energy, conservation is equally appropriate for capital and even labor. Although compelling reasons for conservation include preservation of scarce resources, reduction of pollution, and elegance of design, the principle incentive for industrial energy conservation is profit. In general, resource consumption carries an associated cost, so conservation is potentially a route to increased profit. Of course there are costs associated with conservation as well. In this discussion the methods used to analyze conservation costs and savings will be presented. Technical approaches to industrial energy conservation will be introduced and categorized based on their efficacy, and the analysis of conservation decisions will be demonstrated with several case studies.

The role of the engineer in the design and modification of industrial facilities is to meet a production specification within some type of economic constraint, such as minimum cost or maximum return on investment. Conservation is thus a subgoal or even an artifact of design decisions to meet technical specifications and economic constraints. In the United States the economic constraint on the designer is usually maximization of return on equity. Equity is the capital that must be invested to design, construct, and start up a facility. Return is the cash generated by the facility (i.e., revenue less expenses and taxes). Most design decisions incorporate a compromise between equity and return; by investing more capital, greater revenue or smaller expenses can be obtained. Conservation decisions exemplify the trade-offs between increasing equity and reducing expenses.

A project or facility can be considered an open system surrounded by a boundary or envelope. Dollar, material, and energy streams cross the system boundary. Knowledge of the magnitude and direction of these streams allows overall analysis of the project as a black box with known inputs, transforms, and outputs. The facility can then be analyzed as a subcomponent of a larger system, such as a corporation or a national economy. Design decisions regarding the internals of the system will determine the flows across the system boundaries over its life. Choices among design options are explicitly or implicitly based on estimation of the dollar, material, and energy flows across the project envelope in the alternative designs.

From a financial standpoint equity is put into the system at its inception. For the life of the project, revenues represent cash flowing into the system;

taxes and expenses represent cash flowing out of the system. For a financially viable system, revenues must exceed expenses for enough of the life of the project to generate enough cash to more than repay the equity. From an engineering standpoint raw materials and energy flow into the system; products flow out of the system.

The boundaries of the system are usually clearly defined physically and analytically; thus the decision-making process for the designer is simplified. Note that the designer's decisions are motivated by economic and technical considerations; social considerations such as pollution control or conservation of scarce resources are largely irrelevant unless present in the technical specifications. Of course the designer is influenced by personal ethical and professional standards. Many designers seek an elegant, efficient design and harmony with the environment, but these personal goals may be in direct conflict with the institutional goals of their employer.

This book has introduced the concept of energy return on investment (EROI) as a criterion for evaluation of investment decisions. This concept recognizes the key role that energy plays in underpinning our entire economic structure. However, from the standpoint of a business entity such as a corporation, EROI is not an explicitly used decision-making criterion. Market valuations of material and energy often do not reflect social or thermodynamic values, at least not in the immediate sense. Thus the designer does not seek minimum energy expenditure unless that goal is coincident with maximum return on investment. But, as described in chapter 2, energy quality and especially the EROI for a society's energy sources are a very important determinant of the technological make-up of its economy. Thus engineers and designers are indirectly influenced by EROI to the degree that their choices are constrained by existing technological capabilities. Beyond this legislative and policy-making authorities may create incentives and disincentives that force economic conditions to reflect externalities, social valuations and encourage consideration of broader performance indexes such as EROI.

To summarize, the incentive for industry to conserve is purely economic. The designer strives to meet a technical specification within financial constraints. Conservation by industry is not a goal (despite corporate public relations statements to the contrary) but an artifact of pursuit of financial goals. With an understanding of the motives of industry achieved, this discussion moves to technical approaches to conservation and evaluation of conservation decisions from a traditional financial perspective and from the perspective of EROI.

How does industry conserve energy? Three principal avenues to energy conservation can be discerned. The first approach involves improvement of operating conditions in a plant to reduce needless consumption of energy. Activities of this sort involve turning off pumps, motors, and lights when not needed, fixing broken windows in factories to reduce heat loss, replacing failed steam traps (which are valves that selectively pass condensate, but retain steam), or cleaning heat exchange surfaces to reduce temperature losses in heating and cooling processes. These types of energy conservation techniques are collectively known as *housekeeping* because they reflect mainly good maintenance and operation practices and require little capital expenditure to implement. In essence, housekeeping is simply the elimination of needless energy waste through better use and maintenance of existing facilities. Housekeeping improvements are readily implemented and often result in large energy savings. In the aftermath of the energy price increases of the early 1970s, U.S. industry was very successful in implementing housekeeping improvements to reduce energy consumption. Because of their ease most housekeeping improvements have already been implemented by industry; further conservation gains will have to come from elsewhere.

The second class of energy conservation techniques involves investment of capital to reduce energy consumption during operation. These techniques encompass such improvements as more efficient motors, variable speed pumps and fans, larger heat exchangers to reduce temperature approaches, and installation of equipment to recover otherwise wasted energy. An obvious example which also shows the problems associated with converting a large existing capital base is given in Fig. 14.1. Collectively, these kinds of improvements can be considered *incremental* conservation because small changes are made to the process or the equipment to elicit gains in efficiency. These techniques are all distinguished by the requirement to spend money (and energy) to save energy (and money). The criteria for decisions on whether to implement an incremental energy conservation project is the financial criteria described before. For industry today most energy conservation options are in this class. Unfortunately there are diminishing returns to incremental conservation, resulting in a type of

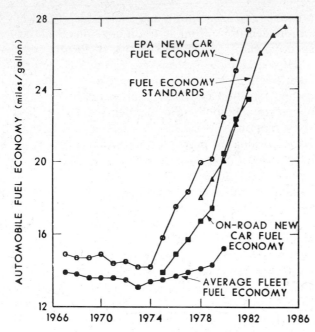

Figure 14.1. Automobile fuel economy estimates. The graph shows federal fuel economy standards from 1978 to 1985, new car fuel economy estimates based on the Environmental Protection Agency test procedures for 1967 to 1980, on-road new car fuel economy as estimated by DOE from 1975 to 1980, and fleet fuel economy as estimated by the Federal Highway Administration from 1967 to 1979. (From Hirst et al., reproduced, with permission, from the *Annual Review of Energy* Volume 8. © 1983 by Annual Reviews Inc.)

negative synergy; with each incremental improvement, subsequent improvements become less attractive. This result will be illustrated in the case studies presented here that outlines details of a few popular incremental conservation approaches and their analytic evaluation.

The final class of energy conservation involves the introduction of new technology to change radically the production process to use less energy. Because of the deep technical changes implied in this type of conservation, it is denoted *structural* conservation. The potential gains are greatest in structural conversion, but large-scale changes and often complete replacement of existing capital stock are required in implementation. Unlike incremental conservation, which has an array of generic approaches applicable to a wide range of facilities, structural conservation is technology intensive and often applicable only to a single production process. Structural conservation is usually associated with research, whereas incremental conservation is typically associated with engineering. After structural conservation has been implemented, further incre-

mental improvements are usually possible. Examples of structural conservation include the Pilkington float process in window glass manufacture, the electric arc furnace in steel production, and the catalytic cracker in petroleum refining. Each of these revolutionized processing and efficiency in its respective industry. Note that these innovations are unique to their process; they are not directly applicable to other energy-intensive production process.

In summary, conservation may be achieved through housekeeping, incremental improvements, or radical process changes. Housekeeping conservation is inexpensive, rewarding, and mostly exhausted. Incremental improvements require trade-offs between investment and return and offer a widely variable potential for savings. Structural improvements offer the greatest long-term potential but also require deep changes and the early retirement of perfectly good capital equipment, and they are difficult to anticipate because they are driven by improvements in technology and science.

The foregoing classification is based on the degree of alteration that conservation imposes on the production process. We also can classify energy conservation based on EROI. Again three broad classes of energy conservation options can be discerned. In the first class, we place energy conservation improvements that are an unqualified success in terms of the energy saved compared to the energy invested. Here we find most housekeeping and structural conservation. These improvements require little or no expenditure of energy to achieve conservation. What these types of improvements require is application of information to energy consumption. Of course the information has an associated energy cost, but this cost is likely to be negligible relative to the energy saved.

Most incremental energy conservation improvements are not unqualified successes. In both EROI and traditional financial terms, investment must be made to be later repaid out of savings. If the savings generated repay the initial investment and provide an excess reward (return), then the incremental improvement is an example of a beneficial trade-off. If the initial investment was not recouped in the savings, then the investment was unjustified, and the modification should not have been implemented. Note that so far we have not specified energy or financial investment and return. From the standpoint of industry financial return is the criterion for a decision. From a policy or social standpoint EROI perhaps should be the criterion for a decision. But these two methods of analysis may give conflicting results. In certain cases a financial incentive for a modification may exist where energy returns are negative, and vice versa. This possibility will be illustrated in the case studies.

In summary, conservation modifications may be unqualified successes, beneficial trade-offs, or unjustified, energy-losing investments. The classification will depend on the method of analysis used for evaluation of investment and return (financial or EROI). Industry will unfailingly use financial criteria to evaluate conservation modifications; social goals may warrant use of EROI to evaluate conservation modifications.

Conservation Case Studies

Several case studies in incremental industrial energy conservation follow. Symbolic representations are used for costs and values to retain generality. The case studies include the addition of a waste heat boiler to a furnace, the installation of sophisticated combustion controls on an industrial boiler, the installation of a heat pump to save steam heating energy, and two examples of coproduction of electrical energy with heat (cogeneration). Each case will be briefly described, and the costs and benefits estimated.

Consider first a hypothetical petrochemical plant that produces several types of plastic with a widely available petrochemical intermediate as its feedstock. The plant purchases natural gas from a local utility for use as fuel at cost per unit energy, g. Assume that the plant is operating close to capacity and that the plant engineer has proposed the evaluation of two energy conservation modifications: the installation of a waste heat boiler on the exhaust from a furnace and the upgrade of the combustion controls on the plant boilers to increase boiler efficiency. Financial and EROI analyses of these modifications in isolation and together are presented here to determine the degree to which they will or will not save money and energy.

In the plant is a furnace where a chemical reaction takes place at high temperature. The products of combustion that leave the furnace are elevated in temperature sufficiently that with the installation of a waste heat boiler, steam could be generated to replace steam that is now generated in boilers fired with natural gas. A vendor of waste heat boilers has proposed a unit that will cost $C1$ dollars and will produce steam at rate $M1$. What are the benefits for

the plant? Should it undertake the installation of the waste heat boiler?

The cost of the installation is known: $C1$. The savings that the installation will produce must be calculated. These savings are in the form of natural gas which will not be burned, since the production of steam from existing gas fired plant boilers will be reduced by $M1$. Thus if h relates steam flow to energy flow, and f is the boiler efficiency, the savings rate $S1$ will be given by

$$S1 = h \times M1 \times \frac{g}{f} \qquad (14.1)$$

For the plant engineer the savings must be balanced against the cost of the modifications. The simplest relationship between cost and savings is termed the simple pay back period, $P1$; it is given by the cost divided by the savings rate and has units of time:

$$P1 = \frac{C1}{S1} \qquad (14.2)$$

When simple pay back is used as a criterion for investment decision, an arbitrary cutoff N is usually chosen which must be met by the project. Thus a project is recommended in the case that the simple pay back is less than the cutoff:

$$P1 < N \qquad (14.3)$$

A more sophisticated, more commonly used analysis incorporating the time value of money and capital, inflation, taxes, and debt financing is provided as an appendix. From an energy standpoint the cost of the project is the energy cost of the new equipment. This must be compared to the savings the project will generate over its life. This term of course is the EROI, here given by

$$\text{EROI} = n \times \frac{S1}{C1 \text{ (energy)}} \qquad (14.4)$$

Here n is the useful life of the project. Great care must be exercised in use of the EROI because energy equivalents do not necessarily reflect equivalence of energy value; that is, the second law of thermodynamics says that all Btus are not created equal, as we are aware from previous considerations of energy quality. Thus we cannot calculate a meaningful EROI based on electricity in and heat out and compare it to an EROI based on fossil fuel in and heat out. This will be demonstrated in a later example. In this case we are concerned with mixed energy investment and fuel energy savings. Note that there is no reason why the EROI and simple pay back should give consistent results, although they might. Note also that a financial equivalent to the EROI could be defined as FROI:

$$\text{FROI} = n \times \frac{S1}{C1} \qquad (14.5)$$

In practice, this term is unused, because it fails to account for the time value of money and capital (i.e., money today is worth more than money tomorrow because today's money can earn interest between now and tomorrow) and the effects of inflation and taxes. These effects are considered in the appendix. Returning to the simple pay back, consider the expanded expression:

$$P = \frac{C1}{S1} = \frac{f \times C1}{h \times M1 \times g} \qquad (14.6)$$

Note that we seek to minimize the pay back period; thus increases in the denominator of the expression encourage the modification, and increases in the numerator discourage the modification. Now consider the other project. This one is for the installation of sophisticated electronic boiler controls to increase the efficiency of the fired boilers in the plant (these controls have no effect on the performance of the proposed waste heat boiler). Call the cost of this modification $C2$. Let the total annual steam requirement of the plant be M. Suppose the new controls result in an efficiency improvement df. Then the savings rate from the modification $S2$ will be given by

$$S2 = \frac{h \times M \times g}{f} - \frac{h \times M \times g}{f + df}$$

$$= \frac{df \times h \times M \times g}{f \times (f + df)} \qquad (14.7)$$

Now consider the effects of implementation of both modifications. First, note that the savings from the waste heat boiler are decreased because of the effect of the increased efficiency of the boilers from which steam is displaced. Note also that the improvement in boiler efficiency from the new controls is less valuable because the boilers are producing less steam; some steam production has been shifted to the waste heat boiler. The total savings $S3$ are given by

$$S3 = \frac{h \times M1 \times g}{f + df}$$

$$+ \frac{df \times h \times (M - M1) \times g}{f \times (f + df)} < S1 + S2 \qquad (14.8)$$

Of course the total cost $C3$ is given by

$$C3 = C1 + C2 \qquad (14.9)$$

Note that the simple pay out of the combined modification is worse than if the savings had been truly additive. This is an example of the diminishing returns or negative synergy of incremental energy conservation modifications. In practice, this is a widespread phenomenon. The same effect is seen with EROI.

Consider now a creamery that produces a variety of products from raw milk. The dairy purchases electricity from a utility at cost e per unit energy and fuel oil from an independent vendor at cost o per unit energy. Inside the creamery, steam is produced in typical industrial boilers fired by fuel oil. The steam is used for heating, sterilizing, and driving refrigeration machines. Electricity is used to drive pumps, fans, mixers, and refrigeration machinery as well as for lighting. To reduce steam consumption, the plant engineer recommends installation of a heat pump for heating hot water. He notes that the current heat exchangers use X steam per year. Based on conversion h from steam to energy, and boiler efficiency f, the annual fuel oil energy consumption is hX/f. But a vendor of heat pumps has guaranteed a heat pump with a coefficient of performance of 2.0; that is, every energy unit of electricity input will result in 2 energy units of heat to the hot water (a typical heat pump performance). The engineer has noted that the unit cost of electricity (\$0.05/kWh = \$11.70/MMBtu) is only 1.8 times the unit cost of steam from fuel oil (\$6.50/MMBtu) at 80% boiler efficiency (\$8.13/MMBtu). This seems like a potential energy and money saver; the plant will consume 1 unit of electricity at a cost of 1.8 where previously the plant consumed 2.0 units of steam at a cost of 2.0—a 10% reduction in energy costs.

From the point of view of both energy and finance this seems like a favorable arrangement for the plant. But consider a wider system boundary, which now includes the power plant where the electricity is generated. Typical thermal power plant efficiencies are around 35%; that is, nearly 3 units of thermal energy derived from fossil fuel combustion must be consumed to produce 1 unit of electrical energy. Recognizing the source of the electricity, we realize now that to provide the 1 unit of electrical energy at the creamery, 3 units of thermal energy are consumed at the power plant. For society, then, the net result is an increase by 50% of the energy that must be used to create the same effect at the creamery. Note that the creamery economics (not atypical numbers!) do *not* reflect the ultimate energy cost of the modification. Two pitfalls in analysis are at work here. The first is the danger of drawing too small a systems boundary and not properly valuing different forms of energy. In this example the engineer considered the electric energy and steam energy as basically equal. The second pitfall is that too often market prices do not accurately reflect relative energy costs. In particular, electricity is often undervalued relative to fuel for various institutional reasons. This encourages decisions at the plant or corporate level that are not in the interests of energy conservation but are motivated by sound financial reasons. Discrepancies like these must be addressed by policymakers and legislators.

Unfortunately for the (real) creamery in this example, a large nuclear power plant was brought on line by the utility at about the time the heat pump was under consideration. Electricity prices increased 40% at that time and plans for the heat pump were shelved. Instead, the creamery engineers began to study the possibility of cogeneration at their site to mitigate the effects of their newly swollen electric bills. Cogeneration is the simultaneous production of heat and power. This coproduction allows greater utilization of fuel energy, at the expense of reduced electrical power production per unit fuel relative to a traditional power plant. Because the creamery had both heat and power requirements, it was a good candidate for fuel and money savings via cogeneration.

Excluding refrigeration, the creamery could identify a steam use of M and an electric power use of E. Refrigeration was excluded in the first analysis because it could be provided by steam or electricity. An engineering consulting firm told the creamery executives that for cost C, they could have a cogeneration system that would completely satisfy their steam use M and produce a fraction x of their power use E. The firm prepared the following analysis which compares the existing creamery and the creamery after installation of the proposed cogeneration system.

To provide the steam for the existing creamery, the current boilers add $h0$ energy per unit steam at efficiency f. Thus the cost of steam CS is currently given by

$$CS = \frac{h0 \times M \times o}{f} \qquad (14.10)$$

The cost of electricity CE currently is given by

$$CE = e \times E \qquad (14.11)$$

Thus the total operating costs for plant utilities $C0$ is given by

$$C0 = CS + CE = \frac{h0 \times M \times o}{f} + e \times E \quad (14.12)$$

In the new cogeneration facility, the same amount of steam is produced but at a higher temperature and pressure, so that before it is used for heating, it is expanded through a turbine to the old temperature and pressure. The expansion of the steam through the turbine generates electricity. Because the steam pressure and temperature are now higher, the energy input per unit steam $h1$ is also higher:

$$h1 > h0 \qquad (14.13)$$

The total cost of utilities $C1$ is now given by

$$C1 = \frac{h1 \times M \times o}{f} + (1 - x) \times e \times E \qquad (14.14)$$

Note that the fuel component of the cost has increased, since $h1 > h0$, and that the electric component of the cost has decreased, since $x < 1$. The net savings S to the creamery are given by

$$S = C0 - C1$$

$$= x \times e \times E - (h1 - h0) \times \frac{M \times o}{f} \qquad (14.15)$$

By virtue of the first law of thermodynamics we know that energy must be conserved, and if extra energy is put into a system, it must be removed somewhere. Comparing the existing system to the proposed system, note that the excess energy put into the steam system is given by

$$Q = (h1 - h0) \times \frac{M}{f} \qquad (14.16)$$

Note that the excess energy that the system produces is simply the electric power cogenerated. This must equal the extra input. So we have

$$Q = x \times E \qquad (14.17)$$

Substituting from (equations 14.16) and (14.17) into (equation 14.15), we can write

$$S = (e - o) \times Q \qquad (14.18)$$

Thus the net savings is given by the difference in unit costs of electricity and oil times the extra energy input to the system. Economic incentive can be readily determined. Because electricity is virtually always more valuable on a unit energy basis than fuel, the savings are almost always positive and likely to remain so. This indicates the financial incentive to cogenerate. Consider now the energy incentive. We have noted earlier the amount of additional energy that must be supplied to the cogeneration system to produce the electricity. Because a heat sink is available for rejected heat from electricity production, the additional heat added to the system is equal to the electricity production. Compare the heat required to make the electricity in a conventional thermal power plant $Q0$ with efficiency r :

$$Q = x \times E = r \times Q0 \qquad (14.19)$$

Thus

$$\frac{Q}{Q0} = r \qquad (14.20)$$

and

$$Q0 - Q = Q0 \times (1 - r) \qquad (14.21)$$

The energy savings is given by the complement of the efficiency; cogeneration under these circumstances is a very attractive approach to energy conservation.

Several qualifying statements are necessary, however. First, note that the comparison with a thermal power plant is not applicable to a hydroelectric power, which has no associated fuel costs. But because hydropower is very limited, incremental power production is nearly always fuel based. Consider that the value of the fuel for a cogeneration plant is likely to be higher than that for a power plant. Cogeneration plants typically burn fuel oil or natural gas; power plants may be nuclear or coal fired. In this case national security considerations (i.e., reduction of imported energy) may suggest consumption of more energy. Also cogeneration plants are less likely to have pollution controls like SO_2 scrubbers that are required for central power stations. Despite these caveats, cogeneration has excellent potential for industrial energy conservation but, as we have shown, very careful calculations must be made before intuitively attractive conservation measures are undertaken.

EROI FOR CONSERVATION

Presumably many conservation practices can have a higher return on investment of both dollars and

kcals than does investment in the expansion of our energy base, while having a positive effect on both employment and the production of goods and services. But most studies that purport to demonstrate large savings do not include a thorough systemwide EROI analysis that examines the energy *costs* of conservation materials and activities. In this section we give an example of such a study that did conclude that conservation had a very high EROI. The study also illustrates in detail the process of determining EROI.

The Cayuga Station versus Increased Insulation Analysis

The New York State Electric and Gas Corporation (NYSEG) service area is a loosely aggregated patchwork of rural counties, suburbs, and small cities encompassing much of central New York state. The region is noted both for its scenic beauty and for its long, harsh winters. As of 1975 the NYSEG service area included 556,000 residential electric customers (or "families") residing in approximately 458,000 houses and apartment buildings (NYSEG, 1974). Of these 458,000 dwellings, about 30,000 (or 6.5%) are heated electrically. The mean electricity use per capita is 2154 kWh (or 1.8×10^6 kcal of electrical energy) per year. Unlike most of the United States, the NYSEG service area is a winter peak load system and is expected to remain so until at least 1990. Furthermore about half the absolute growth in the winter peak demand over 20 years was expected to be due to increased use of electric space heat (NYSEG, 1974). Changeovers from cheaper (and more efficient) gas and oil heat are thought likely owing to possible increased prices and especially restrictions in the availability of oil and gas.

Faced with an expected increase in electricity use, NYSEG developed plans for a series of new power plants. One such facility was the proposed 870 MW$_e$ coal-burning power plant in Tompkins County, New York, the "Cayuga Station" (an alternative site is Somerset, New York, on Lake Ontario, which is currently the preferred site, but the following analysis would apply with trivial corrections to that site). The Cayuga plant would be built next to the 270 MW$_e$ Milliken Station that has been operational since 1955.

Public apprehension over shrinking oil and natural gas supplies for home heating and the consequent anticipated increase in electric heat demand also could be alleviated by increasing the amount of insulation in housing. The following discussion will examine the effectiveness and efficiency of the power plant and the insulation for meeting local energy needs, using energy return on investment methods. The study had three objectives. The first was to apply EROI procedures to the proposed Cayuga electricity-generating station as well as to a comprehensive regional program of insulation. The second was to compare methods and conclusions based on economic analysis versus three available energy-assessment methods. Finally, the study attempted to examine the conclusions of the analyses in light of some institutional constraints that are important components of the decision-making process. The methods given here are, in theory, applicable to any proposed energy facility.

The energy efficiency of a coal-fired power plant traditionally is found by dividing the heat equivalent of the electrical energy generated by the heat content of the coal used to fire the boilers:

$$\text{gross efficiency} = \frac{E_g}{CB} = 30\text{–}40\%,$$

where E_g is the electrical energy generated expressed in thermal equivalents and CB is the total quantity of energy in the coal that is burned to produce E_g. This efficiency is generally between 30 and 40%.

This study redefined the efficiency of coal-burning plants so as to compare the electrical energy with total energy investment required to supply that need. We call this the net or system efficiency of the plant, which is identical to the EROI:

EROI = system efficiency =

$$\frac{E_g - E_{pu} - E_{tl}}{CB + E_m + E_p + E_t + E_{pc} + E_{tfc}}$$

where E_g is as before, E_{pu} is the energy utilized internally by the power plant, E_{tl} is electricity lost in transmission (which is a function of the line voltage and the distance transmitted), E_m, E_p, and E_t are the energy required to mine, process, and transport coal, E_{pc} is the energy required to manufacture and install the generating station and associated equipment, and E_{tfc} is the energy required to make transmission facilities such as line towers, high voltage lines, transformers, and other substation equipment amortized over the nominal life span of the power

plant. A similar procedure was used for determining the energy required to produce insulation. Details of procedures are given in Hall et al. (1979a).

Results of the Comparison

The results of this EROI analysis show that the power plant was less efficient than the insulation program at providing energy to the region by at least a factor of 4 when viewed as energy return per dollar invested and at least a factor of 10 when viewed as energy returned per calorie invested (Figure 14.2).

Energy Return from the Proposed Power Plant

The Cayuga Station is rated at 869.7 MW$_{\text{net electric}}$ (and 924.7 MW$_{\text{gross electric}}$), has a nominal efficiency (of net electricity output over coal input, both expressed in thermal units) of 37.78%, and was expected to be operating an estimated 77% of the time over its 30-year nominal life span (NYSEG, 1974). The gross energy output (E_{go}) of the Cayuga Station over 30 years, calculated as mean output capacity times operating time, is 187 million MWh, equal to 161×10^{12} kcal.

The net output capacity of electricity (E_{no}) is the gross output minus electricity used for internal plant operations. The Cayuga Station is expected to use internally 55 MW$_e$, or about 6% of gross output. Thus, the net lifetime output is 176 million MWh ($= 151 \times 10^{12}$ kcal). The energy output of the Cayuga Station delivered to the consumer (E_d) over 30 years is the net electrical output minus the 3% lost in transmission, or 169 million MWh ($= 146 \times 10^{12}$ kcal $= 0.577$ quad).

Dollar and Energy Investments for Proposed Power Plant

The total dollar investment cost to the consumer of the energy produced by the Cayuga Station, computed as the cost of the energy delivered to the consumer, is $4.9 billion (1975 dollars).

The coal required (CB) to generate 151×10^{12} kcal of electricity, calculated from the mean net output of electricity of the power plant and the efficiency of the plant, was 59.7 million metric tons.

The energy cost of constructing the power plant was derived from an itemized list of components for the construction of the Cayuga plant, obtained from the engineering firm contracted for design (United Engineers, 1975). This list, given in Table 14.3, was combined with four methods of energy-cost-per-unit analysis (Chapter 5) to give energy estimates

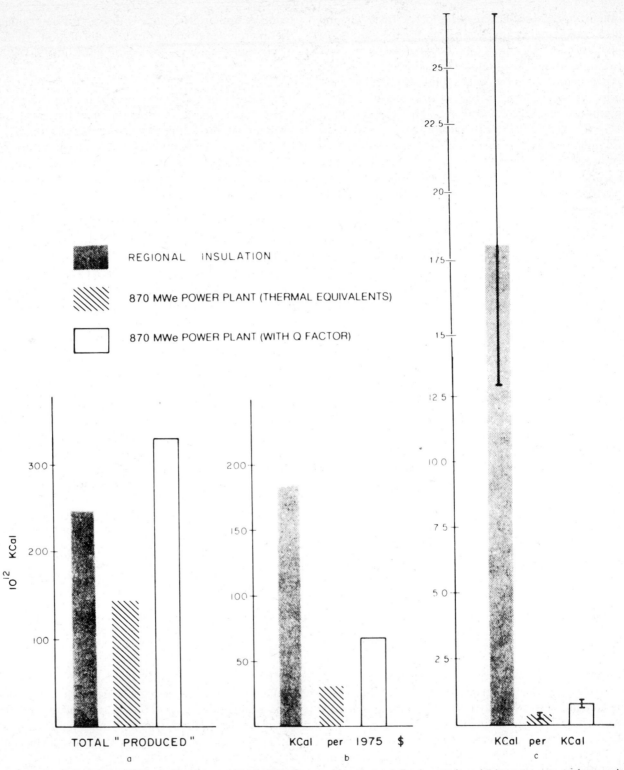

Figure 14.2. Comparison of (*a*) total energy produced, (*b*) energy produced per dollar invested, and (*c*) energy returned per unit energy invested (EROI) for the proposed Cayuga electric power station and the alternative of an extensive home insulation program. (From Hall et al., 1979. Reproduced with permission of Springer-Verlag, New York.)

Table 14.3. Dollar and Energy Cost for Construction of Cayuga Coal-Fired Generating Station

Component	Quantity and Units		Dollar Cost (thousands of 1975 dollars)	Energy Cost (10^9 kcal)		
				Process Energy Method[a,b,c]	Energy Input-Output Method[c]	Input-Output Plus Labor Method
Land and land rights			660	ND[d]	1.57	1.57
Structures and improvements						
Cleaning and grubbing	50	AC	25	0.10	0.14	0.28
Soil testing			50	ND	0.28	0.57
Site grading						
Earth	700000	Yd^3	1400	2.37	7.97	15.88
Rock	1250000	Yd^3	10000	9.19	56.93	113.41
Landscaping			250	ND	1.42	2.84
Roads and paving	50000	Yd^2	350	3.06	16.01	17.99
Fence and gates	15000	Ft	135	0.75	0.95	1.72
Drains and sanitary lines	10000	Ft	500	0.19	3.92	6.75
Lighting	100	Fixt	160	0.75	0.82	1.73
Railroads + 18 turn outs	35000	Ft	1575	6.18	9.23	18.12
Fire protection and potable water	11000	Ft	630	0.21	4.95	8.50
Subtotal			15735.00	22.80	104.20	189.35
Main building-excavate rock	10000	Yd^3	250	3.19	1.77	3.18
Backfill concrete	15000	Yd^3	525	7.81	3.71	6.67
Earth	10000	Yd^3	100	0.03	0.71	1.27
Substructure concrete	17000	Yd^3	2443	8.85	17.25	31.05
Superstructure concrete	2000	Yd^3	400	1.04	2.83	5.08
Structural steel	8500	Tons	8500	57.71	60.03	108.01
Miscellaneous steel	450	Tons	675	3.06	4.77	8.58
Siding	300000	Yd^2	1500	8.78	10.59	19.07
Roof, doors, partitions			1260	7.38	8.90	16.02
Painting	8500	Tons	510	3.21	3.60	6.48
Services			3360	13.18	23.73	42.71
Subtotal			19523.00	114.25	137.88	248.15
Administration	10000	Yd^2	700	6.36	3.99	7.94
Other buildings	20000	Yd^2	800	12.73	4.55	9.07
Subtotal			1500.00	19.09	8.54	17.01
Boiler plant equipment						
Steam generator & equipment			37200	209.80	218.55	428.66
Feed water system			7600	43.87	35.09	78.01
FD and ID fans			3400	13.26	5.70	34.90
Air and gas ducts	3500	Tons	6800	23.76	31.39	69.80
Precipitators			5900	23.14	27.24	60.56
Ash and dust handling			5100	20.01	26.12	54.93
Chimney and foundation	650	Ft	3200	60.45	18.22	36.29
Coal handling structures and equipment			16200	96.50	82.98	174.48
Pulverizer motors and foundations			300	1.58	1.38	3.08
Bunkers and equipment			1400	8.20	9.89	17.79
Fuel oil equipment			3500	16.18	17.93	37.70
Miscellaneous equipment			500	1.96	2.24	5.07
Boiler plant piping			10700	61.76	83.99	144.43
Insulation, paint, etc.			1300	11.51	9.70	17.04

Table 14.3. (*Continued*)

Component	Quantity and Units		Dollar Cost (thousands of 1975 dollars)	Energy Cost (10^9 kcal)		
				Process Energy Method[a,b,c]	Energy Input-Output Method[c]	Input-Output Plus Labor Method
Instruments and controls			3200	10.01	15.21	33.28
Process waste system			1800	7.06	15.69	25.85
Subtotal			108100.00	609.05	611.32	1221.88
Turbine generator units						
Turbine pedestal	6500	Yd³	1450	3.39	8.25	16.44
Turbine mat	5000	Yd³	350	2.18	1.99	3.97
Turbine generator and equipment			37400	210.92	219.73	430.97
Condenser and auxiliaries			5300	46.82	41.60	71.54
Intake structure (including equipment)			2000	ND	11.39	22.68
Circ water pipe			7000	40.41	54.95	94.49
Instrumentation			1500	4.69	7.13	15.60
Paint and insulate			100	0.89	0.57	1.13
Subtotal			55100.00	309.28	345.61	656.82
Accessory electric equipment			9300	40.23	44.20	96.73
Miscellaneous power plant equipment			2300	9.02	10.32	23.31
Station equipment-transmission			3800	14.12	19.53	40.99
Subtotal			15400.00	63.37	74.05	161.03
Subtotal-all direct components			**215358.00**	**1137.84**	**1281.61**	**2494.24**
Indirect components			20600	ND	117.27	233.62
Sales tax			4400	ND	69.98	69.98
Environmental studies			7600	ND	35.09	35.09
Engineering, A.E. costs, owners cost, land, etc.			38040	ND	132.99	132.99
Training, startup, tools, etc.			7000	ND	32.32	32.32
Contingency			34000	ND	193.56	385.59
Interest during construction			100000	ND	349.60	349.60
Subtotal			211640.00	0.00	930.81	1239.19
Grand totals			**426998.00**	**1137.84**	**2212.42**	**3733.43**

Source: From Hall et al. (1979).

[a] Process energy per kilogram were derived from the following references: Barnes and Rankin, 1975; Berry and Fels, 1973; Brevard et al., 1972; Chapman, 1974, 1975; Chapman et al., 1974; Ford Foundation, 1974; Goudarzi et al., 1976; Kline et al., 1977; Lenchek, 1976; Makhijani and Lichtenberg, 1972; Wright, 1974. (See Hall et al. 1979)

[b] Derived from energy per unit mass, where possible, or

[c] Derived from energy per dollar.

[d] Not determined.

for each item on the list. The energy costs for all items were summed according to each of the methods of analysis to give a range of the energy required to construct the Cayuga Station (Table 14.4). The fourth method (the aggregate national ratio applied to total plant cost) was 326 million × 11,816 kcal/dollar, or 3.85 × 10^{12} kcal. Thus the range of the energy required to construct the Ca-

yuga Station is between 1 and 3.9 × 10^{12}, depending on the method and comprehensiveness of the analysis. Similar analyses were done for transmission lines and other components of the power plant system. The importance of the coal energy (for which an unequivocal energy assessment was available and used) with respect to other inputs is clear.

Table 14.4. Summary of Dollars and Energy Used and Saved for Constructing the Cayuga Power Station

Dollar Cost (thousands 1975 dollars)	Cayuga Station (and accessories)	Regional Insulation
Total capital cost	351,000	913,000[a]
Cost of plant or insulation	(326,000)[b]	(879,000)
Cost of powerline or additional furnaces	(25,000)[c]	(34,000)
Cost of coal	1,610,000[d]	—
Salaries, profits, taxes, interest, repair	2,939,000[e]	417,000[f]
(if based on "incremental cost")	(1,570,000)	
Total Dollar Investment		**1,330,000**
Based on mean 1975 price of electricity	**4,900,000**[g]	
Based on "incremental cost"	3,531,000	

Energy Cost (10⁹ kcal)	Low (Direct Process)	Medium (I–O)	High (IO plus Labor)	Low (Direct Process)	Medium (I–O)	High (IO plus Labor)
Total capital cost	1,244	2,762	4,153	7,875	8,588	9,305
Plant	(1,138)	(2,212)	(3,850)			
Powerline	(106)	(550)	(303)			
Coal						
As coal	400,000	400,000	400,000			
Mining, etc.	9,000	20,320	19,030			
Electricity used[h]						
Internally[i]	(9,577)	(9,577)	(9,577)			
In transmission losses[j]	(4,530)	(4,530)	(4,530)			
Subtotal	**410,244**	**423,082**	**423,183**	**7,875**	**8,588**	**9,305**
Salaries, profits, taxes, interest, repair	—	13,572	34,727	—	1,457	4,927
(if based on "incremental cost")	—	(7,250)	(18,551)			
Total Energy Cost (10⁹ kcal)	**410,244**	**436,654**	**457,910**	**7,875**	**10,045**	**14,232**
Based on "incremental cost"	—	(430,332)	(441,734)			
Total energy, less coal	10,244	36,654	57,910	7,875	10,045	14,232
Total energy delivered/saved for 30 years						
Thermal	**146,000**[k]	**146,000**	**146,000**	**244,000**	**244,000**	**244,000**
With quality factor	**328,500**	**328,500**	**328,500**	**244,000**	**244,000**	**244,000**
Return on investment						
kcal/$	**29,790**	29,790	29,790	**183,459**	183,459	183,459
kcal/"incremental $"	41,336	41,336	41,336			
kcal/kcal (thermal)	**0.36**	**0.33**	**0.32**	23.29	18.26	**12.89**
kcal/kcal (w/qual)	0.80	0.75	0.72	23.29	18.26	12.89
kcal/kcal (less coal; w/qual)	32.00	8.96	5.67	23.29	18.26	12.89

Source: From Hall et al. (1979).

Note: Energy costs of insulation is "conservative," that is, high by perhaps 30%.

[a] The total cost of insulating all existing houses to Class VIII was multiplied by 0.75 to represent dollar and energy cost of adding on new insulation to the existing mix of insulation. Dollar and energy cost of insulating all new houses to Class VIII rather than Class VII are $39 million and 340 billion kcal, respectively.

[b] From Table 1.

[c] P. Komar, personal communication.

[d] Computed at $27 per ton, cleaned and delivered.

[e] Obtained by difference between total revenues and capital plus coal costs. Does not count cost of repair to electricity-using devices.

[f] Real interest at 8% for 10 years derived from local bank president. Upon the advice of local contractors and bankers we did not include money for repair to insulation, but did use high estimates for original cost to provide for good-quality installation and to avoid the need for repairs.

[g] 169 × 10⁹ kWh delivered to consumer times 2.90¢ per kWh NYSEG 1977.

[h] Not included in total energy cost.

[i] Calculated at 6% of gross plant output.

[j] Calculated at 3% of plant output (S. Linke, personal communication).

[k] Assumes 100% end-use efficiency.

Energy and Dollar Costs and Benefits of a Regional Insulation Program

We also analyzed the potential energy savings of a comprehensive regional housing-insulation program by combining local demographic, housing-type, fuel-use, and level-of-insulation information with a heat-loss-from-houses model in a computer program of regional heating-energy use. Details are given in Sloane et al. (1979).

In 1977 about half the houses in this region were reasonably well insulated, one-quarter were moderately insulated, and one-quarter were poorly insulated, including 12% that had no insulation at all. Virtually none were insulated to the highest standards we would find—those suggested by the Edison Electric Institute (1977) and HUD (1975). Our computer analysis investigated the dollar and energy costs of raising the insulation levels of the majority of houses in the region to meet these standards (5% were left uninsulated on the assumption that even minimal insulation would be impossible for some houses) and the energy that would be saved by that program. The analysis showed that, for example, a single-family, two-story house would use 116 million kcal each year if uninsulated, 41 million kcal if reasonably insulated, and 28 million kcal if insulated to the highest standards. Our model predicts that the present mix of houses and levels of insulation in the NYSEG region use annually about 21×10^{12} kcal. If 95% of these houses were insulated to the highest standards, the annual use would be about 13.6×10^{12} kcal for a 30-year savings of 244×10^{12} kcal, including savings in new houses (Table 14.6).

Such an insulation program would cost some $900 million, according to our estimates, plus a net cost of $34 million for installing more expensive furnaces to replace the less expensive baseboard heaters. Interest at 8% for 10 years adds another $417 million. The energy cost, derived in detail in Sloane et al. (1978), would be from 8 to 14×10^{12} kcal.

Energy Return on Investment

The Cayuga power plant and associated structures and activities that are required to deliver the electricity would cost $4.9 billion to deliver 146×10^{12} kcal of energy to the user. If we assume that about 40% of the electricity continues to be sold to residential users, then the cost of the new plant is about $3000–$4000 per present customer.

The system efficiency of the power-plant system in supplying electricity to consumers can be calculated using the previously presented formula. We used the middle (input-output) energy cost estimate for this analysis. Using the high and low energy cost estimates (as discussed in Chapter 3 and given in Table 14.4) would change the ratio to 32 and 36%, respectively. We see from this analysis that there would be a return of some 146×10^{12} kcal of energy to society from about 437×10^{12} kcal invested in the Cayuga Station.

Table 14.5. Dollar and Energy Costs of Insulation

| Type of Insulation | R^a | 1975 Dollar Cost | | Energy Cost (kcal/kg) | Energy Cost kcal/ft^3 |
		Material	Labor		
Fiberglass (batt)	3.15	3.5¢/in/ft^2	10¢/ft^2	5477	3726
Cellulose (loose-fill, attic only)	3.7	3.67¢/in/ft^2	2.5¢/in/ft^2	389	406
Ureaformaldehyde^b	4.2	14¢/in/ft^2	Included	48,764	17,292
Storm windows^c	0.72	$2.0/ft^2	tr^d	14,848 kcal/$	na^e
Storm doors^a	1.67	$3.0/ft^2	tr	14,848 kcal/$	na

Source: From J. Sloane, C. Hall, and L. Fisher, 1979, p. 513. Reprinted with permission of Springer-Verlag, New York.

^{a}Resistance to heat flow, per inch of material.
^{b}Based on polystyrene, as estimate for ureaformaldehyde is proprietary information.
^{c}Dollar costs for storm windows and doors are based on local retail prices. Energy costs for storm windows and doors are based on an assumption of one-half fabricated metal products and one-half glass (a high estimate, since there is no correction for retailing costs).
^{d}tr = assumed trivial.
^{e}na = not applicable.

Table 14.6. Type and Amount of Insulating Materials in Houses in the NYSEG Service Area

Insulation Class	Type and Amount of Insulating Materials	Percentage of Houses in 1977[a]	Annual Energy Use by House Style ($\times 10^9$ kcal/year)				
			1	2	3	4	5
I	No insulation	12	108	116	196	357	564
II	Less than 3″ (nominally 2.5″) fiberglass batting in attic	1	58	78	121	243	352
III	As above plus storm windows	4	50	63	100	203	298
IV	As above plus storm doors	4	49	63	99	200	295
V	As above but with 2.5″ fiberglass batting in exterior wall	4	38	44	72	128	198
VI	As above but with more than 3″ (nominally 4.75″) fiberglass batting in ceiling	24	34	41	66	118	179
VII (PSC)	3.5″ fiberglass batting in wall, or urea-formaldehyde foam in walls if existing home does not have insulation; 4.3″ loose-fill cellulose in roof; floors insulated with foil sheet and air gap; single-glazed plus storm windows; 2″ solid wood doors plus storm doors	51	27	34	55	99	147
VIII	Walls filled with 3.5″ ureaformaldehyde foam; roof with 10″ loose-fill cellulose; floors with foil sheet, air gap, and 2″ fiberglass batting, single-glazed plus storm windows; 2″ solid wood doors plus storm doors	0	21	28	43	80	114

Source: From J. Sloane, C. Hall, and L. Fisher, 1979, p. 514. Reprinted with permission of Springer-Verlag, New York.

[a] In NYSEG service area.

[b] 1 = "ranch," 2 = "colonial," 3 = duplex, 4 = small apartment, 5 = large apartment.

The insulation program, on the other hand, would cost about $1.3 billion to deliver approximately 244×10^{12} kcal (Table 14.4), although the actual energy produced will be larger over time, since houses usually last more than 30 years. The energy cost would be from 8 to 14×10^{12} kcal. Thus, if the objective is to provide space heat, to free fossil fuel and/or electricity for other uses, or, most important, to heat with less gas and oil, the insulation program is a better investment by approximately a factor of 4 to 6 when viewed as an economic investment and a factor of from 15 to 60 (depending on methods used) when viewed as energy return on energy investment. The difference in rate of return between economic and energy analysis is due mainly to the low dollar price per kilocalorie of coal compared to the embodied energy. The insulation is a better energy return on energy investment even if the coal is not included in the calculation.

If we assume that an investment in insulation is a much better return on investment than a power plant, why is this not reflected in the marketplace? To some degree it is. Locally, insulation contractors have been booming and electrical demand has consistently fallen well below forecasts over the past 10 years. A principal reason that insulation is not replacing all new electric demand, however, may be related to consumer habits: The capital outlay for insulation is large and must be paid before the energy savings can take place. Utilities, on the other hand, can borrow or raise the capital needed for the new plant, often at lower interest rates than a householder can get. Then utilities can charge consumers relatively small monthly increments for service. This has in effect predisposed consumers toward the construction of generating plants rather than insulation, even though insulation may be far cheaper per dollar than electricity (or other fuels). Although in August 1977 the state legislature passed a bill requiring utilities to provide low interest loans to consumers for the installation of insulation that would be paid back by small increments on monthly utility bills over a period of up to 7 years, voluntary cooperation with the existing New York Plan has been disappointing so far. Perhaps the centralized leadership of state agencies and utilities that are charged with providing energy supplies could be shifted to comprehensive insulation programs.

Such a comprehensive insulation program could easily cut the need for oil and gas heating fuels by one-third in central New York. This would eliminate the NYSEG-assumed need for new electric space heat, at a considerable savings of consumers' money with far less use of national energy resources and, presumably, with less pollution. Such a program would lessen the impact of possible future oil embargoes or coal strikes, and a recent analysis indicates that a dollar spent on installing insulation produces many more jobs than a dollar spent on power plant construction (Bullard, 1978). Nevertheless, despite NYSEG encouragement, consumer response has been slow.

CONCLUSION

The material in this chapter indicates that while the U.S. can and has reduced the energy intensiveness of many of its economic tasks, there is a wide range in effectiveness among conservation activities. The degree of potential energy conservation is not nearly as great as many superficial analyses indicate because energy-saving capital equipment requires energy for its production and use. Simple housekeeping and building insulation appear to have quite favorable EROIs, but many other conservation schemes may have only marginal returns when proper system boundaries are used. In addition many of the easy, high EROI conservation investments have already been made. Thus any proposed conservation scheme, like any energy resource, requires a comprehensive EROI analysis. The identification and implementation of high EROI investments in conservation is an important national task.

APPENDIX 14.1: RETURN ON INVESTMENT CALCULATIONS

A more sophisticated financial analysis goes beyond simple pay back and takes into account several ad-

ditional financial factors. These include the time value of money d, which is simply a discount of future money relative to present money, since present money could earn interest between now and the future. Also included are projections of inflation r in the cost of fuel and any charges for debt service b and interest i if money must be borrowed to complete the project. Finally, provision must be made for tax rate t on energy savings. If a plant modification reduces expenses of operating the plant, this, all other factors being equal, results in increased profit for the operator and hence increased taxes, reducing the benefits of the modification. The effects of taxes on the increased profits are partially mitigated by depreciation rate d of the purchased equipment, which is an expense for tax purposes but does not actually require a cash outflow. Depreciation is the loss of value of capital equipment as it ages and wears out. Finally, the life of the project n must be taken into account to determine the total cash benefit for the plant operator.

For a given year j in the life of the project, the gross savings for that year $S(j)$ are given by

$$S(j) = r(j) \times S \qquad (14.A.1)$$

where S is the savings in the base year. Note that these savings constitute a revenue, which is reduced by taxes and debt service. The tax $T(j)$ can be calculated from the gross profit (which is the gross savings) less the depreciation and interest expense at the tax rate:

$$T(j) = t[S(j) - d(j) - i(j)] \qquad (14A.2)$$

At this point the net profit $P(j)$ can be determined:

$$P(j) = S(j) - T(j) \qquad (14A.3)$$

Note that most operators are not particularly concerned with net profit, however, because it does not actually reflect cash flow. Put differently, net profit includes depreciation, a noncash expense, and does not include principle repayment on debt, which is a negative cash flow. The plant operator thus forms a statement of net cash flow N that includes all actual cash flows and excludes noncash expenses:

$$N(j) = S(j) - B(j) - T(j) \qquad (14A.4)$$

Finally, return on investment can be determined. Return on investment r is defined as the discount rate that results in the savings generated by the investment, balancing the discounted cash flows over the life of the project. In simple terms return on investment is much like interest paid on a deposit at a bank; the return on investment tells the investor what interest rate would be necessary to give the same yield on the investment.

The return on investment is calculated from the expression:

$$C = \text{sum from } j = 1 \text{ to } n \; \frac{N(j)}{(1 + r)^n} \qquad (14\text{A}.5)$$

EDITORIAL

OIL AND GAS AVAILABILITY: A HISTORY OF FEDERAL GOVERNMENT OVERESTIMATION

The citizens of a democratic nation require relevant information to guide business investments, make political choices, and understand other economic and social processes. One function of government is to supply the citizenry and its political representatives with information necessary to make informed decisions, and a certain proportion of tax revenues is diverted toward that end. Generally, relatively little concern is expressed about the accuracy of the data collected and published by the U.S. government and the vast majority of economic and natural resource-related data are supplied by government agencies. Presumably most of these data are reasonably accurate. Various branches of the government also routinely provide us with *predictions* of economic conditions and resource supplies—for example, predictions of GNP, inflation rates, and energy availability. We would like to examine in detail what we believe is a very important subset of government predictions—those of the future availability of liquid and gaseous petroleum.

Predictions of future petroleum availability have been, and still are, very important with respect to influencing government policies toward energy pricing, energy imports and exports, conservation, synthetic fuel development, and other energy-related programs. Yet we believe that until very recently official government estimates of the future availability of petroleum have been grossly in error and, more important, have been received extremely uncritically despite the ongoing availability of much more explicit and more accurate alternative methods of assessment. We do not know why this state of affairs has existed beyond the observation that it is the desire of governmental entities to appear effective, and hence optimistic, although one might argue that for at least the first two-thirds of this century resource optimists have been more or less on target. The following is a more specific history of

the official and unofficial estimates of U.S. petroleum reserves and the procedures used to make them.

METHODS OF ESTIMATING UNDISCOVERED PETROLEUM RESOURCES

There are three general methods for estimating yet to be discovered petroleum resources, although each method has many variations: First are *extrapolations of historic discovery rates or performance patterns*. Davis' (1958) discovery rate curve and Hubbert's (1962) growth curve projections and discovery rate curves (1967) are examples of this type of analysis. Second are *volumetric yield methods* which have been the most popular and widely used methods of assessment, particularly by the U.S. Geological Survey (USGS). Examples of this type of approach are the pioneering work by Weeks (1948) and various official USGS estimates (Zapp, 1962; Hendricks, 1965; Miller et al., 1975). Third are *combined methods* that incorporate both subjective geological evaluations and statistical models. Hubbert (1974) and the most recent USGS estimate (Dolton et al., 1981) are examples of this method.

The two primary methods, extrapolation of past discovery rates and volumetric yield, often produce different estimates even when applied to the same region because different assumptions are employed by each. The discovery rate approach is based on the existing data and behavior of the industry and on the statistical extrapolation of past performances into the future. As such, this approach is most appropriate in mature areas of development and not in frontier areas where geology and economic conditions may be distinctly different from past experience. The volumetric yield approach is more subjectively based in that it involves extrapolating average yields (barrels of oil per cubic volume of sedimentary rock) from known areas to frontier areas that have been determined to be similar to the original area. Obviously, when using this method, there is often a range of opinions of not only what yield ratio is appropriate but also what constitutes a geologically similar region.

In 1956 Hubbert published an estimate that the ultimate recovery of crude oil from the lower 48 states would be about 150 billion bbl (Figure III.5). In 1962 Hubbert estimated about 170 billion bbl of oil, based on the extrapolation of past relations between production, discoveries, and proven reserves, a technique discussed in detail in Chapter 7. Hubbert's analyses were based on the available data for the oil industry and were selected partly on the belief that the domestic oil and gas industry was in a more or less mature stage of development so that the extrapolation of past trends was a reasonable way to approach the problem of resource assessment.

During the same time period that Hubbert was publishing his 1962 and 1967 analyses, a series of official government estimates of future petroleum availability were released, primarily by the USGS (Zapp, 1961, 1962; Hendrick, 1965), which were many times higher than Hubbert's projections (Figure III.5). The USGS method of assessment throughout the 1960s and early 1970s was primarily a form of a volumetric yield model developed by Zapp (1962). The so-called *Zapp hypothesis* is based on the assumption that since oil is discovered only through drilling, exploration for oil would not be complete until *all* potential oil-bearing regions had been drilled intensively enough to reach a well density that would leave virtually no fields undiscovered. Zapp and his colleagues estimated that this would require an overall density of one well per 2 mi^2, drilled to either the basement of the sedimentary rock or 20,000 ft. Implicit in this approach is the important assumption that oil would continue to be found at a constant rate of about 118 bbl per foot of exploratory drilling, the mean rate up to that time. Hence, the validity of Zapp's and the USGS estimates is dependent on the validity of the hypothesis that, on the average, the finding rate for oil would continue to remain relatively constant over time. Based on this model, Zapp estimated that about 590 billion bbl of crude oil would be produced in the United States. It is possible, however, that this high value should not be attributed to Zapp for he died at about the time these estimates were released and hence had no chance to revise or update his original analysis. Thus Zapp, a serious and

scholarly scientist, may have been treated poorly by history because of the actions of others— something we may never know because of his untimely death. The Zapp hypothesis, with slight modifications, was the primary theoretical basis for all USGS estimates until the mid-1970s. The then assistant chief geologist (and later chief) for the USGS, Vincent E. McKelvey, stated in 1962 that "those who have studied Zapp's method are much impressed with it, and we in the Geological Survey have much confidence in his estimates."

A test of the validity of the Zapp hypothesis is to see if in fact oil is found at a constant rate of success per unit of drilling. This test is readily performed with the existing data base of the oil and gas industry. As we have seen already, Davis (1958), Hubbert (1967), and Hall and Cleveland (1981) clearly documented that oil is not discovered at a constant rate per unit drilling through time but rather exhibits a trend over time of decreasing returns per drilling effort (Figure 7.19). This undeniable fact illustrates clearly how the Zapp hypothesis was not even a reasonably good approximation of reality and led to unwarranted optimism for any resource estimate based on it (Figure III.6). The Zapp hypothesis does not take into account the important and inescapable trends in oil and gas exploration described in detail in Chapter 7, namely that the few very large fields that contain most of the recoverable petroleum are discovered first with relatively little drilling effort. As drilling effort accumulates, increasing levels of drilling are required to locate similar volumes of oil because the fields that are found are smaller.

Despite the undeniable observation that oil was being found at much less than the 118 bbl per foot implicit in Zapp's analysis, and even the quite startling fact that Hubbert's original 1956 prediction that U.S. production would peak around 1970 occurred as predicted, the Zapp-type method or other variations of the volumetric yield approach prevailed as the official U.S. estimate of petroleum reserves throughout the 1960s and early 1970s. In 1961 a USGS estimate based on Zapp's methodology put the ultimate recovery of crude oil in the United States at almost 600 billion bbl (Figure III.5). Thirteen years later the USGS still estimated a minimum of about 310 billion bbl of ultimately producible crude oil. As recently as 1975, the USGS estimated between 250 and 300 billion bbl of oil (Miller et al., 1975). The irreconcilability of the estimate made in this Circular 725 with the existing data of the industry was noted by, among others, Hub-

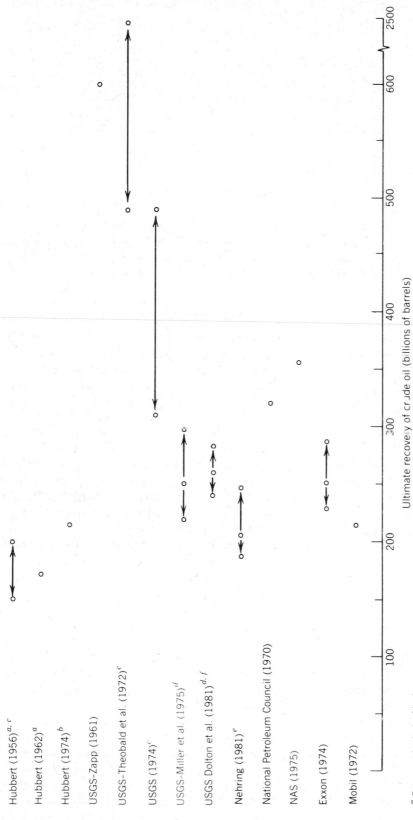

Hubbert (1956)[a, c]

Hubbert (1962)[a]

Hubbert (1974)[b]

USGS-Zapp (1961)

USGS-Theobald et al. (1972)[c]

USGS (1974)[c]

USGS-Miller et al. (1975)[d]

USGS Dolton et al. (1981)[d, f]

Nehring (1981)[e]

National Petroleum Council (1970)

NAS (1975)

Exxon (1974)

Mobil (1972)

100   200   300   400   500   600   2500

Ultimate recovery of crude oil (billions of barrels)

[a] Does not include Alaska.
[b] Including Alaska.
[c] The arrows indicate the range of the maximum and minimum estimates.
[d] The lower value is a 95% probability estimate, meaning that there is a 95% chance that actual ultimate recovery will exceed that value. The upper value is a 5 percent probability estimate, and the middle value is the statistical mean.
[e] The three values are, respectively, the 90%, 50% and 10% probability estimates.
[f] Subsequently reduced by 16 billion barrels.

Figure III.5. Estimates of ultimately producible crude oil in the United States, including outer Continental Shelf waters, made by various authorities in various years.

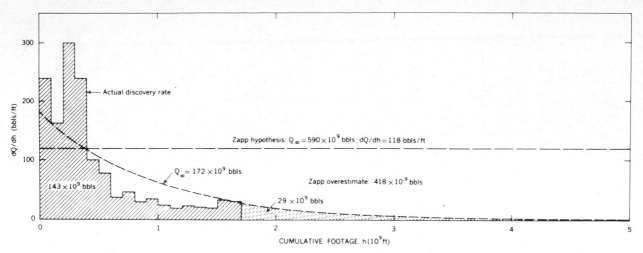

Figure III.6. A comparison of the predictions of the Zapp (1961) hypotheses to the historical record. Based on Zapp's method of a constant oil discovery rate (dQ/dh) of 118 bbl per foot drilled, the USGS predicted ultimate oil recovery (Q_∞) of 590 $\times$ 10^9 barrels. As the record indicates, the actual discovery rate per foot declined substantially, leading to a large overestimate of Q_∞ by the USGS. (From Hubbert, 1974.)

bert (1978) and Nehring (1981). At the conclusion of the most exhaustive empirical analysis of U.S. petroleum resources to date, Nehring (1981) stated:

> . . . we do not believe that anyone could develop a plausible list of geologic prospects . . . in the lower 48 onshore containing an amount of petroleum even approaching the estimates of Circular 725. . . .

As Hubbert (1978) notes, the publication in 1975 of USGS Circular 725 was a "significant historical event," despite the fact that it was still considerably higher than Hubbert's estimate. Circular 725 marked the end of a 14-year period during which the succession of USGS estimates of the ultimate amounts of oil were two to three times higher than could be justified based on existing petroleum industry data on drilling, production, and discoveries. One reason for the dramatic downward revision was the increasing pressure being brought to bear on the unwarranted high USGS estimates of the 1950s, 1960s, and early 1970s brought about in part by the realities of poor drilling success of the 1970s. The new revised USGS estimates were part of the basis for President Carter's new emphasis on the energy problem in general and the need for conservation in particular. Some cynics have suggested that the new low estimates followed Carter's political needs for a national rallying point—an energy crisis—but we believe that reality had finally achieved political respectability and that President Carter had correctly assessed the situation.

In its most recent estimate the USGS's 95 percent probability estimate (meaning there was 95% chance of the amount actually being greater) was 218 billion bbl of ultimately recoverable crude oil (Dolton et al., 1981; Table III.1). The 5% probability, or highest, estimate of the USGS group would lead to about 280 billion bbl of oil, whereas the mean estimate was about 260 billion bbl. The lower figure corresponds rather closely with Hubbert's latest estimate of 213 billion bbl of oil from the lower 48 states, Alaska, and bordering continental shelves, and is consistent with what he has been saying since 1956 (Figure III.5). It is interesting to note that the approach used by Dolton et al. was a combination of exploration history, finding rate studies, petroleum geology, and volumetric yield procedures, and it was the first time that statistical finding rate methods were used explicitly in the USGS methodology, some 25 years after Hubbert had made his initial resource estimates. Meanwhile history continues to substantiate Hubbert's analysis as one of the single most accurate and consistent major economic predictions ever made (Figure III.7).

The use of volumetric yield methods by USGS for its petroleum resource assessments is really not surprising for they are consistent with the attitude that many resource analysts and politicians have held for many years—namely that human ingenuity, both technical and political, has no imaginable limits and that our ever-increasing technical

Table III.1.   The Most Recent Official U.S. Estimates of Remaining Domestic Oil Resources

Area	Cumulative Production[a]	Identified Resources[a]			Undiscovered Recoverable Resources		
		Measured Reserves[b]	Indicated Reserves	Inferred Reserves	Low F_{95}[c]	High F_5[c]	Mean
		Crude Oil (billion bbl)					
Onshore							
Alaska	1.2	8.7	0	5.0	2.5	14.6	6.9
Lower 48 states	111.4	15.9	3.6	16.8	36.1	62.0	47.7
Entire onshore	112.6	24.7	3.6	21.8	41.7	71.0	54.6
Offshore							
Alaska[d]	0.7	0.2	0	0.1	4.6	24.2	12.2
Lower 48 states	7.5	3.0	Negl.	1.4	8.7	25.1	15.8
Entire offshore	8.2	3.1	Negl.	1.5	16.9	43.5	28.0[e]
Entire United States	120.7	27.8	3.6	23.4	64.3	105.1	82.6
		Natural Gas (trillion cu ft)					
Onshore							
Alaska	1.2	30.0	NA	4.4	19.8	62.3	36.6
Lower 48 states	519.3	123.3	NA	132.1	288.6	525.9	390.2
Entire onshore	520.6	153.3	NA	136.5	322.5	567.9	426.8
Offshore							
Alaska[d]	0.6	2.0	NA	1.2	33.3	109.6	64.6
Lower 48 states	56.8	36.3	NA	39.8	66.1	148.2	102.4
Entire offshore	57.5	38.2	NA	41.0	117.4	230.6	167.0
Entire United States	578.0	191.5	NA	177.5	474.6	739.3	593.8
		Natural Gas Liquids (billion bbl)					
Entire United States	19.1	5.7	NA	4.8	NE	NE	17.7

Source: From Dolton et al. (1981).

[a] Cumulative production and reserves are as of December 31, 1979. Production and reserve figures were derived from API and AGA data (American Petroleum Institute, American Gas Association, and Canadian Petroleum Association, 1980), except for California where production and reserve data were taken from the California Division of Oil and Gas (1980) and the U.S. Geological Survey (Kalil, 1980).

[b] Does not include gas in storage.

[c] F_{95} denotes the 95th fractile; the probability of more than the amount F_{95} is 95 percent. F_5 is defined similarly. Fractile values are not additive.

[d] Includes quantities considered recoverable only if technology permits their exploitation beneath Arctic pack ice—a condition not yet met.

[e] Subsequently reduced to 12.0; see p. 349. This would reduce the total to 66.6.

cleverness at performing economic tasks will overcome any implications of increasing resource scarcity, even by providing more and more oil. This attitude is typified by the following statement made by V. E. McKelvey, former director of the USGS, in 1972 (italics added):

> Personally, I am confident that for millennia to come we can continue to develop the mineral supplies needed to maintain a high level of living for those who enjoy it now and to raise it for the impoverished people of our country and the world. My reasons for thinking so are that there is visible undeveloped potential of substantial proportions in each of the *process by which we create resources* and that our experience justifies the belief that these processes have dimensions beyond our knowledge and even beyond our imagination at any given time.

Dr. McKelvey reiterated his cornucopian outlook toward future resource availability in a 1977 speech

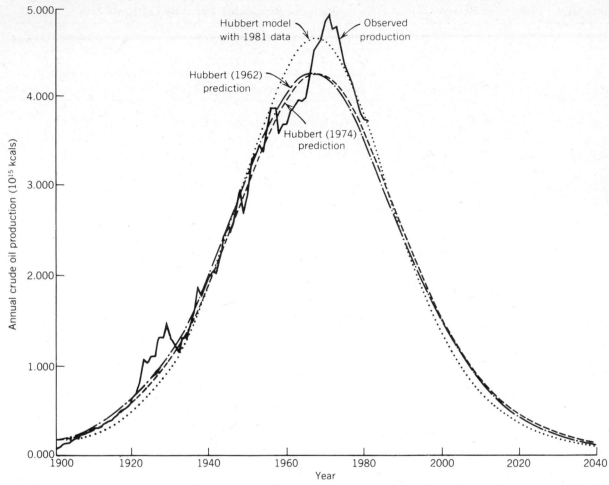

Figure III.7. A comparison of Hubbert's 1962 and 1974 model of the life cycle of annual oil production rates in the lower 48 states with production data through 1981. New economic incentives and, especially the coming on line of the Prudhoe field have resulted in a stabilization of production from 1981 to at least 1985, although nearly all of this production is from existing fields.

in Boston where he stated that natural gas reserves were so large that they amounted to "about 10 times the energy value of all previous oil, gas and coal reserves in the United States combined." No one knows of course how much gas we will eventually find, but this attitude appears to us to be inconsistent with the peaking of gas production in 1973 and the fact that in 1985 proven natural gas reserves are but 12 years of current rates of use. As was described in Chapter 8, the outlook for future domestic gas availability is somewhat more promising than it is for crude oil. It is known that large amounts of gas exist in unconventional formations such as deep gas and geopressured gas. But the energy costs of developing these deeply buried and/ or dilute natural gas sources must be evaluated before we include them in our estimates of future energy supplies. In the case of geopressured gas in the

U.S. Gulf Coast region, preliminary evidence indicates that the average geopressured reservoir may have only a modest EROI, much less than conventional natural gas (Cleveland and Costanza, 1984).

Government attitudes toward future supplies of our natural resources affects all of us directly. Natural gas regulations from the mid-1950s to the late 1970s kept the price of gas artificially low and encouraged its irresponsible use and depletion. The points made in Chapters 2 and 3 suggest that these optimistic attitudes were adopted during a time of abundance of fossil fuels and other resources and perhaps were appropriate for the conditions that existed at that time. During times of decreasing availability of some important natural resources, however, existing attitudes and policies must be scrutinized closely to determine whether they still yield rational and logical policy relative to changing resource conditions. Government subsidies and price regulations cannot always compensate for changing resources realities, and in fact they can exacerbate the problem by giving false promises of future abundance of conventional petroleum.

Despite the increasing awareness of the trends in the quality of our energy resources, many individuals remain very optimistic about our ability to increase our domestic supply of conventional oil and gas. The attitude of the Reagan administration is that increased financial incentives to the oil industry will lead to more intensive searching for petroleum, and ultimately to substantially more oil and gas being found and produced. While this may prove to be true, it is, we believe, more likely that it will not be true at all. In fact virtually all evidence from the oil and gas industry explicitly contradicts this notion. The facts are that the energy crisis that followed the oil embargo in 1973 occurred when the official USGS estimate of that period was that we still had over 500 billion bbl of oil to be found in the United States. Why have we discovered only 11.8 billion bbl of new oil between 1974 and 1980, the equivalent of a little over two years' use, despite unprecedented drilling? We believe that the reasons for this decline are simple and can be found in the physical nature of the resource itself, and not only in the economics of the industry that extracts it. Although it is true that allowing market prices to rise assists in discouraging consumption (an excellent goal, but not without its price in economic growth and material well-being), its effects on encouraging new discovery and production is a two-edged sword. Any present-day production simply hasten the inevitable point in time when the energy cost of domestic production equals the energy gains.

It is possible that large new oil reserves may be found in relatively untested provinces, such as off the coast of Alaska or even Iowa, or that large new gas deposits may be found in deep or unconventional beds, but there is little or no empirical evidence that indicates that the last 10 years of substantially greater price incentives and exploratory efforts is substantially increasing annual finds, especially of new oil. In Texas, for example, 90% of the rather modest 6 billion barrels of oil added to reserves from 1973 to 1984 came from existing fields, rather than from new fields (*Oil and Gas Journal,* July 22, 1985). In the meantime the much larger drilling and other exploration rates are greatly increasing energy costs. Most new provinces that remain will be extremely, perhaps critically, energy-expensive to develop. We should not decrease our search for new petroleum, but neither should we be deluded as to the costs of developing new oil deposits. Alternatively, if future exploratory efforts in frontier regions become more effective, we can reduce our pessimism. At this time we believe such optimism is not warranted, for the fact remains that most of the domestic oil we use in the United States in the 1970s and 1980s is from oil fields discovered before 1940. This reality is catching up with official government estimates. For example, based on the failure of the past 200 offshore wells to find significant quantities of oil the U.S. Dept. of Interior decreased its estimates of oil to be found there from 28 to 12 billion barrels, less than 5 years' use (Norman, 1985).

Surely those who had been listening to Hubbert and Davis all along were neither shocked by the energy crisis nor by our large dependence on foreign sources of oil. On the other hand, the average citizen was quite unaware of Hubbert's analysis and was dependent on official government estimates that we believe were outrageously optimistic. Reputable, although conservative, estimates from some oil industry analysts were discounted by many because they appeared (and may have been) self-serving to an industry trying to raise its domestic oil prices. Even now there is very little recognition or appreciation by the general populace of the precariousness of our petroleum situation and even less government assistance in understanding this problem or dealing with the inevitable future economic and social repercussions.

PART FOUR

ENVIRONMENTAL AND HUMAN HEALTH IMPACTS OF ENERGY EXTRACTION AND USE

OVERVIEW

Both the extraction and use of energy modify natural ecosystems and have adverse effects on human health. In particular, industrial ecosystems often diminish or otherwise alter the storage of natural energies or their flow into the economy. Once a useful natural energy flow like photosynthesis or the provision of clean air or water is disrupted, or human health is compromised, economic energies must be used by human society to replace the services of the lost natural energy or to provide medical care. Alternatively, an economic loss occurs to consumers of that particular natural service or human health degenerates. Unfortunately changes in the quality or availability of natural energy flows, or much of the health loss, are reflected only rarely in either the price of energy or the assessment of the dollar or energy cost of producing and using it. As a result the market price of energy and the EROI evaluations given in earlier chapters are incomplete and normally conservative, for they do not include an assessment of losses of natural energy flows or many of the human health impacts, or the industrial energies that are diverted from the rest of the economy to compensate for these losses. For example, if air pollution reduces crop growth, additional economic energies must be used to produce more fertilizer in order to achieve the same yields, which reduces the energy available to produce other goods and services. If acid rain produced from a power plant kills the fish in a lake, the quality of life is reduced for those who like to fish as is the income for those who once rented boats or sold tackle to the fishermen. If the fishermen wish to continue fishing, they then must drive further to find fish,

raising the energy cost for equivalent personal satisfaction.

ENVIRONMENTAL IMPACTS AND EROI

As noted in the overview of Part Two, energy analysts are just beginning to be able to evaluate empirically the magnitude of natural energy flows used by industrial ecosystems. Despite this incomplete data base, it is important to analyze and, where possible, to incorporate into our EROI calculations the actual amount of natural energy flows disrupted by using fossil, nuclear, and hydro energy so that we can understand and evaluate the true costs and benefits of using a particular fuel.

Most of the impacts of industrial energy use on natural systems are very destructive, while some have little effect. Only a small but significant subset of interactions actually is beneficial to natural environments. Both positive and negative impacts often are diffuse and cryptic, and sometimes counterintuitive. For example, cleaning soot from coal-burning power plants apparently aggravates the problem of acid rain (Likens, 1974; Patrick et al., 1980). Routinely operating coal power plants may release more radiation at the site than routinely operating nuclear power plants (McBride et al., 1978), and one of the nation's largest environmental impacts related to fuel extraction caused by oil operations in southern Louisiana is almost unknown outside of Louisiana (Craig et al., 1977). In addition, because most attempts to mitigate environmental impacts have their own, often large, impacts, it is important to analyze each effect comprehensively. These experiences lead many who deal with them to

351

advocate what is called a *systems approach* to environmental assessment and management. A systems approach is a purposefully nebulous term meaning that considerable effort should be expended to examine not only the most straightforward impacts but also feedbacks, interactions, secondary impacts, energy-use trade-offs, alternate procedures for obtaining the same economic and/or social benefits, and so on. We have brought together some of the philosophy, working tools, and successes and failures of a systems approach in a previous volume (Hall and Day, 1977), and this systems concept has been advocated in a special National Academy of Sciences report (NAS, 1978). Nevertheless, much of energy-environment research and policy development remains fragmented, characterized by the absence of a systems approach. As a result comprehensive calculations for the environmental contributions to the EROI of most fuels are rare, especially for fuels not presently used.

The following chapters examine some of the particular ways in which extracting and using fuel affects environmental systems and give, where possible, preliminary assessments of the dollar and energy cost of such degradation. But first we consider the contributions of natural systems to industrial ecosystems and our economy in more detail.

PUBLIC SERVICE FUNCTION OF NATURE: AN ECONOMIC PERSPECTIVE

So far we have focused on the ways in which economic energies are used to extract and upgrade natural resources to finished goods and services of direct value to humans. Natural energies also contribute large economic benefits to society. Activities of natural systems such as water elevation, regulation, and purification, microclimate regulation, timber production, and provision of recreational opportunities are collectively called the *public service function* of nature (Hall, 1975).

Natural energy flows, although often diffuse, have special value because any economic activity, or even human existence on Earth, would not be possible without a properly functioning global natural environment. Collectively the green plants of the world ensure that free oxygen exists in the atmosphere and help regulate the carbon dioxide concentration of the atmosphere. A forested watershed improves water quality, buffers floods and droughts,

and cleanses polluted water and air. Trees provide microclimates more suitable and pleasurable for human beings, and the forest and its wildlife provide opportunities for recreation and aesthetic pleasure. Foresighted New York City officials wisely purchased and protected the watersheds of its Catskill Mountain reservoirs, and as a result New Yorkers enjoy a water quality almost unknown in other large cities as well as nearby and lovely recreational areas. Although the original land costs were large, subsequent expenditures for drinking water treatment are very small.

All natural services contribute to our economy. All require the continual use of natural energies for their provision, but few are specifically or completely internalized in the price of the good or service that uses or depends on that natural service for its production. When buying lumber, the price the consumer pays has been determined by specifically and directly accounting for the value of the economic energies, including human labor, required to remove a tree from an unmanaged forest. But the consumer pays only indirectly, if at all, for the value of the sunlight, water, and nutrients that grew the tree or for any environmental damage caused by the logging process. One's water bill is certain to cover the cost of pumping and purification of the water at a treatment plant but not the opportunity cost of the solar energy it took to move the water from the ocean to the watershed or of the biotic energy that maintained or even upgraded the quality of that water, for example, by raising the pH of acidic rain. In a cost-benefit analysis of the value of a forest versus the value of a subdivision that might be built in its place, the public service function of the forest may not be included if there is insufficient or no monetary value assigned to the services that the forest was performing, because most such cost-benefit analyses are based on monetary assessments. Not internalizing the costs of degraded or foregone natural services has led us to destroy more of nature than would be the case with a systems-oriented economic assessment.

Externalities and Common Property Resources

Traditional economics is concerned primarily with the interactions between supply and demand that occur in the area in the center and right of Figure 2.1. In this model of the economy all flows of energy, whether they be in the form of fuel, capital, labor, or land, have a clearly defined counterflow of dollars associated with it. The traditional model assumes that if traders have complete information based on the flow of dollars, the market can mediate these transactions in an efficient manner. But if our definition of the boundaries of the economy are expanded to include the natural environment, of which the economy is really a subsystem, there exists another set of equally important interactions between nature and the human economy to which dollar values are not assigned. This information is not included in market transactions. Nature is not recompensed for the solar energy it expended to produce the fossil fuels, transport and clean air and water, fix nitrogen, concentrate minerals, and provide the other environmental services we use. Nor does the natural system in any sense buy from us the noxious by-products of the production process.

The market's inability to assign proper dollar values to these flows leads in many cases to the phenomena we know as pollution. *Externalities* are the costs of lost or degraded environmental (or other) services imposed upon society but not included in the market price of a good whose production contributed to that degradation. Externalities reduce society's wealth by decreasing or degrading the public service functions of nature. Pollution and other externalities are allocations of resources and goods and services that would not occur to the same degree under so-called perfect market conditions in which consumers and/or markets have accurate and complete information.

In terms of environmental impacts, the most important market failure is the lack of a system of well-defined and enforceable *property rights* for most environmental goods and services. Because there are no owners who can bill the users of most environmental goods, a market price cannot be attached to them, leading to suboptimal use of those resources. If full ownership of a good or service exists, whether it be an automobile or a pristine mountain stream, the owner can prevent others from using that resource without first compensating him or her for that use. But most natural resources, including water, air, and some types of land do not, and by their very nature cannot, have clearly defined and enforceable property rights. These are *common property* resources or *public goods*.

Common property and public goods are by definition resources that are freely available to all potential users. It would be difficult to divvy up a river or the atmosphere and assign ownership rights

of the parcels to various individuals. Public goods are those goods whose use by one individual does not diminish the good's potential to be used by others. Examples are national defense, a scenic view, or public street lighting. Common property resources, on the other hand, includes resources like air and water whose future quality and availability can be diminished by current users. Air and water by their physical nature are common property, whereas other services such as highway systems are managed as common property for various economic and political reasons.

Policy in the United States has generally managed common properties according to the principle of *open access,* or "anything goes." As a result, they have been freely available to anyone with few, if any, rules controlling their use. This approach may have made sense in the past because society's demands on common property were relatively small compared to the size of the resource. In the past quarter century, however, rapidly rising population levels, rates of fuel consumption, and economic growth have increased greatly human impact on the natural environment.

A policy of open access usually favors one party over another (Dales, 1968). For example, swimmers using a lake for recreational purposes cannot harm or hinder a steel plant that is using the lake as a waste receptacle. The pollution from a steel plant, on the other hand, may pose a serious threat to swimmers in the lake. Unfortunately, it seems as if those damaging the environment tend to be favored over those who are not. The problem of managing common property resources has received substantial attention in the past decade or so, most notably in Garett Hardin's (1968) "tragedy of the commons" example of how unrestricted common property rights almost always leads to economic, political, and social misuse of natural resources. A short example illustrates his point.

Suppose a productive fisheries industry exists on a river that empties into the ocean. Somewhere upstream oil is discovered and developed, and the extraction industry dumps its unwanted toxic by-products into the river, destroying the fishing industry downstream. If the cost of using the river as a waste receptacle is not reflected in the market price of the oil, the oil industry will probably continue to use the river as a dump because the alternative of reclaiming and reprocessing the waste has a very clear cost to the firm owning the petroleum plant. The *total* cost of that oil to society, however, includes not only the itemized costs to the oil company but also the cost of a less productive or even a destroyed fisheries industry. Society must therefore weigh the benefits of increased chemical production against the cost of the foregone fisheries production when trying to determine what the optimal amount of pollution is and must interfere with the free market to ensure that full costs are borne.

Decisions and laws governing total costs and optimal pollution levels are difficult to make, however, without a means for calculating what the dollar value of such externalities. Estimates of these costs are called *shadow prices*. Several different methods for calculating shadow prices of environmental services currently exist, none of which yet does so completely satisfactorily.

Quantifying the Value of Natural Services

One way to measure the value of nature is to estimate the willingness of people to pay for recreational aspects of natural or seminatural systems, such as national parks. The basic strategy of this approach is to evaluate how much people are willing to pay to visit that particular site (e.g., Kneese et al., 1970; Krutilla and Fisher, 1975), or how much people would be willing to pay if entrance became more expensive. Although this approach satisfies the subjective criterion of what people thought the area was worth to them, this approach misses many other contributions that tie natural systems to the economy and human well-being, contributions that were taken for granted (even though important) or else were not priced because people were not used to attaching a price to them (such as for clean air). Several economic analyses have attempted to measure and include their dollar costs (e.g., Isard, 1972; Dohan, 1977). Nevertheless, it is difficult to determine when a truly comprehensive assessment has been made.

Another approach is to calculate the *industrial equivalent* of a particular service—namely how much it would cost (in dollars or energy) to perform the same service that nature provides using economic energies. For example, swamps and other types of wetlands store and cleanse water. The same functions can be obtained by constructing dams and purification plants. The values of such a swamp to our economic system can be calculated based on how much it would cost to do the same thing with concrete and steel structures. This basic

idea was used by the Northeast region of the Army Corps of Engineers who, in a very enlightened manner, determined that the cheapest way (measured in dollars) to provide flood protection for communities along the Charles River in Massachusetts was to purchase and preserve the many swamps that store excess water during floods. Similarly, Gosselink et al. (1973) estimated the large dollar cost of building a tertiary waste treatment to replace the services performed by wetlands in coastal Louisiana.

Ecoenergetics

Another, more theoretical approach to this problem has been developed by Odum (1971), Gosselink et al. (1973), Odum and Odum (1976), Lavine et al. (1978), Costanza (1982), and others. This approach assumes that natural systems optimize, through millions of years of natural selection, their use of incoming solar energy to provide the essential services required by life (including human beings) and to generate natural resources. According to this ecoenergetics approach, economic energies cannot be invested to produce the same natural processes with greater effectiveness and efficiency because industrial energy systems have not had sufficient time to evolve or develop procedures to perform the same functions as effectively. As a result, the minimum economic energy needed to perform life support functions must be at least as great as the natural energies that presently perform these functions. Consequently, they assume that evaluating the work done by solar energy provides one measure of the energy value of the public service functions of a region.

The ecoenergetics approach is performed as follows. First, the analyst quantifies the natural energy flows for a given natural process (e.g., primary production or the hydrologic cycle). This is often an extremely arduous task, for our knowledge of energy flows in natural system is sparse compared to our knowledge of economic energy flows. Nevertheless, much field work currently is being done in this area, and some preliminary estimates of the solar energy cost of some natural processes have been made. Table IV.1 gives Costanza and Neill's (1981) calculations of the energy cost of various natural services for the biosphere. Lavine and Butler (1982) have attempted to quantify similar energy flows on a broader scale.

Once the energy cost of a particular environmen-

Table IV.1. The Solar Energy Costs of Producing Various Environmental Services

Commodity	Embodied Energy Intensity[a]
1. Manufactured goods and services	191.2 E6 kcal solar/\$ = 17,850 kcal FF/\$
2. Agricultural products	13.9 E3 kcal solar/g = 6.2 E6 kcal solar/lb = \$.03/lb
3. Natural products	39.2 E3 kcal solar/g = 17.7 E6 kcal solar/lb = \$.09/lb
4. Nitrogen	0.63 E6 kcal solar/gN = \$1.49/lbN
5. Carbon dioxide	57.1 E3 kcal solar/gC = \$.13/lbC
6. Phosphorous	1.17 E6 kcal solar/gP = \$2.75/lbP
7. Water vapor	.55 E18 kcal solar/cu · km = \$2.87/cu · m = \$.01/gal
8. Liquid water	.55 E18 kcal solar/cu · km = \$2.87/cu · m = \$.01/gal
9. Fossil fuel	96.4 E3 kcal solar/g FF = 10,711 kcal solar/kcal FF

Source: From Costanza and Neill (1981).

[a]The values above were calculated using standard conversion factors (451 g/lb, 264.2 gal/cu · m, etc) and the intensities calculated by the model for the other commodities (i.e., 191.2 E6 kcal solar/\$). Fossil fuel was converted to kcal using 9 kcal/g FF.

tal process or good is estimated, it is converted to a dollar value using the appropriate conversion factor from the industrial society (dollars per kilocalorie), usually the economic energy/real GNP ratio (see Figure 2.10). In 1982, for example, an average nominal dollar spent in the United States embodied about 7000 kcal economic energy from fossil nuclear and hydro power. Then the natural energy flows are multiplied by this ratio to get a dollar value for natural systems.

Cleveland and Costanza (1983) employed this approach to evaluate the energy costs of developing geopressured gas resources in coastal Louisiana (see p. 217). If the wetland subsides as geopressured fluids are withdrawn, its services as a nursery for young fish and wildlife and for water purification may be lost. This study found that incorporating estimates of the energy value of wetlands lost by geopressurized gas development could eliminate the modest energy gained from most proposed projects.

Although the ecoenergetics approach provides an integrative approach to the problem of considering the public service functions of nature, its methodological tools need substantial refinement. It will be many years, if ever, before the contributions of nature to industrial economics can be understood and measured well enough to enter into routine analysis. Although it is not possible at this time to give a generally accepted methodology for internalizing the contributions of nature for general economic assessments, the functions of nature should be included at least conceptually in comprehensive economic assessments. Meanwhile it gives further qualitative ammunition to other arguments, including aesthetic, economic, recreational, and spiritual perspectives, for protecting many natural ecosystems.

15

IMPACTS FROM EXTRACTION
AND PROCESSING

ENVIRONMENTAL IMPACT OF PETROLEUM SYSTEMS

Extracting, refining, and using petroleum constitute a very large potential for environmental impact. Petroleum supplies about 70% of the primary energy used by the United States, of which about two-thirds is produced domestically. Some attributes of this impact characterize any fossil fuel: the impacts of mining, the emission of carbon dioxide and oxides of sulfur and nitrogen, and the production of heat. On the other hand, the per kilocalorie impact of petroleum, especially gaseous petroleum, is much less than that of coal or shale oil.

The degree of environmental degradation imposed by petroleum extraction depends in large part on whether the drilling site is on dry land or below water. Relatively little has been written about the impact of terrestrial petroleum wells on dry land, and there may be in fact little impact. We have observed pumping rigs in Florida, inland Louisiana, California, and the Rocky Mountains, and they always seem to be quietly going about their business in an unpolluting way. Of course all such rigs are driven by electricity or (rarely) some other energy source, and the burning of the fuel (usually natural gas) to generate the electricity produces various pollutants. Additionally various toxic materials are found in association with oil, and oil may be accompanied by considerable quantities of water. For example, the Big Horn field in Wyoming and many fields in Eastern Ohio extract up to 10 barrels of water with each barrel of oil. If this water can be injected back into the well it presents little problem. On the other hand, if it cannot (it is costly to keep pumping that water again and again), then the disposal of this water, which is often full of salts and toxic materials, may constitute an environmental problem by polluting surface or ground waters. In general, oil rigs take up relatively little space, so they do not take significant quantities of land out of agricultural or natural production. In many cases agriculture continues around the immediate area of the rig. To some the rigs are an eyesore, but to others they are an interesting diversion on the landscape.

The construction of the Alaska pipeline created the most discussion about the potential impacts of terrestrial petroleum extraction. Among these were the possible disruption of migratory routes of animals, such as Caribou, and the potential long-term disruption of the permafrost. Fortunately heavy pressures from environmentalists (e.g., Chiccetti, 1972) and industry response to these pressures greatly reduced these potential impacts.

Unfortunately petroleum extraction in marine or wetland environments has had severe impacts. Oil spilled on land is absorbed by the ground and causes only a local mess, but oil spilled on water spreads out and is carried afar by water currents. The thin film resulting from such a spill produces many problems because, although oil is only a moderately toxic substance directly, it has a propensity to clog appendages, body coverings, and membranes of marine animals. The following section is excerpted from Hall et al. (1978), where we reviewed the environmental impacts of oil in coastal waters. The interested reader is referred to the original article for a more fully developed and documented treatment which also considers the various pathways by which oil enters the marine environment. Similar impacts probably occur where oil is extracted from freshwater environments but, at least in the United States, little oil is extracted from freshwater environments.

The Fate of Spilled Oil

Oil spilled or discharged onto water forms a slick that rapidly spreads away from the original source, generally at a speed of about 1–4 km/hr. The behavior and form of an oil slick varies greatly depending on the composition of the oil, water temperature, and meteorological conditions. Typically a slick consists of relatively thick pools of oil (0.05–1.0 cm thick with diameters of 0.5–10.0 m) floating on a much thinner layer of oil (0.001–0.1 cm thick). Under favorable conditions and a rapid human response, some spilled oil can be contained and cleaned up. But even under optimum conditions most oil is not removed from the water but disappears gradually through a variety of natural processes, including (1) evaporation into the atmosphere, (2) dispersal and dissolution into the water column, (3) incorporation into the bottom sediments, and (4) oxidation by chemical and biotic pathways. The relative importance of each of these pathways in nature is not well known. Most assume that no further damage occurs once the oil passes from sight, but this is not true. If the oil ends up in the water column or in the bottom sediments it will affect biotic communities, perhaps more so than the slick itself. Regardless of where they go, the chemical compounds of the oil remain in the environment, perhaps in an altered form, until they are oxidized completely to carbon dioxide and water.

An editorial in *Science* magazine stated that "a wide variety of animals and microorganisms metabolize or detoxify hydrocarbons." This conclusion is misleading for several reasons. Most work on the microbial degradation of oil has been carried out in the laboratory where the rates of degradation are much higher than rates actually occurring in nature. Nutrient and oxygen availability often limit the rate of microbial degradation in seawater, and in nature other types of carbon sources are always available, many of which are used preferentially by bacteria. But these conditions are not normally reproduced in the laboratory.

Many oils are rapidly oxidized by strong ultraviolet light, such as that found in sunlight at the surface of the ocean. An oil slick absorbing this light forms activated molecules that react rapidly with oxygen, transforming hydrocarbons into more soluble (and toxic) products such as peroxides and phenols. Since these compounds are more soluble, they are likely to end up in the water column, hastening the disappearance of the surface slick but increasing the negative impacts of the original spill.

All molecular weight fractions of oil from a slick can enter bottom sediments quickly. Although oil is less dense than seawater and therefore might be expected to remain afloat, oil sinks via adsorption onto sediment particles, such as clays suspended in the water column, or by sedimentation in oil-laden fecal pellets of zooplankters. One study calculated that a population of healthy calanoid copepods grazing in the vicinity of a surface oil slick could transport three tons of oil per square kilometer to the bottom daily.

Oil in the sediments is very persistent, degrading very slowly over years or even decades. Once in the sediments, oil may migrate laterally affecting biotic communities over a wide area. In the well-documented case of a spill from the oil barge *Florida* off the coast of West Falmouth, Massachusetts, the oil moved slowly about, in, and with the sediments, until it finally came to rest after a considerable period of time in deeper, softer deposits. The same was true of the famous Santa Barbara spill. The oil also can leach slowly from the sediments into the overlying water column, providing a chronic source of pollution. Following the spill of the *Olympia Arrow* in Chedabucto Bay, Nova Scotia, oil was slowly passed back into the water from the sediments for at least 5 years.

Biological Effects of Oil

Impacts of Surface Slicks

The larval and egg stages of fish are especially sensitive to the toxic effects of oil, and a surface slick may have its greatest impact on immature fish. The larvae and/or eggs of many fish float at or near the surface for a long period, typically 4–5 months, while they develop. For example, on Georges Bank, a shallow rich fishery to the east of Cape Cod (Figure 15.1), larvae of haddock, cod, pollock, whiting, red hake, cusk, herring, American dab, yellowtail flounder, gray sole, and sea scallops drift in surface waters, as do the eggs of all of these species except herring. Unfortunately, studies of the effects of spilled oil on fish larvae or eggs in nature are practically nonexistent, for spills tend to occur under very adverse water conditions that make sampling difficult or impossible.

Following the *Argo Merchant* spill in December 1976, however, some preliminary studies were made to determine the fate and biotic impacts of the lost oil. Although the severity of the weather hindered these investigations, the *Argo Merchant* spill

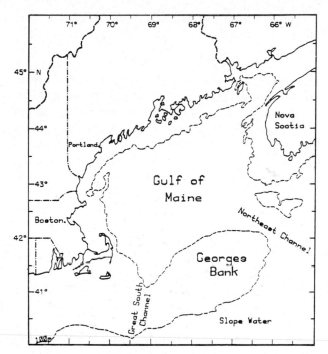

Figure 15.1. Georges Bank off the coast of Massachusetts. During normal years water currents move from North to South, sweeping fish larvae to sea. On some years currents move continuously clockwise around the Bank, keeping fish larvae and their food in this region of high productivity. It is thought that such conditions lead to exceptional years classes of fish. (From Davis, 1982.)

is noteworthy because the slick moved across, and covered, much of Georges Bank. The preliminary biological data from the National Marine Fisheries survey indicated a striking effect on fish eggs and larvae: for example, mortality rates were 46% for pollock embryos and 25% for cod embryos throughout the region covered by the slick; about 90% of the pollock eggs sampled from one site near the spill area (probably just outside the slick) were found to have oil globules adhering to them and were dead or moribund. Since dying fish eggs tend to sink to the bottom (and would therefore not be recovered by surface samples used for investigating fish larvae), actual egg mortality probably was considerably greater.

Populations of most commercially important species of fish are extremely variable—that is, production of new young fish of a particular species is low most years but occasionally is very high, producing a strong *age class* (Chapter 18). The abundance of adult fish generally is higher in years following a strong age class, and successful commercial fishery depends on these strong age classes. Any event producing a year class failure,

particularly simultaneous failures in several species, is very damaging to the fisheries.

The factors that cause these very strong year classes are not clearly understood and vary from species to species, although recent research is beginning to help us understand why some years are so much better for fish than others. Colton and Temple (1961), and increasingly other biologists, believe that strong age classes on Georges Bank result from unusual hydrographic conditions that maintain the same parcel of surface water on the Bank for many months, long enough for the young fish to develop and settle in this highly productive environment instead of in less productive deeper waters. Davis (1982) has shown that water movements that keep the young fish on the bank also retain the zooplankton that are the food for the fish. If this mechanism is correct, these same physical factors would keep spilled or chronically released oil on the Bank. This pollution could destroy otherwise outstanding year classes. Thus it is important to understand the particular conditions of each region to understand the impacts of pollution.

Howarth (1980) estimated that if there was a large oil spill on Georges Bank, and if that impact resulted in the loss of an otherwise outstanding year class of fish, the energy required to replace the lost fish catch with an equivalent quantity of more energy-intensive beef could actually be about as large as the 500 or so million bbl of oil that are thought to be recoverable from Georges Bank. Although such calculations contain considerable uncertainty, Howarth's assessment shows how large environmental impacts can decrease the EROI of an energy resource.

Effects of Surface Slicks on Birds

Marine and coastal birds are extremely vulnerable to oil spills, and such spills are partly to blame for the decline of some auk colonies. In the year following the famous Torrey Canyon spill, breeding populations of puffin, guillemot, and razorbills on the Island of Rouzie off the Brittany coast were reduced by 80–88%. Diving birds are particularly susceptible to damage. At least some species of seabirds may actively seek oil slicks and dive into them, possibly because the slick resembles schooling fish. The specific factors contributing to the great susceptibility of diving birds to oil damage are still not well known.

Effects of Oil Dissolved in the Water Column

Although the impact of oil spills on birds and beaches is most obvious, the invisible, soluble fractions of oil probably cause greater environmental damage. A number of the constituent hydrocarbons of oil are somewhat soluble in water, and of the more soluble fractions, some (e.g., the aromatics) also tend to be the most toxic. Research shows some striking lethal and sublethal effects to a variety of organisms. For example, Mironov (1968) found 40% mortality in plaice eggs treated for a period of days with oil and oil products at a concentration of only 10 parts per billion (ppb). For comparison, the data of Brown et al. (1973) suggest that the concentration of petroleum hydrocarbons in tanker lanes in the western North Atlantic is about 1–20 ppb. An oil spill from an offshore drilling rig raised the concentration of dissolved and suspended oil an order of magnitude to 70 parts per million (ppm) near the rig compared to 1 ppm a mile away.

In experiments with whole planktonic ecosystems enclosed in plastic bags (CEPEX experiments), Lee and Takahashi (1975) observed a change in the structure of algae communities caused by soluble oil concentrations as low as 20 ppb. The oil additions induced the replacement of normally dominant centrate diatoms by much smaller pennate diatoms and microflagellates. Such a change may have serious implications for many adult and larval fishes and other heterotrophs that depend on the susceptible larger diatoms for food, perhaps resulting in the replacement of fish as top carnivores in the ecosystem by ctenophores and other organisms that are not of direct value to people. The species composition of phytoplankton in the North Sea—one of the world's richest fishing grounds—may now be changing in a fashion similar to that found in CEPEX experiments, with larger species being replaced by smaller ones. Low-level oil pollution may be at least partly responsible for this, but we do not know for sure.

Effects of Oil on Benthic Communities

Once oil reaches sediments, it can have a large effect on the benthic (bottom) biotic communities resident there. A relatively small oil spill in the Great Bay Estuary of New Hampshire killed much of the benthic fauna as did a small spill at Portland, Maine. Following the West Falmouth spill described earlier, the bottom fauna were killed, leaving some areas completely azoic. These areas were repopulated fairly quickly by large numbers of the polychaete worm, *Capitella capitata*. This same worm was one of the very few species able to survive and reproduce in the oil-contaminated sediments of Los Angeles Harbor, where in 1957 the oil industry discharged 1.5 million bbl of waste per day. Within a year after the West Falmouth spill, other species started reinvading the oiled sediments. Had the ecosystem recovered? That depends on one's definition of recovered, and we cannot offer an easy answer, for the definition is subjective. That is a common problem for assessing the impacts of pollution, for some species are encouraged while others are eliminated.

Effects of Oil on Salt Marshes

Salt marshes are particularly vulnerable to oil spills, since an oil slick can go no further than the shore. One factor affecting the damage caused by a spill is the degree to which the oil is weathered before it comes ashore. Extensive damage occurred in a British salt marsh following the spill from the Torrey Canyon. But the same oil caused much less damage when it came ashore in France after it had weathered at sea for 14–18 days.

The susceptibility of a marsh to oil-related damage also depends on the depth to which oil penetrates into the sediments, and the rate at which the marsh recovers appears to be correlated to the rate of sediment deposition. If oil sinks deep enough into the sediments to reach the grass rhizomes, recovery is very slow. But if the oil penetration into the sediments is slight and the rhizomes are undamaged, recovery can be rapid. The marshes along the East Coast of the United States, unlike European marshes, are permeated by fiddler crab burrows which probably facilitate the movement of oil into deeper sediments where it is more damaging. This may be an additional explanation why spills in East Coast marshes were more damaging than the spill in the French marsh.

Non-Oil Effects Associated with Petroleum Facilities

Development and operation of petroleum-related energy facilities have impacts on coastal environments other than those caused directly by oil spills. These include (1) fresh water demands of oil refineries, (2) introduction of heavy metals and other toxins into the environment during the oil-drilling operations, (3) excessive sediment disruption—with sometimes heavy increases in turbidity—associated with oil drilling, construction of oil and gas pipelines, and dredging of channels for tankers and drill rigs, and (4) environmental effects of oil production in coastal marshlands.

Environmental Effects of Oil Production in Louisiana Wetlands

Southern Louisiana's coastal region contains 40% of the nation's wetlands (except Alaska) and also has one of the nation's largest commercial fisheries. The washing of dead marsh vegetation into adjacent estuaries and coastal waters is an important food source for the fish. The wetlands of southern Louisiana also support a large hunting and trapping industry. Along with this rich endowment of biotic resources Louisiana also has supplied about 17% of our nation's petroleum resources. Because most of the oil and gas deposits lay beneath the wetlands, the wetlands have been severely affected by oil and gas extraction activity.

Oil and gas extraction and other human industrial activities destroy over 100 km^2 of Louisiana's wetlands every year (Figures 15.2 and 15.3; Gagliano, 1982). Large channels have been dredged to float well-drilling barges and to lay pipelines, and the

spoil normally is deposited next to the channels. These *spoil banks* effectively prevent water from flowing across the marsh surface, thereby depriving the marsh of fresh inputs of sediment and nutrients needed for survival. Without these essential inputs the marsh vegetation dies, and the region is converted to open water. Such conversions reduce the production of the detrital food system that is the cornerstone of the highly productive Gulf of Mexico fisheries.

Craig and Day (1977) estimated that about 1–2.6% of Barataria Basin (approximately 5700 km^2 or 2200 mi^2) has been removed by dredging. If the marsh destroyed by dredge spoils is included, perhaps 10% of the region has been lost from biotic production. The total area adversely affected approaches 20% when secondary effects, such as loss of sheet water flow, are included. Once canals are dug, they tend to widen each year from erosion; thus the effects continue even after dredging or extraction ceases. The report also estimates that some $8–$17 million lost from fisheries each year can be attributed to land modification and related effects. Minor changes in dredge spoil handling could lessen greatly the vegetation and fish losses. The economic and social implications of this environmental disruption of southern Louisiana are analyzed in a series of excellent articles on the first page of the *Wall Street Journal* (October 21–24, 1984).

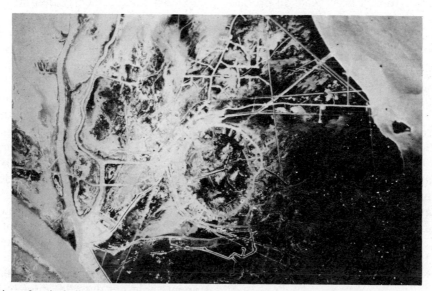

Figure 15.2. Aerial view of typical oil exploration and development impact on Louisiana wetlands. The circular pattern is the Venice oil field, which is 5 miles across. The straight lines are canals dug for oil exploration and development. Thousands of square kilometers are impacted similarly in southern Louisiana. (Courtesy of John Day, Louisiana State University)

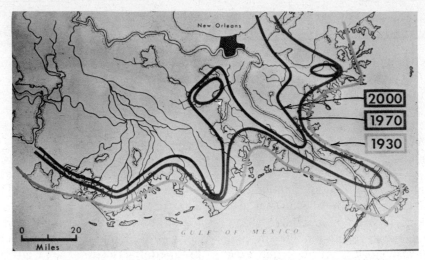

Figure 15.3. Land loss and projected land loss in southern Louisiana. The dark lines are the 50% land-50% water border for different years. About half of this land loss is due to the levying of the Mississippi River, which has eliminated deposition during normal spring flooding. The other half is due to the impacts of dredging and filling, principally in association with exploiting petroleum. (Gagliano, personal communication.)

ENVIRONMENTAL IMPACTS FROM SHALE OIL DEVELOPMENT*

Despite the large deposits of oil shale reserves, their exploitation presents a number of very large problems, including the capital and operating expenses of mining and processing vast quantities of material, waste disposal, water availability, air and water pollution, and problems associated with the rapid development of rural regions. Large-scale oil production from shale would undoubtedly cause large long-lasting changes in the environment.

Many oil industry planners feel that rapid development of oil shale resources can be achieved with a minimum of environmental damage, but Colorado Governor Richard Lamm claimed in November 1979 that a crash oil shale development program "could do irreparable damage to our water supply, to our communities, to our environment." (*Time*, November 19, 1979). Environmental regulations, including the legal disputes over water rights, and the very large capital investments required are major short-term constraints on the growth of the oil shale industry. Longer-term constraints include supply of resources and water and the ability of the region to absorb large population increases.

*Prepared by Chris Neill.

Solid Waste Production

Compared to most other fuels, oil shale contains much less energy per kilogram so that the material processing associated with oil shale development would be enormous. Producing 1 million bbl per day (about 7% of U.S. consumption in the mid-1980s) would require mining, transporting, crushing, and retorting about 1.5–2 million tons of shale per day and disposing of nearly that much spent shale. The operation would have to handle about a billion tons per year. By contrast, U.S. coal production in 1984 was about 890 million tons mined. The underground mine needed to supply a 100,000 bbl per day oil shale processing plant would be 10 times larger than the largest underground coal mine now in operation. Nevers et al. (1978) stated that a 50,000 bbl per day plant requires mining 100,000 tons and disposing of 90,000 tons of shale per day, nearly equal to the largest open pit mine in the world. Strip mining would produce 3 bbl of overburden for every barrel of oil produced, in addition to spent shale. Even modified *in situ* processing would involve mining and moving about 250 million tons per year to support a 1 million bbl per day industry.

A single typical 50,000 bbl per day oil shale processing plant will produce 60,000–90,000 tons of spent shale per day, an order of magnitude greater than what would be produced by a similar sized plant designed to produce liquid fuel from coal. Spent shale will be disposed in canyons, leaving bare, exposed shale on top. Colony Development Operation calculated that in 20 years the spent shale from a 50,000 bbl per day plant would cover approximately 300 ha with nearly 100 m of slag, enough to fill three typical canyons in the region.

Backfilling underground mines with spent shale is a disposal option with advantages and problems. It is possible that backfilling would increase the yield from room and pillar mining by about 15% because thinner pillars could be left to support the mine roof. Unfortunately, although the weight of spent shale after retorting is less than the shale taken from the ground, its volume expands 50% before compaction and 12% after compaction, so other methods of disposal still would be required even if total backfilling was employed. But the industry is reluctant to backfill because lower grade deposits that might be economically recoverable at a later date would be made inaccessible.

The Department of the Interior estimated that a mature oil shale industry would disturb 80,000 ha of land over 30 years. Although it is not yet known how to revegetate spent shale without extensive management, it has been possible to revegetate other disturbed land in similar areas. It is not clear how quickly, if ever, natural vegetative cover can be reestablished on spent shale piles. Some companies have reestablished native grasses after heavy watering, but whether plants can survive the semiarid conditions without intensive and energy-expensive management is not known. So far studies indicate that if native vegetation is to be reestablished, the highly saline spent ore must be leached with large amounts of water, covered by at least 30 cm of topsoil, fertilized with nitrogen and phosphorus, and irrigated regularly. Cook (1979) reported that Paraho-retorted spent shale will not support an adequate stand of vegetation without intensive management, and for other spent shales, "The data at present cannot answer the question of whether fertilizer and water are required to establish diverse plant communities on the disturbed sagebrush and juniper ecosystems of the Piceance Basin." Obviously, even if successful, these are energy-intensive procedures with largely unknown effects on EROI. Another problem is that watering and fertilization tend to favor water- and fertilizer-dependent plant cover which may do poorly when these inputs are removed. Finally, vegetation growing on retorted shale may accumulate toxic levels of boron, and molybdenum may accumulate in some plant species to levels toxic to grazing animals.

Disturbing large areas of land will cause damage to wildlife habitats and fragile soils in the region—primarily grassland, shrub steppes, pinion juniper forest, and rangeland. The area supports a number of rare species, including golden eagles, peregrine and prairie falcons, great sandhill cranes, a rare and endangered plant, the milkvetch *Astragalus lutosus,* and one of the largest herds of migratory deer in the world, 30,000–60,000 mule deer. What the full impact of oil shale development on fish and wildlife in the region would be is not clear; it may be that the increasing human population, and hence recreational pressures, associated with an expanding industry will pose a more serious threat to the environment than the oil shale industry itself, as has sometimes occurred with Wyoming coal development. As with other energy industries, "The trade-offs between 'minimize degradation' and 'maximize oil shale production' perspectives are severe" (Rattien and Eaton, 1976).

Water Requirements

The supply of fresh water for processing, cooling, mining, land reclamation, and a rising population in the oil shale region may place limits on the industry's growth. The Green River formation lies entirely within the Upper Colorado River Basin, where yearly precipitation is low, most surface water already is legally committed to other uses, and water allocations already are a hotly contested, emotional issue. Because oil shale industry would consume large amounts of water to be used as a source of hydrogen for conversion, for evaporative cooling, and for land reclamation, most analysts estimate that a mature oil shale industry would need 2–5 bbl of water for every barrel of synthetic fuel produced. This does not include the substantial quantities of water required to support the work force and their families (Figure 15.4).

Estimates of the amount of water in the Upper Colorado River Basin that could be diverted to the oil shale industry vary and depend heavily on the site and seasonal fluctuations in stream flow, as most of the flow is derived from spring snowmelt. One study concluded: ". . . water availability in the Upper Colorado River Basin may be limited be-

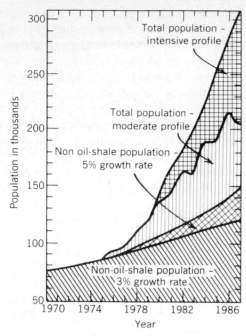

Figure 15.4. Population growth in a three-county region in the Upper Colorado River Basin with and without a developing oil shale industry. (From Ramsey et al., 1978, University of California, Lawrence Livermore National Laboratory and the Department of Energy.)

cause all of the water rights to most of the free flowing water in the basin are already allocated. These water rights would have to be transferred to support additional energy development or water transferred by transbasin diversion" (Gold and Goldstein, 1978).

Water throughout the Western United States is allocated by what is called the prior appropriation doctrine. In this system water rights are assigned by seniority (who "signed up first"). Senior water rights are completely satisfied before junior rights are honored. Generally, the only requirement regarding the use of water once a water right is assigned is the need to put the water to "beneficial use," the definition of which is usually loosely interpreted. Water rights are considered to be property and can be bought and sold as such. One economic limit to the development of the industry is the high price at which water rights would be sold.

Groundwater may also be a source of water for a growing oil shale industry. Estimates of the groundwater available in the Upper Colorado River Basin range from 50 to 115 million acre-ft (60–140 billion m³—enough to produce 15–35 billion bbl of shale oil). This water is replenished at an annual rate of about 4 million acre-ft, enough for about 1 billion

bbl of shale oil per year. All groundwater is subject to the same allocation rules as surface water, is generally judged to be of lower quality, and has many other uses already. In addition it would require a very expensive and complex regional system to exploit it.

Probstein and Gold (1978) estimated that a 1 million bbl per day oil shale industry in the Green River formation would consume 52–86% of the available surface water. They conclude: "a production level of 1×10^6 bbl/day of synthetic crude from oil shale could not easily be attained in the Colorado–Utah area without supplementing the surface water supply by other means such as groundwater use." Even with groundwater use, transbasin diversion, and transferal of existing water rights, the upper limit to oil potential production from shale in the region will probably be about 2 million bbl per day or 0.7 billion bbl per year, about 13% of 1984 oil use (Brown et al., 1977).

Indirect Cost and the EROI of Shale Oil

If water in the oil shale region is diverted to synthetic fuel production there will be a corresponding loss in agricultural production, for most of the available water supplies in Colorado today are devoted to agriculture. If that water is used for shale oil and the agricultural production is replaced by production elsewhere, it inevitably will be on marginal land that requires more energy per unit of agricultural production (see Chapter 20). Water also will be required for municipal use and electrical generation for a population predicted to grow from its 1970 level of 119,400 by 6000 persons for every new 50,000 bbl per day plant that goes into operation (Figure 15.4).

Oil shale production also may contaminate existing water supplies. Mining, particularly *in situ* techniques, may damage underground water systems. Runoff from highly saline spent shale piles and water diversions from the streams in the region may increase the salinity of the Colorado River above and beyond current high levels, and disrupt downstream agriculture. Effluents from oil shale plants may contain potentially carcinogenic hydrocarbons as well as toxic metals.

A large oil shale industry will produce significant amounts of air pollution—particulates, sulfur dioxide, nitrous oxides, carbon monoxide, hydrocarbons, and some potentially carcinogenic polycyclic aromatics (Ramsey et al., 1978). Rattien and Eaton claim that the production of as little as 200,000 bbl of oil shale per day would decrease ambient air quality to the limits set by the state of Colorado. Even lower levels of production would degrade air quality below the standards of the federal Clean Air Act of 1970, which prohibits degradation of unpolluted areas.

Another study concluded: "It will not be possible to make shale oil a major source of liquid fuel in this country until processing techniques are perfected that greatly reduce air emissions compared to those expected with present designs" (Hinman and Leonard, 1977). The other option of course is to change the air quality standards for the oil shale region.

Preliminary calculations suggest that retorting of oil shales from the Green River formation and burning of the product oil could release one and a half to five times more carbon dioxide than burning of conventional oil to obtain the same amount of usable energy (Sundquist and Miller, 1980). Large reserves of oil shale are available, but the EROI is poorly known, and is probably no greater than 7:1; it may be substantially lower if the very large environmental impacts are included.

ENVIRONMENTAL IMPACT OF COAL

Environmental impacts caused by extraction and combustion of coal may be the most important factor influencing increased use of our abundant coal reserves. The various ways that the public, coal-producing companies, the electric utility industry (the primary consumer of coal), and the federal government (the primary regulators of the industry) view these environmental impacts are critical since their combined attitudes largely will determine the future demand for and supply of coal. The importance of public perception in the development of any energy source is evidenced by the fact that, although environmental and human health considerations of nuclear power receive considerable media attention, the *present-day* human health and environmental impacts of coal extraction and combustion are probably much greater than for nuclear power. For example, many coal plants routinely release more radiation than the average nuclear plant (McBride et al., 1978), and an estimated tens of thousands of people a year die from sulfur released by burning coal. There is of course an enormous

potential for adverse affects from the nuclear power cycle. The point is that not only are the impacts themselves important to consider, but it also is important to weigh one energy source against others and examine how the public and private sector perceive and react to those impacts.

Compared kilocalorie to kilocalorie with oil, natural gas, and nuclear power, the use of coal more severely disrupts the environment from which it is extracted and the environments that receive the noxious by-products of its combustion. Every phase of coal use—extraction, processing, transportation, and combustion—affects our land, air, and water resources. An increase in the utilization of coal as an energy source by our nation will depend heavily on our ability to increase our understanding of, and our capacity to deal with, the environmental problems associated with its use.

The specific impacts of coal-related activities vary in severity, are temporal and regional in extent, and depend on a variety of factors: mining method, topography, type of climate, geologic conditions, and demographic factors (Figure 15.5). For example, environmental impacts resulting from underground mining include subsidence, ground and surface water degradation from acid mine drainage and elevated turbidity levels, and mine fires and miner health and safety considerations. Surface mine impacts include erosion, surface water degradation, altered hydrologic patterns, and altered ecological habitats.

An example of the importance of climatic and geologic conditions is the fact that in humid mining regions acid mine drainage is a serious problem associated with surface mining. In humid regions surface or rain water infiltrates the spoil banks and reacts with sulfide minerals such as FeS_2 to produce sulfuric acid (H_2SO_4), which pollutes streams and groundwater resources in the area. Water problems are not as severe in arid regions due to, in part, the scarcity of water and, in part, to the fact that arid soils are more alkaline than soils in humid environments. Erosion rates may be more sensitive to other mining activities such as road building and associated wind erosion (Keller, 1979). Since soils are generally thin and water generally scarce in arid regions, reclamation poses more of a problem there than it does in more humid regions.

The diversity and site specificity of the impacts of coal suggest that their impacts might be dealt with most effectively at the local or state level. In recent years, however, increasing pressure has

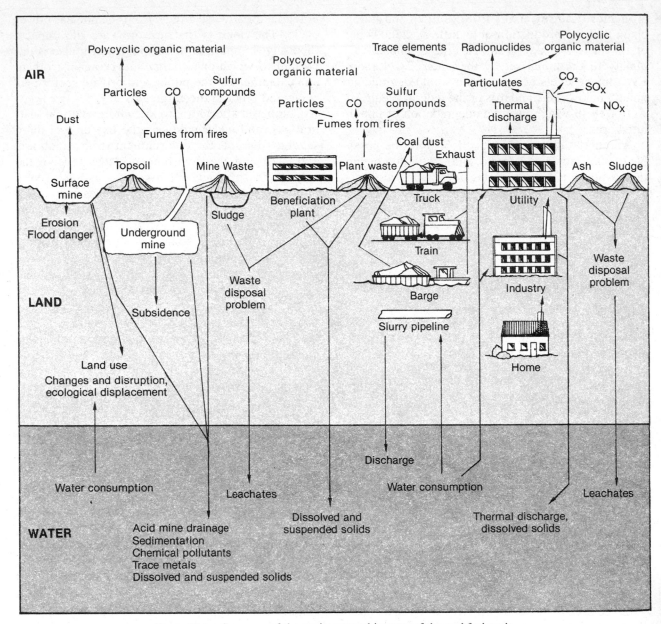

Figure 15.5. Summary of the environmental impacts of the coal fuel cycle.

been placed on the federal government by environmental groups to get federal involvement through the enactment and enforcement of coal-related environmental legislation. Environmentalists argued that in the major coal-producing states the coal industry wielded too much economic and political clout for any effective environmental measures to be passed. The coal industry, on the other hand, argued that only the individual states have the specific information and expertise to deal with the problems and that federal involvement would only

increase bureaucratic red tape and expenses, rather than enhance the quality of the environment.

The result of this debate has been, in part, the passage of control measures such as the Coal Mine Health and Safety Act (1969), the Clean Air Act (1970, amended 1977), and the Surface Mine and Reclamation Act (1977), all of which have had profound influences on coal producers and users since their enactment. The results of these programs have been mixed. Miner safety has increased significantly since 1969, but over half a million hectares of coal lands still need reclaiming, and our ability to understand and control the effects of SO$_2$ emissions from coal combustion is still in the preliminary data-gathering stages. Increased coal production, stimulated by rising oil prices has increased again both coal-related accidents and generation of pollutants. With the administration change in 1980 there has been renewed interest in returning environmental monitoring and treatment of coal-related activities back to the individual states. The resolution of this debate will certainly affect the manner in which all environmental problems associated with coal production and use will be dealt with in the future.

Impacts on Water Systems

Virtually every coal supply technology requires substantial quantities of water at some point (Table 15.1). In addition all coal conversion technologies (e.g., gasification and liquefaction) require water as chemical feedstocks. Finally, both electric generating plants and coal gasification plants require very large quantities of water for cooling purposes. The potential for impacts on the hydrologic cycle therefore is present at every point in the coal fuel cycle.

In 1975 coal production and combustion consumed about 3 percent of the total freshwater intake by all industrial processes (see Table 22.1). In certain areas large water requirements limit implementation of specific coal technology, particularly in Western states where evapotranspiration is sometimes greater than precipitation. For example, water consumption in the Colorado and Missouri River basins was over 20% of total runoff in 1975 (NAS, 1980). Two of the most authoritative studies to date on the subject, Devine et al. (1981) and Harte and El Gasseir (1978) both concluded that future large-scale energy development in the United States may be constrained by inadequate freshwater supplies, particularly in the West. In particular, coal gasification and liquefaction, along with oil from shale, require significant amounts of water, with most required during the conversion stage (see Table 22.2).

The west branch of Susquehanna River is a prime example of a river affected by acid mine drainage. At first glance the Susquehanna is a beautiful river, crystal clear, rapidly flowing, and looking like it should be brimming with trout or bass. The beauty of the river is misleading, however, for the water is clear because virtually no plant or animal life can grow in it, the type of growth that gives biologically productive rivers a more turbid appearance. The reason for the lack of life in the Susquehanna and many other rivers in Pennsylvania, West Virginia, and Kentucky is that they drain the major coal-producing regions of the East where there is sulfur in close proximity to the coal seams. Following coal mining, sulfur makes its way into the

Table 15.1. Water Requirements for Coal System Activities

Coal Activity	Size[b]	Peak Water Consumed by Energy Facility		Peak Water Requirements Associated with Population Increases[a]	
		Acre-ft per Year	Gallons/10^6 Btu in Product	Construction (acre-ft/yr)	Operation (acre/ft/yr)
Surface mine	12 MMtpy	3400	4	71	323
Underground mine	—	0	0	275	1490
Power generation	3000 MW$_e$	29,000	54 (thermal) 157 (electric)	853	260
Lurgi gasification	250 MMscfd	6700	28	1570	350
Synthane gasification	250 MMscfd	9100	38	1570	350
Synthol liquefaction	100,000 bbl/day	17,500	28	1750	1800
Slurry pipeline	25 MMmtpy	18,400	14	Not considered	Not considered

Source: White et al., "Energy from the West," EPA-600/7-77-072a, July 1977.

[a] Assumes 150 gallons per capita per day, a multiplier of 2 to account for added service personnel during construction, and a multiplier of 3.5 to account for families and service personnel during operation.

[b] MMtpy = Million tons per year; MMscfd = Million standard cubic feet per day.

river, leaving it too acidic to support most living creatures.

Impacts on surface water systems occur primarily during the extraction phase of the coal fuel cycle and include acid mine drainage, siltification of streams due to increased erosion, and local changes in hydrology. Acid mine drainage is most severe in underground mines in Appalachia, where the sulfur content of the coal is very high (2–10% by weight) and where soils are poorly buffered. Sulfur in Eastern coals is generally combined with iron in the form of iron disulfate (Fe_2S), which on exposure to air and water forms sulfuric acid (H_2SO_4) and a red iron precipitate. Sulfuric acid that eventually drains into nearby streams often results in the destruction or alteration of aquatic life. It has been estimated that over 16,000 km of surface streams (8000 in Pennsylvania alone) and another 4850 ha of lakes and ponds have been rendered incapable of supporting life due to the annual discharge of over 11 million metric tons of sulfuric acid from underground mines (Lindbergh and Provorse, 1977).

Underground coal mining can disrupt and degrade groundwater systems as well, although usually on a local level. The most common impacts on aquifers are normally by direct contamination from acidic mine water seepage or by physical dislocation that occurs when the depth of the mine reaches below the level of the water table itself. Not only can the water be contaminated and the level of the water table lowered, but water wells in the area also may be affected.

Impact to the Land

Land impacts of the use of coal are caused by mining and waste disposal. Unlike air and water impacts where chemically-induced alterations are the most severe, land impacts usually alter the landscape and remove that land from its ecological function by strictly physical means. The amount of land affected directly by coal use is indicated in Table 15.2. It is interesting to note that Western surface mining disturbs much less land per ton of coal extracted than does Eastern and Central surface mining because many Western coal seams are over 15–20 m thick, allowing extraction of more tons per hectare mined (OTA, 1979). Note also that the impacts of underground mining include all undermined land that is potentially subject to subsidence. Land requirements for disposal include the 16 or more

Table 15.2. Land Requirements for Coal System Activities

Facility	Acres over 30 Years
Mining: 10^{15} Btu per year[a,c]	
Western strip	15,000–96,000
Central strip	216,000
Central room and pillar[b]	525,000
Eastern contour	470,000
Eastern room and pillar[b]	560,000
Gob and refuse disposal	15,000
Combustion/Conversion:[c] Input of 10^{15} Btu/year	
Power plants	13,600
Lurgi gasification	6500
Synthetic oil liquefaction	6300

Source: Calculated from White et al., "Energy from the West," EPA-600/7-77-072a, July 1977.

[a] For a high heat value coal (e.g., Eastern coal with 12,000 Btu/lb) this is equivalent to approximately 42 million tons per year. For a low heat value coal (e.g., Western coal with 8,500 Btu/lb) this is equivalent to approximately 59 million tons per year. For a medium heat value central coal of 10,000 Btu/lb, this is equivalent to 50 million tons per year.

[b] Includes undermined land which is potentially subject to subsidence.

[c] A 1000 MW_e power plant would "consume" approximately 59×10^{12} Btu/year; thus it would take nearly 17 such plants to consume 10^{15} Btu/year. Similarly to use 10^{15} Btu of coal per year would require about 8 Lurgi gasification plants of 250 million cu ft per day and about 3 Synthetic oil liquefaction plants, each producing 100,000 bbl per day.

hectares per year required by an average electric power plant for disposal of solid wastes generated by air pollution controls such as stack scubbers.

In a study on land utilization and reclamation in the mining industry from 1930 to 1971, Paone et al. (1974) concluded that coal mining has historically disturbed more land than any other type of mining activity. Of the 1.56 million ha utilized by the entire mining industry from 1930 to 1971, 667,000 ha or about 42% was due to bituminous and anthracite coal mining activities. Of the land disturbed by coal mining, about 62 percent of it is the actual mined-out area, 32% is used to store wastes from surface and underground mines, and another 4 percent has been affected by subsidence. It is interesting to note that although coal has disturbed the most land, a far greater percentage of coal land has been reclaimed relative to other types of mining industries. Still, strip mining coal disturbs over 800 ha of land every week based on 1977 production rates (Lindbergh and Provorse, 1977), making reclamation a key environmental issue related to coal use. The Surface Mine Control and Reclamation Act (SMCRA) of 1977 mandates that all land disturbed by strip mining must be returned its original contour and ecological function. The effects of this law have been two sided. Although environmental degradation has to an extent been minimized, the cost of surface-mined coal to the consumer has increased due to long delays in the permitting process, the preparation of environmental impact statements, and the expenditure of money for the reclamation procedure itself. Meeting the requirements of the SMCRA is particularly difficult in the West where the water necessary to reestablish vegetation is limited, wind erosion is severe, and little organic matter exists in the soil.

The Council on Environmental Quality (CEQ, 1980) has estimated that of the 1.1 million ha of land disturbed by mining of all types in the United States that needs to be reclaimed, 445,000 ha are mined-out coal land. The states of Pennsylvania, Ohio, Illinois, Kentucky, West Virginia, Alabama, and Missouri all have 20,000 or more hectares of unreclaimed coal land. The CEQ has identified this unreclaimed land as a major cause of nonpoint source pollution in the country.

Land subsidence is another problem caused by underground coal mining that currently plagues many coal mining towns in the East. Subsidence, the sinking of ground after the removal of a solid mineral matter such as coal, usually occurs years after a mine has been closed and abandoned, when the pillars of coal left behind for support begin to crumble and collapse. Because the mine usually has been closed and sealed, treating the problem is very difficult and expensive. When subsidence occurs under populated areas, as it has in Scranton, Pennsylvania, buildings and property damage can occur. Since 1930 about 25,000 ha of land in the United States have been effected by subsidence from abandoned underground coal mines (Paone et al., 1974).

A rather bizarre and potentially lethal type of impact from coal extraction are the mine fires that burn through the coal left in place after the mine has been abandoned. Coal mine fires usually begin in waste piles left behind or from combustible gases that build up in sealed-off mines. In the early 1970s

there were as many as 200 of these fires burning out of control in the Eastern United States. Similar to the problem of subsidence, dealing with these potentially dangerous fires is complicated by the fact that the mine usually has been abandoned for some time and many of the shafts and tunnels are no longer navigable.

The small borough of Centralia, Pennsylvania, with 1200 residents, has been haunted by the same fire for the last 20 years (Seigel, 1981). In December 1979 a resident responding to the cries of one of his children found fumes and smoke rising from his basement floor. The temperature of his basement walls was 180°F, and the fumes turned out to be carbon monoxide and methane. In spring 1981 the ground underneath a young child playing in his backyard suddenly opened up, and he started sliding into the chasm that was spewing hot steam containing deadly carbon monoxide. The child was pulled to safety by another youngster, but the incident brought to a head the townspeople's anger over having to deal with the fire, which by then encompassed almost 60 ha and threatened at least 15 homes and business establishments. Since 1962 more than $3 million has been invested in measures designed to put out the fire, but all have proved futile. The only sure way to stop the fire would be to dig up the mine itself. Excavation would dig up at least 60 ha of land to depths approaching 100 m, and require the relocation of a state highway, natural gas and utility lines, the destruction of over 100 structures, and the relocation of the entire town of Centralia—at a dollar cost approaching $100 million and of course a very substantial energy cost.

Incidents such as these do little to enhance the already poor environmental record of coal use in the United States, a record that must be improved upon if coal is to once again assume a major role in the nation's energy picture. Those who lived in the pre-petroleum era have vivid memories of cities blackened by soot, discolored streams, scarred landscapes, and other effects that coal had on the environment we live in. Long before environmentalism became a popular and powerful social movement, the social costs of extensive coal use were well known to those who had to live with its consequences. In the past three decades we have become spoiled in a sense by the relatively benign effects that the oil and gas fuel cycles have on the environment and on our own health. The reemergence of ''King Coal,'' if it is to occur at all, will depend in large part on not only our technological capability to deal with the severe environmental problems associated with coal use but also on the degree to which a more environmentally aware public will tolerate those problems.

ENVIRONMENTAL IMPACTS OF THE URANIUM FUEL CYCLE*

The use of uranium for electric power production has several environmental impacts, some that characterize the use of any bulk mineral and others specific to uranium's radioactivity. The most important *actual* impacts include the land, water, energy, and material requirements for mining and processing the large quantity of uranium ore used, and the construction and routine operation of nuclear power plants. The most serious *potential* hazards of nuclear energy, however, are releases of radioactivity to the environment from possible large accidents or from long-term accumulations of effluents and wastes. Spent fuel rods and the highly radioactive leach solutions from reprocessed rods are especially worrisome, for as yet no acceptable method for long-term disposal has been found. Even more worrisome is the possibility of a catastrophic malfunction of a power plant, for large quantities of highly radioactive material could be released to the environment. Such releases would be more serious than accidents from the extractive and processing portions of the fuel cycle because most of the radionuclides produced during extraction and processing occur naturally, whereas nuclides from power plants or wastes are much more radioactive after having undergone fission. Of these fissioned nuclides, strontium 90 and cesium 137 are the most dangerous over the short term (half-lives about 30 years), whereas plutonium 239 (half-life 24,400 years), americium 241 (half-life 458 years), and iodine 129 (half-life 17 million years) present a long-term hazard. The half-lives of the naturally occurring nuclides also are often long: radium (1600 years), thorium 230 (77,000 years), and uranium isotopes (160 thousand years).

Consideration of long time scales not only adds a new dimension to the problem of human impact on the biosphere but also poses unprecedented problems for those who wish to evaluate fully all costs and benefits of nuclear power to calculate its EROI.

*Prepared by Juliet Tammenoms-Bakker.

A worst-case accident resulting from the meltdown of a reactor, no matter how unlikely, could release vaporized fission fragments into the atmosphere, long term contamination of large areas, and worldwide fallout of long-lived radioisotopes. Such a large accident could wipe out the positive EROI of a given nuclear power plant or significantly reduce the EROI for the entire nuclear industry because of the large energy investments required to offset the damages. Estimates of the probability of significant malfunctioning at a power plant generally are very small (e.g., Rasmussen, 1977), although a great controversy surrounds such estimates, and recent official estimates have been revised upward (e.g., DOE, 1982). It is incontrovertible, however, that normally operating nuclear power plants emit very low levels of radioactivity and even the rocks of parts of Manhattan emit more radioactivity than most nuclear power facilities.

Presently waste disposal is a greater immediate danger than a potential meltdown. Although few people have been subjected to wastes to any serious degree, considerable quantities of radioactive wastes have leaked from temporary storages into the local environment, especially at Hanford, Washington. The principal concern is that if large quantities of radioactive materials were released into the environment from a large-scale accident at a power plant or from a serious disruption of the fuel cycle, natural ecosystems could be contaminated seriously, exposing human beings and other organisms to radiation which would increase rates of cancer, birth defects, and chromosome damage (see Chapter 17). Thus the relatively low present-day impact must be compared against the largely unknown, but potentially very large, impacts of a future with many operating nuclear plants. The problem has been compounded by the inability of Congress to formulate an acceptable, safe long term waste disposal plan. The ultimate concern is of course that the global spread of nuclear power might accentuate the global spread and possible use of nuclear weapons.

The following sections describe the radiological and non-radiological impacts of the nuclear fuel cycle. The land and water requirements for a 1000 MW light water reactor tend to be less than those for a similar-sized coal-fired plant. The radioactive emissions at each step of the fuel cycle are given in Figure 15.6. Most of these data are derived from Pigford (1974), but the reader is cautioned that other, sometimes divergent, estimates are available. We also discuss some of the most promising waste disposal techniques. Unfortunately there are no assessments yet of the energy costs for preventing or cleaning up the environmental impacts associated with nuclear power. It is obvious, however, that reprocessing and storing wastes and decommissioning power plants requires significant quantities of economic energy and will decrease the EROI to nuclear power (see p. 278).

Environmental Impact of Mining and Milling

Mining and milling processes can be considered together because they have similar environmental impacts. Most of the uranium ore mined in the United States comes from the Western states of New Mexico, Colorado, Texas, Utah, and Washington. The principal deposits are found in stratiform sedimentary rocks that lie parallel to the bedding plane. About 60% of the mines are open pit surface mines, and the remaining 40% are underground or leach mines. Because ore veins often are covered by large quantities of overburden and contain less than 0.1% uranium, large piles of tailings accumulate nearby. These on-site wastes contain more than 95% of each ton of ore mined (ORNL, 1979). A 1000 MW$_e$ light water reactor (LWR) using about 60 metric tons of uranium yearly requires about 186 thousand metric tons of ore mined from about 7 ha, which displaces 2.5 million metric tons of overburden. The ore is milled to produce 180 metric tons of concentrated U$_3$O$_8$ yellowcake while 270,000 metric tons of sludge wastes are formed. These wastes are pumped into impoundments covering a little over 1 ha near the mill. Eventually solids in the wastes settle out and the water is either evaporated or recycled for use in the mill.

Apart from being an eyesore, mine and mill residues are dangerous because they contain the bulk of the ore radionuclides (Table 15.3, Figure 15.6). These nuclides enter surface and subsurface waters through runoff and percolation, infiltrate soils, and contaminate the atmosphere. Radium 226 is of particular concern because it leaches easily and is relatively mobile in groundwater. Radioactive releases to the atmosphere include gaseous radon 222 and particulates of thorium 230, polonium 210, and lead 210. Mill tailings also contain trace amounts of toxic metals such as arsenic, selenium, vanadium, and molybdenum which the mill chemicals mobilize from the ore. Uranium mines are located in the

Southwest, often in areas of native American reservations, where they have had probably their greatest health impact.

Some impacts on groundwater can be circumvented by diverting surface waters from the tailings or by collecting contaminated waters and reusing them during ore processing. At active mine and mill sites the emissions from the piles are monitored so that most impacts arise from unmonitored inactive sites. Although only a few inactive mines exist, there are 23 abandoned mill sites (DOE, 1979). For many years tailing piles at these inactive sites were left unattended, and in Colorado some mill tailings even were used in construction of new houses. Since 1978 the Uranium Mill Tailings Act mandates that the Environmental Protection Agency set management standards for the tailings, and that the Department of Energy control and reclaim both active and inactive sites.

Disposal strategies for tailings piles depend on the volume of wastes and the disposal site chosen. Usually dry tailings are capped with an impermeable clay, concrete, or asphalt layer to stabilize the pile and contain the radon gas. The impermeable cap is covered with soil and reseeded. But land reclamation is hampered by xeric conditions prevalent in the West, where the disposal of tailings has unavoidable and permanent impacts on the biota. Typically, primary production over a 250 ha site is greatly diminished, and seed production and mammal and bird biomass decrease by 90% (Wewerka, 1979). Secondary production in regions studied decreased by 60% (see Table 15.4). Where the tailings are successfully stabilized, contained, and reclaimed, plants and animals take up and accumulate significant quantities of toxic trace elements or radionuclides.

Since prevention is frequently the best cure, an *in situ* leach mining technique, which generates far fewer tailings from both the mining and milling, is increasingly attractive. This technique requires the pumping of solutions that dissolve the minerals into the ore body through an injection well. This solution is channeled to a series of collection wells, and the uranium is recovered at the surface. The leacheate solution is chemically reconstituted and recirculated through the field. Sludge wastes, drilling muds, and cuttings from the injection wells usually are dried and buried onsite. Thus *in situ* leaching disturbs less land at the mine site while reducing the amount of tailings generated. In addition *in situ* leaching brings up less than 5% of the hazardous

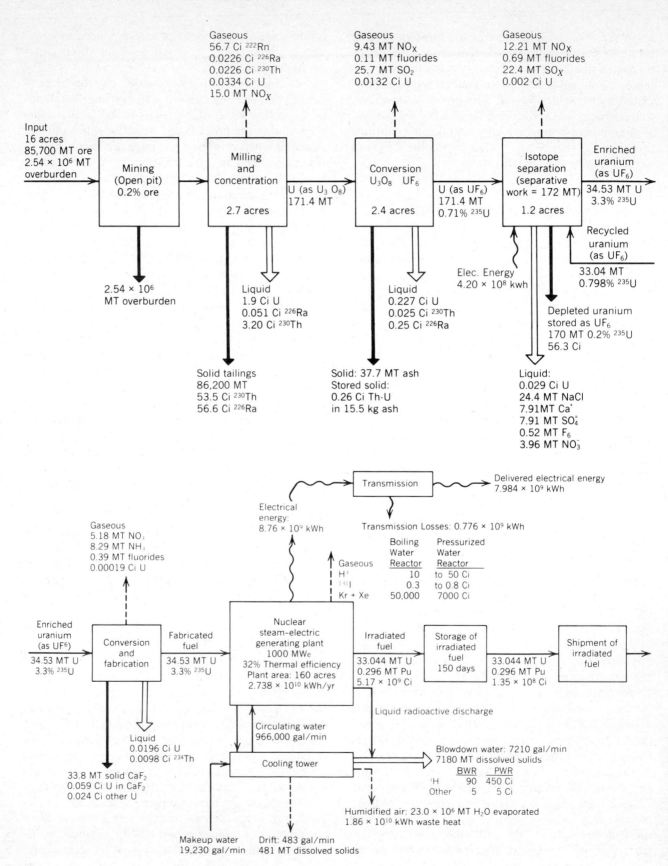

Gaseous
56.7 Ci ^{222}Rn
0.0226 Ci ^{226}Ra
0.0226 Ci ^{230}Th
0.0334 Ci U
15.0 MT NO$_X$

Gaseous
9.43 MT NO$_X$
0.11 MT fluorides
25.7 MT SO$_2$
0.0132 Ci U

Gaseous
12.21 MT NO$_X$
0.69 MT fluorides
22.4 MT SO$_X$
0.002 Ci U

Input
16 acres
85,700 MT ore
2.54 × 10^6 MT
overburden

Mining
(Open pit)
0.2% ore

Milling
and
concentration

2.7 acres

Conversion
U$_3$O$_8$ UF$_6$

2.4 acres

Isotope
separation
(separative
work = 172 MT)

1.2 acres

Enriched
uranium
(as UF$_6$)
34.53 MT U
3.3% ^{235}U

U (as U$_3$O$_8$)
171.4 MT

U (as UF$_6$)
171.4 MT
0.71% ^{235}U

Recycled
uranium
(as UF$_6$)
33.04 MT
0.798% ^{235}U

Elec. Energy
4.20 × 10^8 kwh

2.54 × 10^6
MT overburden

Depleted uranium
stored as UF$_6$
170 MT 0.2% ^{235}U
56.3 Ci

Liquid
1.9 Ci U
0.051 Ci ^{226}Ra
3.20 Ci ^{230}Th

Liquid
0.227 Ci U
0.025 Ci ^{230}Th
0.25 Ci ^{226}Ra

Liquid:
0.029 Ci U
24.4 MT NaCl
7.91 MT Ca$^+$
7.91 MT SO$_4^=$
0.52 MT F$_6$
3.96 MT NO$_3^-$

Solid tailings
86,200 MT
53.5 Ci ^{230}Th
56.6 Ci ^{226}Ra

Solid: 37.7 MT ash
Stored solid:
0.26 Ci Th-U
in 15.5 kg ash

Transmission

Delivered electrical energy
7.984 × 10^9 kWh

Electrical
energy:
8.76 × 10^9 kWh

Transmission Losses: 0.776 × 10^9 kWh

Gaseous
5.18 MT NO$_3$
8.29 MT NH$_3$
0.39 MT fluorides
0.00019 Ci U

Gaseous	Boiling Water Reactor	Pressurized Water Reactor
H^3	10	to 50 Ci
^{131}I	0.3	to 0.8 Ci
Kr + Xe	50,000	7000 Ci

Enriched
uranium
(as UF^6)
34.53 MT U
3.3% ^{235}U

Conversion
and
fabrication

Fabricated
fuel
34.53 MT U
3.3% ^{235}U

Nuclear
steam-electric
generating plant
1000 MW$_e$
32% Thermal efficiency
Plant area: 160 acres
2.738 × 10^{10} kWh/yr

Irradiated
fuel
33.044 MT U
0.296 MT Pu
5.17 × 10^9 Ci

Storage of
irradiated
fuel
150 days

33.044 MT U
0.296 MT Pu
1.35 × 10^8 Ci

Shipment of
irradiated
fuel

Liquid radioactive discharge

Liquid
0.0196 Ci U
0.0098 Ci ^{234}Th

Circulating water
966,000 gal/min

Cooling tower

33.8 MT solid CaF$_2$
0.059 Ci U in CaF$_2$
0.024 Ci other U

Blowdown water: 7210 gal/min
7180 MT dissolved solids

	BWR	PWR
^{3}H	90	450 Ci
Other	5	5 Ci

Humidified air: 23.0 × 10^6 MT H$_2$O evaporated
1.86 × 10^{10} kWh waste heat

Makeup water
19,230 gal/min

Drift: 483 gal/min
481 MT dissolved solids

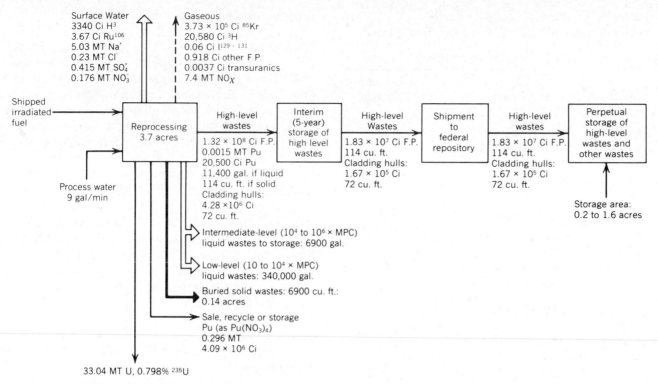

Figure 15.6. Radioactive emissions and other environmental impacts at each step of the nuclear fuel cycle for a 1000 MW$_e$ light water reactor (assuming 100 percent capacity; 1 acre = 0.39 ha, MT = metric tons; Ci = curies; MPC = maximum permissible concentration). (From Pigford, reproduced, with permission, from the *Annual Review of Nuclear and Particle Science* Volume 24. © 1974 by Annual Reviews Inc.)

radium normally brought up by conventional mining techniques (DOE, 1979). The leacheate solution, however, may contaminate nearby aquifers, and the sludge wastes, although not a radioactive hazard, contain trace amounts of toxic metals which may leach out. For these reasons prerequisites to *in situ* leaching include a thorough knowledge of the regional hydrology, the existence of impermeable rock layers above and below the ore body, and constant pressure on the leacheate fluid.

Presently the cleanup of mine and mill tailings is hamstrung by jurisdictional disputes between the Nuclear Regulatory Commission and the Environmental Protection Agency, unresolved litigation between environmentalists and the uranium industry, and the inability of Congress to develop a unified program which is acceptable to industry, environmentalists and the state governments (Crawford, 1985). Meanwhile, 191 million tons of toxic tailings in seven states lay unattended, prompting a congressional aide for one affected state to remark "these tailings are just blowing around New Mexico."

Environmental Impacts of the Conversion Process

The conversion process gasifies the U_3O_8 concentrate into the UF_6 required for enrichment. Only two commercial conversion plants currently operate in the United States: the Kerr-McGee facility in Gore, Oklahoma, and the Allied Chemical plant in Metropolis, Illinois. Together they process 10,000 metric tons of U_3O_8 yearly (180 metric tons of U_3O_8 yields the 270 metric tons of UF_6 necessary each year for a 1000 MW$_e$ LWR (EPA, 1973)).

The Kerr-McGee plant uses a dry hydrofluor process involving reduction, hydrofluorination, fluorination, and refinement of the U_3O_8. Radioactive effluents occur as a gas at Kerr-McGee and a liquid at Allied Chemical. Both plants recover and recycle virtually all of the radionuclides and chemicals present so that uranium is virtually the only nuclide released to the environment (see Figure 15.6). Conversion of the total 10,000 metric tons of U_3O_8 to UF_6 produces about 1000 million tons of solid waste that is shipped to one of three open sites

Table 15.3. Annual Production and Accumulations of Tailings from Uranium Mining

	Volume of Tailings (10^8 cu m)	
Year	Annual[a] Production	Accumulated[b] Material
1979	0.05	0.66
1980	0.04	0.70
1985	0.12	1.11
1990	0.14	1.75
1995	0.14	2.48
2000	0.15	3.20

Source: From DOE (1973).

Note: Amounts are based on the following ore assays and mill recoveries:

　1979–1984 = 0.15 percent ore assay and 92.8 percent mill recovery,

　1985–1989 = 0.139 percent ore assay and 92.0 percent mill recovery,

　1990–2000 = 0.131 percent ore assay and 91.8 percent mill recovery.

[a] Tailings are assumed to be generated in the same year as the uranium is charged to the reactor.

[b] Includes the pre-1979 inventory of tailings at active sites.

for commercial waste disposal at Richland, Washington, Beatty, Nevada, and Barnwell, South Carolina, where they are buried in shallow trenches (EPA, 1973). Since regulations do not require the plants to report the total amount of each radionuclide discharged per year, and since air and water control technologies combine removal of radioactive and chemical materials, it is difficult to estimate radionuclide emission accurately (see Figure 15.6).

Environmental Impacts of Enrichment

Enrichment compresses UF_6 gas through about 1700 semipermeable barriers to concentrate the ^{235}U. Three government gaseous diffusion plants at Oak Ridge, Tennessee, Paducah, Kentucky, and Portsmouth, Ohio, enrich all uranium in the United States. Because these plants supply certain military requirements, specific emission data per 1000 MW plant requirements are not public information. The largest amount of radioactive liquid and solid wastes generated is from cleaning and replacing equipment for recovery of lost uranium. These wastes are less than 1 metric ton per annual LWR requirement and need less than 0.01 acre per year for storage (EPA, 1973). Remaining waste is incinerated, and the small amounts of fluorine and uranium released appear to be well within regulations (see Figure 15.6).

Environmental Impacts of Fuel Fabrication and Scrap Recovery

Ten fuel fabrication plants convert the enriched UF_6 to UO_2 by hydrolyzing UF_6 to uranyl fluoride, converting the fluoride to ammonium diuranate, and calcining the product to ceramic UO_2 pellets. The pellets are loaded into zircalloy fuel rods 3.7 m long and 13 mm wide. Scrap material is dissolved in nitric acid and reprocessed into UO_2. Enriched uranium is the principal radioactive component in air and liquid effluents. Chemical effluents include elemental fluorine, and fluorine and nitrogen compounds. These are mostly vaporized, while the calcium compounds from the neutralizing process are buried on site. About 680 metric tons of calcium

Table 15.4.   Changes in Annual Biotic Production in Areas Affected by Nuclear Fuel Extraction

Ecosystem Characteristic	Ecosystem Characteristics for 250 ha Uranium Mill Sites	Typical Loss from Mill Site
Primary production (2.2×10^7 kcal/ha)	5.4×10^7 kcal	up to 5.4×10^7 kcal
Seed production (5.0×10^7 kcal/ha)	1.3×10^7 kcal	1.2×10^7 kcal
Secondary production (2.5×10^5 kcal/ha)	6.3×10^5 kcal	2.5×10^5 kcal
Small mammal biomass (35–96 g/ha)	9–240 kg	8–220 kg
Bird biomass (161–174 g/ha)	40–44 kg	39 kg
Livestock		5 cows or 25 sheep displaced
Large mammals		3–5 pronghorn displaced

Source: From Wewerka (1979).

fluoride compounds are buried yearly, using about 220 m³ of land per year per 1000 MW LWR (Pigford, 1974).

Environmental Impacts of Reactor Operation

Operating nuclear power plants also release some radioactivity to the environment. Cracks in fuel cladding may permit diffusion of barium, cesium, tritium, krypton, and iodine into the coolant. Consequently small amounts of these gases are released to the environment when the coolant is cleansed. The monthly purge of air inside the reactor building also releases some radioactive gases to the atmosphere. Contaminated reactor equipment may release radionuclides to the environment when disposed in shallow land burial sites (Norman, 1982). None of these releases is considered an important health hazard for a properly operating reactor although that assessment, like most others in the nuclear industry, has been challenged. Finally, the large quantities of heat released by reactors to lakes or streams (1.3×10^{13} kcal per 1000 MW$_e$ per year) may have adverse impacts on the aquatic fauna and flora.

Environmental Impacts of Decontamination and Decommissioning

All facilities associated with the nuclear fuel cycle eventually become obsolete because of either adverse economics or old age. Nuclear power plants become highly contaminated over their 30–40 year lifetime (EPA, 1973) and therefore require decontamination and decommissioning. The alternatives and some of the environmental risks have been considered on p. 280.

Nuclear Wastes

Whereas the environmental impacts of fuel extraction and processing involve releases of mostly naturally occurring radionuclides, wastes generated by operating nuclear power plants are far more hazardous because these materials contain very concentrated amounts of both the natural materials and the more active and long-lived daughter products such as cesium 137, strontium 90, plutonium 239, americium 241, krypton 85, iodine 129, and tritium. The

wastes (including contaminated equipment) produced by the back end of the nuclear fuel cycle can be classified as: spent fuels, high-level wastes (HLW), low-level wastes (LLW), or transuranic wastes (TRU).

Spent fuel includes irradiated fuel rods discharged from the plant once they are unable to sustain nuclear chain reactions at economic levels. This occurs when accumulated fission products poison the process by absorbing the neutrons needed to sustain the chain reaction (Ehrlich et al., 1977). About one-third to one-fourth of the core must be replaced every year. Initially, nuclear planners assumed that these spent fuels would be reprocessed, recovering reusable uranium and plutonium. Accordingly, plants were equipped with cooling ponds where the rods could give off the greater part of their heat and short-lived radioactive emissions over 10 years before being reprocessed. The only U.S. plant to ever reprocess commercial wastes, however, the Western New York Nuclear Services Center in West Valley, New York, operated only from 1966 to 1972. The plant closed in 1972 to increase its capacity, but never reopened because tighter safety regulations made the modifications uneconomical. In 1977 President Carter banned reprocessing of commercially generated waste fuels. President Reagan lifted the ban in 1981, but reprocessing of commercial wastes has not started up as of 1985. Consequently, the nation's commercial cooling ponds contain over 7000 fuel rods, and have become the *de facto* storage solution for spent fuels (Skinner and Walker, 1981).

Since the cooling ponds were designed for smaller loads, the original fuel rod racks have been replaced by new racks with closer spacing in order to increase storage space for the fuel assemblies. This crowding of the rods decreases the safety factor of the plants in several ways. First, there is a greater possibility of a spontaneous chain reaction because the pool is less able to disperse the accumulated heat. Second, the pools were designed with extra capacity for full core storage in case of emergency or plant cleanup. The accumulated fuel rods have encroached on reserve space, reducing this safety factor. Finally, the packing of the pools also decreases operational flexibility because the full core cannot be unloaded in case the reactor needs repair or modifications (NRC, 1978).

In 1977 the U.S. government announced a spent fuel policy in an attempt to resolve the problem of rod accumulation. By this plan the Department of

Energy would accept responsibility for spent fuels from the power plants for a fee to cover transportation and storage in an interim form until a final storage decision is made (DOE, 1980). A national waste terminal storage repository was planned for 1985, but no agreement has yet been reached concerning a suitable location. DOE also is considering building away-from-reactor (AFR) storage basins to supplement the power plant cooling capacities, but this option only postpones an urgently needed final solution and would continue storage in inadequate facilities.

High level wastes (HLW) were generated by reprocessing nuclear fuels at West Valley. This activity involves dissolving the fuel rods in nitric acid and recovering the uranium and plutonium from the acidic liquid. The uranium is converted to UF_6 and recycled into new fuel rods, but the plutonium is presently unusable and is stored at DOE sites. Because reprocessing breaks all the barriers between the irradiated fuel and the environment, it represents the largest potential for serious contamination of the environment from the uranium fuel cycle. For this reason the DOE is directing its efforts toward defining the final waste form and location for disposal of the existing HLW generated at West Valley. The wastes at this site are stored in two underground tanks. The tank containing 454,800 liters of alkaline waste presents a particular problem because an insoluble sludge containing most of the radionuclides has precipitated to the bottom. Since dissolving this sludge for disposal of the waste also would dissolve the tank, it is not clear how the wastes will be moved (DOE, 1982). Disposal options for HLW are discussed in the waste disposal section.

Low level wastes (LLW) include all contaminated materials with radioactive concentrations below 10 μCi/g, such as mildly contaminated equipment, work clothes and boots. Although these wastes generate little heat, they remain radioactive for up to a few hundred years and must be shielded. Current disposal practices for LLW involve burial in shallow trenches about 60–260 m long, 8–20 m wide, and 5–8 m deep. Water has infiltrated several low-level burial sites contaminating nearby groundwater (NAS, 1980). Long term disposal of low-level wastes has also been hampered by political snafus which have prevented Congress from formulating an acceptable plan. Frustration over the absence of a comprehensive disposal plan promised by Congress in 1980 prompted the governors of the 3 states which have the only commercial low-level waste sites to threaten to close the sites if federal legislation is not passed quickly (Norman, 1985).

Transuranic wastes (TRU) are wastes with more than 10 μCi/g, but less activity than HLW. They are usually contaminated with alpha emitters such as plutonium, but they also have low levels of penetrating radiation. These wastes result from reprocessing and generally require little shielding. Presently commercial TRU wastes are buried with low-level wastes at the Richland site in Washington.

Ultimate Disposal of Wastes

Unlike low-level wastes and some transuranic wastes which can be buried relatively safely in shallow land fills, high-level wastes and some TRU wastes must be disposed of carefully. The intense radioactivity of their fission products over hundreds and even thousands of years requires isolation from the biosphere (Klingsberg and Duguid, 1981). Since the radionuclides must not contaminate the soil or groundwater, it is essential that the waste form be solid, chemically stable, and water insoluble.

Options for HLW disposal are burial in geologic repositories on land or below the seabed, injection into the Earth's subduction zones, or shooting into outer space by rockets (Blomeke, 1979a, 1979b). The last alternative seems the least attractive because a rocket might explode on site before getting into space. Also, the energy cost of the space alternative probably is prohibitive.

The disposal technique recommended by the Department of Energy is to solidify the wastes as a borosilicate glass for burial in a geologic repository (DOE, 1980). Already 8600 liters of liquid HLW have been vitrified at DOE's Pacific National Laboratory. Wastes also could be solidified to SYNROC which is more resistant to leaching than borosilicate glass of hollandite, zironolite, and perovskite. Solidified wastes would be encapsulated in canisters made of corrosion resistant alloys, and placed in excavated geologic repositories 300–1000 m below the ground surface. Such disposal is known as multiple-barrier disposal where the waste capsule, the repository, and the host rock form individual barriers contributing to the cumulative effectiveness. If one barrier fails, the others should hold the radionuclides. To ensure the effectiveness of this strategy, it is important that the host rock have high structural strength, adequate thermal conductivity and sorptive capacities, and low thermal expansion rates. There should also be a minimum of undesirable chemical reactions between fluids trapped in the rock and the waste containers (Gonzales, 1982). The Department of Energy is looking for a national waste terminal storage area with such properties and has proposed the Waste Isolation Pilot Plant (WIPP) test repository in New Mexico. This site is located in a salt dome, but other host rocks being considered include bedded salt, granite, basalt, volcanic tuffs, and shales.

Salt formations are receiving special consideration because of their long-term stability, isolation from circulating fluids, self-healing plasticity, low permeability, and high degree of minability (Klingsberg and Duguid, 1981). Salt formations were first considered in the late 1950s and early 1960s with the project "Salt Vault" in an abandoned salt mine in Lyons, Kansas. The Lyons mine was deemed unsuitable, but many of the characteristics of salt were recognized as useful. Some less desirable properties of salt also were discovered, including the ready solubility of salt in the presence of brines entrapped in the salt, the poor sorptive qualities of salt, and a high plasticity that might make it difficult to keep the mine open as the wastes were emplaced.

Granite may be suitable because it occurs in large quantities and is chemically and structurally more stable than salt. In addition, there are few other resources of interest near granite deposits being considered. Granite is impermeable, contains large quantities of secondary minerals which make for a high sorptive capacity, and few dissolved solids or fluids that could corrode the waste containers. On the other hand, granite would have to be fractured, which means that eventually the wastes could be exposed to migrating water. In addition the greater depths required for excavation in granite relative to salt would increase disposal costs by a factor of 2.5.

Basalt occurs most commonly as thick slabs on the oceanic crust and is much more rare on the continental crust. Conveniently a basalt slab is present under the federally owned Hanford Reservation in Washington, where many nuclear wastes are currently stored. Positive attributes of a basalt repository include intermediate thermal conductivity, low permeability, high structural strength, and good sorptive qualities. Negative aspects include problems presented by the frequent columnar joints, the water-bearing porous patches, and the deep drilling depths required.

Shale has some attractive characteristics for waste disposal. It has low permeability and porosity, and like salt, is highly plastic and has good sorptive qualities. Unfortunately, shales also contain significant amounts of mineralized, corrosive waters, and frequently are located near aquifers. Shales also often have associated resources such as coal, petroleum, and groundwater which could not be developed if an adjacent shale bed became a waste repository.

Subseabed disposal would insert waste canisters either in unconsolidated sediments or in basement rocks of abyssal plains where biotic productivity is low (Kelly, 1981). The concept relies on the stability and remoteness of the disposal area, and on the confinement characteristics of the sediments. The most promising substrates are the red abyssal clays. These sediments have stable distributional histories, do not fracture, and heal when disturbed. It is thought that water movement within these sediments is a maximum of only 30 cm and that the sediments are not altered by heat given off by the wastes. An area of 100,000 km^2 could accommodate the nation's HLW through the year 2050 if they were placed on centers 300 m apart (Hollister, Anderson, and Heath, 1981).

Environmental Impacts of Transportation

Most waste disposal techniques require transporting the waste. Such movements raise the possibility of an accident involving the waste-carrying trucks or trains, with subsequent rupture of the tanks and release of the highly toxic wastes to the environment. Because 156 states and local districts have banned the transportation of nuclear wastes through their jurisdictions, the routes to repositories probably will not be the most direct and may be very expensive and energy intensive (Hill et al., 1982). The subseabed disposal alternative may be especially energy intensive as it will require transportation to harbors, special storage buildings at the harbors, and shipping to the disposal area.

Conclusion

Each step of the nuclear fuel cycle requires environmental controls to minimize the release of radioactive elements to the biosphere. These control technologies are complex and expensive. Although no attempt has been made here or, we believe, elsewhere to analyze their energy costs, it is fairly evident that they are energy intensive. Interestingly, existing energy cost analyses of nuclear power generally have not considered the energy costs of fuel disposal; therefore EROI calculations are greater than the actual returns of the nuclear energy system. Including such energy cost assessments is unlikely to change dramatically the EROI for present-day nuclear power (Table 12.4), but that cannot be ascertained until we are actually disposing the wastes. Finally, we believe that although there are no overwhelming technical problems *in principle* to derive a safe nuclear fuel cycle, the recurrence of serious accidents that were not supposed to occur and the long-term failure of the industry and its regulators to derive satisfactory solutions to nuclear waste disposal has seriously eroded public confidence in our nation's ability to realize the potential of nuclear power. Fortunately, whatever waste problems we do face now are much less than if nuclear power had grown enormously, as was once projected.

ENVIRONMENTAL IMPACTS OF SOLAR ENERGY

Although the environmental impact of extracting and burning fossil or nuclear fuels often is readily apparent, many people perceive so-called *alternative* or *soft* energy, like solar or biomass fuels, as essentially environmentally benign, for solar panels or a cornfield do not produce any apparent pollutants. Unfortunately, this is only partly true. Solar convertors, like all energy technologies, cause significant environmental impacts. A comprehensive analysis must include not only the direct impacts but also the environmental impact associated with the other segments of the economy that produce the materials used in any energy convertor, as well as the impact of the extraction and combustion of the energy used to produce those materials. It is important to evaluate the environmental impact of these fuels on a per kilocalorie basis since one reason that pollutants from fossil fuel burning are so prevalent is that the use of these fuels is so large.

Hydroelectric

Hydroelectric power is the most important industrial solar energy convertor, and so we consider its impacts first. In some respects hydroelectric power is extremely clean, for virtually no fuel is burned to convert the potential energy of the elevated water into electricity, and the hydrologic cycle that provides the hydroelectric head is run by clean solar power. And many people view reservoirs as good boating environments and as attractive recreation sites for fishing and camping. Many hydroelectric dams produce cold, clear, and attractive rivers below them that are ideally suited for sport fisheries for fish such as trout that are often more desirable to anglers than are the fish that originally were found there. But constructing large dams and reservoirs to generate significant quantities of hydroelectric power causes large environmental impacts in the region under and below the reservoir. Evaluating the impacts of reservoirs therefore is in large part a question of personal preference.

In regions like New England water power produced from damming rivers was an important, and often the most important, source of energy for early industrialization. One environmental impact is still felt. Many of these rivers (such as the Merrimac and Connecticut) once had very large runs of Atlantic salmon, a prized commercial, food, and sport fish. At one time salmon were so abundant that indentured servants sometimes had clauses in their contracts limiting salmon dinners to no more than twice a week. When dams were erected on spawning streams several years of excellent fishing ensued at the base of the dam where the salmon's upstream migration was stopped. But the fish were eventually eliminated because adults could not get to their spawning grounds. At that time there were no fish ladders or other means for the fish to get around the dams. By 1850 Atlantic salmon were eliminated from all rivers in the United States except for a few rivers in northern Maine. Recent restoration programs have been partially successful in reestablishing salmon runs in a few rivers such as the Penobscot and the Connecticut, but these programs are very expensive and, at least for the present, depend on continual restocking with hatchery-reared fish. Nevertheless, for many of us the return of salmon to many of our Eastern rivers after a hundred or more years of absence is exciting proof that some aspects of environmental quality can be improved by prudent investment of economic energies—if the social will permits.

In recent decades large hydroelectric dams in the Pacific Northwest have reduced substantially the salmon runs on rivers such as the Columbia despite large investments in fish ladders, hatcheries, and other mitigation devices. Dams affect the fish in a number of ways. Most obviously, dams block upstream movements of adults. Conversely, dams provide a rather treacherous roadblock for little salmon moving downstream (called smolts), for the smolts either have to go through the turbines themselves or go over the spillways that are often 50 or 100 m high. Furthermore, as the water passes over the spillways, it often becomes supersaturated with nitrogen, giving fish nitrogen embolism or, as it is more familiarly known to divers, the bends. This disease causes the eyes of the fish to bulge and often leads to death. Another problem is that when large quantities of water are diverted for irrigation, some of the smolts are diverted with the water and end up pathetically stranded in agricultural fields. Fortunately fish often can be kept out of the diverted waters if a piece of bear skin is placed at the entrance of the diversion canal. Many of the fish killed are replaced with hatchery fish, but we are learning slowly and painfully that hatchery fish simply do not have the same genetic resources as wild fish and often are inferior in many respects.

The construction of dams also produce other environmental stresses. Whatever land and river area that originally existed above the dam is obliterated. In a comprehensive review of land use by energy facilities, Smil (1984) found that although hydroelectric facilities produced less than 2% of the industrial energy generated in the United States, such facilities represented nearly a third of all land area devoted to energy production and distribution. Because river valleys tend to be good farmland, large areas of prime farmland are lost. As a result more energy is used elsewhere if that lost production is compensated by increased use of poorer quality farmland. The water backed up behind a dam weighs a great deal. Occasionally dams break. A dam break may produce disastrous results downstream. More frequently the great weight of impounded water causes minor earthquakes, although this process is not yet well understood.

Many hydroelectric dams are also flood control dams that reduce the magnitude of most floods. This, as well as federal flood insurance, encourages people to build residential, commercial, and industrial facilities on flood plains (the often flooded areas adjacent to rivers). Unfortunately, floods inevitably occur that are too large to be handled by flood control devices, and structures on flood plains are lost or damaged. So, as flood control expenditures have increased, so have flood losses. Both flood controls and flood loss replacements are energy intensive.

Finally, the materials (e.g., concrete) used in the construction of dams require considerable amounts of energy, and produce pollution elsewhere. But because the EROI for hydropower is greater than that for many other sources, the general pollution caused per kilocalorie produced by hydroelectric dams is much less than other technologies such as new coal mines or other solar convertors which have a lower return on investment.

Perhaps the most important impact of dams cannot be quantified. Free flowing rivers are emotionally, aesthetically, and recreationally very important to many people. The COANES (1979) report concluded that since free flowing rivers are now relatively rare, damming them produced among the largest environmental impact of any energy source. As a result this committee recommended against large-scale additions to existing hydropower capacity.

Solar Collectors

The environmental impact of solar collectors stems from the low energy density of solar power—solar collectors require a great deal of land and materials to collect the same amount of energy relative to fossil fuel convertors. The factor is probably at least 100 to 1 for solar versus fossil energy. But, if one includes the land required to surface mine coal and dispose of wastes over 30 years, the relation drops to about 6 or 8 to 1 (Smil, 1984). If all industrial energies consumed by the United States in 1980 (19.2×10^{15} kcal) were supplied by solar collectors at 5 percent efficiency, 260 thousand km^2, 3.5% of U.S. land area, would be required for the collectors (Romer, 1976). If this land area is shaded by collectors, obviously it can no longer provide the public service functions produced by the natural systems. On the other hand, if the collectors are located on roof tops or other areas where the natural system already has been displaced, that problem is essentially nonexistent.

The second major impact of solar collectors is their large material requirement per unit of energy delivered. For example, a 1000 MW$_e$ coal-fired plant uses from about 10,000 to 20,000 tons of steel. A thousand megawatt (mean capacity) solar collector located in a Southwestern desert would require some 26 million m^2 (10 mi^2) of collectors with a minimum of 400,000 metric tons of steel (COANES, 1979). Thus the pollution caused by the production of the steel required by collectors would be more than 20 times greater than that required for a coal plant, as well as additional pollutants from the much larger use of glass, concrete, and so forth. An even more comprehensive analysis that examined all the various pollutants produced by manufacturing and moving all the components of any low EROI power plant presumably would show a high ratio of pollutants to energy capacity, especially if that plant used exotic materials in its construction (see p. 427). Finally, if the preliminary values for EROI of Table 2.1 characterize mature solar and biomass industries, the economic energy required will range between one-quarter and one-half of the energy eventually recovered from the system. If there were exponential growth of, say, an active solar industry with a 2 for 1 energy return on investment, each kilocalorie delivered to society could produce at least the same pollution as if that energy were derived directly from fossil or nuclear fuel used to produce the collectors.

Biomass

The environmental impacts of energy from forests are covered in various ways in Chapter 19. Producing energy from crop biomass, which often has a low net energy return, may have large environmental impacts. For example, if the 430 million metric tons of crop residues left in the U.S. fields every year were converted to usable fuel (ethanol, methane, or electricity), they could satisfy nearly 10% of the nation's total energy demand (based on an average heating value of 4.7×10^6 kcal per dry ton of biomass, Pimentel et al., 1981). But the authors of this report concluded that net yield from this potential energy source would be very small because of the large energy costs necessary to offset the environmental impacts associated with the entire energy conversion process. In addition the power density from biomass is so low that huge amounts of land would be needed (Smil, 1974).

Using crop residues as fuel degrades the environment at several stages. Indirect impacts are caused by the production and operation of the large machines that collect and transport the residues. More important, harvesting residues would increase soil erosion and runoff from prime agricultural farmlands in the United States. Crop residues now left lying on the soil surface maintain soil quality by reducing soil erosion and increasing water infiltration rates into the soil. The potential impact of energy production from biomass is heightened by the fact that soil erosion from agricultural land already is a severe environmental problem in the United States. Already an estimated 25 to 30 metric tons of soil per hectare are washed into streams, lakes, reservoirs, or other environments annually, compared to the 3–4 or so tons per year formed.

Harvesting crop residues under natural conditions also affects the environment in ways that are presently difficult to assess but in the long run may be as severe as those outlined here. Sedimentation from agricultural runoff tends to fill in streams, lakes, and reservoirs, bodies of water that often are used as sources of fresh water and recreation by millions of people. Hydroelectric capacity also is lost, and an unknown but significant proportion of the hydroelectric electricity generating potential in the United States has been lost by the siltation of reservoirs. Siltation also may decrease natural energy flows in aquatic ecosystems, for suspended sediments reduce light penetration. Conversely, additional nutrient inputs associated with agricultural runoff may stimulate phytoplankton productivity causing eutrophication in certain lakes and streams. An increased use of forest biomass would cause a similar increase in nutrient loss and aquatic environmental damage.

Topsoil lost through erosion in the United States contains about 9.5 million tons of nitrogen and 1.7 million tons of phosphorus (Larson et al., 1978). Crop residues conserve these nutrients by holding soil and reducing erosion. If these nutrients were lost by erosion of topsoil, farmers would need more fertilizer to maintain soil productivity. In turn fertilizer production by the petrochemical industry is extremely energy intensive and its subsequent use on farmlands causes a complex array of environmental problems. Pimentel et al. (1981) concluded that the EROI from the conversion of corn and other agricultural residues into electricity, ethanol, or biogas would approach the energy break-even point over a 30-year period due to the energy inputs required to offset decreases in crop yields caused by erosion of topsoil laden with valuable nutrients. On the other hand, local management of preexisting farm wastes to supply on farm needs can produce small but locally important sources of energy with minimal environmental disruption (Jewell et al., 1982).

Although the burning of wood does not tend to produce large quantities of sulfur dioxide, wood smoke does contain many other very potent pollutants, including some carcinogens (Cooper, 1980; Deis, 1980; Oglesby and Blosser, 1980; Hewett, 1981). The effects of such pollutants can be enhanced because very often wood burning is done with furnaces that leak smoke into the interior of houses and other structures. At this time a comprehensive analysis of the total human health impacts of burning fuel wood does not exist.

One interesting perspective relating to the land requirements of various energy types is given by Smil (1984). Smil found that in highly industrialized societies roughly as much land is devoted to energy production as is devoted to roads or urban areas. Thus in some sense energy production generates the major environmental change to our landscape after agriculture.

16

GENERAL IMPACTS OF BURNING FOSSIL FUEL

Fossil fuels all contain the same basic elements, principally carbon and hydrogen, with smaller but variable amounts of oxygen, sulfur, and other trace materials (Table 16.1). Because of these similarities, combustion of fossil fuels produces similar by-products, thereby creating similar environmental impacts. Two of the major impacts, the release of carbon dioxide and the acidification of precipitation, are discussed first. Then the effects of trace materials and more complicated by-products of combustion are considered.

THE RELEASE OF CARBON DIOXIDE

Since all fossil fuels are reduced carbon compounds, their combustion releases carbon dioxide. No method is known by which most of the energy contained in the chemical bonds of fossil fuels can be released without producing large volumes of carbon dioxide and water (Table 16.2). Carbon dioxide probably could be removed from stack gases, but the dollar cost is astronomical and the energy cost would be a very large proportion of the energy produced by the original combustion—negating the effects of carbon dioxide removal (Baes et al., 1980; Edwards and Reilly, 1984). This is another example of how energy costs can limit a technologically feasible solution.

Atmospheric Concentration of Carbon Dioxide

The Earth's atmosphere contains trace quantities of carbon dioxide, about 0.035% (or 350 ppm). It is believed that the present concentration of carbon dioxide is about 20–50% higher than it was a century ago. Unfortunately we do not have any direct measurements before 1957. Some new data on CO_2 concentrations of ancient glaciers, presumed to be in equilibrium with ancient atmospheres and stable since then, indicate that the atmosphere may have varied from 150 to 250 ppm CO_2 over the past several thousands of years. These data also suggests that ancient fluctuations of CO_2 were closely correlated with records of temperature (as preserved in isotope ratios) over that time period (Neftel et al., 1982; Shackleton et al., 1983). Since 1958 the concentration of carbon dioxide has been monitored daily at Mauna Loa, Hawaii, a high mountain where well-mixed air passes, air thought reasonably characteristic of the mid Northern Hemisphere where an average parcel of air circles the Earth about once every two weeks. The data show annual fluctuations that are principally a result of net photosynthesis removing CO_2 from the atmosphere during spring and summer and net respiration returning CO_2 to the atmosphere during fall and winter (Figure 16.1; Hall et al., 1975). The record also shows a clear and unequivocal increase from each year to the next, an increase that appears unquestionably a result of human activity.

The Impact of Increasing Carbon Dioxide

At or near current concentrations (350 ppm), carbon dioxide has no known deleterious effects on the physiology of humans or other animals. At high levels (9000 ppm) carbon dioxide depresses the respiratory center of animals, leading to coma and death at about 500,000 ppm (Guyton, 1976). But much smaller changes can effect plants because CO_2 is a plant nutrient. Carbon dioxide concentrations have a positive effect on photosynthetic rates of most plants even at moderately enhanced levels, so that the levels of atmospheric carbon dioxide that are likely to be present in coming decades may

Table 16.1.   Selected Elements Mobilized into the Atmosphere From Weathering of Natural Land Surfaces and the Combustion of Fossil Fuels

Element	Fossil Fuel Concentration (ppm) Coal	Fossil Fuel Concentration (ppm) Oil	Fossil Fuel Mobilization ($\times 10^6$ kg/yr) Coal	Fossil Fuel Mobilization ($\times 10^6$ kg/yr) Oil	River Flows ($\times 10^6$ kg/yr)
Li	65		9		110
B	75	0.002	10.5	0.00	360
Na	2000	2	280	0.33	230,000
Mg	2000	0.1	280	0.02	148,000
Al	10,000	0.5	1400	0.08	14,000
P	500		70		720
S	20,000	3400	2800	550	140,000
Cl	1000		140		280,000
K	1000		140		83,000
V	25	50	3.5	8.2	32
Cr	10	0.3	1.4	0.05	36
Fe	10,000	2.5	1400	0.41	24,000
Co	5	0.2	0.7	0.03	7.2
Ni	15	10	2.1	1.6	11
Zn	50	0.25	7	0.04	720
As	5	0.01	0.7	0.002	72
Se	3	0.17	0.42	0.03	7.2
Sr	500	0.1	70	0.02	1800
Mo	5	10	0.7	1.6	36
Cd		0.01		0.002	
Ba	500	0.01	70	0.02	360
Hg	0.012	10	0.0017	1.6	2.5
Pb	25	0.3	3.5	0.05	10
U	1.0	0.001	0.014	0.001	11

Source: Adapted from Bertine and Goldberg (1971).

Note: Mobilization rates are calculated from 1967 fossil fuel production. The flux of the elements from rivers into the sea is included for comparison.

Table 16.2.   Major Emissions from Combustion of Gas, Oil, and Coal (kg/10^9 kcal)

	Gas	Oil	Coal
Sulfur oxides	1.1	1500–1660	1200–7900
Particulates	9–27	250–1300	110–17000
Carbon monoxide	30–36	72	80–160
Hydrocarbons	1.8–14	13	23–79
Nitrogen oxides	144–1260	230–1370	1200–4400
Carbon dioxide	210,000	288,000	360,000

Source: From Scientists Institute for Public Information (1979).

stimulate the growth of the entire terrestrial biota, which is likely to be beneficial to human economic activities.

Increasing the atmospheric concentration of carbon dioxide, however, may seriously disrupt the Earth's thermal energy budget, changing climates and disrupting present patterns of rainfall, agriculture, and human well-being. The temperature of the Earth is a result of the balance between incoming short-wave solar radiation and the long-wave radiation that is reradiated to space (Figure 16.2). Minute quantities of carbon dioxide and water vapor in the Earth's atmosphere are relatively transparent to incoming short-wave radiation but absorb and reflect back a small proportion of the reradiated long-wave radiation, converting it to sensible heat. As the concentration of carbon dioxide in the atmosphere increases, more long-wave radiation is absorbed and converted to heat, leading to a global warming. This has been labeled the *greenhouse effect* (which is actually a misnomer because greenhouses are warmed principally by their glass roofs limiting upward advective transport of warmed air).

If present trends continue, the concentration of atmospheric carbon dioxide could double from its

ATMOSPHERIC CO$_2$: MAUNA LOA DATA

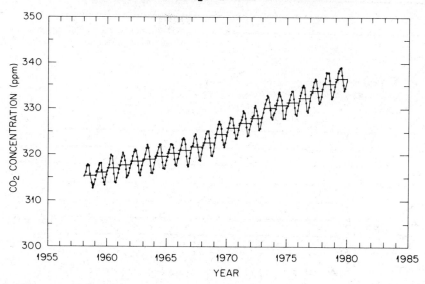

Figure 16.1. Increase of CO$_2$ in atmosphere at Mauna Loa, Hawaii. The wiggles are caused by biotic activity for growing and nongrowing seasons. (From DOE, 1982; see also Hall et al., 1975.)

preindustrial level of about 250–280 ppm to perhaps 560 ppm by the middle of the next century (Figure 16.3). Currently most scientists who study the problem believe that increased carbon dioxide concentrations will increase temperature significantly, especially at the polar regions (Manabe and Weatherald, 1969; Schneider, 1975; Manabe and Stouffer, 1980), and that such changes may be just beginning to be distinguishable from normal variations (Hansen et al., 1981; Gornitz et al., 1982). These climate changes also could shift rainfall distributions significantly, changing global agricultural patterns for better or worse, depending on where you are. Much tropical agriculture probably would benefit, while the American Midwest, the Ukraine, and Scandinavia might suffer. Additionally a rise in temperature could melt parts of the polar ice caps, which, by increasing the sea level, would cause enormous economic damage by flooding coastal cities or, initially, their subways. All of these changes could require very large quantities of energy for adjustment or mitigation, although the increased CO$_2$ and rain (in certain areas) would decrease the energy cost of raising crops. A particular problem with the CO$_2$ buildup is its persistence— even if we stopped the net flux of CO$_2$ to the atmosphere today, the existing CO$_2$ would be reabsorbed only very slowly.

It is important to note that the projected temperature changes are based on models that are quite controversial and still essentially unverified. In particular, there is a great deal of uncertainty pertaining to vertical heat transfer in water vapor and to the role of clouds. Idso (1982) summarizes the position of a minority of scientists who believe that climate warming may be far less than predicted by the models most often used. He also believes that the increased crop productivity that should result from the increased CO$_2$ will be essential for feeding future human population increases. On the other hand, other recent research, summarized in Kerr (1983), suggests that other residuals of industrial society, including methane, nitrous oxide, and chlorofluorocarbons could double the warming resulting from increased CO$_2$.

Human Impacts on the Global Carbon Cycle

At first, the secular increase in the atmospheric concentration of carbon dioxide was thought to be caused simply by industrial combustion of fossil fuels. This hypothesis was confirmed empirically by Revelle and Suess (1957) who measured the changes in the abundance of a radioactive tracer in the atmosphere. A very small percentage (about 1 part in 10^{12}) of atmospheric carbon dioxide contains the isotope ^{14}C, produced by cosmic rays entering the upper atmosphere. Since fossil fuels have been buried, and hence sequestered away from cosmic rays, for long periods of time, virtually all of the radioactive ^{14}C has decayed to stable ^{12}C. Therefore

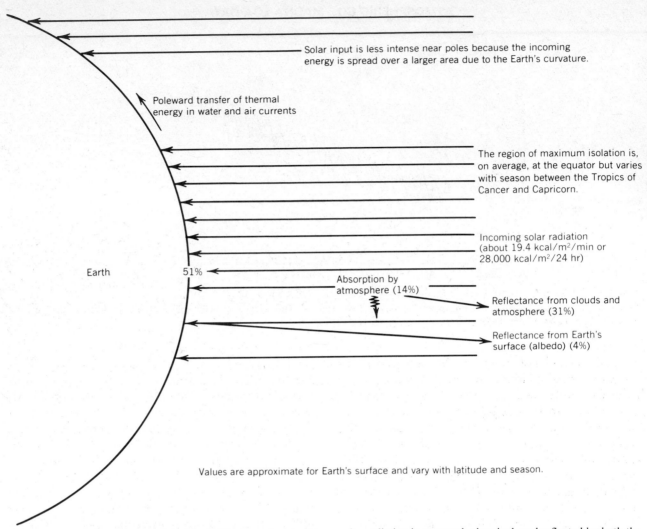

Figure 16.2. Energy balance of the Earth. The incoming shortwave solar radiation is scattered, absorbed, and reflected by both the atmosphere and the ground, and eventually an approximately identical amount is reradiated as long-wave radiation.

burning fossil fuels releases only ^{12}C, diluting the proportion of ^{14}C in the atmosphere. This dilution (the *Suess effect*), confirms the hypothesis that fossil fuel combustion has been a major contributor to the increased concentration of carbon dioxide in the atmosphere.

The Missing Carbon: Further Considerations

Questions concerning the mass balance of this dilution arose when the combustion of fossil fuels and the atmospheric concentration of carbon dioxide were examined quantitatively. Only about 60% of the total carbon dioxide released by fossil fuel consumption could be found in the atmosphere. Where did the rest go? One logical place was the ocean,

since carbon dioxide is forced into seawater as its concentration in the atmosphere increases. Oceanographers agreed, but argued that not all of the missing carbon dioxide could be found in the world's oceans because thermal stratification limits the contact between atmosphere and most of the deeper ocean (Broecker et al., 1979; Siegenthaler, 1980; see Chapter 18). Another suggestion was that the missing carbon dioxide was taken up by the world's terrestrial biota, as photosynthesis increased in response to the enhanced carbon dioxide concentration.

The Global Carbon Cycle

A pictorial summary of the global carbon cycle is given in Figure 16.4. Huge reservoirs of carbon are

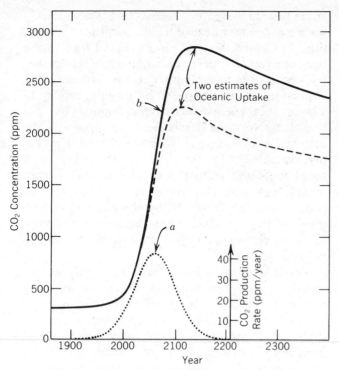

Figure 16.3. (a) potential for fossil fuel burning for the world with assumption of a 2.5 percent increase per year; (b) resultant CO_2 increase in atmosphere thought likely. (From Bacastow and Keeling, 1973.)

stored in the ocean, ocean sediments, and in certain sedimentary rocks of the Earth's mantle. Smaller reservoirs are contained in fossil fuels—those carbonaceous rocks in which the ratio of organically derived carbon to other material is great enough to be mined and burned with an energy and economic gain. In addition small quantities of carbon are stored in the atmosphere, the surface of the ocean, and the biota, especially the terrestrial phytomass and soil.

Interestingly, the smaller reservoirs are the focus of the current debate concerning the global carbon cycle and human activity. The largest reservoirs are sealed off from the atmosphere and therefore have little or no effect on atmospheric concentration of carbon dioxide. Most of the deep ocean is sealed off by a permanent vertical thermocline (steep temperature difference) which prevents advective mixing. This exists from about 50 degrees North to 50 degrees South. The ocean sediments and uneconomic fossil carbon in rocks are buried deep in the Earth's crust and are unreactive with the atmosphere. Thus it is surprising that human activity, which affects the flow of only 5–10 billion metric tons of carbon each year via fossil fuel combustion and impact on storages in the biota, can have such a significant impact on the atmosphere. An important scientific

CARBON INVENTORY: AVERAGE OCEAN

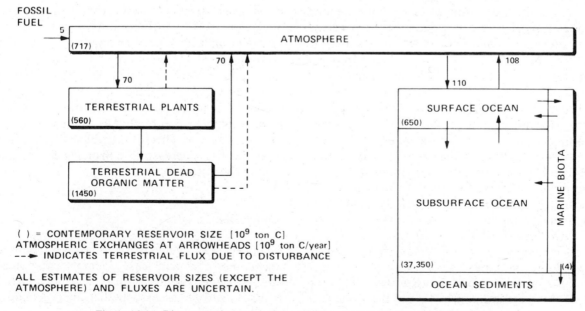

Figure 16.4. Diagrammatic model of the global carbon cycle. (From DOE, 1982.)

question is that it is not possible presently to balance the global carbon cycle—that is, of the 5 billion metric tons released into the atmosphere each year, less than about 60% can be found as increases in atmospheric carbon and only about another 1 billion metric tons are thought to be found in the sea (Siegenthaler and Oeschger, 1978; Broecker et al., 1979). Where is the missing carbon?

The Role of the Earth's Biota

Early analysts of the problem thought that the terrestrial phytomass was a net sink for carbon dioxide, fixing it through increased photosynthetic rates due to enhanced concentrations of carbon dioxide, and thus absorbing the missing carbon. Many now believe, however, that terrestrial phytomass, especially tropical forests, contributes 1 to 4 billion metric tons of carbon per year to the atmosphere, a number not too different from the release from fossil fuels (e.g., Woodwell et al., 1977; Bolin et al., 1979; Detwiler et al., 1985). If land-derived carbon release is large, this makes it even harder to balance the global carbon budget and leads to considerable uncertainty about what, if anything, we should do to ameliorate the problem.

Changing patterns of human land use are the major cause of carbon release from terrestrial systems. These changes are especially important in the tropics, where forests often contain more carbon per hectare than most temperate ecosystems, and where forests are being converted to uses that store much less carbon per unit of land. For example, changing tropical forest in Panama to pastureland reduces the amount of carbon stored from about 170 metric tons per hectare to about 5 metric tons per hectare (Hall et al., 1985). On the other hand, many forests in industrialized temperate areas, such as New England and the American Southeast, are recovering, for more food can be grown on less land due to higher yields in recent decades and because food is imported from other regions (Armentano and Ralston, 1980; Delcourt and Harris, 1980).

The principal cause of the changing patterns of land use and subsequent carbon release in the tropics is human population growth, sometimes combined with the various social structures that encourage and maintain ownership of much of the most fertile land in the hands of a very small percent of the population (Plumwood and Routley, 1982). Because of this population growth and the absence of

an industrial base, the forests are viewed as a source of land for needed local agricultural production. In Central America large areas of forest have been converted to pastureland, much of which supplies beef for U.S. hamburger chains (Myers, 1981). Other forests are destroyed by logging, mostly to support local energy and material demands but also for sale to Northern Hemispheric corporations. The conversion of forests to agricultural land is probably exacerbated by the greatly increasing price of fossil fuels and industrial fertilizers, which often means that more land is needed per unit of food production, but decreased by the planting of high yielding crop varieties. Thus the question of tropical land clearing is a complex and highly charged political issue.

There is considerable debate as of this writing about the rate at which tropical forests are being cut, and thus the rate at which carbon is being added to the atmosphere. Myers (1980) suggests a quite rapid rate, while Lanly and Clement (1982) give lower, although still substantial, rates. Our own results (e.g., Detwiler et al., 1985) suggest releases of roughly a billion metric tons carbon a year, not as large as the largest estimate of biospheric release but still sufficient to make it a mystery that our knowledge of the carbon budget does not balance. A few more recent papers by some geochemists imply that their earlier estimates of CO_2 uptake by the oceans were on the low side. It is possible, but unlikely, that if fossil fuel use continues to grow slowly and if population growth in the tropics continues to increase, there could be more CO_2 release from the biosphere each year than from fossil fuels—at least until the forests are exhausted. There is some evidence that land clearing from 1850 to perhaps 1950 added more CO_2 to the atmosphere than industrial fuel burning (Figure 16.5). Thus it is now fairly well accepted that over the past 100 years both fossil fuel burning and destruction of vegetation have contributed to the increases in CO_2 that we observe. There is not yet any hard evidence that the CO_2 increase is enhancing forest growth and carbon storage.

Conclusion

Many scientists believe that the CO_2 problem is the most important and least tractable component of the energy-environment interface. Although it is now too early to assess whether the greenhouse ef-

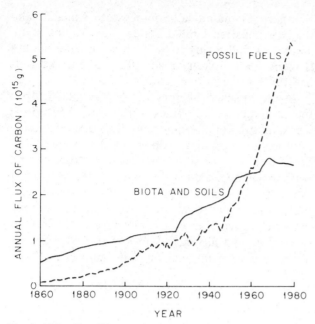

Figure 16.5. Possible contribution of CO_2 from industrial fuel burning and from destruction of plants. (From "Changes in the Carbon Content" by R. A. Houghton, J. E. Hobbie, J. M. Mellilo, B. Moore, B. J. Peterson, G. R. Shaver, and G. M. Woodwell. *Ecological Monographs*, 1983, 53:235–260. Copyright © 1983 by the Ecological Society of America. Reprinted by permission.)

fect will have an overall positive or negative impact on human designs, it almost certainly will have large impacts on human economic and social welfare.

Because the impacts of climate warming from increased CO_2 levels are so large, so rapid, and so irreversible, most of the scientific community (including ourselves) believe it is wise to limit the anthropogenic production of CO_2 as much as possible. To do so, we must regulate human activity. This places severe constraints on, for example, the large-scale use of coal or oil shale, for the carbon dioxide emitted per unit of energy produced is very large (Burnett, 1981). On the other hand, it seems unlikely to us that we will ever voluntarily restrict our use of fuel. Should all fossil fuel burning be replaced by nuclear and/or solar power to avoid the CO_2 buildup? Can the CO_2 buildup be reversed by planting more trees? Does the use of fossil energy to make fertilizers actually decrease the net carbon release by decreasing the amount of new land required for agriculture? What would be the energy costs of replacing structures or moving cities that we have built at the edge of the sea?

The CO_2 problem and its related trade-offs are of

an overwhelming scale that neither science nor society has had to deal with before, and these trade-offs probably have enormous impact on the ultimate wealth society derives from energy use. Unfortunately it would be difficult to make a strong case for undertaking any large-scale social program aimed at alleviating the impacts of carbon dioxide while we cannot balance the global carbon budget. A large coordinated scientific program presently underway is helping to reduce that uncertainty, and within a few years it should give us better answers to the presently unresolved scientific problems. From this information we will be in a better position to decide what, if any, social policies should be implemented.

ACID RAIN*

Many of us who like to fish learned about how to catch trout in the wonderful books of Ray Bergman, who often wrote of the fishing in Adirondack mountain lakes. The high-quality fishing in these lakes in the last century also has been immortalized in paintings by Winslow Homer. Motivated in part by the artistry of Bergman and Homer, the senior author of this book hiked into a number of these lakes in 1971. The lakes were beautiful, but the fishing was very poor—in fact I caught no fish at all, nor did I see any sign of fish. At first I attributed my poor luck to my considerably lesser skill than Bergman, who caught many fish in these same lakes. A few years later I found out the real problem—acid rain. The combustion of fossil fuels releases large quantities of the oxides of sulfur and nitrogen into the atmosphere. These oxides combine with water to form strong mineral acids which then precipitate as rain, often hundreds to thousands of kilometers downwind from the source. This rain increases the acidity of poorly buffered lakes and kills fish. As of the late 1970s over 200 lakes and ponds in the Adirondack Mountains, mostly above 600 m in elevation, no longer contained trout or other fish, apparently because of acid rain. It appears that many lakes and streams in various parts of the industrial world are losing their fish and many other organisms due to acid rain.

Preindustrial rain was, to the best of our knowledge, generally neutral to slightly acidic. We know this because the pH of precipitation prior to the

---

*Written in part by Thomas Butler.

industrial revolution, preserved in glaciers and continental ice sheets, is generally greater than 5.0 (Mateev, 1970; Langway et al., 1965; Rainwater and Guy, 1961). Unfortunately data on rain chemistry in the first half of this century are incomplete, but the presence of relatively large amounts of bicarbonate found in samples of rainwater and snow from Virginia, Tennessee, and New York before 1930 indicate a pH of around 5.6 or greater (Hammer, 1977; Ellet and Hill, 1929; MacIntire and Young, 1923; Delmas and Legrand, 1980; Collison and Mensching, 1973). Furthermore, analyses of contemporary precipitation from relatively unpolluted Greenland and Antarctica indicate a pH of about 5.5. But more recent measurements of rain falling on areas downwind of major industrial centers in Central Europe,

Southern Scandinavia, Southeastern Canada, and the Northeastern United States, yield values of 4.0 and lower, indicating 10 to 100 times greater acidity (Likens, 1976; Likens et al., 1979; Likens and Butler, 1981; Figures 16.6, 16.7).

In the absence of mineral constituents, CO_2 in the atmosphere combines with water to form a weak solution of carbonic acid. This chemical reaction produces precipitation with a pH of about 5.6. Some additions of sulfur and nitrogen from natural sources slightly reduce that natural pH. But when the sulfur, which comprises 1–4% of most fossil fuels, and the nitrogen from the atmosphere combine with oxygen during combustion to form various sulfur and nitrogen oxides (SO_x and NO_x, respectively) very large changes in rainwater pH can

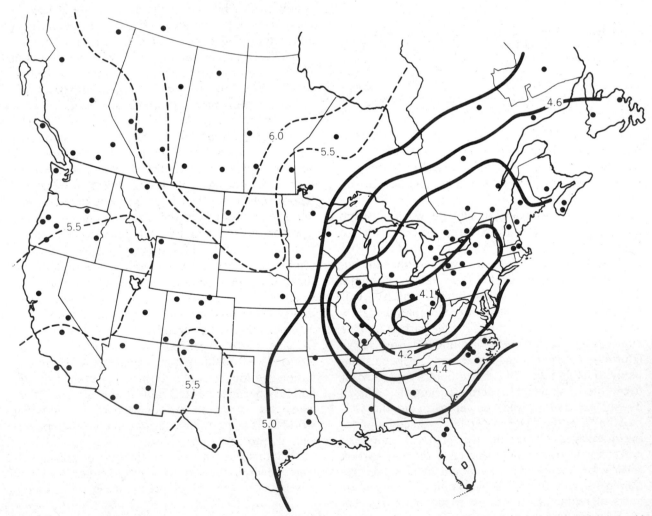

Figure 16.6.   Acid precipitation in the United States. Rain values higher than 2.5 microequivalents per liter (pH 5.6) are defined as acid precipitation.

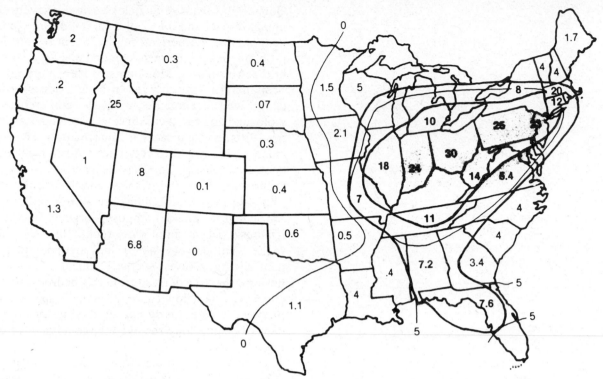

Figure 16.7. SO₂ emission contours in g/m²/yr, for the United States. (From OTA, 1979.)

occur. These oxides are transformed chemically to sulfuric acid aerosols and nitrogen acid vapors over hours to days by processes that are still not well understood. These acid components disassociate strongly, producing a dilute solution of a strong acid. This low pH, highly acidic rainfall is called acid rain, and it characterizes the rain falling over most of the heavily industrialized world. Because the emission of both sulfur and nitrogen oxides has increased in recent decades (Figure 16.8), acid rain is becoming increasingly widespread—for example, in the Southeastern United States (Likens et al., 1979; Likens and Butler, 1981). Although some representatives of the electric power industry claim that most acid rain is natural (see Hanson and Hidy, 1982), this view is challenged, we think quite effectively, in Cogbill et al. (1984).

Acid rain affects both human-made structures and the natural environment. The best-documented effects at this time are the corrosion of building surfaces and a reduction in the pH of many lakes and rivers which kills fish and many other organisms. Limestone buildings are especially susceptible to the corrosive effects of acid rain. A well-known example is the Parthenon in Athens, Greece, which has been damaged more by acid rain and pollutant-

related aerosols in the past few decades than, with the exception of a gunpowder explosion, in all the time since its construction many centuries ago. In Florence, Italy, Michelangelo's magnificent statue *David* had to be moved indoors, with a copy set in its place, because acid rain and related industrial emissions had damaged it. The total effect on municipal and private buildings around the world has not been quantified, to our knowledge, but the economic cost must be large and the artistic loss incalculable. In addition acid rain probably accelerates the corrosion of many metals and may contribute, for example, to the need for more frequent bridge repair.

The best-documented environmental impact of acid rain is the reduced pH of lakes and small rivers in regions with poorly buffered soils. The impact on fish is limited principally to watershed areas having soils with low buffering capacity and includes especially high, old mountain ranges, granitic regions such as parts of Central Norway and Sweden (where many important salmon runs and thousands of lakes have been lost to acid rain) and large parts of the Canadian Shield. David Schindler, of Environment Canada, estimates that fish in about 10,000 lakes are lost each decade due to acid rain in both

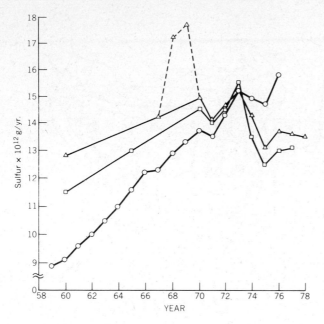

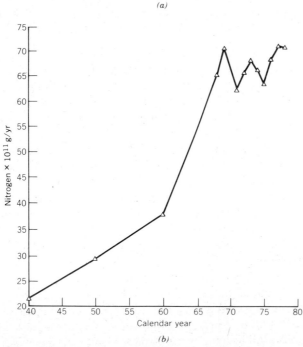

Figure 16.8. Emission trend of primary pollutants by (a) SO_x emissions from industrial fuel combustion as estimated by three authors; (b) oxides of nitrogen emitted from automobile, electric utilities, and industrial fuel combustion. The sharp rise since 1960 is thought due to all three sources. (From McKenzie et al., 1984.)

Quebec and Ontario. Such losses have a severe economic impact and undermine an important component of Native American culture in that region. This problem has been most comprehensively studied in the Adirondack Mountains of New York and in Scandinavia. The changes in New York are especially well documented because comprehensive biological and chemical surveys were undertaken in the 1930s, giving us baseline information. The basic chain of biological effects in the Adirondacks has been unraveled by Carl Schofield and his coworkers (Schofield, 1976, 1977; Cronan and Schofield, 1979). The Adirondack Mountains are an uplifted area of the Canadian Shield, composed largely of old, hard igneous and metamorphic rock. Because these rocks weather very slowly and are poor in acid-neutralizing cations such as calcium and magnesium, the soils derived from this bedrock are thin and have a very low cation exchange capacity. The problem is particularly exacerbated in the Adirondacks because they receive a large amount of orographic rain.

Much of the precipitation that falls on the Adirondacks is derived from air masses passing over industrialized sections of the Midwest, western New York, or the rapidly industrializing Southeast, and it is quite acidic, with average annual pH values of about 4.0 to 4.2. With little neutralizing capacity in the watersheds, stream and lake water acidity have increased over time. Complicating this matter even further is the fact that the Adirondack Mountains receive a large portion of their precipitation as snow. Hydrogen ions (acidity is a measure of hydrogen ion concentration) tend to leach out of the snowpack during the first thaws and thus send a pulse of highly acid snowmelt down the watersheds.

Fish mortality due to increased acidity may occur by several different processes. Canadian research has demonstrated that reduced calcium levels in acid lakes can prevent fish from spawning. In Norway reduced levels of sodium and chloride in the blood caused by high acidity killed fish directly. In the Adirondacks, aluminum, which becomes highly soluble below a pH of 5.0, leaches out of the watersheds of streams and lakes. The resulting high aluminum levels in acidified lakes interfere with the gas exchange of gill tissue surfaces, suffocating the fish (Cronan and Schofield, 1979). In most cases eggs and younger fish are the most susceptible to acidification. A common sign in a lake undergoing progressive deterioration from acid rain is that the fish caught are increasingly older, for no young fish survive.

Acid rain also may affect land plants. Very low pH rain can damage leaves physically, although this probably does not happen on a large scale at present pH levels. Acid rain damages important soil microbes and increases plant uptake of toxic aluminum from the soil. Yet occasionally acid rain can be beneficial to plants, for example, to some sulfur-limited agricultural crops of the Tennessee Valley. The nitrogen in acid rain may be an even more general fertilizer, but positive effects have yet to be shown for a specific crop or species. We expect that on balance the biotic effect of acid rain overwhelmingly is negative. Like most pollutants, however, the mechanisms are complex.

Finally, acid rain may have negative effects on human health. Many toxic metals, including aluminum, lead, arsenic, and mercury, are more soluble in acidified waters, increasing the concentration of these metals in drinking water. Sources of the metals include both reservoir substrates and municipal and domestic plumbing. Quabbin Reservoir, the major water supply for Boston, has already lost most of its ability to neutralize acids, and Massachusetts officials are concerned that the concentrations of dangerous metals may increase rapidly in coming decades. The public expenditures of money and energy to compensate for this energy externality would be enormous.

EFFECT OF AIRBORNE FOSSIL FUEL RESIDUALS ON CROP AND FOREST PRODUCTIVITY*

Chemical Nature of Fossil Fuel Emissions

This section attempts to synthesize a number of existing studies of fossil-fuel-derived air pollutants. First, we provide an overview of the constituents, processes, and geographical patterns of air pollutants, and then we give a qualitative and, to some extent, quantitative estimate of the impact of these pollutants on our economy. At present the impact of pollutants on crops and forest productivity is a substantial but poorly quantified externalized cost to our economy. There is not always an obvious linkage between biological damage and economic loss. For example, pollution-induced crop losses could require additional fossil energy for making up

*Prepared by T. V. Armentano.

crop losses through additional inputs of fertilizers, but the additional cost might be covered by a program of government subsidies. Ideally biological damage and economic loss should be incorporated into a comprehensive assessment of an energy system's EROI. Some suggestions are given on how that might be done, but at this time an assessment cannot be made with a high degree of reliability.

The combustion of all fossil fuels produces pollutants that are, in general, detrimental to natural systems, human health, and human-made materials. As we have seen previously, fossil fuels produce oxides of carbon, nitrogen, and sulfur. Many other pollutants also are emitted, depending on fuel type, environmental conditions, and combustion technology. Coal and petroleum contain many substances that are toxic in the unburned state or as combustion by-products, whereas natural gas (mostly methane) contains relatively few impurities (Tables 16.1, 16.2, 16.3).

Coal contains higher quantities of most toxic trace elements than do oil and gas except for vanadium, molybdenum, and mercury, which occur in highest concentrations in petroleum. Also, some fuel oils are richer in sulfur than are many high-quality Western coals, which may contain one-third

Table 16.3 Distribution of Trace Elements in Coal Combustion Residues

Group I: Equally Distributed in Bottom Ash and Fly Ash			
Barium	(Ba)	Manganese	(Mn)
Chromium	(Cr)	Rubidium	(Rb)
Cerium	(Ce)	Scandium	(Sc)
Cobalt	(Co)	Samarium	(Sm)
Europium	(Eu)	Strontium	(Sr)
Hafnium	(Hf)	Tantalum	(Ta)
Lanthanum	(La)	Thorium	(Th)

Group II: Preferentially Concentrated in Fly Ash			
Arsenic	(As)	Lead	(Pb)
Beryllium	(Be)	Antimony	(Sb)
Cadmium	(Cd)	Selenium	(Se)
Copper	(Cu)	Uranium	(U)
Gallium	(Ga)	Vanadium	(V)
Molybdenum	(Mo)	Zinc	(Zn)
Nickel	(Ni)		

Group III: Discharged as Vapors			
Chlorine	(Cl)	Mercury	(Hg)
Fluorine	(F)		

Source: From OTA (1979).

or less sulfur than Eastern coal. If the United States increases emphasis on coal as a primary energy source, higher emissions of most residual materials can be expected. Therefore increased coal production is likely to lead to greater impacts on the environment and human health. Controlling impacts would require diverting a greater portion of our resources to pollution control and health care.

Toxic constituents of coal and oil include the metals cadmium, selenium, and arsenic, organic compounds such as benzapyrene and other hydrocarbons, and radioactive elements like uranium and thorium. Combustion products such as sulfur dioxide (SO_2), nitrogen oxides (NO_x), or mercury (Hg) enter the atmosphere as gases, and heavy metals or inorganic acids form particulate matter in solid or liquid form.

Bertine and Goldberg (1971) reported that fossil fuel combustion released 51 elements into the atmosphere (Table 16.1). Concentrations of these substances ranged from over 20,000 ppm of sulfur, 1000–10,000 ppm of potassium, chlorine, aluminum, and iron down to less than 0.1 ppm of rare substances like antimony and rhenium. Elemental concentrations may vary widely, however, depending on the region where the fossil fuels were extracted. For example, uranium concentration may range from 0.2 to 43 ppm of coal within the bituminous coal deposits of the interior coal province of the United States, although the average value for all U.S. coal is 1.8 ppm (Fish and Wildlife Service, 1978).

The magnitude of fossil fuel emissions can be compared to the quantities of materials released by the weathering of land surfaces and subsequently transported by the world's rivers (Table 16.1). The release of nickel (Ni) and mercury (Hg) from fossil fuels, for example, appears to be within an order of magnitude of the total river fluxes of these elements. This indicates that fossil energy conversion has become a significant geochemical process in terms of the release and cycling of these elements and that natural cycles have been significantly amplified by industrial activity (Bertine and Goldberg, 1971). Most mobilized elements are released from a relatively small part of the world—the midlatitudes of the Northern Hemisphere where human industrial activity is centered.

The release of nitrogen compounds from fossil fuel combustion (largely as nitrogen oxides) is of special interest because these substances participate in atmospheric reactions forming ozone and acid rain. Annual NO_x emissions from all anthropogenic sources, estimated at 22×10^{12} g of nitrogen, exceed the global production by lightning (National Research Council 1981). Although only about 27 percent of total fuel-derived NO_x emissions originated from power plant combustion in 1975, compared to 45 percent from the transportation sector, utility releases of NO_x are growing more rapidly. Total anthropogenic release has nearly tripled over the last 25 years and significant future increases are forecast as a result of increased coal-burning by both utilities and other industries.

Distribution of Primary and Secondary Emission Products

The exhaust fumes emitted from a tall stack are immediately subject to mixing processes from turbulent air which displaces the fumes vertically and horizontally (Table 16.4). Vertical air movement often is constrained by an inversion layer which may determine when and where the plume touches the ground. While the plume is aloft, primary emission products like SO_2 and NO_x (principally NO_2 and NO) are converted into secondary products such as ozone, and sulfate and nitrate aerosols. These materials, as well as primary and secondary gaseous products, may contact the ground or vegetation surfaces in the dry state (dry deposition) or be dissolved in precipitation.

The rate of conversion of emission gases into secondary products depends on a number of factors such as sunlight intensity, moisture, and the concentration of the reactants. The conversion rate influences how far from the emission source the gases will exist at concentrations high enough to be toxic to organisms. Monitoring of ground-level concentrations of SO_2 and NO_2 near point sources reveals an elliptical deposition area skewed in the direction of the prevailing wind (Figure 16.9), although topographic features like mountains or open water may alter dispersal patterns. At distances of 3–15 mi from the source, gas concentrations decline to natural background levels, as a result of diffusion, removal by vegetation and land surfaces, and chemical transformations. Elevated concentrations of SO_2 and NO_x gases tend to be localized near large industrial sources such as coal-fired power plants or smelters, whereas secondary by-products are subject to long-distance transport to areas extending hundreds of kilometers or more away from the emission source (Ferman et al., 1981).

Table 16.4.   Reactions in the Atmosphere

Reaction Type	Reaction	Catalyst Special Conditions	Comments
Indirect photooxidation	$SO_2 \xrightarrow[\text{radicals}]{\text{Free}} H_2SO_4$	Smog, sunlight, water vapor, $Ho\cdot$, $Ho\cdot_2$, $CH_3O_2\cdot$ radicals	Important reaction rates up to 5 percent per hour giving half-life of SO_2 of $1/2 \rightarrow 2$ days. Rate depends on SO_2, hydrocarbons, NO_x levels, amount of sunlight.
Heterogeneous catalytic oxidation	$SO_2 \xrightarrow{O_2} SO_4{=}$	Liquid water; metal ions	Because of dependency on catalyst concentration, probably important in plumes and polluted urban atmospheres; probably minor in rural atmospheres. Virtually zeroth order in SO_2.
Heterogeneous oxidation by strong oxidants	$SO_2 \xrightarrow[H_2O_2]{O_3,} SO_4{=}$	Water droplets	Importance in dispute, rate estimates vary by a factor of 100.
Heterogeneous oxidation in the presence of ammonia	$SO_2 \xrightarrow{H_2O} H_2SO_4$ $NH_3 + H_2SO_4 \rightarrow$ $NH_4^+ + SO_4{=}$	Water droplets	Rates unknown, dependent on availability of ammonia and pH of droplet.
Surface catalyzed reactions		Soot	Soot has been shown in the laboratory of catalyze oxidation of SO_2. Importance unknown.

Source. From OTA (1979).

Trace Elements

Many trace elements released by fossil fuel combustion are highly toxic to plants and animals. The effects have been documented most clearly near smelters and heavy industrial areas where concentrations are considerably higher than those downwind of most fossil fuel combustion sources. Similar toxic effects have been found where pesticides containing trace elements have been applied heavily for many years. Exposure to high concentrations of pesticide-derived arsenic, fluorine, nickel, copper, and other elements clearly has killed plants and animals. Mercury and lead are potent animal toxins but may not damage terrestrial plants because these metals often are rendered chemically inactive by binding by soils or because of their low solubility in plant and cell fluids (Baumhardt and Welch, 1972). Other metals such as zinc and aluminum, however, can damage absorptive roots of trees after uptake (Jordan, 1975). Boron, molybdenum, zinc, and manganese are trace elements essential to plant growth in small concentrations but toxic at high concentrations.

Smelter operations that release large quantities of trace metals have eliminated nearby vegetation as a result of long-term pollution at several sites (Freedman and Hutchinson, 1980), although some of the observed effects also result from SO_2 releases. Copper, nickel, lead, and zinc commonly accumulate to high levels in soils, sediments, and organisms near smelter sources. Under these heavy doses destruction of plant and animal resources has been extensively documented. Jordan (1975), for example, reports that the denudation of 485 ha of woodland in Pennsylvania is attributable to smelter-derived fire stress and high soil zinc levels (up to 8 percent by weight of the A horizon). Deposition rates associated with fossil fuel combustion are or-

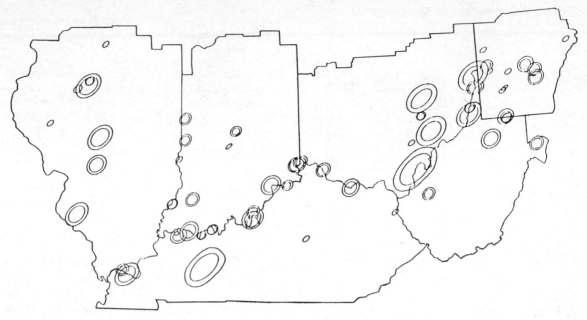

Figure 16.9. Areas around coal-fired power plants in the Ohio River Valley where calculated SO$_2$ concentrations reach either 0.05 ppm (outer ring of ellipse) of 0.10 ppm (inner or single ring of ellipse). Within the ellipses, particularly in the presence of regional ozone enhancement, synergistic effects on sensitive crops and forest plants are likely. (From Loucks et al., 1980.)

dinarily much lower, and there is no evidence of damage to ecosystems attributable exclusively to these emissions. Based on other studies, however, we must assume that subtle adverse impacts occur over much wider areas than where the easily detectable effects are observed.

Several studies show that trace metal emissions frequently are transported to pristine wilderness areas remote from pollution sources (Hirao and Patterson, 1974; Wiersma and Brown, 1980). Accumulation of sulfur and selenium in Greenland ice sheets (Weiss et al., 1971) and vanadium in remote oceanic areas (Duce and Hoffman, 1976), respectively, has been attributed to fossil fuel burning. Accumulation levels in remote locations may far exceed apparent natural background levels. For example, lead in soil from the Great Smoky Mountains National Park, hundreds of kilometers from any significant source, is 7–19 times higher than levels observed in the Olympic National Park where prevailing winds originate over the Pacific Ocean. The higher levels in the Smokies are attributable to industrial and gasoline sources. Although the Olympic sites could be regarded as representative of true background levels, there may be local or trans-Pacific lead sources reaching the Olympics (Wiersma and Brown, 1980). Locating sites that are assumed to be totally free of industrial contamination is becoming

increasingly difficult as the sensitivity of scientific analysis improves.

Some Geographic Dimensions of Air Pollution Effects

Regional Accumulations of Haze and Ozone

Air pollution commonly occurs as a diverse mixture of substances that vary spatially depending on proximity to particular industries, sea coasts, and agricultural areas. In the eastern United States pollution builds up over large areas, taking the form of a dense atmospheric haze. The haze typically is composed of aerosols (fine particles having mean diameters below 10 microns), gases, and water vapor. Most of the particulate matter consists of sulfates, nitrates, hydrocarbons, and trace metals. The gases, particularly those of sulfur and nitrogen, are converted into secondary aerosol forms and into ozone.

Haze causes reduced visibility and diminishes scenic vistas in parks and wilderness areas. An analysis of haze in the mountains of Virginia found that the predominant visibility-reducing component was sulfate particles, and the second most abundant was carbon, mostly in organic form (Ferman et al., 1981). In urban areas air pollution contributes to corrosion of building surfaces, eye irritation, and other effects on human health and welfare (Fennelly, 1976; EPA, 1979; Nautusch and Wallace, 1974). Atmospheric haze has increased significantly in the eastern United States over the past several decades. The pattern of haze buildup correlates well with the spatial and temporal trends of increased coal consumption (Husar et al., 1981). Industrially-derived aerosols have a long residence time in the atmosphere and have been responsible for local climate change in areas near industrial activity (Chagnon, 1968; Hobbs et al., 1974). Climatic effects on a larger scale, perhaps worldwide, are possible but further study is needed (Hobbs et al., 1974; Fennelly, 1976). There is even evidence that visibility degradation in the lower atmosphere of the Arctic is due to aerosols derived from industrial areas of the north temperate zone (Kerr, 1979).

Evidence of the deleterious effects of haze components has been established most clearly for photochemical oxidants. This complex of reactive chemicals was first recognized as a problem soon after World War II in the Los Angeles area. Observations showed that high solar insolation rapidly drives a series of atmospheric reactions that transforms combustion emissions into a complex of photochemical oxidants often called smog. The primary pollutants or precursors, which are transformed into the secondary smog compounds, consist principally of nitrogen oxides, hydrocarbons, aldehydes, and carbon monoxide. Ozone is the principal pollutant, and it is often used as a proxy for the entire photochemical oxidant complex.

Since ozone is formed over perods of several hours or more as the polluted air mass moves downwind of the source area, ozone continues to build up in nonindustrial locations where it may exceed levels found in urban areas (Cleveland and Graedel, 1979). Regional analysis confirms that local air pollution problems are often exacerbated by the influx of pollutants from outside the state or region. For example, ozone concentrations exceeding 0.20 ppm in Connecticut and Massachusetts can be traced to New York City emissions (Cleveland and Graedel, 1979). Ozone concentrations exceeding air quality standards in New York City also were associated with air parcels originating in Ohio 1–2 days earlier (Lioy and Samson, 1979). The Midwest, especially the Ohio River Valley, also contributes significantly to ozone accumulation in the southern Appalachians (Skelly, 1980), and eastern North Carolina (Heagle and Heck, 1980).

The large geographic setting of air pollution episodes directly affects the regulation and the cost accounting associated with the economic effects of air quality degradation. Because of the great separation between primary emissions and downwind location of secondary products, it is difficult to assign the costs of pollutant impacts to specific sources or political units. Often emission sources are not even located in the same state where the high concentrations are observed. State or local governments are thus unlikely to protect successfully their own resource values if pollution control regulations lack procedures for enforcement of cooperative interstate or national air quality regulations.

Air Pollutant Effects on Crop Yields

The effects of air pollutants on plant growth have been studied extensively in the United States (Mudd and Koszlowski, 1975; Heck and Brandt, 1977). Most of this research consists of studies conducted under conditions that are not applicable directly to a practical assessment of pollution effects on agricultural yield. Because of the difficulty in deriving adequate controls in the field, most early pollution research was conducted in greenhouses or growth chambers by exposing plants to pollutants over a 1–8 hr period. Evaluation of effects in these short-duration studies was based on visible leaf injury criteria such as chlorosis (loss of chlorophyll) or necrosis (occurrence of dead tissue) (Tingey et al., 1973). The underlying assumption was that this physical expression of air pollution injury served as an index of reduction in plant growth or yield. A second research approach concentrated on determining concentrations at which physiological processes, particularly photosynthesis, were inhibited on a short-term basis.

Many of these studies have helped elucidate the nature of physiological response to air pollution and the injury mechanisms operating in plants. Each pollutant, for example, tends to induce a characteristic pathological response in leaves that may help identify which pollutant caused the damage in a field situation. But visible injuries have been found to correlate poorly with reductions in most economically important responses like growth or yield (Sprugel et al., 1980). Unfortunately secondary air quality standards still rely heavily on visible injury criteria. For accurate analyses, however, many new field studies of pollutant effects on yield are required in order to estimate the costs of air pollution impacts on productivity.

An advancement in the science of studying pollution effects is the development of large transparent plastic growth chambers with open tops which are placed over a portion of the crop (Heagle et al., 1979). The chambers are equipped with fans for maintaining air turbulence. They have proved to be particularly useful for studying effects of pollutants like ozone that may persist for many days. Another technique for studying pollution effects employs a system of pipes that dispenses pollutants to the crop canopy without using artificial enclosures. It has been used to study the effects of repeated short-term, fumigation episodes characteristic of SO_2 or NO_2 exposures (Sprugel et al., 1980). Both approaches have enabled scientists to expose plants to pollution at exposure levels similar to those occurring in agricultural areas, and they have provided a realistic basis for estimating the economic effect of air pollution.

Effects of Ambient Pollutants

Evidence gathered from these two independent methods of analyzing pollution effects indicates that air pollution in many parts of North America is causing significant reductions in crop yields in areas at great distances from fossil fuel emission sources (Hileman, 1982; Reich et al., 1982). The evidence is strongest for legume crops such as soybeans and alfalfa but has also been observed in potatoes, snap beans, carrots, radishes, and other vegetable crops. Grass crops such as field corn, wheat, and rye appear to be more resistant, although evidence suggests that grass crops are affected in some locations at economically significant levels (Heggestad, 1980; Loucks and Armentano, 1982).

Effects of O_3

Heggestad et al. (1977) reported that ambient ozone in western Maryland reduced the yield of sweet corn, snap beans, and tomatoes by 4–9% compared to yields of plants exposed to filtered air. Yield reductions in four varieties of soybean, the second most important U.S. crop, averaged 20% compared to plants grown in filtered air. Reductions of 10% were noted for certain potato varieties. In northern Illinois seed weight per plant and total seed weight of an important Midwestern soybean variety were reduced 15% or more by exposures as low as 0.065 ppm of O_3 for seven hr/day (Kress and Miller,

1981). These data, plus confirmatory results from similar experiments conducted in North Carolina (Heagle and Heck, 1980), suggest that in much of the eastern United States, fossil fuel residuals (particularly O_3) are increasing the dollar and energy cost of maintaining high crop yields. Reich et al. (1982) provide further evidence for this hypothesis in an analysis of soybeans in New York.

Effects of SO_2

According to long-term studies in northern Illinois, chronic exposure to sulfur dioxide also can reduce yield of soybeans (Sprugel et al., 1980). Soybeans were exposed for two years to SO_2 released from an open air fumigation system at concentrations resembling those found around power plants. Yield reductions ranged from 6% at a mean exposure of 0.095 ppm to 45% at a mean exposure of 0.79 ppm.

There are few data for other crops of the kind generated by Sprugel et al. Thus it is less certain whether effects on other crops occur at moderate to low SO_2 concentrations. At higher SO_2 concentrations close to industrial sources, crop damage has been widely reported. The most comprehensive measurements were made near a SO_2 source at Biersdorf, Germany (Guderian, 1977). Damage to wheat, alfalfa, potatoes, and many other crops occurred on a scale from severe to unmeasurable over a distance from 325 to 6000 m from the source.

The major research question with respect to pollution effects on crops is not whether ambient air exposures in the eastern United States reduce crop yields but how serious are the effects and what are the regional economic impacts. Later we try to answer that question. It should be noted here that in major agricultural areas of California, high ambient oxidant levels have forced major changes in crop distribution and in the choice of varieties grown. But even after variety changes, significant yield losses of tomatoes, alfalfa, citrus, and other crops in southern California still are attributable to annual development of oxidant episodes (Oshima et al., 1977; Oshima et al., 1976; Thompson et al., 1972). Although large investments in crop research have produced crop varieties that are resistant to air pollution, this strategy has not completely offset the losses to farmers. Of greater concern, it reflects the tendency of society to deal with the costs of pollution by making technological adjustments funded by public tax dollars rather than by forcing expensive pollution control measures on industrial sources.

Air Pollution Effects on Natural Ecosystems

The increasing evidence that agricultural yields are being significantly affected by polluted air over rural areas has raised the question of the extent of air pollution effects on natural ecosystems. Quantifying effects of pollutants on natural vegetation, however, is more difficult than for agricultural crops, particularly because most ecosystems contain mixed communities of long-lived perennial flora that differ widely in sensitivity to pollutants. Each species is composed of genetically different populations that vary in sensitivity to environmental perturbations. Natural communities are subject to competition for suboptimal supplies of water and nutrients that can affect their responses to other stresses. Although in managed forests resistant species or genotypes can be planted, in most forest areas little management adjustment to pollution is possible.

Sensitive clones of pine (especially *Pinus strobus* and *P. ponderosa*) and quaking aspen (*Populus tremuloides*) have been found to be among the most sensitive of all plants to O_3 and SO_2. Ozone at 0.05 ppm administered over 4 hr causes depression of photosynthesis rates and formation of leaf lesions in young white pine (*Pinus strobus*). Exposure to SO_2 at the same concentrations has elicited similar responses (Eckert and Houston, 1980). Higher concentrations and longer exposure durations cause more severe effects under experimental conditions and affect a wider range of species. Consequently an important question to be addressed concerns the response of natural forests to chronic air pollution known to occur over large areas.

The first regional-scale evidence that chronic air pollution can reduce growth and survival of entire forests was given in a study of conifer forests of the San Bernardino Mountains located inland from the South Coast Air Basin (SCAB) in California (Miller, 1973). These forests have been exposed repeatedly to photochemical oxidant concentrations reaching 0.40 ppm or more over the last three decades. Effects on ponderosa pine over a 405 km^2 area included leaf epidermal lesions, premature senescence of needles, reduction in total foliage mass, slowed wood growth, and increased mortality. Ongoing studies have revealed that other species are also exhibiting serious stress responses (Taylor, 1980; Miller et al., 1980). In addition to direct effects ozone was found to interact indirectly, by predisposing trees to fatal outbreaks of pests and pathogens.

Serious economic impact could result from increased mortality or reduced wood growth in timber species. McBride et al. (1975) showed that wood production of ponderosa pine was being reduced by up to 75% in the high pollution zones of the San Bernardino Mountains. Whether the lower ozone levels widespread in the eastern United States and elsewhere in California also are reducing wood production over large areas is presently uncertain. In some areas of the United States whole tree populations exhibit reduced growth, apparently influenced by a variety of pollutants that may or may not include ozone and acid rain (Puckett, 1982). Extensive mortality of trees in the German Black Forest as a result of various pollutants made international headlines in 1984.

The observations by Skelly et al. in Virginia, Karnosky in Wisconsin, McLaughlin et al. in east Tennessee, among others, indicate that over wide areas the presence of pollution only moderately elevated above background levels is selectively eliminating sensitive lines of white pine. The extreme sensitivity of the species as a whole suggests that reduced vigor could be occurring throughout Eastern North America due to regional ozone episodes. Increasing evidence indicates that loblolly pine (*P. taeda*) also has strains that vary in pollution sensitivity and may be undergoing differential replacement of sensitive races at prevailing pollution levels (Kress et al., 1982). These impacts may have significant economic costs because pines are important timber species, but no attempt has been made to calculate regional monetary losses.

Community Level Effects of Chronic Air Pollution Stress

Suppression of growth or reproduction of dominant species could affect other species by altering their survival capacity. Miller (1973) has predicted that in the San Bernardino Mountains chronic suppression of ponderosa pine will convert oxidant-damaged forests to a cedar and white fir community. Nearby pine and black oak forests may be shifting to a black oak forest type, and in sites that are suboptimal for oak, shrubs may replace trees as the dominant vegetation (Miller et al., 1980). In the Ohio River Valley changes in forest species composition have occurred along a gradient of chronic air pollution (McClenahen, 1978). Secondary effects of species replacements include alterations in litter quality, forest microenvironment, and production of seeds as food for animals (Kickert and Gemmill, 1980).

Higher environmental stress leads to further reductions in species diversity and community complexity, often in predictable patterns. The responses are strikingly similar regardless of the kind of stress. In Poland and Germany forests located near heavy industrial districts where SO_2 and heavy metal concentrations are high have changed over time into a series of disturbed communities of lower biomass ordered along the stress gradient. Close to the pollution source is a barren zone where no higher plant life survives (Knabe, 1976; Guderian, 1977). Moving away from the source leads sequentially to grass-dominated communities, shrub communities, and tree communities. Similar situations have been observed around large areas of pollution in North America (Figure 16.10) and at Biersdorf, West Germany.

Simulating Community Level Response to Stress

Computer modeling can be used to simulate small changes in species composition over decades or centuries as a function of differential species response to stress. The forest simulation model FORET, for example (West et al., 1980), simulates forest community dynamics based on equations that combine the growth characteristics of each species with treatment of species responses to air pollution. The model was validated by comparing its results with the actual changing populations of Appalachian oak forests over several decades after loss of the American chestnut, a formerly dominant species that was completely killed by a fungal parasite. Simulations of forest response to air pollution stress showed that, as expected, growth of sensitive species was reduced most, leading to either less biomass or total elimination depending on whether pollution exposure was intermittent or continuous.

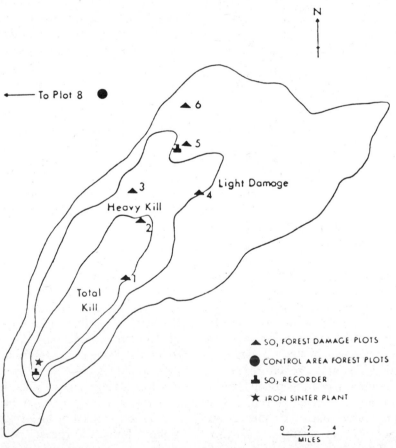

Figure 16.10.   Forest damage survey in the vicinity of the Wawa, Ontario, iron-sintering plant in 1970. Prevailing winds are from the Southwest. The plant emitted 200,000 tons of sulfur annually. There was total kill of higher plants over 108 km^2, heavy kill over 191 km^2, and light damage over 889 km^2. (From Linzon, 1978.)

Tolerant species benefited from the reduced competition and after decades, assumed greater dominance. Species of intermediate sensitivity, however, responded differently depending on their ecological characteristics. Thus in the simulation yellow poplar, a rapidly growing early successional species, responded positively to stress probably by benefiting from reduced competition. But the same stress applied to black oak, another intermediate species, caused a greater biomass reduction than expected. This effect was attributed by West et al. (1980) to the poor competitive ability of black oak.

The overall effect of chronic pollution on the economic or ecological value of a stand depends on both the response of individual species and the entire forest. West et al. note that the growth of the entire stand was reduced overall because reductions in sensitive species were not totally compensated for by stimulation of more resistant species. Even if total growth were unchanged, the loss of black cherry or black walnut, which are valuable lumber species, might not be compensated for by increased growth of economically less desirable species such as red maple or chestnut oak. Second-order interactions of species probably occur in all communities, but responses differ according to a host of community variables such as the stand age, vigor, original species densities, and environmental factors. These effects cannot be estimated accurately. Thus assessments based on short-term impacts are likely to underestimate total costs which emerge only after many decades.

Pollutant Interactions

As previously mentioned, the ubiquitous distribution of O_3 pollution suggests that results of SO_2 studies could be misleading if O_3 were ignored. In most of the eastern United States and in the valleys of the Pacific Coast, crops growing downwind of smelters or power plants will be exposed to pulses of SO_2 in an atmosphere frequently laden with O_3 at concentrations reaching phytotoxic levels.

A recent analysis suggests that the presence of O_3 even at moderately polluted levels (e.g., at 0.06 ppm, twice natural background levels) significantly increases the phytotoxic effect of SO_2. In other words, the gas mixture causes effects that are not elicited by exposure to the gases individually (Reich et al., 1982; Heggestad and Bennett, 1981). Similar synergistic effects occur when other pollutant combinations occur, such as NO_2 and SO_2 (Tingey et al., 1971; Ashendon, 1979), although NO_2 appears

to be less toxic than O_3 or SO_2. The degree of plant injury is dependent on the concentrations of each gas and on cultivation conditions. Because experimental conditions have not been standardized, differences in experimental results are considerable. Pollutant interactions nevertheless add a new dimension to the calculation of the externalized costs of fuel use which have been based largely on the effects of pollutants operating individually.

Pollutants as Growth Stimulants

Modern industrialized agriculture uses crop varieties dependent on high inputs of nitrogen, phosphorous, and potassium in fertilizers. Therefore where rainfall is adequate, crop growth can be limited by elements that are not commonly supplied in fertilizer, such as sulfur. Intensive use of nitrogen fertilizer and the development of high yield (and hence high nutrient-requiring) crop varieties can create a sulfur demand that outstrips soil S supplies, particularly if plant residues that contain S are not retained on the land (Stewart and Porter, 1967). Such cases of increased yields have been observed under controlled exposures to SO_2 (Irving and Miller, 1981). However, Brown et al. (1981) found that in some areas the addition of S to crops did not increase yields, apparently because the soil S is adequate for growth. Intensive studies of tree responses to S pollution in Ontario also have not shown any evidence of beneficial effects (Linzon et al., 1979). Thus growth stimulation by air pollution appears to be a local phenomenon that does not compensate for pollution damage over large areas.

Economic Assessment

The previous evidence suggests that there is little doubt that air pollutants can have deleterious effects on economically valuable plants. Yet quantifying these losses, particularly over large areas, has proved difficult. Part of this difficulty is due to the paucity of data on the effects of ambient air pollutants on properties of plants that are of direct economic value (Adams and Crocker, 1982). This problem has been partially remedied by recent studies, and research continues.

Several early estimates of economic losses from air pollutants effects were made based on visible leaf injury symptoms or on subjective classifications of crop species pollution sensitivity and calculation of exposures from county emissions data. Although these assessments may have been useful because they helped publicize the importance of externalized costs of fossil fuel residuals released to the environment, they probably underestimated costs.

Economic Losses from Air Pollutants

Estimating the economic effects of pollution requires calculating the cost of damage:

$$\text{cost of damage} =$$

$$\frac{\text{percent yield}}{\text{reduction of crop}} \times \frac{\text{total crop}}{\text{production}} \times \frac{\text{unit price}}{\text{of crop}}$$

The last two factors of the formula can be easily estimated for most areas, but considerable uncertainty is associated with estimating yield loss functions. Recent estimations appear to permit placing probable upper and lower limits on air pollution costs to agriculture, although no estimates have been made for losses from tree damage except on a local basis.

For example, Page et al. (1982) converted damage functions for three major crops in the Ohio River Basin into monetary units using the yield loss coefficients for soybeans, wheat, and corn that were calculated for 1976 by Loucks and Armentano (1982). Page et al. estimated that over the period 1976–2000, monetary losses to farmers would approach 12% of the yield under clean air conditions, or over $7 billion. In a similar exercise the Office of Technology Assessment (1983) measured the soybean yield gains associated with reducing ozone to natural background levels. The area of greatest gain was found where crop production is the highest, just north of the most polluted area (Figure 16.11). In this approach the cost of air pollution is interpreted in economic terms as the value gained by reducing damages caused by pollution levels in excess of clean air.

For major crops across the United States, losses attributable to photochemical oxidant exposure probably exceed $1 billion annually. Ozone effects appear to account for 75–95% of national crop losses caused by air pollution (Heagle and Heck, 1980; Ryan et al., 1981). In regions like the Ohio River Valley the proportion rises to 99% (Loucks et al., 1980). A similar percentage probably also applies to the Pacific states. Ryan et al. (1981) calculated that if air quality was improved to meet federal air quality standards, at least $1.75 billion could

be gained by meeting the O_3 standard; the benefit of meeting the SO_2 standard was $34 million.

Various assessments disagree on which crops are affected most. Although Ryan et al. (1981) estimate that greatest monetary losses are associated with corn and tomato (principally in the Pacific states), Heagle and Heck (1980) project that nearly half the national crop loss attributable to ozone would occur from reductions in soybean yield, with little effect on corn. The work by Heggestad (1980) also suggests that soybeans are relatively pollution sensitive and corn relatively tolerant.

Freeman (1979) has suggested that annual crop losses of around $3.6 billion nationally are attributable to ozone, based on average yield losses of 15%. He has noted that crop loss estimates must consider factors other than direct pollutant effects. These include the effect of reduced crop output on food product prices and on the costs to producers, and the costs of switching to more pollution-resistant crop varieties.

The economic impacts of air pollution are concentrated in the mid-Atlantic, East–North Central (Ohio to Wisconsin) and Pacific states where air pollution exposures and the acreage of susceptible crops are highest. Ryan et al. estimates that the East–North Central states would enjoy benefits of $731 million, 41% of the national benefit, by eliminating pollution crop damage caused by exceeding air quality standards. These gains would come principally from protecting yields of corn and soybeans. The Pacific and mid-Atlantic states would gain 43% of the national benefits between them.

Closing Comment

It has become clear in our study of the effects of air pollution on crop and forest productivity that a realistic assessment of pollutant impacts must represent actual conditions for a specific location of interest. Such an analysis of effects must proceed over a long period so that fluctuations in climate, pollutant loadings, and other influencing factors can be averaged. For agricultural crops the average study time should be at least 3 years (Heggestad et al., 1977; Sprugel et al., 1980). For forest effects the average time probably should be significantly longer, except where pollutant levels are high enough to cause severe effects.

Recent research based on these concepts has uncovered pollution effects that were previously un-

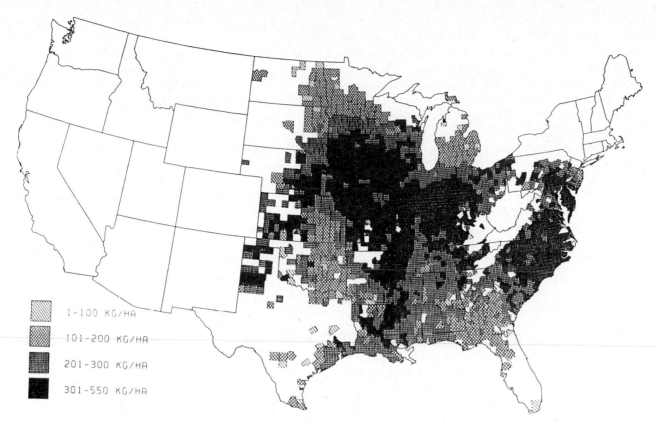

Figure 16.11. Soybean gains (kg/ha) predicted to occur if ozone were reduced to natural background levels. Based on 1978 U.S. Department of Agriculture census and 1978 EPA ozone monitoring data. (From Office of Technology Assessment, 1983.)

recognized. The magnitude of the productivity losses attributable to air pollution has correspondingly been revised upward and now stands at billions of dollars annually. Total costs are still unknown because current understanding of ecosystem effects, particularly long-term and secondary ones, is quite limited. Clearly much of the policymaking of the past decade has not yet incorporated significant recent findings on air pollution.

In a practical sense the maturity or adequacy of assessment research can be gauged by the lowest effects that are statistically demonstrable. One test of maturity would be a retrospective one, namely whether the minimum effects detectable by empirical methods (analogous to the detection limit of an instrument) are lower than the effect level at which economic damage is acceptable to the producer. Such an evaluation suggests that pollution effects research had not reached this level even as recently as a few years ago. The analytical sensitivity of current pollution assessment research may now be approaching the economic acceptability level for crops, although many complex socioeconomic vari-

ables interact to define economic parameters. Nevertheless, it is clear that much additional work is needed on effects of chronic pollution exposure on ecosystems and that additional externalized costs of air pollution are likely to be revealed. All of these costs of course increase the energy cost of producing our food. If Freeman's estimate of a 15% reduction in crop productivity is correct, then we can assume that air pollution has increased the energy cost of producing our national crops by roughly 15% of all of our on-farm energy use, or about 4.10×10^{14} kcal. This figure, equal to the output of 100 large power plants, should be added to the denominator of our national mean EROI assessment.

AQUATIC ENVIRONMENTAL IMPACT FROM THE PRODUCTION OF ELECTRICITY

Some of the most intensively studied, and probably some of the most important, environmental impacts of energy use are those inflicted on aquatic environments by generating electricity. At current rates of conversion efficiency two-thirds of the thermal energy in fuel must be rejected to neighboring environments, and most electrical power plants are located adjacent to water bodies large enough to absorb that heat. Increasingly this means coastal water bodies. The following discussion is derived from a comprehensive review of the literature on the impact of electrical generating systems, with special reference to those located in coastal zones (Hall et al., 1977).

Most of the physical and chemical changes produced by electricity generation cause biotic impacts that at times are severe and easily characterized or, more frequently, modest and cryptic. Because coastal and other aquatic systems are naturally highly dynamic and variable, evaluating these impacts often is difficult. Compounding the problem, power plants occasionally produce positive effects on ecosystems. Nevertheless, many economically and culturally valuable coastal resources are damaged by power plants, including many species of coastal fish that are an important source of protein with a relatively high food energy return on energy invested (Figure 6.10).

Large and expensive environmental studies have attempted to assess the environmental impacts of electricity generation on aquatic environments, often with ambiguous results. Much of this research is of high quality. But, since the research is normally funded by the electric utilities involved, there is often a problem of "assigning the fox to watch the henhouse," and long and bitter debates in the courts have focused on the interpretation of data reported—and not reported. In the case of the Hudson River generating plants, for example, analysis and debate persisted over two decades and cost hundreds of millions of dollars. Even with that large effort it is not clear whether or not power generation has, in general, cumulatively and irreparably stressed certain aquatic systems beyond their normal biotic operating conditions. Research has shown, however, that most impacts are site-specific and depend on specifics of water flow patterns, species' life history and behavior, and details of plant siting and technology. Thus many impacts can be eliminated by intelligent siting.

We review what we feel are the most important impacts on aquatic ecosystems of electric power plant construction and operation, emphasizing coastal systems. Unfortunately most studies ignore the systems approach and instead examine components of the effected ecosystems (i.e., one species of fish or algae) so that, in general, it is difficult to evaluate what the entire impact of the power plant is. More comprehensive views are necessary but generally lacking (see Merriman, 1976, and Kemp, 1977, for truely comprehensive analyses).

Environmental Impact of Construction

The Environmental Protection Agency regards increased turbidity of waters adjacent to construction sites as the most significant aquatic impact caused during the construction of power plants. Dredging to make room for structures, such as intake and effluent canals, docks, and berms, as well as the addition of sediments from building sites, disturbs sediments, and these sediments are transported both vertically and horizontally over long distances in the water, producing a variety of deleterious effects. Turbidity reduces light penetration, decreasing primary productivity of phytoplankton, important sea grasses, and bottom-dwelling algae; this in turn limits secondary production of plant-dependent animal communities and may cause oxygen stress. In addition these sediments may release excessive dissolved plant nutrients or disperse undesirable pollutants throughout the estuary. Finally, many productive bottom communities are removed physically.

Impacts of Operation: Temperature Effects

The cheapest and most common method of dissipating the heat generated by electrical plants is through *once through* cooling systems (Figure 16.12) which pump waste water from a river, lake, estuary, or the ocean through heat converters inside the power plant. Raising the temperature of the adjacent water body is sometimes known as *thermal pollution,* although thermal loading or calefaction is a more appropriate term. There has been a great deal of scientific and media attention paid to thermal pollution, and many very expensive decisions have been made accordingly.

The Federal Water Pollution Control Act of 1972 essentially forbids the addition of heat to natural water bodies and required utilities to incorporate the best available (in terms of environmental protection) cooling system technology into its plants by July 2, 1983. Currently wet or dry cooling towers,

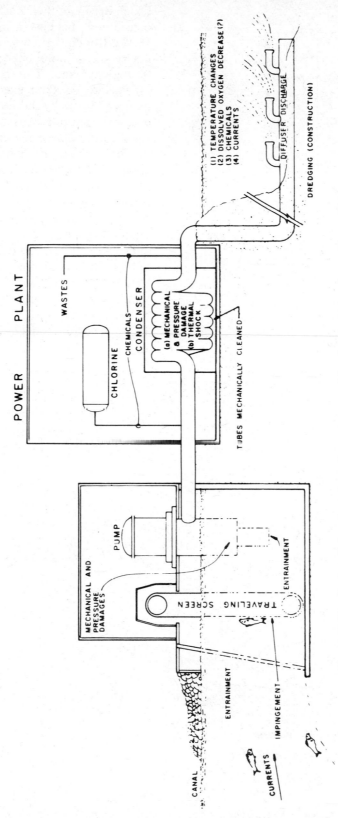

Figure 16.12. Design of once through cooling systems used for electric power plants. (From AEC, 1972.)

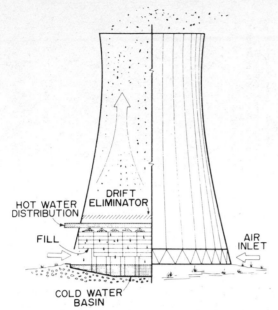

Figure 16.13. Wet cooling tower. (From Clark and Brownell, 1973.)

which dissipate their heat into the atmosphere, are considered the best available technology (Figure 16.13). But constructing and operating cooling towers requires more energy and dollars than once through cooling. This prompted power company officials to seek an exemption from the requirements put forth by the 1972 Act. Officials who could establish that once through cooling would not alter significantly the indigenous ecosystem could use this lower cost alternative while debate continues about impacts of once through cooling systems.

In cases where cooling water discharges are relatively large in comparison to the dissipative capacity of the receiving body, water temperatures often rise over a substantial area. Shallow, enclosed, and/or poorly mixed bodies of water are most vulnerable to heat loading. For example, the first two Crystal River (Florida) power plants raised the temperature of 2 km^2 of the shallow receiving bay by about 5.5°C, and 15 km^2 by about 1°C (Figure 16.14). The additional unit(s) planned would affect a

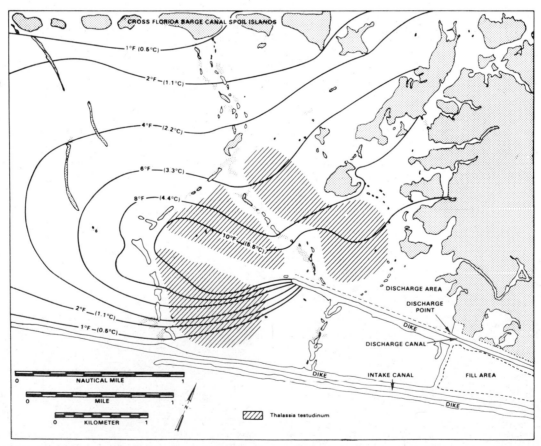

Figure 16.14. Impact of Crystal River, Florida, power plants on water temperatures in discharge region. (From Clark and Brownell, 1973.)

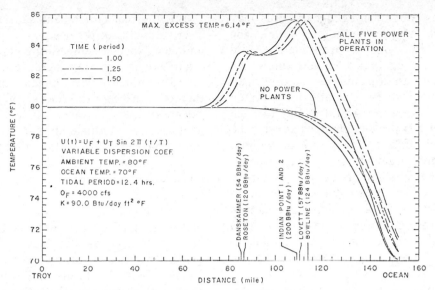

Figure 16.15.   Temperature of the Hudson River from Troy to the ocean at three tidal times with and without five power plants in full operation. (From Clark and Brownell, 1973.)

proportionately larger region. Clark and Brownell (1973) estimate that five generating plants on the mid Hudson River would raise the temperature of a 55 km section of the river by 2–3°C (Figure 16.15).

Two general procedures may be used to evaluate thermally induced stress: controlled laboratory experiments coupled with extrapolations to the field and in-the-field comparison of affected environmentals with unaffected ones. Much more work has been done using the former approach, although the latter or a combination of the two is more useful. Raney et al. (1973) and Krenkel and Parker (1969) have compiled massive bibliographies on the effect of heat stress in aquatic systems.

Impacts on Individuals

Higher water temperatures have their most important impact on individual organisms by interfering with normal physiological processes and/or behavioral responses. This in turn increases their vulnerability to disease, produces embolisms by changing gas solubilities, and increases the impacts of other pollutants by increasing reaction rates of toxicants or other chemicals in the water.

Changing Physiological Processes

A vast body of literature indicates that temperature has a very important effect on the rates, and in some cases even presence, of important biological functions including photosynthesis, respiration, enzyme activity, feeding, and reproduction. The Arrhenius or van t'Hoff law, a broad generality of physiology, states that a 10°C increase in temperature will double or triple a given physiological activity of an organism. Any change in water temperature affects the metabolism of aquatic plants, bacteria, and ectotherm animals, since they are unable to regulate their internal heat balance actively. Such changes may hurt or help organisms.

At some upper limit, frequently not far above the preferred or optimal temperature, organisms die. The mechanisms by which heat kills vertebrates are not well understood but may include the failure of smooth muscle peristalsis, denaturation of proteins in the cells, increased lactic acid in the blood, and oxygen deficit due to increased respiratory activity. A bizarre mechanism has been documented in the laboratory by Laurence (1973), who found that high temperatures increase the metabolism of developing tautogs (blackfish) so that embryos exhaust their energy reserves before a functional mouth develops. Sublethal physiological effects also occur and include a decrease in the overall hatching success of eggs, the inhibition of larval development in some bivalves, erythrocyte degeneration, changes in the normal osmotic balance of fish, and the lowering of respiration rates. Over a year the effects on growth can be ambiguous, since thermal loading could enhance growth in cooler months and retard it in warmer months.

Organisms often recover from short-term temperature increases, but prolonged exposures may induce irreversible shock or direct mortality. Living but severely stressed organisms are sometimes said to have experienced ecological death, since they are vulnerable to predation and serve no normal, functional ecological roles. Aquatic forms often can reduce many of these harmful effects through acclimation. If temperature is changed slowly, the organisms' normal limits can be exceeded without undue harm to the organism until the absolute, species-specific thermal maximum or minimum is approached or exceeded. Even at a particular acclimation temperature the ability of a coastal organism to tolerate temperature change varies according to salinity, variety of food and nutrients present, quantity of dissolved oxygen in water, and presence of pollutants.

As winter temperatures decrease, fish populations may become thermally marooned as they orient and condition themselves to the artificial temperature of a power plant discharge plume. As regional biomass tapers off in the fall, little food is brought to the isolated populations through the discharged water. Sometimes the sport fishing is very good for these hungry and even starving fish. When a generating unit shuts down for scheduled or unscheduled repairs, the sudden temperature drop can kill fish directly or heighten exposure to predation and unfavorable currents. Clark and Brownell (1973) report that 100,000–200,000 menhaden were killed by cold shock in January 1972 at the Oyster Creek Plant in New Jersey when it's warm waste discharge stopped. On the other hand, benthic invertebrate biomass may increase after a shutdown because of reduced predation.

Behavioral Changes

Mobile organisms such as most fish tend to seek areas that they perceive as comfortable and to avoid regions that they find uncomfortable. Animals selectively enter or avoid a thermal discharge depending on whether or not it is compatible with their preferred temperature range. For example, some researchers feared that the operation of the Connecticut Yankee Plant (CYP) would seriously disrupt the spawning behavior of American shad populations in the Connecticut River. But a study of fish movements in the vicinity of the plant using sonic tracers showed that the fish simply migrated up the other side of the river ignoring the warmer plume (Legett, 1976). The shad spawning near the CYP therefore have been unaffected by power generation.

But thermal enrichment can change the reproductive cycles of other aquatic species. For many animals a specific temperature pattern elicits reproductive behavior. Slow increases in temperature often foster gonadal development, and a quick change thereafter triggers the release of gametes. Natural selection has favored spawning temperatures ecologically coordinated with the occurrence of food and favorable water conditions. Premature spawning may be induced in the vicinity of thermally enhanced outfalls, causing fish to spawn before food for the young is plentiful. Rigorous site-specific studies are the only way to determine if there is a need for the considerable effort and expense to correct this problem. For example, in a thermally enriched portion of the Connecticut River female brown bullheads and white catfish developed ovaries prematurely but did not spawn early.

Other behavioral changes, varying widely in ultimate severity to thermally affected populations, include (1) adverse changes in the feeding rates of fish and bivalves, (2) evacuation of protective structures by tube-dwelling amphipods with concommitant exposure to predation, (3) reduction of locomotory behavior in cold-water crabs, and (4) an increased tendency of penaeid shrimp to emigrate (Miller and Beck, 1975). Additionally, increases in water temperature often enhance parasitism in fish.

It is difficult to evaluate the overall significance of the direct thermal effects discussed. Reports of general and massive plant-induced mortality or reports of direct heat and cold shock are rare. Heat shock has been documented in the field, but other factors, often power plant related (e.g., chlorination), complicate the appraisal. We believe that direct heating effects, to date, have had but limited impact on the survival of temperate coastal organisms, but this generality is less applicable to tropical and subtropical waters.

Effects of Calefaction on Aquatic Communities and Ecosystems

The physical, chemical, physiological, and behavioral changes on individual organisms documented in this chapter may combine to affect the relative well-being of groups of species or even entire aquatic ecosystems by changing primary and secondary production, community respiration, species composition, biomass, and nutrient flows. One example of an unexpected ecosystem change began in December 1969 when the Jersey Central Power and Light Company (JCPL) started operating its Oyster Creek Nuclear Generating Station on two adjacent rivers in the town of Forked River, New Jersey. Oyster Creek originally had sufficient land-derived runoff to maintain an essentially freshwater environment in the vicinity of the marinas and small docks located upstream. As a consequence, Oyster Creek was void of marine life over much of its length. In particular, there were no shipworms— one reason the boat facilities were located there. In contrast, Forked River, where cooling water for the power plant was withdrawn was (and still is) saline, and two species of shipworms existed there.

In the summer of 1971, after the power plant was turned on, the shipworm population exploded, as did the tempers of the marina owners on Oyster Creek because their docks and businesses began to collapse. In April 1972 the marina owners presented their grievances to the Atomic Energy Commission (AEC). The marina owners brought suit unsuccessfully in the New Jersey Superior Court during the fall of 1973. Although the court recognized that the power plant probably caused the damage from shipworms, mildew, dry rot, silting, and fog (which also plagued the marinas), it proclaimed that JCPL was not legally responsible for these damages because permission to operate had been granted by the AEC and the New Jersey Public Commission. Late in January 1975 the attorneys for the marina association agreed to a settlement of $2 million with JCPL. By July of that year the marinas left Oyster Creek; the shipworms did not.

Entrainment

Up to 70% of all estuarine fauna have planktonic eggs and larvae (Miller and Beck, 1975), and a large percentage of commerical fish species spend some part of their development in estuaries in one or all of these stages (Clark and Brownell, 1973; Douglas and Stroud, 1971; Smith, Swartz, and Massman, 1966). Common coastal species with planktonic larvae include most of those familiar to sport and commercial fishermen: for example, striped bass, menhaden, tautog, flounders, sea robins, and scup along Northeastern U.S. shores. These tiny juvenile fish, generally measuring a millimeter to a centimeter, use estuarine and coastal waters as a nursery, since coastal waters have a much higher concentration of food than do offshore waters. The fish drift more or less passively with the current but have behaviors that help them seek estuaries. The coastal waters of the Northeastern United States, and probably of most other parts of the world, contain from about 100 thousand to 10 million tiny fish larvae in each million cubic meters of water during much of the summer.

A thousand megawatt nuclear power plant uses about 6 million m^3 of water per day for once through cooling, and a similar sized fossil fuel plant uses about 4 million m^3. A plant located along the coastline pulls (*entrains*) from a half million to as much as a hundred million small fish, and many more invertebrates, through its intake condenser coils and exit structures daily. The number of fish killed this way over the 30-year life span of a coastal power plant is truly staggering, from tens to hundreds of billions. Thus in contrast to the situation of thermal pollution we believe that entrainment has a very large potential for significant environmental impact.

Mortality of Entrained Organisms

Larger free-swimming fish normally are strong enough to escape the influence of the strong intake currents. All fish in cold water, and small fish at all times, however, are much less able to maintain and control their position and are swept or impinged onto the trash racks or intake screens (Figure 16.12). Here larger fish are often either asphyxiated or are killed by removal from the water by the traveling screens. They also suffer external and internal mechanical damage and descaling by screen washing sprays. Smaller fish are pumped through the plant (entrained) and are subject to severe mechanical and thermal stress. Fish not killed but discharged with the effluent in a damaged or disoriented condition are more vulnerable to predation.

Death from all these causes may be immediate or latent, and estimates of the proportion of entrained organisms killed by the combined effects of temperature, mechanical abrasion, pressure changes, chlorination, and predation vary widely. Nearly all zooplankton and planktonic fish entrained by normal operating conditions at the CYP were killed (Merriman and Thorpe, 1976). But Miller and Beck (1975) cite survival rates of 0–94% for organisms undergoing the entire entrainment process and Eiler and Delfino (1975) also found wide-ranging estimates (1.2–91.5%) of zooplankton mortality. Differences in durability of species or individuals are partly responsible for the divergent results.

Specific examples of documented losses and estimates of potential impact from entrainment include the Brayton Point power plant on Mount Hope Bay, Massachusetts, which lost daily to inner plant kills during the summer of 1971 from 7 to more than 160 million menhaden and river herring; the Millstone plant on Niantic Bay, Connecticut, which killed 36 million young fish in 16 days in November 1971 and 150 million fingerling menhaden during 5 days in August 1972; the Connecticut Yankee Plant which was estimated to have killed 179 million fish larvae during 1969 and 1970; the Seabrook power plant in New Hamphire with predicted entrainment losses of 74 million clam larvae killed per day during the plant's initial operation period; and the Storm King pumped-storage plant which could kill up to 3 million young striped bass each day (Clark and Brownell, 1973; Culliver, 1976; Hall, 1977). These numbers are large, but the mortality must be compared to the billions of fish and invertebrates in river and coastal environments not entrained. Estimates of the percentage of a region's plankton population that are entrainment depend on the specific physical and biotic characteristics of both the area and the power plant in question. Massengill (1976) found at the CYP, which extracts 6% of the low flow volume of the Connecticut River, that an average of 4.2% of the potentially entrainable zooplankton passed through the facility. Estimates of the proportion of a year's potential production of Hudson River striped bass killed by entrainment, based on complex computer simulations, vary from about 4 to about 90%, depending on the number of power plants considered and basic assumptions about impact (Hall, 1977; Wallace, 1975). The most recent, and to our mind reasonable, estimates suggest about 20 to 30 percent for the power plants presently operating in the mid Hudson (Barnthouse et al., 1977), or perhaps twice that should all plants once planned be built, but some uncertainty remains. Nevertheless there have been large reductions in the number of striped bass in much of the Northeast that are consistent with earlier computer simulations that suggested large impacts, although other factors probably also have contributed.

The lack of clearly defined boundaries of impact for coastal power plants makes it much harder to estimate the proportion of open coast plankton (e.g., as compared to the Connecticut or Hudson River) that are subjected to entrainment. Entrainment can effect a large proportion of the production of young in enclosed estuaries in or near important spawning or nursery areas, but the impact is less elsewhere. For example, the Shoreham plant, located along the North Shore of Long Island, uses about 3.2 million m^3 of water per day for cooling. Is it justified to compare this volume of water and the number of organisms in it with that of the entire Long Island Sound ecosystem? Although a precise answer remains uncertain, some important biological facts can help resolve this problem.

The Herrod Point Shoals, an important spawning area for fish, is located near the plant intake. Life history studies of some of the Sound's fish showed that many young fish, which spawn in the spring, drift passively for several months with the longshore current, moving from west to east, until they reach a certain size. When large enough, they actively seek out estuarine regions where they spend their first summer and (sometimes) autumn. In a sense these organisms are programmed to search for any shoreward-flowing streams of water. The Shoreham plant probably has a larger impact than that indicated by a simple random sampling of Long Island Sound waters due to the geography and behavior of the fish—some fish spawn close to the plant, and young fish spawned earlier elsewhere seek intake currents that they perceive as suitable nursery areas. Taking these and other considerations into account, the environmental impact statement for Shoreham estimated that the plant would entrain from 11 to 74% of the eastward-flowing nearshore water with its planktonic fish. But many other coastal plants not near important spawning or nursery grounds have a much smaller impact. Thus assessing the impact of electrical generating plants is site specific, and good siting can avoid plant-related problems to a large degree.

Two major strategies can be used to mitigate the impacts of once through cooling systems for operating power plants. Alternative cooling methods, such as cooling towers, can be used, but they are costly in terms of dollars, energy, and other environmental impacts. Since most power plants are part of a grid, a second option is to turn off individual power plants when their local impact is largest, as is now the case for the Hudson River power plants. In this case much of the environmental impact was avoided while saving the energy and dollar costs of cooling towers.

Release of Toxins

Routine operation of most power plants both inadvertently and deliberately releases materials into the aquatic environment that are toxic to many forms of life. These include chlorine (used to control fouling organisms), heavy metals (lost from plumbing in the cooling system), and airborne pollutants. In the case of nuclear reactors, radionuclides can be released into the atmosphere or cooling waters. Coal-fired generators sometimes acidify drainages below stockpiles of coal, and in all plants small quantities of human sanitary wastes are released.

Chlorine

Chlorine is added to the cooling waters of most coastal power plants to prevent and remove the buildup of bacterial slime and other fouling organisms (such as barnacles and clams) that attach to the inside of condenser conduits. Chlorine is added one to several times daily, and dosages vary from 0.1 to 5.0 mg/liter. Since chlorine is added to kill organisms, nontarget species entrained in the coolant and/or residing in adjacent waters also are affected. For example, chlorine concentrations of 0.1–1.2 mg/liter at the Millstone Point Nuclear Plant on northeastern Long Island Sound reduced photosynthetic activity 79% immediately after application (Carpenter et al., 1972). Goldman and Davidson (1977), however, found evidence that the phytoplankton recovered with no apparent permanent effects. Chlorine may have more diverse effects on animals such as larval fish or invertebrates. Eppley et al. (1976) and Goldman et al. (1977) have suggested that the gross effect of chlorine on phyto plankton is minimal compared to effects on organisms higher in the food chain.

Most important, chlorination may provide a path by which persistent chlorinated compounds enter the marine environment. Some of the organochlorine compounds that can be formed are highly refractory, but the ultimate fate and toxicity of most of these chlorinated compounds is virtually unknown. Various alternatives to chlorination have been suggested, including the use of mechanical cleaning devices and antifouling paints, but chlorination remains the principal procedure used.

Heavy Metals

Virtually all power plants with once through cooling use copper in their condensers because of its good heat exchange properties. The toxicity of copper also retards the growth of fouling organisms. As far as we know, few comprehensive studies exist of copper leached from power plants. Copper from power plant discharges can kill fish outright, has been found toxic to aquatic microbiota, and we have heard that copper turned the meat of oysters green in an aquaculture project at Northport, Long Island. Using abrasive sponge balls to clean condenser tubes of fouling bacteria slime in order to avoid the toxic effects of chlorine can release large quantities of copper, as in the Surrey plant in Virginia.

Stack Effluents from Fossil-Fueled Generators

Nearshore coastal sediments frequently contain a complex mixture of hydrocarbons with surface concentrations much higher than those in deeper, older sediments (Farrington et al., 1977). Low levels of these compounds in older sediments probably were originally released by forest fires (Walker et al., 1975). However, dramatic increases in the concentration of pollutant hydrocarbons occurred simultaneously as the rate of fossil fuel combustion grew rapidly. Polycyclic aromatic hydrocarbons (PAH), many of which are known carcinogens, follow this same pattern: concentrations are much higher in sediments laid down since the Industrial Revolution. These PAH in sediments probably originate from fossil-fueled electric generating stations and suggest that hydrocarbons in Buzzards Bay, Massachusetts, could come from as far away as New York City.

Impacts of Operation: Interactive and Synergistic Effects

When two or more factors act in concert to produce an effect greater than that predicted by either action alone, a *synergistic* condition is said to exist. In many cases power plant effects combine with natural events to cause physiological stress and mortality that are difficult to attribute to any one source. For example, excessive heat and chlorine together produce larger adverse effects on organisms than does either alone. An unspecified number of fish were killed in Biscayne Bay by the combined influence of low salinity and chlorine concentration, probably derived from the cooling system of a power station. Aquatic species are more vulnerable to certain toxic metals at elevated temperatures. Furthermore heat can decrease the resistance of microbes to other stresses. Moss et al. (1976) believe that massive fish death rarely is caused by a single factor and feel that relationships between temperature, dissolved oxygen, carbon dioxide, and salinity regulate such events.

In summary, generating of electricity routinely kills millions to billions of coastal and aquatic organisms each day in the United States. Although we understand many of the mechanisms by which this occurs, quantifying a specific impact is difficult. Some of the impact reduces the production of marine commercial fish, which probably is reflected in a small but significant reduction of commercial fish catch and a small but significant increase in the energy cost of fish caught and another increase in the energy cost for our food, since most land-derived protein is more energy intensive than coastal fish (Rochereau and Pimentel, 1977). But the complexity of natural processes, the natural variation in fish stocks, and the ability of other fish to fill the void left by the impacted fish probably reduces the total impact, or at least our ability to measure it.

17

HUMAN HEALTH IMPACTS

Carrie Koplinka-Loehr and Charles A. S. Hall

Throughout most of this book we emphasize energy costs and benefits as important decision-making criteria. We also emphasize the importance and difficulty of including the value of environmental services in such assessments. Perhaps most difficult to evaluate, however, are the impacts of various energy technologies on human health and well-being, for how are we to put a number or value on human life? Economists sometimes measure the value of a human life according to present value of lifetime earnings or net accumulations, but such a definition is inadequate for most of us. We begin this chapter by discussing the variables that complicate assessments of human health impacts caused by extracting and using fuels. Later we evaluate the health impacts of specific energy technologies, dividing them into nonrenewable and renewable energy sources. The health impacts of each technology are split into two main categories: *occupational* (health impacts on workers in that industry) and *social* (health impacts on an entire population). For each category we outline causes of, and statistics for, accidents, illnesses, and deaths according to what we believe are reasonable sources. Nevertheless, we recognize that very different estimates of impacts are available from other sources.

ASSESSMENT: METHODOLOGICAL MUDDLERS

No assessment is complete without warnings concerning the intricacies involved in making that assessment. Some of the problems encountered include (1) defining the system boundaries of a particular technology, (2) dealing with imperfect data, (3) comparing dissimilar impacts of different technologies, and (4) determining appropriate health standards. Finally, it should be remembered

that virtually all employment, and of course life itself, has some risk and that although energy-related health impacts are high, they are not necessarily higher than those from other heavy industries.

Defining System Boundaries

Holdren et al. (1980) describe how difficult it is to define the boundaries of a technology:

> Accidents and disease befalling workers in the energy industries occur at *all stages* of the fuel cycles of the various supply options (e.g., during exploration, harvesting, conversion, and transport) and *at all phases of each stage* (e.g., in construction, operation, maintenance, and decommissioning). A complete accounting of these damages requires identifying the kinds and quantities of labor needed for each phase of each stage—that is, for all the activities connected with supplying energy in usable forms—and associating with each entry in the catalogue of labor inputs the appropriate rates of occupational injury and illness.

In other words, questions such as: Does one count the labor to make the dragline that mines the coal to make the steel to make the (solar) collector? must be answered before the analysis begins. The boundaries chosen for analysis can affect conclusions when, for example, comparing the effects of generating electricity from materials-intensive solar energy systems with fuel-intensive coal plants.

Imperfect Data Base

The data needed to make a scientifically valid assessment are not always complete. For example, the data base used to support claims about health

413

effects of low level radiation is not as unequivocal as it could be for several reasons. First, experiments on laboratory animals produce data that may or may not be applicable to human health. Second, epidemiological studies (i.e., studies of diseases in populations) are needed to validate conjectured cause and effects. Even when such studies have been done, which is rare, they are often incomplete or contain uncontrolled variables. Third, as Holdren et al. (1980) note, injuries rather than illnesses comprise the bulk of occupational health impact statistics, although they may not be the major impact on health. Injuries are easy to correlate to the workplace, whereas illnesses are not, and only 5 percent of the total nonfatal worker-days lost are currently attributed to occupational illnesses. Finally, illnesses listed are predominantly skin disorders—which have clearly visible external symptoms. Workers can develop serious internal disorders in the workplace that might never be correlated with their occupation. So, although identified occupational hazards have been declining in recent years, real undetected occupational hazards may be more prevalent than most people believe. Such has been the case for black lung disease, which often afflicts coal miners only after many years of exposure and incubation.

Incomparable Impacts

Thomas Neff, in *The Social Cost of Solar Energy* (1981), states:

> . . . different technologies have qualitatively different impacts and thus may not be directly comparable. For example, it is difficult to compare the health impact of coal mining on coal miners with the effects on the public of perceptions of risks associated with nuclear wastes. Impacts may be different in character, may affect different groups or institutions in society, and may even affect different generations.

The effects of air pollution on people of different ages are especially difficult to compare. Children are particularly susceptible to air pollution, perhaps because of faster breathing rates, frequent mouth breathing, and habitual respiratory tract infections. Athletes, the elderly, asthmatics, smokers, and persons with preexisting bronchitis or emphysema are other groups that are especially sensitive (Comar and Sagan, 1976).

Setting Standards

Government health standards for human exposure to potentially harmful substances should rarely be viewed as absolute. The difficult task of setting allowable concentrations of hazardous substances in the workplace belongs to the Occupational Safety and Health Administration (OSHA). According to the 1970 act that established it, OSHA must "set the standard 'which most adequately assures . . . on the basis of the best available evidence, that *no employee* will suffer material impairment of health or functional capacity even if such employee has regular exposure to the hazard for the period of his working life'" (Neff, 1981). Numerous factors such as carcinogenicity and synergistic effects must be considered in setting standards. Yet OSHA has few data for the human health effects of the 12,000 toxic chemicals currently in use or the 500 new toxic chemicals introduced annually. Questions such as what the impact will be on a worker who applies a potentially carcinogenic organic solvent to a photovoltaic cell must be answered by OSHA despite limited, if any, data on those effects.

Carcinogens in particular are very difficult to set standards for because as more studies examine the effect of 30–50 years of exposure to low levels of toxic substances, more and more of these substances appear to be correlated with increased incidence of cancer. Although OSHA must assign absolute levels of exposure to carcinogens, it has announced that any level of exposure to carcinogenic agents—even levels that present technology cannot measure—can be responsible for significant health effects (Neff, 1981).

Despite these methodological difficulties, several studies have attempted to assess the impacts of various energy technologies. These are presented in the following pages and sometimes different conclusions are reached. An attempt has been made to approach each technology chronologically. We generally cite the impacts in the order they would occur in producing energy: first exploration, then extraction, transportation, refining, conversion, and finally, transmission.

NONRENEWABLES: PETROLEUM

Occupational Hazards of Petroleum

A broad spectrum of occupational hazards exists in the petroleum industry, ranging from scuba diving in offshore oil exploration to the construction of platforms and pipelines, to dangers associated with high pressure gases from deep within the Earth. These hazards are summarized in Table 17.1.

In offshore exploration or drilling, divers often inspect and maintain the rigs. While doing so, they often are subject to hypothermia, being struck by falling objects, being stranded by communication breakdowns, and drowning. Divers also are susceptible to decompression sickness or the bends, with symptoms ranging from headache to coma or even death. The United Brothers of Carpenters and Joiners of America claim that about 1000 occupation-related fatalities per 100,000 workers per year occur during activities in the oil and gas exploration industry, 10 times higher than that for mining and quarrying fatalities (ILO, 1978).

Drilling wells involves numerous occupational hazards. Mud used for lubrication in the wellhole may contain alkaline additives that irritate or burn skin and mucous membranes. Inhaling large quantities of H_2S bubbling up from underground can cause nausea, dizziness, or death. Blowouts are extremely dangerous because the uncontrolled release of fluids or gas from the wellhole can cause explosions, fire, or high-speed expulsion of hydrocarbons.

Workers in petroleum refining plants may be exposed to more carcinogens than other workers. Thomas et al. (1980) suggest that certain refinery and petrochemical employees had greater mortality rates from cancer of the liver, pancreas, lung, and skin. White males who had been union members for 10 or more years had increased relative frequencies of stomach and brain cancer, leukemia, and multiple myeloma (tumor of the bone marrow).

Current estimates of injuries and illnesses in the petroleum industry as a whole include data from 447,000 employees of 101 U.S. oil and gas companies (Table 17.2). When all phases of the industry are compared based on total injuries and illnesses recorded per total number of hours worked, drilling was the most hazardous, causing 20.3 incidents per 100 workers in 1979. Chemicals ranked second (7.2), and refining was third (6.1). The overall incidence rate for the industry in 1979 was 4.4 (American Petroleum Institute, 1980). The highest fatality rates were in drilling (1 per year per 2534 workers), gas processing, and refining. Fatalities for the entire industry in 1979 were 52 out of 447,040 or 1.16 deaths per 10,000 workers.

Table 17.1. Occupational Hazards in the Petroleum Industry

Activity	Effect
Exploration	
Diving	Hypothermia, "the bends," drowning, injury, falling equipment, miscommunication
Other	Accidents, equipment failure (e.g., helicopters), explosions
Construction	
Platform, pipeline	Accidents: falls, crushing, collision
Welding	Exposure to UV rays, infrared rays, vapors, and ionizing radiation
Drilling	
Inclement weather—cold	Hypothermia
—heat	Heat stress
—storms	Collapse of rigs, drowning
Exposure to H_2S	Nausea, dizziness, death
Exposure to mud	Skin and mucous membrane injury
Vibration	Numbing of limbs
Noise	Hearing impairment, hampered communication
Abnormal working/living conditions (i.e., 12-hr shift)	Insomnia, fatigue, malnutrition, alcoholism
Refining	
Exposure to toxic chemicals, carcinogens	Cancer of the lung, skin, esophagus, stomach, brain; bronchitis
Other	Accidents
Transportation	Accidents
Combustion—power plants	Accidents, bronchitis

Source: Thomas et al. (1980) and ILO (1978).

Table 17.2. Report of Occupational Injuries and Illnesses for 1979 Covering Operations Subject to OSHA Recordkeeping Requirements Only

Function	SIC	Number of Employees	Hours Worked (thousands)	Recordable Cases			Fatalities
				Injuries	Illnesses	Total	
Exploration and production	1311	79,543	161,152	2,777	16	2,793	12
Gas processing	1321	8,082	16,804	347	5	352	2
Drilling	1381	2,534	5,343	541	1	542	1
Chemicals	2800	58,485	104,654	3,679	69	3,748	4
Refining	2911	80,317	164,945	4,860	177	5,037	16
Marketing—wholesale	5171	48,152	99,240	2,330	11	2,341	8
Marketing—retail	5541	35,811	63,926	1,491	8	1,499	2
Pipeline crude and products	4610	14,411	29,008	587	7	594	2
Pipeline gas	4922	18,198	35,475	965	18	983	0
Marine	4400	5,191	19,778	293	5	298	0
Research and development		16,290	31,866	326	7	333	0
Engineering and general service		9,455	18,211	188	3	191	0
Miscellaneous		62,770	122,994	819	25	845	3
Combined functions		12,801	25,341	357	3	360	2
Total		442,040	898,737	19,560	356	19,916	52

Source: From API (1980).

Social Impacts of Petroleum

The per kilocalorie impacts on the health of the general population from petroleum are similar to, but much less than, those from coal, since petroleum is a cleaner fuel. It is virtually impossible to determine whether a given molecule of SO_2 came from coal or oil combustion, however, so we discuss social impacts in the coal section.

COAL

Occupational Hazards

Since 1900 more than 102,000 men and women have died mining coal in the United States. These deaths were due primarily to accidents in the exploration, development, mining, maintenance and repair, and construction phases of the industry. In 1911 these vocations claimed 2656 lives; in 1978 they claimed only 106. The decline in mortality rate during that 67-year period was from 1.9 deaths per million worker-hours to 0.27 deaths per million worker-hours (Figure 11.15).

The 1969 Coal Mine Health and Safety Act (CMHSA), and the Amendments Act of 1977 were two important legislative acts that helped to decrease fatalities in coal mining. The 1969 act established the Mining Safety and Health Administration (MSHA), a sister agency to OSHA. The most significant changes required by the CMHSA were more annual safety inspections of mines, education and training standards for workers, and research on toxic substances produced by the mining (American Conference on Industrial Hygiene, 1979).

Whereas data on mining fatalities due directly to accidents are fairly clear-cut, fatalities from coal miners' diseases are less so, although these diseases may claim as many as 1000 additional lives a year (Ramsay, 1978). Miners' lungs and bronchial tracts are the primary sites of illness. Minute particles of coal dust become lodged in the respiratory system and can lead to or exacerbate bronchitis, emphysema, and coal workers' pneumoconiosis (CWP, or *black lung*). One to 2% of miners develop severe cases of CWP whereby large black solid masses accumulate in the lungs, eventually obstructing or restricting air flow (DHEW, 1977).

Correlations between dust inhalation and mortality have yet to be determined conclusively. A 20-year follow-up study of a group of British miners indicated those with simple CWP had normal mortality patterns (Ramsay, 1978). The problem is complicated by the fact that sensitive individuals who develop symptoms may stop mining (Lave, 1972), or that coal miners may smoke cigarettes at a different rate than the general population.

It is safe to say, however, that miners suffer more respiratory disease symptoms and impairments than other workers, and therefore have higher occupational morbidity (disease) rates than workers in other industries. Lainhart (1969; cf., Lave, 1972) has shown that shortness of breath and persistent cough increased with a miner's years underground. Furthermore, for workers with 15 years of underground experience or more a linear correlation exists between the occurrence of CWP and years worked.

Noise from equipment is another significant hazard. Of 200,000 underground coal miners, 7.2% are exposed to noise above current standards. Of 25,000 surface mine machine operators, 40% are exposed to above-standard noise levels. A survey done by NIOSH (Scheib, 1978) showed that 70% of a group of 60-year-old miners had a hearing impairment, compared to 22% for a control group of 60-year-olds.

Many occupational diseases also may be linked to handling and processing coal. Bituminous coal has been known to be carcinogenic to human skin since at least 1775 when an English surgeon, Sir Pott, observed abnormally high rates of scrotal cancer in British chimney sweeps and linked it with coal soot lodged in the folds of the scrotum (Eckhardt, 1959; cf. Young, 1979). Carcinogens may be particularly important hazards in industries producing synthetic fuels from coal because many of these products are chemically similar to compounds found in chimneys of coal fires.

Other researchers (Doll et al., 1972; cf. Walsh et al., 1981) who have monitored the mortality of coal gas workers over a 12-year period reported increases in lung, bladder, skin, and scrotal cancer. They also found that street pavers and those who worked with asphalt (a by product of petroleum and coal-tar refining) had a lung cancer incidence 64% higher than the general population (Walsh et al., 1981).

Social Impacts of Coal

Occupational hazards of the coal industry are summarized in Table 17.3. Coal is a unique energy resource in that large quantities of it must be transported through densely populated regions on its way from the mine to the consumer.

Table 17.3. Occupational Hazards in the Coal Industry

Mining	
Deep mining	Accidents: Explosions, cave-ins
	Inhalation of dust: shortness of breath, persistent cough, bronchitis, emphysema, pneumoconiosis
	Exposure to low level radiation: cancer, especially lung
	Noise: impairment of hearing
	Exposure to toxic materials, gases
Surface mining	Accidents
	Noise
Transportation	
Railroad	Accidents at crossings
Cleaning and refining	Accidents
	Cancer, especially lung, bladder, skin, scrotal
Combustion—power plants	Accidents
	Inhalation of dust and emissions
	Noise

Source: From Morris et al. (1979); AMA Council on Scientific Affairs, *Combustion* (April, 1979); and Ramsay (1978).

Although coal transportation data are not readily available, autos colliding with coal trains at crossings may cause between 0.1 and 1.5 deaths per coal-fired power plant per year (NRC, 1976). Additionally about 12.2 nonfatal injuries can be attributed annually to coal transport for a 1000 MW$_e$ coal-fired plant (Barrager, 1976).

Numerous trace elements are emitted from coal burning, including mercury, cadmium, selenium, arsenic, fluoride, iodine, beryllium, uranium, thorium, and radon. Although amounts emitted vary according to the type of furnace and coal used, all of these elements are potential health hazards and some are radioactive (McBride et al., 1977). Mercury is especially dangerous in methylated forms when concentrated in biotic food chains (Barrager, 1976).

Stack gases also contain numerous organic compounds, including potentially carcinogenic hydrocarbons. A National Academy of Sciences report (1979) suggests that these compounds may be correlated with increased incidence of lung cancer. The report points out that incidence of lung cancer among urban dwellers is twice that of those living in rural areas. The incidence of lung cancer was even greater where concentrations of fossil fuel products within specific urban communities from industrial usage were high. But because organic compounds are released from sources other than coal combustion, epidemiological evidence for assessing health effects due to coal-derived compounds is limited.

Radioactive substances found in coal are released in both stack gases and fly ash of coal-burning power plants (McBride et al., 1977). A 1000 MW_e plant that burns low sulfur bituminous coal releases each year 2320 g of uranium (0.068 curies) and more than 46000 g thorium and its by-products (1.2 curies) through stack emissions. Additional amounts are released from stored ash. A properly functioning nuclear power plant (BWR) of comparable size annually releases stack gases containing 43 curies of tritium, 10 curies of carbon-14, 0.3 curies of iodine-131, and 7000 curies of crypton and xenon.* The health effects of the radioactive substances from these two sources, however, are not directly comparable. Gaseous radionuclides from nuclear power plants are predominantly noble gases that expose the entire body to radiation but do not combine with biological compounds.

Soluble radionuclides in coal fly ash are long-lived bone seekers—compounds that tend to lodge in bones—and insoluble radionuclides that expose the lungs to radiation. Because the measured radiation doses from coal-fired power plants are below the radiation protection guidelines recommended by the Federal Radiation Council, investigators suggest that these emissions cause negligible social costs. Nevertheless, McBride et al. conclude that in many cases the human health impact from nuclear power plants is less than from coal-burning plants. Thus the major radiological problems of nuclear power occur in other parts of the cycle (see Figure 15.6), although the data base for these trace elements is still limited (Barrager, 1976).

HEALTH HAZARDS TO THE PUBLIC OF EMISSIONS FROM ALL FOSSIL FUELS

When combusted, both oil and coal produce particulates—tiny fragments of solid matter such as dust, fumes, and smoke. Fossil fuel combustion also produces varying quantities of carbon monoxide, sulfur dioxide, and nitrogen oxides. In 1978 electric utilities, industries, and transportation vehi-

---

*In addition other larger emissions come from other parts of the fuel cycle (see Figure 15.6).

cles pumped some 24 million metric tons (mmt) of sulfur dioxide and 23 mmt of nitrogen oxides into the atmosphere. Once in the air, these gases may be converted into sulfates and nitrates. To date, sulfates and particulates are two agents proved to affect human health. Other gases in air pollution, however, are by no means harmless; nitrogen dioxide is a known irritant, and carbon monoxide in small dosages contributes to cardiac disease (Ramsay, 1978).

Several decades of analysis suggest correlations between air pollution and human mortality, although the picture is not completely clear. We have summarized Lave and Seskin (1970) studies linking air pollution to the following specific ailments:

Bronchitis.   Bronchitis is marked by inflammation of the trachea leading to the lungs. Epidemiological studies in England, the United States, and other countries concluded that air pollution is largely responsible for the doubling of bronchitis mortality rate for urban dwellers compared to rural residents. Lave and Seskin note that "if the air in all of Buffalo [N.Y.] were made as clean as the best air, a reduction of approximately 50 percent in bronchitis mortality would probably result."

Total Respiratory Disease.   In studying cases of emphysema (marked by swelling of the lungs and extreme difficulty in breathing), pneumonia, pulmonary tuberculosis, and lung cancer, English scientists have found a significant correlation between air pollution and morbidity rates.

Cancer.   Hagstrom et al. (1969) found the rate of cancer deaths among middle-class Nashville, Tennessee residents to be 25% higher in polluted areas than in nonpolluted areas.

Heart Disease.   Epidemiological studies show that a reduction of air pollution would produce a 10–15% reduction in mortality and morbidity rates for heart disease.

Asthma.   Studies by Sheppard et al. (1980; 1981; cf. *Science* 212:1251) indicate that asthma victims are sensitive to concentrations of sulfur dioxide in the air that are at or even below allowable once-a-year peak levels. Asthmatics who bicycled in an atmosphere with controlled levels of pollution had difficulty breathing and minor throat constrictions.

Numerous other studies have confirmed the correlation between adverse health effects and pollution. The EPA's Community Health and Environmental Surveillance System (CHESS), begun in 1970, concluded that polluted areas were correlated with high incidences of respiratory disease (Eisenbud, 1978). In their Report to the President the Federal Interagency Toxic Substances Committee (1980) determined that air pollution put 35 million U.S. urbanites at special risk to respiratory diseases. The sectors of the population most readily affected by urban air pollution are young children, the elderly, and those persons in ill health. Table 17.4 is an attempt to illustrate the number of fatalities and nonfatal illnesses associated with the emissions from all power plants in 1975.

Exactly what portion of these deaths and illnesses can be attributed to the combustion of oil versus that of coal is difficult to determine. More than twice as much oil is burned as coal, but oil burns more cleanly than coal, and distillate oil, which is used in domestic heating, is especially low in sulfur and ash. Coal is notorious for its variable but significant quantities of sulfur that contribute heavily to emissions and pollution. Table 17.5 is one assessment of the total health impacts of operating a large coal-fired power plant for one year.

NUCLEAR POWER

We live in a world of radiation, for cosmic rays and the Earth's rocks constantly expose us to radioactivity (Table 17.6). On the other hand, in the past

Table 17.4. Health Effects Associated with Sulfur Oxide Emissions

	Remote Location	Urban Location
Cases of chronic respiratory disease	25,600	75,000
Person-days of aggravated heart-lung disease symptoms	265,000	755,000
Asthma attacks	53,000	156,000
Cases of children's respiratory disease	6200	18,400
Premature deaths	14	42

Source: From C. L. Comar and L. A. Sagan, reproduced, with permission, from the *Annual Review of Energy* Volume 1. © 1976 by Annual Reviews Inc.
Note: Chronic refers to both long-term and permanent diseases.

Table 17.5. Summary of Coal Fuel Cycle Effects per 1000 MW$_e$ Plant-Years, 65% Capacity Factor

Fuel Cycle	Effects per 0.65 GW$_e$	
	Deaths	Disease/Injury
Mining		
Public	—	
Workers		
Accidental injury	0.6	42
Occupational disease	0.02–0.4	0.5–1.0
Processing		
Public	Not estimated	
Workers		
Accidental injury	0.05	28
Occupational disease	—	—
Transport		
Public and workers		1.2–5.9
Accidents	0.3–1.3	1.2–5.9
Electricity Generation		
Public		
Air pollution (50-mile radius)	2 (0–10)	10
Air pollution (total U.S.)	10 (0–50)	50
Airborne radionuclides	0–0.19	0–0.19
Workers		
Accidental injury	0.1 (0.02–0.3)	3.3 (2.7–4.0)
Total	11.1–11.6	100–107

Source: From Hamilton et al. (1981).

four decades we have learned how to concentrate the dilute radiation of nature to produce a nuclear technology that augments nature's radiation to a significant degree.

Ionizing Radiation

What is *radiation*? A nucleus of an atom is considered radioactive if it decays to another nucleus, emitting rays or particles in the process. The radiation emitted from reactor fuel is known as *ionizing radiation* because it causes neutral atoms that are hit by it to lose one or more electrons, a process referred to as ionization. In a reactor, radioactivity is emitted primarily from the fuel in three ways. First, fission produces radioactive fragments including isotopes such as krypton, strontium, iodine, xenon, and cesium. Second, nonfission reactions

Table 17.6. Mean Radiation Exposure in the United States for 1970

Radiation Source	Average Dose Rate per Person (mrem/yr)
Natural Sources	
Cosmic rays at ground level[a]	44
Rocks, soil and building materials[b]	40
Sources within the body (largely ^{40}K)	18
Subtotal	102
Artificial Sources	
Fallout from nuclear weapons testing	4
Medical and dental diagnosis and treatment	73
Nuclear power installations[c]	0.003
Occupational exposure	0.8
Miscellaneous[d]	2
Subtotal	80
Total	182

Source: From Romer (1976).

Note: Individual exposures may vary greatly.

[a] Range within the United States: 38–75, with altitude.

[b] Range 15–140 depending on type of soil and building material.

[c] Nuclear power reactors and fuel reprocessing plants; estimated to increase to 0.4 mrem/yr by the year 2000.

[d] Television sets, airplane travel, etc.

(e.g., neutron capture in the first step of the conversion of uranium 238 to plutonium 239) produce the *actinides*: thorium, protactinium, uranium, neptunium, plutonium, americum, and curium. Third, the radioactivity is produced by neutron activation of structural materials and cladding—the material that surrounds the nuclear fuel. The hazards associated with specific radionuclides are summarized in Figure 17.1.

Ionizing radiation occurs in several forms. Alpha particles (α) are high-energy helium nuclei. If unblocked, they will travel less than 1 mm into human tissue. Although alpha particles can be stopped by a sheet of paper or even a few inches of air, inhalation or ingestion of these particles will expose nearby cells to ionizing radiation. Beta particles (β) are high-speed electrons or negatrons that also can ionize tissue; they are capable of traveling several centimeters into flesh but can be stopped by such barriers as a thin metal sheet. Gamma and X-rays are electromagnetic radiation and can pass com-

pletely through the human body, leaving a trail of ionized atoms (Gyorgy et al., 1979).

Radiation deposits energy in human tissue as it passes through it. We measure this amount of energy as a *r*adiation *a*bsorbed *d*ose, or *rad*. A rad is defined as 100 ergs deposited per gram of tissue (Nero, 1979). Whole body doses greater than 25 rad delivered over a period of minutes or hours can result in nausea, vomiting, diarrhea, and loss of hair, with an increasing probability of death as the dose is increased. An exposure of 1000 rad would lead to death within a few weeks. People would be exposed to such high doses only through an extremely serious nuclear power plant accident or nuclear war (Nero, 1979).

The same dose in rads from different radiations may produce different degrees of biological effect. The dose in *rem* (roentgen equivalent man) is the product of the dose in rads and a factor called the quality factor, which depends on the relative biological effectiveness (RBE) of the radiation involved. A rem is that dose from any radiation that produces biological effects in man equivalent to one rad of X-rays. The mean radiation absorbed by a person in the United States from natural sources is roughly 0.09 rem/yr. The maximum dose for the public (in addition to natural and medical doses), recommended by the International Commission on Radiation Protection, is about 0.17 rem/yr or twice the natural amount. The legal limit for workers in the nuclear industry is 5 rem/yr. The maximum allowable dose has changed dramatically in this century, from 100 rem/yr in 1910, to 15 rem/yr in 1948, to 5 rem/yr in 1958 (Gyorgy et al., 1979).

Scientists are uncertain whether even low level radiation is safe. Debate centers on how to extrapolate the easily measured effects of high doses of radiation to low doses. Studies by Sparrow (1979) summarizing very careful experiments indicate that there is no per unit decline in impact at low dose rates, whereas Yarow (1983) states strongly that there is a threshold level below which there is essentially no impact (see discussion in BEIR, 1981).

An incident that emphasized the importance of this controversy began with Dr. Thomas Mancuso in 1965. Mancuso was hired by the Atomic Energy Commission (and then the ERDA) to study the effects, if any, of low-level ionizing radiation on workers in atomic energy facilities. The researchers began to examine the death certificates of 3520 such workers who died from 1944 to 1972. Because cancer may remain latent for several decades, Man-

IONIZING RADIATION

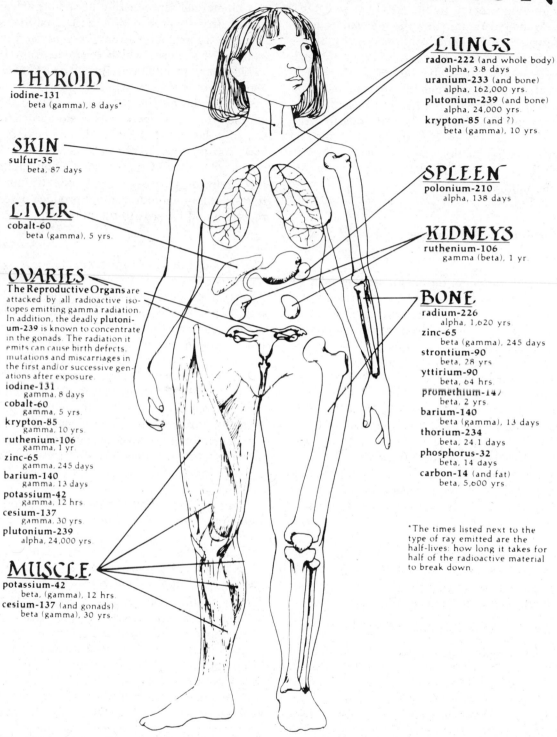

THYROID
iodine-131
beta (gamma), 8 days*

SKIN
sulfur-35
beta, 87 days

LIVER
cobalt-60
beta (gamma), 5 yrs.

OVARIES
The Reproductive Organs are attacked by all radioactive isotopes emitting gamma radiation. In addition, the deadly plutonium-239 is known to concentrate in the gonads. The radiation it emits can cause birth defects, mutations and miscarriages in the first and/or successive generations after exposure.
iodine-131
gamma, 8 days
cobalt-60
gamma, 5 yrs.
krypton-85
gamma, 10 yrs.
ruthenium-106
gamma, 1 yr.
zinc-65
gamma, 245 days
barium-140
gamma, 13 days
potassium-42
gamma, 12 hrs.
cesium-137
gamma, 30 yrs.
plutonium-239
alpha, 24,000 yrs.

MUSCLE
potassium-42
beta, (gamma), 12 hrs.
cesium-137 (and gonads)
beta (gamma), 30 yrs.

LUNGS
radon-222 (and whole body)
alpha, 3.8 days
uranium-233 (and bone)
alpha, 162,000 yrs.
plutonium-239 (and bone)
alpha, 24,000 yrs.
krypton-85 (and ?)
beta (gamma), 10 yrs.

SPLEEN
polonium-210
alpha, 138 days

KIDNEYS
ruthenium-106
gamma (beta), 1 yr.

BONE
radium-226
alpha, 1,620 yrs.
zinc-65
beta (gamma), 245 days
strontium-90
beta, 28 yrs.
yttirium-90
beta, 64 hrs.
promethium-147
beta, 2 yrs.
barium-140
beta (gamma), 13 days
thorium-234
beta, 24.1 days
phosphorus-32
beta, 14 days
carbon-14 (and fat)
beta, 5,600 yrs.

*The times listed next to the type of ray emitted are the half-lives: how long it takes for half of the radioactive material to break down.

Figure 17.1. Human organs affected by ionizing radiation. (From Gyorgy et al., copyright © 1979 Anna Gyorgy, with permission of South End Press.)

421

cuso's results after 12 years did not show a clear pattern of increased mortality when he was pressed to publish his findings. He refused to publish; in 1977 his funding was cut by the ERDA (now DOE) because, Mancuso claims, they did not want to hear his final results. An independent analysis by Steward and Kneale (cf. Gyorgy et al., 1979) in the final year of the Mancuso study, however, showed a definite relationship between the development of bone marrow cancers and radiation doses. Similarly, arguments have ensued about the possible impacts of bomb testing in Nevada in the 1950s. In summary, whether or not there is a health impact from low levels of ionizing radiation currently depends on whom you ask.

Occupational Hazards from Nuclear Power

Mining

As in the coal industry, the major portion of fatal and nonfatal occupational diseases for present day nuclear industry occurs in uranium mining (Ramsay, 1978). Particles inhaled from hard-rock mining produce silicosis—massive fibrosis of the lungs characterized by shortness of breath.

Uranium mines are radioactive as well. In addition to being exposed to whole-body radiation, miners who inhale small dust particles—which include radioactive radon gas and its products—receive an especially high dose of radiation in the lungs. Several studies have shown correlations between uranium mining and both respiratory problems and lung cancer. Furthermore, exposure to radon gas in mines may lead to cancer of the lymphatic system (Ramsay, 1978). Uranium miners have a 10% chance per year of having a nonfatal disabling injury, and their fatality rate appears somewhat higher than that of coal miners (Hamilton, 1981). Health impacts associated with the nuclear fuel cycle are summarized in Table 17.7.

Enrichment and Reprocessing

Three gaseous diffusion enrichment plants have operated in the United States for a combined total of 80 reactor-years. They have handled large quantities of uranium fluoride with no identifiably significant environmental or employee health consequence. Experience with centrifugal enrichment plants is less extensive, but their safety record also is good. Events such as earthquakes or airplane crashes on the buildings are considered to be the only possible causes of a major failure of the process equipment or storage facilities at enrichment plants.

Commercial reprocessing of spent nuclear fuels is not currently practiced in the United States. The single U.S. commercial reprocessing plant in West Valley, New York, which operated from 1966 to 1972, met with numerous problems; between 100 and 150 workers were exposed to rates of radiation exceeding the maximum allowed annual dose. In addition the high level waste storage tanks, designed to last only 30 years, already have begun to leak, posing serious environmental and human health dilemmas. If reprocessing commercial fuels is revitalized, releases of krypton-85, tritium, strontium-90, and radioiodine could produce external irradiation and hazards to the bones and lungs.

The Public: Effects of Emissions

The average dose to any person in a reasonably large population group surrounding a properly operating nuclear power plant is less than 0.001 rem/yr. One estimate is a total of 5 person-rem/yr to the population within 50 mi, or the equivalent of the maximum allowable dosage per year for one worker (Nero, 1979). The composition of these emissions depends on the type of reactor. Operating boiling water reactors release primarily gaseous radioactivity, generally in the form of noble gases, and small amounts of radioiodine. Pressurized water reactors release primarily liquid radioactive waste (Lave et al., 1972).

Estimates as to how the public is affected by this radiation, and to what degree, are still highly controversial. At the upper extreme, Gofman and Tamplin (1971) calculated that if the world population were exposed from birth to 0.17 rad/yr (the permissible dose), there could be an additional 32,000 cancer deaths each year, and from 150,000 to 1.5 million extra genetic deaths per year. Another large-impact assessment was derived by the Advisory Committee of the Biological Effects of Ionizing Radiation (BEIR), created by the National Academy of Science in 1970. The 1972 BEIR Report estimated that radiation releases meeting the U.S. permissible dose could cause from 3000 to 15,000 cancer deaths per year; their most likely estimate was 6000 deaths (Gyorgy et al., 1979). This estimate has been widely criticized, and some consider the potential impacts in the BEIR Report to be

Table 17.7. Nuclear Fuel Cycle Effects Summary per 1000 MW$_e$ Plant-Years with a 65% Capacity Factor

Fuel Cycle	Effects per 0.65 GW$_e$/yr	
	Deaths	Disease/Injury
Mining[a]		
Public	0.086	0.047
Workers		
Radiation-induced cancer	0.095	0.019
Nonradiation health effects		
Occupational disease	0.07	0.14–2.8[b]
Occupational accidents	0.17	6.4
Subtotal	0.42	6.6–9.3
Processing		
Public	0.002	0.002
Workers		
Radiation-induced cancer	0.034	0.034
Occupational accidents	0.004	1.3
Subtotal	0.04	1.3
Electricity Generation		
Public, routine radionuclide emissions	0.013	0.011
Workers		
Radiation-induced cancer	0.04	0.045
Occupational accidents	0.011	1.13
Catastrophic accidents	0.1	—
Subtotal	0.16	(1.2)
Waste Management		
Public	4.1×10^{-5}	4.0×10^{-5}
Workers	5.3×10^{-4}	5.9×10^{-4}
Subtotal	6×10^{-4}	6×10^{-4}
Transport		
Public, routine exposure	3.0×10^{-4}	3.0×10^{-4}
Workers		
Radiation-induced cancer	3.0×10^{-4}	3.5×10^{-4}
Occupational accidents	0.01	0.10
Catastrophic accidents		
Cancers	8.3×10^{-5}–7.1×10^{-4}	8.3×10^{-5}–7.1×10^{-4}
Prompt deaths	2.1×10^{-7}–9.3×10^{-5}	
Subtotal	0.01	0.10
Decommissioning		
Public	4.2×10^{-9}	3.6×10^{-9}
Workers		
Radiation-induced cancer	2.0×10^{-3}	2.7×10^{-3}
Occupational accidents	8.0×10^{-4}	0.07
Subtotal	2.8×10^{-3}	0.073
Total	0.63	9.3–12[b]

Source: From Hamilton et al. (1981).

A 1000 MW$_e$ power plant operating with an average capacity factor of 65% produces 0.65 GW/yr or 1.94×10^{13} Btu (2.05×10^{16} J), electrical energy in a year.

Note: Some more recent estimates are higher for radiation-induced cancers.

[a]Assumes 60% underground, 40% surface mining; GESMO p. IVF16 (NRC, 1976).
[b]Based on ratio of occupational disease/death in coal miners.

excessive. Hamilton (1981) states that the low level doses given off by nuclear power plants at a low rate may not induce cancer deaths at all. No wonder the public is confused.

What is our assessment of the "safe dosage" controversy? Unfortunately, scientists, despite their wealth of knowledge in this area, still are reinterpreting old data to arrive at new conclusions. A case in point is the reappraisal of genetic effects of children born to survivors of the Hiroshima and Nagasaki bombings. The parents in the study received small but statistically significant exposure to radiation, yet analysis of their reproductive success suggests that these low levels of exposure may have caused some genetic damage (Schull et al., 1981). Although at present we see little compelling evidence that indicates that operating nuclear plants routinely subject the general public to large health risks, and overall health risks from nuclear power appear less than those from coal combustion, future research, longer health records, or new accidents could invalidate this statement.

Mining Wastes

The mining of uranium transfers rocks that contain long-lived radioactivity from underground sites to the surface. Milling separates high-grade uranium from mildly radioactive residues which persist as tailings. These tailings have a relatively low specific activity when compared with many other low-level radioactive wastes, but they could contribute measurably to the total dosage that might be delivered to humans over very long periods of time (NEA, 1978). Pohl (1976) makes the analogy that failing to recognize the cumulative health effects of the radon-222 that comes from the decay of thorium-230 in tailings, and that is released to the atmosphere and water table, is like quoting the annual installment payment of an object instead of the full price. He estimated that the health impact from radon-222 emanating from mill tailings left on the ground up until the time the isotope decays, which would take millions of years, may approach 400 eventual lung cancer deaths per GW_e per year of power plant operations.

Transportation

To date no transportation accidents resulting in significant radiation exposure to humans have been reported, probably because a great deal of care is taken in building uranium transportation vessels. Brobst (1973; cf. Lave et al., 1972) estimates that a truck driver involved in a transportation accident while carrying spent reactor fuel is much less likely to be injured from radiation exposure than from nonradiological crash effects. Data on accidents involving transportation of uranium, however, are limited because national accident rate statistics are not completely differentiated according to materials transported. In addition, because the United States does not have a nuclear waste processing program, spent nuclear fuel is normally stored at the reactor site. When a reprocessing program is finally undertaken, large quantities of spent fuel will be transported.

Gyorgy et al. (1979) give transportation impacts less of a clean bill of health. They say that despite the fact that casks for containing the high level wastes are tested adequately for strength, resilience, and so forth, ordinary steel drums are often used for the actual transportation of many low level radioactive wastes.

Spent Fuel Storage

The principal long-term health dangers from spent fuel are caused by the alpha particles emitted by transuranic elements. Although these heavy particles cannot penetrate skin, they can be carcinogenic if inhaled; the particles may come in contact with lung cells or be transported in the blood stream and affect other living tissue. Quantities as small as ten-millionths of a gram are potentially harmful; one pound of plutonium, if deposited in the lungs of everyone on Earth, would kill the entire population (Gyorgy et al., 1979).

But the major problem in assessing the potential hazard from spent fuel storage is *time*. Such hazards will remain dangerous for from 25,000 to a million years due to long half-lives of the nucleides. Humans can barely conceptualize the eons required for a dangerous isotope to decay fully. Even worse, we have only a weak idea of the type of vessel required to contain a hazardous substance during its most radioactive period. Some nuclear waste repositories are already beginning to generate very difficult problems (see p. 280). Thus any successful waste storage program will have to be effective for a longer time than any human structures have yet lasted. Advocates of particular burial schemes are convinced, however, that since the geologic areas selected for storage have been inert for even longer periods than it takes for virtually all decay to take place, they will be a safe place for storage.

Accidents

Perhaps the greatest hazard arising from nuclear reactors is the possibility that something could go wrong in a big way, for the large inventories of highly radioactive materials make nuclear reactors potentially very dangerous. Light water reactors cannot detonate like a bomb, but they can release very hazardous materials. Several attempts have been made to predict the probability of nuclear reactor accidents and the subsequent deaths and diseases that would result. One of the most comprehensive, but also the most controversial, was the WASH-1400 Reactor Safety Study (RSS, 1975) directed by Norman Rasmussen (Table 17.8). This study attempted to estimate the risk of a large reactor accident by assessing which components would need to fail and then combining the failure probabilities (which are known) of the various valves, pumps, and other hardware. The RSS estimated very low probabilities: i.e. that 0.02 "early fatalities and total cancer deaths" per reactor-year would result from possible LWR accidents over a 30-year period.

This estimate, derived by multiplying the probability of an accident times the assumed human health consequences, has since been shown to be both misleading and inaccurate. Nero (1979) points out that Rasmussen's number for *latent* cancer death must be multiplied by 30 (years) to obtain the actual overall risk of 2 deaths per reactor-year, making long-term effects much more important than acute effects. Holdren (1982) states that subsequent reviews of the Rasmussen document estimate the loss of life at between 0.0002 and 20 deaths per reactor-year. Hamilton's estimate (1981), based on estimates of the Ad Hoc Risk Assessment Group (Lewis et al., 1978; cf. Hamilton, 1981) and ERRI (1979; cf. Hamilton, 1981) is 0.1 deaths per GW_e/yr, or about 0.1 death per reactor-year (Table 23.7). Another problem is that our experience with less serious accidents such as Three Mile Island suggests that mechanical failures are not always independent, as the Rasmussen report assumed, but may even cause one another.

Another risk associated with the use of nuclear power is that of deliberate sabotage, especially for

Table 17.8. Approximate Annual Average Societal and Individual Risk Probabilities from Potential Nuclear Power Plant Accidents

Consequences	Societal	Individual
Early fatalities[a]	3×10^{-3}	2×10^{-10}
Early illness[a]	2×10^{-1}	1×10^{-8}
Latent cancer fatalities[b]	7×10^{-2}/yr[c]	3×10^{-10}/yr
Thyroid nodules[b]	7×10^{-1}/yr[c]	3×10^{-9}/yr
Genetic effects[d]	1×10^{-2}/yr	7×10^{-11}/yr

Source: From WASH (1975), quoted in Nero (1979).

Note: Based on 100 reactors at 68 current sites.

[a] The individual risk value is the per capita impact based on 15 million people living in the general vicinity of the first 100 nuclear power plants.

[b] This value is the rate of occurrence per year for about a 30-year period following a potential accident. The individual rate is based on the total U.S. population.

[c] To obtain the net risk of these delayed effects, for comparison with early effects, these entries should be multiplied by 30 (see note b). This yields an annual risk of 2 cancer fatalities and 20 thyroid nodules.

[d] This value is the rate of occurrence per year for the first generation born after a potential accident; subsequent generations would experience effects at a lower rate. The individual rate is based on the total U.S. population.

political purposes. This risk is unquantifiable, to say the least. Writes Holdren (1982): "Sabotage . . . may well prove more important than the intricate mechanical failures more easily studied."

In conclusion, the health impacts of present nuclear power are neither as large as presented in some of the antinuclear media nor as insignificant as proposed by some proponents of nuclear power. But many very important questions are unanswered, so that it can be argued that despite 30 years of commercial nuclear power we simply do not know what the health inpacts of a large scale nuclear program would be. The impact today of this uncertainty is less than what might have been the case if nuclear power had grown as much as was once projected. A remaining, unanswered question is the degree to which a large accident or sabotage could dramatically change the risk assessment numbers. In the meantime the continued occurrence of small and moderate accidents, engineering blunders, unexpected material failures, and the failure of finding a good way to dispose of waste has increased the distrust and concern of many about nuclear power. Such subjective attitudes are probably more important than objective statistics in the long run in determining the future of nuclear power in America.

RENEWABLE RESOURCES

Just as the actual health impacts of certain energy technologies may be exaggerated in the public eye, the impacts of other technologies such as solar are insufficiently appreciated. The following material is taken directly from the review by Holdren, Morris, and Mintzer (1980) except where supplemented by referenced materials.

Direct Solar Heating and Cooling

Many of the health impacts of solar heating and cooling designs are associated with the insulation needed to make solar heating work rather than the solar technologies directly. Some examples are (1) falls during installation, (2) chemical pollutants, (3) indoor air pollution due to reduced ventilation rates, and (4) poor choices of insulation materials which could lead to fire hazards and accumulation of outgassed toxic substances.

Solar devices also have problems. Flat-plate collectors for space heating and water heating in residential and commercial buildings usually are mounted on roofs, increasing the probabilities of injury during installation and maintenance. Water in such collectors may contain toxic additives to inhibit rust and prevent the growth of organisms. Fluids other than water are frequently hazardous and can pose pollution hazards when discarded or if they leak from the collector, the associated plumbing, or storage. Leaks through heat exchangers into the potable water supply could be particularly hazardous. Rock heat storage bins provide excellent growing conditions for allergenic molds and fungi, which can be transferred to living quarters by forced-air heating systems. Even normal operation may give off potentially hazardous compounds via outgassing from the increased insulation, or from the plastic compounds, epoxies, and selective coatings often associated with solar heating systems. Overheating for one reason or another increases outgassing and severe overheating can cause fires. Fires of any origin can mobilize the toxic materials embodied in the collector or working fluid.

Both Holdren et al.'s and our review of the literature revealed little quantitative analysis concerning the extent or severity of the impacts of passive solar systems. We think it likely that assuming sensible choice of materials, the use of passive solar design in architecture will produce fewer and less significant health effects than most fossil fuel or nuclear technologies.

Other Collectors

Optical concentration and sun-tracking mounts usually employ more exotic working fluids, such as fluorocarbons or liquid sodium, than do flat-plate collectors. In addition the higher temperatures attained may increase the fire hazard. Adsorption air conditioners and some technologies for solar cogeneration of electricity and process heat may require evaporative cooling towers, with their associated water use and pollution from anticorrosion and antifouling additives. The working fluids for heat engines associated with saline ponds, parabolic trough collectors, or Offshore Thermal Electric Technologies (OTEC) (e.g., ammonia, fluorocarbons) can present hazards in the event of leaks and upon disposal.

Power-tower and parabolic-trough approaches have particular hazards. For example, workers can fall, aircraft can collide with the tower, and the concentrated beams may cause human or animal burns. Additional hazards may be associated with the large amount of thermal storage needed in both approaches to achieve high capacity factors. The storage medium may be rocks (with oil as heat-transfer fluid), a molten salt, or a liquid metal.

Terrestrial Photovoltaics

Photovoltaic technologies and production processes are far from mature, resulting in a lack of information on precisely how these technologies may be used and a lack of data on what impacts actually occur. Nevertheless, preliminary health and environmental analysis of generic technological choices can play an extremely important role in decisions made concerning photovoltaic technologies. Indeed, analysis of present processes may help in determining which suboptions are most attractive from a social cost standpoint, or which process steps need improvement before an option may be considered acceptable. (Neff, 1981)

Other Impacts

In addition, the power-conditioning equipment used to boost voltage and convert the electricity from photovoltaic cells to alternating current contains many toxic substances that may leak or be released in fires. Storage of electricity in lead-acid or other batteries poses chemical, fire, and explosive hazards.

Our summary, based mainly on Neff's (1981) review, includes the potential health impacts of the three main lines of photovoltaic (PV) research and development:

Large-crystal silicon (Si) wafer cells in flat-plate arrays.

Cadmium sulfide (CdS) cells in flat-plate arrays.

Gallium arsenide (GaAs) cells in the flat-plate or concentrator arrays.

Silicon

The process of mining silicon dioxide (SiO_2) from sand or quartzite and reducing it with heat to pure silicon (SiO) is essential for most silicon-based photovoltaic technologies. SiO or SiO_2 particulates and fly ash from coke released during refining are the principal agents of direct occupational health problems. As in uranium mining, silica particulates can cause silicosis or be translocated to the kidneys or other sensitive organs. Synergistic effects involving other pollutants also are likely. For example, particulates may provide catalytic surfaces for reactions involving chemical pollutants or a transport mechanism into the lung; impairment of lung function may increase the likelihood of other diseases. The presence of SiO (which is more reactive than SiO_2) in emissions, and its role in health effects, are of particular interest. Conventional refinement processes may involve hazards from silicon dust or oils used in cutting up the silicon sheets.

Several of the steps for cell fabrication and array assembly involve toxic substances. When a cell surface is *doped*, small amounts of impurities such as boron, zinc, and selenium are added to provide the excess or deficiency of electrons needed to make the cell a semiconductor. The substances used at this stage are highly toxic and are applied as fine sprays or vapors which are readily inhaled.

During the final stage in cell manufacture a fine metal conductive grid is applied which will collect electrical energy when the cell is used. This step may expose workers to heavy metal or organometallic vapors. The process may use other hazardous substances such as cyanides.

When the completed cells are assembled, soldered, embedded, and covered with a coating, po-

tentially toxic or carcinogenic chemical agents, including organic solvents, are used in some of the steps. Heavy metal vapors or particulates from soldering may be a particular problem.

Cadmium Sulfide

The toxicity and carcinogenicity of cadmium and cadmium sulfide is still uncertain. The industrial use of cadmium in electroplating, alloys, batteries, and paint, for example, has an unhappy health history. Chronic exposures of workers, largely through inhalation, and public health damage, through deposition on food crops or into rivers that supported commercial fisheries, are real but went unnoticed for decades. Cadmium metal and its oxide—fumes and dust—are now classified as acutely and chronically toxic materials. Inhalation of fumes, which is possible during refining or compounding, is most dangerous; inhalation of dust, and ingestion less so. Acute doses (fatal pulmonary edema) have occurred at exposures of 2500 mg/m^3 of air.

Chronic exposures to very low levels of cadium, lower than 100 $\mu g/m^3$, may cause emphysema and increase gastrointestinal disorders, kidney damage, and bone mineralization. Increased cancer incidence and hypertension also have been linked to low but extended exposure. The long residence time of cadmium in the body (half-life – 20 years) and the delay in appearance of health effects make the determination of dose–response relationships difficult.

Gallium Arsenide

Arsenic and its inorganic compounds are toxic and carcinogenic. Exposures at levels of 100 $\mu g/m^3$ have been reported to cause skin irritation, and higher levels lead to vomiting, nausea, diarrhea, inflammation and ulceration of mucuous membranes and skin, and kidney damage. The effects of chronic arsenic poisoning include increased pigmentation of the skin, dermatitis, muscular paralysis, visual disturbances, fatigue, loss of appetite, and cramps. Liver damage and jaundice may result, as well as kidney degeneration, edema, bone marrow injury, and nervous system disorders.

As more has been discovered about these effects, and as epidemiological data linking arsenic exposure to cancer incidence has been developed, regulatory standards for exposure have decreased precipitously. Occupational limits dropped from 500 $\mu g/m^3$ in 1978 and further reductions . . . are possible. There are as yet no public exposure standards. (Neff, 1981)

Orbiting Photovoltaics

One potentially large human health effect of an orbiting photovoltaic system (see p. 317) is the possible adverse impact of the 2.45 gigahertz (GHz) microwave beam that would transmit electrical energy from the orbiting cells to antennae on Earth. This beam may be hazardous to workers in the orbiting units, to workers at or near the 150 km^2 receivers on the ground, to members of the public in the vicinity of the receivers, and to aircraft passengers, birds, pollinating bees, and other insects flying through the beam.

Microwaves are not the only environmental hazard of satellite power systems. Occupational and safety problems in space will include cosmic rays, effects of weightlessness, and hazards associated with malfunctions of rockets and life support systems. In addition, placing 50 million kg of components per satellite into orbit would require about 200 flights of a large launch vehicle, the exhaust of which will pollute both trophosphere and atmosphere. Such a large program could change weather patterns of the trophosphere, deplete the ozone layer, and/or change the vertical structure in the stratosphere, with an unknown impact on the climate (Holdren et al., 1980).

Ocean Thermal Energy Conversion

Preventing the heat exchangers of OTEC systems from being fouled with algae and bacteria probably will require the use and discharge of considerable quantities of biocides. The manufacture and degradation of heat-exchanger materials will routinely emit metals, especially copper, nickel, and aluminum. Leaks of working fluids (including large leaks if a ship should collide with the plant) also could cause pollution problems.

Fuels from Biomass

Most dry forms of biomass can be burned directly to produce heat, steam, or electricity (see Chapters 13 and 16). Combusting biomass produces low to moderate levels of particulates and other pollutants, and in particular, combustion of fuel wood in wood-burning stoves may produce substantial indoor as well as regional air pollution. Some of these pollutants are carcinogenic. New wood stove catalysts seem to reduce wood smoke considerably and could presumably decrease pollution from burning wood if installed on a massive scale.

Another environmental problem is that the fire hazards of wood heat exceed those of gas, oil, and electricity, producing severe human health impacts and necessitating large energy investments to replace burned structures. Finally, lumbering is traditionally a very dangerous industry. Although we do not know of any specific estimate of accidents from cutting firewood, extrapolation of occupational hazards from the lumber industry indicates that there would probably be many accidents and deaths in an expanded fuelwood industry. Holdren et al. (1980) were not able to find information to derive meaningful rates of accidents per unit of energy derived from firewood, solar collectors, or windmills to allow comparisons with other energy production technologies.

COMPARISON OF TOTAL IMPACTS

The following excerpt is from Holdren et al., 1980:

In 1978 and 1979, widespread attention was drawn to a study by Inhaber of the Atomic Energy Control Board of Canada, which claimed to have tallied up consistently and comprehensively the occupational damages of building and operating five conventional energy systems (electricity from coal, oil, natural gas, nuclear light-water reactors, and hydropower) and six unconventional ones (electricity from solar thermal, photovoltaic, wind and ocean thermal systems, space heat from flat-plate collectors, and methanol from wood). Based on sums of his estimates of occupational and public damages, Inhaber concluded that the "health risks" associated with renewables are much greater than those of natural gas and nuclear power and in many cases comparable to those of coal and oil. Scrutiny of the work by a variety of reviewers . . . revealed, however, that these conclusions rested on gross inconsistencies in the treatment of the conventional and unconventional technologies, serious misreadings of literature, multiple calculational errors, and a variety of conceptual confusions. In late 1979, the Atomic Energy Control Board cancelled a contemplated revision of the report and declared all earlier versions out of print. With repeated emphasis on the care with which such figures must be interpreted, we present in [Table 17.9] our own estimates of occupational health effects of renewable energy sources.

Table 17.9.   Estimate of Occupational Health Risks from Various Energy Transformation Technologies

Energy System	Employment[a] $(10^3 WYr/10^{18}J)$	Fatalities $(D/10^{18}J)$	Fatalities $(D/10^3 WYr)$	Lost Work Days[b] $(10^3 WDL/10^{18}J)$	Lost Work Days[b] $(WDL/10^3 WYr)$
Heat					
Solar flat plate collectors	56–310	16–86	0.28–0.29	41–220	710–730
Chemical Fuel					
Fuels from biomass	5.6–61	1.3–15	0.23–0.25	3.2–41	570–670
Snyfuels from coal	20–30	17–25	0.83–0.85	28–42	1400
Refined oil products	3.5–3.8	1.1–1.2	0.31–0.32	1.8–2.1	510–550
Electricity					
Solar thermal electric	65–110	15–24	0.22–0.23	44–95	680–860
Solar photovoltaic	44–120	11–21	0.18–0.25	39–90	750–890
Wind	41–60	8.8–14	0.21–0.23	26–43	630–720
Conventional coal	42	35	0.83	57–58	1400
Nuclear LWR	18	4.7	0.26	13–14	720–780

Source: From J. Holdren et al., reproduced, with permission, from the *Annual Review of Energy* Volume 5. © 1980 by Annual Reviews Inc.

[a] WYr = Worker years.
[b] WDL = Worker days lost.

Because the renewable energy systems are more materials- and labor-intensive, the occupational effects generated by their construction will likely be higher than those experienced for conventional technologies [*but*] our calculations suggest that [only in the most extreme and unlikely cases will] the total occupational effects for all stages of the renewable energy system fuel cycles conceivably . . . be equal to the total occupational effects of obtaining an equivalent amount of energy from coal. [Authors note: See Inhaber (1985) for a rebuttal by Inhaber and some very interesting correspondence.]

In conclusion, the energy technology best for human health requires us to choose the lesser of many evils: *all* energy technologies involve at least *some* substantial risk to human health. The risks associated with coal and oil are fairly well understood, except perhaps for possible future climatological changes from CO_2 that could affect the health and well-being of billions of people by changing rainfall distribution and agricultural production. Nuclear power appears to have less day-to-day impact but has the potential for very large-scale impacts on human health.

SUMMARY

Clearly more studies need to be done regarding the hazards of energy technologies for humans. Yet we know enough to set guidelines for the future, and we know enough to recognize that there are no simple solutions that would eliminate major health impacts. A recent summary (Travis and Etnier, 1983) found major health impacts from all energy technologies and concluded that occupational impacts per unit energy were maximum for coal power plants, less for unconventional technologies, and least for nuclear power. They suggest that, although nuclear power has not yet killed directly a member of the public, any new technology is generally thought less safe than well-known technologies. And of course genuine uncertainties about potential large accidents still remain. Thus the public perception of a technology may be just as important as actual data on that technology's performance.

The health impacts of energy production could be decreased by tightening control immediately on fossil fuel emissions, by restricting the volume of nitrous oxides, sulfates, and particulates emanating from utility stacks, and by maintaining stricter operating standards and developing waste disposal policies for nuclear plants. Satisfactory means to dispose of wastes definitely need to be developed. But substantial human health impacts associated with any energy production will never be eliminated, no matter what technology is developed, if for no other reason than the provision of energy is a large, extensive business that employs many people.

EDITORIAL

TRADE-OFFS AND ECONOMIC OBJECTIVES

The preceding chapters have shown that all energy technologies have important environmental impacts. As a consequence, virtually all economic activities have environmental impacts associated with their energy costs. The types of fuels used may differ however, as do their environmental effects. For many reasons petroleum, especially natural gas, is relatively benign because smaller amounts of CO_2 and sulfur are released per kilocalorie burned compared to most alternatives (except hydropower and nuclear) and because extracting petroleum has a smaller impact on local environments than most alternatives. In terms of the criteria given in Chapters 15 and 16 as well as *known* human health impacts (Table 17.9), the *present* generation of nuclear power probably produces the least disruption per kilocalorie. Yet this point is debatable, for it ignores the potentially catastrophic results of a major nuclear accident and potential hazards associated with the absence of satisfactory technologies for long-term nuclear waste disposal. On a global scale some have argued that the possible impacts of disruption caused by increased CO_2 concentrations in the atmosphere from fossil fuels could be greater than even a large nuclear accident. A dependence on petroleum produces less immediate impact than coal or perhaps nuclear, but possible environmental impacts of large scale military activity related to insuring supplies could be very large. Unfortunately science currently is unable to weigh unambiguously the total environmental disruption of one energy system compared to another, especially when the probabilities of large disasters are unknown.

Large environmental problems are not new to the Earth; large, even catastrophic, environmental changes appear to have occurred in the past and will undoubtedly occur in the future. Some have been slow, like the onset of an ice age. Others have been more rapid, the most spectacular being the hypothesis that on at least one occasion the Earth was hit by a giant meteor whose debris blocked sunlight for years. This caused the massive extinction of dinosaurs and many other species as seen in the fossil record, such as at the end of the Cretaceous geological period (Alverez and Alverez, 1980). An alternative agent for the same environmental disruption has been postulated to be a very large volcano (see Officer and Drake 1983). Clearly today the only human environmental disruption that would be as catastrophic for all life would be an all out nuclear war between the super powers (Toon et al., 1983). There are, however, legitimate concerns that future growth in energy consumption (e.g., Figure 16.3) could diminish human welfare substantially while causing extinction of many other species. In the meantime day-to-day use of energy kills untold millions of organisms, causes widespread, generally deleterious, changes in the atmosphere and water, and more or less permanently disrupts large land areas. Most of the costs are not internalized in the price of energy, but they are real, important, and affect our economic, social and moral well-being.

ENVIRONMENTAL TRADE-OFFS

One problem facing even the best-intentioned environmental plans is that procedures designed to mitigate one environmental problem often cause other environmental impacts that may or may not be as severe as the problem they originally were designed to eliminate. Pollution abatement technologies are often energy intensive, so either directly or indirectly they diminish fuel reserves and increase the environmental impacts associated with any energy use. The next section considers several examples of this general problem.

Coastal Power Plants

As developed in Chapter 16, electricity generation normally requires large quantities of water for cooling. A number of states responded to concerns about thermal pollution in the 1970s by passing laws that specified that the effluent water from power plants could not be heated more than a given number of degrees above the ambient temperature. To satisfy these laws, engineers increased the quantity of water cycled through the cooling loops, so that the heat load was spread out over a greater quantity of water. As a result the number of organisms entrained increased, which probably caused greater damage then the original thermal pollution.

An alternative plan designed to reduce thermal loading and entrainment is to construct cooling towers. These towers often are found, paradoxically, adjacent to large bodies of water potentially useful for cooling, since the legislation requiring towers was passed *after* plants were sited on large water bodies that originally were to be used for cooling. These cooling towers have a considerably greater energy cost than once-through cooling and therefore increase the environmental impacts associated with fuel use and material extraction and refining.

It is difficult to compare the environmental impacts of the once-through cooling to those of cooling towers, for the impacts are both qualitatively and quantitatively different and tend to affect different species. One rather remarkable effort to compare their total impact was done by Kemp et al. (1977, 1979) who compared the total environmental impacts of once-through cooling with those arising from the use of cooling towers for the Crystal River power plant complex in Florida. These investigators also conducted detailed studies of the trophic relations of the organisms in the environment affected by the power plants so that each species could be assigned a *quality rating* based on their place in the food chain (Figure IV.1)—that is the loss of a kilocalorie of top carnivore was weighted more heavily than the loss of a kilocalorie of herbivore or plant. Assessments also were made of the impact of changes in physical energy flows such as water currents. The results indicated that contrary to the original assumptions of many environmentalists, the total environmental impact of constructing and operating the cooling towers was greater than that of operating once through cooling systems, mostly because of the energy and other resources used to build and operate the enormous cooling towers. Although that conclusion is in part a function of the many assumptions that the authors used to quantify impacts, the large difference in the final results probably makes such assumptions un-

important. Although cooling towers were ultimately built at Crystal River because of political considerations, they were avoided in a similar case for the Hudson River by turning off different power plants up and down the river when their impact on the fish was thought to be greatest (see Hall, 1977; *New York Times*, December 20, 1980, p. 1). In this case an enormous amount of energy was saved, and environmental impacts were decreased.

There have been other instances of corrective measures designed to protect the environment that have had negative or ambivalent results. For example, tall smoke stacks are designed to clean up local pollution, which they do. But they also allow pollutants to remain in the air for longer periods of time, turning serious local problems into serious regional and even international problems (Likens, 1974). Precipitators that remove particles from smoke stacks appear to exacerbate the problem of acid rain, since the particulates formerly left in stack gases combine with acid radicals in the smoke plumes to produce neutral salts rather than sulfur oxides (MacIntire and Young, 1973; Likens, 1974). Finally, strict guidelines for strip mine reclamation that require restoring the land to the original contours necessarily limit the possibilities for creatively recontouring the lands for wildlife management (Harold Bergman, personal communication).

All of this is not to suggest that environmental regulations are always, or even often, counterproductive. Strong, inflexible laws are sometimes needed to avoid unscrupulous consequences. These examples do suggest, however, that a comprehensive and flexible systems approach to both energy and environmental problems sometimes can be very useful, and that good energy and good environmental solutions need not be mutually exclusive.

IS THE OBJECTIVE OF ECONOMIC GROWTH DESIRABLE?

If large-scale economic growth continues to be a major political and social goal, major new fuel sources will have to be developed both to offset expected shortfalls of domestic petroleum and to power additional economic production, if that is possible. Domestic fossil energy reserves of coal and, perhaps, oil shale could provide the fuel required for our economy to grow for at least several decades—if we as a society are willing to divert an ever larger proportion of our fuel budget from con-

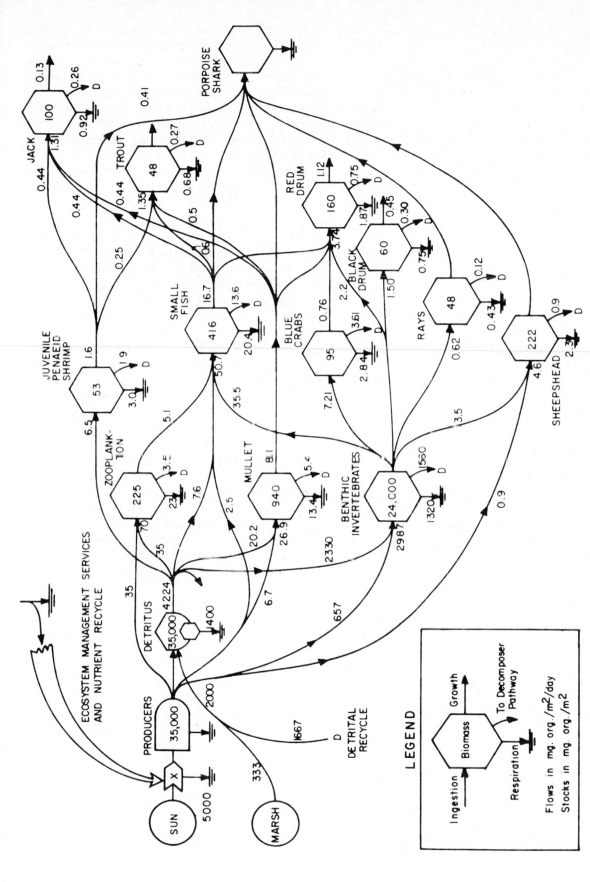

Figure IV.1. A diagrammatic representation of the food web for the Crystal River estuary illustrating the convergence of different trophic pathways to larger complex animals that have "management" roles in the ecosystem. This diagram is based on extensive quantitative field sampling and stomach analysis by Mark Homer which was used to calculate energy quality indexes for organisms lost from entrainment and impingement. (From *Ecosystem Modeling in Theory and Practice*, C. Hall and J. Day (eds.). Reprinted by permission of John Wiley & Sons.)

433

sumption to the production of new fuel resources, and if social factors (availability of trained workers, lack of widespread environmental opposition to expanded coal use, miners strikes, etc) allow the physically available resources to be utilized. If we do continue to use domestic fuels at present or increased rates, the environment will suffer, as will human health, for the fuels remaining that are abundant (coal, oil shale, and nuclear-with-breeders) exact a greater environmental and health cost per kilocalorie than the relatively clean liquid and gaseous petroleum and fission power that they would replace.

The Objectives of Growth

Compared to many other nations, the *majority* of the people living in the United States are reasonably well fed and well housed. In a sense much of the economic growth that has occurred in the past several decades, and that may occur in the future, is for luxury items, or for once-luxury items that have now become necessities such as new automobiles, homeowning, consumer gadgets and expensive vacations. Ironically, at least in the eyes of some social observers (e.g., Packard, 1959, 1960), much of the desire to accumulate wealth is not for the purpose of achieving the wealth *per se* but rather as a means of buying status, of seeking a ready metric of success, or as a sublimation of other desires. How much of the great depletion of American petroleum reserves and the environmental degradation that occurred during the 1950s, '60s, and '70s was simply buying unneeded and even unwanted goods and services simply to purchase status in the eyes of others who may not even have cared? How much of our future desecration of the American West's natural and cultural environment will be for the same purpose? How many people think about the fact that for most purchases each dollar spent depletes our petroleum reserves by more than 5000 kcal while producing, somewhere, about a kilogram of various waste products?

A second disquieting social observation is from the economist Galbraith (1958) who argues that much of the demand that exists for goods and services is artificial in that it would not exist without advertising. A depressing statistic is that roughly half of the psychology majors of many of our major universities go to work for advertising agencies in order to help those agencies figure out how to get the rest of us to buy something we otherwise wouldn't, and that we would be perfectly happy without. Presumably without advertising our demand for goods and services would be less, leading to less of the accompanying pollution and resource depletion. But, would not that lead to less economic activity, less jobs, and so on? Probably, but one solution, suggested by Bertrand Russell, is to cultivate leisure activities that are not resource intensive, so that we all work fewer hours, earn less, consume less, and enjoy life more. The frenetic general commercialization and drive for economic success is to a large degree an oddity of American life, one that seems peculiar when returning from many other cultures. Is our total life more enjoyable for our headlong rush after affluence? That is hard to answer, but it is clear that much of the pollution and depletion that has occurred has not been for especially good reasons.

PART FIVE

ENERGY AND THE MANAGEMENT OF RENEWABLE NATURAL RESOURCES

Natural resources are the raw stuff from which human wealth ultimately derives following, as we have pointed out, the proper application of energy and the technology to use that energy. Renewable resources, such as fisheries, forests, and rangelands, might at first not seem to have a nonrenewable component. The present scale of exploitation, however, and the technological approaches we use, makes present-day exploitation of most renewable resources very energy and material intensive. Additionally many management practices are destructive to the resource base itself so that many renewable resources are in effect mined.

The exploitation of natural resources can be done with greater or lesser efficiency and effectiveness, however, and this most certainly applies to the ways that we use energy to exploit other resources. The following chapters discuss the nature of several resource bases, various schemes for management that have been used, or proposed, to manage those resources, and the interaction of these management procedures with energy use. We believe that as high quality fossil fuel resources are increasingly depleted, it will become necessary to reconsider many traditional energy-intensive management procedures. More important, we believe that it is essential to begin the process of understanding the various ways that cheap petroleum has allowed us to mismanage various resources without fully realizing the consequences.

The following chapters examine energy use in various renewable resource industries in the United States. The different chapters also provide different approaches and philosophies related to resource problems, reflecting in part the different interests and training of their authors. For example, Chapter 18 emphasizes interactions of theory and management, Chapter 19 provides a very data-intensive analysis of time trends, and Chapter 21 focuses on needed changes in policy. Collectively we believe they provide a thorough perspective on the fascinating field of natural resource management.

18

FISHERIES

Fisheries refers to the fish, the people who fish for a living, their boats, canneries, markets, and so forth, and all other agents and products associated with the catching, processing, and marketing of fish and shellfish. Modern fisheries differ from modern agriculture and animal husbandry in that, for the most part, they rely on the natural environment, more or less unmodified by people, for production. Nevertheless, fishing, like agriculture, is a very energy-intensive process. Some fisheries appear to use considerably less energy per kilogram of protein produced than is the case for raising livestock, whereas other fisheries are much more energy intensive (see Figure 6.11). The reasons for the wide range in the energy cost per unit of fish caught are numerous and are related to the productivity of the natural environment from which the fish are caught, the distance of the fishing grounds from ports or markets, and, of particular interest here, the intensity of fishing pressure and the philosophies and effectiveness of the policies with which we manage fish. We consider a high quality fisheries to be, as usual, one in which the economic energy required to catch a kg of fish is relatively low. One thing of particular importance is that very often the energy intensity of the fisheries depends to a large degree on the management strategy. This chapter reviews some of the historical and other characteristics of fisheries and some of the management concepts that have evolved over time, and then compares a number of them with field data on fish populations. Finally we suggest a few ways by which energy analysis can and cannot be used to improve the effectiveness of fisheries management.

In earlier times much fishing pressure was directed toward *anadromous* fish, fish that lived in the sea but spawned in fresh water. This allowed the relatively large productive resources of the ocean to be harvested with a relatively small investment of human energy, since the fish were concentrated periodically in river mouths. For most of the history of fishing harvests were based almost entirely on the use of human energy resources and simple raw materials. About three millennia ago wind energy was harnessed to power fishing boats, allowing fisheries to exploit more distant fish stocks. Modern fossil fuel-intensive fisheries in Europe evolved from the steam tugs that once pulled wind-powered fishing boats out of North Sea ports during calm weather. It did not take the people long to realize that more reliable fishing could be obtained by eliminating the middleperson, so to speak—that is, the sailboat. Similarly, in the United States it was found in the latter part of the last century that far more menhaden could be caught with fossil fuel powerboats than in sailboats. From these relatively humble beginnings evolved today's modern fishing fleet, which is very capital and energy intensive. Commercial fishing is normally done with nets, dredges, or traps of some kind or with long strings of baited fishhooks (Figure 18.1). Fossil energy is used to build gear and boats, travel to and from fishing grounds, tow nets, and preserve the catch. Many high seas fishing boats weigh about the same when they leave port as when they return, the weight of fuel used being roughly equal to the weight of fish brought back.

HOW MANY FISH CAN WE CATCH?

There is a myth that more or less still remains as a romantic legacy of the ocean—that fisheries will feed the world's starving billions. In reality most of the oceans are a virtual desert, devoid of sufficient plant nutrients, primary production, or fish to make exploitation worthwhile. Some relatively small regions of the ocean are more productive, and these areas have been exploited heavily for decades or centuries, but the protein production of even the richest of these areas is low compared to what can be obtained from the land.

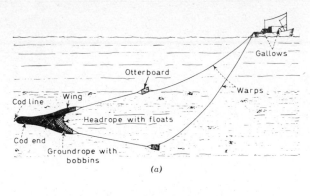

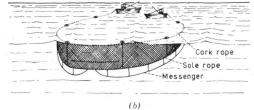

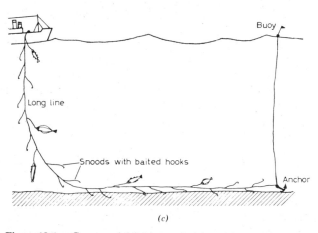

Figure 18.1. Commercial fishing techniques. (*a*) otter trawl, (*b*) purse seine, and (*c*) long line fishing. (From Tait, 1968, reproduced with permission of Plenum Publishing Corp.)

The basic reason that most of the world's oceans are a biotic desert is their relatively low primary production (i.e., rate of fixation of energy and carbon by plants). Plants need two principal resources to grow: sunlight for energy and mineral nutrients to synthesize the necessary chemicals of life. Plants also need carbon for their basic cellular structure, but that is generally abundant in the sea. The quirk is that, for the most part, the nutrients are where the light is not, and vice versa, because the vast bulk of the ocean is thermally stratified, meaning that, in general, where sunlight is intense there is a sharp temperature differential between surface and bottom waters. Because of the differences in density

warm surface waters do not mix with cold deep waters, as illustrated recently by the finding that radioactive debris from early 1960s nuclear bomb testing penetrated no deeper than the surface 300–500 m of the world's oceans even more than a decade later (Ostlund et al., 1976). There is a constant rain of material from surface waters to deep waters via biotic debris, most notably the fecal pellets of the small crustaceans that are the most important grazers of the world's seas (Figure 18.2). This downward flux of material constantly drains the surface waters of phosphorus, nitrogen, silica, and other critical nutrient elements. As a result these nutrients are removed from sunlit waters, thereby restricting plant growth. There are some regions of the ocean, however, where nutrients are resupplied by various processes (Figure 18.3). These regions tend to be where important fisheries are located (Figure 18.4). For example, productive fisheries are often found in temperate and boreal oceans North and South of about 30°–40° latitude. Here surface waters cool sufficiently so that the water is mixed all the way to the bottom during winter, and the nutrients lost during the summer are resupplied to the surface waters.

Photosynthesis, and hence fish production, also tends to be high in coastal regions. Nutrients are supplied to coastal regions by river flows, by river-driven inmoving deep salt waters (see Day et al., in press), and by wind-driven coastal upwelling processes, which occur along the west coasts of all continents. In addition coastal regions often are sufficiently shallow so that nutrients that fall to the bottom are regenerated (mineralized) by biotic activity rather than lost to the depths (Nixon, 1981).

Although most of the world's fish catch comes from salt water, the life cycles of many of these salt water fish are tightly linked to estuarine or nearshore ecosystems (e.g., see Smith, 1966; Stroud, 1967; Day et al., in press). Estuaries are highly productive regions due to the river and tidal energies that continually resupply nutrients to the euphotic (well-lit) zone. Fisheries also exist in natural fresh waters, although (with the major exception of mainland China) freshwater fisheries are not important on a global scale. In managed fresh- or saltwater fish ponds very high yields are possible through the intensive use of fertilizers and human labor.

Thus we view commercial fishing as the energy-intensive exploitation of relatively small parts of the ocean's surface, where fish tend to be concentrated at hundreds to thousands of times the density of the

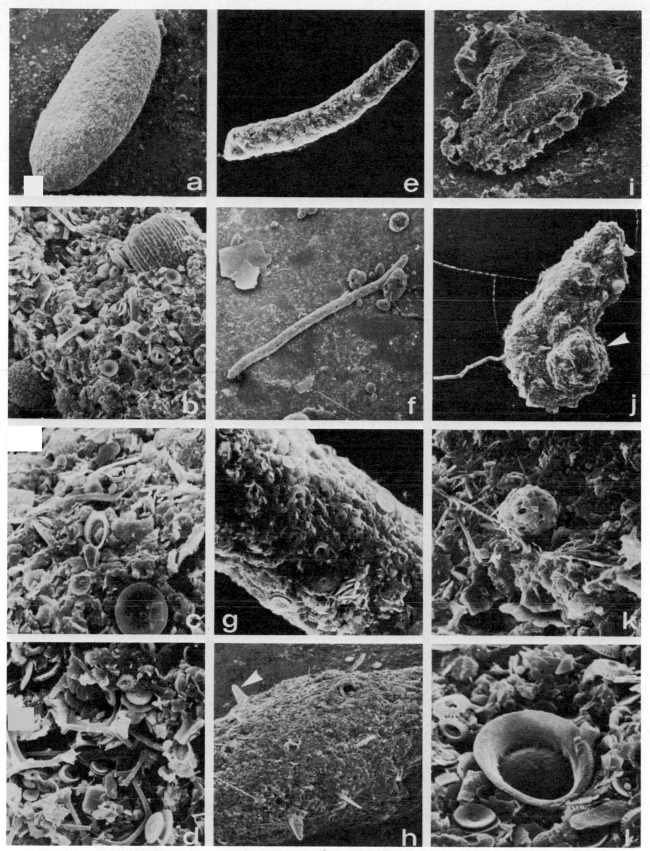

Figure 18.2. Zooplankton fecal pellets. (Honjo, personal communication.)

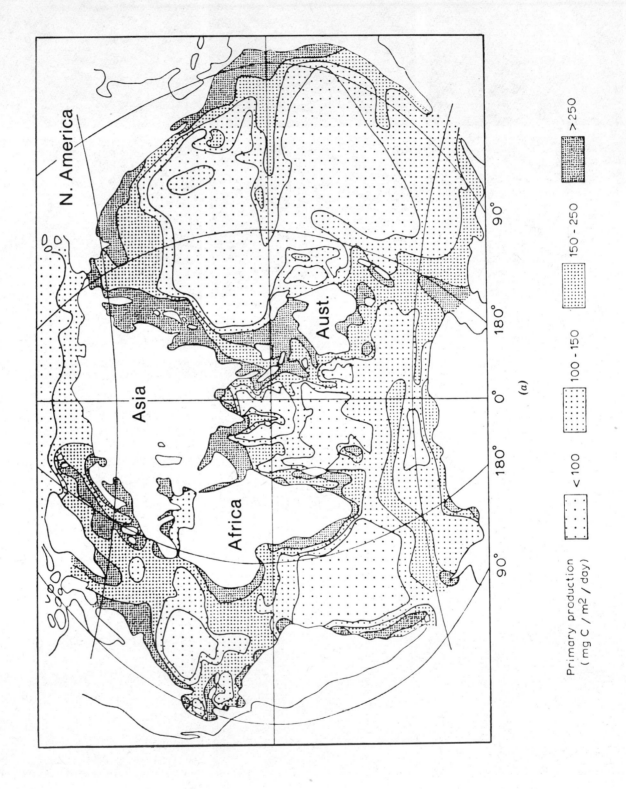

Primary production
(mg C / m² / day)

< 100   100 - 150   150 - 250   > 250

(a)

440

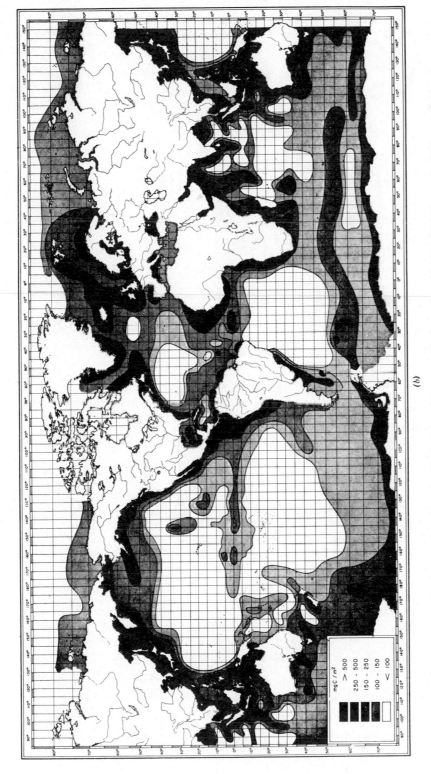

Figure 18.3.(*a*) Distribution of primary production in the world oceans. (Redrawn from Koblentz-Mishke et al., 1970, and reprinted with permission from Parsons and Takahashi, *Biological Oceanographic Processes*, copyright 1973, Pergamon Press, Ltd.) (*b*) Distribution of zooplankton in the world oceans (From Unesco, no date).

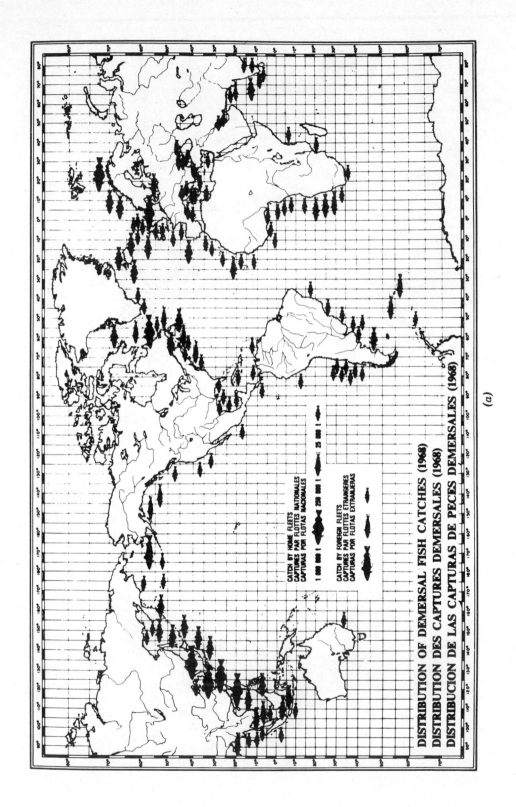

DISTRIBUTION OF DEMERSAL FISH CATCHES (1968)
DISTRIBUTION DES CAPTURES DEMERSALES (1968)
DISTRIBUCION DE LAS CAPTURAS DE PECES DEMERSALES (1968)

(a)

442

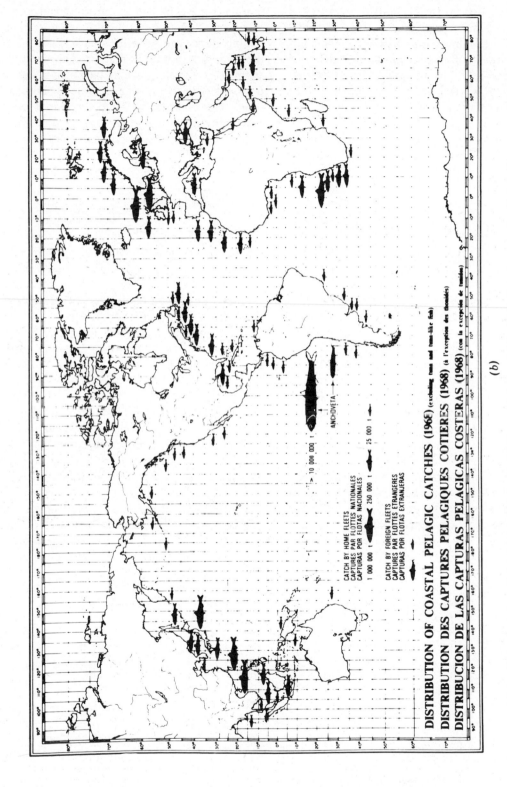

DISTRIBUTION OF COASTAL PELAGIC CATCHES (1968) (excluding tuna and tuna-like fish)

DISTRIBUTION DES CAPTURES PELAGIQUES COTIERES (1968) (à l'exception des thonidés)

DISTRIBUCION DE LAS CAPTURAS PELAGICAS COSTERAS (1968) (con la excepción de túnidos)

(b)

Figure 18.4. Location of global oceanic fisheries (a) Demersal (bottom) fish, (b) Relagic (water column) fish. (From UNESCO no date.)

ocean as a whole. For example, a trawl can catch hundreds of tons per hour in the Northeast Atlantic (Cushing, 1975). But a trawl hauled in tropical oceans might catch only 60 kg of fish per hour, and a typical catch for longline tuna fishing in the mid-Pacific is only one to five fish per hundred hooks (spread out over 5 to 10 km) in several day's fishing. Thus it is quite inappropriate to extrapolate catch rates of our best fishing areas to the ocean as a whole.

Ryther (1969) was among the first to derive clearly a theoretical limit to the sustainable quantity of fish obtainable from the sea. He calculated the quantity of organic material fixed annually by the plants of the ocean (which had been estimated in the recent past using the then-new ^{14}C method) and estimated the efficiency with which this organic material was converted to fish flesh. The results of his calculations (Table 18.1) indicated that the 1960s world *catch* of sea fish (about 60 million tons) was not much lower than his estimated annual *production* of sea fish (about 100 million tons), and suggested that it was unlikely that we could increase the world's catch of sea fish by very much.

Other work (e.g., Alverson et al., 1970; Gieskes et al., 1979) indicated that the radioactive carbon (^{14}C) method used as a basis for Ryther's calculations may underestimate total oceanic plant production, perhaps by as much as a factor of 10, and that other reasonable estimates of food chain efficiencies could be used. As a result modern revisions of Ryther's estimate are somewhat higher. Independently Gulland (1972) estimated a potential yield of about 100 million tons per year, based on several independent analyses of individual fish species. This is much less than some earlier estimates but still greater than the actual catch. The reality of the

situation, however, is that the world fish catch has not increased a great deal recently despite a very large increase in fishing effort (Figure 18.5). And our high technology, high fossil-energy-intensive fishing fleets are so effective at catching fish that most of our traditional major fisheries are either severely overfished or are in danger of being so (Cushing, 1982). Most contemporary fisheries experts (e.g., Hennemuth, 1979) believe that we are approaching the limits of the ocean to provide people with fishes, with the possible exception of Antarctic Krill.

ENERGY AND FISHING

One factor limiting the increase in the world's fish catch is that most new fishing grounds are more energy intensive to exploit than traditional areas, so that expanding our fisheries often requires as much or more energy per kilogram of protein produced than most land-based animal systems and much more energy than protein from plants (Table 18.2). Thus present capital-intensive fishing methods are not an especially efficient way of producing human food from fossil fuels. Another factor increasing the energy cost of fish is that many industrial nations, such as Japan, East Germany, and the Soviet Union, send marine fleets thousands of kilometers to exploit the fish off other countries' shores. Finally fish are energy intensive to catch because we often mismanage fisheries. In short, the appropriate question about the future size of the world's fish catch may not be How many fish can we catch? but rather Do we have the petroleum to catch a given amount of fish, and would it be an efficient use of that petroleum? One might argue that consumers, in

Table 18.1.   Estimates of Primary Productivity and Potential Fish Yield for the World's Oceans (from Ryther, 1967)

Province	Primary Production of Organic Carbon (tons)	Trophic Levels	Efficiency (%)	Fresh Fish Production (tons)
Oceanic	16.3×10^9	5	10	16×10^5
Coastal	3.6×10^9	3	15	12×10^7
Upwelling	0.1×10^9	1.5	20	12×10^7
Total				24×10^7

Source: From J. H. Ryther, *Science* Vol. 166, pp. 72–76, Table 3, 1969. Copyright 1969 by the AAAS.

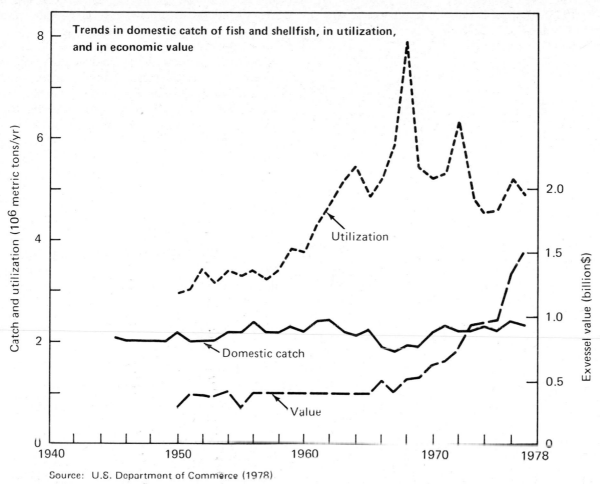

Figure 18.5.   Trends in catch of marine fish and in catch per capita for the U.S. (From Brown and Lugo, 1981.)

their market decisions, in large part determine whether fish are efficient in that sense. But such decisions may not be based on complete information, and market pressures can operate to make fisheries less energy efficient than they otherwise would be.

It is particularly interesting to ask why fisheries use energy inefficiently due to poor management. Unfortunately managing most fisheries effectively is difficult because (1) natural populations are inherently difficult to understand or predict, (2) the theories of management most often used in the past were too often based on ecological theories that were not applicable for real fish populations and that were not adequately tested in the field anywhere, (3) even when effective management schemes are available, they often are at variance with pressure from those looking for short-term economic gains, or they may require data that are

difficult or expensive to obtain, (4) there is no clear agreement as to what constitutes the optimal management goal, (5) even when a given plan for management is agreed to be in the best interests of managers, fish, and fisheries, political implementation may be difficult or costs of implementation large. Therefore a clear understanding of how energy is used to manage resources such as fish requires a careful consideration of the concept of management and the social limitations of applying good management.

Where fish are abundant, as in coastal and upwelling regions, the most desirable fish species are exploited very heavily so that, on average, an adult fish is about as likely to be caught in a fish net as to die of old age or predation. This is especially true for the productive fisheries grounds that are near human population concentrations, such as the fish of the North Sea and those off the coast of Massa-

Table 18.2. Ratios of Fossil Fuel Energy Inputs Per Unit of Protein Energy of Protein-Producing Systems

Protein-Producing System	Fossil Fuel Input to Protein Energy Output (kcal/kcal)
Domestic fishing industry (1975) (excluding boats under 5 GRT)	14
Domestic fishing industry (1975) (including an estimate for boats under 5 GRT)	37
Florida fishing industry (1975)	49
Feedlot beef (1974)	20–44
Feedlot beef (before 1974)	78
Rangeland beef	10
Broiler chicken	22
Eggs	13
Milk	36
Vegetable crops	2–4
Grain crops	3–4

Fishery	Year	
Northeast fishery trawlers	1974	4.1
Washington-otter trawl	1971–72	6.1
Oregon groundfish-otter trawl	1977	6.8
Oregon salmon-purse seine	1977	19.0
Oregon salmon-gillnet	1977	18.0
Oregon shrimp and crab	1977	3.0
Oregon shrimp	1977	3.6
Florida Gulf of Mexico shrimp	1975	198
Florida Spanish Mackerel	1975	7.8
Florida snapper-grouper	1975	20
Florida commercial fishery	1975	48.6

Source: From Brown and Lugo (1981).

chusetts. Often as new fisheries are developed, such as for the Pacific perch or the international fishing for West Atlantic herring, the catch for the total fishery increases dramatically as successful fishing techniques and new markets are developed (Figure 18.6). Very often, however, the fishery eventually will decline severely or even collapse, in many cases due to severe reduction of the spawning stock as a result of intense fishing pressure.

Interestingly, total pounds of fish caught each year by U.S. fisheries remained about constant from 1950 to 1978 despite wild fluctuations in land-ings of individual species and despite an estimated three- to fourfold increase in the quantity of economic energy used for fishing. Thus the catch per effort (and the food energy returned per unit of fossil energy invested) has declined considerably (see Figure 18.13). The basic problem is that it often makes sense for an individual to enter the fishery, since he or she often can make a profit even when the catch for the fisheries as a whole is not increased by the additional effort. Thus two chronic problems in fisheries, and in resource management in general, are overcapitalization and overharvesting—especially with respect to the most desired species—both of which reduce the yield per effort while in many cases actually reducing total harvest. There does not seem to be a solution to these problems within the workings of conventional unregulated markets, and the problem has become larger as the use of more and more energy has increased our ability to exploit living resources.

THE CONCEPT OF MANAGEMENT

The concept of management—that is, human intervention in the production or exploitation of natural living resources—is an ancient one, practiced by nearly all civilizations, but especially the ancient Chinese. It acquired particular status with the silviculturists of Middle European forests during recent centuries. Management was relatively unimportant in the United States until recently, for resources were abundant and could always be found anew a bit to the West. This frontier philosophy remains an important part of our national consciousness and manifests itself today in a deep suspicion by many people of government intervention in many areas, including resource management.

Fisheries management can mean many things, including fish culture in hatcheries or artificial ponds, removal of predatory or competing fishes, and, especially, regulation of catch through various limits on time spent fishing, quantities of fish kept, or gear used. The need for management of fish was at first a hotly contested issue, in both this country and Great Britain. Nielsen (1976) documents some of the controversy that surrounded the development of regulation and management in the 19th century. A particularly lively debate in Britain initially resulted in the abolishment of all laws restricting fishing, based on the concept that natural mortality was much larger than fishing mortality. This view was reversed, however, at a formal debate held at an International Fisheries exhibition in London in 1883, where fisheries scientists who believed that fish were, in some sense, exhaustible and who favored regulation won the debate and set the stage for later regulation of the industry.

The U.S. federal government first recognized the need to manage fisheries when a number of fisheries scientists, among them Spencer Baird, convinced Congress to establish the U.S. Fisheries Commission. Unfortunately, since records kept at that time on the catch of various fisheries were not very comprehensive or accurate, it is difficult to get good documentation on the perceived decline in fisheries noted at that time or even to judge whether things are much worse (or better) now.

Initially managers constructed fish hatcheries to supplement natural reproduction of stocks that were depleted by overfishing. It soon became apparent, however, that hatcheries had little effect, especially for marine fish, or (more frequently), that it was impossible to determine whether a given fish even came from a hatchery. Managers then sought increasingly to regulate effort, especially by limiting the length of fishing seasons, in an attempt to conserve and rebuild supposedly depleted fish stocks.

One of the most commonly heard comments about fisheries is that such and such a body of water, or species, or whatever, is overfished. Although this often means that *others* are catching too many fish, the regulation of fish mortality caused by fishing remains an essential component of modern fisheries management philosophy—that is, it is desirable to regulate the amount of fish caught so that enough remain for breeding and for efficient fishing. As we shall see, modern fisheries management has some rather interesting variations on that theme; there are some quite sophisticated procedures for ensuring (hopefully) that various species are not overfished. Nevertheless, despite years of fisheries research and fisheries management many, perhaps most, U.S. commercial fish stocks are chronically overfished (Brown and Lugo, 1980). In such cases fishing pressure itself decreases resource availability by decreasing fishing success and increasing the energy cost of catching a fish.

The Evolution of Mathematical Procedures for Estimating Desirable Levels of Catch

An important concept in the development of a theoretical rationale for management is *catch per unit effort* (CPUE)—that is, how many fish are caught per person per hour, or per some other unit of effort. Although the concept was familiar (certainly it is the most familiar component to an individual who fishes for a living), the cumulative effect of fishing effort was not generally appreciated until it became obvious that the catch per effort for North Sea fishes had increased following the lull in fishing activities during World War I (Figure 18.7). This was the first clear indication that reducing fishing effort could increase yield, and that fishing effort affected fish populations.

Although some important fisheries management concepts were developed in Russia in the 1930s, (e.g., Baranov, 1918; Borisov, 1960), the first formalized equation for managing fish populations was developed by Russell (1942). Russell summarized changes in stock numbers with the following equation:

$$P_2 = P_1 + (R + G) - (M + F)$$

where P_1 is the stock biomass in a given year, P_2 is the stock in the next year, G is the annual addition to the stock from growth, R is addition due to recruitment of a new age class, M is loss due to natural mortality, and F is losses due to fishing mortality. Although Russell's equation identified explicitly the factors that caused fish stocks to change, none of these parameters could be estimated at that time beyond already measurable fishing mortality, nor was that equation useful in estimating desirable and/or sustainable catch levels.

The only protection that heavily exploited fish had at that time was that at some point (often too late) it became uneconomical to fish for them. Theoretical research in fisheries over the last 50 years has sought a good rationale for restricting catch rates before that point and has attempted to convince those who fished for a living, legislators, and other investigators that only so many fish should be caught. Initially fisheries scientists had no rationale for determining what specific rate of exploitation would be optimal. Interestingly some experimental and theoretical developments within the emerging science of ecology provided a foundation for deriving the needed rules.

In general, this new approach applied mathematical analysis to fish population dynamics. Calculations were based on the most easily obtained information—the number of fish caught by the fisheries each year (these data were easy to get because the fishermen did all the sampling work)—and the approach assumed that catch per effort was directly proportional to the number of adult fish, which is

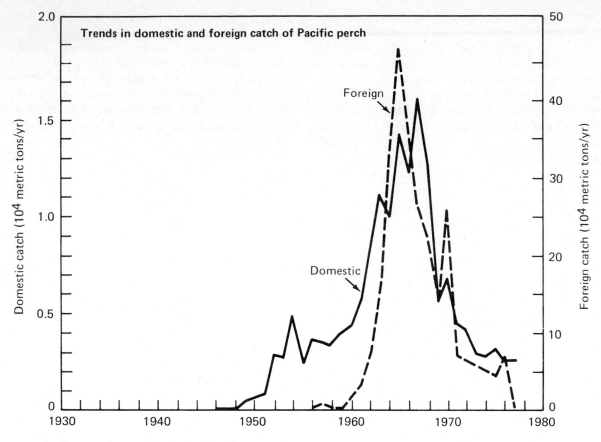

Sources: Prior to 1970, Wise (1974); after 1970, U.S. Department of Commerce (1950-1975, 1978).

(a)

Figure 18.6.   Some examples (there are many) of the development, overdevelopment and apparent destruction of a fishery. (Part *a* from Brown and Lugo, 1981; part *b* from D. H. Cushing, *Fisheries Biology: A Study in Population Dynamics,* © 1981 by The Board of Regents of The University of Wisconsin System; used by permission of the University of Wisconsin Press.)

known as *stock*. Mathematics were used because the problem had a quantitative component and because there was at that time the model of research in physics that showed how successful certain mathematical approaches could be in predicting scientific phenomena. Another reason was that although the behavior of biological organisms was, in general, very difficult to understand, the consequences of certain mathematical assumptions and relations were relatively easy to understand and predict. This mathematical approach was further encouraged by the remarkable early success of some equations for predicting population growth rates in certain laboratory studies.

In 1838 the Belgian mathematician Pierre-Francois Verhulst developed the *logistic curve* to describe population growth (Figure 18.8). The mathematics were simple enough so that they could be applied and understood by even those ecologists

without extensive formal training in mathematics, yet they were complex enough when later applied to the discipline of ecology to give a veneer of elegance to an academic discipline that for too long had been overshadowed by the elegant mathematical advances in other disciplines.

The concept is simple and is based on the logistic, or S-shaped, curve described on p. 118. Many laboratory experiments with fruit flies (reported in Pearl, 1925), *Daphnia* (Terao and Tanaka, 1928), yeast cells (Pearl, 1925), and flour beetles (Gause, 1934) showed that populations that live in a simple homogeneous environment, such as fruit paste (for fruit flies) or a bag of flour (flour beetles), would grow according to the logistic curve. Some field data also seemed to corroborate the lab studies. For example, deer in Utah (Davis and Golley, 1963) and bees in Italy (Bodenheimer, 1937) appeared to show patterns of population growth consistent with the

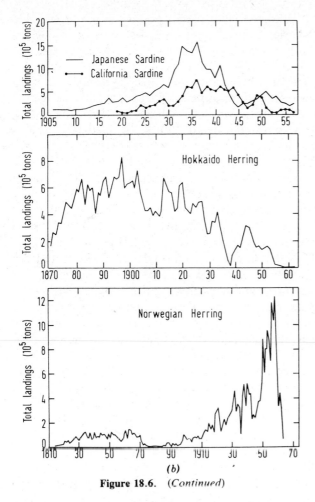

Figure 18.6. (Continued)

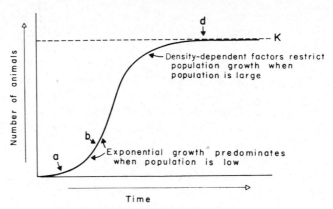

Figure 18.8. Logistic curve, a theory for population growth.

Schaefer (1957). In particular, Ricker developed very elegant and influential techniques using concepts of density compensation through reproduction and early survival to develop a theoretical relation between the number of adults (stock) and the number of young produced for the next generation (*recruitment*). This in turn gave impetus to a clever scheme for managing fisheries. Ricker assumed that year to year variations in spawning success due to density-independent factors (e.g., weather) were small compared to density-dependent effects, and he assumed that both stock and recruitment could be measured either directly or by the yield per unit of effort.

According to the theory behind the logistic equation and its correlates that form the basis for the Ricker approach a fish population would grow rapidly when it was well below its *carrying capacity* (Figure 18.8) but much less so as it approached its carrying capacity. As a result the number of recruits depends in large part on the number of spawning adults. When the stock was low, the total number

logistic curve. For a further description of the theory and the basic mathematics involved see Krebs (1978) or Hutchinson (1978).

The idea of density compensation and the logistic curve was applied to fisheries management by Ricker (1954), Beverton and Holt (1957), and

POPULATION DYNAMICS OF FISHES

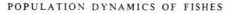

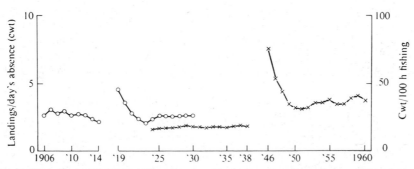

Figure 18.7. Catch of ground fish in the North Sea, showing the effect of a reduction in fishing effort during the world wars. Two different procedures were used for estimating abundance as indicated; CWT means hundreds of pounds. (From D. H. Cushing, *Marine Ecology and Fisheries,* 1975, reprinted with permission of Cambridge University Press.)

of young produced is small (since the number of spawners is small), but the per capita survival of the young would be high because density-dependent mortality is low (point *a* on Figure 18.8). As a result the populations grow. Larger populations produce more young, but the survival rate of each individual decreases due to an increased probability of mortality from density-dependent factors. At the stock's carrying capacity only as many young survive as are necessary to maintain a stable population over time (point *d* on Figure 18.8, assuming adults spawn only once in their lifetime). If the adult population exceeds the carrying capacity, density-dependent factors increase juvenile mortality and reduce recruitment, shrinking the population (Figure 18.8, point *e;* Figure 18.9).

One concept behind Ricker's approach to management was to maximize population growth by depressing the stock population to the point of maximum reproduction (Figure 18.8*b*) through fishing pressure—in a sense having your cake (increasing recruitment) while eating it too (catching the fish). The original data base used to derive this relation (British Columbia salmon) seemed consistent with the model, and the concept was used widely to regulate catch in many fisheries. Increasingly sophisticated corrections were made in the basic model to account for age structure, fishing mortality, and so on. Although we later criticize the Ricker curve approach to management, Ricker himself has been among the most persistent and intelligent critics of the uncritical application of his theory. Nevertheless, we are impressed at the large number of fisheries investigators and managers who used and still use, sometimes uncritically, some variant of the Ricker curve approach (e.g., Walters, 1975; Clark, 1985).

The Ricker relation was readily adapted by fisheries biologists who desired a rationale for analyzing and managing fish populations. The theory was logical, and it formed the basis for much of fish management over decades. And since actual recruitment was difficult to measure accurately in most fisheries, there was little opportunity to disprove the method.

Are the Logistic Curve and Ricker Analysis Accurate Representations of Nature?

Unfortunately the validity of the Ricker curve method for managing fish is not readily apparent. As Laurence (1981), a keynote speaker in a symposium on fish and fisheries, has said, "The ability to predict recruitment would be an invaluable fisheries management tool, but our knowledge is still unrefined. The traditional stock-recruitment principle has not been validated and is likely incorrect for most situations." Although the theoretical base and the empirical validation of the Ricker model are perhaps of dubious validity, the theory was extremely useful politically in that it gave a rationale for preventing excessive overfishing, perhaps preventing the collapse of certain fisheries.

Why shouldn't the Ricker equation work, if the underlying principles make such apparent sense? There are at least three reasons. First, the theoretical base (the logistic equation) does not describe the biology of most real populations. Although the logistic curve does seem to describe adequately the growth of some populations in the laboratory or perhaps even new environments, it has little relevance for most populations that have been existing in a region for long periods of time or where the carrying capacity varies widely over time. Second, each fish species competes for food and spatial resources with other species. If one species is depressed by intensive fishing pressure, interspecific competition may prevent remaining members of the first species of interest to find the resources supposedly freed. In other words, the higher population growth rate in Figure 18.8, point *b*, may be illusionary in a multispecies environment. A third element, long advocated by some biologists, often under the generic term of *density-independent factors,* is that there exist environmental elements, most notably the weather, that have an important effect on populations. *Density-independent* and *density-dependent* arguments have waged back and forth, often acrimoniously, over the years among terrestrial ecologists. Such arguments historically were of less interest to fisheries managers, for it seemed unlikely that the weather had too much effect under the surface of the ocean. Later we show that new research has given strong support to the importance of density-independent factors in the ocean.

Our own critical analysis of the many supposed field examples of logistic growth has failed to locate one good example of logistic growth occurring in nature, despite supposed examples in nearly every ecology textbook and numerous and often very mathematically complex manifestations of the theory appearing in the literature. The Kaibab deer story, once thought an excellent example, is based on poor data, and probably wishful thinking (see Colinvaux, 1975 for a debunking of that example).

Nearly all the examples we have found where the logistic equation represents population growth accurately are from the laboratory. Supported examples from nature (e.g., see Hutchinson, 1978, for a review of many possible examples) all fail as adequate examples because they are (1) from human-managed situations (e.g., sheep in Tazmania), (2) from organisms with reproductive cycles that act, over a season, to give the appearance of logistic growth (e.g., barnacles on rocks), (3) based on either no data (just a nice smooth curve with appropriate axes) or examples where the logistic curve drawn in does not fit the data any better than some other curve, such as a straight line (Mediterranean bees grown in Maryland, Argentinian antlions), or (4) use a semilog scale that gives a misleading appearance of logistic growth (ring-tailed dove in England). A general critique of such use and misuse of mathematical models in ecology is found in Hall and DeAngeles (1985).

These misapplications do not disprove the concept of density dependence as a regulator of many natural populations; clearly density-dependent factors operate occasionally or subtly to keep populations from going extinct or increasing without limit. Although portions of real population data sets resemble logistic growth, that equation certainly does not describe most real populations over time. Thus the data available at this time are in no way sufficient to validate the logistic curve as an adequate representation of year to year variation in natural populations, although they are often presented as if they did. By contrast, real long-term data on populations in nature often show irregular fluctuations (e.g., Watt, 1968). Recruitment in real fish populations may vary by two to six orders of magnitude for a given stock size (Figure 18.10). Thus we should not be surprised if fish populations do not respond as the logistic-based models predict. After all, we have been managing a number of fish stocks for decades in a fashion (reducing stock through harvesting) that (due to stock reduction and supposed density-dependent positive response) should have *increased* yield regularly over time. This does not seem to have occurred consistently for any major U.S. or Canadian fishery. Nevertheless, approaches to fishery management based on theoretical stock-recruitment curves—whether or not the curves actually do represent the real fish populations—have served an important political role in limiting fishing effort which is generally in the best interests of both fish and fisheries.

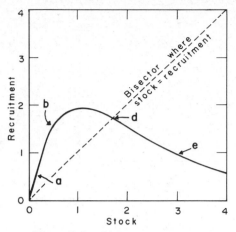

Figure 18.9. Ricker curve theory.

Schaefer Curves

Since it was rather difficult to determine species-specific relations between stock and recruitment that would optimize the number of fish caught and/or the production of progeny, other means of calculating optimum fishing effort have been sought. Schaefer (1967) developed an interesting and reasonably successful approach that relied on information from the fishery itself to derive the optimum catch in a given year. His approach also uses the logistic curve in part for its justification, although the empirical nature of the analysis greatly restricts the importance of the logistic curve.

Schaefer, working initially with Pacific yellowfin tuna, determined the annual amount of effort (i.e., boat-days spent fishing) and the reported catch of the boats in that fishery. He next plotted catch per unit effort (CPUE) and total catch as functions of effort (Figure 18.11). Schaefer found that as annual effort increased, the catch per effort declined, presumably due to depletion of the stock. In other words, the first boats out found it easy to catch fish, for there were many fish to be caught, but in later years, as more boats joined the fleet, the catch per effort declined (Figure 18.11a). The total catch is equal to the catch per effort times the effort, and it at first increased with increasing effort but eventually peaked and declined as the decline in catch per effort became more important than the increase in effort. The new data analyzed by Schaefer fit the model rather well, except that there was insufficient information about the behavior of the fishery at high effort (i.e., on the right-hand side of the curve). In addition problems may occur when the average boat size or type of fishing gear changes (as when

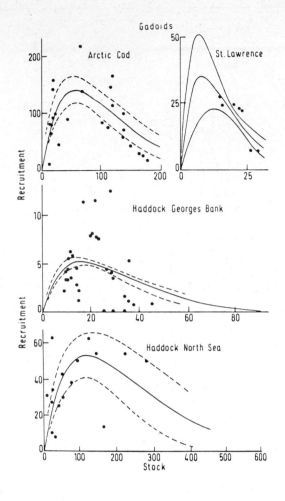

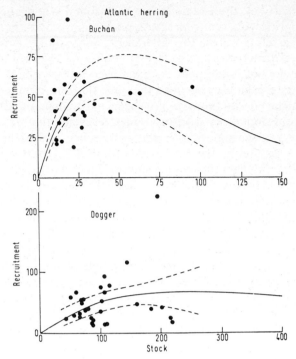

Figure 18.10. Ricker curve reality—many other lines fit the data as well as the Ricker assumption which is the curves drawn in. Although the Ricker concept is a very useful one for the analysis of fisheries, and it does seem to hold for Arctic cod, unfortunately most fisheries do not show a clear Ricker stock-recruitment relation. (From D. H. Cushing, *Recruitment and Parent Stock*, 1973, reproduced with permission of Washington Sea Grant Communications.)

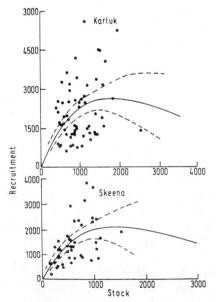

the Pacific tuna fleet changed from using rods and chum to using purse seines) so that intercalibrations must be made. Nevertheless, the Schaefer approach seems to work well for a number of oceanic fish species (Walters and Hilborn, 1976). Without any other justification the Schaefer approach probably is as good as any for managing fish, especially if enough data are available to see if the data fit the model.

If empirically-derived data for fish populations behave in a manner similar to a Schaefer plot, one can calculate catch levels that optimize the fishery according to particular criteria. A common objective is *maximum sustained yield* (MSY), the level at which the greatest sustained catch can be made (Figure 18.11*b*). MSY has served as a management objective in many fisheries for a number of decades, based on the rationale that fisheries should produce as many fish as possible for as long as possible. Presumably the maximum sustained yield for the Schaefer approach also should be more or less where the maximum sustained yield is found for the Ricker approach, that is, at some moderate level of fishing effort. But at this time the concept of maximum sustained yield is being questioned from many quarters (e.g., Nielsen, 1975; Larkin, 1977; Brown and Lugo, 1981). Edwards and Hennemuth (1975) point out that the influence of exceptionally large, but rare, age classes in certain stocks can greatly distort the data and conclusions related to determining MSY.

Other possible management objectives include the best economic return on investment—that is, where the quantity of fish caught per unit of effort (either in terms of dollars or energy) is highest (e.g., Clark, 1985). This would reduce fishing levels to considerably less than MSY. On the other hand, Pacific halibut have been deliberately overfished in Canada for years because it is cheaper in the short run for the province of British Columbia than paying unemployment. Because of this policy it requires about twice as much energy to catch a halibut compared to more restrictive regulations. Finally, conservative catch restrictions might be considered useful from a conservation view or to protect recreational fisheries (e.g., Nielson, 1975).

Toward an Integration of Ecosystem and Population-Level Information

Given the failure of most straightforward Ricker analyses and the obvious importance of both density-independent factors and the interactions of different species, it became necessary to develop a more comprehensive approach to the relation of stock and recruitment (e.g., McHugh, 1959). For example, Cushing (1975) integrated information about the relation between environmental conditions and the survival of young fish, in part by examining the magnitude of the seasonal cycles in the abundance of primary and secondary producers (essentially phytoplankton and zooplankton, the food of many young fish) in aquatic environments. He, and others, have focused on the variation in the number of many phytoplankters in temperate environments over the year and from year to year—frequently by factors of 10 to 1000. Similarly zooplankton abundance varies by (typically) factors of 10 to 100 over the year. Peak abundances of both phytoplankton and zooplankton varied typically much less from one year to another than within any given year. Fish reproduction and migration ap-

pears to be closely geared to the timing and abundance of plankton blooms (Lasker, 1975; Lasker and Smith, 1977; Hall and English, 1985; Day et al., in press). Understanding of the food chain cycles and the ecosystem level factors that drive them should lead to a better understanding of the processes that produce good fisheries. We particularly recommend Cushing's book as a synthetic approach to understanding the processes affecting fish populations by going one step beyond a stock-recruitment approach to a more ecosystem-oriented approach. But neither Cushing's book nor, to date, any other has established clearly what the mechanisms are by which environmental conditions are translated into year to year differences in fish recruitment, nor have they given any particularly better prescription for management than had been available previously. Cushing's most recent book (1982; see also review by Hall, in press) begins that process by synthesizing the growing body of evidence pertaining to the role of environmental factors in regulating the population dynamics of most fish, but a synthesis remains elusive.

Ecosystem Factors and Fish Population Dynamics

The basic idea that environmental factors are much more important in determining reproductive success than strictly density dependent factors has been around for a long time (Hjort, 1916). More recent and sophisticated research, however, has shown that density-independent factors can act in more subtle ways than was once generally appreciated, ways that make a great deal of difference to the fish populations. Hennemuth et al. (1980) have analyzed time series of recruitment for 18 fish stocks and found that recruitment varies enormously in ways that appear to reflect very large environmental effects. Many fisheries researchers are beginning to agree that the key to understanding and predicting fish stocks for any given year depends on a thorough understanding of the events that produce high or low survival during the critical days, weeks, or months after the fish are born. Yet a further problem persists even if we are successful at understanding and prediction: How do we translate a knowledge of fish abundance into management?

Laurence (1976) has shown that growth and survival of young haddock depends, in large part, on food supply. Lasker (1975) and Lasker and Smith

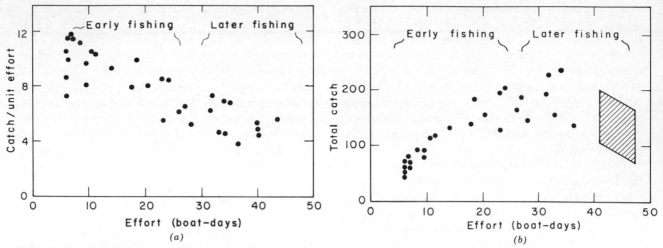

Figure 18.11. Schaefer analysis showing (*a*) a continuous decline in catch per unit effort with increased effort and (*b*) a rising catch with initial increase in effort followed by an asymptotic and, perhaps, eventual decline in catch. In many fisheries total catch declines with greater effort as represented by the shaded area. Dots are data for Pacific yellow-fin Tuna fisheries, 1950s–1960s.

(1977) have shown similar relations for Pacific coast species—the higher the concentrations of plankton in the water the greater the survival rate of young anchovies. To manage fisheries, however, we must, ideally, be able to predict the factors that determine such plankton abundance and how these factors affect fish populations, or at least be able to respond quickly through regulations once such patterns are determined.

One interesting attempt at prediction is from Sutcliffe and his associates (1972; 1973; 1977) who attempted to assess environmental conditions and integrate them with ecosystem biology to predict fish catch. They hypothesized that those years with a large freshwater discharge into the Gulf of Saint Lawrence have associated large nutrient inputs (from both the river- and freshwater-driven, deeper inflowing salt water) to the estuarine zone, raising survival levels of the young of various species. They observed a very strong correlation between river discharge and fish catch in subsequent years, as well as with Sutcliffe's prediction of fish catch based on earlier years' river discharge (Figure 18.12). Thus the abundance of fish in the future and the allowable catch appears easier to predict now, at least where and when such relations are found.

A systems approach also has been used successfully to manage sockeye salmon originating from Great Central Lake, Vancouver Island, British Columbia (LeBrasseur et al., 1979). This is a particularly interesting case because Canadian salmon have been managed intensively for decades accord-

ing to population concepts in a way that should have increased the catch. Nevertheless the catch has been decreasing. Sockeye salmon spawn in the fall in small streams running into lakes or occasionally along lake shores. The young emerge from the spawning bed the next spring and move more or less immediately into the open lake, where they spend about a year feeding on zooplankton. The following spring they migrate down rivers into the sea where they undertake extensive migrations (see Manzer et al., 1964; Royce, et al., 1968; Hall and English, 1985).

LeBrasseur and his colleagues knew that, in general, the larger the salmon were when they went to sea, the higher the probability they would return to spawn several years later. They also knew that the lakes where they grew up were deficient in the element phosphorus, the lack of which decreased the production of phytoplankton in the lake and hence the growth of salmon. Some investigators thought this phosphorus deficiency was due in part to heavy fishing pressure on the adult sockeye salmon. According to stock-recruitment models, enough fish escaped the fishery to produce a suitable number of spawning adults, but the large catch deprived the lake of a major source of phosphorus—the dead bodies of spawned-out adults. In other words, the fish themselves brought phosphorus (in their bones) uphill from the sea. LeBrasseur and his colleagues suggested that this missing phosphorus could be resupplied by adding commercial phosphorus to the lake, and such an experiment was undertaken in Great Central Lake. The results have been extremely encouraging, for the commercial catch has increased from about 50,000 to about 245,000 fish per year while more fish have escaped the fishery and returned to spawn. The economic return on fertilizing lakes was about $2.8 for $1, with a large portion of that budget allocated to scientist's salaries. About 200,000 salmon weighing 2–3 kg, worth at that time about $4/kg, could be produced for less than $200,000 worth of fertilizer. Presumably the energy return on investment would show a similar high return, although the energy cost of catching the fish (which is not large, as they swim right into a net at the river mouth) would have to be figured in. At least in this case a systems approach that used information about the interaction of fish and their environment was more successful than the population approach alone for managing salmon. All such approaches require careful thought and experimentation, however, for as other lakes were fertilized in an attempt to duplicate the success of the Great Central Lake experiment at least one of the lakes produced only millions of commercially useless sculpins and few additional salmon. But many other lakes were successfully fertilized, and it is hoped that this approach will be able to reverse the long-term downward trend of salmon for at least some salmon stocks. An added bonus is that the genetic makeup of the salmon is largely unchanged, which is not the case with hatchery fish.

How Are Fisheries Managed?

Although several theories for managing fisheries are available, most U.S. fisheries are not managed at all or, at most, barely managed. Anyone who wants to fish can do so and may remove as many fish as he or she can or wants to. When catch quotas are initiated, as in the Northwest Atlantic fisheries in the early 1970s, they often are greatly exceeded, for it is hard to enforce the regulations. As a result many of our national fisheries are strongly cyclic as fisheries develop, overcapitalize, overfish, greatly reduce stocks and decline (e.g., Figure 18.6). A consequence of this free-market philosophy is that as overfishing reduces profitability, increases in fishing effort will be discouraged. Sometimes this is true, and sometimes the fish recover after their human predators go elsewhere. But sometimes they do not, and often they fluctuate strongly due to reasons that are unrelated to the fishing. Often fishing pressure directed at other species will continue to exert substantial impact on the overfished stock. One would think that those who fish would prefer a more stable fishery. Such stability can be achieved to some degree by restricting the number of fish that can be caught, as has been done sometimes in Canada where the fishing industry is more tightly regulated. Such regulations have worked on occasion in the United States (as in the tuna fisheries), when the fisheries realized that regulations worked in their own interests. Unfortunately even many Canadian fisheries that are tightly regulated are apparently overfished and declining, for example, nonhatchery Pacific salmon (Larkin, 1979). Whether this is due to the failure of the concept of regulation or the general failure of fisheries scientists to have their proposed regulations enforced as strictly as they would prefer is not clear. Whatever our national management rationale, or lack thereof, it is clear that our fish yield per unit of energy invested has declined a great deal (Figure 18.13). The relation of yield to energy investment has almost never been used explicitly in management philosophies, although many economists recognize that the overcapitalization of many fisheries is a poor economic investment (e.g., Clark, 1985).

The New Challenge to Fisheries Managers

Given the partial or complete failure of many traditional fisheries management schemes, such as the unmodified Ricker analysis, and the difficulties in successfully applying theoretical population methodology to fisheries management, what options are available for management? Although it is impossible to give a simple prescription, a number of promising approaches can be used, with caution, for various fisheries. These approaches may complement, or occasionally contradict, the older single-population concepts. It is important to use data from the fishery as much as possible to help determine which approach makes the most sense for the case in hand.

The first approach, already mentioned, is the Schaefer curve approach (e.g., Figure 18.11), which examines the impact of effort on catch and determines what level of regulation is optimal for whatever economic, social, or environmental goals are chosen. This approach is useful only for well-developed single-species fisheries that have both a relatively lengthy data record, including information from periods of very different fishing intensity,

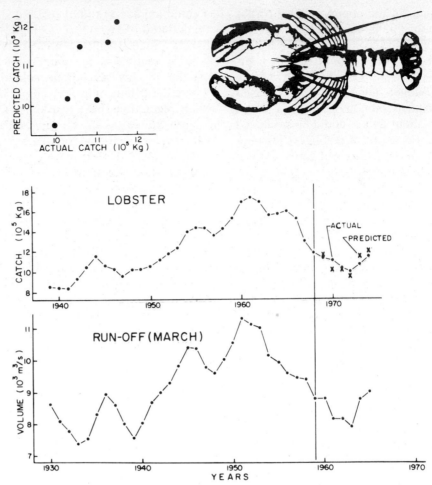

Figure 18.12.   Annual regional lobster catch and March discharge of the St. Lawrence River 10 years earlier. The change in the size of the catch reflects the earlier changes in freshwater discharge and accompanying injection of more or less nutrient-rich salt water into the fish nursery areas. (After Sutcliffe, 1977.)

and well-behaved fish—that is, fish that plot relatively nicely on the axes given in Figure 18.11. Hilborn (1979) has shown that many fisheries are not especially well behaved, but Walters and Hilborn (1976) have given six examples of fisheries where the Schaefer curve does seem to work well. Pitcher and Hart (1982) suggest the method is useful where only catch and effort data are available, which is normally the case. In addition the Schaefer curve method is useful for determining fishing strategies relative to energy criteria. Energy use is closely related to the level of effort and the yield of fish can be measured by edible calories. The leftmost points on Figure 18.11*a* and *b* would be the maximum fish return per unit fossil energy investment, and the middle to rightmost points would be the point of maximum yield (conceptually the same as the point of *maximum power* for the fishing system, p. 63).

Fishing at a greater level of effort would be a waste of energy, for the same number of fish could be caught at about one-third the energy investment.

A second approach replaces standard *stock-recruitment* approach with a *larvae-recruitment* approach (Laurence, 1981). This approach recognizes that density-independent factors that affect the survival of small fish are more important than density-dependent factors, and it replaces the use of estimates of spawning adults with estimates of juvenile fish abundance as a means of estimating the number of fish that can be harvested. At present it is too early to know whether this approach will be generally applicable, although it has shown some success, such as in predicting the year class strength of California sardines (Lasker and Smith, 1977).

A third approach to management is based on predictions of year to year abundance of fish from the functioning of the entire ecosystem rather than from population characteristics alone. We have discussed previously examples based on Sutcliffe's and LeBrasseur's work.

A fourth approach, one that can be combined with others, uses what is known as *adaptive management* (Walters and Hilborn, 1976; Holling, 1978). The term is purposely vague but refers to a general set of procedures for updating decisions based on new information as it becomes available. The flexibility and acknowledgment of natural variation implicit in this approach is essential as we learn more about the degree to which many fisheries fluctuate, both naturally and in response to fishing pressure (e.g., Cushing, 1982; Hennemuth et al., 1982). This approach should be especially effective as the most recent information is usually the most useful. These principles are especially effective for limiting fishing seasons. Instead of making fixed decisions about the length of a salmon fishing season, as is done traditionally, open-ended decisions are made. If more fish come in than expected from predictive formulas (e.g., Ricker analyses), a larger catch is allowed. If, on the other hand, the catch is smaller than anticipated, the fishing season can be shortened. This approach utilizes and protects the fishery more effectively but is tough on those who fish, since it is difficult to plan for how many days' provisions and other items will be needed. But fishing always has been a risky business.

Limited access is a fifth important idea that restricts the number of people who can fish a region. Sophisticated approaches exist for calculating this number (e.g., Rettis and Ginter, 1978), and all can improve yield per effort, thereby reducing the energy cost of catching fish. It is still too early to tell whether any or all of the foregoing possibilities will work generally or even for a specific fishery. Although examples of success can be found, failures are less likely to be recorded in the scientific press. At worst most of these procedures share with the Ricker-type analysis the basic virtue of discouraging overfishing, which all too often has proven disastrous to the fish and to the economic well-being of fishermen.

Presently no well-developed and empirically tested body of theory for optimizing the catch from a fishery exhibiting highly variable recruitment is available for managers. If, for example, we predict the strength of a year class from either the abundance of larval fish or from environmental factors (e.g., Figure 18.12), we still do not know whether it is in our best interest to fish the strong year class especially hard—since there are many, perhaps a surplus, of fish—or not as hard—since the resulting greater abundance of fish might contribute more effectively to rebuilding a (presumably) depleted fish stock. And even if scientific solutions can be found, economic, cultural, and sociological factors may disrupt a management program because fishermen do not like to have their boats sitting idly at the dock in the off years. It is possible that a very complicated management scheme might be devised that would trade off fishing for one species with another depending on the abundance or lack of abundance of each species, so that there would always be something to fish for. In practice, this probably would require an information base very difficult to achieve.

The Remaining Social Problem

Unfortunately even if we develop a more predictive science in fisheries based on increasingly sophisticated knowledge of the relations of stock, environment, fisheries, effort, and catch, and are able to decide what level of fishing is optimal for fishing, social factors make it extremely difficult to implement whatever regulatory conclusions might be reached. If, for example, a fishery shows a reliable and consistent Schaefer relation (Figure 18.11), the most desirable goal can be controversial: Maximum sustained yield? Maximum economic return on investment (Clark, 1985)? Or, highest return on energy (or dollars) invested? Clark (1976) once suggested that it may be optimal to catch all the fish you can right away, so that the maximum money will be available for investment, but his more recent book gives a very sophisticated set of mathematical procedures for optimizing economic gain consistent with other objectives. From our perspective, limiting effort is very desirable for we can often catch about the same number of fish with much less energy invested. Many people who fish, however, would be excluded from their livelihood, and it may be socially undesirable to provide additional unemployment compensation, to suffer the consequences of failing to provide that compensation, or to employ those individuals in some other activity that might be even more fuel intensive while providing, little or no social benefit.

One possible way out of this dilemma is to define the management goal as maximizing the fish return per unit of *energy* invested and then to establish

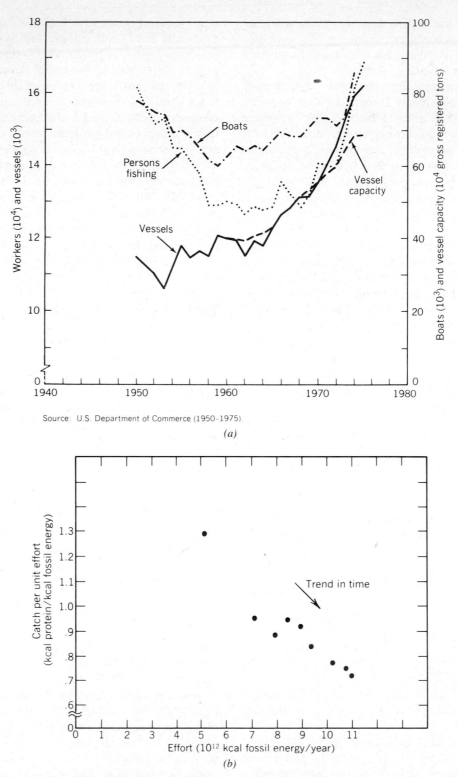

Figure 18.13. Basic statistics of U.S. fishing industry, (*a*) trends in the domestic fishng fleet working on boats, on vessels, and onshore, and (*b*) catch per unit of energy invested, 1950–1980. (From Brown and Lugo, 1981, and Brown, personal communication.)

deliberate regulations that make each person who fishes use less energy and hence become *less* efficient at catching fish. Ideally in this way a fishery could be regulated at the left to middle of Figure 18.11 while maintaining or even expanding employment by substituting human labor-intensive procedures for oil-intensive procedures. It probably would not be necessary to return to a *Captain's Courageous* style of fishing to achieve the goal, but the concept does offer the possibility of avoiding several presently difficult social problems. Perhaps a given quantity of fuel could be allocated to each of 1000 captains for the pursuit of, say, haddock. The fuel would be allocated each year in such a way as to produce about the desired level of fishing intensity, given the current stocks, fishing fleet and technology. Each captain would determine for him or herself the ways to catch the largest number of fish with that quantity of fuel, a process that ideally would lead to a labor-intensive fishery. Although such an approach might seem illogical in an industry that has become increasingly technical and energy intensive, it might be as fair a way of limiting effort as some schemes that are presently in effect, and it may some day be forced on us.

Several other strategies are available to decrease the energy cost of fish. Less than one-quarter of the fish we bring into port are eaten directly by people. The rest are fed to animals or used for fertilizer (Pimentel and Pimentel, 1979). A related problem is that about half of the fish caught are not commercially desirable species. These fish take as much energy to catch as the desired fish (with which they are generally caught) but take up valuable cargo space and ice and may not fetch a good enough price to justify taking them back to port, either at all or else in a condition that allows for human consumption. Trawling is a very energy-intensive process, and if the nontarget species could be marketed, then about twice the poundage of fish could be taken per kilocalorie invested in the trawling process (Fig. 18.13). One way to market these fish, which often are just as tasty as the traditional species we eat, is simply to change their name. *Eelpout* does not sound too tasty, and they are not pretty fish. Consumer acceptance of this species was much greater when they were marketed as *ocean perch*. Goosefish are among the ugliest fish that live on the bottom of the sea in the New England area, but they are very abundant. We once bought at a premium price, and enjoyed, fresh *monkfish* in central New York without realizing that it was just ugly old goosefish. Clearly clever marketing has its role in this game.

In conclusion, there can be no one obvious answer to the problems that plague the fishing industry, and that will increasingly plague the industry as the price of fuel inevitably goes up. Flexible management not tied to outmoded theory certainly will help offset these effects. We hope that integrating ecosystem level concepts (e.g., the examination of factors that produce strong year classes), population concepts (e.g., stock-recruitment relations), and energy analysis will give managers better information with which they can make decisions. But despite such scientific progress we may never be able to eliminate the ambiguities and conflicts of differing social objectives.

19

FOREST RESOURCES AND ENERGY USE IN THE FOREST PRODUCTS INDUSTRY

Jonathan Chapman and Charles A. S. Hall

The *forest products industry,* which includes all U.S. industries that use wood directly to manufacture their products (Table 19.1), harvests between 320 and 410 million m³ of timber each year. These products range from standard items such as 2 × 4's and tissue paper to giant paperboard boxes and mobile homes. The two main sectors of the forest products industry are the lumber and wood products manufacturers and the paper and allied products manufacturers (Figure 19.1). The forest products industry uses 58% of all timber felled annually. Households use 4% as fuel wood, and 38% remains on site as branches and uneconomic wood burned or left to decompose (OTA, 1980). Sixty-three percent of forest products industry harvests are used to manufacture structural products, and 35% are consumed to manufacture fiber-based goods (Figures 19.1, 19.2b). The remaining 2% is used as fuel or as a source for chemicals (Jahn and Preston, 1976).

By some estimates the forest products industry is the fourth largest in the nation. It consumes 15% of the nation's raw materials (metals, fuels, timber, etc.), employs 5% of the work force, and produces 5% of the total gross national product (Pingrey, 1976; Cheremisinoff, 1980). In addition to producing goods from domestic sources for domestic consumption, the industry affects the national economy's balance of trade by importing and exporting large quantities of materials (Table 8.1, Figure 19.2b). Finally, the forest products industry is a producer as well as consumer of energy, producing about 328 trillion kcal (about 1.7% of the energy used in the United States).

In this chapter we assess the supply and quality of the U.S. forest resources and the energy and labor costs of processing these resources into useful goods and services. It is a data-intensive chapter designed in part to show by example the methods by which the energy performance of any industry can be evaluated.

THE EXTENT AND NATURE OF THE RESOURCE BASE

Of roughly 300 million ha of forest land in the continental United States (32.7 percent of the total land mass), 66% (197 million ha) is considered commercially productive, defined as ecosystems producing at least 1.4 m³ of wood per hectare per year (20 ft³/acre/yr; Spurr and Vaux, 1976). Some readers may be surprised that a majority of U.S. forestland (58–60%) is owned by nonindustry private individuals (USFS, 1978; Figure 19.3). Only 14% of commercially productive forestland belongs to forest products companies, which represents 9% of total forestland.

Softwoods account for nearly 80% of harvests from U.S. forests, of which the Douglas fir and the Southern pines are most important. Hardwoods account for the remaining 20%, 48% of which are oaks (Table 19.2). Most of U.S. timber production occurs in the Southeast and Northwest (Figure 19.2a). Alaska also contributes a significant portion, but is far from the major domestic markets for wood and fiber products which are centered in the Midwest and Northeast. Canadian sources are closer and their materials cheaper to transport, so that much of Alaska's timber—as well as that of Oregon and Washington—is exported to Japan as logs and pulp chips (Bethel and Schreuder, 1976). The domestic forest products industry imports a significant amount of raw and intermediate materials to satisfy

Table 19.1.   Subindustries of the Forest Products Industry, with Energy Costs and Value Added in Manufacture for 1980

Industry	Purchased/ Generated Electricity (10^{12} kcal)	Purchased Fuels (10^{12} kcal)	Approximate Energy Cost of Purchased Materials (10^{12} kcal)	Value Added in 1972 Dollars (10^6 \$)
Lumber and wood products	37.94	37.62	213.0	9031
Logging camps and log contractors	0.93	8.14	39.0	1564
Sawmills and planing mills	15.54	10.66	62.7	2622
Millwork	2.16	1.11	20.4	884
Hardwood veneer and plywood	1.17	1.21	7.4	233
Softwood veneer and plywood	5.74	4.31	18.7	589
Particle board	2.13	1.06	2.2	108
Paper and allied products	150.86	279.44	316.7	13507
Paper mills excepting building paper	73.40	126.5	78.5	3507
Pulp	10.06	20.82	13.0	588
Paperboard	41.75	104.4	44.5	1740
Total	189.	317.	530.	22538

the domestic demand for wood and fiber commodities. Softwoods comprise the vast majority of imports, and virtually all of these come from Canada. Approximately 33 million m³ of softwood lumber is imported into the East and Midwest, mainly from British Columbia (Bethel and Schreuder, 1976), as is newsprint and plywood from Central and Eastern Canada. Much of the material imported from Canada could be supplied from domestic sources, but energy and dollar costs of transporting the raw and finished materials from the production centers (the Northwestern and Southeastern United States) to factories or market centers (mostly Midwestern and Northeastern United States) would be higher.

Hardwoods comprise a small fraction of wood

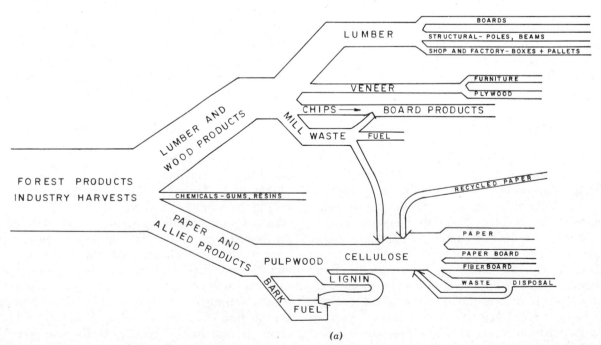

(a)

Figure 19.1.   (a) Flow diagram of the forest products industry and representative final products; (b) energy used and value added in subsectors of the U.S. forest industry.

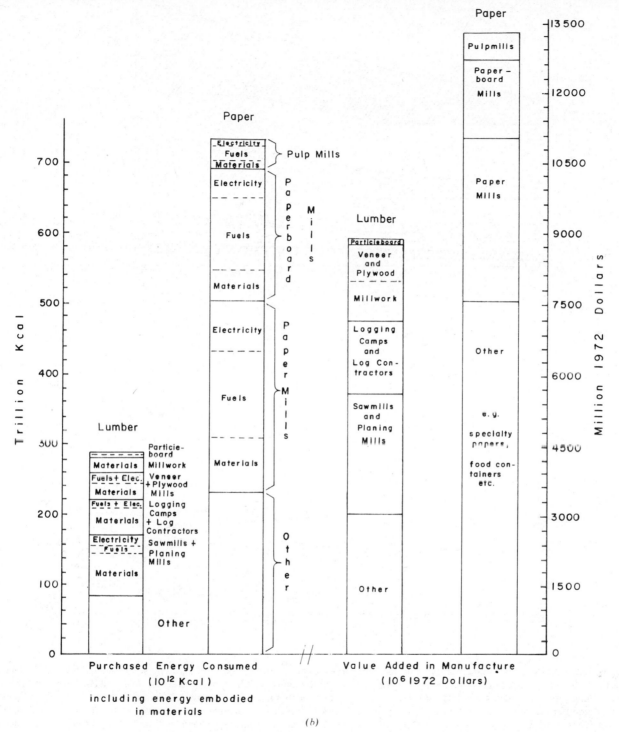

(b)

Figure 19.1. (*Continued*)

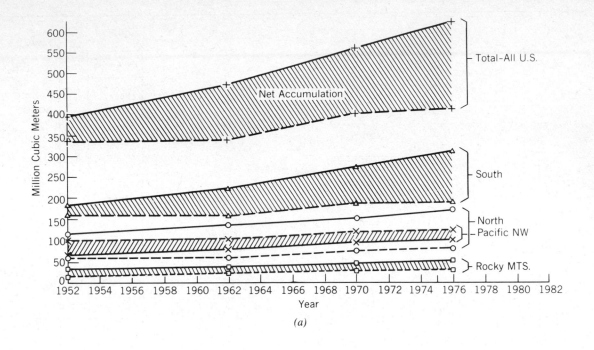

(a)

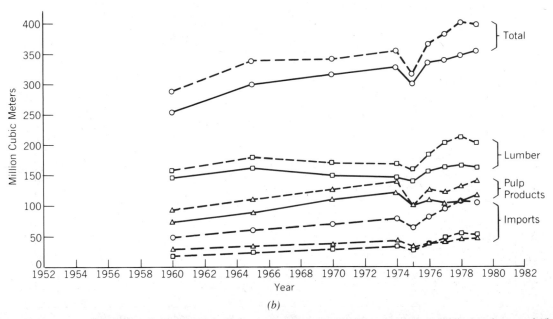

(b)

Figure 19.2. Various rates and efficiencies for the U.S. forest products industry. (*a*) Forest growth (solid line) and removals (broken line) in U.S. commercial forests (million m³/yr). (*b*) Domestic production (solid line), imports (dashed line) and consumption (broken line) (million m³/yr). Squares are data for lumber and triangles are data for pulp. (*c*) Energy efficiency (value added in 1972 dollars/ million kcal) in the two major sectors of the forest products industry. The ×'s include energy from process wastes. National mean energy/GNP ratios were used to assess upstream energy costs. All electricity values multiplied by 3. (*d*) Labor efficiency (value added 1972 dollars/worker-hour) in the two major sectors of the forest products industry.

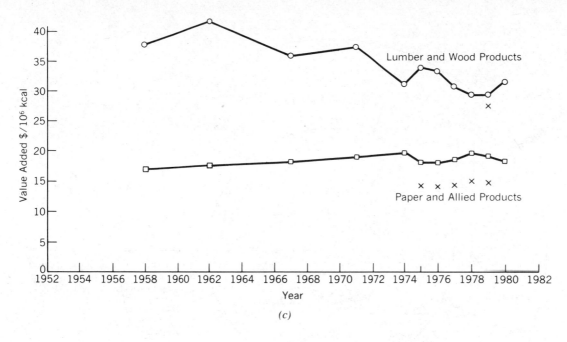

(c)

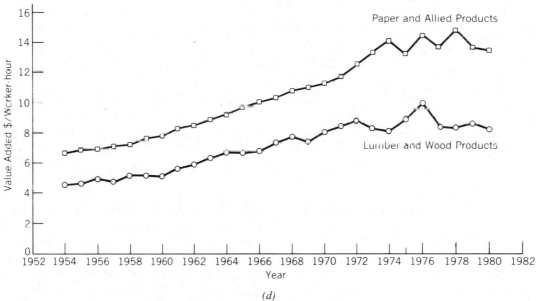

(d)

Figure 19.2.   *(Continued)*

imports, most of which is tropical hardwood and veneer that cannot be grown domestically. Bethel and Schreuder (1976) suggest that the United States could become self-sufficient in hardwoods (except for luxury hardwoods), for the volume of hardwood grown each year exceeds the amount harvested (Figure 19.2a). We are, however, far from self-sufficient in domestic softwood production at this time.

ENERGY USE IN THE FOREST PRODUCTS INDUSTRY

Estimates of total process fuel (including electricity) consumed directly by forest product industries range from 600 to 693 trillion kcal, equivalent to between 18.6 and 21.5% of fuel used by all U.S. manufacturing industries and between 3.0 and 3.5% of total U.S. fuel consumption (Duke and Fudali,

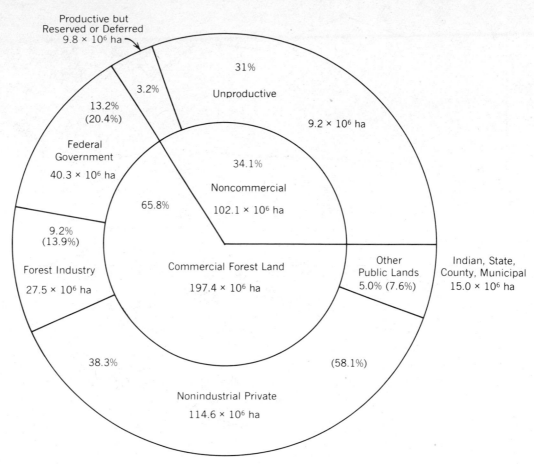

Figure 19.3. Forestland in the United States by ownership. (Percentages without parentheses refer to total land; those within parentheses to commercial land.) (From USFS, 1978.)

Table 19.2.   Production by Important Species in the United States in 1979

Species	Percent
Douglas fir	22.4
Southern yellow pine	21.3
Ponderosa pine	10.5
Hemlock	6.6
White fir	5.6
Redwood	2.4
Cedar	2.2
All other softwoods	9.6
Oak	9.2
Yellow poplar	1.8
All other hardwoods	8.3

Source: From Statistical Abstracts 1982–1983.

1976; Mitre, 1979). Between 85 and 92% of this energy was used to manufacture pulp and paper (Duke and Fudali, 1976; Arola, 1976). Fuels used by forest products companies vary and depend on availability, price, quality, and specific manufacturing requirements. In 1979 fossil fuels supplied 42.2% of the energy used to manufacture pulp, paper, and paperboard, forest-derived fuels supplied 41.1%, hydropower supplied 0.4%, and 15.9% came from purchased electricity. Oil was the predominant fossil fuel, contributing 17.2% of total energy consumption (American Paper Institute [API], 1981; Table 19.3; Table 19.4).

There is general consensus among our sources that forest-derived fuels account for between 277 and 328 trillion kcal (1.1. to 1.3 quad) of process energy in the forest products industry (OTA, 1981; Mitre, 1979). Paper and pulp manufacturing consumes 77–83% of that. Another 50–202 trillion kcal

Table 19.3. Fuels Consumed in the Pulp, Paper, and Paperboard Industry in 1979

Fuel	Amount (10^{12} kcal)	Percent
Forest fuels	251.2	41.1
Spent liquor	201.6	33.0
Hogged fuel	25.3	4.2
Bark	24.3	4.0
Fossil fuels	257.8	42.2
Oil	105.3	17.2
Natural gas	97.2	15.9
Coal	55.1	9.0
Propane	0.4	0.1
Electricity	99.9	16.3
Purchased[a]	97.6	16.0
Self-generated (hydro)	2.3	0.4
Other	7.8	1.3
Energy sold	(5.3)	0.9
Total	611.3	100.1

Source: Based on a sample of 90.6% of manufacturers, reported in American Pulp Institute, 1981.

[a]Multiplied by a quality factor of 3.

are consumed as fuel wood in private homes (0.2 to 0.8 quad; OTA, 1980; Lipfert and Dungan, 1983).

The next section examines the types of operations for the major forest product sectors, as well as their energy consumption, and the manufacturing efficiencies for all steps of production from harvest to final products. More detailed analyses are given for several subindustries.

Energy Use in Harvest and Transport

The harvest and transport of timber require approximately 12% of the fuel used to produce wood-based manufactured goods (Koch, 1976). Diesel fuel and gasoline provide most of this harvesting energy and therefore would be difficult to replace with al-

Table 19.4. Degree of Energy Self-Sufficiency in the Forest Products Industries

Industry	1972	1976
Lumber	30	30
Plywood	50	50
Pulp and paper	40	45

Sources: From USFS (1976); Duke and Fudali (1976).

ternative fuels. In most present-day harvest systems several laborers cut and delimb trees with chainsaws and skid the logs to a landing with the aid of a large rubber-wheeled skidder. Some of the larger harvest operations are accomplished with feller bunchers, which are large self-propelled machines that rapidly cut tree trunks with steel jaws and wrap them with several other trees in bundles. These machines are more often used to clear-cut large areas where tree species, wood quality, and tree age are unimportant. Although feller bunchers require more energy to manufacture and operate than do chainsaws, and can operate only on relatively smooth terrain, they harvest trees much more rapidly than loggers with chainsaws, thereby increasing productivity (volume per time) dramatically.

Once at a landing, logs produced by either method are loaded onto trucks at *tree length* (about 5 m) using a truck-mounted mechanical arm, or are chipped and fed directly into large vans. Logs and chips are driven to a mill where they are used as raw material for lumber, board products, pulp, or fuel. Trucks and vans travel an average of 80 km to and from the pulping mill or sawmill (Smith and Corcoran, 1976; Tillman, 1978). For purpose of analysis we assume a similar value for fuel wood too.

Energy Required and Energy Return on Investment for Fuelwood

We examine first the energy return on investment (EROI) for both chainsaw and feller-buncher-based fuel wood operations—systems that consume fossil fuel energy in order to derive heat energy from fuelwood. Our analysis is based on data collected by Smith and Corcoran (1976) who estimate direct and embodied fossil fuel used per ton of wood harvested.

Values of heat content per cord of wood (but not per kilogram; 1 cord = 3.6 m^3) differ significantly among tree species (Table 19.5), so we used paper birch, a wood of intermediate heat content (5.91 $\times$ 10^6 kcal/cord, or 3521 kcal/kg) for our calculations. Oak or beech would deliver more energy per cord but about the same per kilogram, although the energy costs of securing and trucking a cord of these denser woods would be slightly greater. The converse is true for less dense pine or hemlock.

Based on these assumptions (Table 19.6) we calculate that a feller buncher consumes approximately 13% of direct and 14% of all fuel used for

Table 19.5. Wood Heating Values

Species	10^6 Btu per cord	10^6 kcal per cord	lb/cord	kg/cord	kcal/kg
Hickory	27	6.80	4200	1907	3566
Hophornbeam	27	6.80	4200	1907	3566
White oak	26	6.55	4000	1816	3607
Beech	25	6.30	3800	1725	3652
Sugar maple	24	6.05	3700	1680	3601
Red oak	24	6.05	3700	1680	3601
Yellow birch	23	5.80	3700	1680	3450
White ash	23	5.80	3600	1634	3550
Paper birch	21	5.29	3300	1498	3521
Black cherry	20	5.04	3200	1453	3469
Red maple	20	5.04	3200	1453	3469
Grey birch	20	5.04	3100	1407	3582
American elm	19	4.79	3000	1362	3517
Aspen	15	3.78	2300	1044	3621
Red pine	17	4.28	2500	1135	3771
Hemlock	16	4.03	2400	1090	3697
White pine	14	3.53	2100	953	3704

Source: From Morrow and Gage (1977).

harvests, or roughly double the fuel used by a chainsaw. Transportation consumes about 39% of the total direct fuel used in conventional lumbering operations and accounts for 37% of total embodied energy in timber.

The energy cost of the standard harvest method and the feller buncher (whole-tree chip) system are somewhat different (153 kcal/kg vs. 137 kcal/kg). The standard method requires more fuel per kilogram of wood because logs must be handled individually, whereas the more automated whole-tree chip system uses the chipper as a loader and dumps the chips to unload them. The EROIs for the two systems are 29:1 (standard method) and 32:1 (feller buncher chipper system). Both have quite favorable energy returns on investment, based on Smith and Corcoran's assumptions. Returning to the calculation in Chapter 13 of the national potential for wood as an energy source (p. 308), we see that if forest growth were fully utilized, fuel wood could provide a moderate net contribution to our long-term national energy requirements. Wood presently accounts for up to 11 percent of all energy inputs to space heating nationally (202 trillion kcal, or 0.8 quad) but the proportion of heat actually delivered probably is significantly lower because many woodstoves and all fireplaces are notoriously inefficient (Lipfert and Dungan, 1983).

Energy Requirements in Lumber and Paper Production Processes

The Production Process

Once at a mill, unchipped logs are debarked if they are to be used for lumber, pulp, or plywood, or chipped whole if they are to be used directly as fuel. Sawlogs are sawn into smaller pieces called bolts and sawn again into yet smaller pieces such as 2 × 4's. The ends are trimmed, and the boards planed and sanded if necessary. Each step produces wastes such as shavings and sawdust, which may be burned as fuel or used later in fiber or panel board products. Veneer logs for plywood (or furniture) are spun against a sharp blade to peel thin sheets which can be glued to form plywood, or laminated to less attractive woods to manufacture furniture. Wood is chipped for pulp and board products after debarking, and in the case of pulp either ground or cooked in liquor containing sulfate or sulfite-based pulping chemicals to separate the cellulose fibers and to remove lignin. To manufacture chipboard or particle board, chips are screened for size and glued and pressed into panels.

After being cooked in liquor, pulp is screened, washed, bleached (to remove coloring materials), beaten to break down further and to hydrate the fibers, spread on drying and pressing screens and

Table 19.6. Energy Use per Unit of Fuelwood in the Manufacture and Operation of Timber Harvesting Equipment

Process	Energy for Manufacture[a]		Fuel Used in Equipment Operation		Total Embodied Energy		Percent of Energy Cost for Each Harvest Method
	kcal/ 10^3 kcal[b]	kcal/kg[c]	kcal/ 10^3 kcal	kcal/kg	kcal/ 10^3 kcal	kcal/kg	
Felling with							
Chainsaw	0.002	0.001	2.11	9.17	2.11	9.18	6.0
Feller-buncher	0.66	2.88	3.82	16.58	4.48	19.46	14.2
Transport to Landing with Skidder	0.67	2.89	5.66	24.58	6.33	27.47	16.0 (chainsaw-based) 20.0 (feller-buncher based)
Loading with truck-mounted loader	0.25	1.08	2.78	12.1	3.03	13.17	8.6
Transportation							
Small truck	0.43	1.86	24.31	105.6	27.43	107.46	34.6 (chainsaw-based)
Large truck	0.21	0.92	11.96	52.1	12.17	52.86	38.6 (feller-buncher based)
Chipping by							
whole tree chipper	0.61	2.64	4.19	18.19	4.80	20.83	13.6 (chainsaw-based) 15.0 (feller-buncher based)
Auxiliary management vehicles	0.06	0.28	3.68	16.0	3.75	16.28	
Totals for each harvest method							
Standard method		9.90	33.17	144.20	35.22	152.96	100
Whole tree chip system		9.60	29.31	127.43	31.53	136.90	100

Source: Adapted from Smith and Corcoran (1979).

[a] Based on required replacement rate.

[b] Kcal required per thousand kcal of wood produced.

[c] Kcal required per kilogram of wood produced.

cylinders, calandered (to improve surface smoothness), and rolled. Depending on its final use, the paper also may be sized (to improve wettability), colored, impregnated with resins, laminated, corrugated, or creped. The majority of wastes produced by pulp processes are black, lignin-laden liquor and bark, whereas the by-products of chipboard or particle board manufacture consist mostly of bark. Both bark and black liquor are used for fuel, but the latter must first be evaporated to about 60% solids. An additional waste product from pulp and paper processes is tainted water, which must (or should) be cooled and filtered to remove suspended solids and treated to remove dissolved chemicals and compounds before it can be returned to rivers. According to the *Annual Survey of Manufacturers* (1978) the paper products industry produces 6.68 million m³ of waste water annually. A large mill may use up to 23 thousand m³ of water per day. Linehan and Post (1971) reported that of the 715,000 liters of water (190,000 gal) required to produce a ton of Kraft (sulfate process) paper, 550,000 liters were recycled for reuse. K. V. Sarkanen (personal communication) reports that water use has been greatly reduced since 1971. Waste water is collected from log debarkers and flumes, liquor recovery evaporators, bleach plant washer filtrates, paper machines, and wash water used on pulp screens. Water may then be recycled or pumped to sludge pools where bacteria and protozoans clarify it. Collectively, wastes from structural and fiber manufacture are called *process wastes*.

Approximately 60 percent of pulp is produced in pulp mills integrated with paper mills, and the remainder in mills that produce *market pulp* for sale to paper mills (Parker, 1974). These market pulp mills are especially energy intensive because they require radiant heat or steam energy to dry the pulp in preparation for transport, whereas integrated plants do not. In paper-related industries any drying process is generally one of the most energy-intensive steps. Factors affecting the consumption of fuel for heat in paper industries include age of capital equipment, efficiency of boilers and furnaces, type of wood used, the pulping and paper-making processes used, and the grade of final product (Parker, 1974). Other energy-intensive subindustries include those making paperboard, particle board, fiberboard (and their variants), and laminated lumber from veneers. Many of these subindustries substitute energy for high-grade raw materials, thus allowing the use of lower-grade inputs (e.g., woodchips and shavings) to make high-grade products. Likewise these more energy-intensive subindustries utilize a greater proportion of their timber inputs than do the less energy-intensive (but materials-costly) subindustries like lumber and plywood mills which have 31 and 45% conversion efficiencies of timber to products, respectively (Koch 1976).

Past and Current Use of Harvest Residues and Process Wastes in the Production of Energy

Prior to the Clean Air Act of 1967 process wastes were generally incinerated (Tillman, 1977). Following the enactment of this legislation, as well as modest price increases for natural gas and fuel oil, many more waste-fired boilers were installed to use the energy in process wastes while simultaneously eliminating the smoke-producing incinerators. This trend has continued ever since, although with spurts and lags. One important increase resulted from the passage of the Clean Water Act in 1972 (later revised in 1977; Figure 19.4) which requires industries to reduce greatly the amount of pollutants in effluents. The principal goal of the act is to end the discharge of pollutants by encouraging industries to reclaim and reuse their chemicals (Banks and Dubrowski, 1982). The provisions of this act are an excellent example of how industrial and environmental concerns can be mutually beneficial. Although compliance has been costly due to large investments in equipment, it has provided benefits to many pulp and paper companies. By recycling spent pulping liquor, these plants recover valuable pulping chemicals (e.g., sulfite) as well as the energy-rich lignins which are a significant fuel source (Tillman, 1978; Table 19.3). Sarkanen (per-

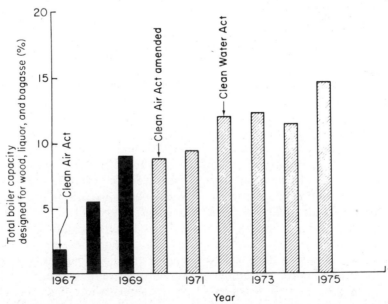

Figure 19.4. Trends in the installed capacity of wood, spent liquor, and bagasse fired boilers for industry as well as some important legislation influencing that capacity. (From Tillman, 1977).

sonal communication) reports that pulp chemicals have always been recycled in Kraft (sulfate) plants because of process requirements, but it is also true that the proportion of sulfite to Kraft pulp production was much greater before the passage of clean water legislation than now (Myers and Nakamura, 1978). Although increasing energy prices encouraged investments in equipment to utilize process wastes, they were not a significant factor in the major increase in wood fuels utilization; "energy cost played only a modest role in [the] resurgence of wood [based fuels] behind increased costs of waste disposal and pollution control" (Tillman, 1978).

The Use of Solid Wastes

Before their recent resurgence as fuels, process wastes from lumber manufacturing were viewed as nuisances by most forest products companies. Unlike liquid wastes of the pulp and paper industry, which could be dumped into a body of water and forgotten, solid wastes had to be buried or incinerated. "It was not unusual to see an oil-fired boiler installed next to a wood-residue-destroying incinerator" (Tillman, 1977). Similarly residues from harvest were left on the ground to decompose, or were piled and/or burned. Purchased fossil fuels were still relatively cheap, and environmental regulation was not a national concern, so there was little incentive to recover and utilize these combustible materials (Table 19.7).

Harvest residues consist of tree limbs and tops, crooked and damaged stems, and rotten, diseased, or otherwise economically useless timber. Process wastes are those parts of the logs that are removed or discarded as a result of carving regular fixed-size right-angle solids out of irregular, roughly conical ones, as well as from shaping, smoothing, and finishing operations (Christensen, 1976). As mentioned before, bark waste and spent pulp liquor containing combustible plant resins are process wastes of the pulp and paper industry.

As a result of recent economic incentives, logging residues now are often harvested along with usable stems and are chipped or shredded. Bark, slabs, edgings, and trim ends from lumber manufacture, along with veneer cores and waste, are chipped (*hogged*), and combined with planar shavings and sander dust. When process wastes (and other fuels) are burned, the heat produced is used to generate steam. This high-pressure steam may be used first to turn turbines that provide mechanical energy, and/or to turn electricity-producing turbines, and later (in the case of many paper plants) to dry pulp spread on drying screens and cylinders (Parker, 1974).

In recent years a new trend has developed in which wood products (but not paper) manufacturers sell their excess wood wastes to manufacturers of pulp and wood-based board products. A 1977 survey of wood-fuel-fired electric power generating plants (Pingrey and Waggoner, 1979) indicated that a significant number of wood products companies sold their wastes to pulp and paper manufacturers, rather than fully utilizing their in-house electrical power generating systems. Although wood wastes are valuable substitutes for fossil fuels, they have

Table 19.7.   Percentages of Product and Preliminary Wastes in Commodity Manufacture

Commodity	Percentage of Each Ton of Logs Converted into Product	Percent Preliminary Waste
Softwood lumber	35	65
Hardwood lumber	28	72
Sheathing plywood	45	55
Structural particle board	70	30
Lumber laminate from veneer	46	54
Softwood pulp chips	89	11
Hardwood plywood	30	70
Underlayment particle board	89	11
Medium density fiberboard	77	23
Insulation board	89	11

Source: From KOCH/CORRIM (1976).

greater economic value in pulp and board production than in energy production (Christenson, 1976). Thus, like some fossil fuels used as chemical feedstocks, some wood residues are too valuable to burn.

ENERGY AND LABOR USED TO MANUFACTURE FOREST PRODUCTS

In this section we calculate the embodied energy in several of the major products of the forest products industry, analyzing these trends over time. Numerous methods exist for calculating manufacturing efficiency, of which three are most useful (see p. 108). The first method we employ uses data from the *Annual Survey of Manufacturers* (1958–1961, 1963–1966, 1968–1971, 1974–1976, 1978–1980), *Census of Manufactures* (1962, 1967, 1977), and inflation corrections from the *Statistical Supplement to the Survey of Current Business* (1979) and *Statistical Abstracts* (1982–1983) to calculate *energy efficiency* as inflation-corrected *dollars of value added per million kilocalories* of energy consumed directly and indirectly in the manufacture of wood and paper products (Figure 19.2c; Table 19.8).

A second estimate of industry efficiency was made using *mass* of product output (kilogram) *per million kilocalorie consumed.* Again, we used data from *ASM, CoM, SCB* (1979), and *SA* (1982–1983), as well as API (1981), to determine the industrial energy used per unit of output (Tables 19.9, 19.10). This second method is more direct and is independent of the somewhat unreliable corrections for inflation. Unfortunately, this method too is subject to some (unknown) error as product mix changes from year to year. Also, data for output in mass units are not available for every year. Results calculated by the two methods can be compared to assess the magnitude of the uncertainty associated with each methodology and to compare the ratios for manufacturing process efficiency calculated by Koch (1976; see also Table 19.8).

Finally, we assess *labor efficiency* (productivity), in inflation-corrected *dollars of value added per worker-hour,* using data for production worker hours (which we refer to as worker-hours) and value-added dollars from the *ASM* and *CoM* (1958–1980). We estimated labor productivity in forest products industries by dividing inflation-corrected (1972) dollars by worker-hours. Labor productivity is a common measure of efficiency, but when used without a measure of energy consumption, it is a misleading measure of efficiency (see Chapter 2, p. 42).

Patterns and Trends of Energy Efficiency

Results calculated using the first and second methods indicate that in 1979 the paper and allied products industry yielded $18.94 (1972 dollars) of value and produced 81.5 kg of paper goods per million kilocalorie of purchased energy used (Table 19.8). If we include estimates of self-generated power from process wastes (API, 1981), the estimates are reduced to $14.50 of added value and 62.4 kg (Table 19.9). If all manufacturing energy could be supplied from wood, an average of 283 kg of wood (e.g., white birch at 100% efficiency) would have been required to supply the energy used to produce $14.5 (or 62 kg) of paper products. Thus the embodied energy in a kilogram of paper is about 4.5 times greater than its chemical energy.

Similarly in 1979 the harvest and manufacture of lumber and wood products added $29.1 of value and produced 121.5 kg of products per million kilocalorie of purchased energy invested ($27.2 and 88.6 kg per million kilocalorie if the energy from process wastes is included in the denominator) (Table 19.8, Figure 19.2c). More specifically, softwood lumber and particle board manufacture produced 1238 kg and 490 kg, respectively, of products per million kilocalorie of process energy invested (Table 19.10). Comparing the dollar figures to the national energy-GNP ratio of 121.3 value-added dollars per million kilocalorie (see p. 55), we find that both paper and wood products manufacturing are much more energy intensive than the average national commercial process, as was found for various extractive industries by Costanza (1980)—although the output of these industries is not normally final demand goods.

We calculated that over the 22-year period for which *ASM* and *CoM* data are available, the real value added per million kilocalories in the lumber and wood products sector decreased 17% from $37.75 to $31.28 (1972 dollars) per million kilocalorie. On the other hand, real value added in paper and allied products manufacturing rose 13% from $16.08 to $18.07 (1972 dollars) per million kilocalorie (Tables 19.8 and 19.9; Figure 19.2c). No data for waste-supplied energy were available for years prior to 1975, but when estimates for the pe-

Table 19.8. Total Energy Cost of Manufacture in the Lumber and Wood Products and Paper and Allied Products Industries

Year	Purchased Electricity (×10⁹ kcal)	Generated Electricity (×10⁹ kcal)	Purchased Fuel (×10⁹ kcal)	Energy Cost of Materials (×10⁹ kcal)	Total Energy Cost (×10⁹ kcal)	Value Added in Manufacture (×10⁶ 1972 $)	Mass of Product (10⁶ Kg)	Total Embodied Energy Cost of Manufacture[a] (kcal/$)	($/10⁶ kcal)	kg/10⁶ kcal
				Lumber						
1958	8783	830	19,782	103,592	132,989	5019	—	26,493	37.75	—
1962	11,575	873	25,803	104,028	142,280	5738	—	24,796	40.33	—
1967	18,829	580	41,026	140,141	200,578	7176	—	27,948	35.78	—
1971	24,034	934	49,283	132,292	206,546	7682	—	26,884	37.20	—
1974	38,164	430	55,648	199,477	293,720	9088	37,031	32,316	30.94	126.07
1975	37,119	321	45,413	168,200	251,055	8446	34,902	29,721	33.65	139.02
1976	41,693	269	45,511	198,050	285,525	9441	38,836	30,242	33.07	136.01
1977	41,608	256	43,469	240,060	325,395	9904		32,854	30.44	124.57
1978	43,009	271	43,898	253,677	340,856	9917	40,949	34,370	29.10	120.14
1979	41,454	234	42,386	247,814	331,888	9657	40,315	34,365	29.10	121.47
1980	37,871	90	37,623	213,081	288,667	9030	—	31,965	31.28	—
				Paper						
1958	32,351	14,602	192,490	177,181	416,626	6701	—	62,174	16.08	—
1962	43,999	18,204	219,411	186,013	467,628	8241	—	56,744	17.68	—
1967	66,720	19,964	269,211	253,949	609,845	11,061	—	55,131	18.14	—
1971	90,308	21,833	301,465	224,011	637,618	12,031	—	52,997	18.87	—
1974	105,457	23,552	300,261	296,161	725,432	14,168	—	51,201	19.53	—
1975	100,942	21,014	271,275	273,657	666,890	11,939	47,924	55,860	17.90	71.86
1976	112,385	21,924	290,178	304,689	729,175	13,024	54,880	55,987	17.86	75.26
1977	114,978	23,552	292,295	302,582	733,407	13,486	56,523	54,383	18.39	77.07
1978	117,691	23,112	288,666	298,542	728,009	14,143	58,425	51,476	19.43	80.25
1979	119,110	23,318	287,834	311,654	741,916	14,053	60,490	52,794	18.94	81.53
1980	128,200	22,658	279,442	316,794	747,096	13,500	—	55,340	18.07	—

Source: From ASM and COM (various years); SCB (1979); SA (1982–83).

[a] Does not include waste generated electricity.

Table 19.9. Total Energy Cost of Manufacture Including Waste Energy in the Paper and Allied Products Industry

Year	Purchased Electricity (×10⁹ kcal)	Waste Power[a] (×10⁹ kcal)	Purchased Fuel (×10⁹ kcal)	Energy Cost of Materials (×10⁹ kcal)	Total Energy Cost (×10⁹ kcal)	Value Added in Manufacture (×10⁶ 1972 $)	Mass of Product (×10⁶ kg)	Total Embodied Energy Cost of Manufacture (kcal/$)	($/10⁶ kcal)	kg/10⁶ kcal
1975	100,942	204,817	271,275	273,658	850,693	11,939	47,924	71,256	14.03	56.34
1976	112,385	232,510	290,178	304,689	939,762	13,024	54,880	72,157	13.86	58.40
1977	114,978	244,955	292,295	302,582	954,810	13,486	56,523	70,800	14.12	59.20
1978	117,691	252,377	288,666	298,542	957,275	14,143	58,425	67,687	14.77	61.03
1979	119,110	250,409	287,834	311,654	969,007	14,053	60,490	68,954	14.50	62.42

Source: From ASM and COM (various years); SCB (1979); SA (1982–83); API (1981).
[a]Data for the pulp, paper and paperboard industry, a major subset of the paper and allied products industry consuming approximately 80% of purchased fuels and electricity.

Table 19.10. Summary of Process Energy for Various Wood Commodities (kcal/kg)

Commodity	Gross Manufacturing Energy — Logging and Preparation	Electricity	Heat	Transport	Total Energy	Manufacturing Energy Potentially Supplied by Residue Energy	Available Residue Energy	Net Total Fossil Fuel Requirement for Manufacture[b] (kcal/kg)	(kg/10⁶ kcal)
Softwood lumber	262	214	1128	546	2150	1342	(2309)[a]	808	1238
Laminated lumber from veneer	206	32	1790	546	2574	1822	983	1591	629
Oak flooring	298	235	1346	549	2427	1581	(3164)	847	1181
Softwood sheathing plywood	207	41	1868	578	2694	1909	1027	1667	599
Hardwood plywood paneling	289	68	2777	549	3683	2845	(2953)	838	1193
Underlayment particle board	1282	696	1555	333	3866	2251	1827	2039	490
Structural flake board	266	161	1926	365	2717	2087	(2393)	631	1585
Insulation board	173	1367	1561	345	3446	2928	185	3261	307
Wet formed hardboard	206	2756	2706	319	5887	5462	222	5765	173
Medium density fiberboard	217	1042	1543	319	3121	2585	761	2360	424

Source: Adapted from KOCH/CORRIM (1976).
[a]Parentheses indicate surplus.
[b]Assuming supplementary requirements are met with fossil fuels.

riod 1975 to 1979 were used to adjust paper products' energy efficiency, a somewhat larger increase in efficiency (3.3 vs. 1.0%) is calculated for this time period (Figure 19.2c). No yearly estimates for lumber industry waste utilization are available, so a comparable correction for that industry is not possible. Trends during the last decade for both lumber and paper industry yields in kilograms per million kilocalories corroborate the results for value added. Physical measures of energy efficiency in manufacturing show a 4% decline in the lumber industry between 1974 and 1979 and an 11% increase in the paper industry between 1975 and 1979.

Results for Labor Efficiency

As with other economic sectors, rising labor productivity has been achieved in large part by subsidizing workers with increasing amounts of energy; therefore we would expect sectors with high labor productivity rates to have relatively low energy efficiency, and visa versa. In 1979, $13.64 (1972 dollars) of value were added per worker-hour in paper production, compared to $8.62 (1972 dollars) in the lumber industry. This 60% difference probably was not achieved by superior workers in the paper industry but instead reflects a greater energy subsidy per worker in the more automated paper industry. For every worker-hour of paper manufacture in 1979, 638 thousand kcal of purchased and internally generated energy were expended, versus 96 thousand kcal per worker-hour in lumber production.

Over the period 1954 to 1980, labor efficiency increased in both the lumber and paper industries. Labor productivity in the lumber industry rose 81% overall but has declined 38% from its all-time high of $9.98 value added per worker-hour in 1976 (Figure 19.2d). The 27-year trend for labor productivity in the paper industry was similar: it rose 101% but has declined 20% from the 1978 high of $14.80 per worker hour (Figure 19.2d). Presumably these declines reflect the impact of increasing energy prices, and they are consistent with the theory of labor productivity given in Chapter 2.

Lumber Industry Efficiencies

Greber and White (1982) attribute much of the long-term rise in labor productivity in the lumber and wood products industry to technical and institutional change. Their analysis of the industry between 1951 and 1973 indicates that technological changes, such as the replacement of chainsaw-toting loggers with large mechanical devices, have served to displace labor and to increase labor's productive efficiency. They cite similar changes in materials handling in mills, and also the decline of small labor-intensive sawmill and logging firms in the late 1950s and 1960s as other factors that increased the industry's overall efficiency. But the lumber industry appears to have been having difficulty since 1972 in sustaining the momentum of its increase in labor efficiency due, we believe, to the large increase in the real price of fuels, which made the implementation of new technologies less economically attractive, and the associated decrease in fuel consumption.

Even though the lumber and wood product industry has been able to meet approximately 30% of its energy needs with internally generated power, it has failed to produce proportionately more goods with the additional energy it uses (Figure 19.2c). The only change in this trend occurred in 1975 when the industry experienced increases in both labor and energy efficiency during a slowdown in production. In the following 2 years, however, as production increased again, both gains were negated. This is the same trend as seen in copper (Figure 4.14).

Rising productivity rates observed during the 1975 recession, followed by declining rates during the subsequent recovery lends some credence to the hypothesis that extractive industries operate more efficiently at low rates. As manufacturers decrease production during a recession, they usually abandon the marginal or lowest quality resources first because they are less economic. As a result, manufacturers concentrate on higher quality resources and use proportionately less energy. When production expands again, as in 1976, the higher *plus* the lower quality forests are exploited, requiring proportionately more energy, thus decreasing energy efficiency. This is what we observed in the lumber industry as it expanded between 1975 and 1978 and contracted again between 1979 and 1980. Energy requirements for various products are given in Table 19.10.

Trends in Paper Industry Efficiencies

During the 27-year period we studied, both labor productivity and energy efficiency rose in the paper industry. This is in contrast to the lumber industry, but it indicates that in some industries it is possible to have an absolute increase in energy efficiency. These improvements have occurred despite an increasing ratio of domestic pulp production to paper and paperboard production and a decline in the proportion of recycled paper used as a raw material. According to Myers and Nakamura (1978), the use of "wastepaper or imported pulp requires less energy consumption than production based on pulpwood." These authors cite growth in the use of the energy-intensive Kraft (sulfate) pulping process as increasing the amount of energy consumed, although they point out that this process also has made possible the extensive use of pulping liquors for fuel.

Meanwhile several technical factors contributed to an improvement in energy efficiency: the increased use of lumber industry process wastes, the recycling of pulping chemicals and water—and the capture of the heat they contain—and the implementation of continuous pulping digesters, vapor-recompression concentration of pulping liquor, new bleaching techniques, and high-pressure paper drying presses (Ross, 1981).

As in the lumber industry, the paper and allied products industry has had difficulty sustaining the momentum of increasing labor and energy productivity (Figure 19.2d). Labor productivity had improved steadily for at least 19 years prior to the 1973 oil embargo but since 1973 has remained relatively stable. Energy efficiency also had steadily improved until 1975, at which time increases in dollar costs of fuel and equipment, in conjunction with economic recession, mandated investments in pollution control, and, possibly, a decline in resource quality, worked together to reverse the paper industry's improvements in productivity. An 11% reduction in production coupled with a 1% increase in capacity resulted in an 8% reduction in capacity utilization, a factor that Myers and Nakamura (1978) believe also increased the purchased energy to output ratio. The relation between energy efficiency and labor efficiency observed in the lumber industry in the years following the 1975 recession also seems to have held true in the paper industry. Energy efficiency declined slightly in 1976 while labor efficiency rose dramatically; when energy efficiency rose in 1977, the labor efficiency dropped as quickly as it had risen. The following years' efficiencies, however, have not exhibited this relation.

THE FUTURE OF THE FORESTRY INDUSTRY: WILL FOREST BIOMASS CONSUMPTION INCREASE, AND IF SO, WHAT WILL BE ITS EFFECTS?

Forest biomass has recently been examined as a potentially significant energy source. A number of authors (Stanturf, 1979; Cheremisinoff, 1980) have stated that American forests have great potential for decreasing our dependence on imported fossil fuels, whereas others (Pimentel, et al., 1981; Tillman, 1978; and Morrow, 1979) caution against overexpectation and overexploitation (see p. 380). The U.S. Forest Service (1976) estimated that national forests contain 907 million metric tons (one billion oven-dry tons, ODT) of noncommercial (economically useless) timber in rough, diseased, and dead trees, which if harvested and burned could yield 3.0 quadrillion kcal or 16.1% of 1984 national energy consumption. In addition, they estimated that annual harvests in forests produce 109 million metric tons (120 million ODT) of wood and bark residue, 7.26 million metric tons (8 million ODT) of bark, and 18.1 million metric tons (20 million ODT) of fiber from timber stand improvements, the sum of which could yield 0.48 quadrillion kcal or 2.5% of national consumption in 1984. Harvest residues and waste wood could provide between 0.214 quadrillion and 0.416 quadrillion kcal if collected and burned (Pimentel, 1981; Zerbe, 1978). Although some estimates of energy potential from residues and noncommercial timber are encouraging, it is important to consider the energy and dollar cost of collecting, transporting, and handling them. Residues usually are dispersed, so their collection is likely to be costly, and transportation alone can use almost 50% of the total energy for harvest (using traditional methods), even over distances as short as 80 km (Smith and Corcoran, 1979). Perhaps more important, much of these combustible wood residues could instead be used as raw materials for board and fiber products, a use that is usually of greater value to forest products manufacturers. Finally, removing all old-growth, crooked, and diseased timber is neither feasible nor necessarily desirable. Much forested land has multiple uses, such as for watershed protection, wildlife habitat, recreation, or as scenic or wilderness areas, in addition to growing timber. If they are to continue to be useful for these other purposes, they cannot all be harvested.

Methods of Increasing Production

Fuel wood plantations are another potential source of biomass receiving much attention. As discussed in Chapter 13, these use fertile land to grow wood for energy production. According to Tillman (1978) plantations would use up to 80% of the energy delivered because of the costs of management and conversion needed to grow and utilize the fuel products (Table 19.11). Intensively managed plantations may degrade soil as nutrients are removed during harvest, and therefore would require large applications of fertilizer presumably made from fossil fuels (Wells et al., 1975; Henry, 1979). Because plantations require large land areas (11,500 ha cited as a minimum by Henry, 1979), they may divert land from timber production and/or agriculture (Holdren et al., 1980). Thus preliminary calculations for the EROI of plantations (1.2:1) are much smaller than for naturally grown wood (Tillman, 1978; Table 19.11).

As in biomass plantations, whole-tree harvest and complete residue collection remove soil nutrients, and the energy cost of offsetting this loss may be another factor limiting residue collections. Wells et al. (1975) report that whole-tree removal may deplete three times the quantity of nutrients removed in ordinary stemwood harvests while yielding only one-third more biomass. This is because young trees (generally less than 7 years old; Hansen and Baker, 1979) contain a larger percentage of nutrients than older trees, for young trees have a larger ratio of branches, bark, and leaves to stem. If trees are harvested whole and/or residues are collected and burned for fuel, forest soils may have to be refertilized. Morrow (1977) suggests that wood ash be returned to forests in order to recycle needed nutrients.

In addition to nutrient loss in harvested trees, intensive harvests may degrade soil through erosion and reduction of the soil organic matter (Pimentel et al., 1981). In summary, indirect environmental costs of biomass fuels from either plantations, whole-tree harvest, or residue collection severely limit the sustainable use of biomass as a large scale energy source. Fuel wood will, however, continue to play an important and increasingly large part in small scale, local applications, especially for home heating.

Thinning crowded forest stands is one method for collecting biomass that appears especially attractive because it preserves environmental integrity. Historically forest products companies have managed only about 14% of commercial U.S. forestland (i.e., their own holdings), thereby providing just enough timber to maintain a constant supply. If they were to initiate thinning and timber stand improvement (TSI) operations on a portion of the other 86 percent (roughly 30 million ha), however, they could increase national fiber and fuel yield as well as the growth rate of commercially valuable crop trees. This could double net forest production in several decades (PAPTE, 1973 in Bethel and Schreuder, 1976). Initiating such practices in the nonindustrial forests could be delayed by institutional difficulties such as small holdings, adverse local tax systems, multiple ownership, and owner displeasure with changed aesthetics in recently harvested areas.

Thinning and TSI operations may have the greatest impact in the eastern United States, where hardwoods are numerous. Harvesting hardwood thinnings could enlarge the flow of valuable fiber for fuel and for pulp or board manufacture.

Table 19.11. Detailed Energy Analysis of the Wood-Based Energy Farm Producing Liquid Fuels (in 10^{12} kcal)

Total energy to be grown	1000
Wood-growing energy costs	
Cultivation and planting	0.5
Fertilization	32.5
Subtotal	33.0
Extraction energy costs	
Felling and bunching	4.4
Skidding	6.2
Chipping	4.7
Transport to stockpile	6.2
Auxiliary	3.7
Subtotal	25.2
Conversion energy costs[a]	667.0
Transportation energy costs[b]	21.5
Final combustion energy costs[c]	50.0
Net energy delivered	203.3
EROI	1.20
Efficiency if used for electricity production	1.25

Source: From Tillman (1978).

[a] Assumes hydrogenation.

[b] Assumes putting liquid fuel in national energy transportation system; therefore, uses 2.15 percent energy budget.

[c] Losses are taken only from liquid fuel produced.

Additionally the OTA (1980, as reported in Ahokas, 1982 (unpublished manuscript) suggested that enlarged harvests achieved by improved management could relieve industrial pressure to harvest timber in scenic and wilderness settings with aesthetic, recreational, and ecological value. Although many of these areas are relatively inexpensive to harvest because of their large old-growth trees, harvesting would disrupt their alternative uses, often permanently. Therefore thinning operations on other, already disturbed, lands may be beneficial for more than purely supply-oriented reasons.

Factors Limiting Increased Forest Products Use

A major factor affecting the future of the forest products industry as well as its role in our nation's economy is our great dependence on imported softwood lumber and newsprint from Canada. Bethel and Schreuder (1976) state that "the United States is currently so dependent on Canada as a source of softwood lumber and newsprint that any major restriction in price or volume could create a major materials supply problem for this country." Presently these imports are processed in areas where transportation costs render domestic products economically noncompetitive, and a significant price increase for imported softwood would be required to make the production and transportation of domestic softwood products from domestic sources economically feasible. Even if such transport were feasible, it is doubtful that domestic supplies could meet the new demand without a considerable time lag.

Another element regulating the supply and consumption of domestic hardwoods and softwoods is the rate of investment and its effect on manufacturing capacity. For a number of years (e.g., 1974 to 1976; also 1980 to 1982) the rates of investment in new production equipment were less than normal because of the recession (Myers and Nakamura, 1978). Bethel and Schreuder (1976) suggest that the industry's growth has been hindered in recent years by compliance with expensive pollution control regulations. They suggest that approximately half of new capital investments by wood conversion plants during the past several years have been in pollution control equipment. American Paper Institute data (1981) do not corroborate this, showing instead that between 1971 and 1979 on average 26% of capital expenditures by the pulp and paper industry was invested in "environmental protection equipment."

Although investments in pollution abatement equipment do not enlarge production capacity in the near term, they can benefit the industry. As mentioned before, process wastes have become valuable sources of energy and raw materials, and equally important, using them has reduced pollutants and helped to maintain (and in some cases improve) environmental quality and the image of the paper companies in the eyes of the public.

Factors That May Lead to Increased Use of Forest Products

The gradual replacement of softwoods by hardwoods may help to alleviate a possible softwood shortage. Today increasing quantities of hardwoods are used to manufacture traditionally softwood-based commodities such as pulp, plywood, and particle board. Substituting wood types is a relatively new trend, but it is slow to occur because the relatively lower harvest and manufacturing costs of softwoods discourage this switch. Hardwood particle board is presently taking a significant share of the particle board market, and because U.S. demand for particle board is increasing faster than that of any other structural forest commodity, hardwoods may eventually replace softwoods in board manufacture. Unfortunately, although less energy is required to produce a ton of pulp from hardwood than from softwood, it makes weaker paper because of hardwood's shorter fibers. Where reuse is important short fibers are a great disadvantage because hardwood-derived paper cannot be recycled as many times as paper from softwoods. Nevertheless, because of a looming shortage of softwoods in Maine—the result of repeated ravages by spruce budworms on young monocultures of that tree species—several companies have initiated research to find an economical means of producing high-quality papers from hardwoods (Lovell, 1983). Although hardwoods will never completely replace softwoods in all markets, the use of more hardwoods may free domestic softwood supplies enough to decrease our dependence on imports from foreign sources (Figure 19.2b). As mentioned before, there is a surplus of hardwood growth over harvests, and there is a potential to double hardwood production through proper management. Therefore an increased utilization of hardwoods is a promising prospect for a stable future in the forest products industry and in turn the national economy.

One last aspect of the industry that will have

increasing economic influence in future years is our national rate of paper and paperboard consumption. In 1825 U.S. citizens consumed less than 1.5 kg of paper per person per year (*Historical Statistics of the United States, 1976*). Today Americans annually consume nearly 215 kg of pulp products per person (*Statistical Abstracts, 1982–1983*). If consumption continues to grow at the 1979 rate, per capita consumption could become 250 kg within a decade. At such levels it would be difficult, if not impossible, to increase harvest of forests for energy, and the many environmental problems associated with the production and disposal of paper would be exacerbated.

We cannot do without paper products, but we can use them more efficiently. By reducing excessive packaging, by recycling, and by burning any paper that cannot be recycled economically, we may avoid unnecessary waste. Americans currently recycle about 23% of all paper produced domestically, but according to Jahn and Preston (1976) recycling could reach 30%. Above this point, Jahn and Preston believe that the dollar cost of collection and separation becomes prohibitive. It is not clear at exactly what point more energy would be used for recycling than would be saved. Gunn and Hannon (1983) suggest that we are near that point now, in agreement with the economic finding of Jahn and Preston.

Burning wastes to generate energy is an alternative solution to the solid waste problem. In the past burning merely redistributed waste from solid debris to air pollution. Today generating energy from burning, coupled with pollution control measures, is being contemplated and sometimes even employed in both municipal and factory power systems in order to replace fossil fuel. But even this strategy is not without costs. Combustion decreases energy costs of transportation and recycling, but it destroys fiber that required approximately 4.5 times as much energy to produce as is released by incineration.

The price of imported fossil fuels is not likely to decline in the long run, nor is total energy consumption likely to decrease in the forest products industry as paper and lumber consumption increase. As a result of probable increases in both energy prices and energy consumption, the forest products industry will seek new ways to squeeze more products and more fuels from the raw materials it consumes. Thus the forest products industry probably will not become a significant source of energy for the rest of society, since its general approach is to use energy to produce forest products, rather than the converse.

Because fuels are needed for transportation and harvest, the forest products industry probably will never be totally energy self-sufficient, although it may come very close. The paper industry has increased its efficiency of producing products from purchased energy by about 0.6–0.8% per year from 1958 to 1980. Further, an increasing percentage of its energy now comes from previously incinerated wood waste and previously dumped pulp liquor. Through increased but regulated use of harvest residues and process wastes, increased management of forest stands, and improvements in operations efficiency, the forest products industry has the potential to become the first of the giant U.S. industries to approach energy self-sufficiency while still preserving the health and productivity of its materials sources—America's forests. Whether or not it achieves such a goal, however, will depend on the leadership and vision of the industry itself and those who regulate the industry, especially where such long-term goals conflict with short-term profits.

20

CROPLAND

Charles Staver, with the assistance of Melinda Dower

The diverse patterns of soils, vegetation, topography, and water that comprise the Earth's surface have developed and continue to change in response to geological forces, direct solar inputs, air, water, heat fluxes, and biotic forces. Many of these features influence how human society uses the land. In turn, human use has become a significant force in shaping the Earth's surface.

One of the most fundamental human uses of land now and in the past is agriculture. Chapter 6 examined the development of energy use in agriculture from its solar- and labor-powered beginnings to its current fossil energy dependence. This chapter expands the information of Chapter 6 by considering the use of land as it relates to soil and other land qualities, fossil energy requirements, and the prospects for sustaining and increasing agricultural production. We look first at the characteristics of land relevant to its use in agriculture and relate land use to land quality. In the second section we examine the quality, quantity, and distribution of U.S. land resources. Finally, we project the role of land quality in terms of the adequacy of the land base and possible changes in the use of fossil energy inputs.

From this discussion we conclude that the pattern of modern human use of a renewable resource such as agricultural land follows a pattern similar to that of the nonrenewable resources examined in Chapters 7–12. Four points will emerge: First, land quality is determined essentially by differences in soil, climate, landscape, and location. Human standards of land quality, however, have changed with agricultural technology and human settlement patterns. Second, high-quality agricultural lands require smaller energy investments to utilize than their lower-quality counterparts. High-quality agricultural lands are also more resistant to degradation of their productive capacity by human use. Third, over time crop production in the United

States has become concentrated increasingly on higher-quality lands where returns to economic energy investments are greater. As these lands are lost from agricultural production due to soil erosion, salinization, and conversion to other uses, or as the agricultural land base needs to expand, lower-quality lands must be brought under cultivation, increasing per hectare energy inputs. Fourth, technical change in modern agriculture has increased crop yields per hectare and farm output per laborer. As a result total agricultural output also has increased, even though total labor use and land area under cultivation have declined. This has been achieved with an ever-increasing expenditure of fossil energy inputs to save labor and increase land productivity.

CHARACTERISTICS OF LAND RELEVANT TO AGRICULTURAL USE

The basis of agriculture is the fixation of solar energy by plants for subsequent use by humans or animals (Spedding, 1971). The quality of agricultural land must be evaluated in relation to soil and climate, both attributes of land fundamental to the growth and productivity of plants. *Soil* is the support medium for roots and the repository of nutrients and water. *Climate* is the distribution of light, heat, and precipitation that controls the plant growth cycle and determines the level of productivity. Soil and climate quality affect the quantity of fossil energy used and the response of the crop to different inputs and types of management.

Landscape, which is composed of diverse patterns of soils and relief, is a third characteristic of land important for agriculture. Landscape influences the spatial organization of agricultural production, the type and scale of technology used, and the levels of production attained. A single technol-

481

ogy or management will not be appropriate for all types of land nor will it result in the same yields.

A final aspect of land that influences and is influenced by patterns of human land use is its *location,* a characteristic of its spatial relation to other types of land and land uses. In the following sections we examine these characteristics individually and relate them to an overall view of land quality and of land and energy use.

Soil

Soil consists of mineral particles, organic matter, water, and air in more or less distinct horizontal layers. As a medium for plant growth, soil provides physical support for the plant and receives and stores water and nutrients. The soil also is the site for the physical, chemical, and biological processes that convert mineral and organic matter into more soil (for a basic text on soils, see Brady, 1974).

Soil formation is influenced by five interdependent factors:

1. Climate (particularly temperature and precipitation).
2. Living organisms (especially native vegetation as well as microorganisms and larger fauna).
3. Parent material (texture and structure, chemical and mineralogical composition).
4. Topography.
5. The time that parent materials are subjected to soil-forming processes.

Soil can be considered an open system that evolves in response to the energy of geologic, biotic, hydrologic, and meteorologic processes. These energies transform less desirable raw materials (i.e., parent materials) into an economically useful product.

Crops and forages can be grown on a wide range of soils depending on the plant species, variety, and soil management techniques. Soils vary in fertility, water-holding capacity, rooting depth, possible susceptibility to erosion, and internal drainage. These features in large part determine the potential crop yield, the quantity of fossil energy needed to achieve such yield, and the sustainability of such yield. The variability of these factors is related to soil texture, cation exchange capacity, depth, and stratification which are relatively permanent features of a soil, and soil structure and organic matter levels which are altered through human use.

Soil texture is determined by the relative proportions of sand, silt, and clay that constitute soil minerals. Mineral matter (and its texture) influence the nutrient and water retention capacity of a soil while supplying many essential nutrients such as calcium, magnesium, and potassium. A soil dominated by sand is loose, well aerated and drained, and easy to work but has little capacity to hold water and nutrients. A large clay fraction holds water and nutrients well but requires more energy to plow, and it may have poor drainage and aeration. A loam soil with similar proportions of sand, silt, and clay has intermediate characteristics in workability, aeration, drainage, and moisture retention and is well suited for the growth of many crops.

The capacity of a soil to retain cations through ionic bonding is known as *cation exchange capacity* and is a function of the clay and decomposed organic matter of the soil. This quality is important because calcium, magnesium, and many other nutrients are held in the soil and released slowly through exchange reactions for uptake by plants. In soils with little organic matter and a large fraction of sand, cation exchange capacity is small and the reserve of nutrients held in the soil is limited. The productivity of soils of this type can be upgraded by applying fertilizer or animal manures, but if these added nutrients are not taken up immediately by plants, they are likely to be leached from the soil and lost into the groundwater.

Soil pH influences plant growth via direct effects of hydrogen ions and indirect effects on nutrient availability and toxicity. Soil at the proper pH permits maximum crop response to nutrients from fertilizer and from decomposing organic matter. Soils with pH's unsuitable for agricultural production can be altered by the use of economic energy. Lime, for example, is used on acid soils to increase the pH and enhance the availability to the plant of phosphorus, calcium, magnesium, and molybdenum and to reduce the availability of iron, aluminum, and magnesium which are toxic to crop plants under very acidic conditions. The pH of alkaline soils can be reduced to enhance crop production by adding gypsum.

Soil *stratification* refers to the internal arrangement of layers of different textural, chemical, and structural characteristics. These strata are referred to as A, B, and C horizons. The A horizon is the surface horizon that accumulates organic matter and is leached of minerals and clays. Directly below the A horizon is the B horizon, which is composed of weathered material and minerals leached from the A horizon. The C horizon is unconsolidated and unweathered material below the B horizon and is relatively unaltered by soil-forming processes. The depth to the B and C horizons can vary greatly from one soil to another, so that some soils may have fairly uniform characteristics throughout the entire rooting zone and others have differently textured materials or different minerals in the different strata. For example, a soil may accumulate clay in the B horizon, which restricts the movement of water and the penetration of roots. Other soils may be only a thin layer over bedrock so that even low rates of soil erosion may render them completely unproductive.

Soil *structure* characterizes the aggregation of mineral particles into crumbles or granules. Soil structure influences water infiltration, soil erosion by wind and water, root penetration, and aeration. Soil structure (in the surface layer) is termed *tilth* and is improved by alternating periods of wetting and drying, freezing and thawing, and by the physical action of roots, soil fauna, and decaying organic matter. Tilling can have both positive and negative effects on soil structure through stirring action and compaction caused by the passage of heavy machinery, respectively.

Although organic matter in many soils may be only 3–5% by weight, it strongly influences the potential of the soil for agricultural use. Partially decayed plant and animal residues have a high capacity to hold moisture and nutrients and bind individual particles into soil structure. Many important nutrients such as nitrogen, sulfur, and phosphorus are released as plant residues or manures decay. This process occurs throughout the growing season, releasing a steady supply of nutrients for the crop. Because of its importance the level and type of soil organic matter often is used as a measure of the adequacy of management in maintaining soil productivity.

In summary, a good soil is a deep, well-drained loam with medium to high organic matter content. This soil type is able to retain enough water to satisfy crop requirements, even with uneven rainfall, and also nutrients for later release. A crumbly, granular surface structure allows good water infiltration and resists wind and water erosion. Although soils with other characteristics can be highly productive, they require more precise timing of tillage, weeding, and harvesting operations and also higher levels of economic energy in the form of fertilizer, water, and fuel for machinery. Many other soils are so deficient in one or more of these characteristics that the large energy costs necessary to upgrade their quality prohibit their use for large-scale agricultural production.

Climate

Climate is obviously important for plant growth, for sun and water are prerequisites for any life form. The amount of sunlight, temperature, and rainfall determine the periods of the year when crops can be grown without irrigation. In colder regions these factors also determine if and when a particular crop can be grown at all. In the Northern United States the frost-free season, the period when temperatures do not drop below 0°C, determines the cropping schedule. The longer frost-free season or mild winter temperatures in Southern areas may permit crop production for Northern winter markets, double cropping, or the growth of longer season crops. Nevertheless, cold winter temperatures can be very beneficial for agriculture as the cold weather destroys many competing weeds, and the short nights of temperate summers reduces crop energy losses to nighttime respiration compared to universally long tropical nights.

The quantity and distribution of rainfall influences the crop species and their productivity. Corn, for example, is grown in the relatively wet Midwest, although even there yields may vary up to 25% from year to year due to rainfall variation. Crops requiring less water, like sorghum and wheat, are grown farther west where rainfall is less abundant and more variable in its distribution. In many other areas crop production is impossible without irrigation, a practice that increases crop yields at considerable energy cost.

Landscape

Agricultural land quality and land use patterns also are influencd by factors other than soil and climate. The term *landscape* embraces both the relief and pattern of the mosaic of soils and slopes in a given region.

Relief refers to differences in elevation within an area and affects water drainage and soil erosion. Water moves down a slope, either as surface runoff or as groundwater within the soil itself, and eventually enters a stream or river. The proportions of rain moving internally or as runoff depends on soil texture, soil structure, slope, and type of surface vegetation (USDA, 1951). The length and the steepness of the slope influence the amount and rate of water runoff, which in turn largely determines the amount and rate of soil erosion. Internal movement of water ranges from excessive in the case of porous soils on steep slopes to very poor in soils with a high water table or in soils at the foot of slopes which receive drainage water from higher elevations. Crops grown on soils with excessive drainage often will suffer water shortages. Conversely, crops grown on poorly drained soils can do poorly because of excess water.

Landscape pattern refers to the spatial complexity of different soils and degrees of relief. A landscape with one or two soil types over large areas with a simple, uniform slope can be managed as a unit, whereas a landscape of many soil types in a hilly or mountainous area with complex slopes requires many variations in management. Pattern and relief influence field size, potential farm size, production technologies, the need for land improvement measures such as drainage, terracing, and crop rotations, and permanent crops. All of these in turn influence the energy used to produce a crop. An example of contrasting regional land use patterns due to landscape differences is the homogeneous agriculture of the relatively flat Midwest compared to the mixed agriculture with diverse management practices of the Appalachian Mountains of Kentucky and West Virginia.

Location

The aggregate pattern of land use, including agricultural, residential, urban, forest or wetland uses, also is related to land quality. The range of possible uses of a given parcel of land is influenced by other types of uses within the immediate area and the relative location of other factors like markets or processing plants. Small parcels of flat, well-drained loam soils at the urban fringe, for example, often are not suitable for agriculture due to the absence of agriculture-related support services, diseconomies of scale caused by using expensive farm machinery on small parcels of land, or restrictions due to zoning ordinances. On the other hand, relatively poor quality land located near a major city

may be used profitably for fresh milk production due to lower transportation costs.

As technology, demand, and surrounding land use change, locational factors change. Major fresh-vegetable-producing areas, for example, which were located near major urban centers before the advent of refrigeration and processing technology, are now concentrated in California, Florida, and Texas. This concentration has been powered in large part by high quality fossil fuels used to transport crops rapidly over long distances to consumption centers. There also has been a shift away from the consumption of such nonperishable vegetables as cabbage and carrots to more perishable vegetables such as lettuce, tomato, and broccoli, crops that require different soils and climate for optimum growth.

Land Quality: A Summary

One attempt to measure land quality from an energy perspective was based on the solar energy embodied in soils (Odum et al, 1981). Another was by Regan (1977) who quantified the energy contributed by each soil-forming factor—climate, biological organisms, topography, and parent material—over the period needed for soil development. Regan also looked at the energy stored in the soil in the form of organic matter, available ion exchange bonds, and water-holding potential. These calculations, however, are incomplete measures of agricultural land quality because they ignore the structure and organization of the soil, especially stratification and the patterns of soil distribution in the landscape.

A more widely held view is that land quality cannot be measured in absolute terms, but rather it is a relative concept expressing the capacity or potential of land to support a given use at a given time with given technological conditions or management levels. Defined in this way, land quality changes with changes in technology, management skills, the availability of other resources, and the surrounding land use (Barlowe, 1978).

In a study of changes in agricultural land use during 100 years in two counties in Wisconsin, Auclair (1976) found that thickness of the B horizon, the percentage of sand in the B horizon and the depth to bedrock had become important characteristics of the land under cultivation due to erosional impact and especially technological change. In contrast elevation, soil pH, and the percentage of clay in the surface horizon were no longer important characteristics.

In summary, the relevant characteristics of agricultural land quality are soil, climate, landscape, and location—most of which are derived from natural energy processes and all of which contribute to the ultimate measure of land quality as the quantity and value of the yield. Under a given set of technical and economic conditions, land with certain physical and locational characteristics is limited to specific uses. But as new technology is developed, as economic conditions change, and as the surrounding land use changes, the same land may have a wider or narrower range of uses.

THE QUALITY AND USE OF U.S. LAND RESOURCES

The most widely used land quality classification system in the United States is the Soil Conservation Service Land Capability Classification system for which abundant data are available (Klingebiel and Montgomery, 1961). This system consists of eight classes and four subclasses as shown in Table 20.1. It is based on an interpretation of the effects of soil type, climate, slope, and other permanent land features on productive capacity. It also considers the effects of these variables on limitations of land use, on soil management requirements, and on the risks of land degradation. The classification is designed primarily for agricultural land use and assumes the use of contemporary agricultural technology, including low labor inputs and high levels of fuel-intensive inputs such as mechanization, fertilizer, irrigation, and drainage.

Land quality is measured by the Soil Conservation Service system according to the range of possible uses for a land class, yield levels attainable with given amounts of inputs, the need for additional inputs to attain a given yield level, and the susceptibility of the land to permanent loss of productivity under specified management systems. The first criterion of land quality is the range of possible uses for each capability of class (Figure 20.1). As seen in the figure, Class I can be used safely for everything from intensive cultivation to forestry, but Class VIII land is suitable only for wildlife habitat. Class I would therefore be considered of higher quality for agricultural purposes.

Table 20.1. Soil Conservation Service Land Capability Classes and Subclasses

Capability Classes

Lands Suited for Cultivation and Other Purposes

I   *Soils with few limitations that restrict their use*
These lands have deep, well-drained soils with good water and nutrient-holding capacities. They are nearly level and are not subject to erosion.

II *Soils with some limitations that reduce the choice of plants or require moderate conservation practices*
These lands may have one or a combination of the following limitations: (1) gentle slopes, (2) moderate susceptibility to erosion, (3) less than ideal soil depth, (4) somewhat unfavorable soil structure and workability, (5) possible flooding, (6) wetness which may be corrected by drainage.

III *Soils with severe limitations that reduce the choice of plants or require special conservation practices*
These lands may have one or more of the following limitations: (1) moderately steep slopes, (2) high susceptibility to erosion or severe past erosion, (3) frequent flooding, (4) slow permeability of subsoil, (5) some wetness even after drainage, (6) shallow root zone, (7) low moisture holding capacity, (8) fertility problems not easily corrected.

IV *Soils with very severe limitations that restrict choice of plants or require very careful management*
The limitations occurring singly or in combination of these lands are (1) steep slopes, (2) high susceptibility to erosion, (3) severe past erosion, (4) shallowness of soil, (5) excess wetness even with drainage, (6) severe crop damage due to flooding.

Lands Limited in Use

V   *Soils with little or no erosion hazards, but with other limitations that restrict their use to pasture, rangeland, forest, or wildlife food or cover*

VI *Soils with severe limitations that make them generally unsuited to cultivation and restrict their use largely to pasture, rangeland, woodland, or wildlife food and cover*

VII *Soils with very severe limitations that make them unsuited to cultivation and that restrict their use largely to grazing, woodland, or wildlife*

VIII *Soils with limitations that preclude their use for commercial plant production and that restrict their use to recreation, wildlife, water supply, or to aesthetic purposes*

Capability Subclasses

e   Susceptibility to erosion or past erosion damage is the principal problem in their use.

w   Excess water is main problem limiting use of these lands.

s   Soil factors such as shallowness, stoniness, low moisture-holding capacity, low fertility, or salinity are main hazards in use.

c   Soil for which unfavorable climate (temperature or moisture) is main problem in use.

Source: From Sampson (1981).

Another indicator of land quality is the amount of economic energy needed to upgrade land for a given use or to obtain a given level of output from the land. Corn, for example, can be grown in Iowa on soils from all five subclasses, I to IIIw, although with slightly decreasing yields (see Table 20.2). This pattern indicates the declining quality of the different subclasses, and more energy-intensive management practices are needed on the lower subclasses to obtain even those lower yields while minimizing land degradation. The best lands require only those practices to maintain soil organic matter levels that tend not to be especially energy intensive. The same management would be necessary for corn production on Subclass IIe, but these lands also require additional costs for contour plowing, terracing, and strip cropping to prevent soil erosion. Most Subclass IIIe land would require many energy-intensive erosion control measures. For IIw and IIIw lands, dollar and energy investment in drainage would be necessary—in addition to certain erosion control practices—to produce corn.

Land class also reflects soil erosion and especially the rising energy costs for preventing soil ero-

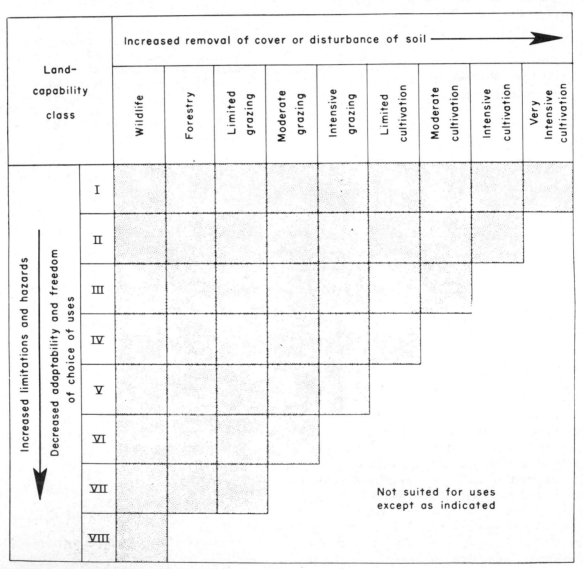

Figure 20.1. Intensity with which each land capability class can be used with safety. Note the increasing limitations on the uses to which the land can safely be put as one moves from Class I to Class VIII. (From Hockensmith and Steele, 1949.)

Table 20.2.   Predicted Average Corn Yields from Soil Survey of Guthrie County, Iowa

Capability Class/Subclass	Percentage that is Prime Land	Average Yield Expected (kg/ha)
I	100	7105
IIe	87	6856
IIw	86	6290
IIIe	26	5975
IIIw	43	5472

Source: From K. O. Schmude. Published originally in the *Journal of Soil and Water Conservation,* Volume 32, pp. 240–242. Copyright © 1979 Soil Conservation Society of America.

sion (Table 20.3). The average annual sheet and rill erosion rate increases from 6.3 metric tons/ha on Class I land to 11.7 metric tons/ha on Class III land to 25.2 metric tons/ha on Class VI land. These rising rates of erosion require concurrent investments of economic energy to reduce erosion rates on lower quality lands below the Soil Conservation Service's "acceptable" limit of 11.25 metric tons/ha. Without these measures the productivity of poorer quality land would decline rapidly as the nutrient-rich topsoil is eroded.

Another example of the increased economic energy costs of using lower quality land is the use of agricultural machinery in New York State. The

Table 20.3.   Annual Erosion Rates from Total U.S. Cropland by Capability Class

Capability Class	Percentage of Total Cropland	Erosion Rate (Sheet and Rill) (Metric tons/ha/yr)	Percentage of Total Cropland Erosion
I	8	6.3	5
II	45	8.1	35
III	32	11.7	35
IV	11	14.8	15
VI	3	25.2	8
VII	1	31.9	2

Source: From D. E. McCormack and W. E. Larson. Published originally in the *Journal of Soil and Water Conservation,* pp. 392–406. Copyright © 1981 Soil Conservation Society of America.

amount of fuel oil, grease, repairs, and maintenance energy required per hectare was 5% greater on Classes III and IV land compared to Classes I and II, and 10% greater for Classes VI and VII. Part of the energy costs can be attributed to increased machinery wear and tear and greater power requirements on lands with steeper slopes and/or poorer drainage (Knoblauch and Milligan, 1981). Although these differences per hectare are not great, they would be much greater expressed as energy per bushel of corn, for crop yields are much lower on Classes VI and VII land.

Current and Past Land Use

The most recent inventory of uses of nonfederal rural lands, which encompass 51% of the total U.S. land area, was carried out by the Soil Conservation Service in 1977 (USDA, 1979). Of these nonfederal rural lands, 44% is Classes I through III land, 12% Class IV land, and 44% Classes V through VIII land. Approximately 50% of this total cropland is limited by susceptibility to erosion, 25% by excess water, 10% by a soil limitation in the root zone, and 6% by climate (USDA, 1980). Table 20.4 indicates land use by capability class. Not surprisingly cropland is concentrated on the better lands, whereas pasture, rangeland, and forests are found primarily on land of intermediate and poor quality. Agricultural land use patterns exhibit the same trend as fuel and mineral resources. The highest quality land is used preferentially and is the most productive. Lower quality lands can be brought under cultivation only with significantly greater energy costs.

Table 20.5 shows the distribution of prime farmland by farm production regions. Prime farmland, according to the Soil Conservation Service, is land that gives the highest crop yields with the least inputs of energy and money and that causes the least damage to the environment. Prime farmland includes all Class I lands, 80% of the Class II lands, and about 30% of the Class III lands (Schmude, 1977). The Corn Belt and Northern and Southern Plains contain 60% of the total prime farmland, whereas the Northeast and Southeast states and Appalachia contain only 17%.

From about 1930 to 1970 total area in cropland declined after a long period of expansion (Table 20.5; Table 20.6) (Barlowe, 1978). More important, during that period crop production became concen-

Table 20.4.   Use of Rural Nonfederal Lands in 1977 by Capability Class

	Classes I and II	Classes III and IV	Class V	Class VI	Classes VII and VIII	Total	%
	(million ha)						
Cropland	89	70	1	5	2	167	29
Pasture and Rangeland	21	66	3	67	62	219	37
Forestland	17	45	7	34	44	150	26
Other	5	8	1	2	29	45	8
Total	132	192	12	108	137	581	100
Percent	23	33	2	19	23	100	

Source: From Sampson (1981).

trated on the higher-quality lands. In 1975, 55% of the total cropland was Classes I and II land, an increase of 8% over 1950. At the same time cropland on Classes V through VIII dropped from 9 to 4%. This shift, coupled with the overall decline in cropland area, suggests that much of the decline in total cropland area has resulted from the conversion of marginal cropland to pasture and forests.

Trends in the use of other factors in agricultural production are shown in Figure 20.2. Energy inputs

Table 20.6.   Trends in the Use of Nonfederal Lands for Cropland by Capability Class

Class	1950[a]	1958[b]	1967[c]	1975[d]
I and II	47%	49%	51%	55%
III and IV	44%	45%	44%	41%
V–VIII	9%	6%	5%	4%
Total (million ha)	190.7	181.0	176.9	161.9

[a] From SCS (1953).   [c] From USDA (1971).
[b] From USDA (1965).   [d] From SCS (1977).

Table 20.5.   Regional Distribution of Prime Farmland in the United States, 1975

	Total Prime Farmland (million ha)	Percent	Prime Farmland Cropped (1975) (million ha)	Total Lost (1967–1975) (million ha)	Percent	Remaining Reserve[a] (million ha)	Percent	Lost as Percentage of Reserve
Appalachia	10.7	7	5.3	0.31	10	1.7	18	18
Corn belt	31.1	20	24.8	0.58	18	1.2	12	48
Delta states	12.0	8	6.2	0.2	6	0.7	7	29
Lake states	15.5	10	12.3	0.24	7	0.4	4	60
Mountain states	7.3	5	6.4	0.1	3	0.1	1	100
Northeast states	7.2	4	3.6	0.38	12	0.2	2	190
Northern plains	29.2	19	23.5	0.21	6	1.5	16	14
Pacific states	5.6	4	3.6	0.25	7	0.3	3	83
Southeast states	9.6	6	3.6	0.65	20	1.2	12	55
Southern plains	27.2	17	11.8	0.34	11	2.4	25	14
Total	155.5	100	101.1	3.25	100	9.7	100	34

Source: From Sampson (1981).
[a] Prime farmland estimated in 1975 to have high potential for conversion to cropland under 1974 economic conditions.

in the form of fertilizer and irrigation have increased. The energy of human labor and animal power has been replaced in large part by energy embodied in mechanical convertors like tractors. This has reduced the total number of farm laborers while increasing labor productivity (see Figure 20.3). Spedding (1979) has calculated that one-third of the fossil energy used in agriculture is to increase yields through inputs like fertilizer, irrigation, and improved crop varieties. The other two-thirds replaces and subsidizes labor inputs, especially through mechanization. As Figure 20.3 indicates, crop production per hectare has increased steadily since 1930. Thus, even though total crop area has declined, total agricultural production has increased

through the use of progressively larger amounts of fossil energy inputs on high-quality land.

Farmland Conversion and Losses in Productivity

As we have already described, Class I land can be used easily for many competing economic purposes. In the case of crop, pasture, and forestland the use for one purpose does not necessarily preclude its use for another in the future. Unfortunately high-quality land for agriculture also is often the best land for urban development. Characteristics of poor agricultural lands, such as poor drainage, high water table, shallow depth to bedrock,

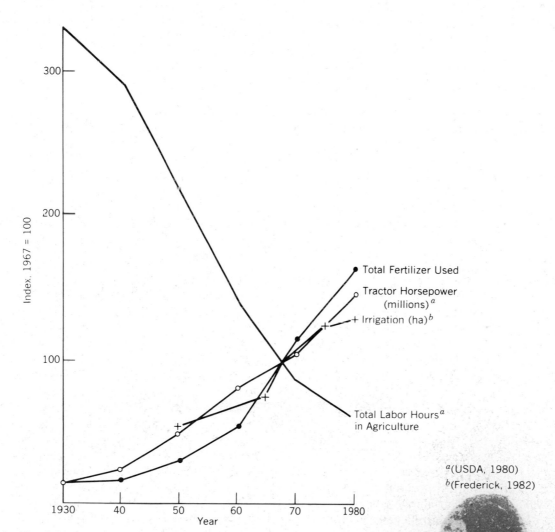

Figure 20.2. Trends in the use of factors of production in U.S. agriculture, 1930–1980. (Data from USDA, and Frederick, 1982.)

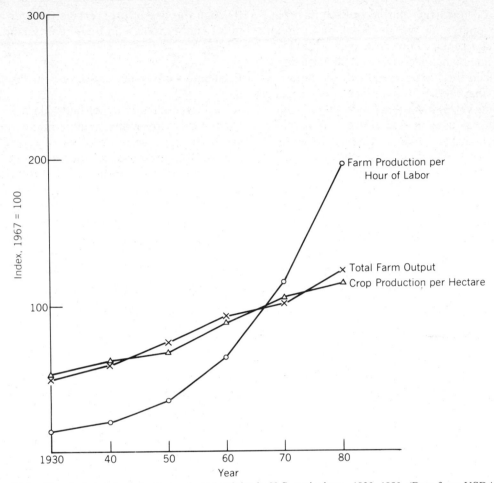

Figure 20.3.   Trends in total production and productivity in U.S. agriculture, 1930–1980. (Data from USDA, 1982.)

and steep slopes increase the costs for housing developments and road construction and limit site potential for septic tank seepage fields and waste disposal (Olson, 1981). As a consequence high-quality land in many areas of the United States is being converted for urban uses which for all practical purposes preclude the reconversion of that land to agricultural use. Agricultural land also is lost permanently through conversion to reservoirs and mines, and through erosion.

From 1967 to 1977 the net agricultural land resource base lost was approximately 0.72 million ha/yr, with 2.1 million ha lost and 1.38 million ha gained by conversion to cropland. Of this net change 39% was lost to urban development, and the remainder was converted to pasture and forestland (Dideriksen et al., 1979). Of the amount lost to urban uses and water, 34% was prime farmland and another 34% was nonprime cropland.

The severity of such loses varies across regions as shown in Figure 20.4. The Corn Belt and Southeast lost the largest quantities, but the Northeast and Mountain states lost a larger percentage of their remaining reserve. Although the total loss of prime farmland during this decade was only 2% of total prime farmland, the impact in certain regions is far greater than in others. In addition, effective losses may be greater, since for every hectare lost directly to urban development, at least one more hectare is indirectly affected in ways that make intensive agriculture difficult or impossible (Sampson, 1981). These include reducing field and farm size to a degree too small for efficient machinery use, the destruction of regional agriculture-related services, and the implementation of land use restrictions through zoning.

Land lost due to energy production includes 161,940 ha/yr of land disturbed for mining, 80,970

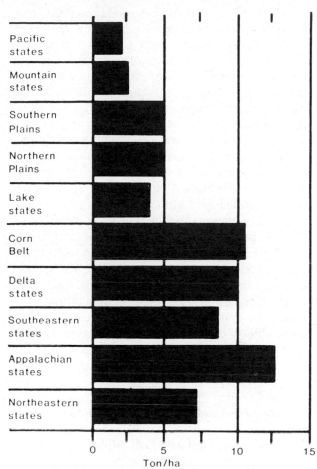

Figure 20.4. Average sheet and rill erosion by cropland region in 1977. (From Larson et al., *Science* Vol. 219, pp. 458–465, Figure 3, 1983. Copyright 1983 by the AAAS.)

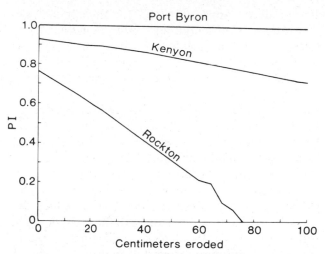

Figure 20.5. The productivity index (PI) of three soils plotted against the centimeters of soil removed through erosion. (From Larsen et al., *Science* Vol. 219, pp. 458–465, Figure 7, 1983. Copyright 1983 by the AAAS.)

ha/yr for reservoirs, and an unquantified amount for roads, transmission lines, and power generating plants (Sampson, 1981). Some of this land may be restored after use, but in many cases its productive potential has been diminished.

Soil erosion alters the land base largely by decreasing productivity, although in extreme cases land may be abandoned for crop production due to erosion. Productivity losses vary from soil to soil depending on the soil depth and type of subsoil. Three soils from the Midwest are compared in Figure 20.5. Productivity of the Port Byron soils are not altered, even with the loss of 100 cm of soil, whereas the productivity of the Rockton soil drops to zero with the loss of 75 cm of soil. In other regions the loss of 10 cm of soil may decrease drastically the land's productivity.

To establish a tolerable limit for annual soil ero-

sion rates, the Soil Conservation Service estimated soil formation rates for the A horizon, the layer from which soil is lost during erosion. If soil is being lost at less than the formation rate, there is no loss of productivity due to erosion. This approach is not entirely satisfactory because soil formation rates vary in response to the five factors mentioned earlier in this chapter. In addition, the soil formation rate for the total rooting depth (weathering of parent rock or deeper soil horizons) occurs at a much slower rate. Nevertheless, 11.25 metric tons/ha/yr has been established as the tolerable limit for soil erosion; this corresponds to a general approximation of the formation rate of the A horizon (McCormack and Larson, 1981).

Sixty-six percent of total cropland loses less than 11.25 metric tons/ha/yr, 23% loses between 11.25 and 22.5 metric tons/ha/yr, and 11% loses more than 22.5 metric tons/ha/yr (Sampson, 1981). These rates of soil erosion vary greatly among regions (Figure 20.4). The Appalachian Region, the Corn Belt, and the Delta States average more than 10 metric tons of soil lost per hectare per year, whereas the more arid regions have much lower rates of erosion due to low rainfall. Mississippi, Missouri, Tennessee, and Iowa have the highest erosion rates, exceeding 22.5 metric tons/ha/yr (USDA, 1980). Sampson (1979) has attempted to quantify total soil loss due to erosion on a hectare-equivalent basis. Sampson calculated that if 1460 metric tons of soil are considered equivalent to a hectare with a layer of soil

10 cm deep, approximately 1.2 million hectare-equivalents of potential agricultural land are being lost due to erosion each year. Most of the soil lost originates from Classes III through VII, but losses from Class II are also significant in absolute terms (Table 20.3).

Agricultural technology can reduce productivity and land quality in ways other than erosion. Heavy equipment used when soil moisture content is not optimum can cause soil compaction. This restricts root penetration leaving crops less responsive to fertilizer and more susceptible to drought. Intensifying crop production also may reduce the level of organic matter, leading to the breakdown of soil structure. Hay and pasture generally are considered soil building crops that improve soil structure and increase organic matter. Conversely, cultivating crops in rows exposes the soil to greater temperature extremes, more frequent tillage, and the destructive impact of falling rain, leading to the more rapid oxidation of organic matter and the disintegration of soil aggregates. Between 1967 and 1977 the area in row cropping increased 27%, while the area in rotation hay and pastures declined 40% (Sampson, 1981). Much of the water used to irrigate land has high levels of sodium, which subjects the land to salinization that reduces productivity and may force land abandonment. Frederick (1982) indicates that 25–35% of the irrigated lands in the West have salinity problems. In areas of the Southern Plains some irrigated lands have been abandoned as groundwater has become scarce or too energy intensive to pump (Frederick, 1982).

FUTURE USE OF THE LAND

Future demand for and supply of agricultural land are a complex function of (1) changes in technology that increase yield potential and decrease land degradation, (2) changes in fossil energy costs and availability, (3) changes in demand for agricultural products both domestically and abroad, and (4) changes in demand for land for other uses. Managed properly, land is a renewable or flow resource that can be used again and again without loss of potential for future use. If mismanaged, agricultural land must be viewed as a stock resource available for use in limited quantities because it is replaced only very slowly by natural energies compared to losses caused by human misuse. In the United States the agricultural land base is declining in both size and quality, even as agricultural production has continued to rise. This apparent paradox had been made possible only by increasing inputs of economic energy per hectare of cropland and per kilocalorie of crop produced. How should we in the United States manage our land resources for agriculture in anticipation of a future of rising costs, increasing scarcity of energy, and a rising demand for output? Will we need to expand the agricultural land base again to include lower-quality lands or will we continue to increase the use of economic energies on the highest-quality land? Can we develop effective high yield technologies that are also increasingly energy efficient and sustainable?

In this final section we examine how land quality will affect land use in the future. Two aspects are fundamental: (1) the adequacy of the land base and possible competing uses for agricultural land given present trends, and (2) changes in agricultural technology that could increase energy efficiency and/or sustainability of agricultural production or that might alter current use patterns or land capability standards (see Chapter 6, p. 134).

Increasing the Land Base for Crop Production

The area of rural land with high or medium potential for conversion to cropland has been estimated by the Soil Conservation Service at 51.4 million ha, of which 36.4 million ha is presently pasture and rangeland and 12.6 million ha is forestland (Sampson, 1981). This represents a potential 30% expansion of current cropland. Of the land with high potential for conversion, 49% is Class I or II land. Land with medium potential is only 22% Class I or II land (Heady, 1982). Fifty-five percent of current cropland is Class I or II land. Thus, the pool of available reserve cropland is of lower quality than the cropland in use. We also should point out that potential for conversion was based on 1974 prices, conditions that no longer apply because of rising energy prices and declining crop prices. Converting large areas of pasture and rangeland to cropland also would reduce the land base for grazing sheep, goats, and cattle with several possible consequences: (1) a shift to more fossil-energy-intensive feed production systems, (2) a reduction in meat production, or (3) an increase in adverse environmental impacts caused by intensified production on current rangelands (see p. 502).

A significant proportion of reserve prime farmland is located in the Northern and Southern Plains where quantity and distribution of rainfall are highly variable, and in Appalachia where the landscape pattern is highly complex and erosion is a serious threat to land productivity (Table 20.5). These areas may be less useful as cropland because they require more fossil energy per unit output than croplands currently used.

The conversion of cropland to urban, commercial, and industrial uses, transportation, and reservoirs will absorb 3.2% of all cropland by the year 2000 if present trends continue (Heady, 1982). An important question is whether per capita urban land use will rise as it did from 1930 (0.07 ha/capita) to 1975 (0.10 ha/capita) (Heady, 1982). The shift in population from industrialized regions of the Northeast and Midwest to the Sun Belt and the shift from urban to rural lands as primary sites for residential development suggest that the trend may continue. This would diminish the amount of land available for crop production. On the other hand, Ruttan (1982) listed several moderating influences on land conversion from agricultural to nonagricultural uses. The interstate highway system is nearly complete, the potential sites for water impoundment are nearly exhausted, travel and commuting costs have been increasing, and the country's population on the average is becoming older.

Two factors cloud projections of future demand for cropland: the level of food exports and the amount of land devoted to biomass production for conversion to energy. Land devoted to producing agricultural exports increased from 20% of total cropland in 1960 to 31% in 1977 (Cotner et al., 1981). This represents a large buffer that could be shifted to domestic consumption should domestic demand increase or the need to export food decline. Potential demands for land devoted to biomass production for energy are discussed in Chapter 19.

The U.S. land base for crop production appears adequate to satisfy future domestic demand and to permit substantial exports. But the uncertainty of energy supplies and prices, increased urban expansion in certain regions, and a possible leveling off of per hectare yields suggests that conservation of prime farmland is the prudent strategy to ensure present and future use. Prime lands which require less fossil energy and are more resistant to resource degradation should be used now in a way compatible with agricultural use in the future. How this will be achieved by a free-market system in which the economic returns from land in urban uses surpass the returns from land in agriculture is a difficult and complex question. If prime farmland is not preserved for agricultural use, the agricultural land base will expand onto lower-quality land, thereby increasing the economic energy used to grow food. For a further discussion of these issues see Sampson (1981), Schnepf (1979), NAS (1975), and Crosson (1982).

Land Quality and the Use of Fossil Energy Inputs

We have described how crop production has shifted to the highest-quality land while using progressively larger amounts of fossil energy. Will this trend continue in the future, or will lower-quality lands be brought under cultivation in order to expand production? DeWit (1979) and Frissel (1977), suggest that both agricultural yields and energy efficiency can be increased by concentrating fertilizer and energy use *intensively* on as small an area of high-quality land as possible. Economic energy inputs would be used to control all possible biological aspects of production and upgrade the physical and chemical properties of already high-quality land. This could have gains for forest and wildlife because more undisturbed land would be available. Odum (1971), on the other hand, proposed that fossil energy should be used *extensively* to complement, redirect, and concentrate natural energy and material flows so that fossil and solar power could be combined to optimize both total resource production and efficiency, thus allowing the sustained economic exploitation of all landscapes. According to this view natural systems provide support services to both the urban and agricultural sectors, and the uses of high- and low-quality lands should not be judged in isolation from each other, but rather as they are interconnected by nutrient cycles, energy flows, and economic interactions.

At the present time data are not available to reconcile this important debate, but we will examine various options.

The increased use of appropriate capital investments can maintain or increase crop yields while increasing energy efficiency. Capital investment in permanent land improvements can emphasize long useful life with low operation and maintenance costs in contrast to the currently used criterion of the minimization of initial costs. Irrigation can be used to illustrate this point. Energy costs for installing and operating irrigation systems vary (Table

Table 20.7.   Energy Inputs for the Irrigation of a 65-ha Farm

Type of Irrigation	Energy Cost (10^6 kcal)			
	Installation	Pumping	Labor[a]	Total
Surface with runoff recovery	180	48	0.3	228
Trickle	530	468	0.1	998
Hand-moved sprinkler	160	804	4.8	968
Center pivot	388	864	0.1	1252
Traveler sprinkler	289	1569	0.4	1858

Source: From USDA (1977).
[a] Direct energy used by labor.

20.7). The traveler sprinkler, center pivot, and hand-moved sprinkler have low initial costs but much higher pumping (operating) costs than the trickle system. Which type of system should the farmer use? Much will depend on the soil texture, stratification, and degree of slope of the land to be irrigated, but high pumping costs make these systems less viable alternatives should there be an energy-scarce future. Unfortunately discounting procedures used for economic cost–benefit analyses (p. 146) tend to favor investments with low upfront capital costs, so that mostly cheap but inefficient systems are installed. If energy costs rise substantially as they have in the past, such decisions will be economically, as well as environmentally, inappropriate.

Society could encourage energy investments in capital improvements that increase energy efficiency while decreasing the amount of recurrent inputs such as fertilizer and pesticides. For example, installation of capital-intensive drainage and water capture procedures on agricultural lands permit optimal plant response to fertilizer. Certain types of agricultural research can be viewed as a capital investment that has a major impact on energy efficiency and crop productivity. Although research is expensive in dollars and energy, the increases in basic knowledge and applied technology are permanent improvements for human society and serve as the basis for future advances in agricultural yield and efficiency. Long-term advances in basic research, such as in plant growth regulators, genetic engineering, and water and mineral behavior in soil, may allow us to design highly efficient crop production systems. Currently the development of integrated pest management, pest resistant and tolerant crop varieties, soil fertility management, and weather prediction are practical tools that may en-

able farmers to optimize crop response to fossil energy inputs. Shoemaker (1977) has used computer models of alfalfa insect pests to predict the periods of greatest pest vulnerability. Pesticide applications could be reduced by concentrating them during these periods. Bouldin and Lathwell (1968) have estimated that carefully timed applications of nitrogenous fertilizer can reduce total use by 25–30%. On the other hand, such changes in management would intensify the seasonal demand for nitrogen fertilizer, and for higher investments in manufacturing, storage, and distribution facilities for fertilizer. In this, as in all such complex issues, comprehensive systems analyses are needed to evaluate genuine dollar and energy savings to the economy and society.

A recent Office of Technology Assessment report (OTA, 1983a) suggests that the increased use of both capital and recurrent inputs, including fertilizer, pesticide, and improved crop varieties, has masked a gradual deterioration of land productivity that has resulted principally from wind and water erosion. Nevertheless, the report also states that economically viable technology is available that achieves high levels of crop production and conserves long-term productive potential of the land. The most important such approach is no-till agriculture, or, more accurately, limited tillage, in which soil disturbance for planting is reduced sharply. Such an approach maintains protective crop residues on the soil surface and thus reduces soil erosion and fuel use for machinery operation, while increasing yields in areas of limited moisture. Using no-till, farmers can plant steeply sloped lands in row crops and thus contain erosion without major investments in erosion control structures. Soil organic matter levels stabilize at higher levels under no-till than under conventional tillage and may reach higher levels than under undisturbed natural vegetation (Palmer, 1983). This short-term improvement in soil quality results from increased amounts of crop residues due to high levels of fertilizer use and to proper crop residue management. The system does, however, require more herbicides to suppress the weeds controlled by plowing under conventional tillage. No-till methods also lead to an increase in pests.

The area under limited tillage in the United States increased from 1.5 to 13.4 million ha from 1963 to 1973, and about one-half of the total cropland area may be managed by some form of reduced tillage by 1990 (Allen et al., 1976). (See OTA, 1983a and Geisler et al., 1981 for a discussion of research, governmental policy, and government programs. These papers include a discussion of social issues that relate to, and sometimes discourage, the use of no-till and other productivity-conserving techniques in agriculture.) The use of living mulches (living plants that serve to protect the soil without competing with the crop) is another practice that also improves land productivity with little recurrent cost and no capital expenditure (Nicolson, 1983). Cover crops require only a small investment for seed and planting and act to protect the soil from water and wind erosion and to hold nutrients not taken up by the main crop so that these nutrients can be returned to the soil when the cover crop is plowed under (Cornell University, 1983).

The approaches mentioned in the preceding paragraphs lower levels of both capital and recurrent inputs. They provide the best energy return on energy invested when applied to lands with deep, well-drained soils with high water and nutrient retention capacity. They are less effective on lands with soils characterized by shallow depth, low nutrient- and water-holding capacity, poor internal drainage, or other impediments to root penetration. If yields increase on higher-quality lands, lower-quality lands of this sort may be returned to pasture and forest.

Patterns of land use affect energy flow and nutrient cycling. Present food production depends on one way nutrient flows powered by fossil energy inputs. Nutrients are purchased in the form of chemical fertilizer, applied to the best land, lost to erosion or removed in the crop and sent to the city where they pass through humans and are dumped into the rivers. These cycles could be closed in several ways. First, if nutrients could be recycled from urban areas back to areas of agricultural production or into those natural systems that provide important services to the urban or agricultural sector. The high energy costs of transporting wastes suggest that in the long run human settlement patterns may have to be dispersed to reduce the distance between urban and agricultural centers, although toxic metals and disease vectors must be removed from wastes produced by human society.

Second, nutrients could be recycled between crop and animal production systems more effectively. For example, commercial vegetable production needs large inputs of organic matter. Animal feedlots currently have trouble disposing of their immense quantities of organic wastes and nutrients in an environmentally benign manner. The two systems could be integrated. Another possibility is to encourage the intensive grazing of pastures on intermediate quality land (rather than mechanical harvesting of crops for consumption by confined animals) to reduce the fossil energy cost of harvesting forage while enhancing the recycling of nutrients from animals to crops and back again.

Third, marginal lands could be incorporated into production through the design of integrated systems that promote solar-powered energy flows and nutrient cycles. Agricultural production on these lands is neither economically nor energetically viable under current fossil-energy-intensive technology. However, these lands may be important or useful in flood prevention, groundwater recharge, waste disposal and water filtration, pest predator reproduction, and as the seasonal source for forage, forest products, and high value products such as nuts or medicinal herbs requiring labor-intensive harvesting. Research to understand better the role of these lands in maintaining the productivity of intensive agriculture and their response to selectively used energy and nutrients will facilitate the design of minimum fossil energy input technology and management strategies for their use. Legislation based on regional land use planning also may be necessary to achieve the linkage of marginal lands and high quality agricultural lands under different ownerships.

MORE LAND OR MORE FOSSIL ENERGY INPUTS—WHERE DO WE STAND?

At the present time the United States produces a surplus of wheat, milk, corn, cheese, soybeans, and many other agricultural products and has an abundance of agricultural land. In the long run high-quality land for agriculture probably will continue to be managed with increasing intensity. In the past decades this trend has required more and more fossil energy (Figure 20.2) and such may be the case in the future. However, research advances in plant growth regulators, time release fertilizers, management of soil microflora and fauna, precision soil tillage, planting, and weed control, and the integrated control of insects and diseases have the potential to increase yields, while decreasing energy use. Similarly advances in the sciences of animal physiology and reproduction have the potential to increase the efficiency of milk, egg, and meat production. Our increased scientific understanding of global and local energy flows and nutrient cycles as a result of research in climatology, geology, ecology, and biology could serve as a basis for linking agricultural, urban, and natural systems and lead to a long-term reduction in fossil energy use. The incentive, however, for increasing the energy-efficient use of prime croplands and an energy-efficient total land management approach is only likely to occur under extreme energy restrictions.

21

RANGELANDS

Beverly I. Strassmann

As the quality of many of our natural resources deteriorates, or as yields are increased, production techniques tend to become increasingly energy intensive. As described earlier, increasing crop production by either bringing low quality land into cultivation or improving yields on intensely farmed croplands often increases the energy required for each bushel of grain harvested (see also Pimentel et al., 1976). In the United States agriculture appears to have reached the point of diminishing marginal returns where increasing energy investments produce smaller increases in farm output (Steinhart and Steinhart, 1974). Likewise increasing expenditures of fossil fuel in U.S. commercial fisheries have not led to commensurate increases in fish catches—particularly, as we increasingly exploit waters of marginal productivity or overfished zones (Rochereau and Pimentel, 1978; Brown and Lugo, 1981; Chapter 18).

This chapter examines the energy costs of intensive rangeland management. Will intensive management of Western rangelands lead to a parallel increase in beef production or will it only raise the energy required per kilogram of beef produced? If intensive management increases beef production substantially, will the increased energy subsidy be cost effective? These two questions are critical because the federal government has adopted a major program of intensive management to restore the productivity of federal rangelands (USBLM, 1974, 1975; CBO, 1978; USFS, 1981; USGAO, 1982). Livestock overgrazing reduced as much as four-fifths of the Western rangelands to fair or poor condition by 1966 (Pacific Consultants, 1970; Box, 1979) and was a major cause of the desertification that has essentially removed 91 million ha from production (Dregne, 1977; Sheridan, 1981).

This chapter will evaluate the energy and dollar costs of intensive management of Western range-lands, and compare the federal government's intensive management program for federal lands with an alternative approach that places greater emphasis on short-term stocking reductions. Criteria for comparing these approaches will include their potential for: (1) restoring rangeland productivity, (2) protecting the environment, (3) sustaining beef production, and (4) minimizing energy use.

THE WESTERN RANGELANDS

Rangelands are defined as areas where the native vegetation is predominantly grasses, grasslike plants, forbs (herbaceous flowering plants other than grasses, sedges, and rushes), or shrubs suitable for grazing or browsing by native ungulates or livestock (SCS, 1976). Western rangelands, which stretch from the Great Plains to the Pacific Coast, consist of three major biomes: grassland, woodland-brushland, and desert (Figure 21.1). *Forest-range* expands the term *rangelands* to include grazing lands that may also be found in the coniferous forest biome (Johnston, 1978; Pimentel et al., 1980). Ninety-nine percent of the 263 million ha of U.S. rangelands outside of Alaska and Hawaii is in 17 Western states (USFS, 1981).

Western rangelands are characterized by low annual rainfall and periodic drought (Johnston, 1978). With local exceptions, the average annual precipitation is less than 51 cm (Figure 21.2). Such aridity limits net primary production (Odum, 1971; Lauenroth, 1979) keeping the carrying capacity for livestock low. Perry (1978) has aptly described rangelands as "areas which are too dry, too hot, too cold, too shallow, too steep, or too infertile for more intensive forms of agriculture . . . we are dealing with the residue of land after the more productive lands have been selected for more intensive land use."

The 102 million ha of federal rangeland in the

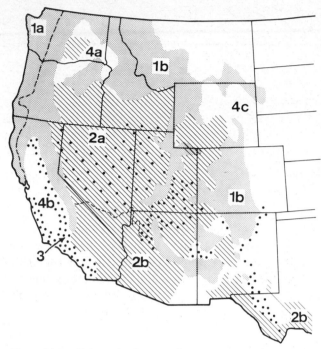

Figure 21.1. Schematic diagram of the Western biomes (boundaries approximate). Subcategories (i.e., *a* vs *b*) are separated by dashed lines. (Adapted from BLM, 1974, based on Küchler, 1980.)

1 = Coniferous forest biome
    1*a* = Northwest coastal forest
    1*b* = Montane Forest
2 = Desert
    2*a* = Cold desert
    2*b* = Hot desert
3 = Woodland-brushland biome
4 = Grassland biome
    4*a* = Palouse Prairie
    4*b* = California annual grassland
    4*c* = Other grassland

contiguous states comprise about 39% of the total rangeland area (Figure 21.3). The U.S. Bureau of Land Management (USBLM) in the Department of the Interior administers 69 million ha and the U.S. Forest Service (USFS) in the Department of Agriculture administers 20 million ha. Other agencies in the Department of the Interior and the Department of Defense administer 13 million ha of rangeland (USFS, 1981). Grazing of rangelands is measured in animal unit months (AUMs). An AUM is the amount of forage needed to sustain a 450 kg cow or five sheep for 30 days. In 1976 rangelands of the coterminus states provided 217 million AUMs of

livestock grazing: 29 million on federal lands and 188 million on nonfederal lands (Cutler, 1979).

In fiscal year 1981 the USBLM issued 22,000 permits to individuals and agribusiness corporations for grazing livestock on approximately 68 million ha of grazing districts in 10 Western states. The estimated actual use of these lands by livestock totaled 12 million AUMs (USBLM, 1982). In calendar year 1982 the USFS issued 15,000 permits for grazing 42 million ha of forest-range in the National Forest System in 36 states (USFS, 1982). Authorized grazing on these lands totaled 9.9 million AUMs (USFS, 1982). In fiscal year 1980 the U.S. Fish and Wildlife Service issued 1400 permits for livestock grazing and haying (largely for livestock fodder) on 981,954 ha at National Wildlife Refuges (Strassmann, 1983).

DEMAND FOR RANGELANDS

Although the productivity of the Western rangelands is naturally low, the USBLM has proposed an intensive management program to increase AUMs of grazing on public lands from 12.7 million in 1970, to 16.2 million in 2000 and 17.9 million in 2010 (USBLM, 1974). In 1974 a federal district court judge in *Natural Resources Defense Council et al.* v. *Morton et al.* found that the USBLM violated the National Environmental Policy Act by failing to base this program on site-specific environmental impact statements. As a result of this finding, the USBLM has been obligated to prepare such statements before proceeding with intensive management (USBLM, 1980). The U.S. Forest Service also has proposed intensive management and reports that grazing in the United States can thereby be increased by an additional 152 million AUMs by 2030 (USFS, 1981). The USFS projects most of this increase for USBLM and private lands. A key assumption of the USFS is that forage-based rangeland beef production consumes less fossil fuel energy per kilogram of beef produced than do grain intensive feedlot systems and that, as fossil fuel becomes less available, rangelands will play a more important role in U.S. beef production (USFS, 1981).

The USFS's 1981 assessment for Congress of forest and rangeland resources (USFS, 1981) states:

> Range livestock production is a relatively low consumer of energy compared to production systems

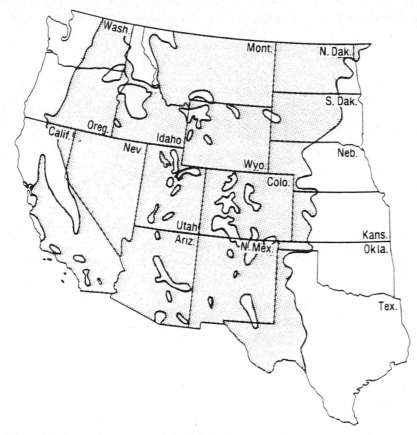

Figure 21.2. Areas with less than 51 cm of average annual precipitation. (From Geraghty et al., 1973, as adapted by Sheridan, 1981.)

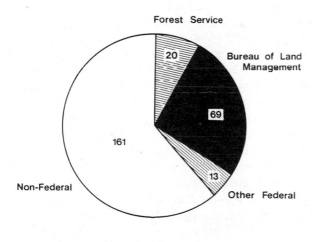

Forest Service

Bureau of Land Management

Non-Federal

Other Federal

Total Rangeland (263)

Figure 21.3. Ownership of rangeland in the contiguous states in millions of hectares, 1976. (From USFS, 1981.)

using large quantities of grains. Forage production on range is largely a function of natural processes using energy from the sun, whereas grain production depends on cultivation activities using high-cost fossil fuel energy.

The USFS's analysis overlooks the large quantities of fossil fuel energy that are presently consumed by the average ranching operation for running pickup trucks and machinery, pumping water, and constructing fences. Table 21.1 shows the energy used by an average ranching operation in the intermountain Great Basin and the Central Great Plains. In the Great Basin considerable quantities of energy (in descending magnitude) are used for pickup trucks, machinery, barbed wire fences, and feed. In the Central Great Plains the sequence is slightly different: machinery, pickup trucks, feed, and barbed wire fences.

Despite the large quantities of energy expended by ranching operations, the belief that feedlot beef production is more energy intensive than rangeland beef production is widespread (e.g., Heady et al.,

Table 21.1.   Average Cultural Energy Expended for Cattle Ranches (250 cows)

Item	Intermountain Great Basin		Central Great Plains	
	Quantity	kcal	Quantity	kcal
Labor (hr)	3170	7925	2737	6842
Machinery (hr)	1200	54,000	1744	78,480
Pickup trucks used in herding	20,000	83,660	15,000	62,745
Transportation of materials (kg × km)	573	1662	7	20
Feed (kg)	20,089	22,500	30,134	23,490
Barbwire fences (km)	80	34,856	30	13,071
Water (AUM)	3000	13,815	3000	16,327
Total		218,418		200,975
Market weight or gain (kg)	108.8		184.8	
Calf crop (%)	93		92	
Total energy in carcass (Mcal)[a]	49,498		50,578	
Number marketed	233		233,230	
Live weight marketed (kg)	42,127		42,509	
Cultural energy/food energy	4.41		3.97	
Mcal cultural energy/kg of live weight	5.18		4.73	

Source: From Pimentel (1980).
[a] Mcal = 1000 kcal.

1974; Steinhart and Steinhart, 1974; Ward et al., 1977; Hoveland, 1980). But a recent analysis of energy consumption by cow-calf operations of rangelands and feedlots in all regions of the United States suggests that feedlot beef is, in fact, less energy intensive (Pimentel et al., 1980). Cow-calf operations used an average of 120×10^3 kcal of fossil energy per kilogram of beef produced, and feedlots used 101×10^3 kcal/kg (Pimentel et al., 1980). Thus, at the present time there is insufficient basis for concluding that greater U.S. dependence on rangeland beef will reduce the energy costs per unit of beef produced.

Other factors that could result in greater U.S. dependence on rangeland beef are complex and will not be explored in this book. Chief among them is the rate of beef consumption which in turn is dependent on population growth, disposable income, and the cost of other products (Ward, 1977). The National Cattleman's Association reports that unless there are substantial gains in real incomes, future growth or even maintenance of the beef industry at present levels will depend on (1) domestic population growth, (2) future increases in beef exports, (3) new forms and kinds of beef products, (4) increased consumer acceptance of beef, and (5) more competitive prices for beef relative to other products (SAC, 1982). Although it is not clear how high the future demand for beef will be, it is clear that federal rangelands have a limited capacity to provide for

any increases in production. Federal rangelands in the 17 Western states provide only 13% of the total U.S. AUMs of livestock grazing (Cutler, 1979), and they provide less than 8% of the feed consumed by all U.S. beef cattle.

In competition with the demand for rangelands for beef production is the increasing interest in the scenic, recreational, archaeological, historical, wildlife, and wilderness values of federal rangelands (USBLM, 1980; USFS, 1981). These competing uses and values of federal rangelands are protected by several statutes, most notably, the Federal Land Policy and Management Act (FLPMA) of 1976. This act mandates that federal rangelands administered by the U.S. Forest Service and the Bureau of Land Management be managed for multiple use—meaning that wildlife, recreational, and economic values all must be supported. Both FLPMA and its 1964 precursor, the Classification and Multiple Use Act, represented a shift from the Taylor Grazing Act of 1934 which emphasized the value of rangelands for livestock production. The extent to which the multiple uses of rangelands can be accommodated harmoniously will depend in large part on rangeland quality.

RANGELAND QUALITY

Rangeland quality is protected by FLPMA, which requires that management of USFS and USBLM lands must prevent "permanent impairment of the productivity of the land and the quality of the environment." Rangeland quality is usually assessed by a technique called the range site and condition evaluation (Dyksterhuis, 1949; Bell, 1973). Range sites are delineated according to the composition and productivity of the climax vegetation, which is influenced by a combination of edaphic, climatic, topographic, and biotic factors (Dyksterhuis, 1949; Ibrahim, 1975). The condition evaluation includes an estimate of the current plant species composition of a range site relative to that which would be present at the climax stage. Plant species are grouped into ecological categories based on their response to grazing by livestock (Figure 21.4). Climax species that initially become more abundant when grazed, but decrease in abundance with continued grazing pressure, are termed increasers. Decreasers, which are more palatable or are physiologically less resistant to grazing, are climax species that become less abundant when grazed. Invaders are species that were not present in the native climax vegetation of a particular range site or, if present, were restricted to disturbed areas like prairie dog towns. As increasers and decreasers are gradually removed by

grazing, invaders become abundant. To calculate range condition, the percentages of all increaser and decreaser species in a site's existing vegetative cover are compared to the percentages of these species in the site's presumed climax composition. Different agencies use somewhat different classifications (e.g., USFS, 1981), but according to the basic scheme, a range site with 85–100% of the climax composition is in excellent condition. Fifty to 75% would signify good condition, 25–50% would signify fair condition, and 0–25% would signify poor condition (Dyksterhuis, 1949; Bell, 1973; Stoddart et al., 1975). Other variables such as plant production, ground cover, and soil erosion also factor into condition evaluations and modify their outcome (Stoddart et al., 1975).

Like all indexes the range site and condition evaluation has both advantages and disadvantages. Among the advantages of this index are (1) it provides the best quantitative means presently available for comparing the quality of rangelands which differ in their resource characteristics and resource uses (cf. Dyksterhuis, 1949; Bell, 1973), and (2) it usually provides a good assessment of rangeland quality for livestock production because the climax

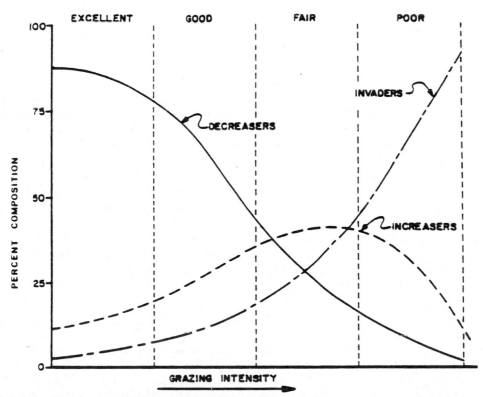

Figure 21.4. The relationship between grazing intensity and range condition, and the relative percentages of decreaser, increaser, and invader plants in each condition class. (From Sims and Dwyer, 1965, as adapted by Stoddart et al., 1975.)

plants are generally more palatable and nutritious and therefore contribute more to livestock weight gains than do earlier successional species (Bell, 1973). (The annual grasslands of California are among the many exceptions to this rule because they can provide excellent grazing although their condition is far from climax; Dyksterhuis, 1949.) Among the disadvantages of this index are (1) the difficulty of determining the composition of the climax vegetation (Stoddart et al., 1975), (2) the potential irrelevance of the climax composition, particularly in the case of artificial seedings or rangelands that would revert to forest under climax conditions (Dyksterhuis, 1949), and (3) the failure of range site and condition evaluations to assess important determinants of the quality of wildlife habitat such as availability of cover, food, and water (Sharpe, 1979).

The Western Range (U.S. Senate, 1936) was published two years after passage of the Taylor Grazing Act and is the first quantitative report on the condition of the Western ranges. The categories of range depletion reported in this study are not based on the same techniques presently employed in range site and condition evaluations. Therefore they are not comparable to recent evaluations but are nevertheless informative (Box, 1979). *The Western Range* concluded that 57.5% of the public ranges were in a condition of severe or extreme depletion due to overgrazing. Only 16% were in reasonably good condition. No other major evaluation of range conditions appeared before 1970 when Pacific Consultants completed an inventory for the Public Land Law Review Commission. This inventory concluded that overgrazing had reduced as much as 80% of the ranges to fair or poor condition (Pacific Consultants, 1970). Based on the inventory data, Box et al. (1976) concluded that 75% of the Western ranges were producing less than half the climax vegetation of the native range (cf. Wagner, 1978a).

In 1975 the USBLM prepared the "Range Condition Report" (USBLM, 1975). This report, which used five categories for range condition, found that 83% (54.6 million ha) of the land managed by the USBLM was in fair, poor, or bad condition. Only 17% (11.3 million ha) was in good or excellent condition. According to the USBLM a range in excellent condition is "producing all or nearly all of the forage that it can and has a fully productive stable soil." A range in bad condition has "only a sparse stand of low value plants, mostly annuals or un-palatable shrubs. Topsoil is largely gone. The remaining soil is exposed to wind and water and active erosion occurs." Assuming no changes in management practices, the report's prognosis for the future stated:

> Public rangeland will continue to deteriorate; projections indicate that in 25 years productive capability could decrease by as much as [an additional] 25 percent—a decline that could be reflected in the possible loss of livestock grazing privileges . . . Other losses will be suffered in terms of erosion, water quality deterioration, downstream flooding, loss of wildlife and recreation values, and decline in basic productivity.

The report's evaluation of soil erosion concluded that overgrazing was causing 91% of the public lands to suffer some soil loss and that for 10% of the lands this erosion was critical or severe. Thirty-five percent of the lands were suffering moderate soil loss—defined as erosion during heavy storms with an associated loss of productivity and a heightened susceptibility to rapid deterioration. Forty-six percent were sustaining slight soil loss and, although vulnerable, were not suffering immediate damage (USBLM, 1975). The lands in the moderate, severe, and critical categories, making up 45% of the total, were considered unsatisfactory by the USBLM.

More recent analyses by the USBLM and the U.S. General Accounting Office (USGAO) suggest that the 1975 report greatly underestimated the poor quality and the rate of deterioration of the public rangelands (USGAO, 1977; Sheridan, 1981). Although the condition of large areas of federal rangeland is worsening, better quantitative data on the trend in range condition for most areas remain a critical management need.

ECOLOGICAL EFFECTS OF LIVESTOCK OVERGRAZING

To understand why overgrazing can reduce rangeland productivity, it is necessary to consider the various ecological processes that are responsible, including their effects on vegetation, soil, streams, and wildlife.

Vegetation

Livestock graze selectively (Bell, 1973). Therefore one of the first effects of overgrazing is that the palatable species—usually perennial grasses—do not have the opportunity to recover. This process gives a competitive edge to the unpalatable, often woody, species such as sagebrush, snakeweed, and rabbitbrush, allowing them to proliferate (Bell, 1973; Wagner, 1978a; Day, 1979; Sheridan, 1981). Day (1979) estimated that livestock grazing caused a third of the range and pasturelands in the United States to become infested with undesirable woody plants. According to Day:

> The single overwhelming ecological fact of range management is that grazing animals selectively harvest forage plants and leave nonforage species to prosper undamaged. The consequence is the decline of forage species and increase in unpalatable, poisonous, and noxious species . . . There are numerous ways to soften the impact of selective harvest, most of them having to do one way or another with conservative utilization.

Soil

The relatively widely spaced woody species that take over after more palatable species are selectively grazed do not hold the soil as well as would a continuous cover of grasses. Bare areas left unprotected by vegetation are then vulnerable to erosion. Although soil erosion is a natural process that is particularly important on steep slopes or fragile soils, excess stocking is the major cause of soil erosion on rangelands (Johnston, 1978). Soil erosion reduces rangeland productivity by removing nutrients needed for plant growth. Low rates of plant growth result in even less vegetative cover, thereby enhancing the vulnerability of the soil to further erosion (cf. Gifford and Whitehead, 1982).

Soil water is an important, if not the foremost, factor limiting plant growth in arid Western rangelands. Increased levels of surface runoff therefore reduce rangeland productivity (cf. Powell et al., 1978; Thomas, 1978; Gifford and Whitehead, 1982). Soils with a good protective cover usually have high infiltration rates and water percolates through them to the water table. Where the protective cover has been lost through overgrazing, surface runoff increases (Lusby, 1970; Meehan and Platts, 1978). Trampling by livestock also contributes to runoff by compacting soils which reduces water infiltration rates (Meehan and Platts, 1978; Lusby, 1979). Lusby (1979) found that excluding livestock from a salt-desert rangeland in Colorado reduced surface runoff by 40% between 1953 and 1973.

Streams

Overgrazing of watersheds degrades the quality of streams in two major ways. First, runoff water discharges into streams erratically and consequently is less able to maintain stream flow during droughts than is the slow, steady release of groundwater. Second, runoff water carries a high sediment load which accumulates in streams where it increases water turbidity and lowers water quality (Meehan and Platts, 1978; Platts, 1981). Sedimentation of streambeds reduces stream cover for fish, kills fish embryos by silting over the gravel beds used for spawning, and clogs interstices in aquatic insect habitat which in turn reduces the prey available to fish (Platts, 1981). Lusby (1979) reported that excluding livestock for 20 years reduced sediment yield by 63%.

Livestock overgrazing also can damage riparian areas directly. In fact several investigators report that overgrazing by livestock is the most pervasive cause of the deterioration of riparian ecosystems on public lands (Carothers, 1977; Cope, 1979; Knoppf and Cannon, 1982). Riparian vegetation is especially prone to overgrazing because cattle do not graze steep slopes or travel very far to water and so concentrate their grazing in riparian environments (Van Vuren, 1982).

The principal results of overgrazing of riparian ecosystems include (1) trampling which causes erosion and hence increased water turbidity and changes in channel morphology (Platts, 1978, 1981), and (2) loss of willows and other vegetation important for providing shade (e.g., Davis, 1982; Rickard and Cushing, 1982). As a result, stream water temperatures rise and dissolved oxygen levels decrease (Platts, 1978, 1981; Davis, 1982; Rickard and Cushing, 1982). In the arid and semiarid West, riparian ecosystems are potentially the most productive (Platts, 1978; Sharpe, 1979). Damage to them therefore can have major effects on wildlife.

Wildlife

Livestock grazing is the single most important factor limiting wildlife production in the West (Smith, 1977; Platts, 1978). Its major effects include proliferation of invader and increaser species at the expense of climax plant communities on which wildlife like pronghorn antelope (*Antilocapra americana*) and sage grouse (*Centrocercus uro-*

phasianus) depend (Wagner, 1978a), depletion of the key browse plants of critical wintering areas (Mackie, 1978; Wagner, 1978a), reduction of the nesting cover for waterfowl and upland birds (Clark, 1977; Braun et al., 1978; Kirsch et al., 1978; Kessler, 1982), reduction of fawn survival through removal of vegetative cover (McNay and O'Gara, 1982), reduction of wild ungulate populations through competition for forage (Mackie, 1978; Wagner, 1978a), and degradation of habitat for fish and other wildlife through increased water turbidity and elevated water temperatures (Bue et al., 1964; Platts, 1981; see Strassmann, 1983, for a more detailed discussion of the effects of livestock grazing on wildlife with particular reference to National Wildlife Refuges).

The change in the use of Western ranges by wild ungulates and cattle has been roughly estimated by Wagner (1978b) (Figure 21.5). American bison (*Bison bison*), bighorn sheep (*Ovis canadensis*), pronghorn antelope (*Antilocapra americana*), and elk (*Cervus canadensis*) populations have declined a great deal while cattle populations have soared. The pre-Columbian populations of climax species—bison, bighorn sheep, and pronghorn antelope—have declined by about 95%, and elk populations have declined by about 75% (Wagner, 1978a; 1978b). Data do not exist to document the historic use of Western ranges by nongame wildlife such as smaller mammals, songbirds, reptiles, and various predatory species, but all of the climax fauna have probably been greatly reduced (Wagner, 1978b). Deer, which prefer intermediate successional vegetation, have become more abundant following increases of brush in grassland areas (Wagner, 1978a) and the attention they receive from game management programs.

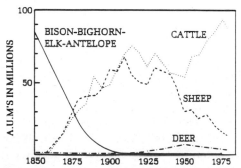

Figure 21.5. Conjectured AUM's of grazing by wild ungulates and livestock on rangelands of the 11 Western states. (From Wagner, 1978b.)

INTENSIVE MANAGEMENT OF RANGELANDS

Previous sections have provided an overview of the Western rangelands including the ecological effects of livestock overgrazing, rangeland quality, and the demand for rangelands. The remainder of this chapter will evaluate the federal government's intensive management program for federal rangelands and compare it to an alternative approach that places greater emphasis on short-term stocking reductions. As stated earlier, criteria for comparison will include (1) restoration of rangeland productivity, (2) environmental protection, (3) sustainable beef production, and (4) energy use.

Range Improvement Techniques

To restore the quality of the public lands for both livestock and wildlife production, the USBLM has developed a list of range improvements needed (see Table 21.2). Range improvements can increase forage production, but they also can have some adverse environmental effects.

Many of the range improvement techniques aim to control the vegetation that is unpalatable to cattle and that has a competitive edge under grazing pressure. Removal of unpalatable forage can increase the water and nutrients available for growth of palatable forage and raise the carrying capacity for livestock (Sheridan, 1981). Unwanted vegetation has been removed by the USBLM and the USFS using rootplows, bulldozers, brush-cutters, and other machinery. Chaining is a popular technique for crushing pinyon-juniper stands and sagebrush. A heavy anchor chain is dragged across the land between two tractors (Paulsen, 1975). In sagebrush steppe, prescribed burning has been found to be an effective and cheap means of controlling sagebrush (Bradshaw and Bartlett, 1978). For example, Greene and Aldworth (1959) observed a 69% increase in forage production 3 years after a controlled burn. Vegetative removal practices can cause erosion if they are attempted on steep slopes or unstable soils (Wagner, 1978a; Sheridan, 1981).

Herbicides also are used to discourage the growth of unwanted species and are presently a major focus of intensive management programs. The EPA lists about eight herbicides for use on rangelands. The compounds 2,4-D and 2,4,5-T are used on over 85% of the area treated (Williams, 1979). The targets of 2,4-D are sagebrush and annual

Table 21.2. Range Improvements on Public Lands

Range Improvement Structures and Techniques	Existing Quantity in 1974[a]	Additional Quantity Needed[a]	Quantity Accomplished in FY 1975–1982[b]
Fences (km)	45,000	159,400	6,624
Water developments			
Springs	5,470	5,300	818
Pipelines (km)	6,900	26,600	420
Wells	2,700	6,500	309
Reservoirs	16,000[c]	25,000	986
Vegetation removal (ha)			33,127
Chemical (ha)	690,400	1,053,400	
Mechanical (ha)			
Pinyon-juniper chaining (ha)	182,100	277,600	
Brush removal (ha)	81,300	216,500	
Plowing (ha)	819,900	645,500	
Prescribed burning (ha)	95,900	214,500	
Seeding (ha)	1,297,500	1,173,600	149,942

[a] From BLM (1974).
[b] From BLM, *Public Land Statistics,* various years.
[c] In FY 1974, 890,806 reservoirs were built on public lands according to *Public Land Statistics.* Presumably, most of these were built after the total of 16,000 tallied here.

broadleaf weeds; the targets of 2,4,5-T are annual broadleaf weeds, cacti, mesquite, and oaks (Williams, 1979). Because 2,4,5-T is contaminated with dioxin, a compound associated with such human impacts as stillbirths, use of this herbicide on rangelands has been a subject of environmental litigation (Hinkle, 1979). The Dow Chemical Company found dioxin to be present at 3–4 ppt in three of eight samples of beef fat from a ranch in Iron County, Texas. The pasture was sprayed with 2.7 kg of 2,4,5-T per hectare, and the cattle grazed the treated pasture for 30 days (Hinkle, 1979). The use of 2,4-D, like the use of 2,4,5-T, has raised disturbing questions about consequences for human health because 2,4-D may be a mutagen and a teratogen (Sheridan, 1981). Herbicides usually are applied to rangelands by aerial spraying rather than by selective application (Williams, 1979)—further aggravating their health and environmental risks (Hinkle, 1979).

Initial increases in forage yields following treatment with herbicides are common. For example, Bartel et al. (1973; cited in Bradshaw and Bartlett, 1978) reported a 363 kg/ha increase in forage production following sagebrush control with 2,4-D, and a 540 kg/ha increase in forage production following two applications of a 2,4,5-TP and picloram mixture three years apart. On the other hand, Thilenious et al. (1974) sprayed an alpine meadow with 2.2 kg/ha

of 2,4-D and virtually eliminated all forbs. Furthermore, production dropped from 1348 kg/ha to 1269 kg/ha, and the protein and carotene content of the plants was reduced. When herbicides are used to increase forage production, periodic maintenance spraying and other measures such as periodic rest from grazing are necessary to prevent the eventual reinvasion of brush (Vallentine, 1980).

Another intensive management technique aimed at increasing forage for livestock is plowing the land and seeding it with an exotic perennial grass, typically crested wheat grass. The resulting monocultures may produce more forage (Bradshaw and Bartlett, 1978), but they are relatively barren of wildlife (Wagner, 1978a). Seeding often is accompanied by fertilization and the resulting yield is then highly dependent on the inherent productivity of the site (Pimentel et al., 1980). For example, 84 kg/ha of nitrogen increased the yield of pensacola bahiagrass in the humid Southeast by as much as 2900 kg/ha (Beaty et al., 1974), whereas in arid foothill range in Utah, 45 kg/ha increased the yield of crested wheat grass by 585 kg/ha (Cook, 1965). The USFS concluded that the use of fertilizers in native range is unlikely to become widespread because of "continuously escalating costs of inorganic fertilizers in relation to benefits" (USFS, 1980). The USFS further concluded: "However, fertilization does offer limited opportunities to increase forage, especially on private lands with high productivity and where livestock can be very intensively managed.''

Because livestock production on Western rangelands is limited by the accessibility of water to grazing areas (Cooley et al., 1978), water developments such as wells and reservoirs are a major part of intensive management efforts. Water developments are vital for expanding arid land beef production, but they increase competition between cattle and native arid-adapted wildlife (Campbell and Knowles, 1978; Nielson, 1978).

Irrigation of pastures is a range improvement technique that is limited by water scarcity, particularly in the arid West. Impending depletion of some important groundwater reserves suggests that water may become the greatest constraint on U.S. food production (Bertrand, 1979). Moreover competition for scarce water resources among competing uses may lower the proportion of water available for agricultural use (Gertel and Wollman, 1960).

Grazing Systems

In addition to the plans for range improvements, the federal land managing agencies have begun a widespread program to implement grazing systems on federal lands (CBO, 1978; Currie, 1978; Pieper et al., 1978; Wagner, 1978a; Van Poollen and Lacey, 1979). Grazing systems are defined as the manipulation of livestock to accomplish a desired result (Range Term Glossary Committee, 1964; Soil Conservation Society of America, 1976). By contrast, range improvements do not involve the direct manipulation of livestock although they may indirectly influence the location of grazing. The multiplicity of grazing systems has been classified into 14 categories by Lacey and Van Poollen (1979). The essential feature of the diverse grazing systems being implemented by the federal land managing agencies is that rangelands are checkerboarded into subunits separated by barbed wire fences. On a variety of schedules, cattle are rotated between the various subunits to give each area a periodic rest interval. Total AUMs typically are not reduced, thus stocking intensity in the subunits being grazed increases.

The purpose of the federal agencies' rotational grazing systems is to accomplish better livestock distribution, greater herbage and livestock production, and improved range condition without reducing grazing (CBO, 1978; Pieper et al., 1978). Early studies, for example, those of Hormay and Talbot

(1961) and Merrill (1954), reported that such results can be achieved. More recent studies often have failed to demonstrate that rotational grazing systems are advantageous. Nonsignificant forage responses, reductions in livestock production, and cost increases from rotational grazing systems when compared to continuous grazing, have been reported (Currie, 1978; Pieper et al., 1978; Van Poollen and Lacey, 1979). In riparian and aquatic habitats, Platts and Martin (1980) found that the effects of rotational grazing systems are little better than those of yearlong grazing. They rated rotational grazing as poor to fair and yearlong grazing as poor. The varied results on the effectiveness of grazing systems stem partly from differences in the experimental and environmental conditions of each study (Pieper et al., 1978). It will be difficult to conclude that federal agencies can use rotational grazing systems effectively before site-specific research demonstrates which systems will work where. In the past the tendency has been for blanket implementation of a few model systems, and it is not clear that funding for site-specific research will be forthcoming.

There are other problems with rotational grazing systems. For example, if overgrazing has already occurred, plants may require many years to recover in arid rangelands (e.g., Sheridan, 1981; Wagner, 1978a). The rest periods in rotational grazing systems, however, usually are not longer than one grazing season or one year, and in the nonrest periods stocking intensity is often increased. Thus time spans for recovery often are inadequate. In addition the thousands of miles of barbwire fences erected for rotational grazing systems have harmed wild ungulates, especially pronghorn antelope, and numerous birds (Wagner, 1978a; Braun et al., 1978).

Energy Costs of Intensive Management

Pimentel et al. (1980) reported that rangeland beef production in the United States is more energy intensive than feedlot beef production. As management intensifies through such techniques as mechanization, herbicide use, fertilization, development of water supplies, irrigation, and fencing, the energy costs of rangeland beef production increase still further (Ward et al., 1977). The importance of this fact has been recognized by Baumer (1978), who stated: "Today energy is one of the most important limiting factors to the increase of rangeland uses, as it is no longer cheap or abundant." Whether intensive management techniques will succeed will depend on their dollar and energy costs relative to how much they increase production.

Pimentel et al. (1980) have calculated the labor and energy inputs needed to increase the total digestible nutrients (TDN) of forage produced under current extensive management (low management inputs per hectare) and under intensive management (high management inputs per hectare). They concluded that due to the aridity, low productivity, poor condition—in short, marginal quality—of much of the grazed forest-range, large energy and labor inputs would be needed to achieve a relatively small increment in total forage yield. Specifically, to increase the current yield of total digestible nutrients on 338,052,000 ha of grazed forest-range from 47.3 to 68.0 million kg would require an increase in labor from 48.8 to 276.1 million hours. Energy inputs would have to increase dramatically from 1.0 $\times$ 10^{12} kcal on current grazed forest-range to 220.0 $\times$ 10^{12} on improved grazed forest-range (Table 21.3). Thus Pimentel et al. found that intensive management of forest-range requires more than a 200-fold increase in energy to double output.

Klopatek and Risser (1982) compared the fossil fuel subsidies for beef production on native rangelands, an extensively managed system, and improved pastures, an intensively managed system, in Oklahoma. They found that based on fossil fuel expenditures, rangelands are two to three times more efficient producers of beef than improved pastures, although rangeland beef production is considerably lower per hectare. As beef production increased from the rangeland to the pasture systems in Oklahoma, the efficiency of fossil fuel use decreased (Figure 21.6). This is consistent with a pattern shown for nearly all resources examined in this book—namely a diminishing yield per energy input as inputs are increased.

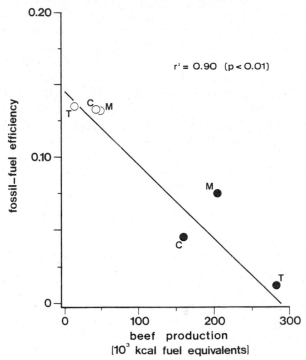

Figure 21.6. The relationship between fossil fuel efficiency (kcal of beef per kcal of fossil fuel) and beef production on pastures and rangelands in Oklahoma: ● = pastures; ○ = rangelands; C = Cleveland County, average annual precipitation = 87.6 cm; M = McCurtain County, average annual precipitation = 192.0 cm; T = Texas County, average annual precipitation = 45.0 cm. (From Klopatek and Risser, 1982.)

Klopatek and Risser's comparison of improved pastures in three counties with varying precipitation averages (Texas County—45 cm per year; Cleveland County—87.6 cm/yr; and McCurtain County—192 cm per year) indicated that upgrading rangelands to pasturelands in the driest county required the greatest fossil fuel subsidies. Future research should attempt to define more precisely the relationship between fossil fuel subsidies required for intensive management and average annual precipitation, an important and easily quantifiable

Table 21.3. **Current and Potential Yield of Total Digestible Nutrients (TDN) and Protein per Year from Current Grazed Forest Range and Improved Grazed Forest Range, and the Required Inputs of Energy, Labor, and Land**

Grazed Forest Range	TDN (kg $\times$ 10^6)	Protein (kg $\times$ 10^6)	Energy (kcal $\times$ 10^{12})	Labor (hr $\times$ 10^6)	Land (ha $\times$ 10^3)
Current	47.271	7.643	1.00	48.8	338.1
Improved	68.042	11.003	220.05	276.1	338.1

Source: From Pimentel et al. (1980).

measure of resource quality. Such information would help ranchers and agencies evaluate the potential benefits of range improvements. The Oklahoma study suggests that as precipitation decreases, the fossil fuel subsidy required for intensive management will increase. This conclusion is consistent with the linear relationship previously found between net primary production and mean annual precipitation (except in mesic areas) (Odum, 1971; Lauenroth, 1979). It is also consistent with generalizations that abound in the literature. For example, Thomas (1978) concluded:

> Most of this rangeland [the World's] is not presently adapted to intensive land use because of rough topography, severe temperatures, or poor soils; but the predominant limiting factor in production is usually lack of moisture.

Lack of nutrients is another factor that should be considered. In the Southern Sahel, low levels of nitrogen and phosphorous limit production more than does low rainfall (Bremen and DeWit, 1983).

The USBLM manages approximately 69 million ha that are characterized by both depleted soils and low moisture (Figures 21.2, 21.7). These features should escalate the energy subsidies required for the USBLM's intensive management program and raise its dollar cost. A U.S. General Accounting Office (USGAO) report entitled "Public Rangeland Improvement—A Slow, Costly Process in Need of Alternate Funding" (USGAO, 1982) confirms that the USBLM's program will be expensive. The USGAO report states that the USBLM estimates that the cost of its intensive management program will be at least $183 million. If delayed, inflation will raise the cost beyond $183 million.

Although the Public Rangelands Improvement Act of 1978 authorized special funding for range improvements totaling at least $15 million annually in fiscal years 1980 to 1986, and $20 million annually in fiscal years 1987 to 1999, the full sums have not actually been appropriated (USGAO, 1982). In fiscal year 1980 Congress appropriated $5.6 million in special funds, and in fiscal year 1981 it appropriated $9.4 million. No special funds were appropriated in fiscal year 1982, none were even requested by the USBLM in fiscal year 1983, and none are anticipated for fiscal year 1984 (USGAO, 1982). The USGAO concludes: "In view of efforts to control federal spending, large requests are not practicable at this time." Thus, although the USBLM is relying on range improvements to restore the productivity

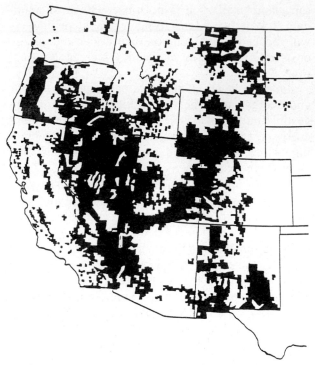

Figure 21.7. Public lands administered by the Bureau of Land Management. (From BLM, 1974.)

of the federal rangelands, the special funding for these improvements is at a standstill. As shown in Table 21.2, few improvements have actually been implemented compared to the USBLM's estimates of improvements needed. In the late seventies the USBLM's obligation to complete environmental impact statements (EISs) might have delayed range improvements. But scores of EISs are now finished, and it is clearly funding that is delaying a large-scale range improvement program.

Although the required funding has not materialized, it is worth examining how cost effective it might be. The Vale project, which was the USBLM's first major range rehabilitation effort, is instructive. In 1962 Congress authorized $16 million for range improvements following the political aftermath of some proposed stocking reductions for eastern Oregon. Most of the funding was never appropriated, but some funds were invested over the next seven years in such practices as seeding, brush control, fencing, and construction of pipelines, reservoirs, springs, and wells (Stevens and Godfrey, 1972). These techniques increased the forage in the Vale District about 25% above the 1964 level. Allowing for a project life of 30 years, the value of this extra forage to Vale area households totaled $1.0 to

$1.5 million (Stevens and Godfrey, 1972). The total project cost to the USBLM (and ultimately the taxpayers), however, was $2.9 million (Stevens and Godfrey, 1972). Stevens and Godfrey conclude that this outcome demonstrates the "inefficiency of public range investments in both the production of forage and the redistribution of income in the case of the Vale project."

Bartlett and Clawson (1978) evaluated the trade-offs between maximizing red meat production, minimizing the use of fossil fuel energy, and maximizing profits at a ranch in northern California. This ranch had adequate water supplies, including irrigation, so it had the potential to respond well to range improvements (Bartlett and Clawson, 1978). Costs of range improvements were borne by the ranch. The study found that when meat production was maximized by intensifying management, fossil fuel use was highest and profits were lowest (Table 21.4). Bartlett and Clawson concluded that efficient use of energy will become a necessity if maximum profits are to be attained.

In light of this it is not surprising that many of the USBLM's proposed range improvement projects have not cleared the Office of Management and Budget's cost-effectiveness rules. Sharpe (1979), of the Izaak Walton League of America, wrote:

> Since the 1940's, the dream has been that we could have more forage and more cattle, if only enough money were available for range improvements. But the dream has been a trap; the promised investments have not materialized; and the range has continued to deteriorate under the pressure of grazing livestock—too many, too concentrated, turned out too early, and left on too long.

As the energy costs of range improvements—particularly on unproductive USBLM lands—continue to increase, an intensive range management program that will restore the quality of these areas will continue to elude us.

SHORT-TERM STOCKING REDUCTIONS

The pessimistic outlook for intensive management of federal rangelands suggests an alternative: stocking reductions. Presently stocking reductions are given minimal consideration by the federal land-managing agencies (cf. USBLM, 1974, 1982; USFS, 1981). The Congressional Budget Office's 1978 evaluation of the quadrennial authorizations for the USBLM concluded:

> [US]BLM designs grazing programs so that, if possible, total AUM's are not reduced. In effect, public capital subsidies and investments are substituted for a reduction in grazing (permitted) . . . It seems likely that in many cases, grazing will have to be reduced until range condition improves.

Stocking reductions would improve range conditions by allowing a gradual growth of the later successional, or decreaser, plant species (see Figure 21.4) which provide better-quality forage. What effect would stocking reductions have on the quantity of forage produced? Studies overwhelmingly indicate that lower stocking intensities result in greater forage yields over the long term than do higher stocking intensities. For example, a mixed grass prairie in South Dakota yielded a 5-year average of air-dry herbage from heavily (3.5 ha per AUM), moderately (5.7 ha/AUM), and lightly (7.7 ha/AUM) grazed pastures of 1527, 1700, and 2067 kg/ha, respectively. Over the subsequent 4 years average yields on the heavily, moderately, and lightly grazed pastures (after a 2%, 9.7%, and 13.7% increase in stocking level, respectively) were 1274, 1835, and 2577, respectively. Thus heavy grazing reduced average yields from the first period to the second by 253 kg/ha, moderate grazing approximately maintained yields, and light grazing led to a 510 kg/ha increase in yields (Johnson et al., 1951). The nine-year average herbage yield from the mod-

Table 21.4. Trade-offs Between the Alternative Goals of Maximizing Meat Production, Minimizing Fossil Fuel Requirements, and Maximizing Profits at a Ranch in Northern California

Item	Maximize Meat Production	Minimize Fossil Fuel Requirement	Maximize Profit
Meat production (kg)	451,460	131,680	262,600
Fossil fuel need (liter)	649,040	51,820	246,150
Ranch profit ($)	15,880	23,610	120,310

Source: From Bartlett and Clawson (1978).

erately grazed pastures was 20 percent higher than the yield from the heavily grazed pastures and 30 percent lower than the yield from the lightly grazed pastures.

During the first year of the study the range condition in the heavy, moderate, and lightly grazed pastures was rated as 70, 72, and 73% of climax. In the last year of the study the ratings were 51, 73, and 81%, respectively. Thus relative cover of decreaser plants was reduced markedly under heavy grazing, maintained at former levels under moderate grazing, and increased under light grazing.

The reduction in the quantity and the quality of forage under heavier grazing intensities was reflected in a decline in the carrying capacity of the range for livestock production. In the spring of the last year two heavily grazed and one moderately grazed pasture had to be temporarily rested because their grass was too meager to support grazing. The summer was dry, and grazing had to cease on the heavily grazed pastures 45 days early. The experiment was terminated although there was still sufficient grass on the other pastures to have carried the cattle longer.

When the cattle were removed, there was a difference of 87 kg in the weight of cows on the heavily grazed vs. lightly grazed ranges. During the last year of the study the cows on the lightly grazed ranges produced 62 kg more calf per cow than did those on the moderately grazed ranges. The cow and calf gains per hectare were also greatest on the lightly grazed pastures in the last year.

On a shortgrass range in Colorado, Klipple and Costello (1960) found that the three-year average yields of air-dry blue grama and buffalo grass herbage from heavily (greater than 50% of current growth consumed) moderately (31–51%) and lightly (0–30%) grazed pastures were 749, 733, and 662 kg/ha early in their experiment. Six years later the 3-year average yield from the heavily grazed pasture was reduced significantly to 527 kg/ha, but the corresponding yield from the moderately grazed pasture was nearly the same as before at 749 kg/ha while the yield from the lightly grazed pasture had increased slightly to 736 kg/ha. In the following decade midgrass herbage yield from the lightly grazed pastures was significantly greater than from the moderately grazed pastures in every year but one (Klipple and Bement, 1961).

Over 11 years the range condition of the four heavily grazed pastures deteriorated. Two of the four moderately grazed pastures maintained the same range condition as before, and the other two improved somewhat. All four lightly grazed pastures improved in range condition.

Cattle on the heavily grazed areas produced significantly smaller weight gains per head than did cattle on the moderately and lightly grazed areas. Under heavy grazing, cattle gained weight through September but usually lost weight during October. Under moderate grazing, cattle usually maintained their weight during October, and under light grazing cattle were able to make small October weight gains. Average weight gains per hectare were usually greatest in the heavily grazed pastures due to the larger number of animals present, but in the twelfth year of the study, average gains per hectare under moderate use surpassed those under heavy use (Klipple and Costello, 1960).

On a shortgrass range in Kansas, Launchbaugh (1957) reported that over eight years the average gain per head under heavy (0.8 ha/head), moderate (1.4 ha/head), and light (2.1 ha/head) grazing was 55 kg, 85 kg, and 98 kg, respectively. Over an 11-year period steers on average continued to gain weight at the end of the grazing season on the lightly grazed range, whereas steers on the moderate and heavily grazed range were losing weight. After the first 7 years the heavily grazed pastures provided only 154 days of grazing because there was a "complete lack of forage to carry (the cattle) further," while the moderate and lightly grazed pastures maintained the original 180-day grazing season. After 2 more years of grazing the heavily grazed pasture provided an all-time low of only 92 days of grazing, whereas the other pastures still maintained the 180 day grazing season.

Once the carrying capacity of a range has been reduced through overgrazing, it can be restored to a higher sustainable level through stocking reductions (Klipple and Costello, 1960; Klipple and Bement, 1961). In a study first reported by Klipple and Bement (1961), the Rocky Mountain Forest and Range Experiment Station compared the effects of light grazing on deteriorated shortgrass vegetation with those of moderate grazing on shortgrass vegetation in good condition. At the start of the study, yields of palatable grasses in kilograms of air-dry herbage per hectare averaged 628 on the deteriorated ranges and 897 on the good condition ranges. Eight years later light grazing enabled average yields on the deteriorated pastures to increase to 913 kg/ha, a 45% increase, but moderate grazing slightly reduced yields on the good condition ranges to 829 kg/ha, a 7% decrease. The increased herbage production of the deteriorated range was maintained after six subsequent years of moderate use. Klipple and Bement (1961) concluded that most of the improvement in herbage production from the deteriorated range occurred within 5 to 7 years after light grazing was instituted.

In addition to increasing the long-term carrying capacity of deteriorated ranges, stocking reductions can increase beef production both per animal and per hectare over a single grazing season. For example, in a Colorado study of the seasonal response of growing heifers to stocking rates of 9.9, 4.9, 3.2, and 2.5 ha/AUM on native gambel oak rangeland, Rittenhouse et al. (1982) found that the average daily gain per head per day in kilograms at each successive stocking level was 1.05, 0.89, 0.79, and 0.60 in 1978 and 1.32, 1.22, 1.19, and 0.80 in 1980. Rittenhouse et al. reported that the cumulative gain per animal over the course of the summer was greater at the lower stocking rates—particularly in the dry summer of 1978 when forage was more limited. During this summer the heifers at the higher stocking levels stopped gaining weight altogether due to insufficient forage. Increasing the stocking rate to 3.2 ha/AUM increased cumulative weight gain per hectare, but at this point cumulative weight gain peaked and was reduced by additional increases in the stocking rate.

Jones and Sandland (1974) showed that there is a linear decrease in animal gain with increasing stocking rate which they expressed by the equation: $y_a = a - bx$, where y_a is the gain per animal, x is the stocking rate, and a and b are constants whose values must be measured in the field for each range type and class of grazing animal. At very low grazing intensities, however, forage is superabundant and stocking rate does not influence gain per animal (Jones and Sandland, 1974; Hart, 1978; Rittenhouse, 1982). The relationship between gain per hectare, y_h, and stocking rate can be expressed by the quadratic equation: $y_h = ax - bx^2$, where x, a, and b are defined as before. Thus gain per hectare increases with increasing stocking rate to an optimum level, peaks, and then decreases (Jones and Sandland, 1974; Hart, 1978; Rittenhouse, 1982). Gain per hectare and gain per animal cannot be maximized simultaneously (Jones and Sandland, 1974; Stoddart et al., 1975). Figure 21.8 which is based on data from a continuously grazed tropical grass-legume pasture illustrates these relationships (see p. 63).

It should be emphasized that the optimal stocking level based on gain per hectare is not necessarily the optimal stocking level from the standpoint of sustaining productivity over the long term because it does not show future animal weight gains and does not contribute any information on the response of vegetation and soil on which animal weight gains ultimately depend (Rittenhouse, 1982).

From the perspective of economic returns, moderate stocking intensities, as long as they are sustainable, are generally considered most advantageous (Bement, 1969; Stoddart et al., 1975). Heavy stocking intensities incur the following costs: (1) in-

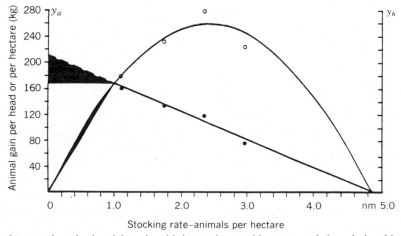

Figure 21.8. The linear decrease in animal weight gain with increasing stocking rate, and the relationship between weight gain per hectare and stocking rate, which is a quadratic function. The shaded triangle indicates that at low stocking rates, in this example below 1 animal per hectare, animal weight gain is expected to be independent of stocking rate. (From Jones and Sandland, 1974).

vestment in additional animals, (2) reduced weight gains per animal, (3) reduced calf crops, (4) reduced price per kilogram for lower quality beef, and (5) increased death loss (Launchbaugh, 1957; Stoddart et al., 1975). Due to these factors Launchbaugh (1957) found that the lowest stocking rate in his study had a net return comparable to that of the moderate stocking rate. The heavy rate of 4.9 ha/head led to a net loss of $9.02/ha, the moderate rate of 8.4 ha/head led to a net gain of $12.28 per hectare, and the light rate of 12.6 ha/head led to a net gain of $12.87/ha.

Klipple and Bement (1961) demonstrated that when the objective is not simply to sustain the quality of the forage resource, but actually to improve it, light grazing is an economically feasible means toward this end.

By how much would stocking reductions increase forage production compared to forage increases under grazing systems? Van Poollen and Lacey (1979) reviewed the literature on vegetation responses to grazing systems and stocking reductions and found that grazing systems, which intensify management by requiring fencing, additional water developments, and frequent relocation of cattle, are less effective than stocking reductions. Specifically, mean annual herbage production increased by an average of 13% across 15 studies when grazing systems were implemented and stocking intensity was held constant at moderate use. Under continuous grazing, however, reducing livestock use from heavy to moderate increased mean annual herbage production by an average of 35% across 11 studies, and reducing moderate use to light use increased mean annual herbage production by an average of 27% across 9 studies.

Van Poollen and Lacey concluded that land managers should place more emphasis on proper stocking intensity and less on grazing system implementation. For example, if a land manager simultaneously implemented a grazing system and reduced livestock use from heavy to moderate, on average a 48% (13 + 35%) increase would be expected in mean annual herbage production. The grazing system would be responsible for 27% of the increase in herbage production, and the reduction in livestock use would be responsible for 73% of the increase. Note that these figures are averages and that the efficacy of stocking reductions at a particular site would depend on such factors as range type, climate, and weather.

Stocking reductions are especially important on low quality, unproductive rangelands where the potential benefits of range improvements are particularly limited. For example, in a depleted salt-desert rangeland in Colorado, reducing grazing intensity was the only viable method for increasing forage production (Bradshaw and Bartlett, 1978). In other areas stocking reductions are important to prevent range deterioration in the first place or, if the point of no return through natural recovery has been reached, they must be combined with range improvements (Klipple and Bement, 1961). For example, where invader species such as mesquite have become established, vegetative removal practices may be necessary (Paulsen and Ares, 1961; Wagner, 1978a). It is not desirable, however, to delay stocking reductions pending funding for range improvements. Such delays simply increase the task at hand once funding is secured.

Stocking reductions do not require any capital investments (Klipple and Bement, 1961); therefore, compared to intensive management, they are a cheap and energy-efficient way to restore the productivity of federal rangelands. In the short term stocking reductions would slightly decrease beef production. It should be noted, however, that only 13% of the total AUMs of livestock grazing in the United States are derived from federal lands (see Cutler, 1979) and grazing (both range and nonrange) presently supplies only about 64% of the feed consumed by all U.S. beef cattle (USFS, 1981). If federal lands of the West supported both range and nonrange grazing, federal lands would provide the feed for 8% ($0.13 \times 0.64 = 0.08$) of all U.S. beef cattle. Because federal lands support only range grazing, however, they must actually provide the feed for less than 8% of all U.S. cattle. Stocking reductions of, for example, 25% across all federal lands, would therefore cut total beef production in the short term by less than 2% ($0.08 \times 0.25 = 0.02$). Thus the short-term impact of federal stocking reductions on U.S. beef production as a whole would be slight.

It has commonly been assumed that stocking reductions on federal lands would decrease beef production on private lands because many ranchers have insufficient forage on their own lands to graze all of their cattle on them year-round. These ranchers therefore graze the federal rangelands for one or more seasons a year (Heady et al., 1974). Godfrey (1978) evaluated this assumption for counties where federal stocking cuts exceeded 25% and compared such counties to similar ones without significant stocking reductions. His results indicate that the cattle industry was still able to grow in each county affected by livestock reductions on federal lands, even though the reductions were not followed by increases in federal grazing permitted. In the areas with the most extensive adjustments in the use of federal lands, however, the rate at which cattle numbers increased tended to be "somewhat smaller."

In the long term stocking reductions on federal lands would result in increased forage production and thereby would increase the carrying capacity of these areas for livestock and wildlife. Because 75% of the Western ranges were producing less than half the climax vegetation of the native range by 1966 (Pacific Consultants, 1970; Box et al., 1976), stocking reductions have the ecological potential approximately to double palatable forage production and hence double the carrying capacity of forage for livestock and wildlife.

Due to new selective pressures (more domestic herbivores relative to native herbivores, fewer fires, etc.), the plant communities that will result from stocking reductions in overgrazed areas will not necessarily resemble historical climax communities. They may, however, be fully as productive (Stoddart et al., 1975). Vegetative succession will occur most slowly in areas where grazing stress remains high, soil erosion is severe, precipitation is low, invader species have become established, and few late successional native species remain to provide seed (Stoddart et al., 1975). These constraints also retard range improvement under intensive management.

If stocking reductions are not instituted, the productivity of federal rangelands will remain at its present impoverished level or, still worse, will continue to deteriorate. Consequently the long-term carrying capacity for livestock will remain considerably below potential. Thus, for restoring rangeland productivity, sustaining beef production, and minimizing costs, stocking reductions hold more promise than does intensive management.

A long-term increase in the carrying capacity for livestock resulting from stocking reductions would benefit the rangeland livestock industry. But those individual ranchers forced to cut their operations on federal lands would either have to liquidate or relocate some of their livestock (Sheridan, 1981). Because federal grazing fees are lower than those for private and state lands (see Stoddart et al., 1975; USDA, 1977; CBO, 1978; Sheridan, 1981; US-

GAO, 1982), relocation of cattle often would reduce income. Thus the reduction in short-term forage consumption would be a valuable investment in future forage production when considered from the perspective of the public as a whole or the ranching industry but not necessarily from the perspective of the individual rancher. Agencies responsible for managing federal lands can play an important role in resource protection through maintaining higher levels of sustained yield than is economically feasible for lands under private ownership (Stevens and Godfrey, 1972).

CONCLUSION

Western rangelands are inherently low in productivity, but their quality has deteriorated still further due to failure to regulate livestock use to sustainable levels. Livestock overgrazing reduced 80% of the Western ranges to fair or poor condition by 1966 and has been a major cause of the desertification that has essentially removed 91 million ha from production.

The intensive management programs adopted by the federal land-managing agencies to restore the productivity of Western rangelands will require enormous fossil fuel subsidies. As management intensifies, these fossil fuel subsidies will have to be larger and larger for smaller and smaller new contributions to beef production. Such energy inefficiency makes intensive management programs very expensive, and adequate funds to implement them successfully have not been appropriated. While the federal agencies await funding for range improvements, these areas will be vulnerable to further degradation, and there is no hope of restoring productivity to former levels.

As an alternative to intensive management, stocking intensity should be reduced so that it does not exceed light use over all federal lands needing improvement in range condition. Stocking intensity on other rangelands should not exceed moderate use. The magnitude of the stocking intensities representing moderate and light use will vary with local rainfall and other determinants of rangeland quality.

Stocking reductions do not require energy and capital investments and are more effective than grazing systems at increasing forage yields. By restoring the productivity of Western rangelands, stocking reductions have the ecological potential approximately to double the carrying capacity of forage for both livestock and wildlife. As an added benefit, stocking reductions do not cause adverse environmental effects as do certain range improvement techniques. Thus from the perspective of (1) restoring rangeland productivity, (2) environmental protection, (3) sustaining beef production, and (4) minimizing energy costs, stocking reductions have more to offer than an approach that emphasizes intensive management.

22

WATER FROM AN ENERGY PERSPECTIVE*

There are similarities and differences between the ways that water and energy interact in natural and industrial systems. Both systems need water to support life. Cooling is an essential water-aided process in both systems. For example, water is transpired to maintain leaf temperatures at ambient levels that enable continued energy upgrading from photons to organic states. Similarly, water is necessary to maintain power plant temperatures that allow energy to be upgraded from fossil fuel or nuclear forms to electricity. In natural systems water transports weathered rocks, leaving behind valleys and canyons and making deltas while transporting nutrients and other biotically important solutes. Similarly in industrial systems water is used for transport, for coal slurry conveyance, in manufacturing most items and to wash coal, gravel, crops, and other essential materials. Whether making canyons or washing industrial materials, more work can be done when the water is purer, for water with less solutes has more Gibbs free energy. Thus, for most human-directed uses of water, purer water is of higher quality.

Large quantities of water, mostly of high quality, are necessary for the maintenance of a high standard of living in the industrialized countries. Very often it requires large amounts of energy to supply this water, and in turn large amounts of water are required to supply energy. The extent and pervasiveness of the water requirements for energy can be seen from the following quote from Harte and El-Gasseir (1978):

> Energy technologies use water resources in numerous ways. For example, the cooling of electric generating plants or coal gasification and liquefaction plants may consume freshwater. Coal and oil shale conversion processes require water as a chemical feedstock. Coal mining and land reclamation subsequent to surface mining require water. Solar bioconversion plantations are likely to require irrigation water. Hydroelectric power consumes water in the sense that artificial lakes enhance evaporation losses. In fact, nearly every imaginable energy system demands water. On the other hand, water is one of the few non-energy materials whose supply can produce energy. In other words, water supply systems frequently have associated hydropower installations.

WATER AVAILABILITY

Superficially there appears to be more than enough fresh water, in most parts of the globe, but the critical problem is that it is not distributed evenly or regularly in either time or space (Table 22.1). Water is recycled constantly through the environment in the hydrological cycle, which is driven by the evaporation of about 453,000 km³ of water annually by the sun. Of this water about 38,000 km³ per year falls on the inhabited continents and is available for human use as runoff in rivers (Penman, 1969; Falkenmark, 1976; Figure 22.1). Although there is much more rain than present or anticipated human needs most of the total runoff occurs as flood waters so that year-round availability is only 30–40% of runoff. Evaporation, transpiration, and droughts also cut the amount of runoff that can be depended on for use (Figure 22.2). In the United States about 5.8 trillion m³ falls each year on the continental land mass (USWRC, 1979). Only about one-third of this precipitation is available for potential human use since nearly two-thirds is converted back into water vapor by transpiration by plants and evaporation (Nace, 1976; Gartska, 1978; Pimentel et al., 1982). But because of periodic low flows only about 675 bgd (billion gallons per day; 1 gal = 0.003785 M³) of surface and subsurface flow is relied upon for society's use (USWRC, 1979).

As a result in many watersheds there are insufficient supplies of good quality water for at

*The authors thank Larry Dryer and Neo Martinez for contributions to this chapter.

Table 22.1. Water Resources in Different Parts of the World

Continent	Runoff, (cu km per year)			Stable Runoff as Percentage of Total Runoff
	Total	Stable Portion[a]	Unstable Portion	
Africa	4225	1905	2320	45
Asia, except USSR	9544	2900	6644	30
Australia	1965	495	1470	25
Europe, except USSR	2362	1020	1342	43
North America	5960	2380	3580	40
South America	10,380	3900	6480	38
USSR	4384	1410	2974	32
All continents, except polar areas	38,820	14,010	24,810	36

Source: From Falkenmark (1976).

[a]Derived from groundwater or regulated by lakes or reservoirs.

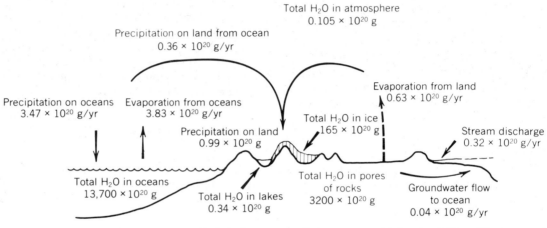

Figure 22.1a. The global water cycle. One $M^3 = 10^6$ g $= 264.2$ gal. (Reproduced with permission, from *Chemical Cycles and the Global Environment* by R. Garrels, F. T. MacKenzie, and C. Hunt. Copyright © 1975 by William Kaufmann, Inc. All rights reserved.)

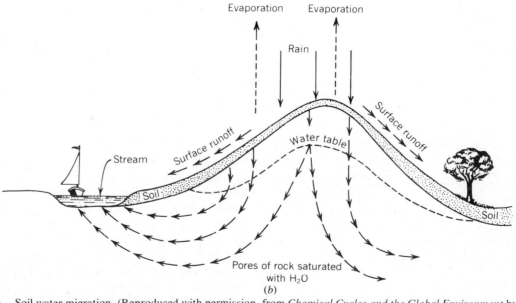

Figure 22.1b. Soil water migration. (Reproduced with permission, from *Chemical Cycles and the Global Environment* by R. Garrels, F. T. MacKenzie, and C. Hunt. Copyright © 1975 by William Kaufmann, Inc. All rights reserved.)

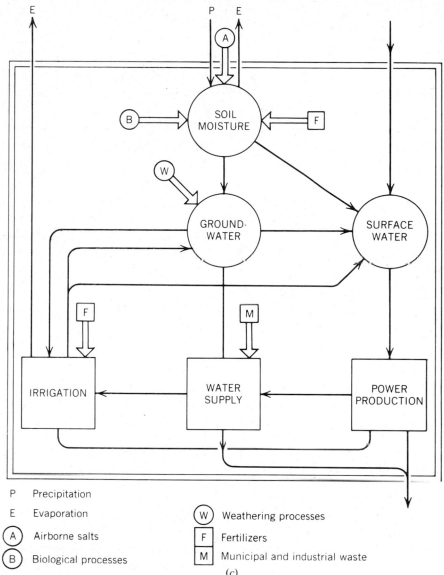

P   Precipitation

E   Evaporation

Ⓦ   Weathering processes

Ⓐ   Airborne salts

F   Fertilizers

Ⓑ   Biological processes

M   Municipal and industrial waste

(c)

Figure 22.1c.   Schematic description of a generalized runoff module (water resource unit) taking into consideration the balance of water and of substances dissolved in same. Large circles denote natural reservoirs; squares denote human-made reservoirs. Small circles denote natural sources of dissolved substances; small squares denote human-made sources of dissolved substances. (From Falkenmark, 1976.)

least some months of some years. Energy invested in new water projects can alleviate some water supply problems by increasing the storage of flood waters, although most of the good reservoir sites in the United States already have been developed (Nace, 1976). The damming of rivers to create reservoirs also has its associated water costs. For example, the evaporation from human-made reservoirs represents 12% of the total fresh water consumption in the United States, about twice the total municipal use including domestic and commercial use (Table 22.2).

Types of Water Use

Water use can be classified into two categories. *Withdrawal* means that water is diverted from its normal flow pattern to human uses. The water is often returned, and thus can be reused, although the quality of the water is usually degraded. The withdrawn water normally picks up dissolved substances that may render it unfit for certain end uses, for example, for drinking water, but this may not affect its suitability for uses such as irrigation (Table 22.2). Water that is completely lost to the

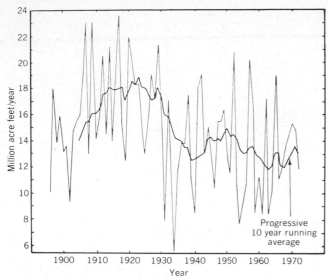

Figure 22.2. The annual discharge of the Colorado River at Lee Ferry, Arizona, shows the uncertainty and changes over time of water availability. Variation within the year increases the uncertainty. (From Moss, 1967.)

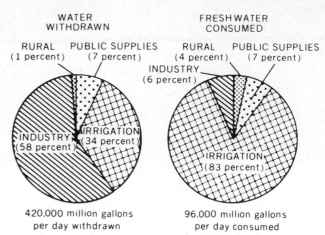

Figure 22.3. Water withdrawal and consumption in the United States in 1975. (After Murray and Reeves, 1977.)

runoff system, as by evaporation, is considered *consumed,* although it may rain back locally or in a distant watershed. The quantities of water withdrawn and consumed for various purposes in the United States are given in Table 22.2 and Figure 22.3. The greatest withdrawal and consumption of water is for irrigation, livestock and miscellaneous rural use, which together account for 76% of the United States total (Table 22.2). Energy facilities both withdraw and consume substantial amounts of water. The largest withdrawal is for electric plant cooling, after which the water is returned more or less unaffected except for being of higher temperature and containing some chemicals (see p. 403). Consumptive uses include reservoir evaporation, water lost in energy processing and water used to encourage vegetative growth on reclaimed mining land.

Energy Costs of Making Water Available

There are many costs associated with making water available. The most obvious are energies required for pumping, building dams and other capital structures, and purification. More subtle costs include those associated with the disruption of aquatic and wetland ecosystems (Harte and El-Gasseir, 1978; Ward and Stanford, 1979; Brooker, 1981) and the fact that present groundwater withdrawal increases future energy costs by lowering water tables.

Regional Water Availability

It is important to view water availability from a regional perspective (Figure 22.4). Table 22.3 compares the availability and use of water for 18 regions of the coterminous United States plus Alaska and Hawaii. In general, more people live in wetter than drier regions, in part due to the effects of water availability on the desirability of a region for human

Table 22.2. Fresh Water Use in the United States (in km³/yr)

Use Category	Withdrawal	Consumption
Municipal use including domestic and commercial	40.0	9.2
Industrial mining and manufacturing	52.0	5.8
Thermal electric power plant cooling	180.0	2.6
Irrigation, livestock, and rural use	200.0	115.0
Evaporation from human-made reservoirs	18.0	18.0
Total	490.0	151.0

Source: From Harte and El-Gasseir (1978).

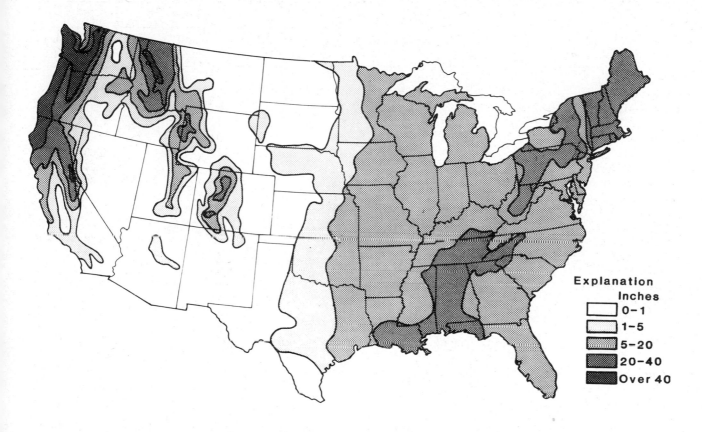

Explanation

Inches

☐	0-1
☐	1-5
▨	5-20
▨	20-40
■	Over 40

Figure 22.4.   Average annual precipitation for the United States.

residence and agriculture. Undeveloped energy resources, however, tend to be found in drier regions.

The principal sources for most of the water we use are continuously recharged through rain. One especially important resource, groundwater, however, is not, even though there is about 10 times more fresh water stored underground as above the ground. For example, the Ogallala aquifer, which supplies water for irrigation in Texas, New Mexico, Oklahoma, Colorado, Kansas, and Nebraska, is being pumped out at the rate of nearly 2 m/yr. It recharges at the rate of about 1 cm or so a year and seems to be running dry rapidly. The use of the Ogallala aquifer is akin to an underground strike of gold, silver, oil, or other valuable resource which is being mined out. This is forcing some farmers in that region to shift to growing dry-climate wheat, rather than more lucrative corn, for although there is enough water, the dollar cost of the energy required to pump the deep water has become prohibitive in some cases. In the United States, in general,

about 25% more water is withdrawn from underground aquifers than is replenished by rain (USWRC, 1979). A total of about 113 billion m³ of groundwater are withdrawn annually compared to a recharge of about 84 billion m³ (Pimentel et al., 1982), although this varies greatly by region.

Figure 22.5 indicates where major water-requiring energy deposits are found; those areas in the West overlap considerably with regions where water is chronically in short supply. Table 22.4 gives estimates of the quality of water required for various energy extraction and production industries, and Table 22.5 shows where water is used within each energy industry.

ENERGY AND WATER

As indicated previously, energy and water interact in two fundamental ways, for they are interdependent. First, virtually all water management requires

Table 22.3.   Regional Runoff, 1975 Consumption, per Capita Runoff, and Consumption per Unit Runoff

Region	Mean Annual Runoff (cu km/year)	Data for 1975 Consumption (cu km/year)	Per Capita Runoff (10^3 cu m/person/year)	Consumption/Mean Annual Runoff
New England	93.0	0.61	7.9	0.0066
Mid-Atlantic	120.0	2.2	3.0	0.018
South Atlantic Gulf	270.0	5.1	10.2	0.019
Great Lakes	100.0	1.5	4.5	0.015
Ohio	170.0	1.7	8.0	0.01
Tennessee	57.0	0.39	17.0	0.0068
Upper Mississippi	90.0	1.3	4.6	0.014
Lower Mississippi	100.0	7.6	17.0	0.069
Souris-Red-Rainy	8.6	0.17	12.0	0.016
Missouri	75.0	24.0	8.4	0.32
Arkansas	100.0	16.0	16.0	0.16
Texas Gulf	44.0	13.0	4.2	0.30
Rio Grande	6.9	6.0	3.5	0.87
Upper Colorado	18.0	3.4	40.0	0.19
Lower Colorado	4.4	10.0	1.7	2.3
Great Basin	10.0	5.5	7.0	0.55
Pacific Northwest	290.0	18.0	44.0	0.062
California	86.0	34.0	4.1	0.40
Alaska	800.0	0.0077	2000.0	9.6×10^{-6}
Hawaii	18.0	0.77	22.0	0.043
United States	2471.0	151.0	11.0	0.060
United States excluding Alaska and Hawaii	1653.0	150.0	7.8	0.091

Source: From "Energy and Water," J. Harte and M. El-Gasseir, *Science* Vol. 199, pp. 623–634, Table 1, 1978. Copyright © 1978 by the AAAS.

substantial quantities of energy, although sometimes hydroelectric energy can be gained through water works. Second, virtually all energy production and use is water intensive, including petroleum refining, coal mining, shale oil development, uranium enrichment, biomass production, and especially electric power production from any fuel. Much of this water use degrades the quality of the water used in one way or another, so that water availability limits energy production and in turn energy production degrades water availability and quality. This second aspect of energy and water has been covered in Chapter 15, so the rest of this chapter considers the energy cost of water. It is important to note, however, that the large-scale development of Western coal, shale oil, or synthetic fuels would greatly aggrevate an already tense series of conflicts about the distribution of a resource already scarce in that region (see Harte and El-Gasseir, 1978, for projections).

The Energy Cost of Water Management

Water is used in the United States economy principally for the following purposes: (1) to meet human physiological and sanitary requirements, (2) to maintain natural supportive ecosystems, (3) to grow food crops and livestock, (4) to provide process water for a myriad of industrial processes, and (5) to cool electric generating plants and many other industrial facilities. The principal energy cost of desalinizing seawater and lifting it to the elevation of most human endeavors is of course provided by the sun, and as a result the dollar and fossil energy cost of delivering water to the majority of our farmlands and cities is relatively small per unit of water. If our total economic activity were small, then most of our water needs could be met principally by solar energy alone. But the problem arises with larger economic production because (1) that production requires increasing use of drier farmlands, (2) greater

Figure 22.5.   Relation of major U.S. water-intensive energy deposits to drainage systems. (Courtesy of D. Pimentel.)

levels of industrial and human waste production overtax the natural cleansing capacity of waterways, and (3) the forthcoming exhaustion of our highest grades of petroleum will mean increased reliance on much more water-intensive synthetics, from coal or perhaps shale oil, for fluid energy supply.

Three water-provision management problems illustrate how the industrial energy requirements have been increasing for supplying water that could once be met more or less completely by natural, low investment means. The first is for supplying groundwater for irrigation, the second is for supplying municipal water for Southern California, and the third is for flood protection.

Irrigation

In the United States in 1980 about 24 million ha were irrigated, about 20% more than in 1974 (Slogett, 1982). This land produced at least 29% (in dollars) of U.S. agricultural sales, and the direct energy required for pumping was 23% of U.S. on-farm energy use. Although the irrigation cost was nearly $2 billion in 1980, this was not a large proportion of the total energy or dollar cost of the food produced (Slogett, 1982).

The most important ways in which energy is used in irrigation are (1) to manufacture equipment, (2) to lift water from the water table in the ground to the point of application, and (3) to apply pressure to the

Table 22.4.   Water Used by Industry in Producing Various Items

Industrial Product	Unit	Gallons of Water Used per Unit	Cubic Meters of Water Used per Unit
Automobile	1 auto	12,000–16,000	45–60
Steel	1 ton	1400–65,000	5.3–244
Glass bottle	1 bottle	120–650	0.45–2.4
Synthetic rubber	1 ton	30,000–600,000	113–250
Pulp and paper	1 ton	20,000	75
Corn	1 bushel	6000	23

Source: From Chanlett (1979).

Table 22.5. Water Required for the Production of Energy from Conventional and Synthetic Fuels

			Category of Use				
	Mining[a]	Reclamation[b]	Transport by Slurry Pipelines[c]	Conversions[d]	Associated Urban[d]	Total with Slurry Pipelines	Total without Slurry Pipelines
Low Btu Gas							
Eastern coal							
Surface mined	0.0028–0.0035	0.0–0.030	0.045–0.057	0.083–0.058	0.018	0.15–0.69	0.10–0.63
Deep mined	0.0062–0.0078	0.0	0.045–0.057	0.083–0.058	0.018	0.15–0.66	0.11–0.61
Western coal							
Surface mined	0.0028–0.0070	0.0028–0.14	0.045–0.11	0.083–0.058	0.018	0.15–0.86	0.11–0.74
Deep mined	0.0062–0.010	0.0	0.045–0.11	0.083–0.058	0.018	0.15–0.72	0.11–0.61
High Btu Gas							
Eastern coal							
Surface mined	0.0035–0.0042	0.0–0.036	0.057–0.069	0.083–0.58	0.049	0.19–0.74	0.14–0.67
Deep mined	0.0078–0.0095	0.0	0.057–0.069	0.083–0.58	0.049	0.20–0.71	0.14–0.64
Western coal							
Surface mined	0.0035–0.0085	0.0036–0.17	0.057–0.14	0.083–0.58	0.049	0.20–0.95	0.14–0.81
Deep mined	0.0078–0.012	0.0	0.057–0.14	0.083–0.58	0.049	0.20–0.7	0.14–0.64
Syncrude							
Eastern coal							
Surface mined	0.0031–0.057	0.0–0.048	0.051–0.093	0.11–0.74	0.029	0.19–0.92	0.14–0.82
Deep mined	0.0070–0.013	0.0	0.051–0.093	0.11–0.74	0.029	0.20–0.88	0.15–0.78
Western coal							
Surfaced mined	0.0031–0.011	0.0032–0.23	0.051–0.19	0.11–0.74	0.029	0.20–1.2	0.14–1.0
Deep mined	0.0070–0.017	0.0	0.051–0.19	0.11–0.74	0.029	0.20–0.98	0.14–0.79
Oil from Shale							
Surface technology							
Surface mined	0.0040–0.0056	0.033–0.053	na	0.030–0.044	0.0069–0.0092	na	0.074–0.11
Deep mined	0.0041–0.0056	0.032–0.056	na	0.030–0.044	0.0082–0.011	na	0.074–0.12
In situ technology							
Modified in situ	0.0019–0.0026	0.014–0.030	na	0.027–0.047	0.0087–0.010	na	0.052–0.090
True in situ	na	0.0–0.0077	na	0.0–0.044	0.0088–0.010	na	0.009–0.062

Source: From "Energy and Water," J. Harte and M. El-Gasseir, *Science* Vol. 199, pp. 623–634, Fig. 1, 10 February 1978. Copyright © 1978 by the AAAS.

Note: Data are expressed as km³/10¹⁸ J of synthetic fuel product. All calculations are based on coal energy content of 28, 22, and 14×10^6 J/kg of bituminous, subbituminous, and lignite coals and on the conversion efficiencies of 67 to 85%, 55 to 67%, and 41 to 75% for low and high Btu gasification and liquefaction, respectively.

[a] In the East, surface and deep mining consume 2.3 and 5.2 m³/10¹² J of coal mined. In the West, consumption is 2.3–4.7 and 5.2–6.8 m³/10¹² J mined, respectively.

[b] In the East, land disturbance is 22–65 m²/10² J of coal mined and annual water consumption is 0–0.015 m³/m² over a 1- to 2-year period. In the West, the corresponding figures are 3.9–31 m²/10¹² J of coal mined and 0.30–0.61 m³/m² over 2 to 5 years. The shale estimates include consumption for revegetation as well as processed shale disposal.

[c] Slurry pipelines consume 38 and 37–76 m³/10¹² J of coal mined in the East and the West, respectively.

[d] For coal and shale conversion factors see Harte and El Gasseir.

water to spray or otherwise distribute the water. The first energy cost is calculated as in Chapter 5. The second energy cost, E, can be estimated as

$$E = MDg\,\frac{1}{e}$$

where M = mass lifted, D equals meters of lift, g is the acceleration due to gravity, and e is the efficiency of a pump (about 60% on average). Thus between 10 and 20 kcal of petroleum are required to lift 1 m³ of water 1 m in altitude. Electric pumps are more efficient but may use more total fossil energy when conversion efficiencies are figured in. Pumping costs generally increase as the level of the water table drops. The energy required for distribution depends on the volume of water distributed, the pressure used (which depends on the device), and the efficiency of the device. In 1980 roughly 37 × 10¹² kcal were used to irrigate some 12% of U.S farmland, about two-tenths of 1% of total U.S. energy use, and the use of irrigated land is expected to increase in the future (see Slogett, 1982, and Roberts and Hagan, 1975, for more detailed estimates).

The energy cost of irrigation can be reduced substantially by installing drip irrigation or certain other capital-intensive systems (Pimentel et al., 1982). The dollar and energy cost of the required capital equipment is large, discouraging its use, but the savings can be large in the long run. In addition drip irrigation, for example, uses much less water than spray irrigation and therefore slows the depletion of groundwater, thereby reducing energy use in future years (see Blackwelder, 1984, for innovative alternatives to traditional water development schemes that save water and energy).

Water Supplies in Southern California

The public works by which water is supplied to Southern California is an excellent example of the general principle given in this book that the least energy-intensive supplies of a resource tend to be developed first, so the energy cost of resources tends to increase over time. The groundwater resources originally used were not especially energy intensive (Figure 22.6). The first major water supply project, the Owens Valley project, was a net source of energy to the region because of the hydroelectric power generated. But this huge project presently

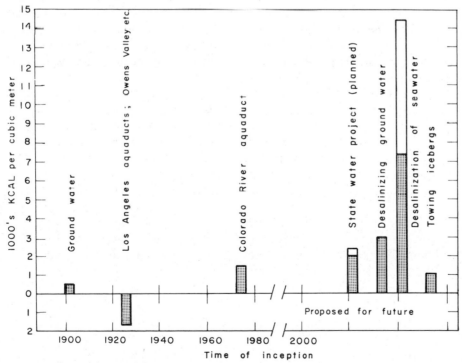

Figure 22.6. Energy costs of supplying municipal water to the Los Angeles area have been increasing as the best resources are developed. Times of project inception are approximate. (Data from Roberts and Nagon, 1975.)

supplies only about 80% of the region's growing needs, so a portion of the Colorado River has been diverted to make up the deficit. In this case instead of an energy gain there was a considerable energy cost. Future designs to provide additional water are even more energy intensive. For example, water can be *supplied* by cleaning up dirty water using industrial processes. It takes about 139 kcal to give the 1 m^3 of water primary and secondary treatment and required tertiary treatment to give higher quality water would cost roughly 5–10 times more energy per cubic meter (Figure 22-6). The California water plan proposes the use of brackish groundwater and the purification of municipal wastes, both of which would cost much more energy than the Colorado project while delivering water of substantially less quality. One large-scale scheme might be somewhat less energy intensive—the towing of icebergs from Antarctica. In any case it is clear now that supplying water to Southern California is becoming an increasingly energy-intensive proposition as the highest quality alternatives have been utilized.

Flood Protection and River Management

One of the larger public works projects in the United States has been the national effort to provide flood protection. The theory for federal support for flood protection seems simple and humane, for every year we learn of the large human and financial toll that results from acts of nature seemingly beyond the control of human beings. Compassion would seem to argue for some sort of flood insurance and for flood protection by building dams. But all such events related to large-scale flooding, including both the destruction (and replacement) of property and the construction of flood control projects, have had large associated energy components that are largely unquantified.

Unfortunately flood control and insurance are not as effective as it might seem. For example, although federal flood control expenditures increased dramatically in each decade from the 1920s to the 1960s, so too did flood damage. The problem was that flood control devices did indeed reduce the damage from the *average* flood. But, as a result, more people and businesses moved onto flood plains adjacent to rivers. Then, when a flood larger than what the flood protection devices can handle comes along—and that always happens—there is much more property to be impacted by the flood. Some flood control structures even increase flood damage in other regions. For example, the levees built along the middle Mississippi River keep floodwaters in the channel (rather than spreading out on the flood plains) which creates higher floods in the channel itself and causes new floods in unleveed tributaries, some of which actually flow *upstream* in response to the levees in the main channel.

A similar problem is faced by engineers who try to maintain large rivers in particular configurations. A particular example is the lower Mississippi River. Over the last ten thousand or so years the Mississippi has changed its bed about 10 times in response to new energy gradients produced from the prograding of existing deltas, the net effect of which is to decrease the gradient of the river to the sea and encourage the river to take a new steeper path to the sea. At present the river is attempting to switch back to the steeper gradient of the Atchafalaya Basin, and it would have done so during the great 1978 flood except for a very large energy-intensive effort by the Army Corps of Engineers. This would have been disastrous for the city of New Orleans whose economy is largely linked to its role as a port.

The lesson to be learned from these analyses is that eventually rivers will flood, switch, or otherwise do what the natural energy regimes dictate. Human intervention can delay but not stop these changes, and such delaying tactics are extremely energy intensive if they are on a large scale. If industrial energy does become less available, it will become increasingly important to understand and accept the changes that inevitably will occur and to design human infrastructure in accordance with, and not in opposition to, nature.

FINAL EDITORIAL

HOW IMPORTANT IS ENERGY?: MATERIALISM VERSUS HUMANISM

The formation, occurrence, and distribution of natural resources, and the resource transformation technologies by which they are used for human survival and material well-being, provide a common thread by which we can understand both natural and industrial ecosystems. The models and data presented throughout this book concerning the relation between resource (especially fuel) quality and a variety of ecological and economic parameters substantiate the view that an explicit assessment of natural resource conditions are essential to a complete understanding of all living systems, including human societies. Such a view might seem intuitive in light of the material presented in this book, but not all economic analysts agree that physical principles require explicit consideration in economic models. There is a wide spectrum of opinion on what the role of natural resources is in the economic process and how that role should be represented in our models of the economy. At the poles of this philosophical continuum are what Daly (1984) calls the *economic imperialists* and *energy reductionists*. Economic imperialists assume that subjective tastes and preferences are the driving force behind the economy and that individual wants are the origin of all economic value. Accordingly, any constraints imposed by nature can be overcome by technological change, the vehicle by which new ideas are used to increase human control over the flows of energy and matter in nature. Energy reductionists, on the other hand, tend to reduce all ecological and economic phenomena into energy units and assume that physical laws in large part determine the laws of the market and other social institutions. Energy reductionists equate economic value with the direct and indirect energy required to produce a commodity.

Neither extreme view is a complete and accurate picture of the economic process, because each searches for a single principle by which to explain all economic phenomena. In doing so, each side explicitly rejects or implicitly ignores credible components of the opposing theory. What is needed is a dualist position that accounts for the limitations imposed on human aspirations by resource and other environmental conditions. Both Frederick Soddy, an early energy analyst, and Herman Daly, a contemporary critic of the growth-oriented neoclassical economic paradigm, recognized the need for a dualist economic theory. Daly (1984, p. 25) stated

> A dualist position recognizes that (economic) value has roots both in the physical world and in the mind. Its physical roots are finitude and entropy. Finitude keeps all wants from being satisfied and imposes the necessity of choice. Low entropy is the physical quality of matter-energy that enables us to satisfy our wants, that can only be rearranged, but never created by human activity. The mental roots (of the dualist position) are subjective tastes and preferences of objective value (i.e. a moral principle for evaluating the relative value of subjective tastes).

In a similar vein, Soddy (1922, p. 26) emphasized that economics occupies the middle ground between the physical and metaphysical world.

> In each direction possibilities of further knowledge extend *ad infinitum*, but in each direction diametrically away from . . . the problems of life. It is in this middle field that economics lies, unaffected whether by the ultimate philosophy of the electron or the soul, and concerned rather with the interaction, with the middle world of life of these two end worlds of physics and mind in the commonest everyday aspects, matter and energy on the one hand, obeying the laws of mathematical probability or chance as exhibited in the inanimate universe, and, on the other hand, with the guidance, direction and willing of these blind forces and processes to predetermined ends.

A dualist position recognizes that neither energy nor subjective tastes and preferences alone are necessary *and* sufficient conditions for economic value.

Energy is a necessary condition, for without energy no economic process could operate. But consideration of energy divorced from consideration of human purpose leaves us in the noneconomic world. Similarly, the human contribution to economic progress is, by itself, insufficient for understanding such progress, for only when human aspirations become physical realities through the use of energy do those aspirations serve to better the conditions of human existence. As such, all economic activities are constrained to varying degrees by the physical laws governing energy and matter transformations.

While acknowledging the importance of human ideas, we have emphasized the unique contribution of energy and other natural resources to the economic process. We chose this tack because the humanist approach has dominated formal economic theory for the last 200 years to the almost complete exclusion of physical laws and natural resource constraints. The exclusion of natural resources from neoclassical models of economic growth is due in large part to the abundance and low cost of resources relative to capital and labor throughout the early stages of the industrial revolution. The cost of natural resources was insignificant relative to capital and labor, so resources were simply ignored or taken as a given. Natural resource abundance, however, does not mean freedom from resource constraints. Natural resources have always been the basis of material wealth regardless of their relative abundance and cost. Nature no longer affords us the luxury of ignoring or downplaying the role of natural resources. It is time to account for the biophysical origins of economic wealth in our models of the economic process. Studying the flow of energy in society is a means by which such a goal can be partially realized. In the final analysis, the quantity and quality of energy available to human society sets broad but distinct limits on what is and is not economically possible at any given time. Ignoring such limits leads to the euphoric delusion that the only limit to the economic expansion of the human race exists in our own minds.

Our view may be dismissed by those who believe humans are so unique that no physically based model can ever accurately explain aspects of human behavior. We have neither proposed nor advocated a strict deterministic relation between energy use and all economic behavior. Although we know of no logically or empirically consistent model that supports a universal cause and effect relation between energy use and all aspects of economic behavior,

we agree with Cottrell (1955) that available energy sources and energy transformation technologies largely determine the general structure of society. The validity and usefulness of our models does not hinge on whether or not "everything human [is] reducible to physical principles wrapped up in mechanical models" (Daly, 1981). Instead, we have tried to show how the natural resource base influences the direction and rate of social change by constraining the economic choices presented to us. Natural resource conditions also influence which technological and social choices will be successful and prosperous.

An ecological analogy may help clarify this point. Developing fully testable models of natural selection is difficult because the theory of evolution does not and cannot rely on direct causality. The size and shape of the beaks on Darwin's finches—small birds in the Galapagos Islands that were first studied by Charles Darwin—were not *caused* by the size and shape of seeds available to the birds (Costanza, 1981). Rather, seed size and myriad other environmental factors exerted subtle yet pervasive pressure on the finches to adapt to physical conditions of their environment. An individual finch was free to eat whatever seeds it chose, and there is no reason to not believe that various individual finches eat smaller or larger seeds. Nevertheless, certain size seeds were more abundant and certain size beaks were more effective at exploiting the more abundant seed types. The end result was a type of beak and other resource-exploiting traits that enabled the finches to develop the resources available to them at a better rate and efficiency than once-competing species that no longer exist.

In a similar vein an energy-based theory of economic development does not rely on direct causality. Rather, the natural resource base exerts the same subtle yet pervasive influence on humans. The "seeds" available to a society (fuels, minerals, air, water, etc.) influence the type of resource transformation technologies adopted by that society in the same way that the resources available to the finches modified physical and behavioral aspects of the finch society. Of various alternative technologies, those that allow societies to control and direct more energy than others give an advantage in dealing with other competitors for the same limited supply of resources. Some technologies are selectively eliminated because they in some way reduce their users' competitive position. Over time the successful and prosperous societies develop technologies that generally increase their control of energy while not adversely affecting their chances of survival in other ways.

We have also emphasized that energy is not the only factor in the evolution of human behavior. There are important dissimilarities between human and natural systems. There is no a priori reason to believe that the same principles that guide nonhuman evolution also guide all aspects of human evolution. Environmental conditions certainly direct, to a degree, the course of biological evolution by reinforcing certain resource transforming behaviors and selectively eliminating others. Humans, however, physically alter the environment they live in to a much greater degree than other organisms. Indeed, industrial ecosystems have become dominant forces shaping the conditions under which all other living systems must operate in many regions of the biosphere. In doing so, humans alter significantly the conditions under which natural selection operates. Humans can, so to speak, change to a degree the rules of the game in which they and other species are playing.

A biophysical perspective cannot explain why a particular individual prefers a Ford to a Chrysler. But it can help explain how and why technologies were developed that produced automobiles and other petroleum-based technologies, and how social institutions might change when the resources underlying that technology decline in quality. The goals of a biophysical perspective are not therefore to suggest cause-and-effect relations between energy and a particular behavior but rather to identify the constraints that the laws of energy and matter place on our economic aspirations. Ultimately, the criterion by which a biophysical approach should be judged is the degree to which it helps us understand our economic choices. When compared vis-à-vis the dubious track record of standard economic models of the past decade that emphasize the primacy of subjective consumer preferences and omnipotent technological change, a biophysical perspective for assessing future economic options provides an important and much needed approach.

RESOURCE OPTIMISM VERSUS RESOURCE PESSIMISM

The environmental movement that began in earnest in the late 1960s and the energy events of the 1970s thrust natural resources onto the national political agenda and back into the economic vocabulary. The question many seem to be asking is, will the supply of natural resources be adequate to satisfy a growing population that has come to expect an ever-expanding material standard of living? In Chapter 4 we described in detail two contrasting philosophies of this important issue. The analytic responses have run the gamut from Chicken Little predictions of imminent or eventual economic and environmental collapse (Meadows et al., 1972) to myopic extrapolations of past trends of mineralogical abundance (Simon, 1981) and pie-in-the-sky predictions of economic rhapsody (Kahn, 1983). The former group, the so-called resource pessimists, argue that many of society's vital fuel, mineral, and biological resources are being depleted rapidly, with resulting inevitable shortages combined with expanding population causing a cessation or even a reversal of economic growth. The so-called resource optimists either disavow the possibility of resource shortages or concede that resource shortages loom on the horizon but argue that technical change will postpone their arrival or develop substitutes by the time shortages arrive. An extreme case of the latter type of reasoning, which could be termed the omnipotent technology hypothesis, maintains that society can maintain historic rates of growth indefinitely with an ever declining resource base.

Both sides of the resource scarcity issue base their arguments on extrapolations of past trends. The bleak forecasts made by Meadows et al. (1972) in the *Limits to Growth* were based on computer simulation models that extrapolated population growth and rising rates of per capita use of finite stocks of nonrenewable and renewable resources. Shortages were the logical conclusion of their models in which rapidly expanding demands were placed on rapidly dwindling stocks of nonrenewable resources. Predictions of resource shortages were not invented with the computer. W. Stanley Jevons, one of the founders of neoclassical economics, predicted in 1865 that within the near future England would face severe economic difficulties due to the depletion of its domestic coal resources. Jevon's predictions proved incorrect because he foresaw neither technical advances that enabled English miners to dig coal from deeper mines nor the general switch from coal to petroleum. Resource optimists today argue that the modern version of resource pessimism exemplified by the *Limits to Growth* model is incorrect because technical advances will enable society to develop petroleum deposits currently too expensive to use and alternative energy technologies that will supply the same or greater amount of fuel at similar prices.

As described in Chapter 4, resource optimists believe that resource depletion automatically gives rise to solutions for the problems it creates: increasing resource scarcity leads to higher resource prices which induce research and development of new technologies and substitutes that circumvent the shortage. As Barnett and Morse (1963, p. 236) would have us believe,

> . . . a strong case can be made for the view that the cumulation of knowledge and technological progress is automatic and self-reproductive in modern economies, and obeys a law of increasing returns.

Since there is no a priori reason to believe that the rate of new ideas generated by humans will decline in the future, resource scarcity would be a moot issue if the situation were the way Barnett and Morse depict it, for all the minerals in the entire Earth's crust would be extractable at reasonable cost. Indeed, as we discussed in Chapter 4, some analysts assert that the vast quantities of minerals that exist at concentrations at or near their crustal abundances will someday be extracted at costs not significantly higher than the price we now pay for the extremely rare and concentrated deposits we currently mine (Goeller, 1979; Simon, 1981). Brooks and Andrews (1974) summarize this position succinctly when they assert that there is no reason to believe that we won't mine average country dirt to meet our future mineral needs. Given their underlying assumptions (i.e., self-generating technological change), the reassuring conclusions of the omnipotent technology model are logically consistent if difficult to validate empirically. If economic growth were limited only by the rate of advance of human ingenuity, there would be no need to debate our economic future.

The simple extrapolations represented by the extreme cases of both resource optimists and pessimists do not accurately reflect the conditions confronting industrial society. Clearly, using the methods of Meadows et al. of dividing known reserves by annual rates of use leads to overly pessimistic assessments of resource availability. Technological possibilities are too numerous and societies' tastes are too adaptable to tie human welfare in toto to a single resource. On the other hand, the seductive notion of self-correcting, self-generating technical change is too simplistic, especially in view of the fact that a large component of all technological change to date has relied on increased per capita use of fossil fuels. We must come to grips with the fact that there may not be a technical solution to all our future economic problems, or that the answers to these problems will necessarily arrive before a growing population expands beyond the limits of a shrinking resource base which itself is the taproot of technology.

THE LESSONS OF HISTORY

History teaches us that unquestioning faith in the market mechanism, technological change, or other human institutions to maintain our level of material wealth indefinitely and political power is an unwise and potentially dangerous delusion. In the past, changing resource conditions combined with nonadaptive social institutions and growing population led to the downfall of powerful nations. The Egyptian and Roman empires declined in large part due to overexpansion of their economies and social infrastructures relative to growing population pressures, mismanagement of certain natural resources, and an overall decline in quality of their natural resource base. These crises contributed to civil unrest and eventually social revolution, a restructuring of social values, and opportunities for invading armies. In more modern times, the rise of the British empire was fueled by that nation's ability to expand and control international trade using wind power via the sailing ship, and later when it was the first nation to fully realize the economic opportunities made possible by coal-driven steam engines. Like Egypt and Rome, however, the British empire faded due in part to an overextension of its economic and political systems relative to a declining domestic resource base which created a range of social pressures at home. Such trends are documented by Cottrell (1955), Cook (1976), Ryan (1980), Adams (1982), and others. In each of the instances cited above, an early warning sign of a nation reaching the zenith of economic expansion and political power was a growing dependence on foreign sources of natural resources. This occurs when a nation cannot support a growing standard of living on domestic resource supplies alone. When this occurs, a nation grants greater control of its internal affairs to outside agents. The dependence of the United States and other nations on OPEC countries for petroleum is an obvious and painful example of this predicament.

A prudent course of action for the United States would be to not presume that the future of its material wealth and political power are unconditionally guaranteed. If questioned about their economic future, the average Egyptian and Roman were probably optimistic. Yet their societies declined in prominence due to social changes induced by changes in their natural resource bases, and energy technologies in particular.

Framing the resource scarcity debate as we have is useful in highlighting the distinct differences with which people view our economic future. The resource optimist–pessimist dichotomy, however, paints the misleading picture of a world in which technology races against resource depletion, trying to keep economic growth one step ahead of economic collapse. We seem to be faced with either economic utopia or a return to the Stone Age. Polarizing our possibilities in this way glosses over the intimate connection between resource quality and technical change highlighted by a biophysical perspective. It ignores the vast middle ground between the two extremes where resource limitations preclude the possibility of 3–4% annual growth rates in GNP but where technical change still allows a comfortable standard of living for a large number of people. Such a society would place a premium on economic *development* rather than economic *growth*. Such a society would need to achieve population levels commensurate with existing resource conditions. It also would require modification of existing social values that measure progress in terms of annual rates of growth in GNP. Economic growth models based on the assumption that technical change is independent of resource conditions must be tempered with models that incorporate the natural resource foundation of technology and that increasingly give importance to non-material indices of human welfare.

We believe that the trend of using increasing quantities and higher-quality forms of fossil fuel to implement new technologies differentiates present and forthcoming resource shortages from those that confronted society in Jevons' time. While past resource shortages were often averted by developing new technologies, most of those technologies used greater quantities of higher-quality, higher EROI fuel. This pattern is reflected by the historic shift from wood to coal to petroleum to electricity. But what of the next 50 years? Despite research efforts

that are massive relative to earlier efforts, no alternative fuel technology presently under research and development appears capable of supporting a new technological burst such as the one made possible when we tapped $1 per barrel crude oil from Texas and Oklahoma in the 1930s at an EROI of greater than 100:1. Many are quick to reply that we may not be able to foresee the energy technology of the future that will replace high EROI petroleum without the economy skipping a beat, just as Jevon's did not foresee petroleum replacing coal in 19th century England. To this we reply that our technical knowledge of the Earth's resources is far greater than it was in Jevons' time. The science of thermodynamics was in its infancy when Jevons wrote *The Coal Question* in 1865. Today we have knowledge not only of Jevons' miscalculations and countless others to learn by but also extensive knowledge of how energy is used in the economy and the needed characteristics of alternative energy resources if economic growth is to be maintained. Most important, the long lead times associated with the development of large-scale energy systems means that any alternate fuel must be proved feasible very soon if we are to avoid energy shortfalls. We certainly cannot predict to the letter what energy technologies will be feasible in 50 years, but neither is it likely that we will be surprised by some completely unforeseen technology. It is highly likely that the energy technologies supplying the bulk of our energy needs 50 years from now are identified and/or used to some degree today.

ENERGY AND TECHNOLOGICAL CHANGE

The apparent lack of an adequate replacement for petroleum threatens to undermine a basic economic strategy of the past 150 years: ameliorating the drudgery of economic tasks accomplished predominantly by human mechanical energy (often in slavery systems) by developing technologies that empower or supplement labor with increasing quantitites of fossil fuel, especially petroleum. These fuel subsidies reduced the labor time necessary to extract natural resources even as those resources were becoming physically scarcer. We do not dispute the empirical analyses of Barnett and Morse (1963) and Simon (1981) which document the decline in labor costs per unit output in the extractive sectors. These analyses are consistent with our findings presented in Chapters 2 and 4 which showed a strong correlation between energy use per laborer and labor productivity in the U.S. economy. We do emphasize, however, as did Georgescu-Roegen (1975), that the explanations of Barnett and Morse and Simon are incomplete. Attributing the decline in labor costs to self-generating technical change leaves unexplained the origin and mode of implementation of new technologies. Without increasing fuel subsidies per worker in the future, technical advance in the extractive sectors is unlikely to keep pace with resource depletion, and resource prices will rise. In Chapters 4 and 7 we discussed the analyses of Norgaard (1975) and Cleveland and Costanza (1983) which documented how technical change in the oil and gas industry offset only part of the decline in petroleum resource quality, with the result being higher discovery costs and declining EROI over time. Because the omnipotent technology model assumes the origins of technical change to be solely a function of human ideas, it is not surprising that the natural resource events of the past decade represent a significant departure from the historic trends in the models of Barnett and Morse and Simon. Their rosy view of our resource future becomes less reassuring when it is recognized that the physical conditions underlying the historic trend of declining resource prices have been irrevocably altered. Fuels with large EROI are no longer cheap or abundant.

The physical basis of technical change is not widely appreciated. Most economic models of technical change cling tenaciously to the notion that technical progress is powered by advances in human ingenuity alone. It is not surprising that these models have gone through increasingly difficult acrobatics in trying to explain current economic conditions. Labor productivity stagnation does not stem from a stagnation of human ideas but rather from the increased cost of energy that discouraged the implementation of new energy-using ideas. The model of self-correcting resource shortages seems weak at best, at least in the case of domestic petroleum resources. The real price of petroleum quadrupled between 1972 and 1981, expenditures on petroleum exploration and development increased over 500%, but domestic petroleum production and discoveries continued their general downward trend.

Properly used, a biophysical perspective can increase our understanding of our current economic problems. The problem is not a sudden lack of human ingenuity, for clearly society's understanding of its relation with the physical environment expands every day. New ideas of how to transform resources through economic tasks are being generated and tested daily. The issue is whether new technologies can be developed that maintain historic rates of labor productivity and economic growth with a declining natural resource base. With a physical basis of technical change clearly identified, it becomes incumbent on the resource optimists to demonstrate specifically how technical change can offset the deleterious effects of resource depletion as it did in the past when resources were cheap and abundant.

THE FADING BEACON OF ECONOMIC GROWTH?

As the EROI for fossil fuels declines, common sense dictates that we consider all plausible scenarios of our economic future. One such scenario is that energy limitations may force industrial society into a zero growth (steady-state) or even declining economy. The case for a steady-state economy—in which all stocks, including material artifacts and population levels, are maintained at a constant level and resource throughputs managed so as to maintain these constant stocks—has been forcefully and cogently presented by Daly (1977) and Boulding (1966). Based on our knowledge of our energy and resource prospects, we cannot conclude unequivocally that a steady state would be necessary and desirable in the foreseeable future. We are equally unsure, however, about the feasibility and desirability of maintaining a rapidly expanding economy into the indefinite future. It is imperative that we investigate the implications of both possibilities, for to exclude one at the expense of the other would preclude a smooth transition to that economic future should it become a reality.

A steady-state view of the future is not a philosophy of gloom and doom as it is sometimes depicted by the advocates of indefinite economic growth. The economic and social consequences of trying to force economic growth under resource conditions that cannot physically support such growth would be far more injurious than those associated with policies aimed at development commensurate with existing resource conditions. As Daly (1977) emphasizes, steady-state economies need not be technically or intellectually stagnant. Technical change would be oriented more toward economic development, as opposed to economic growth, and toward diversification of economic and social opportunities. Natural resources needed to implement new technologies would be diverted from existing uses rather than enlarging total resource use.

It is important to evaluate a range of energy and hence economic futures because of the influence energy systems have on society as a whole. Energy technologies do much more than simply supply new quantities of energy. They are important determinants of many other technologies and of the structure of the social systems using those technologies. Alteration of a society's basic energy system leads to fundamental changes in traditional institutions and values. The social and cultural impacts of the agricultural and industrial revolutions, the two most significant shifts in energy systems in human history, are examples of such impacts. As Cottrell (1955) emphasized, resistance to social change induced by a change in energy systems is often severe. Those in power (i.e., those who control the energy transformation technologies) resist a shift to any energy source which they do not control, or will try to secure control over the new energy technology. The Pharoahs of ancient Egypt secured and maintained their political power by diverting most of their society's surplus energy (in the form of surplus human labor power) into the construction of the Great Pyramids. As Egypt's dominion expanded and it controlled increasing quantities of resources, the size and diversity of its social structure also increased. The merchant and military class rose in power and challenged and eventually overcame the power of the Pharoahs. In a similar fashion, the British transition to a fossil fuel economy gave increasing wealth and power to the emerging industrial class and placed them at odds with the aristocracy, which had ruled Britain for centuries by virtue of their control over the primary energy source of their time—agricultural land. It is not surprising, therefore, that there is substantial resistance to even a consideration of a transition to a steady-state economy. Such a transition implies a redistribution of the economic products of natural resources, and therefore a redistribution of economic and social power as well.

Economic growth has been used to postpone debate on social issues such as the equitable distribution of income and political power, a luxury less materially prosperous nations have not been afforded. As Daly (1980, p. 23) stated:

growth takes the edge off distributional conflicts. If everyone's absolute share of income is increasing, there is a tendency not to fight over relative shares, especially since such fights may interfere with growth and lead to a lower absolute share for all. But these problems cannot be held at bay forever, because growth cannot continue indefinitely.

Without a growing pie, one group's demand for a larger slice must be taken from another group's slice. The mineralogical and energetic bonanza of the past 40 years offered the genuine prospect of upward mobility for many Americans. No one likes to be told that his or her material standard of living might be lower in the future. Such a possibility is unpalatable to politicians because they would not be able to take credit for last quarter's GNP growth and would instead have to take time to explain to the less well off their plans for improving economic conditions.

We must resist the temptation to believe that all our economic and social problems can be solved by political means. The unprecedented magnitude of material wealth achieved by the United States has been made possible primarily by large quantities of high-quality fuel and other resources used to empower the effort of human labor in the production process. Although the exploitation of these resources has been encouraged by our democratic political philosophy, other nations with quite different political philosophies have become as rich or richer than the United States. Thus, the international relation between per capita fuel use and per capita economic output is much stronger than the relation between political ideology and economic output. Of course the quantity, method, and efficiency with which energy is used in the production process are not independent of political philosophy. But within developed nations it is likely that specific political decisions affect principally *who* gets the end products of energy use and only secondarily what the absolute level of material wealth will be. Such limits are more a function of the natural resources available to the nation. Thus, the ability of a political party or system to make good its promise of increased production is dependent principally on the continued or increased availability of high-quality energy and raw materials, and only secondarily on the ways in which those factors are used to empower labor through technologies or political action. On the other hand, there certainly are resource-rich underdeveloped nations that lack the political wherewithal to mobilize those resources.

THE CHOICES BEFORE US

Our students and other readers of early drafts of this book often asked for guidelines concerning energy policy recommendations. Like other energy or resource analysts, we have our own ideas concerning what we think our economic future will be like, as well as opinions as to what we would like the future to be like and how we should prepare for it. Since the purpose of this book is more analytical than prescriptive we did not try to provide an energy blueprint. Because there are myriad complex choices before us, any single best answer is almost *a priori* suspect. We do, however, see the utility of making some general recommendations concerning our energy future based on the information we have presented within the constraints of what we see as viable energy alternatives. We can also provide suggestions as to some of the desired characteristics of plausible technical and institutional solutions to the problems we have raised.

We must resist the temptation to base our energy and economic policies on the belief that the stage of depletion of domestic conventional oil and gas resources is a temporary and artifically contrived situation, resolvable by proper economic or regulatory incentives. We have read many claims that substantial quantities of conventional oil and gas remain to be discovered in the United States through continued price deregulation, oil import tariffs and/or quotas, and other regulatory policies aimed at increasing the incentives to the petroleum industry. Most of these claims, however, have no basis in reality and rely on the fervently held notion that a new wave of drilling or technological change will overcome the current petroleum shortage. Since 1973 there has been all the economic incentive one can imagine for increased oil petroleum production and discoveries, yet no dramatic turnaround has occurred. Instead, huge quantities of wealth have been transferred to the major oil companies as a result of oil price increases.

The analyses of domestic petroleum availability presented earlier by Davis (1958), Hubbert (1956, 1974), Menard and Sharman (1975), Nehring (1981), Hall and Cleveland (1981), and Cleveland and Costanza (1983) were based on the empirical record of the industry and therefore embody the effects of past technological change on petroleum discovery and production rates. These analyses indicate that conventional domestic petroleum resources cannot be relied upon much longer to supply 70–75% of total energy needs as they have for most of the past two decades. The outlook for future natural gas production is possibly brighter than the prospects for crude oil. Natural gas price regulation imposed by federal regulations has led to disincentives to explore for and develop certain types of natural gas deposits. When these price regulations are removed, there may be a modest surge in the discovery and production of natural gas. On the other hand the production declines in the Hugoton-Panhandle Field in Oklahoma and the Louisiana gas fields (Figure 7.9) may be a more accurate indicator of the future.

Because oil and gas depletion is real, the United States needs to look for replacements. This book has explored many of the possibilities. We believe that *all* potential energy transformation technologies should be researched aggressively, while explicitly and comprehensively assessing their EROI. Such assessments should include identification and quantification of all direct and indirect energy costs to the largest degree possible. Such an approach is required by federal laws 93-577 and 96-294 (Title II) and has been endorsed as a practical and beneficial energy policy tool by the Government Accounting Office (1982). Despite such endorsements, net energy analysis remains essentially unknown to the general public and is ignored by most policy makers. Net energy assessments can help identify those technologies that can provide society with the greatest quantities of net fuel, and thereby with the largest potential to produce economic output. This would allow us the greatest number and variety of economic and social choices.

The energy return on investment should not be the sole criterion by which we select the next generation of fuels. The employment, environmental, and national security impacts of every energy technology should also be considered. Energy technologies differ in their relative ability to preserve those values and institutions that ensure society's survival and well-being. As Cottrell (1972, p. 25) stated:

> . . . the conditions necessary to convert increasing energy flow may in some cases (as, for example, in setting fires to burn off unwanted vegetation) impose only limited conditions on the society that uses it. In others, (for example, controlled nuclear fission or fusion) they may be extremely narrow and the technology and social organization required to operate them be extremely precise. Here, if the [energy] flow is to be secured and maintained, most

of the arrangements are dictated by the facts discovered by science to be necessary.

Even if breeder reactors can produce the same amount of net energy at the same EROI as a combination of, for example, renewable fuels and conservation, the social impacts of the hard versus soft paths would differ radically. A breeder system would of necessity be highly centralized, and the general level of security in the United States would need to be increased to prevent large quantities of plutonium from falling into the wrong hands. On the other hand the soft path would decentralize energy systems, but with such disbursement, consumer responsibility and participation also would increase. Such differences have been forcefully presented by Lovins (1977), Jungk (1979), and others.

Three controversial fuel sources—imported oil, coal, and conservation will figure prominently in our near-term energy future. The continued importation of oil has some strong advantages and disadvantages. The strongest argument against such a policy is that it leaves us dependent on sometimes politically unstable and unfriendly governments in the Middle East. The economic impacts of oil price shocks such as we experienced in the 1970s is another obvious disadvantage. Oil importation is not without its advantages, however. Most important, imports slow the rate at which the United States depletes domestic oil reserves. Since oil is already the preferred fuel for much of our capital stock, imports could postpone the inevitable investment we will have to make in new energy converters. Perhaps the most important advantage is that petroleum is a very high-quality, flexible, and relatively clean-burning fuel.

We believe that coal, in one form or another, will be the major fuel used by the United States in 50 years. We don't particularly like coal, but its abundance relative to petroleum is unquestionable. The EROI for coal will probably be favorable relative to alternative fuels for many years to come, even with its substantial environmental and human health impacts. Substantial investments in new technologies will be required to minimize such impacts. Removing the sulfur from coal is a must, and much remains to be learned about improving the efficiency of coal combustion technologies. The carbon dioxide problem will remain, and by the time we determine whether increased carbon dioxide is good or bad it may be too late to do anything about it. Despite all its drawbacks, we will probably have to learn once again how to power our economy on substantial quantities of coal.

We have emphasized the strong link between energy use, economic output, and social development. We must not ignore, however, the opportunities that exist for true energy conservation improvements that increase the quantity of economic output and social welfare that can be produced from a given quantity of energy. When considered as energy transformation technologies, conservation and improved efficiency must be given the highest priority, except where their EROIs are demonstrably low.

As of this writing, the future of the nuclear power industry is at best uncertain, as the last sections of Chapter 13 describes. Due to decreased demand for electricity, construction and safety-related problems, the impact of new nuclear plants on electricity price increases, and adverse reaction to nuclear plants in capital markets, the costs of constructing new nuclear facilities have skyrocketed. Chapter 13 highlights the recent history of the nuclear power industry in the United States, a history dominated by severe cost overruns and questions of reactor safety and security. The situation is a far cry from that in the not-too-distant past when utilities depicted nuclear power as being "too cheap to meter" and predicted up to 1000 plants to be operating by 2000. Instead, by the end of 1984 there were 81 reactors licensed for operation, and no new reactor units had been announced in 6 years. In addition, there is concern that rapidly declining uranium ore grades could reduce the already low EROI for nuclear plants.

Breeder and fusion power represent an energy panacea to many futurists just as conventional light water reactors did to some 30–40 years ago. Nevertheless, our experience with conventional nuclear reactors suggests that it would be unwise to assume a priori that either fusion or breeders will have a large EROI. The scientific and engineering problems that remain to be solved are formidable and may never be solved. Nuclear power, in general, always has the potential to go askew in a very big way, implying large potential social costs that are difficult to incorporate in any cost–benefit analysis. This is especially true for breeders, where the fuel cycle would be one of the most pervasive technologies in society, and the chances of plutonium falling in unwanted hands thereby considerably greater. On the other hand, if breeders work, they could have a very large EROI.

Imported oil, conservation, and coal cannot sustain industrial society forever. If humans are to survive as an industrial species, we will have to develop renewable resources to their fullest potential. We are optimistic about the future of solar energy while acknowledging its limitations, especially its present low EROI. Technical improvements are certain to improve the EROI for many solar energy technologies, as evidenced by the four- to fivefold increase in the EROI for photovoltaics in recent years. Solar energy can probably provide 10–30% of energy needs in the medium range future if society encourages the development of appropriate solar technologies such as water heating, passive building heating and cooling, some biomass fuels, and certain industrial applications. At the present time capital-intensive solar electricity generating technologies such as the solar power satellite appear to have very marginal EROI ratios. If government subsidies to the nuclear industry were reduced and if the social costs of coal combustion were internalized, solar energy probably could contribute moderate quantities of energy in the beginning of the next century. At that time we would have the experience to assess whether it will be possible to maintain a highly industrialized society based predominantly on solar energy transformation technologies.

THE FINAL QUANDARY: PASCAL'S WAGER

The United States and other industrialized nations are faced with a remarkable quandary not previously encountered in human history. Natural resource shortages and energy transitions have come and gone throughout the history of civilization but never before has a society with such a sophisticated physical infrastructure and a high material standard of living been so dependent on a finite store of low-entropy matter and energy. The question before us is a simple one: Can human technologies circumvent the declining quality of fossil fuels as it did in many (but not all) past resource shortages, and thereby allow economic growth to continue at rates comparable to those we've experienced since the Industrial Revolution?

No definitive answer is available at this time. We have presented evidence in this book sufficient to make defensible the proposition that economic production cannot double every 25 years as it has in the past. It may even decline. In the meantime, population levels will continue to increase which could decrease per capita wealth substantially. Yet we cannot rule out the possibility that technical breakthroughs might make such growth possible. There is then a certain degree of risk involved in implementing energy and economic policies based on one of these two scenarios. Despite our uncertainty of future conditions, the United States must formulate policies based on the best information presently available.

The relative risk and return of the two strategies can be analyzed by applying the logic behind Pascal's wager (Daly, 1977). We can adopt the omnipotent technology hypothesis and later find it to be false, or we can reject it and later find out that the necessary technical breakthroughs do in fact exist. Which error would have the least deleterious economic and social impacts? If society assumes technology will mitigate all resource scarcities and sets its sights on rapid economic growth, it could easily overshoot its carrying capacity. By the time society realized the necessary technology had not arrived its biological support systems could be damaged irreparably. On the other hand, if we planned for slower growth and later found that technical capabilities permitted higher production rates, some individuals would have lost some portion of a higher standard of living. Nevertheless, once the error was discovered, the option of adjusting our economic course could still be possible.

It appears that it would be wise to at least plan for the contingency that omnipotent technology will not be the savior of infinite economic growth. This does not mean that we should accept such limits passively. We must realize that the remarkable social achievements of the past 150 years could not have occurred without empowering labor with greater quantities of energy, particularly fossil fuels. In the future genius must work toward replacing the bludgeon of fossil fuel with the rapier of knowledge. We encourage you to think of your professional interests from a biophysical perspective, where it is appropriate, and to utilize such information in your professional endeavors. There is much for us to do.

BIBLIOGRAPHY

Aage, H. 1984. Economic arguments on the sufficiency of natural resources. *Cambridge J. Econ.*, **8**:105–113.

Adams, F. G. and P. Miovic. 1968. On relative fuel efficiencies and the output elasticity of energy consumption in Western Europe. *J. Ind. Econ.*, **17**: 41–56.

Adams, R. M. and T. D. Crocker. 1982. Dose–response information and environmental damage assessments: An economic perspective. *J. Air Pollut. Control Assoc.*, **32**:1062–1067.

Adams, R. N. 1982. *Paradoxical Harvest: Energy and Explanation in British History, 1870–1914.* Cambridge University Press, New York.

Adelman, M. A., R. E. Hall, K. F. Hansen, J. H. Halloman, H. D. Jacoby, P. L. Jaskow, P. W. MacAvoy, H. P. Meissner, D. C. White, and M. B. Zimmerman. 1974. Energy self-sufficiency: An economic evaluation. *Tech. Rev.*, **76**:23–58.

Aerospace Corporation. 1977. Aerospace Corporation characterization of the U.S. transportation system. Aerospace Report ATR-77(7398)-1 Vol. II, El Segundo, CA.

Air Quality and Automobile Emission Control. 1973. Prepared for the Committee on Public Works, U.S. Senate. National Academy of Engineering. Vol. 1, No. 93-24.

Akarca, A. T. and T. V. Long. 1980. Dynamic modeling using advanced time series techniques: Energy–GNP and energy–employment interactions. In *Energy Modelling II*, pp. 501–516. Institute of Gas Technology, Chicago.

Alessio, F. J. 1981. Energy analysis and the energy theory of value. *Energy J.*, **2**:61–84.

Alexander, J. F., D. P. Swaney, R. J. Rognstad, and R. K. Hutchinson. An energetics analysis of coal quality. 1980. In A. Green (Ed.), *Coal Burning Issues*. University of Florida Press, Gainesville, pp. 49–70.

Alfven, H. 1972. Energy and environment. *Bull. Atomic Scientists*, **23**(5):5–8.

Alfven, H. 1974. Fission energy and other sources of energy. *Bull. Atomic Scientists*, **30**:4–8.

Alvarez, L. W., W. Alvarez, F. Asaro, and H. Michel. 1980. Extraterrestrial cause for the Cretaceous–Tertiary extinction. *Science*, **208**:1095–1108.

Alverson, D. L., A. R. Longhurst, and J. A. Gulland. 1970. How much food from the sea? *Science*, **168**:503–505.

American Association of Petroleum Geologists. 1981. Annual review of drilling in North America. *AAPG Bull.*, various years.

American Petroleum Institute, American Gas Association and Canadian Petroleum Association. 1980. Reserves of crude oil, natural gas liquids, and natural gas in the United States and Canada. API, Washington, D.C.

Andersen, L. C. and J. L. Jordan. 1968. Monetary and fiscal actions: A test of their relative importance in economic stabilization: Review. Federal Reserve Bank of St. Louis, St. Louis, MO.

Anderson, J. R. and J. Q. Jones. 1980. Uranium production, Supply Branch, Supply Analysis Division, Grand Junction Office, U.S. Department of Energy.

Andrews, R. M. 1979. Reproductive effort of female *Anolis limifrons* (Sauria: Iguonidae). *Copeia*, **1979**:620–626.

Andrews, R. M. and T. Asato. 1977. Energy utilization of a tropical lizard. *Comp. Biochem. Phys.*, **58A**:57–62.

Angel, J. L. 1975. Paleoecology, paleodemography and health. In S. Polgar (Ed.), *Population, Ecology and Social Evolution*. Mouton, The Hague, pp. 167–190.

Annual Survey of Manufacturers and Census of Manufacturers. U.S. Department of Commerce, Bureau of the Census.

Anonymous. 1979. Tapping the riches of shale. *Time*, **114**:62–63 (Nov. 19).

Anonymous. 1980. The watering down of energy proposals. *Business Week*, **2625**:52–53 (Feb. 25).

Anonymous. 1981. *Business Week*, **2698**:11 (July 27).

Anonymous. 1982. Steam tubes further stay TMI-1 restart. *Sci. News*, **121**:283.

Argonne National Laboratory. 1976. Health and Ecological Effects of Coal Utilization. Draft Report, U.S. Nuclear Regulatory Commission.

Armentano, T. V. and C. W. Ralston. 1980. The role of temperate zone forests in the global carbon cycle. *Can. J. For. Res.*, **10**:53–60.

Arola, R. A. 1976. Wood fuels—How do they stack up? In *Energy and the Wood Products Industry*. The Forest Products Research Society, Madison, WI.

Arrow, K. J. 1981. The response of orthodox economics. In H. E. Daly and F. Umana (Eds.), *Energy, Economics, and the Environment*. Westview Press, Boulder, CO., pp. 109–114.

Ashenden, T. W. 1979. The effects of long-term exposures to SO_2 and NO_2 pollution on the growth of *Dactylis glomerata L.* and *Poa pratensis L. Environ. Pollut.*, **18**:249–258.

Asinof, R. 1983. Deluge. *Environ. Action*, **14**(7):10–15.

Aspey, W. P. and S. I. Lustick (Eds.). 1983. *Behavioral Energetics. The Cost of Survival in Vertebrates*. Papers from a colloquium, Columbus, Ohio, October 1980. Ohio State University Press, Columbus.

Attanasi, E. D., L. J. Drew, and J. H. Schuenemeyer. 1980. Petroleum resource appraisal and discovery rate forecasting in partially explored regions: An application to supply modeling. U.S. Geol. Surv. Professional Paper 1138-c. U.S. Government Printing Office, Washington, D.C.

Averitt, P. 1975. Coal Resources of the United States: January 1, 1974. *U.S. Geol. Surv. Bull.* 1412. U.S. Government Printing Office, Washington, D.C.

Avineri, Shlomo. 1968. *The Social and Political Thought of Karl Marx*. Cambridge University Press, Cambridge.

Ayers, E. and C. Scarlott. 1952. *Energy Sources—The Wealth of the World*. McGraw-Hill, New York.

Ayres, R. U. 1978. *Resources, Environment, and Economics*. Wiley-Interscience, New York.

Ayres, R. U. and I. Nair. 1984. Thermodynamics and economics. *Phys. Today*, **November**:62–71.

Baes, C. F., Jr., S. E. Beall, D. W. Lee, and G. Marland. 1980. Options for the collection and disposal of carbon dioxide. Oak Ridge National Laboratory, Oak Ridge, Tenn. ORNL-5657.

Baily, M. N. 1982. Economic models under challenge. *Science,* **216**:859–862.

Baker, J. G. 1981. Sources of deep coal mine productivity change, 1962–1975. *Energy J.*, **2**(2):95–106.

Balcomb, J. D., J. C. Hedstrom, and R. D. McFarland. 1977. Simulation analysis of passive solar heated buildings—preliminary results. *Solar Energy,* **19**(3-E):277.

Baranov, F. I. 1918. On the Question of the Biological Basis of Fisheries. Nauchn. Issled. Ikhtiologicheskii Inst., *Izv.,* **1**(1):81–128.

Barish, N. and S. Kaplan. 1978. *Economic Analysis for Engineering and Managerial Decision Making*. McGraw-Hill, New York.

Barlowe, R. 1978. *Land Resource Economics*. Prentice Hall, Englewood Cliffs, N.J.

Barnes, D. and L. Rankin. 1975. The Energy Economics of Building Construction. *Build. Int.,* **8**:31–42.

Barnett, H. J. and C. Morse. 1963. *Scarcity and Growth: The Economics of Natural Resource Availability*. Johns Hopkins University Press, Baltimore.

Barnett, H. J. 1979. Scarcity and growth revisited. In V. K. Smith (Ed.), *Scarcity and Growth Reconsidered*. Johns Hopkins University Press, Baltimore, pp. 163–217.

Barnthouse, L. W., J. B. Cannon, S. G. Christensen, A. H. Evaslan, J. L. Harris, K. H. Kim, M. E. LaVerne, H. A. McLain, B. D. Murphy, R. J. Raridon, T. H. Row, and R. D. Sharp. 1977. A Selective Analysis of Power Plant Operation on the Hudson River with Emphasis on the Bowline Generating Station. ORNL/TM-5877/Vols. I & II. Oak Ridge National Laboratory, Oak Ridge, Tenn.

Baron, S. 1978. Solar energy: Will it conserve nonrenewable resources? *Public Utilities Fortnightly,* 31–36.

Barrager, S. M., B. R. Judd, and D. W. North. 1976. The Economic and Social Costs of Coal and Nuclear Electrical Generation. Prepared by Decision Analysis Dept., Stanford Research Institute, Stanford, CA.

Bartholomew, G. A. 1982. Energy metabolism. In M. S. Gordon, G. A. Bartholomew, A. D. Grinnell, C. B. Jorgensen, and F. N. White (Eds.), *Animal Physiology: Principles and Adaptations*. Macmillan, New York, pp. 46–93.

Bartlett, E. T. and W. J. Clawson. 1978. Profit, meat production or efficient use of energy in ranching. *J. Anim. Sci.,* **46**(3):812–818.

Batelle Columbus Laboratories. 1977. Survey of the applications of solar thermal energy to provide industrial process heat. Columbus, Ohio.

Batelle Laboratories. 1975. *Interim Reports on Energy Use Patterns in Metallurgical and Non-Metallic Mineral Processing*. U.S. Bureau of Mines. Open File. Report 80-75. U.S. Government Printing Office, Washington, D.C.

Baumer, M. C. 1978. Environmental impacts of rangeland uses. In D. N. Hyder (Ed.), *Proceedings of the First International Rangeland Congress*. Society for Range Management, Denver, CO, pp. 17–20.

Baumhardt, G. R. and L. F. Welch. 1972. Lead uptake and corn growth with soil applied lead. *J. Environ. Qual.,* **1**:92–96.

Baumol, W. J. and E. N. Wolff. 1981. Subsidies to new energy sources: Do they add to energy stocks? *J. Pol. Econ.,* **89**:891–913.

Bayley, S. and J. Zucchetto. 1977. Energetics and systems modeling: A framework study for energy evaluation of alternative transportation modes. Report submitted to U.S. Army Engineer Institute for Water Resources, Fort Belvoir, VA.

Beaty, E. R., Y. C. Smith, and J. D. Powell. 1974. Response of Pensacola bahia grass to irrigation and time of N fertilization. *J. Range Manage.,* **27**:394–396.

Bell, H. M. 1973. *Rangeland Management for Livestock Production.* University of Oklahoma Press, Norman, OK.

Bell, J. F. 1953. *A History of Economic Thought.* Ronald Press, New York.

Bell, J. N. B., A. J. Rutter, and J. Relton. 1979. Studies of the effects of low levels of sulphur dioxide on the growth of *Lolium perenne L. New Phytol.,* **83**:627–643.

Bement, R. E. 1969. A stocking-rate guide for production on blue-grama range. *J. Range Manage.,* **22**:83–86.

Benjamin, B., P. R. Cox, and J. Peel (Eds.). 1973. *Resources and Population.* Academic Press, London.

Berger, M. 1981. Energy Return on Energy Invested for Presently Operating and Future Nuclear Power Systems. Honors Thesis, Cornell University, Ithaca, NY.

Berndt, E. R. 1978. Aggregate energy, efficiency, and productivity measurement. *Ann. Rev. Energy,* **3**: 225–273.

Berndt, E. R. 1982. From technocracy to net energy analysis: Engineers, economists, and recurring energy theories of value. Massachusetts Institute of Technology, Studies in Energy and the American Economy, Discussion Paper No. 11.

Berndt, E. R. and D. O. Wood. 1975. Technology, prices, and the derived demand for energy. *Rev. Econ. Stat.,* **57**:259–268.

Berndt, E. R. and D. O. Wood. 1979. Engineering and economic interpretation of energy capital complementarity. *Am. Econ. Rev.,* **69**:342.

Berner, W., H. Oeshger, and B. Stauffer. 1980. Information on the CO_2 cycle from ice core studies. *Radiocarbon,* **22**:227–235.

Berry, R. S. and M. F. Fels. 1973. The energy cost of automobiles. *Bull. Atomic Scientists,* **27**:11–19.

Bertine, K. K. and E. D. Goldberg. 1971. Fossil fuel combustion and the major sedimentary cycle. *Science,* **173**:233–235.

Bertrand, A. B. 1979. *Rangeland Policies for the Future.* Proceedings of a Symposium January 28–31, Tucson, Arizona. USDA, USDI, CEQ. U.S. Government Printing Office, Washington, D.C. GTR WO-17.

Bethel, J. S. and G. F. Schreuder. 1976. Forest Resources: An Overview. In P. Abelson and A. Hammond (Eds.), *Materials: Renewable and Nonrenewable Resources,* Vol. 4. American Association for the Advancement of Science.

Beverton, R. J. H. and S. J. Holt. 1957. On the dynamics of exploited fish populations. *Fish. Invest. Ser.,* **219**, 533 pp.

Bigelow, H. B. and W. C. Schroeder. 1953. Fishes of the Gulf of Maine. Fishery Bulletin 47 of the Fish and Wildlife Service, Volume 53. U.S. Government Printing Office, Washington, D.C.

Blackwelder, B. Unpubl. Alternatives to traditional water development. Presented at the Annual Meeting of the American Association for the Advancement of Science, May 24–29, 1984. New York.

Bliss, R. 1978. Why not build them right in the first place? In R. H. Williams (Ed.), *Toward a Solar Civilization.* MIT Press, Cambridge, MA.

Blomeke, J. O. 1979a. Origin, magnitude, and treatment of radioactive wastes. Oak Ridge National Laboratories and the Office of Nuclear Waste Management. Washington, D.C. CONF 791185-1.

Blomeke, J. O. 1979b. Disposal of radioactive wastes. Oak Ridge National Laboratories. CONF 791185-2.

Bodenheimer, F. S. 1937. Population problems of social insects. *Q. Rev. Biol.,* **12**:393–430.

Boes, E., et al. 1976. Distributions of direct and indirect solar radiation availabilities for the USA. Sandia Labs. Report No. SAND76-0411.

Boretsky, M. 1975. Trends in U.S. technology: A political economist's view. *Am. Sci.,* **63**:70–82.

Borgstrom, G. 1969. *Too Many: A Study of the Earth's Biological Limitations.* Macmillan, New York.

Borisov, P. G. 1960. *Fisheries Research in Russia. A Historical Survey.* U.S. Dept. of Commerce, Office of Technical Services, Washington, D.C.

Boshier, J. F. 1978. Can we save energy by taxing it? *Tech. Rev.,* **80**(8):62–71.

Botkin, D. B. and E. A. Keller. 1982. *Environmental Studies: The Earth as a Living Planet.* Merrill, Columbus, OH.

Bottrell, D. R. 1979. *Integrated Pest Management.* Council on Environmental Quality. U.S. Government Printing Office, Washington, D.C.

Boulding, K. E. 1966. The economics of the coming spaceship earth. In *Environmental Quality in a Growing Economy.* Resources for the Future, Washington, D.C.

Box, T. W. 1979. The American rangelands: Their condition and policy implications for management. In *Rangeland Policies for the Future.* Proceedings of a Symposium January 28–31, Tucson, Arizona. USDA, USDI, CEQ. U.S. Government Printing Office, Washington, D.C. GTR WO-17.

Box, T. W., D. D. Dwyer, and F. H. Wagner. 1976. The public range and its management. Unpublished report to the Council on Environmental Quality.

Boynton, W. R. 1975. Energy Basis of a Coastal Region: Franklin County and Apalachicola Bay, Florida. Ph.D. Thesis. University of Florida.

Bradshaw, K. R. L. and E. T. Bartlett. 1978. *Expected*

Benefits from Range Improvements in Colorado Ecosystems. Colorado State University Range Science Department, Science Series No. 30.

Brady, N. C. 1974. *The Nature and Properties of Soils.* Macmillan, New York.

Braun, C. E., K. W. Harmon, J. A. Jackson, and C. D. Littlefield. 1978. Management of national wildlife refuges in the U.S.: Its impact on birds. *Wilson Bull.,* **90**(2):309–321.

Bravard, J. C., H. B. Flora, and C. Portal. 1972. *Energy Expenditures Associated with the Production and Recycling of Metals.* Oak Ridge National Laboratory Report ORNL-NSF-EP-24.

Bredehoeft, J. D. and T. Maini. 1981. Strategy for radioactive waste dispersal in crystalline rocks. *Science,* **213**:293–296.

Breman, H. and C. T. DeWit. 1983. Rangeland productivity and exploitation in the Sahel. *Science,* **221**:1341–1347.

Brett, J. R. and T. D. D. Groves. 1979. Physiological energetics. In S. Hoar, D. J. Randall, and J. R. Brett (Eds.), *Fish Physiology,* Vol. 8. Academic Press, New York, pp. 280–352.

Brewer, M. and R. Buxley. 1982. The potential supply of cropland. In P. Crosson (Ed.), *The Cropland Crisis: Myth or Reality?* Johns Hopkins University Press, Baltimore.

Brobst, D. A. 1979. Fundamental concepts for analysis. In V. K. Smith (Ed.), *Scarcity and Growth Reconsidered.* Johns Hopkins University Press, Baltimore, pp. 106–142.

Brobst, D. A. and W. Pratt (Eds.). 1973. *United States Mineral Resources.* U.S. Geol. Surv. Prof. Paper 820. U.S. Government Printing Office, Washington, D.C.

Brobst, W. A. 1973. Transportation accidents: How probable? *Nucl. News,* **16**(5):48–54.

Brooks, D. B. and P. W. Andrews. 1974. Mineral resources, economic growth and world population. *Science,* **185**:13–19.

Brown, A., M. I. Schauer, J. W. Rowe, and W. Hely. 1977. *Water Management in Oil Shale Mining,* Vol. 1. Golder Associates, Kirkland, WA.

Brown, E. R., J. J. Hazdra, L. Keith, I. Greenspan, J. B. Kwapinski, and P. Beamer. 1973. Frequency of fish tumors found in a polluted watershed as compared to nonpolluted Canadian waters. *Can. Res.,* **33**:189.

Brown, G. M. and B. C. Field, Jr. 1978. Implications of alternative measures of natural resource scarcity. *J. Pol. Econ.,* **86**:229.

Brown, H. 1967. *The Next Ninety Years.* California Institute of Technology, Pasadena.

Brown, J. R., W. O. Thom, and L. L. Wall, Sr. 1981. Effects of sulfur application on yield and composi-

tion of soybeans and soil sulfur. *Commun. Soil Sci. Plant Anal.,* **12**:247–261.

Brown, L. R. 1978. Carrying capacity, biological systems, and human numbers. In *Key Issues in Population and Food Policy.* University of America Press, Washington, D.C.

Brown, L. R. 1981. World population growth, soil erosion, and food security. *Science,* **214**:955–1002.

Brown, S. and A. E. Lugo. 1981. *Management and Status of U.S. Commercial Marine Fisheries.* Council on Environmental Quality, Washington, D.C.

Bryson, R. A. and B. M. Goodman. 1980. Volcanic activity and climatic changes. *Science,* **207**:1041–1044.

Budiansky, S. 1980. Bioenergy: The lesson of wood burning. *Environ. Sci. Tech.,* **14**(7):769–771.

Budnitz, R. J. and J. P. Holdren. 1976. Social and environmental costs of energy systems. *Ann. Rev. Energy,* **1**:553–580.

Bue, I. G., H. G. Uhlig, and J. D. Smith. 1964. Stock ponds and dugouts. In J. P. Linduska (Ed.), *Waterfowl Tomorrow,* U.S. Government Printing Office, Washington, D.C., pp. 391–398.

Bullard, C. W. and R. A. Herendeen. 1975. Energy costs of goods and services. *Energy Policy,* **3**:263–278.

Bullard, C. W., P. S. Penner, and D. A. Pilati. 1978. Net energy analysis handbook for combining process and input–output analysis. *Resources and Energy,* **1**:267–313.

Bupp, I. C. 1979. The nuclear stalemate. In R. Stobaugh and D. Yergin (Eds.), *Energy Future.* Random House, New York, pp. 127–166.

Burton, T.M. and G. E. Likens. 1975. Energy flow and nutrient cycling in salamander populations in the Hubbard Brook Experimental Forest, New Hampshire. *Ecology,* **56**(5):1068–1080.

Butti, K. and J. Perlin. 1980. *A Golden Thread: 2500 Years of Solar Architecture and Technology.* Cheshire Books, Van Nostrand Reinhold, New York.

Campbell, R. B. and C. J. Knowles. 1978. Elk–hunter–livestock interactions in the Missouri River Breaks. Unpublished *Job Final Rept.* 150 pp.

Caputo, R. (no date). Projection of distributed-collector solar thermal electric power plant economics to 1990–2000. Jet Propulsion Laboratory, Pasadena, CA.

Carnot, S. 1824. *Reflexions sur la Puissance Motrice du Feu et sur les Machines.* (Reflections on the Motive Power of Fire and on Machines Fitted to Develop that Power.) Chez Bachelier, Libraire, Paris.

Carothers, S. W. 1977. Importance, preservation, and management of riparian habitats: An overview. In R. R. Johnson and D. A. Jones (Eds.), *Importance, Preservation, and Management of Riparian Habitat:*

A Symposium. USDA Forest Service General Technical Report, Tucson, Arizona. RM-43.

Carpenter, E.J., B. B. Peck, and S. J. Anderson. 1972. Entrainment of marine plankton through the Millstone Unit 1. *Mar. Biol.,* **16**:37–40.

Carver, T. N. 1935. *The Essential Factor of Social Evolution.* Harvard University Press, Cambridge, MA.

Chambers, R. S., R. A. Herendeen, J. J. Joyce, and P. S. Penner. 1979. Gasohol: does it or doesn't it produce positive net energy? *Science,* **206**:789–795.

Chanlett, E. T. 1979. *Environmental Protection,* 2nd ed. McGraw-Hill, New York.

Chapman, P. F. 1974. The energy costs of producing copper and aluminum from primary sources. *Met. Miner.,* **8**(2):107–111.

Chapman, P. F. 1975a. The energy costs of materials. *Energy Policy,* **3**:47–57.

Chapman, P. F. 1975b. Energy analysis of nuclear power stations. *Energy Policy,* **3**:285–294.

Chapman, P. F. and N. D. Mortimer. 1974. Energy inputs and outputs of nuclear power. Open University Energy Research Group Research Report, Milton, Keynes, England.

Chapman, P. F., G. Leach, and M. Slesser. 1974. The energy cost of fuels. *Energy Policy,* **2**:231–243.

Cheng, C. 1982. Steam generator tube experience. Office of Nuclear Reactor Regulation, Nuclear Regulatory Commission. NUREG-0886.

Cheremisinoff, N. P. 1980. *Wood for Energy Production.* Ann Arbor Science Publishers, Ann Arbor, MI.

Cherry, Robert D. 1980. *Macroeconomics.* Addison-Wesley, Reading, MA.

Christensen, G. W. 1976. Wood Residue Sources: Uses and Trends. *FPRS Wood Energy/Energy Proceedings.* Forest Products Research Society, Madison, WI, p. 75-93.

Cicchetti, C. J. 1973. The wrong route. *Environment,* **15**(5):4–11.

Clair, A. 1976. Ecological factors in the development of intensive management ecosystems. *Ecology,* **57**:431–444.

Clark, C. 1985. *Bioeconomic Modeling and Fisheries Management.* Wiley-Interscience, New York.

Clark, C. E. and D. C. Varisco. 1975. Net energy and oil shale. *Q. Colo. Sch. Mines,* **70**(3):3–20.

Clark, J. 1977. Effects of experimental management schemes on production and nesting ecology of ducks at Malheur National Wildlife Refuge. M.S. Thesis, Department of Fisheries and Wildlife, Oregon State University.

Clark, J. R. and W. Brownell. 1973. *Electric Power Plants in the Coastal Zone: Environmental Issues.*

American Littoral Society Special Publication no. 7, Highlands, NJ.

Clark, W. 1974. *Energy for Survival.* Anchor Press, New York. 652 pp.

Clarke, G. L. 1946. Dynamics of production in a marine area. *Ecol. Monogr.,* **16**:321–335.

Cleveland, C. J. And R. Costanza. 1986. Ultimately recoverable hydrocarbons in Louisiana: A net energy approach. *Energy Policy,* in press.

Cleveland, C. J. and R. Costanza. 1984. Net energy analysis of geopressured gas resources in the U.S. Gulf Coast Region. *Energy,* **9**(1):35–51.

Cleveland, C. J., R. Costanza, C. A. S. Hall, and R. K. Kaufmann. 1984. Energy and the U.S. economy: A biophysical perspective. *Science,* **225**:890–897.

Cleveland, W. S. and T. E. Graedel. 1979. Photochemical air pollution in the northeast United States. *Science,* **204**:1273–1278.

Cochrane, A. L. 1973. Relation between radiographic categories of coal-worker's pneumoconiosis and expectation of life. *Br. Med. J.,* **2**:532–534.

Coffin, T. 1981. Water: the sun belt's catch 22. *The Washington Spectator,* **7**:4.

Cogbill, C. V., G. E. Likens, and T. J. Butler. 1984. Uncertainties in historical aspects of acid precipitation: Getting it straight. *Atmos. Environ.,* **18**:2261–2270.

Cole, L. C. 1951. Population cycles and random oscillations. *J. Wildl. Manage.,* **15**:233–251.

Cole, L. C. 1954. The population consequences of life history phenomena. *Q. Rev. Biol.,* **29**(2):103–137.

Colinvaux, P. A. 1973. *Introduction to Ecology.* Wiley, New York.

Colinvaux, P. A. 1975. An ecologist's view of history. *Yale Rev.,* **1975**(Spring):357–369.

Collision, R. C. and J. E. Mensching. 1932. Lysimeter Investigations: II. Composition of Rainwater at Geneva, N.Y., for a 10-Year Period. Bull. No. 193. New York Experimental Station, Geneva.

Colten, J. B. and R. F. Temple. 1961. The enigma of Georges Bank spawning. *Limnol. Oceanogr.,* **6**:280–291.

Comar, C. L. and L. A. Sagan. 1976. Health effects of energy production and conversion. *Ann. Rev. Energy,* **1**:581–600.

Committee on Nuclear and Alternative Energy Systems. 1979. *Energy in Transition 1985–2010.* National Research Council. National Academy of Sciences, Washington, D.C.

Committee on Renewable Resources for Industrial Materials (CORRIM). 1976. Report of the Committee on Renewable Resources for Industrial Materials (CORRIM). *J. Soc. Wood Sci. Tech.,* **8**(Nov).

Commoner, B. 1976. *The Poverty of Power*. Bantam Books, New York.

Commoner, B. 1979. *The Politics of Energy*. Knopf, New York.

Congressional Budget Office. 1978. The proposed four-year authorization for the Bureau of Land Management: An evaluation of issues and alternatives. Staff draft analysis. In *Quadrennial Authorizations for the Bureau of Land Management*. Energy and Natural Resources Committee Hearing, U.S. Senate, 95th Congress, Pub. No. 95-125. U.S. Government Printing Office, Washington, D.C.

Cook, C. W. 1965. Plant and livestock responses to fertilized rangelands. Bull. No. 455, Utah State Agricultural Experiment Station.

Cook, C. W. 1979. Rehabilitation potential and practices of Colorado oil shale lands. Progress report for period June 1, 1978–May 31, 1979. Colorado State University, Ft. Collins, CO, COO 4018-3.

Cook, E. 1976a. Limits to the exploitation of nonrenewable resources. *Science*, **19**:677–682.

Cook, E. 1976b. *Man, Energy, Society*. W. H. Freeman, San Francisco.

Cook, E. 1982. The consumer as creator: A criticism of faith in limitless ingenuity. *Energy Exploration and Exploitation*, **1**(3):189–201.

Cook, R. E. (Ed.). 1962. *How Many People Have Ever Lived on Earth?* Population Bulletin. Vol. 18, no. 1. Population Reference Bureau, Washington, D.C.

Cooley, K. R., G. W. Frasier, and K. R. Drew. 1978. Water harvesting: An aid to range management. In D. N. Hyder (Ed.), *Proceedings of the First International Rangeland Congress*. Society for Range Management, Denver, CO, pp. 292–293.

Cooper, J. A. 1980. Environmental impact of residential wood combustion emissions. *J. Air Pollut. Control Assoc.*, **30**(8):855–861.

Cope, O. B. (Ed.). 1979. *Proceedings of the Forum—Grazing and Riparian/Stream Ecosystems*, November 3–4. Trout Unlimited, Inc., Denver, CO.

Cornell University. 1983. Cornell recommendations for commercial vegetable production. Vegetable Crops Department, Ithaca, NY.

Costanza, R. 1980. Embodied energy and economic valuation. *Science*, **210**:1219–1224.

Costanza, R. 1981. Embodied energy, energy analysis, and economics. In H. E. Daly and A. F. Umana (Eds.), *Energy, Economics, and the Environment*. Westview, Boulder, CO, pp. 119–146.

Costanza, R. 1982. Economic values and embodied energy. Reply to D. A. Huettner. *Science*, **215**:1143.

Costanza, R. and C. Neill. 1981. The energy embodied in the products of ecological systems: A linear programming approach. In W. J. Mitsch, W. Bosser-

man, and J. M. Klopatek (Eds.), *Energy and Ecological Modelling*. Elsevier Scientific, Amsterdam, pp. 661–670.

Costanza, R. and R. A. Herendeen. 1984. Embodied energy and economic value in the U. S. economy: 1963, 1967, and 1972. *Resources and Energy*, **6**:129.

Cotner, M. L., N. L. Bills, and R. F. Boxley. 1981. An economic perspective of land use. In W. E. Jeske (Ed.), *Economics, Ethics, Ecology*. Soil Conservation Society of America, Ankeny, Iowa, pp. 27–50.

Cottrell, F. 1955. *Energy and Society*. McGraw-Hill, New York. (Reprinted by Greenwood Press, Westport, CT.)

Cottrell, F. 1972. *Technology, Man, and Progress*. Merrill, Columbus, OH.

Council of Economic Advisors. 1981. Economic Indicators. Prepared for the Joint Economic Committee by the Council of Economic Advisors.

Council of Economic Advisors. 1983. *Economic Report of the President*. U.S. Government Printing Office, Washington, D.C.

Council of Environmental Quality. 1980. *Annual Report on the Environment*. U.S. Government Printing Office, Washington, D.C.

Coutant, C. C. 1971. Effects on organisms of entrainment in cooling waters: Steps toward predictability. *Nucl. Saf.*, **12**:600–607.

Coutant, C. C. 1974. Evaluation of entrainment effect. In P. L. Jensen (Ed.), *Proceedings of the 2nd Entrainment and Intake Screening Workshop*. Cooling Water Research Project Report No. 15. Johns Hopkins University, Baltimore, pp. 1–11.

Cowles, R. B. and C. M. Bogert. 1944. A preliminary study of the thermal requirements of desert reptiles. *Bull. Am. Mus. Nat. Hist.*, **83**:261–296.

Cox, G. and M. Atkins. 1979. *Agricultural Ecology*. W. H. Freeman, San Francisco.

Craig, N. J. and J. W. Day. 1977. Cumulative Impact Studies in the Louisiana Coastal Zone: Eutrophication and Land Loss. Final Report to Louisiana State Planning Office, 30 June, 1977.

Cram, I. (Ed.). 1971. Future petroleum provinces of the United States. *Am. Assoc. Pet. Geol. Mem. 15*, **1**.

Cranstone, D. A. and H. L. Martin. 1973. Are ore discovery costs increasing? In *MR 137*, Mineral Resources Branch, Department of Energy, Ottawa, Canada, pp. 5–13.

Crawford, M. 1985. Mill tailings: A four billion dollar problem. *Science*, **229**:537–538.

Crittenden, P. D. and D. J. Read. 1979. The effects of air pollution on plant growth with special reference to sulphur dioxide, III. Growth studies with *Lolium multiflorum Lam.* and *Dactylis glomerata L. New Phytol.*, **83**:645–651.

Crosson, P. 1978. Long run costs of production in American agriculture. Unpubl. manuscript. Resources for the Future, Washington, D.C.

Crosson, P. 1979. Agricultural land use: A technological and energy perspective. In M. Schnepf (Ed.), *Farmland, Food, and the Future*. Soil Conservation Society of America, Ankeny, Iowa, pp. 99–111.

Crosson, P. 1982. Future economic and environmental costs of agricultural land. In P. Crosson (Ed.), *The Cropland Crisis: Myth or Reality?* Johns Hopkins University Press, Baltimore.

Crosson, P. and S. Brubaker. 1982. Resource and Environmental Effects of U.S. Agriculture. Johns Hopkins University Press, Baltimore (for Resources for the Future, Washington, D.C.).

Culbertson, W. C. and J. K. Pitman. 1973. Oil shale. In D. A. Brobst and W. P. Pratt (Eds.), *United States Mineral Resources*. U.S. Geol. Surv. Prof. Paper 820. U.S. Government Printing Office, Washington, D.C.

Cummings, R. G. 1974. *Interbasin Water Transfers*. Johns Hopkins University Press, Baltimore.

Currie, T. O. 1978. Cattle weight gain comparisons under season long and rotation grazing systems. In D. N. Hyder (Ed.), *Proceedings of the First International Rangeland Congress*. Society for Range Management, Denver, CO, pp. 579–580.

Curzon, F. L. and B. Ahlborn. 1975. Efficiency of a Carnot engine at maximum power output. *Am. J. Phys.,* **43**:22–24.

Cushing, D. H. 1968. *Fisheries Biology*. University of Wisconsin Press, Madison.

Cushing, D. H. 1973. *Recruitment and Parent Stock in Fishes*. Washington Sea Grant Program, WSG 73-1. University of Washington Press, Seattle.

Cushing, D. H. 1975. *Marine Ecology and Fisheries*. Cambridge University Press, Cambridge.

Cutler, M. R. 1979. Committment to the future. In *Rangeland Policies for the Future: Proceedings of a Symposium,* January 28–31, Tucson, Arizona. USDA, USDI, CEQ. U.S. Government Printing Office, Washington, D.C., pp. 1–5, GTR WO-17.

Daly, H. E. 1973. Introduction. In H. E. Daly (Ed.), *Toward a Steady-State Economy*. W. H. Freeman, San Francisco, pp. 1–29.

Daly, H. E. 1977. *Steady State Economics*. W. H. Freeman, San Francisco.

Daly, H. E. 1979. Entropy, growth, and the political economy of scarcity. In V. K. Smith (Ed.), *Scarcity and Growth Reconsidered*. Johns Hopkins University Press, Baltimore, pp. 67–94.

Daly, H. E. 1981. Postscript. In H. E. Daly and A. F. Umana (Eds.), *Energy, Economics, and the Environment*. Westview Press, Boulder, CO, pp. 165–185.

Daly, H. E. 1983. The circular flow of exchange value and the linear throughput of matter energy: A case of misplaced concreteness. Faculty Working Paper, Economics Department, Louisiana State University, Baton Rouge.

Darmstadter, J., J. Dunkerley, and J. Alterman. 1977. *How Industrial Societies Use Energy*. Johns Hopkins University Press, Baltimore.

Davis, C. S., III. 1982. Processes controlling zooplankton abundance on Georges Bank. Ph.D. Thesis, Boston University.

Davis, D. E. and F. B. Golley. 1963. *Principles in Mammology*. Reinhold, New York.

Davis, J. W. 1982. Livestock vs. riparian habitat management—there are solutions. In J. M. Peek and P. K. Dalke (Eds.), *Proceedings of the Wildlife–Livestock Relationships Symposium,* Coeur d'Alene, Idaho, April 20–22. Forest, Wildlife and Range Experiment Station, University of Idaho, Moscow.

Davis, W. 1958. Future productive capacity and probable reserves of the U.S. *Oil Gas J*. **56**:105–119.

Dawson, F. G. 1976. *Nuclear Power: Development and Management of a Technology*. University of Washington Press, Seattle.

Day, B. E. 1979. Range management, an ecological art. In *Rangeland Policies for the Future*. Proceedings of a Symposium, January 28–31, Tucson, Arizona. USDA, USDI, CEQ. U.S. Government Printing Office, Washington, D.C., GTR WO-17, pp. 91–92.

Day, J. W., C. A. S. Hall, M. Kemp, and A. Yanez. In press. *Estuarine Ecology*. Wiley-Interscience, New York.

DeWit, C. T. 1975. Agriculture's uncertain claim on world energy resources. *SPAN,* 2–4.

DeWit, C. T. and H. D. J. van Heemst. 1976. Aspects of agricultural resources. In W. T. Koetsier (Ed.), *Proceedings of the Plenary Sessions of the First World Congress on Chemical Engineering,* Amsterdam, June 28–July 1, 1976. Elsevier Scientific, Amsterdam, pp. 125–145.

Deevey, E. S. 1960. The human population. *Sci. Am.,* **203**:195–204.

Deis, R. 1980. Where there's wood there's smoke. *Environ. Action,* **12**(6):4–9.

Dekkers, W. A., J. M. Lange, and C. T. DeWit. 1974. Energy production and use in Dutch agriculture. *Neth. J. Agric. Sci.,* **22**:107–118.

Delcourt, H. R. and W. F. Harris. 1980. Carbon budget of the Southeastern U.S. biota: Analysis of historical change in trend from source to sink. *Science,* **210**:321–323.

Delmas. R. J., J. M. Ascencio, and M. Legrand. 1980. Polar ice evidence that atmospheric CO_2 20,000 yr BP was 50% of present. *Nature,* **284**:155–157.

Delmas, R. J. and M. Legrand. 1980. The acidity of polar precipitation: A natural reference level for acid rains. International Conference on The Ecological Impact of Acid Precipitation. Sandefjord, Norway.

Demaison, G. J. and G. T. Moore. 1980. Anoxic environments and oil source bed genesis. *Am. Assoc. Pet. Geol. Bull.*, **64**:1179–1209.

Denison, E. F. 1974. *Accounting for United States Economic Growth 1929–1969*. Brookings Institution, Washington, D.C.

Denison, E. F. 1979. Explanations of declining productivity growth. *Surv. Curr. Business*, **79**:1–24.

Detwiler, P., C. A. S. Hall, and P. Bogdonoff. 1985. Land use change and carbon exchange in the tropics: II. Estimates for the entire region. *Environ. Manage.*, **9**:335–344.

Devarajan, S. and A. Fisher. 1982. Exploration and Scarcity. *J. Pol. Econ.*, **90**:1279–1290.

Devine, M. D., S. C. Ballard, and others. 1981. *Energy From the West*. University of Oklahoma Press, Norman.

Devine, W. D. 1982. An historical perspective on electricity and energy use. Institute for Energy Analysis, Oak Ridge Universities.

Dewhurst, J. F. and Associates. 1955. *America's Needs and Resources*. Twentieth Century Fund, New York.

Dickson, D. 1980. Controversy over $20 billion synfuels programme. *Nature*, **284**:200–201.

Dickson, D. 1982a. Europe's fast breeders move to a slow track. *Science*, **218**:1094–1097.

Dickson, D. 1982b. Breeders and bombs. *Science*, **218**:1095.

Dickson, E. M., et al. 1976. *Impacts of Synthetic Liquid Fuel Development: Assessment of Critical Factors*, Vol. 2. U.S. Government Printing Office, Washington, D.C., ERDA-129.

Dideriksen, R. I., A. R. Hidlebaugh, and K. O. Schmude. 1979. Trends in agricultural land use. In M. Schnepf (Ed.), *Farmland, Food and the Future*. Soil Conservation Society of America, Ankeny, Iowa, pp. 13–28.

Dochinger, L. S., F. W. Bender, F. L. Fox, and W. W. Heck. 1970. Chlorotic dwarf of eastern white pine caused by an ozone and sulphur dioxide interaction. *Nature*, **225**:476.

Dohan, M. R. 1977. Economic values and natural ecosystems. In C. Hall and J. Day (Eds.), *Ecosystem Modeling in Theory and Practice*. Wiley-Interscience, New York.

Dolton, G. L., K. H. Carlson, R. R. Charpentier, A. B. Coury, R. A. Crovelli, S. E. Frezon, A. S. Kahn, J. H. Lister, R. H. McMullin, R. S. Pike, R. B. Powers, E. W. Scott, and K. L. Varnes. 1981. Estimates of undiscovered recoverable conventional resources of oil and gas in the U.S. USGS Circular 860, Washington, D.C.

Douglas, P. A. and R. H. Stroud (Eds.). 1971. *Symposium on the Biological Significance of Estuaries*. Sports Fishing Institute, Washington, D.C.

Dregne, H. 1977. Desertification of arid lands. *Econ. Geogr.*, **53**(4):322–331.

Drew, L. J., J. H. Schuenemeyer, and D. M. Root. 1980. Petroleum resource appraisal and discovery rate forecasting for partially explored regions. USGS Prof. Paper 1138. Washington, D.C.

Drew, L. J., J. H. Schuenemeyer, and D. M. Root. 1980. Petroleum resource appraisal and discovery rate forecasting in partially explored regions: An application to the Denver basin. USGS Prof. Paper 1138-a. U.S. Government Printing Office, Washington, D.C.

Drucker, P. F. 1982. Towards the next economics. In D. Bell and I. Kristol (Eds.), *The Crisis in Economic Theory*. Basic Books, New York, pp. 4–10.

Duce, R. A. and G. L. Hoffman. 1976. Atmospheric vanadium transport to the ocean. *Atmos. Environ.*, **10**:989–996.

Duke, J. M. and M. J. Fudali. 1976. *Report on the Pulp and Paper Industry's Energy Savings and Changing Fuel Mix*. American Paper Institute, New York.

Duncan, D. C. and V. E. Swanson. 1965. Organic rich shales of the United States and world land areas. U.S. Geol. Surv. Circ. 523. U.S. Government Printing Office, Washington, D.C.

Dunkerley, J. 1980a. *Trends in Energy Use in Industrial Societies*. Resources for the Future, Washington, D.C.

Dunkerley, J. (Ed.). 1980b. *International Energy Strategies*. Oegeschlager, Gunn, & Hain, Cambridge, MA.

Durham, W. H. 1976. Resource competition and human aggression, Part I: A review of primitive war. *Q. Rev. Biol.*, **51**:385–414.

Dyksterhuis, E. J. 1949. Condition and management of rangeland based on quantitative ecology. *J. Range Manage.*, **2**:104–115.

Eckardt, R. E. 1959. *Industrial Carcinogens*. Grune and Stratton, New York. (As cited in R. J. Young and J. M. Evans, Occupational health and coal conversion. *J. Occup. Health Saf.*, Sept. 1979.)

Eckert, R. T. and D. B. Houston. 1980. Photosynthesis and needle elongation response of *Pinus strobus* clones to low level sulfur diozide exposures. *Can. J. For. Res.*, **10**:357–361.

Edwards, R. and R. Hennemuth. 1975. Maximum yield: Assessment and attainment. *Oceanus*, **18**(2):3–9.

Ehrlich, P., A. Ehrlich, and J. P. Holdren. 1977. *Ecosci-

ence: Population, Resources, Environment. W. H. Freeman, San Francisco.

Eisenbud, M. 1978. *Environmental Technology and Health: Human Ecology in Historical Perspective.* New York University Press, New York.

Ellett, W. B. and H. H. Hill. 1929. Effect of lime materials on the outgo of sulfur from Hagerstown silt loam soil. *J. Agric. Res.,* **38**:697–711.

Eppley, R. W. and J. D. M. Strickland. 1968. Kinetics of marine phytoplankton growth. *Adv. Microbiol. Sea,* **1**:22–63.

Ehrlich, P. R. and J. P. Holdren. 1972. One-dimensional ecology. *Bull. Atomic Sci.,* **28**:16–27.

Etnier, E. L. and A. P. Watson. 1981. Health and safety implications of alternative energy technologies: II. Solar. *Environ. Manage.,* **5**(5):409–425.

Evans, L. S. 1982. Biological effects of acidity in precipitation on vegetation: a review. *Environ. Exp. Bot.,* **22**(2):155–169.

Evans, L. S., N. F. Gmur, and D. Mancini. 1982. Effects of simulated acid rain on yields of *Raphanus sativus, Lactuca sativa, Triticum aestivum* and *Medicago sativa. Environ. Exp. Bot.,* **22**(4):445–453.

Ewel, J., C. Berish, B. Brown, N. Price, and J. Raich. 1981. Slash and burn impacts on a Costa Rican wet forest site. *Ecology,* **62**(3):816–829.

Exxon, 1978. As reported by J. D. Langston. A new look at oil and gas potential. Paper presented at 16th Annual Institute on Petroleum Exploration and Economics, Dallas, March 10.

Falkenmark, M. 1976. *Water for a Starving World.* Westview Press, Boulder, CO.

Farrington, J. W., W. M. Frew, P. M. Gschwend, and B. W. Tripp. 1977. Hydrocarbons in cores of Northwestern Atlantic coastal and continental margin sediments. *Estuarine Coastal Mar. Sci.,* **5**:793–808.

Fearon, J. G. and M. G. Wolman. 1982. Shale oil: Always a bridesmaid? Forty years of estimating costs and prospects. *Energy Syst. Policy,* **6**:63–96.

Feeny, P. 1970. Seasonal changes in oak leaf tannins and nutrients as a cause of spring feeding by winter moth caterpillars. *Ecology,* **51**(4):565–581.

Fennelly, P. F. 1976. The origin and influence of airborne particulates. *Am. Sci.,* **64**:46–56.

Ferguson, G., J. Hughes, and K. Brown. 1983. Food availability and territorial establishment of juvenile *Sceloporus undulatus.* In R. Huey, E. Pianka, and T. Schoener (Eds.), *Lizard Ecology,* Harvard University Press, Cambridge, MA.

Ferman, M. A., G. T. Wolff, and N. A. Kelly. 1981. The nature and sources of haze in the Shenandoah Valley/Blue Ridge Mountains Area. *J. Air Pollut. Control Assoc.,* **31**:1074–1082.

Field, B. C. and C. Grebenstein. 1980. Capital–energy substitution in U.S. manufacturing. *Rev. Econ. Stat.,* **62**:207–212.

Fisher, A. C. 1979. Measures of natural resource scarcity. In V. K. Smith (Ed.), *Scarcity and Growth Reconsidered.* Johns Hopkins University Press, Baltimore, pp. 249–275.

Fisher, A. C. 1981. *Resource and Environmental Economics.* Cambridge University Press, Cambridge.

Fisher, N. S. 1976. North Sea phytoplankton. *Nature,* **259**:160.

Flint, R. F. and B. J. Skinner. 1974. *Physical Geology.* Wiley, New York.

Ford Foundation Energy Policy Project. 1974. *A Time to Choose: America's Energy Future.* Ballinger, Cambridge.

Forest Service. 1981. *An Assessment of the Forest and Range Land Situation in the United States.* Prepared by the Forest Service, U.S. Department of Agriculture, for submission to Congress. Forest Resource Rept. No. 22. U.S. Government Printing Office, Washington, D.C.

Forest Service. 1982. *Annual Grazing Statistical Use Summary.* U.S. Government Printing Office, Washington, D.C.

Foster, P. and M. Bailey. 1974. Population growth, property taxes and public debt in Maryland. Cooperative Extension, Symans Hall, University of Maryland. 4 pp.

Frabetti, A. J. 1975. *A Study to Develop Estimates of Merit for Selected Fuel Technologies.* Development Sciences, Inc., East Sandwich, MA.

Franklin, B. 1784. Meteorological imaginations and conjectures. In J. Bigelow (Ed.), *The Complete Works of Benjamin Franklin.* Putnam, New York, pp. 486–488.

Frederick, K. 1982. Irrigation and the adequacy of agricultural land. In P. Crosson (Ed.), *Cropland Crisis: Myth or Reality?* Johns Hopkins University Press, Baltimore, pp. 117–159.

Freedman, B. and T. C. Hutchinson. 1980. Long-term effects of smelter pollution at Sudbury, Ontario on forest community composition. *Can. J. Bot.,* **58**: 2123–2140.

Freeman, A. M., III. 1979. The benefits of air and water pollution control: A review and synthesis of recent estimates. Report prepared for the Council on Environmental Quality.

Friedman, M. 1974. Monetary correction: A proposal for escalation clauses to reduce the costs of ending inflation. Institute of Economic Affairs Occasional Paper No. 41. London.

Frissel, M. J. 1977. Mineral nutrient cycling in agroecosystems. *Agro-Ecosystems,* **4**:1–354.

Fuschman, C. H. 1980. *Peat: Industrial Chemistry and Technology.* Academic Press, New York.

Fuss, M. and L. Waverman. 1975. The demand for energy in Canada. Working paper, Institute for Policy Analysis, University of Toronto.

Gagliano, S. M., K. J. Meyer-Arendt, and K. M. Wicker. 1981. Land loss in the Mississippi River deltaic plain. *Trans. Gulf Coast Assoc. Geol. Soc.,* 31:295–300.

Galbraith, J. K. 1958. *The Affluent Society.* Houghton Mifflin, Boston.

Galbraith, J. K. 1971. *The New Industrial State.* Houghton Mifflin, Boston.

Galloway, J. N., G. E. Likens, W. C. Keene, and J. M. Miller. 1982. Composition of precipitation in remote areas of the world. *J. Geophys. Res.,* 87(11):8771–8786.

Ganapathy, R. 1980. A major meteorite impact on the earth 65 million years ago: Evidence from the Cretaceous–Tertiary boundary clay. *Science,* 209: 921–923.

Gardner, G. M. 1977. Comparison of Energy Analysis of Oil Shale. In H. T. Odum and J. Alexander (Eds.), *Energy Analysis of Models of the United States.* University of Florida, Gainesville, pp. 196–232.

Garmon, T. 1981. The box within a box within a box. *Sci. News,* 120:396–399.

Garrels, R. M., F. T. Mackenzie, and C. Hunt. 1973. *Chemical Cycles and Global Environment.* William Kaufmann, Los Altos, CA.

Gause, G. F. 1934. *The Struggle for Existence.* Williams and Wilkins, Baltimore.

Geisler, C. C., J. T. Cowan, M. R. Hattery, and H. M. Jacobs. 1981. Sustained land productivity: Equity consequences of alternative agricultural technologies. Dept. of Rural Sociology Bulletin No. 119, Cornell University, Ithaca, New York.

General Accounting Office. 1982. Actions being taken to help reduce occupational radiation exposure at commercial nuclear powerplants. GAO/EMD-82-91.

Georgescu-Roegen, N. 1971. *The Entropy Law and the Economic Process.* Harvard University Press, Cambridge.

Georgescu-Roegen, N. 1975. Energy and economic myths. *Southern Econ. J.,* 41:347–381.

Georgescu-Roegen, N. 1977. The steady-state and ecological salvation: A thermodynamic analysis. *Bioscience,* 27:266.

Georgescu-Roegen, N. 1979. Energy analysis and economic evaluation. *Southern Econ. J.,* 45:1023–1058.

Geraghty, J. J., D. W. Miller, F. Van der Leeden, M. Pinther, and F. L. Troise. 1973. *Water Atlas of the United States.* Water Information Center, Inc., Port Washington, NY, 122 plates.

Gertel, K. and N. Wollman. 1960. Price and assessment guides to western water allocation. *J. Farm. Econ.,* 42(5):1332–1344.

Gieskes, W. W. C., G. W. Kraay, and M. A. Baars. 1979. Current ^{14}C methods for measuring primary production: Gross underestimates in oceanic waters. *Neth. J. Sea Res.,* 13(1):58–78.

Gifford, G. F. and J. M. Whitehead. 1982. Soil erosion effects on productivity in rangeland environments: Where is the research? *J. Range Manage.,* 35(6):801–802.

Gill, F. B. and L. L. Wolf. 1975. Economics of feeding territoriality in the golden-winged sunbird. *Ecology,* 56(2):333–345.

Gill, F. B. and L. L. Wolf. 1979. Nectar loss by golden-winged sunbirds to competitors. *The Auk,* 96:448–461.

Gilliker, J. P. 1979. State Energy Prices by Major Economic Sector from 1960 through 1977. U.S. Department of Energy.

Gilliland, M. W. 1975. Energy analysis and public policy. *Science,* 189:1051–1056.

Gilliland, M. W. 1982. Embodied energy studies of metal and fuel minerals. In M. J. Lavine and T. J. Butler (Eds.), *Use of Embodied Energy Values to Price Environmental Factors: Examining the Embodied Energy/Dollar Relationship.* Report to National Science Foundation.

Gilliland, M. W., J. M. Klopatek, and S. G. Hildebrande. 1981. Net Energy of Seven Small-Scale Hydroelectric Power Plants. Oak Ridge National Laboratory. ORNL/TM-7694.

Ginzberg, E. and G. J. Vojta. 1981. The service sector of the U.S. economy. *Sci. Am.,* 244(3):48–55.

Godfrey, E. B. 1978. Private adjustments to changes in grazing on public lands. Unpublished final report submitted to the Forest Service.

Goeller, H. E. 1979. The age of substitutability. In V. K. Smith (Ed.), *Scarcity and Growth Reconsidered.* Johns Hopkins University Press, Baltimore, pp. 143–159.

Gofman, J. W. 1979. *Irrevy: An Irreverent, Illustrated View of Nuclear Power.* Committee for Nuclear Responsibility, San Francisco.

Gofman, J. W. and A. R. Tamplin. 1971. *Poisoned Power: The Case Against Nuclear Power Plants.* Rodale Press, Pennsylvania.

Gold, H., and D. J. Goldstein. 1978. *Water Related Environmental Effects in Fuel Conversion,* Vol. I. Water Purification Associates, Cambridge, MA.

Golding, E. W. 1955. *The Generation of Electricity by Wind Power.* Philosophical Library. Reprinted by Halsted Press, New York, 1976.

Goldman, J. C., and J. A. Davidson. 1977. Physical model of marine phytoplankton chlorination at coastal power plants. *Environ. Sci. Technol.,* 11:908–913.

Golley, F. B. 1968. Secondary productivity in terrestrial communities. *Am. Zool.,* **8**:53–59.

Golley, F. B. 1983. Nutrient cycling and nutrient conservation. In F. B. Golley (Ed.), *Tropical Rain Forest Ecosystems: Structure and Function in Ecosystems of the World 14A.* Elsevier Scientific, Amsterdam, pp. 137–156.

Golley, F. B. and E. G. Farnworth. 1974. *Fragile Ecosystems.* Springer-Verlag, New York.

Gonzales, S. 1982. Host rocks for radioactive-waste disposal. *Am. Sci.,* **70**:191–200.

Gorshkov, V. G. and V. R. Dol'nik. 1980. Energetics of the biosphere. *Usp. Fiz. Nauk,* **131**:441–478. [Translation: *Sov. Phys. Usp,* **23/7**, American Institute of Physics.]

Gosselink, J. G., E. P. Odum, and R. M. Pope. 1973. The value of the tidal marsh. Center for Wetlands Research. Louisiana State University, Baton Rouge.

Goudarzi, G. H., L. F. Rooney, and G. L. Schaeffer. 1976. Supply of Nonfuel Minerals and Materials for the United States Energy Industry, 1975–1990. Geological Surv. Prof. Paper 1006-B.

Government Accounting Office. 1982. DOE Funds New Energy Technologies Without Estimating Potential Net Energy Yields. Washington, D.C.

Greber, B. J. and D. E. White. 1982. Technical change and productivity growth in the lumber and wood products industry. *For. Sci.* **28**:(1)135–147.

Greene, G. E. and W. R. Aldworth. 1959. Sagebrush or grass? *J. Soil Water Conserv.,* **14**(4):170–172.

Griffin, J. M. and P. R. Gregory. 1976. An intercountry translog model of energy substitution responses. *Am. Econ. Rev.,* **66**:845–857.

Griliches, A. and D. W. Jorgenson. 1967. Explaining productivity change. *Rev. Econ. Studies,* **34**:249–283.

Grimmer, D. 1978. Solar energy breeders. Los Alamos Scientific Laboratory, Los Alamos, NM. Report No. LA-UR-78-2973.

Groet, S. S. 1976. Regional and local variations in heavy metal concentrations of bryophytes in the northeastern United States. *Oikos,* **27**:445–456.

Guderian, R. 1977. *Air Pollution: Phytotoxicity of Acidic Gases and Its Significance in Air Pollution Control.* Springer-Verlag, New York.

Gulland, J. A. 1970. *The Management of Marine Fisheries.* University of Washington Press, Seattle.

Gulland, J. A. (Ed.). 1977. *Fish Population Dynamics.* Wiley-Interscience, New York.

Gunn, T. and B. Hannon. 1983. Energy conservation in the paper industry. *Resources and Energy,* **5**:243–260.

Gyorgy, A. 1979. *No Nukes: Everyone's Guide to Nuclear Power.* South End Press, Boston.

Hafele, W. (Ed.). 1981. *Energy in a Finite World.* Ballinger, Cambridge, MA.

Hainsworth, F. R. 1974. Food quality and foraging efficiency: the efficiency of sugar assimilation by hummingbirds. *J. Comp. Physiol.,* **88**:425–431.

Hairston, N. G., F. E. Smith, and L. B. Slobodkin. 1960. Community structure, population control and competition. *Am. Nat.,* **94**:421–425.

Hall, C. A. S. 1972. Migration and metabolism in a temperate stream ecosystem. *Ecology,* **53**(4):585–604.

Hall, C. A. S. 1975. The biosphere, the industriosphere and their interactions. *Bull. Atomic Scientists,* **31**: S11–21.

Hall, C. A. S. 1977. Models and the decision making process: The Hudson River power plant case. In C. A. S. Hall and J. Day (Eds.) *Models as Ecological Tools: Theory and Case Histories.* Wiley-Interscience, New York, pp. 37–48.

Hall, C. A. S. In press. Review of D. H. Cushing, Climate and Fisheries. *Environ. Manage.*

Hall, C. A. S., C. Ekdahl, and D. Wartenberg. 1975. A fifteen-year record of biotic metabolism in the northern hemisphere. *Nature,* **255**:136–138.

Hall, C. A. S. and J. Day (Eds.). 1977. *Ecosystem Modeling in Theory and Practice.* Wiley-Interscience, New York.

Hall, C. A. S., R. Howarth, B. Moore, and C. Vorosmarty. 1978. Environmental impacts of industrial energy systems in the coastal zone. *Ann. Rev. Energy,* **3**:395–475.

Hall, C. A. S., M. Lavine, and J. Sloane. 1979. Efficiency of energy delivery systems: Part I. An economic and energy analysis. *Environ. Manage.,* **3**(6):493–504.

Hall, C. A. S., E. Kaufman, S. Walker, and D. Yen. 1979. Efficiency of energy delivery systems: II. Estimating energy costs of capital equipment. *Environ. Manage.,* **3**(6):505–510.

Hall, C. A. S. and C. J. Cleveland. 1981. Petroleum drilling and production in the United States: yield per effort and net energy analysis. *Science,* **211**:576–579.

Hall, C. A. S. and C. J. Cleveland. 1981. Oil exploration. *Science,* **213**:1448–1450.

Hall, C. A. S., C. J. Cleveland, and M. Berger. 1981. Yield per effort as a function of time and effort for United States petroleum, uranium, and coal. In W. J. Mitsch, R. W. Bosserman, and J. M. Klopatek (Eds.), *Energy and Ecological Modelling.* Elsevier Scientific, Amsterdam, pp. 715–724.

Hall, C. A. S. and K. English. 1984. Simulation of the spatial and temporal food-chain and salmon dynamics of the Northeast Pacific Ocean. Manuscript.

Hall, D. C., J. V. Hall, and D. X. Kolk. 1981. Scarcity and growth: An update. Unpublished manuscript. West Virginia University, Morgantown, WV.

Halvorsen, R. 1975. Residential demand for electric energy. *Rev. Econ. Stat.,* February.

Hamilton, J. D. 1985. Historical causes of postwar oil shocks and recessions. *Energy J.,* **6**:95–116.

Hamilton, L. 1981. Comparative risk assessment in the energy industry. In C. R. Richmond, P. J. Walsh, and E. D. Copenhaver (Eds.), *Health Risk Analysis.* Proceedings of the Third Life Science Symposium, Health Risk Analysis, Gatlinburg, TN, October 27–30, 1980. The Franklin Institute Press, Philadelphia, PA.

Hammer, C. U. 1977. Past volcanism revealed by Greenland ice sheet impurities. *Nature,* **270**:482–486.

Hannon, B. 1981a. Analysis of the energy cost of economic activities: 1963 to 2000. Energy Research Group Document No. 316, University of Illinois, Urbana.

Hannon, B. 1981b. The energy cost of energy. In H. E. Daly and A. F. Umana (Eds.), *Energy, Economics, and the Environment.* Westview Press, Boulder, CO.

Hannon, B. 1982. Analysis of the energy costs of economic activities: 1963 to 2000. *Energy Syst. Policy J.,* **6**:249–278.

Hannon, B. and H. Perez-Blanco. 1979. Ethanol and methanol as industrial feedstocks. Report to Argonne National Laboratories, Argonne, IL. Contract No. ANL 31-109-38-5154.

Hannon, B., R. A. Herendeen, and T. Blazeck. 1981. Energy and labor intensities for 1972. Energy Research Group Document No. 307, University of Illinois, Urbana.

Hannon, B. and S. Casles. 1982. Calculation of the energy and labor intensities of U.S. commodities, 1974 and a comparison to 1963, 1967, and 1972. Energy Research Group Document No. 315, University of Illinois, Urbana.

Hansen, J., D. Johnson, A. Lacis, S. Lebedeff, P. Lee, D. Rind, G. Russell. 1981. Climatic impact of increasing atmospheric carbon dioxide. *Science,* **213**:957–966.

Hanson, E. A. and J. B. Baker. 1979. Biomass and nutrient removal in short-rotation intensively cultured plantations. In *Proceedings of Impact of Intensive Harvesting on Forest Nutrient Recycling.* State University of New York (SUNY) College of Environmental Science and Forestry, Syracuse, NY, pp. 130–149.

Harbert, H. P. and W. A. Berg. 1978. *Vegetative Stabilization of Spent Oil Shales. Vegetation, Moisture, Salinity and Runoff—1973–1976.* U.S. Government Printing Office, Washington, D.C.

Harris, M. 1979. *Cultural Materialism.* Random House, New York.

Harris, C. K. and D. E. Rogers. 1979. Forecast of the sockeye salmon run to Bristol Bay in 1979. Circular 79-2, Fisheries Research Institute, University of Washington, Seattle.

Harris, D. P. and B. J. Skinner. 1982. The assessment of long term mineral supplies. In V. K. Smith and J. Krutilla (Eds.), *Explorations in Natural Resource Economics.* The Johns Hopkins University Press, Baltimore, pp. 247–326.

Hart, R. H. 1978. Stocking rate theory and its application to grazing on rangelands. In D. N. Hyder (Ed.), *Proceedings of the First International Rangeland Congress.* Society for Range Management, Denver, CO, pp. 547–550.

Harte, J. and M. El-Gasseir. 1978. Energy and water. *Science,* **199**:623–634.

Hatcher, R. D., Jr. 1982. Hydrocarbon resources of the eastern overthrust belt: A realistic evaluation. *Science,* **216**:980–982.

Hay, K., J. D. L. Harrison, R. Hill, and T. Riaz. 1981. A comparison of solar cell production technologies through their economic impact on society. *IEEE,* **15**:267–272.

Hayes, E. T. 1979. Energy resources available to the United States, 1985 to 2000. *Science,* **203**:233–239.

Heady, E. O. 1982. The adequacy of agricultural land: A demand-supply perspective. In P. R. Crosson (Ed.), *The Cropland Crisis: Myth or Reality?* Resources for the Future, Baltimore.

Heady, H. F., et al. 1974. Livestock grazing on federal lands in the 11 Western states. *J. Range Manage.,* **27**(3):174–181.

Heagle, A. S., D. E. Body, and G. E. Neely. 1974. Injury and yield responses of soybean to chronic doses of ozone and sulfur dioxide in the field. *Phytopathology,* **64**:132–136.

Heagle, A. S., R. B. Philbeck, H. H. Rogers, and M. B. Letchworth. 1979. Dispensing and monitoring ozone in open-top field chambers for plant-effects studies. *Phytopathology,* **69**:15–20.

Heagle, A. S. and W. W. Heck. 1980. Field methods to assess crop losses due to oxidant air pollutants. In P. S. Teng and S. V. Krupa (Eds.), *Crop Loss Assessment,* Proc. of E. C. Stakman Commem. Symp. Misc. Pub. No. 7. Agricultural Experiment Station, University of Minnesota, Minneapolis.

Heck, W. W. and C. S. Brandt. 1977. Effects of vegetation: Nature, crops, forest. In A. C. Stern (Ed.), *Air Pollution,* Vol. 2. *The Effects of Air Pollution,* 3rd ed. Academic Press, New York.

Heggestad, H. E. 1980. Field Assessment of Air Pollution Impacts on Growth and Productivity of Crop Species. The 73rd Annual Meeting of the Air Pollution Control Association, Montreal, Canada.

Heggestad, H. E., R. K. Howell, and J. H. Bennett.

1977. The effects of oxidant air pollutants on soybeans, snap beans and potatoes. U.S. Environmental Protection Agency, EPA/600/3-77-128, Corvallis, Oregon.

Heggestad, H. E. and J. H. Bennett. 1981. Photochemical oxidants and potential yield losses in snap beans attributable to sulphur dioxide. *Science*, **213**:1008–1010.

Heichel, G. H. 1976. Agricultural production and energy resources. *Am. Sci.*, **64**:64–72.

Heilbroner, R. L. and L. C. Thurow. 1981. *The Economic Problem*. Prentice-Hall, Englewood Cliffs, NJ.

Heinrich, B. 1979. *Bumblebee Economics*. Harvard University Press, Cambridge, MA.

Hendricks, T. A. 1965. Resources of oil, gas, and natural gas liquids in the U.S. and the world. USGS Circular 522. U.S. Government Printing Office, Washington, D.C.

Hendrickson, T. A. 1975. *Synthetic Fuels Data Handbook*. Cameron Engineers, Inc., Denver, CO.

Hennemuth, R. C. 1979. Man as predator. In G. P. Patil and M. L. Rosenzweig (Eds.), *Contemporary Quantitative Ecology and Related Ecometrics*. International Co-operative Publishing House, Fairland, MD, pp. 507–532.

Hennemuth, R. C., J. E. Palmer, and B. E. Brown. 1980. A statistical description of recruitment in eighteen selected fish stocks. *Norw. Atl. Fish. Sci.*, **1**:101–111.

Henry, J. F. 1979. Silvicultural energy farm in perspective. In K. Sarkanen, V. Shafizadeh, and D. A. Tillman (Eds.), *Thermal Uses and Properties of Carbohydrates and Lignins*. Academic Press, New York.

Herendeen, R. A. and C. W. Bullard. 1974. Energy Cost of Goods and Services, 1963 and 1967. Document 140, Center for Advanced Computations, University of Illinois, Urbana.

Herendeen, R. A. and C. W. Bullard. 1975. Energy costs of goods and services: 1963 and 1967. *Energy Policy*, **3**:268.

Herendeen, R. A. and C. W. Bullard. 1976. U.S. energy balance of trade, 1963–1967. In *Energy Systems and Policy*, Vol. I, No. 4. Crane, Russak and Co.

Herendeen, R. A. and R. Plant. 1979. Energy analysis of geothermal electric systems. Energy Research Group Doc. No. 272. University of Illinois, Urbana.

Herendeen, R. A., T. Kary, and J. Rebitzer. 1979. Energy analysis of the solar power satellite. *Science*, **205**:451–454.

Hewett, C. E., C. J. High, N. Marshall and R. Wildermuth. 1981. Wood energy in the United States. *Ann. Rev. Energy*, **6**:139–170.

Hewett, D. F. 1929. Cycles in metal production. In *Transactions of American Institute of Mining and Metallurgical Engineers*. American Institute of Mining and Metallurgical Engineers, New York.

Hileman, B. 1982. Crop losses from air pollutants. *Environ. Sci. Technol.*, **16**:495A–499A.

Hill, A. C. and J. H. Bennett. 1970. Inhibition of apparent photosynthesis by nitrogen oxides. *Atmos. Environ.*, **4**:341–348.

Hill, D., B. L. Pierce, W. C. Metz, M. D. Rowe, E. T. Haefele, F. C. Bryant, and E. J. Tuthill. 1982. Management of high-level waste repository siting. *Science*, **218**:859–864.

Hinds, J. 1950. Colorado River Aqueduct. The Metropolitan Water District of Southern California.

Hinga, K. R., G. R. Heath, D. R. Anderson, and C. D. Hollister. 1982. Disposal of high-level radioactive wastes by burial in the sea. *Environ. Sci. Technol.*, **16**:28A–37A.

Hinkle, M. K. 1979. Ramifications of vegetative manipulation—vegetative considerations. In *Rangeland Policies for the Future: Proceedings of a Symposium*, January 28–31, Tucson, Arizona. USDA, USDI, CEQ. U.S. Government Printing Office, Washington, D.C., GTR WO-17.

Hinman, G. and E. Leonard. 1977. Air quality and energy development in the Rocky Mountain West. Los Alamos Scientific Laboratory, Los Alamos, New Mexico. LA-6674.

Hinton, C. and R. V. Moore. 1959. Nuclear propulsion of ships. *Br. Nucl. Energy Conf. J. London*, 4(1):39–50.

Hirao, Y. and C. C. Patterson. 1974. Lead aerosol pollution in the high Sierra overides natural mechanisms which exclude lead from a food chain. *Science*, **184**:989–992.

Hirst, E., R. Marlay, D. Greene, and R. Barnes. 1983. Recent changes in U.S. energy consumption: What happened and why. *Ann. Rev. Energy*, **8**:193–245.

Hirth, H. F. 1963. The ecology of two lizards on a tropical beach. *Ecol. Monogr.*, **33**:83–112.

Hjort, J. 1916. Fluctuations in the great fisheries of Northern Europe viewed in the light of biological research. *Rapp. P.-V. Cons. Int. Explor. Mer*, **20**:1–228.

Hobbs, P. V., H. Harrison, and E. Robinson. 1974. Atmospheric effects of pollutants. *Science*, **183**:909–915.

Hogan, W. W. and A. S. Manne. 1977. Energy–economic interactions: The fable of the elephant and the rabbit? In C. J. Hitch (Ed.), *Modeling Energy-Economy Interactions: Five Approaches*. Resources for the Future, Washington, D.C.

Holdren, J. P. 1982. Energy hazards: What to measure, what to compare. *Technol. Rev.*, **85**(3):32–38, 74–75.

Holdren, J. P. and P. Herrera. 1971. *Energy*. Sierra Club, San Francisco.

Holdren, J. P., K. Anderson, P. H. Gleick, I. Mintzer, G. Morris, and K. P. Smith. 1979. Risk of Renewable Energy Sources: A Critique of the Inhaber Report. Energy and Resources Group Report. ERG 79-3. University of California, Berkeley.

Holdren, J. P., G. Morris, and I. Mintzer. 1980. Environmental aspects of renewable energy sources. *Ann. Rev. Energy,* **5**:241–291.

Holling, C. S. (Ed.). 1978. *Adaptive Environmental Assessment and Management.* Wiley, New York.

Hollingsworth, T. H. 1969. *Historical Demography.* Cornell University Press, Ithaca, NY.

Hollister, C. D., D. R. Anderson, and G. R. Heath. 1981. Subseabed disposal of nuclear wastes. *Science,* **213**:1321–1326.

Hopkinson, C. S., Jr., and J. W. Day. 1980. Net energy analysis of alcohol production from sugarcane. *Science,* **207**:302–303.

Hormay, A. L. and M. W. Talbot. 1961. Rest-rotation grazing—a new management system for perennial bunchgrass ranges. USDA Prod. Res. Rep. 51.

Horsfall, J. 1975. *Agricultural Production Efficiency.* National Academy of Sciences, Washington, D.C.

Hoskins, W. N., R. P. Upadhyay, J. B. Bills, and C. R. Sandberg III. 1976. A technical and economic study of candidate underground mining systems for deep, thick oil shale deposits. *Q. Colo. Sch. Mines,* **71**(4):199–234.

Hoveland, C. S. 1980. Energy inputs for beef cattle production on pasture. In D. Pimentel (Ed.), *CRC Handbook of Energy Utilization in Agriculture.* CRC Press, Boca Raton, FL, pp 419–424.

Howard, H. E. 1920. *Territory in Bird Life.* Murray, London.

Howarth, Richard and C. A. S. Hall. The energy return on investment of nuclear power in the U.S.: Past trends and future projections. (In preparation.)

Howarth, Richard, C. A. S. Hall, and R. Kaufmann. The energy/economic link: An international assessment. (In preparation.)

Howarth, Robert W. 1981. Fish versus fuel: A slippery quandary. *Technol. Rev.,* **Jan.**:68–77.

Howarth, Robert W. and C. A. S. Hall. 1978. What do you want to do with your last 27,000 gallons of oil? *Human Ecol. Forum,* **8**(3):2–11.

Hubbert, M. K. 1949. Energy from fossil fuels. *Science,* **109**:103.

Hubbert, M. K. 1956. Nuclear energy and the fossil fuels. In *Drilling and Production Practice.* American Petroleum Institute, New York, pp. 7–25.

Hubbert, M. K. 1962. *Energy Resources.* A Report to the Committee on Natural Resources, Pub. No. 1000-D, National Academy of Sciences, Washington, D.C.

Hubbert, M. K. 1967. Degree of advancement of petroleum exploration in the United States. *Am. Assoc. Pet. Geol. Bull.,* **51**:2207.

Hubbert, M. K. 1969. Energy resources. In *Resources and Man, National Academy of Sciences.* W. H. Freeman, San Fransisco, pp. 157–242.

Hubbert, M. K. 1974. U.S. energy resources, a review as of 1972. Senate Comm. Interior Insular Affairs Ser. No. 93-40.

Hubbert, M. K. 1978. U.S. petroleum estimates 1956–1978. In Annual Meeting Papers, Production Dept., American Petroleum Institute, Washington, D.C.

Hubbert, M. K. 1980. Techniques of prediction as applied to the production of oil and gas. National Bureau of Standards Special Publication 631. U.S. Government Printing Office, Washington, D.C.

Huettner, D. A. 1976. Net energy analysis: An economic assessment. *Science,* **192**(4235):101–104.

Huettner, D. A. 1981. Energy, entropy, and economic analysis: Some new directions. *Energy J.,* **2**:123–130.

Huettner, D. A. 1982. Economic values and embodied energy. *Science,* **216**:1141–1143.

Hunt, L. P. 1976. Total energy use in the production of silicon solar cells from raw materials to finished product. *IEEE,* 347–352.

Husar, R. B., J. M. Holloway, D. E. Patterson, and W. E. Wilson. 1981. Spatial and temporal pattern of eastern U.S. haziness: A summary. *Atmos. Environ.,* **15**:1919–1925.

Hutchison, G. E. 1978. *An Introduction to Population Ecology.* Yale University Press, New Haven.

Hyman, E. L. 1980. Net energy analysis and the energy theory of value: Is it a new paradigm for a planned economic system? *J. Environ. Syst.,* **9**:313–324.

Ibrahim, K. 1975. *Glossary of Terms Used in Pasture and Range Survey Research, Ecology and Management.* Food and Agriculture Organization of the United Nations, Rome.

Incze, L. S. and A. J. Paul. 1983. Grazing and predation as related to energy needs of stage I zoeae of the tanner crab *Chionoecetes bairdi* (Brachyura, Majidae). *Biol. Bull.,* **165**(1):197–208.

Iglehart, C. F. 1972. North American drilling activity in 1971. *Bull. Am. Assoc. Pet. Geol.,* **56**:1147–1175.

Intertechnology Corporation. 1977. An analysis of the economic potential of solar thermal energy to provide industrial process heat. Warrenton, PA.

Irving, P. M. and J. E. Miller. 1981. Productivity of field-grown soybeans exposed to acid rain and sulfur dioxide alone and in combination. *J. Environ. Qual.,* **10**:473–478.

Isard, W., C. L. Choguill, J. Kissin, R. Seyfarth, and R.

Tatlock. 1972. *Ecologic–Economic Analysis for Regional Development.* The Free Press, New York.

Jahn, E. C. and S. B. Preston. 1976. Timber: More Effective Utilization. In Abelson and Hammond (Eds.), *Materials: Renewable and Non-renewable Resources,* Vol. 4. American Association for the Advancement of Science, Washington, D.C.

Janzen, D. H. 1973. Tropical agroecosystems. *Science,* **182**:1212–1219.

Jenkins, D. M., T. A. McClure, and T. S. Reedy. 1979. *Net Energy Analysis of Alcohol Fuels,* Pub. No. 4312. American Petroleum Institute, Washington, D.C.

Jevons, W. S. 1871. *Theory of Political Economy.* MacMillan, London.

Jewell, W. J., B. A. Adams, B. T. Eckstrom, K. J. Fanfoni, R. M. Kabrick, and D. F. Sherman. 1982. The feasibility of biogas production on farms. Report prepared for USDOE, SERI No. XB-0-9038-1-10.

Johnson, L. E., L. A. Albee, R. O. Smith, and A. L. Moxon. 1951. Cows, calves, and grass. South Dakota Agricultural Experiment Station Bulletin 412.

Johnson, M. H., F. W. Bell, and J. T. Bennett. 1980. Natural resource scarcity: Empirical evidence and public policy. *J. Environ. Econ. Public Policy,* **7**:256–271.

Johnston, A. 1978. Panorama of the rangelands of North America. In D. N. Hyder (Ed.), *Proceedings of the First International Rangeland Congress.* Society for Range Management, Denver, CO, pp. 37–41.

Jones, R. J. 1974. The relation of animal and pasture production to stocking rates on legume based and nitrogen fertilized subtropical pastures. *Proc. Aust. Soc. Anim. Prod.,* **10**:340–343.

Jones, R. J. and R. L. Sandland. 1974. The relation between animal gain and stocking rate. Derivation of the relation from the results of grazing trials. *J. Agric. Sci., Camb.,* **83**:335–342.

Jordan, C. F. and R. Herrara. 1981. Tropical Rain Forests: Are nutrients really critical? *Am. Nat.,* **117**:167–180.

Jordan, M. J. 1975. Effects of zinc smelter emissions and fire on a chestnut oak woodland. *Ecology,* **56**:78–91.

Jorgenson, J. R., C. G. Wells, and L. J. Metz. 1975. The nutrient cycle: Key to continuous forest production. *J. Forestry,* **73**:400–403.

Joyce, J. 1979. Energy. Technical Memo 120, Energy Research Group, University of Illinois, Urbana.

Jungk, Robert. 1979. *The New Tyranny.* Grossett and Dunlap, New York.

Jurik, T. W. 1983. Reproductive effort and CO_2 dynamics of wild strawberry populations. *Ecology,* **64**(6):1329–1342.

Justus, C. G. 1978. Wind energy statistics for large arrays of wind turbines (New England and central U.S. regions). *Solar Energy,* **20**(5):379–386.

Kacelnik, Alejandro, Alasdair I. Houston, and John R. Krebs. 1981. Optimal foraging and territorial defence in the Great Tit (Parus major). *Behav. Ecol. Sociobiol.,* **8**:35–40.

Kahn, E. 1978. The reliability of wind power from dispersed sites: A preliminary analysis. Lawrence Livermore Berkeley Laboratory, LBL 6889.

Kahn, H. 1983. *The Coming Boom.* Simon and Schuster, New York.

Kahn, H., W. Brown, and L. Martel. 1976. *The Next 200 Years.* William Morrow, New York.

Kakela, P. J. 1978. Iron ore: Energy, labor and capital changes with technology. *Science,* **202**:1151–1157.

Karnowky, D. F. 1980. Changes in southern Wisconsin white pine stands related to air pollution sensitivity. In P. R. Miller (Tech. Coord.), *International Symposium on Effects of Air Pollutants on Mediterranean and Temperate Forest Ecosystems,* Riverside Cal. U.S. Forest Service.

Katzenberg, R. and B. Wolfe. No date. *The Federal Wind Program, A Proposal for F.Y. 1979 Budget.* The American Wind Association, Washington, D.C.

Kaufmann, R. K. A statistical analysis of the energy/real GNP ratio in the U.S.: 1929–1981. In preparation.

Kaufmann, R. K. and C. A. S. Hall. 1981. The energy return on investment of imported petroleum. In W. J. Mitsch, R. W. Bosserman, and J. M. Klopatek (Eds.), *Energy and Ecological Modeling.* Elsevier Scientific, Amsterdam, pp. 697–701.

Kellogg, H. H. 1974. Energy efficiency in the age of scarcity. *J. Metals,* **26**:25–29.

Kelly, J. and C. Shea. (no date). The subseabed disposal program for high-level radioactive waste. Draft 5. The Center for Complex Systems, Univ. of New Hampshire, Durham, NH.

Kelly, H. 1978. Photovoltaic power systems: A tour through the alternatives. *Science,* **199**:634–643.

Kemp, W. M., W. H. B. Smith, H. N. McKellar, M. E. Lehman, M. Homer, D. L. Young, and H. T. Odum. 1977. Energy cost–benefit analysis applied to power plants near Crystal River, Florida. In C. A. S. Hall and J. Day (Eds.), *Ecosystem Modelling in Theory and Practice.* Wiley-Interscience, New York, pp. 507–543.

Kemp, W. M., W. R. Boynton, and K. Limburg. 1981. The influence of natural resources and demographic factors on the economic production of nations. In W. J. Mitsch (Ed.), *Energy and Ecological Modeling.* Elsevier, Amsterdam, pp. 827–839.

Kemp, W. M., W. R. Boynton, and A. J. Hermann. 1982. A simulation modeling framework for ecological research in complex systems: The case of submerged vegetation in upper Chesapeake Bay. In *Marine Ecosystem Modeling,* Proceedings From a Workshop Held April 6–8, 1982. U.S. Department of Commerce, Washington, D.C., pp. 131–158.

Kennedy, D., T. Blazeck, B. Hannon, and B. Illyes. 1981. Energy Research Group I-O Model Database: 1963, 1967, and 1972. Energy Research Group Document no. 317. Urbana, Illinois.

Kerr, R. A. 1979. Global pollution: Is the arctic haze actually industrial smog? *Science,* **205**:290–293.

Kerr, R. A. 1981. How much oil? It depends on whom you ask. *Science,* **212**:427–429.

Kerr, R. A. 1983a. Trace gases could double climate warming. *Science,* **220**:1364–1365.

Kerr, R. A. 1983b. Another oil resource warning. *Science,* **223**:382.

Kessler, W. B. and R. P. Bosch 1982. Sharp-tailed grouse and range management practices in western rangelands. In J. M. Peek and P. K. Dalke (Eds.), *Proceedings of the Wildlife-Livestock Relationships Symposium,* Coeur d'Alene, Idaho, April 20–22. Forest, Wildlife, and Range Experiment Station, University of Idaho, Moscow. pp. 133–146.

Kickert, R. N. and B. Gemmell. 1980. Data-based ecological modeling of ozone air pollution effects in a southern California mixed conifer ecosystem. In P. R. Miller (Tech. Coord.), *Proceedings of a Symposium on Effects of Air Pollutants on Mediterranean and Temperate Ecosystems.* U.S. Forest Service, Washington, D.C., GIRPSW-43.

Kinchal, S. K. 1975. Energy requirements of an oil shale industry: Based on Paraho's direct combustion retorting process. *Q. Colo. Sch. Mines,* **70**(3):21–29.

Kirsch, L. M., H. F. Duebbert, and A. D. Kruse. 1978. Grazing and haying effects on habitats of upland nesting birds. In K. Sabol (Ed.), *Transactions of the 43rd North American Wildlife and Natural Resources Conference.* Wildlife Management Institute, Washington, D.C., pp. 486–497.

Klein, G. L., L. Masin, J. Rosen, and A. Wyatt. 1977. A Feasibility Study on Development of a Small Scale Cellulose Insulation Industry in Tompkins County, N.Y. National Center for Appropriate Technology, Montana.

Klepper, R., W. Lockeretz, B. Commoner, M. Gertler, S. Fast, D. O'Leary, and R. Blobaum. Economic performance and energy intensiveness of organic and conventional farms in the corn belt: A preliminary comparison. *Am. J. Agric. Econ.,* **59**:1–12.

Klingebiel, A. A. and P. H. Montgomery. 1961. Land capability classification. *Agricultural Handbook 210.* U.S. Department of Agriculture, Washington, D.C.

Klingsberg, C. and J. Duguid. 1982. Isolating radioactive wastes. *Am. Sci.,* **70**:182–190.

Klipple, G. E. and D. F. Costello. 1960. Vegetation and cattle responses to different intensities of grazing on shortgrass ranges of the Central Great Plains. U.S. Dept. Agric. Tech. Bull. 1216.

Klipple, G. E. and R. E. Bement. 1961. Light grazing—is it economically feasible as a range-improvement practice. *J. Range Manage.,* **14**(2):57–62.

Klopatek, M. and P. G. Risser. 1982. Energy analysis of Oklahoma rangelands and improved pastures. *J. Range Manage.,* **35**(5):637–643.

Knabe, W. 1976. Effects of sulfur dioxide on terrestrial vegetation. *Ambio,* **5**:213–218.

Kneese, A. V., R. V. Ayres, and R. D'arge. 1970. *Economics and the Environment.* Resources for the Future, Washington, D.C.

Knoblauch, W. A. and R. A. Milligan. 1981. Economic profiles for corn, hay, and pasture. 1980 and five year average 1976–1980. Cornell University, Dept. of Agricultural Economics. A.E. Ext. 81–23. Ithaca, New York.

Knopf, F. L. and R. W. Cannon. 1982. Structural resilience of a willow riparian community to changes in grazing practices. In J. M. Peek and P. K. Dalke (Eds.), *Proceedings of the Wildlife–Livestock Relationships Symposium,* Coeur d'Alene, Idaho, April 20–22. Forest, Wildlife, and Range Experiment Station, University of Idaho, Moscow. pp. 198–207.

Koch, P. 1976. Materials balances and energy required for manufacture of ten wood commodities. In *Energy and the Wood Products Industry,* Proceedings Number P76-14. The Forest Products Society, Madison, WI.

Kozlovsky, Daniel G. 1968. A critical evaluation of the trophic level concept. *Ecology,* **49**:48–60.

Krebs, C. J. 1972. *Ecology: The Experimental Analysis of Distribution and Abundance.* Harper and Row, New York.

Krenkel, P. A. and F. L. Parker. 1969. Biological aspects of thermal pollution. In *Proceedings of the National Symposium on Thermal Pollution,* Portland, Oregon, 1968. Vanerbilt University Press, Nashville, Tenn.

Kress, L. W. and J. E. Miller. 1981. Impact of ozone on soybean yield. Radiolog. and Environ. Research Div. Ann. Rept. Ecology, Jan.–Dec., 1980. Argonne National Laboratory. ANL-80-115. Part III.

Kress, L. W., J. M. Skelly, and K. H. Hinkleman. 1981. Growth impact of O_3, NO_2, and/or SO_2 on *Pinus taeda. Environ. Monit. Assess.,* **1**:229–239.

Kress, L. W., J. M. Skelly, and K. H. Hinkleman. 1982. Relative sensitivity of 18 full-sib families of *Pinus taeda* to O_3. *Can. J. For. Res.,* **12**:203–209.

Krutilla, J. V. and A. C. Fischer. 1975. *The Economics of Natural Environments: Studies in Valuation of Commodity and Amenity Resources.* Johns Hopkins University Press, Baltimore.

Kuchler, A. W. 1980. Natural vegetation. In *Goode's World Atlas*, 15th ed. Rand McNally, Chicago.

Kunchal, S. K. 1975. Energy requirements in an oil shale industry: Based on Paraho's direct combustion retorting processes. *Q. Colo. Sch. Mines*, 70(3):21–29.

Kuuskraa, V. A., J. P. Brashear, T. M. Doscher, and L. E. Elkins. 1978. *Enhanced Recovery of Unconventional Gas*. National Technical Information Service, Springfield, VA.

Kylstra, C. and K. Han. 1975. An energy analysis of the U.S. nuclear power system. Nuclear Engineering Sciences, University of Florida. ERDA C Grant E-(40-1)-4398.

Lacey, J. R. and H. W. Van Poollen. 1979. Grazing system identification. *J. Range Manage.*, 32(1):38–39.

Lack, D. L. 1954. *The Natural Regulation of Animal Numbers*. Oxford University Press, New York.

Lainhart, W. S. 1969. Prevalence of coal workers Pneumoconiosis in Appalachian bituminous coal miners. In C. W. Lainhart et al. (Eds.), *Pneumoconiosis in Appalachian Bituminous Coal Miners*. Public Health Service, Cincinnati, OH, pp. 31–60.

Lambert, T. A. 1980. Energy and entropy in American agriculture and rural society; a new paradigm for public policy analysis. *Cornell J. Soc. Relations*, 15(1):84–97.

Landsberg, H. H. 1982. Relaxed energy outlook masks continuing uncertainties. *Science*, 218:973–974.

Landsberg, H. H. and S. H. Schurr. 1960. *Energy in the United States*. Random House, New York (for Resources for the Future).

Landsberg, H. H. and J. M. Dukert. 1981. *High Energy Costs—Uneven, Unfair, Unavoidable?* Johns Hopkins University Press, Baltimore (for Resources for the Future).

Langway, C. C., H. Oeschger, B. Alder, and B. Renaud. 1965. Sampling polar ice for radiocarbon dating. *Nature*, 206:500–501.

Lappe, F. M. and J. Collins. 1977. *Food First, Beyond the Myth of Scarcity*. Houghton Mifflin, Boston.

Larea, T. J. and J. Darmstadter. 1982. Energy and consumer-expenditive patterns: Modeling approaches and projections. *Ann. Rev. Energy*, 7:261–292.

Larkin, P. A. 1977. An epitaph for the concept of maximum sustained yield. *Trans. Am. Fish. Soc.*, 106(1):1–11.

Larson, W. E., F. J. Pierce, and R. H. Dowdy. 1983. The threat of soil erosion to long-term crop production. *Science*, 219:458–465.

Lasker, R. 1975. Field criteria for survival of anchovy larvae: The relation between inshore chlorophyll maximum layers and successful first feeding. *Fish. Bull.*, 73(3):453–462.

Lasker, R. and P. E. Smith. 1977. Estimation of the effects of environmental variations on the eggs and larvae of the Northern anchovy. *Calif. Coop. Oceanic Fish Invest. Rep.*, 19:128–137.

Lasky, S. G. 1950. How tonnage and grade relations help predict ore reserves. *Eng. Mining J.*, 151:81–85.

Lauenroth, W. K. 1979. Grassland primary production: North American grasslands in perspective. In N. R. French (Ed.), *Perspectives in Grassland Ecology: Results and Applications of the US/IBP Grassland Biome Study*. Springer-Verlag, New York, pp. 3–24.

Launchbaugh, J. L. 1957. The effect of stocking rate on cattle gains and on native shortgrass vegetation in west-central Kansas. Kansas Agricultural Experiment Station Bulletin 394.

Laurence, G. C. 1973. Influence of temperature on energy utilization of embryonic and prolarval tautog, *Tautoga onites*. *J. Fish. Res. Bd. Can.*, 30(3):435–442.

Laurence, G. C. 1981. Overview. Modelling—an esoteric or potentially utilitarian approach to understanding larval fish dynamics. *Rapp. P.-V. Reun. Cons. Int. Explor. Mer.*, 178:3–6.

Laurence, G. C. and C. A. Rogers. 1976. Effects of temperature and salinity on comparative embryo development and mortality of Atlantic cod (*Gadus morhau L.*) and haddock (*Melanogrammus aeglefinus L.*). *J. Cons. Int. Explor. Mer.*, 36:220–228.

Lave, L. B. and E. P. Seskin. 1970. Air pollution and public health. *Science*, 169:723–733.

Lave, L. B. and E. P. Seskin. 1973. An analysis of the association between U.S. mortality and air pollution. *J. Am. Statist. Assoc.*, 68:284–290.

Lave, L. B. and E. P. Seskin. 1977. *Air Pollution and Human Health*. Johns Hopkins University Press, Baltimore.

Lave, L. B. and L. C. Freeburg. 1972. Health effects of electricity generation from coal, oil and nuclear fuel. Unpublished Sierra Club conference report.

Lave, L. B. and L. C. Freeburg. 1973 (revised). Health Effects of Electricity Generation from Coal, Oil, and Nuclear Fuel. Paper presented at Sierra Club Conference, January 1972, Johnson, Vermont.

Lavine, M. J. and A. H. Meyburg. 1976. Toward environmental benefit/cost analysis: Measurement methodology. Prepared for National Cooperative Highway Research Program, Transportation Research Board, National Research Council. 20-11A. Unpublished.

Lavine, M. J. and T. J. Butler. 1981. Energy analysis and economic analysis: A comparison of concepts. In W. J. Mitsch, R. W. Bosserman, and J. M. Klopatek (Eds.), *Energy and Ecological Modelling*. Elsevier, New York.

Lavine, M. J., A. H. Meyburg, and T. J. Butler. 1982. Use of energy analysis for assessing environmental impacts due to transportation. *Transp. Res.,* **16A** (1):35–42.

Lawton, J. H. 1973. The energy cost of food-gathering. In B. Benjamin, P. R. Cox, and J. Peel (Eds.), *Resources and Population.* Academic Press, New York, pp. 59–76.

Le Brasseur, R. J., C. D. McAllister, and T. R. Parsons. 1979. Additions of nutrients to a lake leads to greatly increased catch of salmon. *Environ. Conserv.,* **6**(3):187–190.

Lee, R. B. 1969. !Kung bushman subsistance: An input–output analysis. In A. P. Vayda (Ed.), *Environment and Cultural Behavior: Ecological Studies in Cultural Anthropology.* University of Texas Press, Austin, pp. 47–79.

Lee, R. B. 1979. *The !Kung San: Men, Women, and Work in a Foraging Society.* Cambridge University Press, New York.

Lee, R. F. and M. Takahashi. 1975. The fate and effect of petroleum in controlled ecosystem enclosures. In *Petroleum Hydrocarbons in the Marine Environment,* ICES Workshop 65, Aberdeen, Scotland.

Leggett, W. C. 1976. The American shad (*Alosa sapidissima*) with special reference to its migration and population dynamics in the Connecticut River. In *The Connecticut River Ecological Study,* Monograph No. 1. The American Fisheries Society, pp. 169–225.

Lehninger, A. L. 1965. *The Molecular Basis of Biological Energy Transformations. Bioenergetics.* W. A. Benjamin, New York.

Lekachman, R. 1976. *Economists at Bay.* McGraw-Hill, New York.

Lem, P. N., H. T. Odum, and W. E. Bolch. 1974. Some considerations that affect the net yield from nuclear power. Health Physics Society 19th Annual Meeting, Houston, TX.

Len, R. R., B. A. Steward, and P. W. Unger. 1976. Conservation, tillage, and energy. Proceedings of the 31st Annual Meeting of the Soil Conservation Society of America, Minneapolis, MN.

Lencheck, T. E. 1976. Energy expenditures in a solar heating system. *Alternate Sources of Energy,* **21**:13–18.

Leontief, W. W. 1941. *The Structure of the American Economy, 1919, 1929: An Empirical Analysis of Equilibrium Analysis.* Harvard University Press, Cambridge, MA.

Leontief, W. W. 1982. Academic economics. *Science,* **217**:104–107.

Lewis, A. E. 1973. Nuclear *in situ* recovery of oil from shale. Lawrence Livermore Laboratory, Lawrence, California. UCRL 51453.

Lewis, J. D. 1982. Technology, enterprise, and American economic growth. *Science,* **215**:1204–1211.

Lewontin, R. 1980. Economics down on the farm. (Book review of Don Paarlberg's *Farm and Food Policy: Issues of the 1980s,* University of Nebraska Press, 1980.) *Nature,* **287**:661–662.

Lieberman, M. A. 1976. United States uranium resources—an analysis of historical data. *Science,* **192**:431–436.

Likens, G. E. and F. H. Bormann. 1974. Acid rain: A serious regional environmental problem. *Science,* **184**:1176–1179.

Likens, G. E., R. F. Wright, J. N. Galloway, and T. J. Butler. 1979. Acid rain. *Sci. Am.,* **241**(4):43–51.

Likens, G. E. and T. J. Butler. 1981. Recent acidification of precipitation in North America. *Atmos. Environ.,* **15**:1103–1109.

Lind, C. G. and W. J. Mitsch. 1981. A net energy analysis including environmental cost of oil shale development in Kentucky. In W. J. Mitsch, R. W. Bosserman, and J. M. Klopatek (Eds.), *Energy and Ecological Modelling.* Elsevier Scientific, Amsterdam, pp. 689–696.

Lindbergh, K. and B. Provorse. 1977. *Coal: A Contemporary Energy Story.* Scribe, New York.

Lindeman, R.L. 1942. The trophic–dynamic aspect of ecology. *Ecology,* **23**(4):399–418.

Linehan, H. T. and H. A. Post. 1971. Water pollution control and the pulp and paper industry. *J. Forestry,* **62**(12):861–864.

Linzon, S. N. 1978. Effects of air-borne sulfur pollutants on plants. In Jerome Nriagu (Ed.), *Sulfur in the Environment. Part II: Ecological Impacts.* Wiley, New York.

Linzon, S. N., P. J. Temple, and R. G. Pearson. 1979. Sulphur concentrations in plant foliage and related effects. *J. Air Pollut. Control Assoc.,* **29**:520–525.

Lioy, P. J. and P. J. Samson. 1979. Ozone concentration patterns observed during the 1976–1977 long-range transport study. *Environ. Int.,* **2**:77–83.

Lipfert, F. W. and J. L. Dungan. 1983. Residential firewood use in the United States. *Science,* **219**: 1425–1426.

Lockeretz, W., R. Klepper, B. Commoner, M. Gerter, S. Fast, and D. O'Leary. 1976. Organic and conventional crop production in the corn belt: a comparison of economic performance and energy use for selected farms. Center for Biology of Natural Systems, St. Louis, MO. Report AE-7.

Lotka, A. J. 1914. An objective standard of value derived from the principle of evolution. *Washington Acad. Sci.,* **4**:409–418.

Lotka, A. J. 1922. Contributions to the energetics of evolution. *Proc. Natl. Acad. Sci.,* **8**:147–151.

Loucks, O. L. and A. D'Alessio. 1975. *Energy Flow and*

Human Adaptation. Office of Ecosystems Studies, The Institute of Ecology.

Loucks, O. L., T. V. Armantano, R. W. Usher, W. T. Williams, R. W. Miller, and L. T. K. Wong. 1980. *Crop and Forest Losses Due to Current and Projected Emissions from Coal-Fired Power Plants in the Ohio River Basin.* Prepared for Ohio River Basin Energy Study (ORBES) under Prime Contr. EPAR805588. U.S. Environmental Protection Agency, Washington, D.C.

Loucks, O. L. and T. V. Armentano. 1982. Estimating crop yields effects from ambient air pollutants in the Ohio River Valley. *J. Air Pollut. Control Assoc.,* **32**:146–150.

Lovell, J. 1983. Exploring the "new" forest—Ravages of budworm spur paper industry research. *Maine Sunday Telegram,* October 9.

Lovering, T. S. 1969. Mineral resources from the land. In National Academy of Sciences, *Resources and Man.* W. H. Freeman, San Francisco, pp. 109–134.

Lovins, A. B. 1977. *Soft Energy Paths.* Friends of the Earth International, Ballanger, Cambridge, MA.

Lovins, A. B. and J. H. Price. 1980. *Non-Nuclear Futures: The Case for Ethical Energy Strategy.* Harper-Colophon Books, Harper & Row, New York.

Lovins, A. B. and L. H. Lovins. 1980. *Energy/War: Breaking the Nuclear Link.* Friends of the Earth, San Francisco, CA.

Lovins, A. B. 1985. The electricity industry. *Science* **229**:914.

Lowenstein, R. 1983. Firms spend billions hoping to find oil under Alaskan Sea. *Wall Street Journal,* Sept. 15.

Lugo, A. E. 1978. Stress and ecosystems. In J. H. Thorp and J. W. Gibbons (Eds.), *Energy and Environmental Stress in Aquatic Ecosystems.* DOE Symposium Series, CONF. 77114, pp. 61–101, Oak Ridge, Tenn.

Lusby, G. C. 1970. Hydrologic and biotic effects of grazing vs. nongrazing near Grand Junction, Colorado. *J. Range Manage.,* **23**:256–260.

Lanly, J. P. and J. Clements. 1982. Tropical forest resources. FAO Forestry Paper 30. FAO, Rome.

MacElroy, R. D. and M. M. Averner. 1978. Space Ecosynthesis: An Approach to the Design of Closed Ecosystems for Use in Space. NASA Technical Memorandum 78491. Ames Research Center, Moffett Field, California.

MacIntire, W. H. and J. B. Young. 1923. Sulfur, calcium, magnesium and potassium content and reaction of rainfall at different points in Tennessee. *Soil Sci.,* **15**:205–277.

Mackie, R. J. 1978. Impacts of livestock grazing on wild ungulates. In K. Sabol (Ed.), *Transactions of the 43rd North American Wildlife and Natural Resources Conference.* Wildlife Management Institute, Washington, D.C., pp. 462–476.

Makhijani, A. B. and A. J. Lichtenberg. 1972. Energy and well being. *Environment,* **14**(5):10–18.

Makinen, G. E. 1977. *Money, the Price Level, and Interest Rates.* Prentice-Hall, Englewood Cliffs, NJ.

Malthus, T. 1778. *An Essay on Population.* Ward, Lock, and Company, London.

Manabe, S. and R. T. Weatherald. 1975. The effects of doubling the carbon dioxide concentration on the climate of a general circulation model. *J. Atmos. Sci.,* **32**(1):3–15.

Mansfield, E. 1980. Research and development, productivity, and inflation. *Science,* **209**:1091–1093.

Manzer, J. I. 1964. Preliminary observations on the vertical distribution of Pacific salmon (genus Oncorrhynchus) in the Gulf of Alaska. *J. Fish Res. Bd. Can.,* **21**(5):891–903.

Marland, G. 1979. Shale oil: U.S. and world resources and prospects for near-term commercialization in the United States. Oak Ridge Associated Universities, Oak Ridge, Tennessee. ORAU/IEA 79-8(R).

Marland, G., A. M. Perry, and D. B. Reister. 1978. Net energy analysis of *in situ* oil shale processing. *Energy,* **3**:31–44.

Marshall, E. 1979. Synfuels program born in confusion. *Science,* **205**:1356–1357.

Marshall, E. 1981a. Problems continue at Three Mile Island. *Science,* **213**:1344–1345.

Marshall, E. 1981b. Reactor safety and the research budget. *Science,* **214**:766–768.

Marshall, E. 1982a. Reactor mishap raises broad questions. *Science,* **215**:877–878.

Marshall, E. 1982b. NRC reviews brittle reactor hazard. *Science,* **215**:1596–1597.

Marshall, E. 1982c. Using experience to calculate nuclear risks. *Science,* **217**:338–339.

Marshall, E. 1982d. Synfuels program runs out of projects. *Science,* **218**:1098–1099.

Marshall, E. 1982e. Brittle reactors: NRC has a plan. *Science,* **218**:1290–1291.

Marshall, E. 1983a. Ultrasafe reactors, anyone? *Science,* **219**:265–267.

Marshall, E. 1983b. The Salem case: A failure of nuclear logic. *Science,* **220**:280–282.

Marshall, E. 1983c. NRC faults utility, delays reactor start-up. *Science,* **220**:484.

Marshall, E. 1983d. NRC relents on Salem, clears plant for restart. *Science,* **220**:698.

Marshall, E. 1983e. DOE's mixed forecast. *Science,* **220**:800.

Marshall, E. 1983f. How LOFT stayed afloat at DOE. *Science,* **221**:34.

Marx, Karl. 1906. *Capital.* The Modern Library, New York.

Maryland Power Plant Siting Program. 1975. Power Plant Cumulative Environmental Impact Report, PPSP-CEIR-1. Department of Natural Resources, Annapolis, MD. Sept. 1975. (As cited in Ramsay, 1978).

Mason, B. 1958. *Principles of Geochemistry*. Wiley, New York.

Massengill, R. R. 1976. Entrainment of zooplankton at the Connecticut Yankee Plant. In D. Merriman and L. M. Thorpe (Eds.), *The Connecticut River Ecology Study: The Impact of a Nuclear Power Plant*. Am. Fish. Soc. Monograph No. 1, pp. 55–59. Allen Press, Laurence, KS.

Mattice, J. S., and M. E. Zittel. 1976. Site-specific evaluation of power plant chlorination. *J. Water Pollut. Control Fed.*, **48**:2284–2308.

Matveev, A. A. 1970. Chemical hydrology of regions of Eastern Antarctica. *J. Geophys. Res.*, **75**:3686–3690.

Mau, D., S. Bayley, and J. Zucchetto. 1978. Net energy analysis of electrical transmission lines. *Can. Elec. Eng. J.*, **3**(4):9–13.

Mazria, E. 1979. *The Passive Solar Energy Book*. Rodale Press, Emmaus, PA.

McBride, J. R., V. P. Semion, and P. R. Miller. 1975. Impact of air pollution on the growth of ponderosa pine. *Calif. Agric.*, **Dec.**:8–9.

McBride, J. P., R. E. Moore, J. P. Witherspoon, and R. E. Blanco. 1978. Radiological impact of airborne effluents of coal and nuclear plants. *Science*, **202**:1045–1050.

McClenahen, J. R. 1978. Community changes in a deciduous forest exposed to air pollution. *Can. J. For. Res.*, **8**:432–438.

McCormack, D. E. and W. E. Larson. 1981. A values dilemma: Standards for soil quality tomorrow. In W. E. Jeske (Ed.), *Economics, Ethics, Ecology*. Soil Conservation Society of America, Ankeny, Iowa, pp. 392–406.

McDonald, R. J. 1977. Estimate of national hydroelectric power potential at existing dams. U.S. Army Corps of Engineers Institute of Water Resources, July 20.

McFarland, W. N., F. H. Pough, and J. B. Heiser. 1979. *Vertebrate Life*. Macmillan, New York.

McGraw-Hill. 1980. *Encyclopedia of Energy*. McGraw-Hill, New York.

McHugh, J. L. 1959. Can we manage our Atlantic coastal fishery resources? *Trans. Am. Fish. Soc.*, **88**:105–110.

McKelvey, V. E. 1972. Mineral resource estimates and public policy. *Am. Sci.*, **60**:32–40.

McLaughlin, S. B., R. K. McConathy, D. Duvick, and L. K. Mann. 1982. Effects of chronic air pollution stress on photosynthesis, carbon allocation and growth of white pine trees. *For. Sci.*, **28**:60–70.

McNab, B. K. 1980. Food habits, energetics, and the population biology of mammals. *Am. Nat.*, **116**(1):106–124.

McNay, M. E. and B. W. O'Gara. 1982. Cattle-pronghorn interactions during the fawning season in northwestern Nevada. In J. M. Peek and P. K. Dalke (Eds.), *Proceedings of the Wildlife–Livestock Relationships Symposium*, Coeur d'Alene, Idaho, April 20–22, 1981. Forest, Wildlife, and Range Experiment Station, University of Idaho, Moscow. pp. 593–606.

Meadows, D. H., D. L. Meadows, J. Randers, and W. W. Behrens. 1972. *The Limits to Growth*. Universe, New York.

Meehan, W. R. and W. S. Platts. 1978. Livestock grazing and the aquatic environment. *J. Soil Water Conserv.*, **33**(6):274–278.

Menard, H. W. 1981. Toward a rational strategy for oil exploration. *Sci. Am.*, **244**:55–65.

Menard, H. W. and G. Sharman. 1975. Scientific uses of random drilling models. *Science*, **190**:337–343.

Merrill, L. B. 1954. A variation of deferred-rotation grazing for use under Southwest range conditions. *J. Range. Manage.*, **7**:152–154.

Merriman, D. and L. M. Thorpe. 1976. Introduction. In D. Merriman and L. M. Thorpe (Eds.), *The Connecticut River Ecology Study: The Impact of a Nuclear Power Plant*. Allen Press, Laurence, KS.

Merrow, E. W. 1978. *Constraints on the Commercialization of Oil Shale*. Rand Corporation, Santa Monica, CA.

Merrow, E. W., S. W. Chapel, and C. Worthing. 1979. *Review of Cost Estimates in New Technologies: Implications for Energy Process Plants*. Rand Corporation, Santa Monica, CA.

Metz, J. 1974. Oil shale: A huge resource of low grade fuel. *Science*, **184**:1271–1275.

Metz, W. D. and A. L. Hammond. 1978. *Solar Energy in America*. American Association for Advancement of Science, Washington, D.C.

Meyerhoff, A. A. 1976. Economic impact and geopolitical implications of giant petroleum fields. *Am. Sci.*, **64**:536–541.

Michael, A. D., C. R. van Raalte, and L. S. Brown. 1975. Long term effects of an oil spill at West Falmouth, Massachusetts. In *Conference on Prevention and Control of Oil Pollution*. American Petroleum Institute, Washington, D.C., pp. 573–582.

Miernyk, W. H. 1982. *The Illusion of Conventional Economics*. West Virginia University Press, Morgantown.

Miller, B. M., H. L. Thomsen, G. L. Dolton, A. B. Coury, T. A. Hendricks, F. L. Lennartz, R. B. Powers, E. G. Sable, and K. L. Varnes. 1975. Geological estimates of undiscovered recoverable oil and gas resources in the U.S. Geol. Surv. Circ. 725. U.S. Government Printing Office, Washington, D.C.

Miller, D. C. and A. D. Beck. 1975. Development and application of criteria for marine cooling waters. In *Environmental Effects of Cooling Systems at Nuclear Power Plants.* Proceedings of Symposium on the Physical and Biological Effects on the Environment of Cooling Systems and Thermal Discharges at Nuclear Power Stations, Oslo, 26–30 August. International Atomic Energy Agency, Vienna, pp. 639–657.

Miller, P. L. 1973. Oxidant-induced community change on a mixed conifer forest. In Naegle (Ed.), *Air Pollution Impact to Vegetation.* Advances in Chemistry Series 122, Chap. 9. American Chemistry Society, Washington, D.C.

Miller, P. R., G. J. Longbotham, R. E. Van Doren, and M. A. Thomas. 1980. Effect of oxidant air pollution exposure on California black oak in the San Bernadino Mountains. In *Proceedings of the Symposium on Ecology, Management and Utilization of California Oaks.* Pacific Southwest Forest and Range Experiment Station, Berkeley, California. U.S. Forest Service, Washington, D.C., GTR:PSW-4.

Mine Safety and Health Administration (MSHA). 1979. *Annual Report on Coal Mine Injuries and Fatalities.* U.S. Government Printing Office, Washington, D.C.

Mintz, J. 1983. How the engineers are sinking nuclear power. *Science,* 4(5):78–82.

Miranov, O. G. 1968. Hydrocarbon pollution of the sea and its influences on marine organisms. *Helgol. Wiss. Meeresunters.,* 17:335–339.

Montgomery, E., J. V. G. A. Durnin, and J. E. Ellis. 1975. *Energy Consumption.* Workshop Volume manuscript. The Institute of Ecology, Madison, WI.

Monthly Energy Review. U.S. Department of Energy. Washington, D.C.

Monthly Energy Review. December, 1983. Energy Information Administration, Office of Energy Markets and End Use. U.S. Department of Energy, Washington, D.C., DOE/EIA-0035/83/12.

Morowitz, H. J. 1968. *Energy Flow in Biology, Biological Organization as a Problem in Thermal Physics.* Academic Press, New York. (See review by H. T. Odum, *Science,* 164:683–684, 1969.)

Morowitz, H. J. 1980. *Entropy for Biologists.* Academic Press, New York.

Morrow, R. R. 1979. Potential losses associated with harvesting forest biomass for energy. In Conservation Circular 17, no. 9, pp. 7–11. Department of Natural Resources, Cornell University, Ithaca, NY.

Morrow, R. R. and T. E. Gage. 1977. *The Use of Wood for Fuel.* Conservation Circular 15, no. 1. Depart-

ment of Natural Resources, Cornell University, Ithaca, NY.

Moss, (Senator) F. E. 1967. *The Water Crisis.* Praeger, New York.

Mudd, J. B. and T. T. Kozlowski (Eds.). 1975. *Responses of Plants to Air Pollution.* Academic Press, New York.

Mulligan, H. F., A. C. Mathieson, G. E. Jones, A. C. Borror, T. C. Loder, P. J. Sawyer, and L. G. Harris. 1974. Impact of an oil refinery on the New Hampshire marine environment. In *The Impacts of an Oil Refinery Located in Southeastern New Hampshire: Preliminary Study.* University of New Hampshire, Durham, NH, Chap. X.

Murdoch, W. W. (Ed.). 1975. *Environment: Resources, Pollution, and Society,* 2nd ed. Sinauer Assoc., Sunderland, MA.

Murdoch, W. W. 1980. *The Poverty of Nations: The Political Economy of Hunger and Population.* Johns Hopkins University Press, Baltimore.

Myers, J. G. and L. Nakamura. 1978. *Energy Consumption in Manufacturing.* (Report to the Energy Policy Project of the Ford Foundation.) Ballinger Publishing, Cambridge, MA.

Myers, N. 1981. The hamburger connection: How Central America's forests become North America's hamburgers. *Ambio,* 10(1):2–8.

MITRE Corporation, 1979. Near Term Potential of Wood as a Fuel. McLean, VA, USDOE EG-77-C-01-4101.

Nadkarni, N. M. 1981. Canopy roots: Convergent evolution in rainforest nutrient cycles. *Science,* 214:1023–1024.

National Academy of Sciences. 1975. *Agricultural Production Efficiency.* National Academy of Sciences, Washington, D.C.

National Academy of Sciences. 1975. *Mineral Resources and the Environment.* National Academy of Sciences, Washington, D.C.

National Academy of Sciences. 1979. Controlled nuclear fusion. In *Energy in Transition 1985–2010.* W. H. Freeman, San Francisco.

National Academy of Sciences. 1979. Nuclear power. In *Energy in Transition 1985–2010.* W. H. Freeman, San Francisco.

National Academy of Sciences. 1979. *Energy in Transition 1985–2010.* W. H. Freeman, San Francisco.

National Academy of Sciences/National Research Council, Division of Medical Sciences, Advisory Committee on the Biological Effects of Ionizing Radiations (BEIR). 1972. *The Effects on Populations of Exposure of Low Levels of Ionizing Radiation.* National Academy of Sciences, Washington, D.C.

National Academy of Sciences, Committee on Mineral Resources and the Environment (COMRATE). 1975.

Mineral Resources and the Environment. National Academy of Sciences, Washington, D.C.

National Coal Association. 1980. *Coal Facts.* National Coal Association, Washington, D.C.

National Petroleum Council. 1970. *U.S. Energy Outlook: Oil and Gas Availability.* National Petroleum Council, Washington, D.C.

National Research Council. 1981. Atmosphere–Biosphere Interactions: Toward a Better Understanding of the Ecological Consequences of Fossil Fuel Combustion. Comm. on Atmos. and Bios., Bd. on Agric. and Renew. Resour., Comm. on Nat. Res.

Nautusch, D. E. S. and J. R. Wallace. 1974. Urban aerosol toxicity: The influence of particle size. *Science,* **186**:695–699.

Neff, T. L. 1981. *The Social Costs of Solar Energy: A Study of Photovoltaic Energy Systems.* Pergamon Press, New York.

Neftel, A., H. Oeschger, J. Schwander, B. Stauffer, and R. Zumbrunn. 1982. Ice core sample measurements give atmospheric CO_2 content during the past 40,000 yr. *Nature,* **295**:220–223.

Nehring, R. 1978. *Giant Oil Fields and World Oil Resources.* Prepared for the Central Intelligence Agency. Rand, Santa Monica, CA.

Nehring, R. 1981. *The Discovery of Significant Oil and Gas Fields in the United States.* Rand Corporation, Santa Monica, CA.

Nehring, R. 1982. Prospects for world oil resources. *Ann. Rev. Energy,* **7**:175–200.

Nero, A. V., Jr. 1979. *A Guidebook to Nuclear Reactors.* University of California Press, Berkeley.

New York State Energy Research and Development Authority. 1978. NYSERDA Special Report. (West Valley).

New York State Energy Research and Development Authority. 1978. Assessment of hydropower restoration and expansion in New York State. NYSERDA Report 78-6.

Nicolson, G. 1983. Screening of turfgrass and legumes for use as living mulches in vegetable production. M.S. Thesis. Vegetable Crops Department, Cornell University, Ithaca, NY.

Nielsen, L. A. 1976. The evolution of fisheries management philosophy. *Mar. Fish. Rev.,* **38**:15–23.

Nielson, L. A. 1978. The effects of rest-rotation grazing on the distribution of sharp-tailed grouse. M.S. Thesis. Montana State University, Bozeman.

Nixon, S. W. 1980. Between coastal marshes and coastal waters: A review of twenty years of speculation and research on the role of salt marshes in estuarine productivity and water chemistry. In P. Hamilton and K. MacDonald (Eds.), *Estuarine and Wetlands Processes.* Plenum Press, New York.

Nixon, S. W. 1981. Remineralization and nutrient cycling in coastal marine ecosystems. In B. J. Neilson and L. E. Cronin (Eds.), *Estuaries and Nutrients.* Humana Press, Clifton, NJ, pp. 111–138.

Nordhaus, W. and J. Tobin. 1973. Is growth obsolete? In M. Moss (Ed.), *The Measurement of Economic and Social Performance.* National Bureau of Economic Research. Columbia University Press, New York.

Norgaard, R. B. 1975. Resource scarcity and new technology in U.S. petroleum development. *Nat. Res. J.,* **15**:265–282.

Norman, C. 1982. Long-term problems for the nuclear industry. *Science,* **215**:376–379.

Norman, C. 1985. Low level waste deadline looms. *Science,* **229**:448–449.

Nuclear Regulatory Commission. 1975. Reactor safety study: An assessment of accident risks in U.S. commercial nuclear power plants. NRC Report WASH 1400.

Nye, P. H. and D. J. Greenland. 1960. The soil under shifting cultivation. Technical Communication 51. Commonwealth Bureau of Soils, Harpenden, London.

Occupational Injuries and Illnesses in the U.S. by Industry, for 1978. U.S. Dept. of Labor and U.S. Bureau of Labor Statistics. 1980. GPO. no. 2078.

Odum, E. P. 1971. *Fundamentals of Ecology,* 3rd ed. W. B. Saunders, Philadelphia.

Odum, E. P. and A. E. Smalley. 1959. Comparison of population energy flow of a herbivorous and a deposit-feeding invertebrate in a salt marsh ecosystem. *Proc. Natl. Acad. Sci.,* **45**:617–622.

Odum, H. T. 1957. Trophic structure and productivity of Silver Springs, Florida. *Ecol. Monogr.,* **27**:55–112.

Odum, H. T. 1960. Ecological potential and analogue circuits for the ecosystem. *Am. Sci.,* **48**:1–8.

Odum, H. T. 1967a. Biological circuits and the marine systems of Texas. In T. H. Olson and F. J. Burgess (Eds.), *Pollution and Marine Ecology.* Wiley-Interscience, New York, pp. 99–157.

Odum, H. T. 1967b. Work circuits and systems stress. In H. E. Young (Ed.), *Symposium on Primary Productivity and Mineral Cycling in Natural Ecosystems.* University of Maine Press, Orono, ME, pp. 81–138.

Odum, H. T. 1971. *Environment, Power, and Society.* Wiley-Interscience, New York.

Odum, H. T. 1977. Energy, value, and money. In C. A. S. Hall and J. W. Day, Jr. (Eds), *Ecosystem Modeling in Theory and Practice.* Wiley, New York.

Odum, H. T. 1983. Maximum power and efficiency: A rebuttal. *Ecol. Modelling,* **20**:71–82.

Odum, H. T. and R. C. Pinkerton. 1955. Times speed regulator, the optimum efficiency for maximum output in physical and biological systems. *Am. Sci.,* **43**:331–343.

Odum, H. T. and E. C. Odum. 1976. *Energy Basis for Man and Nature.* McGraw-Hill, New York.

Odum, H. T., J. F. Alexander, F. Wang, M. Brown, M. Burnett, R. Costanza, P. Kangas, D. Swaney, S. Leibowitz, and S. Lemlich. 1979. Energy basis for the United States. Report to USDOE, EY-76-S-05-4398.

Odum, H. T., F. C. Wang, J. Alexander, and M. Gilliland. 1981. Energy analysis of environmental values. In H. T. Odum, M. J. Lavine, F. C. Wang, M. A. Miller, J. Alexander, and T. Butler (Eds.), *A Manual for Using Energy Analysis for Plant Siting. Report to the Nuclear Regulatory Commission,* Contract NRC-04-77-123 Mod3. Energy Analysis Workshop. Center for Wetlands. University of Florida, Gainesville.

Office of Technology Assessment. 1978a. *A Technology Assessment of Coal Slurry Pipelines.* Washington, D.C. OTA-E-60.

Office of Technology Assessment. 1978b. *Applications of Solar Technology to Today's Energy Needs.* Washington, D.C.

Office of Technological Assessment, 1980. *Energy from Biological Processes.* Volume II, *Technical and Environmental Analyses.* U.S. Congress, U.S. Government Printing Office, Washington, D.C.

Office of Technology Assessment. 1983a. Industrial energy use. U.S. Congress, Washington, D.C., OTA E-198.

Office of Technology Assessment. 1983b. The regional implications of transported air pollutants: An assessment of acidic deposition and ozone. Draft report to U.S. Congress.

Oglesby, H. S. and R. O. Blosser. 1980. Information on the sulfur content of bark and its contribution to SO_2 emissions when burned as a fuel. *J. Air Pollut. Control Assoc.,* **30**(7):769–772.

Oil and Gas Journal. 1985. Fisher sees stable Texas output. **83**:50–51.

Olson, G. W. 1981. *Soils and the Environment.* Chapman and Hall, New York.

Ortmeyer, C. E., J. Costello, W. K. C. Morgan, S. Sweker, and M. Peterson. 1974. The mortality of Appalachian coal miners, 1963–1971. *Arch. Env. Health.,* **29**:67–72. (As cited in Ramsay, 1978.)

Osborne, D. 1984. America's plentiful energy resource. *The Atlantic,* **March:** 86–102.

Oshima, R. J., M. P. Poe, P. K. Braegelmann, D. W. Baldwin, and V. Van Way. 1976. Ozone dosage–crop loss function for alfalfa: A standardized method for assessing crop losses from air pollutants. *J. Air Pollut. Control Assoc.,* **26**:861–865.

Oshima, R. J., P. K. Braegelman, D. W. Baldwin, V. Van Way, and O. C. Taylor. 1977. Reduction of tomato fruit size and yield by ozone. *J. Am. Soc. Hort. Sci.,* **102**:287–293.

Ostlund, H. G. and R. A. Fine. 1979. Oceanic distribution and transport of Tritium. In *Behavior of Tritium in the Environment,* Proceedings of a Symposium. International Atomic Energy Agency, Vienna, pp. 303–314.

Ostwald, W. 1907. Modern theory of energetics. *The Monist,* **17**(4):480–515.

Ostwald, W. 1911. Efficiency. *The Independent,* **71**:867–871.

Pacific Consultants. 1970. The forage resource. A study produced for the Public Land Law Review Commission. University of Idaho, Moscow.

Packard, V. 1959. *The Status Seekers: An Exploration of Class Behavior in America and the Hidden Barriers that Affect You, Your Community, Your Future.* D. McKay Co., New York.

Packard, V. 1960. *The Waste-Makers.* D. McKay Co., New York.

Page, N. J. and S. C. Creasy. 1975. Ore grade, metal production, and energy. *J. Res. U.S. Geol. Surv.,* **3**:9–13.

Page, T. 1977. *Conservation and Economic Efficiency: An Approach to Materials Policy.* Johns Hopkins University Press, Baltimore.

Page, W. P., G. Arbogast, R. G. Fabian, and J. Ciecka. 1982. Estimation of economic losses to the agricultural sector from air-borne residuals in the Ohio River Basin region. *J. Air Pollut. Control Assoc.,* **3**:151–154.

Palmer, L. 1983. New corn yield increases ahead. *Farm J.,* **107**(5):19–21.

Paone, J., J. L. Morning, and L. Giorgetti. 1974. Land utilization and reclamation in the mining industry 1930–1971. Bureau of Mines Info. Circular 8642. U.S. Government Printing Office, Washington, D.C.

Park, T. 1962. Beetles, competition and populations. *Science,* **138**:1369–1375.

Parker, P. A. 1974. Pulp, paper and paperboard mills. In *Energy Consumption in Manufacturing.* The Conference Board of the Ford Foundation. Ballinger, Cambridge, MA.

Parker, R. L. 1967. Composition of the earth's crust. U.S. Geol. Surv. Prof. Paper 440-D. U.S. Government Printing Office, Washington, D.C.

Patterson, J. 1978. Possible role of shale in uranium supply. Office of Uranium Resources and Enrichment. U.S. Department of Energy.

Paulsen, H. A. 1975. *Range Management in the Central and Southern Rocky Mountains.* U.S. Forest Service, Washington, D.C., RM-154.

Paulsen, H. A. and F. N. Ares. 1961. Trends in carrying capacity and vegetation on an arid southwestern range. *J. Range Manage.,* **14**:78–83.

Pearl, R. 1925. *The Biology of Population Growth.* Knopf, New York.

Peng, T. H., W. S. Broecker, H. D. Freyer, and S. Trumbore. 1983. A deconvolution of the tree ring based 13C record. *J. Geophys. Res.,* **88**(C6):3609–3620.

Penman, H. L. 1970. The water cycle. *Sci. Am.,* **223**:98–108.

Penner, P., J. Kurish, and B. Hannon. 1980. Energy and labor cost of coal electric fuel cycles. Energy Research Group Doc. No. 273. University of Illinois, Urbana, IL.

Perry, H. 1983. Coal in the United States: A status report. *Science,* **222**:377–384.

Perry, R. A. 1978. Rangeland resources: Worldwide opportunities and challenges. In D. N. Hyder (Ed.), *Proceedings of the First International Rangeland Congress.* Society for Range Management, Denver, CO, pp. 7–9.

Peterson, I. 1982a. The Ginna problem: Isolated or generic? *Sci. News,* **121**:68.

Peterson, I. 1982b. Nuclear shutdown: Tubular woes. *Sci. News,* **121**:105–110.

Peterson, I. 1982c. Ginna atomic plant scrapped by steel? *Sci. News,* **121**:277.

Pieper, R. D., G. B. Donart, E. Parker, and J. D. Wallace. 1978. Livestock and vegetational response to continuous and 4-pasture, 1-herd grazing systems in New Mexico. In D. N. Hyder (Ed.), *Proceedings of the First International Rangeland Congress.* Society for Range Management, Denver, CO, pp. 560–562.

Pigford, T. H. 1974a. Environmental aspects of nuclear energy production. *Ann. Rev. Nucl. Sci.,* **24**:515.

Pigford, T. H., M. J. Keaton, B. J. Mann, P. Cukor, and G. Sessler. 1974b. Fuel cycles for electric power generation, Parts I and II. Rep. No. EEED-103/106 for EPA. Teknekron, Inc., Berkeley, CA.

Pilati, D. 1977. Energy analysis of electricity supply and energy conservation options. *Energy,* **2**:1–7.

Pimentel, D. (Ed.). 1980. *CRC Handbook of Energy Utilization in Agriculture.* CRC Press, Boca Raton, FL.

Pimentel, D., 1981. Biomass Energy: Report of the Energy Research Advisory Board Panel on Biomass.

Pimentel, D., L. E. Hurd, A. C. Bellotti, M. J. Forster, I. N. Oka, O. D. Sholes, and R. J. Whitman. 1973. Food production and the energy crisis. *Science,* **182**:443–449.

Pimentel, D., W. R. Lynn, W. K. MacReynolds, M. T. Hewes, and S. Rush. 1974. Workshop on research methodologies for studies of energy, food, man and environment Phase I. Center for Environmental Quality Management. Cornell University, Ithaca, NY.

Pimentel, D., W. Dritschilo, J. Krammel, and J. Kutzman. 1975. Energy and land constraints in food protein production. *Science,* **190**:754–761.

Pimentel, D., E. C. Terhune, R. Dyson-Hudson, S. Rochereau, R. Samis, E. A. Smith, D. Denman, D. Reifschneider, and M. Shepard. 1976. Land degradation: Effects on food and energy resources. *Science,* **194**:843–848.

Pimentel, D., D. Nafus, W. Vergara, D. Papaj, L. Jaconetta, M. Wulfe, L. Olsvig, K. Frech, M. Loye, and E. Mendoza. 1978. Biological solar energy conversion and U.S. energy policy. *BioScience,* **28**(6):376–382.

Pimentel, D. and M. Pimentel. 1979. *Food, Energy and Society.* Wiley, New York.

Pimentel, D., P. A. Oltenacu, M. C. Nesheim, J. Krummel, M. S. Allen, and S. Chick. 1980. The potential for grass-fed livestock: Resource constraints. *Science,* **207**:843–848.

Pimentel, D., M. A. Moran, S. Fast, G. Weber, R. Bukantis, L. Balliett, P. Boveng, C. J. Cleveland, S. Hindman, and M. Young. 1981. Biomass energy from crop and forest residues. *Science,* **212**:1110–1115.

Pimentel, D. (Chairman) and Biomass Panel of the Energy Research Advisory Board. 1983. Biomass energy. *Solar Energy,* **30**:1–31.

Pimentel, D., C. Fried, L. Olson, S. Schmidt, K. Wagner-Johnson, A. Westman, A. M. Whelan, K. Foglia, P. Poole, T. Klein, R. Sobin, and A. Bochner. 1983. Biomass energy: Environmental and social costs. Environmental Biology, Department of Entomology & Section of Ecology and Systematics, NYS College of Agriculture and Life Sciences, Cornell University, Ithaca, NY, Report 83-2.

Pindyck, R. S. 1979a. Interfuel substitution and the industrial demand for energy: An international comparison. *Rev. Econ. Stat.,* **61**:169–179.

Pindyck, R. S. 1979b. The characteristics of the demand for energy. In J. C. Sawhill (Ed.), *Energy: Conservation and Public Policy.* Prentice-Hall, Englewood Cliffs, NJ.

Pingrey, D. W. 1976. Forest Products Energy Overview. In Energy and the Wood Products Industry. The Forest Products Research Society, Proceedings Number P76-14, Madison, WI.

Pingrey, D. W. and N. E. Wagonner, 1979. *Wood Fuel Fired Electric Power Generating Plants, Summary and Report,* Vol. 1. U.S. Department of Energy, Seattle, WA.

Pinker, T. R. and A. Delman. 1983. On the precipitation–vegetation–ozone connection in the Washington, D.C. area. *Atmos. Environ.,* **17**:221–226.

Platts, W. S. 1981. Streamside management to protect bank-channel stability and aquatic life. In D. M. Baumgartner (Ed.), *Interior West Watershed Management,* Proceedings of a Symposium held April 8–11, 1980. Spokane, Washington. Cooperative Extension, Washington State University, Pullman, WA., pp. 245–255.

Platts, W. S. and S. B. Martin. 1980. Livestock grazing and logging effects on trout. In *Proceedings of Wild Trout II,* September 24–25, 1979, Yellowstone National Park, Wyoming. Trout Unlimited and Federation of Fly Fishermen.

Platts, W. W. 1978. Livestock interactions with fish and aquatic environments: problems in evaluation. In K. Sabol (Ed.), *Transactions of the 43rd North American Wildlife and Natural Resources Conference.* Wildlife Management Institute, Washington, D.C., pp. 498–504.

Pohl, R. 1976. Radioactive Pollution. *ASHRAE J.,* **September**: 47.

Popp, M. 1982. German energy technology prospects. *Science,* **218**:1280–1285.

Porter, W. P., J. W. Mitchell, W. A. Beckman, and C. B. DeWitt. 1973. Behavioral implications of mechanistic ecology; thermal and behavioral modelling of desert ectotherms and their microenvironment. *Oecologia,* **13**(1):1–54.

Porter, W. P. and F. C. James. 1979. Behavioral implications of mechanistic ecology II: the African rainbow lizard, *Agama agama. Copeia,* **4**:594–619.

Pough, F. H. 1973. Lizard energetics and diet. *Ecology,* **54**(4):837–844.

Pough, F. H. 1980. The advantages of ectothermy for tetrapods. *Am. Midland Naturalist,* **115**:92–112.

Powell, J., F. R. Crowl, and D. G. Wagner. 1978. Plant biomass and nutrient cycling on a grazed, tallgrass prairie watershed. In D. N. Hyder (Ed.), *Proceedings of the First International Rangeland Congress.* Society for Range Management, Denver, CO, pp. 216–220.

President's Advisory Panel on Timber and the Environment (PAPTE). 1973. *Report to the President's Advisory Panel: Timber and the Environment.* U.S. Government Printing Office, Washington, D.C.

President's Commission on Coal. 1981. U.S. Government Printing Office, Washington, D.C.

Probstein, R. and H. Gold. 1978. *Water in Synthetic Fuel Production: The Technology and Alternatives.* MIT Press, Cambridge, MA.

Puckett, L. J. 1982. Acid rain, air pollution and tree growth in southeastern New York. *J. Environ. Qual.,* **11**:376–381.

Pulliam, R. H. and F. Enders. 1971. The feeding ecology of five sympatric finch species. *Ecology,* **52**(4):557–566.

Pyke, G. H., H. R. Pulliam, and E. L. Charnov. 1977. Optimal foraging: A selective review of theory and tests. *Q. Rev. Biol.,* **52**:137–154.

Rainwater, F. H. and H. P. Guy. 1961. Some observations on the hydrochemistry and sedimentation of the Chamberlin Glacier Area, Alaska. U.S. Geol. Surv. Prof. Paper 414-C, pp. 1–14.

Ramsay, W. 1979. *Unpaid Costs of Electrical Energy: Health and Environmental Impacts from Coal and Nuclear Power.* Johns Hopkins University Press, Baltimore.

Ramsey, W. J., I. Y. Borg, and R. L. Thornton. 1978. Institutional Constraints and the Potential for Oil Shale Development. Lawrence Livermore Laboratory, Livermore, California.

Raney, E. C., B. W. Menzel, and C. E. Weller. 1973. *Heated Effluents and Effects on Aquatic Life with Emphasis on Fishes—A Bibliography.* Ichthyological Associates Bulletin No. 9 and A.E.C. Tech. Inf. Services Rep. No. TID-3918. Oak Ridge, Tenn.

Range Term Glossary Committee. 1974. *A Glossary of Terms Used in Range Management.* Society for Range Management, Denver, CO.

Rappaport, R. A. 1967. *Pigs for the Ancestors: Ritual in the Ecology of a New Guinea People.* Yale University Press, New Haven.

Rattien, S. and D. Eaton. 1976. Oil shale: the prospects and problems of an emerging energy industry. *Ann. Rev. Energy,* **1**:183–212.

Redfield, A. C. 1958. The biological control of chemical factors in the environment. *Am. Sci.,* **46**:205–221.

Regan, E. J. 1977. Energy analysis and models of soil formation. In H. T. Odum and J. Alexander (Eds.), *Energy Analysis of Models of the United States.* Systems Ecology and Energy Analysis Group, University of Florida, Gainesville.

Reich, P. B., R. G. Amundson, and J. P. Lassoie. 1982. Reduction in soybean yield after exposure to ozone and sulfur dioxide using a linear gradient exposure technique. *Water, Air Soil Pollut.,* **17**:29–36.

Reiners, W. A., R. H. Marks, and P. M. Vitousek. 1975. Heavy metals in subalpine and alpine soils of New Hampshire. *Oikos,* **26**:264–275.

Rettig, B. R. and J. J. C. Ginter (Eds.). 1978. *Limited Entry as a Fishery Management Tool.* Proceedings of a National Conference to Consider Limited Entry as a Tool in Fishery Management, Denver. A Washington Sea Grant Publication. University of Washington Press, Seattle.

Revelle, R. and H. E. Suess. 1957. Carbon dioxide exchange between atmosphere and ocean and the question of an increase of atmospheric CO_2 during the past decade. *Tellus,* **9**:18–27.

Ricardo, David. 1891. *The Principles of Political Econ-*

omy and Taxation. G. Bell and Sons, London. (Reprint of the 3rd ed., pub. 1821.)

Richards, P. W. 1952. *The Tropical Rain Forest*. Cambridge University Press, Cambridge.

Rickard, W. H. and C. E. Cushing. 1982. Recovery of streamside woody vegetation after exclusion of livestock grazing. *J. Range Manage.*, **35**(3):360–361.

Ricker, W. E. 1954. Stock and recruitment. *J. Fish Res. Bd. Can.*, **11**(5):559–623.

Ricker, W. E. 1958. *Handbook of Computations for Biological Statistics of Fish Populations*. Fish. Research Board of Canada, Ottawa, Canada.

Ricklefs, R. E. 1973. *Ecology*. Chiron Press, Newton, MA.

Ringwood, T. 1982. Immobilization of radioactive wastes in SYNROC. *Am. Sci.*, **70**(2):201–207.

Rittenhouse, L. R., L. E. Bartel, and A. H. Denham. 1982. Seasonal response of growing heifers to four stocking rates. Colo. Exp. Sta. Prog. Rept. No. 6.

Roach, C. H. and L. L. Smith. 1980. 1980 uranium assessment report, results, and future NURE plans. 1980 National Uranium Resource Evaluation Report. Office of Resource Assessment, Grand Junction Office. U.S. Department of Energy.

Roberts, P. C. 1982. Energy and value. *Energy Policy*, **10**:171–180.

Rochereau, S. and D. Pimentel. 1978. Energy tradeoffs between northeast fishery production and coastal power reactors. *Energy*, **3**:575–589.

Roff, D. A., and D. J. Fairbairn. 1980. An evaluation of Gulland's method for fitting the Schaefer model. *Can. J. Fish. Aq. Sci.*, **37**:1229–1235.

Rogers, W. L. 1975. The Oil Shale Environmental Advisory Panel: The environment and the Federal program—past, present, and future. *Q. Colo. Sch. Mines*, **70**(4):1–18.

Romer, R. H. 1976. *Energy: An Introduction to Physics*. W. H. Freeman, San Francisco.

Root, D. H. 1980. Historical growth of estimates of oil and gas field size. Paper presented at USDOE and National Bureau of Standards Symposium on Oil and Gas Supply Modeling, June 19. Washington, D.C.

Root, D. M. and L. J. Drew. 1979. The pattern of petroleum discovery rates. *Am. Sci.*, **67**:648–652.

Root, D. M. and J. M. Schuenemeyer. 1980. Petroleum resource appraisal and discovery rate forecasting in partially explored regions: Mathematical foundations. USGS Prof. Paper 1138-B. U.S. Government Printing Office, Washington, D.C.

Ross, M. H. and R. H. Williams. 1977. Energy and economic growth. Senate Subcommittee on Energy of the Joint Economic Committee, 91–952. U.S. Government Printing Office, Washington, D.C.

Ross, R. 1981. Energy consumption by industry. *Ann. Rev. Energy*, **6**:379–416.

Rothman, A. J. 1975. Research and development on rubble *in situ* extraction (RISE) at Lawrence Livermore Laboratory. *Q. Colo. Sch. Mines*, **70**(3):159–178.

Rothman, A. J. and A. E. Lewis. 1975. Rubble *in situ* extraction (RISE): A proposed program of recovery of oil from shale. Lawrence Livermore Laboratory, Lawrence, California, UCRL-51768.

Rouston, R. C. and R. M. Bean. 1977. Ground disposal of oil shale wastes: A review with an indexed annotated bibliography through 1976. Battele, Pacific Northwest Laboratory, Richland, Washington, PNL 2200.

Rowe, R. 1978. *Land Resource Economics*. Prentice-Hall. Englewood Cliffs, NJ.

Royce, W. F., L. Smith, and A. C. Hartt. 1968. Models of oceanic migration of Pacific salmon and comments on guidance mechanisms. *USFWS Fish Bull.*, **66**(3):441–462.

Russell, E. S. 1931. Some theoretical considerations on the "Overfishing" problem. *J. Cons. Int. Explor. Mer.*, **6**(1):3–20.

Ruttan, V. W. 1982. Discussion. In P. Crosson (Ed.), *The Cropland Crisis*. Resources for the Future, Baltimore, pp. 57–62.

Ryan, C. J. 1980. The choices in the next energy and social revolution. *Technological Forecasting and Social Change*, **16**:191–208.

Ryan, J. W., and others. 1981. An estimate of the non-health benefits of meeting the secondary national ambient air quality standards, final report. Prepared for the National Committee on Air Quality, Washington, D.C. SRE International Project No. 2094.

Ryther, J. H. 1969. Photosynthesis and fish production in the sea. *Science*, **166**:72–76.

Sampson, R. N. 1979. The ethical dimension of farmland protection. In M. Schnepf (Ed.), *Farmland, Food, and the Future*. Soil Conservation Society of America, Ankeny, Iowa, pp. 89–98.

Sampson, R. N. 1981. *Farmland or Wasteland*. Rodale Press. Emmaus, PA.

Samuelson, P. 1976. *Economics*. McGraw-Hill, New York.

Sarkanen, K. V. 1983. Personal communication, September.

Saxby, J. D. and M. Shibaoka. 1974. Depth of oil origin and primary migration. *Am. Assoc. Pet. Geol. Bull.*, **59**:721–722.

Schaefer, M. B. 1957a. Fishery dynamics and present status of the Yellowfin Tuna population of the Eastern Tropical Pacific Ocean. *Bull. Inter-Am. Trop. Tuna Comm.*, **2**(6):247–285.

Schaefer, M. B. 1957b. A study of the dynamics of the

fishery for Yellowfin Tuna in the Eastern Tropical Pacific Ocean. *Bull. Inter-Am. Trop. Tuna. Comm.,* **2**:245–285.

Scheib, Thomas. 1979. Industrial hygiene for mining and tunneling. Proceedings of symposium, Nov. 1978. American Conference of Governmental and Industrial Hygenics, Inc. Cincinnati.

Schipper, L. 1977. Raising the productivity of energy utilization. *Ann. Rev. Energy,* **1**:455–517.

Schipper, L. and A. Lichtenberg. 1976. Efficient energy use and well-being: The Swedish example. *Science,* **194**:1001–1013.

Schlichter, T., C. A. S. Hall, A. Bolanos, V. Palmieri, and L. Levitan. In preparation. Energy and Central American Agriculture.

Schmude, K. O. 1979. A perspective on prime farmland. *J. Soil Water Conserv.,* **32**(5):240–242.

Schneider, S. H. and C. Mass. 1975. Volcanic dust and temperature trends. *Science,* **190**:741–746.

Schoener, T. W. 1982. Simple models of optimal feeding-territory size: A reconciliation. *Am. Nat.,* **121**:608–629.

Schrödinger, E. 1935. Science and the Human Temperament. George Allen and Unwin, London.

Schroeder, J. L., Jr. 1973. Modern mining methods—underground. In S. Cassidy (Ed.), *Elements of Practical Coal Mining.* American Institute of Mining, Metallurgical, and Petroleum Engineers, New York.

Schull, W., M. Otake, and J. Neil. 1981. Genetic Effect of the atomic bombs: A reappraisal. *Science,* **213**:1220–1227.

Schumacher, E. F. 1973. *Small is Beautiful: Economics as if People Mattered.* Harper & Row, New York.

Schurr, S. H. 1982. Energy efficiency and productivity efficiency: Some thoughts based on American experience. *Energy J.,* **3**:3.

Schurr, S. H. and B. Netschert. 1960. *Energy in the American Economy, 1850–1975.* Johns Hopkins University Press, Baltimore.

Schurr, S. H., J. Darmstadter, H. Perry, W. Ramsey, and M. Russell. 1979. *Energy in America's Future.* Published for Resources for the Future by Johns Hopkins Univeristy Press, Baltimore.

Scientists' Institute for Public Information. 1979. Primer on Natural Gas and Methane. A Report of the Task Force on Natural Gas and Methane.

Segre, E. 1970. *Enrico Fermi, Physicist.* University of Chicago Press, Chicago.

Seyle, H. 1956. *The Stress of Life.* McGraw-Hill, New York.

Shackleton, N. J., M. A. Hall, J. Line, and Cang Shuxi. 1983. Carbon isotope data in core V19-30 confirm reduced carbon dioxide concentration in the ice age atmosphere. *Nature,* **306**:319–322.

Shannon, C. E. and W. Weaver. 1963. *The Mathematical Theory of Communication.* University of Illinois Press, Urbana.

Sharpe, M. 1979. Rangeland condition. In *Rangeland Policies for the Future,* Proceedings of a Symposium, January 28–31, 1979, Tucson, Arizona. USDA, USDI, CEQ. U.S. Government Printing Office, Washington, D.C., GTR WO-17.

Sheppard, D., A. Saistro, J. Nadel, and H. Boushey. 1981. Exercise increases SO_2-induced bronchoconstriction in asthmatic subjects. *Ann. Rev. Resp. Dis.,* **123**:486–491.

Sheppard, D., W. S. Wong, C. F. Vehara, J. Nodel, and H. Boushey. 1980. Lower threshold and greater bronchomotor responsiveness of asthmatic subjects to SO_2. *Annu. Rev. Resp. Dis.,* **122**:873–878.

Sheridan, D. 1981. *Desertification of the United States.* Council on Environmental Quality, U.S. Government Printing Office, Washington, D.C.

Sherry, E. V. 1979. Hard and soft—each to his own taste. *Public Utilities Fortnightly,* Nov. 8.

Shoemaker, C. A. 1977. Pest Management Models for Crop Ecosystems. In C. A. S. Hall and J. Day (Eds.), *Ecosystem Modeling in Theory and Practice.* Wiley-Interscience, New York, pp. 545–574.

Shurcliff, W. A. 1978. *Solar Heated Homes in North America.* Brick House Publishing, Harrisville, NH.

Siegel, B. 1981. *Baton Route Morning Advocate,* April 15, pp. 11-C.

Simon, C. A. 1975. The influence of food abundance on territory size in the iguanid lizard *Sceloporus jarrovi. Ecology,* **56**:993–998.

Simon, J. L. 1981. *The Ultimate Resource.* Princeton University Press, Princeton, NJ.

Sims, P. L. and D. D. Dwyer. 1965. Pattern of retrogression of native vegetation in North Central Oklahoma. *J. Range Manage.,* **18**:20–25.

Singer, D. A., D. P. Cox, and L. A. Drew. 1975. Grade and tonnage relationships among copper deposits. U.S. Geol. Surv. Prof. Paper 907-A. U.S. Government Printing Office, Washington, D.C.

Singer, S. F. 1974. Domestic resources can satisfy the energy needs of the United States. *Energy Syst. Policy,* **1**:81–103.

Sissenwine, M. P., B. E. Brown, and J. Brennan-Hoskins. 1978. Brief history and state of the art of fish production models and some applications to fisheries off the Northeastern United States. In *Climate and Fisheries,* Proceedings from a Workshop held March 29–31, 1978. Center for Ocean Management Studies, pp. 25–48.

Skelly, J. M. 1980. Photochemical oxidant impact on

Mediterranean and temperate forest ecosystems: Real and potential effects. In P. R. Miller (Tech. Coord.), *Proceedings of the Symposium on Effects of Air Pollutants on Mediterranean and Temperate Ecosystems,* U.S. Forest Service, Washington, D.C., GTR:RSW-47.

Skelly, J. M., S. Duchelle, and L. W. Kress. 1979. Impact of photochemical oxidants to white pine in the Shenandoah, Blue Ridge Parkway and Great Smoky Mountains National Parks. In Second Conference on Scientific Research in the National Parks, November 26–30, 1979. San Francisco, CA.

Skinner, B. J. 1976a. A second iron age ahead? *Am. Sci.,* **64**:258.

Skinner, B. J. 1976b. *Earth Resources.* Prentice-Hall, Englewood Cliffs, NJ.

Skinner, B. J. and C. H. Walker. 1982. Radioactive wastes. *Am. Sci.,* **70**:180–181.

Skog, E. and I. A. Waterson. 1984. *Residential Fuelwood Use in the United States: 1980–1981. J. of Forestry,* **82**:742–747.

Slansky, F., Jr. and P. Feeny. 1977. Stabilization of the rate of nitrogen accumulation by larvae of the cabbage butterfly on wild and cultivated food plants. *Ecol. Monogr.,* **47**(2):209–228.

Slesser, M. 1978. *Energy in the Economy.* St. Martin's Press, New York.

Sloane, J., C. Hall, and L. Fisher. 1979. Efficiency of energy delivery systems. III. Assessing the potential energy savings of a comprehensive insulation program. *Environ. Manage.,* **3**(6):511–515.

Slobodkin, L. 1964. The strategy of evolution. *Am. Sci.,* **52**:342–357.

Smith, Adam. 1937. *An Inquiry into the Nature and Causes of the Wealth of Nations.* Modern Library, New York. (Reprinted from 5th ed.)

Smith, G. C. 1971. Ecological energetics of three species of ectothermic vertebrates. *Ecology,* **57**:252–264.

Smith, N. and T. Corcoran. 1979. *The Energy Analysis of Wood for Fuel Applications,* American Chemical Society Proceedings, Fuel Division. Vol. 21, no. 2. American Chemical Society, Washington, D.C.

Smith, R. 1966. *A Symposium on Estuarine Fishes.* American Fisheries Society, Special Publication No. 3.

Smith, R. J. 1977. Conclusions. In *Proceedings of a Seminar on Improving Fish and Wildlife Benefits in Range Management.* U.S. Fish and Wildlife Service, Biol. Serv. Prog., Washington, D.C., FWS OBS-77/1.

Smith, R. J. 1981. A seismological shoot-out at Diablo Canyon. *Science,* **214**:528–529.

Smith, V. K. 1978. Measuring natural resource scarcity: Theory and practice. *J. Environ. Econ. Manage.,* **5**:150–171.

Smith, V. K. 1979a. Natural resource scarcity: A statistical approach. *Rev. Econ. Stat.,* **61**:423–427.

Smith, V. K. (Ed.). 1979b. *Scarcity and Growth Reconsidered.* Johns Hopkins University Press, Baltimore.

Smith, V. K. 1980. The evaluation of natural resource adequacy: Elusive quest or frontier of economic analysis? *Land Econ.,* **56**:257–298.

Sneva, F. A. 1978. Nitrogen and sulfur impacts on the cold desert biome. In D. N. Hyder (Ed.), *Proceedings of the First International Rangeland Congress.* Society for Range Management, Denver, CO, pp. 678–680.

Soddy, F. 1912. *Matter and Energy.* Henry Holt, New York.

Soddy, F. 1922. *Cartesian Economics.* Hendersons, London.

Soddy, F. 1926. *Wealth, Virtual Wealth and Debt.* E. P. Dutton and Company, New York.

Soil Conservation Service. 1953. Land Facts. Tech Paper 123. USDA. Washington, D.C.

Soil Conservation Service. 1976. *National Range Handbook.* SCS, USDA, Washington, D.C.

Soil Conservation Service. 1977. Potential cropland study. Statistical Bulletin 578. USDA, Washington, D.C.

Soil Conservation Society of America. 1976. *Resource Conservation Glossary.* Soil Conservation Society of America, Ankeny, IA.

Solow, R. M. 1974. The economics of resources or the resources of economics. *Am. Econ. Rev.,* **64**:1–14.

Solow, R. M. 1978. Resources and economic growth. *Am. Econ.,* **2**:5–11.

Sonnerblum, S. 1978. *The Energy Connections.* Ballinger, Cambridge, MA.

Sparks, F. L. 1974. Water prospects for the emerging oil shale industry. *Q. Colo. Sch. Mines,* **69**(2):93–101.

Spaulding, M. L., M. Reed, S. B. Saila, E. Lorda, and H. A. Walker. 1982. Oil spill–fishery impact assessment modeling: Application to Georges Bank. In *Marine Ecosystem Modeling,* Proceedings From a Workshop Held April 6–8, 1982. U.S. Department of Commerce, Washington, D.C., pp. 101–130.

Special Advisory Committee. 1982. The future for beef. Beef Business Bull. May 5. National Cattleman's Association.

Spedding, C. R. W. 1979. *An Introduction to Agricultural Systems.* Applied Science Publishers, London.

Spencer, H. 1880. *First Principles.* Appleton, New York.

Sprugel, D. G., J. E. Miller, R. N. Muller, J. J. Smith, and P. B. Xerikos. 1980. Sulfur dioxide effects on yield and seed quality in field-grown soybeans. *Phytopathology,* **70**:1129–1133.

Spurr, S. H. and H. J. Vaux, 1976. Timber: Biological and economic potential. *Science,* **191**:752–756.

Stan, C. 1979. *Current Issues in Energy: A Collection of Papers.* Pergamon, New York.

Stanturf, J. 1979. Wood as an Energy Resource. In Conservation Circular, Vol. 17, no. 9. Department of Natural Resources, Cornell University, Ithaca, NY.

Stark, N. and C. F. Jordan. 1978. Nutrient retention by the root mat of an Amazonian rain forest. *Ecology,* **59**:434–437.

Starr, C. 1978. Energy use: An interregional analysis with implications from international comparisons. In J. Dunkerley (Ed.), *International Comparisons of Energy Consumption.* Resources for the Future, Inc. Washington, D.C.

States, J. B. 1975. Quantitative baseline definition for terrestrial ecosystems at oil shale tract C-a. *Q. Colo. Sch. Mines,* **70**(4):135–141.

Statistical Abstracts. Various years. U.S. Dept. of Commerce, Bureau of Economic Analysis, Washington, D.C.

Statistical Supplement to the Survey of Current Business. Various Years. U.S. Dept. of Commerce, Washington, D.C.

Steinhart, J. S. and C. E. Steinhart. 1974a. Energy use in the U.S. food system. *Science,* **184**:307–316.

Steinhart, C. E. and J. S. Steinhart. 1974b. *Energy Sources: Use and Role in Human Affairs.* Duxbury Press, Duxbury.

Steinhart, J. S. and B. J. McKellar. 1982. Future availability of oil for the U.S. In L. O. Ruedisili and M. W. Firebaugh (Eds.), *Perspective on Energy.* Oxford University Press, New York, pp. 156–186.

Stenslund, G. J. and R. G. Semonin. 1982. Another interpretation of pH trends in the United States. *Bull. Am. Meteor. Soc.,* **63**(11):1277–1284.

Stephens, M. M. and O. F. Spencer. 1957. *Petroleum and Natural Gas Production.* Pennsylvania State University Press, University Park.

Sternglass, E. 1981. *Secret Fallout: Low-level Radiation from Hiroshima to Three Mile Island.* McGraw-Hill, New York.

Stevens, J. B. and E. B. Godfrey. 1972. Use rates, resource flows, and efficiency of public investment in range improvements. *Am. J. Agric. Econ.,* **54**(4):611–621.

Stewart, B. A. and L. K. Porter. 1967. Nitrogen–sulphur relationships in wheat (*Triticum aestivum L.*), corn (*Zea mays*), and beans (*Phaseolus vulgaris*). *Agronomy J.,* **61**:267–271.

Stiglitz, J. E. 1979. Neoclassical analysis of resource economics. In V. K. Smith (Ed.), *Scarcity and Growth Reconsidered.* Johns Hopkins University Press, Baltimore, pp. 36–66.

Stobaugh, R. and D. Yergin, 1979. *Energy Future.* Random House, New York.

Stoddard, L. A., T. W. Box, and A. R. Smith. 1975. *Range Management,* 3rd ed. McGraw-Hill, New York.

Strassman, B. I. 1983. Grazing programs on Federal rangelands and refuges: Consequences for wildlife, beef production, and fossil fuel use. M. S. Thesis, Cornell University, Ithaca, NY.

Strout, A. M. 1962. Market trends in mineral fuels, 1951–60. Fuels Symposium, American Society of Mechanical Engineers, June 5–7, 1962. ASME Paper No. 62-Fus-4.

Stuiver, M. 1976. The ^{14}C distribution in the Atlantic Ocean. *J. Geophys. Res.,* **85**:2711–2718.

Sun, M. 1981. Diablo Canyon license suspended. *Science,* **214**:1102–1103.

Sundquist, E. T. and G. A. Miller. 1980. Oil shales and carbon dioxide. *Science,* **208**:740–741.

Survey of Current Businesses. U.S. Department of Commerce, Bureau of Census. Nov. 1969, Feb. 1974, Feb. 1979.

Sutcliffe, W. H., Jr. 1972. Some relations of land drainage, nutrients, particulate material, and fish catch in two eastern Canadian bays. *J. Fish. Res. Bd. Can.,* **29**:357–362.

Sutcliffe, W. H., Jr. 1973. Correlations between seasonal river discharge and local landings of American lobster (*Homerus americanus*) and Atlantic halibut (*Hippoglossus hippoglossus*) in the Gulf of St. Lawrence. *J. Fish. Res. Bd. Can.,* **30**(6):856–859.

Sutcliffe, W. H., Jr. 1976. Fish production and its relationship to climate and oceanographic variation. In M. P. Latremouille (Ed.), *Biennial Review, 1975/76.* Bedford Institute of Oceanography, Dartmouth, Nova Scotia, Canada, p. 100.

Tait, R. V. 1968. *Elements of Marine Ecology.* Plenum Press, New York.

Tatom, J. A. 1979. The productivity problem. *Rev.—Fed. Res. Bank St. Louis,* **61**(9):3–16.

Taylor, O. C. (Ed.). 1980. Photochemical oxidant air pollution effect on a mixed conifer forest ecosystem. Final Report. Corvallis Env. Res. Lab. Office of Research and Development, U.S. Environmental Protection Agency, EPA-600/3-80-002.

Teller, E. 1979. *Energy From Heaven and Earth.* W. H. Freeman, San Francisco.

Terao, A., and T. Tanaka. 1928. Influence of temperature upon the rate of reproduction in the water-flea *Moina macrocopa Strauss.* *Proc. Imper. Acad. (Japan),* **4**:553–555.

Theobald, P. K., S. P. Schweinfurth, and D. C. Duncan. 1972. Energy resources of the U.S. U.S. Geol. Surv. Circ. 650, Washington, D.C.

Thilenius, J. F., R. Smith, and F. R. Brown. 1974. Effect of 2,4-D on composition and production of an alpine plant community in Wyoming. *J. Range Manage.*, **27**(4):140–142.

Thirring, H. 1958. *Energy for Man: From Windmills to Nuclear Power.* Harper and Row, NY.

Thomas, A., P. DeCoufle, and R. Moure-Eraso. 1980. Mortality among workers in petroleum refining and petrochemical plants. *J. Occup. Med.*, **22**:2.

Thomas, G. W. 1978. Rangeland—the unrecognized resource in world food production systems. In D. N. Hyder (Ed.), *Proceedings of the First International Rangeland Congress.* Society for Range Management, Denver, CO, pp. 77–78.

Thomas, R. B. 1973. Human adaptation to a high Andean energy flow system. Occasional papers in Anthropology, no. 7. The Pennsylvania State University, University Park, PA.

Thomas, R. G. and F. H. Pough. 1979. The effect of rattlesnake venom on digestion of prey. *Toxicon*, **17**:221–228.

Thomas, R., L. Donoven, and M. Richard. 1978. Large wind turbine generators. NASA Lewis Research Center, DOE/NASA/1059/78/1.

Thompson, C. R., G. Kats, and E. Hensel. 1972. Effects of ambient levels of ozone on navel oranges. *Environ. Sci. Technol.*, **6**:1014–1016.

Thompson, L. 1979. Climate change and world grain production. Unpublished paper for Council on Foreign Relations, Iowa State Univ., Ames, Iowa.

Tillman, D. A. 1977. Uncounted energy: The present contribution of renewable resources. In D. A. Tillman, K. V. Sarkanen, and L. L. Anderson (Eds.), *Fuels and Energy from Renewable Resources.* Academic Press, New York.

Tillman, D. A. 1978. *Wood as an Energy Resource.* Academic Press, New York.

Tillman, D. A. and K. V. Sarkanen (Eds.). 1979. *Progress in Biomass Conversion,* Vol. 1. Academic Press, New York.

Time. November 19, 1979. **114**(21):62–63.

Tingey, D. T., R. A. Reinert, J. A. Dunning, and W. W. Heck. 1971. Vegetation injury from the interaction of nitrogen dioxide and sulfur dioxide. *Phytopathology*, **61**:1506–1511.

Tingey, D. T., R. A. Reinert, J. A. Dunning, and W. W. Heck. 1973. Foliar injury responses of eleven plant species to ozone sulfur dioxide mixtures. *Atmos. Env.*, **7**:201–208.

Tinkle, D. W. and N. F. Hadley. 1974. Lizard reproductive effort: Caloric estimates and comments on its evolution. *Ecology*, **56**:427–434.

Tiratsoo, E. N. 1967. *Natural Gas.* Plenum Press, New York.

Tissot, B. P. 1979. Effects on prolific petroleum source rock and major coal deposits caused by sea level changes. *Nature,* **277**:463–465.

Tissot, B. P. and D. H. Welte. 1978. *Petroleum Formation and Occurence.* Springer-Verlag, New York.

Tosi, J. A. 1974. Some relationships of climate to economic development in the tropics. In *International Union for Conservation of Nature and Natural Resources,* Proceedings of The Use of Ecological Guidelines for Development in the American Humid Tropics, February 20–22, 1974, Caracas, Venezuela. Morges, Switzerland.

Townsend, C. R. and P. Calow. 1981. *Physiological Ecology.* Sinauer Associates, Sunderland, MA.

Travis, C. and E. Etnier (Eds.). 1983. *Health Risks of Energy Technologies.* American Association for the Advancement of Science.

Triplett, G. B. and D. M. Van Doren. 1977. Agriculture without tillage. *Sci. Am.*, **236**(1):28–33.

Trombe, F., et al. 1977. Concrete walls to collect and hold heat. *Solar Age,* **August**:13.

Tryon, F. G. 1927. An index of consumption of fuels and water power. *J. Am. Stat. Assoc.*, **22**:271–282.

Turvey, R. and A. R. Nobay. 1965. On measuring energy consumption. *Econ. J.*, **75**:784–793.

TRW, Inc. 1978. *Trace Elements Associated with Oil Shale and Its Processing.* TRW, Inc., McLean, VA.

TRW, Inc. 1979. *Oil Shale Data Book.* TRW, Inc. McLean, VA.

Uldin, D. and D. Lathwell. 1968. Timing of application is key to crop nitrogen use. *N.Y. Food Life Sci.*, **1**(1):9–12.

United Nations. 1968. Hydro-electric potential of Europe's water resources: The present state of assessment in Europe. New York, ST ECE EP 39.

United Nations, Dept. of International Economic and Social Affairs. 1983. *Yearbook of World Energy Statistics.* United Nations, New York.

U.S. Bureau of Land Management. 1974. *Final Environmental Impact Statement.* Livestock Grazing Management on National Resource Lands, Washington, D.C.

U.S. Bureau of Land Management. 1975. *Range Condition Report.* Prepared for the Senate Committee on Appropriations. U.S. Government Printing Office, Washington, D.C.

U.S. Bureau of Land Management. 1980. Four-year Authorization for Fiscal Years 1982–1985. Report to Congress.

U.S. Bureau of Land Management. 1982. *Public Land Statistics.* U.S. Government Printing Office, Washington, D.C.

U.S. Bureau of Mines. *Minerals Yearbook.* Various years.

U.S. Department of Agriculture. 1951. *Soil Survey*

Manual. Agricultural Handbook 18. Agricultural Research Administration, USDA, Washington, D.C.

U.S. Department of Agriculture. 1965. *Soil and Water Conservation Needs. A National Inventory 1958*. Msc. Pub. 971. USDA, Washington, D.C.

U.S. Department of Agriculture. 1971. *Basis Statistics—National Inventory of Soil and Water Conservation Needs. 1967*. Statistical Bulletin 461. USDA, Washington, D.C.

U.S. Department of Agriculture. 1977. *Guide to Energy Savings for the Field Crops Producer*. Federal Energy Administration, Washington, D.C.

U.S. Department of Agriculture. 1977. Study of fees for grazing livestock on Federal lands. A report from the Secretary of the Interior and the Secretary of Agriculture. U.S. Government Printing Office, Washington, D.C., 248-888/6624.

U.S. Department of Agriculture. 1979. SCS National Resource Inventories 1977. Final Estimates. USDA, Washington, D.C.

U.S. Department of Agriculture. 1980. Resources Conservation Act Appraisal 1980. Review Draft. USDA, Washington, D.C.

U.S. Department of Agriculture. 1981. *Agricultural Statistics*. Statistical Reporting Service. U.S. Government Printing Office, Washington, D.C.

U.S. Department of Agriculture. 1982. Economic Indicators of the Farm Sector: Production and Efficiency Statistics, 1980. Statistical Bulletin No. 679. USDA, Washington, D.C.

U.S. Department of Commerce. 1976. Historical statistics of the United States. Washington, D.C.

U.S. Department of Commerce, Bureau of Census. Various years. *Census of Manufacturers*. U.S. Government Printing Office, Washington, D.C.

U.S. Department of Commerce, Bureau of Census. Various years. *Census of Mineral Industries 1939, 1954, 1958, 1963, 1967, 1972, 1977*. U.S. Government Printing Office, Washington, D.C.

U.S. Department of Commerce, Bureau of the Census. Various years. *Highlights of U.S. Export and Import Trade*. U.S. Government Printing Office, Washington, D.C.

U.S. Department of Commerce, Bureau of the Census. Various years. Statistical Abstracts of the United States.

U.S. Department of Commerce, Bureau of Economic Analysis. Various years. Survey of Current Business.

U.S. Department of Commerce, Bureau of Economic Analysis. 1973. Long term economic growth, 1860 to 1970.

U.S. Department of Commerce, Bureau of Economic Analysis. 1980. The Biennial Supplement to the Survey of Current Business.

U.S. Department of Energy. Various years. *Monthly Energy Review*.

U.S. Department of Energy. 1976. *Energy from Coal*. U.S. Government Printing Office, Washington, D.C.

U.S. Department of Energy. 1979. *Bituminous Coal and Lignite Production and Mine Operations—1977*. U.S. Government Printing Office, Washington, D.C.

U.S. Department of Energy. 1980. *Demonstrated Reserve Base of Coal in the United States on January 1, 1979*. U.S. Government Printing Office, Washington, D.C.

U.S. Department of Energy. 1980. *Final Environmental Impact Statement—U.S. Spent Fuel Policy*, Vols. 1–5. U.S. Government Printing Office, Washington, D.C.

U.S. Department of Energy. 1980. *Report of a Workshop on the Role of Temperate Zone Forests in the World Carbon Cycle*. T. V. Armentano and J. Hett (Eds.). USDOE, Indianapolis.

U.S. Department of Energy. 1980. Report of the Energy Research Advisory Board on gasohol. Prepared by the Gasohol Study Group of the Energy Research Advisory Board, U.S. Dept. of Energy.

U.S. Department of Energy. 1983. An analysis of concepts for controlling atmospheric carbon dioxide. Prepared by M. Steinberg, Brookhaven National Laboratory, DOE/CH/00016-1

U.S. Department of Energy. 1983. Statistical data of the uranium industry. GJO-100(83), Grand Junction Office.

U.S. Department of Energy. 1984. *Annual Review of Energy: 1983*. U.S. Government Printing Office, Washington, D.C.

U.S. Department Health, Education and Welfare. 1977. *Occupational Diseases*. U.S. Government Printing Office, Washington, D.C.

U.S. Department of Labor, Bureau of Labor Statistics. Various years. *Monthly Labor Review*.

U.S. Department of Labor, Bureau of Labor Statistics. 1980. Occupational injuries in 1978: Summary. U.S. BLS, Report No. 586.

U.S. Department of Labor, Bureau of Labor Statistics. 1980. Occupational Injuries and Illnesses in the United States by Industry. 1978. Bulletin No. 2078, U.S. Government Printing Office, Washington, D.C.

U.S. 83rd Congress, 2nd Session. 1954. *Colorado River Storage Project*. U.S. Government Printing Office, Washington, D.C.

U.S. Environmental Protection Agency. 1973. *Environmental Analysis of the Uranium Fuel Cycle*. Part 1. *Fuel Supply*. Office of Radiation Programs, Washington, D.C.

U.S. Environmental Protection Agency. 1975. Tamano Oil Spill in Casco Bay: Environmental Effects and Cleanup Operations. Rep. No. 439/9-75-018. Environmental Protection Agency, Washington, D.C.

U.S. Environmental Protection Agency. 1979. Protecting Visibility. An EPA Report to Congress. Office of Air Quality, EPA 450/5-79-008. Washington, D.C.

U.S. Fish and Wildlife Service. 1978. Impacts of Coal-fired Power Plants on Fish, Wildlife and their Habitats. Biological Services Program, U.S. Dept. Interior, FWS/OBS-78/29.

U.S. Forest Service. 1976. *Feasibility of Utilizing Forest Residues for Energy and Chemicals.* National Science Foundation, Washington, D.C.

U.S. Forest Service. 1978. Forest Statistics of the United States, Review Draft. Washington, D.C., U.S. Dept. of Agriculture.

U.S. General Accounting Office. 1977. *Public Rangelands Continue to Deteriorate.* U.S. Government Printing Office, Washington, D.C., CED-77-88.

U.S. General Accounting Office. 1982. *Public Rangeland Improvement—A Slow, Costly Process in Need of Alternative Funding.* U.S. Government Printing Office, Washington, D.C., RCED-83-23, B-204997.

U.S. Geological Survey. 1974. USGS releases revised oil and gas resource estimates. News Release, March 26, Washington, D.C.

U.S. Geological Survey. 1976. Resource/reserve classification scheme. U.S. Geol. Surv. Bull. 1450-A.

U.S. Geological Survey. 1980. Principles of a resource/reserve classification for minerals. Circular 831. U.S. Government Printing Office, Washington, D.C.

U.S. Government Accounting Office. 1982. DOE funds new energy technologies without estimating potential net energy yields. Report to Congress, July. Washington, D.C.

U.S. Nuclear Regulatory Commission. 1975. Reactor safety study: An assessment of accident rate risks in U.S. commercial nuclear power plants. 9 vols. U.S. NRC report WASH-1400, NUREG-75/014. National Technical Information Service.

U.S. Nuclear Regulatory Commission. 1978. Generic Environmental Impacts Statement on Handling and Storage of Spent Light Water Reactor Fuels. Washington, D.C. NUREG-0404. Vol. 1.

U.S. Office of Congressional and Public Communications, Office of Technology Assessment. Impacts of Technology on U.S. Cropland and Rangeland Productivity.

U.S. Senate. 1936. The Western range. Senate document 199, 74th Congress, 2nd Session. U.S. Government Printing Office, Washington, D.C.

U.S. Water Resources Council. 1974. *Water for Energy Self-Sufficiency.* U.S. Government Printing Office, Washington, D.C.

U.S. Water Resources Council. 1979. *The Nation's Water Resources, 1975–2000,* Vols. 1–4. Second National Assessment United States Water Resources Council. U.S. Government Printing Office, Washington, D.C.

Uri, N. D. and S. A. Hassanein. 1982. Energy prices, labor productivity, and causality. *Energy Econ.* 4(2):98–104.

Vallentine, J. F. 1980. *Range Development and Improvements,* 2nd ed. Brigham Young University Press, Provo, UT.

Van Poollen, H. W. and J. R. Lacey. 1979. Herbage response to grazing systems and stocking intensities. *J. Range Manage.,* 32(4):250–253.

Van Valen, L. 1976. Energy and evolution. *Evol. Theory,* 1:179–229.

Van Vuren, D. 1982. Comparative ecology of bison and cattle in the Henry Mountains, Utah. In J. M. Peek and P. K. Dalke (Eds.), *Proceedings of the Wildlife–Livestock Relationships Symposium,* Coeur d'Alene, Idaho, April 20–22. Forest, Wildlife, and Range Experiment Station, University of Idaho, Moscow.

Vayda, A. P. (Ed.). 1969. *Environment and Cultural Behavior.* University of Texas Press, Austin.

Vogt, W. 1948. *Road to Survival.* William Sloane Associates, New York.

Wagner, F. H. 1978a. Livestock grazing and the livestock industry. In P. Brokaw (Ed.), *Wildlife and America: Contributions to an Understanding of American Wildlife and its Conservation.* Council on Environmental Quality, U.S. Government Printing Office, Washington, D.C., pp. 121–145.

Wagner, F. H. 1978b. Western rangeland: Troubled American resource. In K. Sabol (Ed.), *Transactions of the 43rd North American Wildlife and Natural Resources Conference.* Wildlife Management Institute, Washington, D.C., pp. 453–461.

Waldrop, M. M. 1982. Rethinking the future of magnetic fusion. *Science,* 217:1235–1236.

Wall Street Journal. 1979. 26 November, p. 1.

Wall Street Journal. 1981. 3 February, p. 1.

Wall Street Journal. 1984. 22–25 October, p. 1.

Walsh, J. 1980. What to do when the well runs dry. *Science,* 210:754–757.

Walsh, J. J., G. T. Rowe, R. L. Iverson, and C. P. McRoy. 1981. Biological export of shelf carbon is a sink of the global CO_2 cycle. *Nature,* 291:196–201.

Walsh, P. J., B. L. Etnier, and A. P. Watson. 1981. Health and safety implications of alternate energy technologies. III. *Fossil Energy Env. Mgmt.* 5(6):00.

Walters, C. J. 1975. Optimal harvest strategies for salmon in relation to environmental variability. *J. Fish. Res. Bd. Can.,* 32:1777–1784.

Walters, C. J. and R. Hilborn. 1976. Adaptive control of fishing systems. *J. Fish. Res. Bd. Can.*, **33**:145–159.

Walters, C. J. and R. Hilborn. 1978. Ecological optimization and adaptive management. Institute of Resource Ecology, University of British Columbia, Vancouver. Working Paper W-25.

Walters, R. F. 1971. Shifting cultivation in Latin America. FAO Forestry Development Paper 17. FAO, Rome.

Walton, I. (Ed.). 1982. Acid rain: The view from across the border. *Outdoor Am.*, **47**(5):16.

Ward, G. M., P. L. Knox, and B. W. Hobson. 1977. Beef production options and requirements for fossil fuel. *Science*, **198**:265–271.

Ward, J. V. and J. A. Stanford (Eds.). 1979. *The Ecology of Regulated Streams*. Plenum Press, New York.

Warren, C. E., in collaboration with P. Doudoroff. 1971. *Biology and Water Pollution Control*. W. B. Saunders, Philadelphia.

Watt, K. E. F. 1969. A comparative study on the meaning of stability in five biological systems: insect and furbearer populations, influenza, Thai hemorrhagic fever, and plague. In H. H. Smith (Ed.), *Diversity and Stability in Ecological Systems*. Brookhaven Symposia in Biology: Number 22. Brookhaven National Laboratory, BNL 50175 (C-56), pp. 142–150.

Watt, K. E. F. 1976. The Relationship Between Resource Use and Economic Growth: Forces Determining the Future. Marine Sciences Distinguished Lecture Series, No. 76-008. Louisiana State University Center for Wetland Resources, Baton Rouge.

Webb, M. and D. Pearce. 1977. The economics of energy analysis reconsidered. *Energy Policy*, **5**:158.

Weeks, L. G. 1948. Highlights on 1947 developments in foreign petroleum fields. *Am. Assoc. Pet. Geol. Bull.*, **32**:1093–1160.

Weinberg, A., I. Spiewak, D. Phung, and R. Livingston. 1985. The second nuclear era: A nuclear renaissance. *Energy*, **10**:661–680.

Weiss, H. V., M. Koide, and E. D. Goldberg. 1971. Selenium and sulfur in a Greenland ice sheet: Relation to fossil-fuel combustion. *Science*, **172**:261–263.

Wells, H. G., J. S. Huxley, and G. P. Wells. 1939. *The Science of Life*, Book 6, Part V. Garden City Publishing, Garden City, NY.

West, D. C., S. B. Maclaughlin, and H. H. Shugart. 1980. Simulated forest response to chronic air pollution stress. *J. Environ. Qual.*, **9**:43–49.

Wewerka, E. M. 1979. The Disposal and Reclamation of Southwestern Coal and Uranium Wastes. Los Alamos Scientific Laboratories. Proceedings of the Environmental Technology Training Conference, Arizona. LAUR 7916746.

White, L. A. 1949. *The Science of Culture*. Farrar, Straus, and Co., New York.

White, L. A. 1959. *The Evolution of Culture*. McGraw-Hill, New York.

Whitney, J. W. 1975. A resource analysis based on porphyry copper deposits and the cumulative copper metal curve using Monte Carlo simulation. *Econ. Geol.*, **70**:527–537.

Whittaker, R. H. and P. P. Feeny. 1971. Allelochemics: Chemical interactions between species. *Science*, **171**:757–770.

Wiersma, G. B. and K. W. Brown. 1980. Background levels of trace elements in forest ecosystems. In P. R. Miller (Tech. Coord.), *Proceedings of the Symposium on Effects of Air Pollutants on Mediterranean and Temperate Forest Ecosystems*, U.S. Forest Service GTR-PSW3.

Williams, C. S. 1979. Ramifications of vegetative manipulation. In *Rangeland Policies for the Future*, Proceedings of a Symposium January 28–31, Tucson, Arizona. USDA, USDI, CEQ. U.S. Government Printing Office, Washington, D.C., GTR WO-17.

Willrich, M. and T. B. Taylor. 1979. Nuclear theft: Risks and safeguards. In A. R. Norton and M. H. Greenberg (Eds.), *Studies in Nuclear Terrorism*. G. K. Hall and Company, Boston, pp. 59–84.

Wilson, T. 1961. *Inflation*. Harvard University Press, Cambridge.

Wiorkowski, J. J. 1981. Estimating volumes of remaining fossil fuel resources: A critical review. *J. Am. Stat. Assoc.*, **76**(375):534–548.

Wishart, R. S. 1978. Industrial energy in transition. *Science*, **199**:614–618.

Wolf, M., H. M. Goldman, and A. C. Lawson. 1978. Evaluation of options for process sequences. International Electronic and Electrical Eng. P. V. Special Conference, Washington, D.C., pp. 271–280.

Wolff, G. T., P. J. Lioy, and G. D. Wight. 1980. Transport of ozone associated with an air mass. *J. Environ. Sci. Health*, **A15**:183–199.

World Bank. 1980. *World Tables*. Johns Hopkins University Press, Baltimore.

Wright, D. J. 1974. Goods and services: An input–output analysis. *Energy Policy*, **2**:307–315.

Yergin, D. 1979. Conservation: The Key Energy Source. In R. Stobaugh and D. Yergin (Eds.), *Energy Future*. Random House, New York.

Young, R. J. and J. M. Evans. 1979. Occupational health and coal conversion. *J. Occup. Health Saf.*, Sept.

Zapp, A. D. 1961. World petroleum resources. In V. E. McKelvey (Ed.), Domestic and World Resources of Fossil Fuels, Radioactive Materials, and Geothermal Energy. Unpublished report by USGS for Federal Science Council.

Zapp, A. D. 1962. Future petroleum producing capacity of the U.S. U.S. Geol. Surv. Bull. 1142-H.

Zerbe, J. I., R. A. Arola, and R. M. Rowell. 1978. Opportunities for greater self-sufficiency in energy requirements for the forest products industry. In American Institute of Chemical Engineers no. 177, vol. 74.

Zucchetto, J. and S. Brown. 1977. Comparison of the fossil fuel energy requirements for solar, natural gas, and electrical water heating systems. *Res. Rec. Conserv.*, **2**:283–300.

Zucchetto, J., S. Bayley, L. Shapiro, D. Mau, and J. Nessel. 1980. The direct and indirect energy costs of coal transport by alternative bulk commodity modes. *Res. Rec. Conserv.*, **5**:161–177.

Zucchetto, J. and R. Walker. 1981. Time-series analysis of international energy-economic relationships. In W. J. Mitsch (Ed.), *Energy and Ecological Modeling,* Elsevier, Amsterdam, pp. 767–772.

Zucchetto, J., E. Titus, S. Blanco, C. Graizbord, and R. Walker. 1982. Global Energy/Economic Relationships. Working Papers in Regional Science and Transportation #64, Univ. of Pennsylvania, Philadelphia, PA.

INDEX

Abundant elements, 91
Acid mine drainage, 365, 367–368
Acid rain, 389–393
  aluminum, 392
  breeder reactor, 265
  coal, 231
  iron, 365
  power plants, 351
Adirondak Mountains, 352, 389–392
Aerosols:
  haze, 396
  nitrogen, 394
  sulfur, 394
Agricultural surplus, 496
Alaska:
  coal, 234
  natural gas, 204, 206, 209–211, 214
  oil, 162, 166, 168–177
  oil pipeline, 357
  oil shale, 221
  timber, 461
  water, 520
Alpha particles, 420, 424
Alpha t reaction, 264, 267
Aluminum:
  acid rain, 392
  energy cost, 100, 105
  environmental impact, 395
  fossil fuel combustion, 384
  ocean thermal energy conversion, 428
  soil, 482
Anthracite, 229, 234, 241–242, 368
Anticipated transient without scram, 281
Appalachia:
  coal bed methane, 216
  eastern gas shale, 216
  erosion rates, 491
  landscape, 484
Aquafer, see Groundwater
Argo Merchant, 358–359
Ash, 393, 418
Associated gas, 202
Atoms for Peace, 267–268
Autotroph, assimilation efficiency, 11

Backfilling, 363
Bagasse, 312
Beef, 129–130, 133
Benthic community:
  oil spill, 360
  power plant, 404
  thermal pollution, 407
Beta particles, 420

Biomass, energy potential, 308–309
Biomass energy:
  forest residues, 466–467
  harvest residues, 471
  land use, 493
Birds:
  oil spill, 359–360
  overgrazing, 504
  solar satellite, 428
  territoriality, 19
Bituminous coal, 229, 233–234, 240, 368
Black lung, 416
Boiling water reactor, 422
Brazil, 311
Breeder reactor, 261, 265, 523

Cadmium, 306, 427
Calefaction, see Thermal pollution
Cancer, see Carcinogens
Candu reactor, 267
Capital:
  depreciation, 106–107
  embodied, 71
  energy cost, 72, 106
  profit, 71
Capital cost, 45
  coal fired electricity, 333
  coal liquifaction, 254
  concentrating collector, 297
  decommisioning, 280
  fisheries, 444
  flat plate collectors, 291
  free piston engine, 305
  fusion, 265
  gas centrifuge, 277
  insulation, 338–339
  land improvement, 493
  no-till agriculture, 495
  oil shale, 226
  petroleum exploration, 178, 188
  photovoltaic cells, 304–305
  pollution control, 478
  rangeland improvements, 508
  salmon, 455
  solar hot water, 287
  tracking solar collector, 301–302
  underground mine fire, 369
  windmills, 307
Carbon cycle, 385–388
Carbon dioxide:
  atmospheric concentration, 383
  breeder reactors, 265
  fossil fuel combustion, 384

  missing, 386–389
  oil shale, 364–365
  photosynthesis, 383, 388
  physiological effect, 383
  release from biota, 388, 389
Carbon : hydrogen ratio:
  coal, 160, 231, 251–252
  kerogen, 156
  petroleum, 160
Carbon monoxide, 253
Carcinogens:
  biomass, 428
  coal, 417–418
  fossil fuel, 418
  fuelwood, 380–381
  herbicides, 505
  ocean sediments, 411
  petroleum industry, 415
  radiation, 422, 424–426
  safety standard, 414
  silicon, 427
  uranium mining, 422
Carrier bed, 157
Carrying capacity:
  density dependent, 116
  fish, 449
  human, 118
  management, 118
  rangelands, 497, 504
  stocking reductions, 510, 513
Carter, James, 311, 321, 346, 375
Catch per unit effort, see Yield per effort
Cation exchange capacity, 482
Causality, 50, 526–527
Cayuga station analysis, 333
Centralia, 369
Chaining, 504
Char, 252–253
Chattanooga shale, 271, 273, 277
Chemical defence, 20–21, 123
Chemical industry, 325
China:
  agriculture, 123, 130, 140
  population, 115, 117
Chlorine:
  environmental impact, 410–411
  synergistic environment, 411
Chlorosis, 397
Classical model:
  exchange value, 69
  use value, 70–71
Classification and Multi, 500
Clay, 357, 482

Clean Air Act:
  coal, 231, 239, 244, 250–251
  forest product industry, 470
  oil shale, 365
Clean Water Act, 470
Climate:
  carbon dioxide, 387, 389
  crops, 483
  definition, 481
  soil formation, 482
  solar satellite, 428
Climax ecosystem, 501, 503–504, 513
Clinch river, 265
Coal gasification, water, 367
Coal liquifaction, water, 367
Coal Mine Health & Safety Act, 238, 239,
    246–248, 366
Coal seams, 257
Coal workers' Pneumocon, see Black lung
Cogeneration, 326, 331
Coke, 229, 232, 241–242, 244
Colorado River Basin, 363–364, 367
Common property resource, 353–354
Community structure:
  air pollution, 399–400
  oil spills, 360
  thermal pollution, 408
Competition:
  economic, 16, 74–75, 526–527
  human, 74
  interspecific, 400–401, 450
  maximum power principle, 63–65
  surplus value, 71–72
Confinement, fusion, 265
Consumer tastes & preferences, 27, 72–76,
    84–85
Continuous mining machine, 237, 246,
    255–256
Contour mining, 240
Cooling:
  once through system, 404–405
  photovoltaic cell, 305
  water supply, 515
Cooling tower, 405–406, 432
Copper:
  coal, 393
  environmental impact, 395, 410–411
  Gibbs free energy, 95–96
  grade-tonnage relation, 92–94
  minerological threshold, 94–95
  ocean thermal energy conversion, 428
  ore quality, 86–87, 99, 101
  recovery, 97
  substitution, 84
  uranium by-product, 271
Corn:
  energy cost, 46, 136, 195, 197
  exports, 197
  gasahol, 311–312
  irrigation, 519
  labor costs, 70–71
  land quality, 486, 487

pollution, 402
pollution induced losses, 135, 398, 402
production, 135
water cost, 521
Cracking, 328
Critical thermal maximum, 17–19, 406–407
Crystal river, cooling tower, 405–406, 432
Cutting rates, 388
Cycle:
  carbon, 384–389
  fission reactor, 268
  nickel, 394
  nutrient, 121, 123
  hydrological, 515–517

Dams:
  electrical production, 308
  environmental impact, 315, 378–379
  evaporation, 517
Darrieus wind machine, 307
Decommissioning, 278–282
Deep gas, 207
Deforestation:
  coal mining, 241
  Greece, 285
  smolter stress, 395
  sulfur, 400
  tropics, 388
Deindustrialization, 324
Density dependent mortality, 449–451
Density independent mortality, 450, 453
Depression, 59–62, 101, 149
Detritus, oil spill, 14–15, 361
Deuterium, 265
Development, 52, 191
Diablo Canyon reactor, 281
Diet:
  caloric content, 107, 125
  mortality, 120, 131–132
Diminishing returns:
  incremental conservation, 327, 330
  rangeland improvement, 507
  see also Yield per effort
Discounting:
  irrigation, 523
  money, 340–341
  rate, 146–147
Dismantlement, fission reactors, 279
Domestic animals:
  conversion efficiency, 134
  food, 120
  work, 43, 72, 124–125
Dry gas, 202–204

Efficiency:
  agricultural production, 493–495
  assimilation, 11
  automobile, 328
  autotrophs, 11
  breeder reactors, 265
  Carnot, 6
  coal liquifaction, 253

coal transport, 250
cogeneration, 326, 331
concentrating collector, 297
deindustrialization, 324
domestic animals, 134
electrical transmission, 250, 276
electricity, 202
electricity from coal, 333
electric power plant, 7, 66
endothermy vs. ectothermy, 19
energy, 53–59
energy prices, 325
farm size, 490
fisheries, 444
flat plate collectors, 291
flue gas desulfurization, 251
gas, 202
heat, 64–65
heterotrophs, 11
inflation, 61–62
labor-product conversion, 471, 478
maintenance respiration, 129–133
maximum power, 63–66
natural energies, 354–355
oil, 202
oil exploration, 181
oil extraction, 165–166, 225
open pit mining, 225
Pareto optimality, 73
petroleum production, 129–130, 133
photovoltaic cell, 303–304, 306
power plant, 66
rangeland vs. pastureland, 507
room & pillar mining, 225
substitution, 45–46, 54–55
surface mining, 235
synthetic gas, 252
trophic energy transfer, 13, 19
trucking, 105
underground mining, 235
vessel embrittlement, 279
waste heat boiler, 329
water pump, 523
wood stoves, 468
Elasticity:
  efficiency, 57–58
  energy, 45
Electric power plant:
  coal use, 240–245
  efficiency, 7, 66
  energy cost, 109
Energy break even point:
  oil exploration, 182
  oil imports, 198
  oil shale, 225
Energy budget:
  animals, 12
  Peruvian highlanders, 121–122
Energy cost:
  agriculture, 120
  aluminum, 105
  average material, 194–197, 318

bauxite, 100
beef, 129–130, 133
capital, 106
coal fired boiler, 335
coal fired electricity, 333–336
copper, 97, 98, 100
corn, 135, 136, 195, 197
crops, 129–133
decommissioning, 278
economic production, 35, 37
electric power plant, 109
enrichment, 275
fertilizer, 124, 127–128
fish, 101, 444
food, 107
forest product, 472–476
fuelwood, 467–469
geopressurized gas, 218
goods & services, 51, 58
household fuel purchases, 56
hunter-gatherer, 119–120
insulation, 33
iron, 887, 99–100
irrigation, 493–494, 521–523
labor, 106
land quality, 487
manufactured goods, 194–197
measurement error, 110
migration, 19–20
mineralogical threshold, 94
minerals, 91, 96–101
natural resources, 79
natural selection, 16
nitrogen, 124
oil exploration, 179–186
oil refining, 156
oil spill, 359
petroleum, 106
precipitation, 335
production, 105
protein, 446
pulp, 470
railroads, 335
rangeland management, 506–509
rangeland vs. feedlot beef, 498–500
rice, 195, 197
roads and paving, 335
silicon, 317
taconite, 99–100
tractors, 318
treatment, 524
truncation problem, 109–113
water, 518
wheat, 195–197
world agriculture, 139–140
Energy end use:
 U.S., 301
 U.S. industry, 324–326
Energy/GNP ratio:
 international comparisons, 321–324
 product mix, 322–323
 U.S., 53–59

Energy gradient:
 ecosystem, 15
 heat, 7
 redox, 6
Energy opportunity cost:
 biological, 16
 definition, 8–9
 economic, 29
 natural resources, 79, 84–85, 95, 146
Energy payback period:
 passive solar, 301
 photovoltaic cell, 317
 solar water heater, 315
Energy prices:
 ceiling, 198
 coal, 243, 246–248
 conservation, 325–341, 344
 discount rate, 146–147
 domestic crude, 161, 164
 efficiency, 325
 energy/GNP ratio, 322
 EROI, 183, 196–198
 exploration, 170
 forest product technology, 475
 fossil fuel, 208
 free market, 198–199, 207
 gas, 169–170
 household fuel purchases, 56
 irrigation, 136 137, 402
 labor productivity, 44, 479
 land conversion, 492
 natural gas, 201, 205 207, 215
 petroleum exploration, 178–180
 recycling forest product, 471
 substitution, 44–46
 world oil, 190, 195
Energy quality:
 alternative fuels, 104
 biomass, 432–433
 coal, 55–56, 103–104, 160
 electricity, 276, 331–332
 EROI, 145
 gas, 103
 heat, 5, 55, 145
 natural gas, 201, 202
 oil, 103, 197–198
 petroleum, 55–56
 primary electricity, 55–56
 see also Efficiency
Energy return on investment:
 alcohol production, 198
 biomass energy, 380
 carbon dioxide, 383
 Cayuga station, 338–340
 Chattanooga shale, 277
 coal, 254–257
 coal liquification, 145
 decision criterion, 327–330, 532
 decommissioning, 281–282
 defence from predation, 20–21
 definition, 28–29
 discount rate, 146–147

domestic oil, 196
energy prices, 195–197
energy quality, 145
fishing, 456, 458, 466
food, 124–127, 130–133
food capture, 16–17
fuel quality, 83
fuelwood, 467–468
fusion, 533
geopressurized gas, 217–219
housekeeping conservation, 328
hunter-gatherer, 119
hydroelectric power, 315
imported gas, 195, 198
imported oil, 193–197
incremental conservation, 328
insulation, 332
Louisiana gas, 183
Louisiana oil, 183–186
migration, 19–20
natural energies, 351
nonfuel resources, 84, 146
nuclear power, 275–277, 281, 370–371
offshore oil, 183
oil reserves, 164, 171
oil shale, 198, 224–227, 364–366
petroleum production, 166, 180–183
photovoltaic cells, 316–317
solar, 380
solar satellite, 317
solar water heaters, 315–316
structural conservation, 328
substitution, 198–199
synthetic fuels, 254
territorial defence, 20
U.S. agriculture, 128
waste heat boiler, 329–330
wood plantations, 477
Energy storage:
 absorption air conditioning, 303
 conventional fuels, 290–291
 hot water, 301
 hydropower, 313
 solar, 199, 290, 312–313
 technologies, 313
 wind, 314
Energy theory of value, 53, 524
 criticism, 147–148
 prices, 74–75
Engine:
 electric, 326
 free piston, 305
 Rankine, 302–303
 solar, 288
 sterling, 305
Enhanced oil recovery, 165, 176
Enrichment:
 energy costs, 275
 gas centrifuge, 277
 health impact, 422
 uranium, 260–265, 275–276
Enthalpy, energy end use, 301

Entropy, economic systems, 7–8, 34–36, 39
Environmental protection, 404, 419
Erosion:
  agriculture, 481
  biomass energy, 380
  coal, 365
  land productivity, 485–487, 491
  landscape, 484
  loss rate, 137, 140
  nitrogen, 137, 380
  rangelands, 502–503
  wood plantations, 477
  yields, 137
Estuary:
  fish life cycle, 410, 438
  oil formation, 155–156
Exploration:
  cumulative effort, 175
  natural gas, 204
  petroleum, 155, 158, 162, 167, 169
  random, 177
  total effort, 179
  uranium, 271, 274
  wildcatting, 165
Exponential growth, population, 116
Exports:
  agricultural land, 493
  timber, 461
Externalities:
  coal mining, 246, 248
  crop production, 195
  energy, 59
  fission, 259
  shadow prices, 354
  see also Natural energies
Extinction, meteor theory, 431

Fallow, rangelands, 506
Famine, 115–117
Fatalities:
  cadmium, 427
  coal, 416
  energy technologies, 429
  low level radiation, 422
  nuclear power accident, 425
  petroleum industries, 415
Federal Land & Policy Management Act, 500
Federal lands, 488, 512
Federal Power Commission, 206–207
Federal Water Pollution, 404–405
Fertilizer:
  agricultural production, 102, 125, 488–489
  energy cost, 128
  rangelands, 505
  wood plantations, 477
Financial return on investment, 341
  Cayuga station, 334
  conservation, 326–328
  definition, 328–329

rangeland improvement, 508–509
  stocking reductions, 511–512
Fiscal policy, 60, 62, 148–150
Fish:
  acid rain, 389–392
  age class, 359
  chlorine, 410
  effect of overgrazing, 503
  energy cost, 101
  entrainment, 408–410
  oil shale, 363
  oil spill, 358–359
  thermal pollution, 406–407
Fisheries:
  catch, 443
  common property resource, 353–354
  location, 442
Fission:
  externalities, 259
  plant cancellation, 259–260, 265
  public confidence, 259–260
Flaring, 206
Flat plate collectors, 291
Flood, 354, 379, 524
Flue gas desulfurization, 251
Fluidized bed combustion, 254
Food web:
  crystal river, 433
  ecosystem, 14–15
Forest land:
  classification, 488
  resource base, 461
Forest stand improvement, 310, 477–478
Free market:
  agriculture, 137
  conservation, 339
  energy prices, 198–199
  fisheries, 455
  land use, 493
  myth, 27, 150
  natural gas, 207
  oil exploration, 188
Frontier philosophy, 446. See also Management
Frontier region:
  Alaska, 175–176
  definition, 170
  natural gas, 215
  oil exploration, 180
Fuel wood:
  energy potential, 311, 476, 479
  heat content, 468
  household use, 461, 467
  plantations, 477
Fusion, 265, 533

Gallium, 306, 427
Gamma ray, 420
Gasohol, energy potential, 311–312
Gas cap drive, 206
Gas centrifuge, 277
Gas : oil ratio, range, 209–210

Geology:
  mineral formation, 87–96
  natural gas formation, 202
  nutrient cycles, 121, 123
  oil formation, 190
Georges Bank, 358–359
Giant gas fields, 209–210
Giant oil fields, 164, 169, 170–171, 177
Gibbs free energy:
  copper, 95–96
  entropy, 7
  water, 515
Grade-tonnage relation, copper, 92–94
Granite:
  acid rain, 391
  nuclear waste disposal, 377
Green revolution, 102, 128, 138–140
Greenhouse effect, 384–385
Green River Formation, 216, 221, 223, 363
Gross National Product:
  definition, 47–49
  energy use, 50–53
  externalities, 59
  inadequacy, 49, 529
  measure, 39
  underground, 59
Groundwater:
  oil shale, 364
  overgrazing, 503
  supply, 517, 519
  uranium, 373
Growth, 432–434, 527–528
  biomass, 447
  neoclassical economics, 41–42
  projections, 530–534

Half life:
  cadmium, 427
  radioactive materials, 370
  photoxidants, 395
  spent fuel, 424
Hanford Storage Center, 370, 377
Hard path, 324
Hardwoods:
  energy cost, 474
  forest thinning, 477
  harvest investment, 478
  product conversion, 471
  technical change, 478
  timber harvest, 461, 463, 465
Hatcheries, 447
Haze, 396–397
Heating:
  electric, 332
  end use, 297
  passive solar, 285, 299–300
  solar, 299–301
Heat pump 330
Herbicides:
  human health, 505
  rangelands, 504–505
  see also Pesticides

Index 573

Heterotroph(s):
  assimilation efficiency, 11
  definition, 14
  herbivory *vs.* carnivory, 17
Household fuel use:
  coal, 241, 243
  energy/GNP ratio, 58–59
  international comparison, 322
Housekeeping, 327
Hubbert production cycle, 166
  coal, 167
  Hubbert unit, 175, 177
  Louisiana oil, 183–187
  natural gas, 214
  oil, 172
Hudson River, 405, 407, 409–410, 432
Hugoton-Panhandle field, 164, 209, 532
Hydrological cycle, 515–517

Imports:
  forest products, 461–462
  growth, 528–529
  petroleum, 189–199
  softwoods, 478
Incremental conservation, 327
  cogeneration, 331
  heat pump, 330
  system boundaries, 331
  waste heat boiler, 329–330
Industrial equivalent:
  cooling tower, 432
  crop loss, 403
  human life, 412
  Louisiana land loss, 361
  natural energies, 354
  ozone, 398, 402
  pollution, 402–403
  salmon, 455
  sulfur, 402
Industrial Fuel Use Act, 244
Inflation:
  biophysical economic model, 61–62
  embodied energy, 195
  neoclassical model, 60–61
  stagflation, 33
  management, 141, 343
  oil exploration, 179
In situ techniques:
  coal, 253
  oil shale, 224–225
  uranium, 371
Insulation, 332–340
  energy cost, 335
  energy saving, 297, 324, 338
  health effect, 426
Invader species, 501, 503–504
Iron:
  acid rain, 365
  coal use, 241–242
  energy cost, 87, 99–100
  soil, 482

Irrigation:
  agricultural production, 489
  cost, 136–137
  energy cost, 493–494, 521–523
  energy prices, 492
  rangelands, 506
  salmon, 379
  solar power, 288, 302–303
  U.S. water balance, 517
  wind power, 307

Kaibab deer, 450
Kerogen:
  oil shale, 221
  petroleum formation, 155–156
  tar sands, 223
Kerr-McGee, 373
Kinetic energy, 3–5
Krill, 444

Labor, 69
  biological energy cost, 107
  embodied, 70–71
  energy costs, 28, 40, 56, 106, 256–25
  horsepower equivalent, 43
  subsistence wage, 107–108
  surplus value, 71
  wage, 45, 71
  work, 43
Labor productivity, 43–44
  agriculture, 490
  coal industry, 43
  energy prices, 44
  energy supplies, 530
  farmer, 124, 126, 133–134
  forest products, 465, 472–476
  resource extraction, 85–86
  resource price, 83
  surface coal mining, 239
  underground coal mining, 238
  uranium mining, 371
Land:
  area cropped, 487–488
  biomass energy, 493
  coal mining, 368–370
  cutting rates, 358
  dams, 379, 491
  energy equivalent, 40
  mining, 490
  no-till agriculture, 495
  oil exploration, 361–362
  oil shale, 363
  quality, 485–487
  rent, 71
  resource quality, 480
  solar collectors, 379
  technology, 494–495
  uranium cycle, 370–375
  use pattern, 487–496
Landscape, 481

Laws of thermodynamics:
  biological systems, 9–10
  economics, 27, 28, 34–40, 143–144
  first law, 5, 33, 36
  fourth law, 144–145
  second law, 5–6, 35
  value, 74
Leach mining, 371
Lead, 87, 396
Leibig's law of the minimum:
  energy, 144
  resource quality, 102
  yield, 127–128
Light water reactor, 260–261, 263, 267–268
Lignite, 230, 233–234, 249–250
Liquified natural gas, 202
Load curve, 303, 312–313
Logging, 310, 388
Logistic curve, 448–450
Longwall mining, 235–237
Louisiana:
  loss of wetlands, 361
  natural gas, 209
  oil, 181–183, 351
  unconventional natural gas, 215–216
Low interest loans, 340

McKelvey resource class, oil, 164
Maintenance respiration:
  capital, 106–107
  conversion efficiency, 129–133
  economics, 36
  ecosystem, 152
  endothermy *vs.* ectotherm, 17–19
  global, 383, 385
  homeostasis, 10–11
  thermal pollution, 406
Management:
  carrying capacity, 118
  definition, 120
  fertilizer application, 494
  fisheries, 446
  flood control, 524
  frontier philosophy, 446
  grazing systems, 506
  pesticides, 494
  rangelands, 497, 504–506
  wood, 446
Mancuso, Dr. Thomas, 420, 422
Manhattan project, 266
Marketing, 434, 459
Marshes, 360
Material standard of living, 27, 62, 71–72
Mauna Loa, 383
Maximum power:
  economic systems, 63–65, 74
  fisheries, 456
  foraging strategy, 17
  stocking rate, 511
Maximum sustainable yield, 452, 457, 459

Mercury:
  acid rain, 393
  coal, 394, 417
  environmental impact, 395
  grade-tonnage relation, 94
Metallurgical coal, *see* Coke
Michelis-Menten:
  crop yield, 127–128
  resource quality, 102
Migration:
  Alaska pipeline, 357
  oil formation, 157
  plankton blooms, 453
  response to temperature, 408
  salmon, 19–20
Milling, 260, 273–275
Mill residues, energy source, 309–310
Mine mouth power plant, 231, 248–250
Mineralogical threshold:
  copper, 94–95
  definition, 89–91
  energy cost, 94
Mining Safety & Health Act, 416
Molybdenum:
  coal, 393
  environmental impact, 395
  soil, 482
  uranium, 371
Monetary policy, 61–62, 149–150
Monoculture, 124
Mukluk field, 177

National Energy Act, 244
National Environmental Protection Agency,
  498
Native Americans, 371, 393
Natural energies:
  definition, 59
  effect of sulfur, 399
  efficiency, 493
  EROI, 195
  hydrological cycle, 515–517
  industrial equivalent, 354
  land use, 495
  nutrient cycles, 438
  oil refining, 156
  pollution, 351
  public service function, 351–353
  soil formation, 482
Natural gas liquids, 202
Natural resources:
  economic opportunity cost, 103
  energy opportunity cost, 79, 84–85
  prices, 78, 80–81
  substitution, 79, 84–85
  technology, 79–80
  wealth, 526
Natural selection:
  adaptive strategies, 13
  economic systems, 74
  energy cost, 16
  finches, 526
  fitness, 11–13

  goal, 15–16
  humans, 526–527
  pesticide resistance, 128
  reproduction, 408
Necrosis, 397
Neoclassical economics:
  exchange value, 69, 72–73
  failures, 33–34
  growth model, 41–42
  inflation, 60–61
  natural resources, 78
  oil exploration, 188
  Pareto efficiency, 53
  production, 36
  substitution, 44–46
  technical change, 42
Nickel:
  environmental impact, 395
  human impact on cycle, 393
  ocean thermal energy conversion, 428
  zooplankton, 438
Nitrogen:
  acid rain, 389–391
  coal, 160, 231, 394–395
  crop yields, 102, 127–128, 131
  emission level, 392
  energy cost, 124
  erosion, 380
  fertilizer, 135, 206, 401
  fossil fuels, 394, 418
  natural gas, 202
  Redfield ratio, 102
  secondary product, 394–395
  soil organic matter, 483
  uranium fuel fabrication, 374
  wheat yields, 138
  zooplankton, 438
Nonassociated gas, 202, 209, 211–213
No-till agriculture, 494–495
Nutrients:
  crop yield, 138
  energy cost, 124, 127
  fish yields, 453–455

Occupational health effect:
  coal, 238, 246, 248, 416–417
  definition, 413
  nuclear power, 422
  petroleum industry, 414–415
Occupational Safety Health Act, 414,
  416
Ocean:
  carbon dioxide, 386–389
  oil, 358–360
  productivity, 437–438, 440, 441, 444
Offshore oil, 170
  discoveries, 176
  environmental impact, 357–361
  EROI, 183
  occupational health risk, 414–415
  production, 184
  reserves, 347, 349
Ogallala aquafer, 519

Oil spill:
  evaporation, 358
  fish mortality, 358–360
  slicks, 358
OPEC, 161
Open access policy, 353
Open pit mining:
  coal, 240
  oil shale, 224–225
  uranium, 273
Organic matter:
  no-till agriculture, 494
  soil, 481–483
Overfishing, 44, 446–447
Overgrazing, 502, 503, 506, 509–513
Oxidation, oil spills, 358
Oyster Creek Plant, fish kill, 407–408
Ozone:
  environmental impact, 398–399
  formation, 395–397
  industrial equivalent, 402
  solar power satellite, 428
  synergistic environment, 401

Parabolic collectors, 287, 291–296
Particulates:
  black lung disease, 416
  fossil fuel combustion, 384, 418
  precipitators, 432
  silicosis, 422
Pascal's wager, 534
Pasturelands, 488
Pennsylvania, 369
Pesticides, 124
  coal, 394
  management, 494
  resistance, 128, 131, 140–141
Ph:
  precipitation 389–390
  soil 482
  see also Acid rain
Phosphorus:
  erosion, 380
  salmon production, 435–455
Photons, photovoltaic electricity, 305–306
Piceance basin:
  oil shale, 221, 223
  tight gas, 216
Pine:
  effect of ozone, 399
  heat content, 467–468
  timber harvest, 461
Pipelines, coal slurry, 249–250
Pittsburg coal bed, 233
Plankton:
  fish production, 452–453
  oil spill, 360
Plasma, 265
Plutonium:
  breeder reactors, 263
  fission, 261
  half life, 370
  nuclear weapons, 264–265

production, 267
toxicity, 264
Population:
China, 115, 117
density, 120–121
Egypt, 115, 117
exponential growth, 116
land use patterns, 388
oil shale, 364
threat, 138–139
U.S., 40–41
world, 115, 116
Potential energy, 3–5
Precipitators, 432
Pressurized water reactors, 422
Price Anderson Act, 268
Prices:
historical trend, 78
natural resources, 70–71, 80–81, 83
unit extraction cost, 80–81
Primary productivity:
commercial forests, 464
natural ecosystems, 14
rainfall, 508
Prime farmland, 487–488, 490
Prior appropriation doctrine, water, 364
Process wastes:
food, 312
timber industry, 469
Production:
biological, 65
biophysical economic model, 40–41
measures, 47–48
neoclassical model, 36
Productivity:
erosion, 491
estuaries, 155
fisheries, 437
Georges Bank, 359
oceans, 437–438, 440, 441, 444
rangelands, 498–499
Profit:
biophysical perspective, 75
conservation, 326
decision criteria, 340–341
energy intensity, 509
surplus value, 71
Property rights, 353
Protein, 129–133, 444, 446
Public goods, definition, 353. *See also*
Externalities
Public Rangelands Improvements, 508

RAD, definition, 420
Radiation:
alpha particle, 420
beta particle, 420
cancer, 422–424
coal, 351, 365, 418
gamma ray, 420
half life, 370
ionizing, 419
low level health effects, 414, 420–422

nuclear power accidents, 425–426
safety standard, 420
Radioactive wastes, 268, 277, 280, 431
away from reactor, 376
disposal, 376–377
high level, 376
low level, 376
salt mines, 377
shale, 377
spent fuels, 375
subseabed disposal, 376–377
transport, 377
transuranic, 376
Radium, 371
Radon, 371
Railroad:
coal, 241, 248–250
energy costs, 335
Rainfall, crop production, 483
Rank, coal, 229–231
Rasmussen report, 280–281, 370, 425
Rationing, fishing, 459
Reagan, Ronald, 349, 367, 375
Recycling:
forest products, 470–472, 476
hardwood *vs.* softwood, 478
matter, 145
nutrients, 121, 123, 405
paper products, 479
water, 524
wood wastes, 471–472
Redfield ratio, 102
Regulation, natural gas, 201, 206, 217
REM, definition, 420
Rent, 70–71, 80, 83
Reprocessing, 263, 280
human health, 422
spent fuel, 375–376
Reserves:
coal, 231–233
definition, 88–90
development drilling, 165
domestic petroleum, 161–164, 167–168,
172–176, 343–349
energy prices, 215
extensions & revisions, 165, 173, 178
frontier region, 180
geopressurized gas, 217
giant gas fields, 209–210
Louisiana natural gas, 209
natural gas, 201, 204, 207, 209–215, 347
oil shale, 221–222
prime farmland, 488
tar sands, 223
unconventional natural gas, 216–217
uranium, 268–273
volumetric yield method, 343–349
world oil, 190, 194
Resource optimism, 527–528
Resource quality:
copper, 86–87
definition, 85
demand, 71

energy opportunity cost, 95
instantaneous demand, 101
land, 484–485
lumber industry, 475
metal, 86–87
Michelis-Menten, 102
minerological threshold, 88–90
pollution, 77
rangelands, 502
rent, 83
reserve, 88–90
technology, 85, 87, 91
Respiratory disease:
coal miners, 416–417
fossil fuel combustion, 418
uranium miners, 422
Revegetation, 363, 369
Rice:
energy costs, 195, 197
exports, 197
farming methods, 123
yield, 102, 127–128
Ricker curve, 449–450, 454
River, nutrients, 438
Room & pillar mining:
coal, 236
oil shale, 224–225
Rural Electrification Act, 307

Safe storage, fission reactors, 279
Safety standard:
erosion, 487, 491
gallium, 427
radiation, 420
Salinization:
land loss, 492
oil production, 357
oil shale, 363–364
Salmon:
acid rain, 391
dams, 378–379
EROI, 446
phosphorus, 435–455
Ricker curve, 450
Salt mines, waste disposal, 377
Santa Barbara oil spill, 358
Sawah, 123
Scarce elements, 91–92
Schaeffer curves, 451–452, 455
Scrubbers, *see* Flue gas desulfurization
Seabrook Nuclear Plant, 409
Seam, coal, 231–233
Shadow prices, 354. *See also* Energy
opportunity cost
Shippingport reactor, 267, 280
Shoreham nuclear plant, 416
Shortwall mining, 237
SIC, definition, 105
Silicon:
human health, 427
photovoltaic cell, 306
processing, 316–317
zooplankton, 438

Silicosis, 422, 427
Slag, 363, 368, 373–374
Sludge, 371–373
Smelter stress, 395
Social health effects:
    coal, 417–418
    definition, 413
    petroleum, 416
    uranium mining, 422–424
Soft path, 324
Softwood:
    energy cost, 474
    harvest investment, 478
    product conversion, 471
    timber, 461–462
Soil:
    compaction, 492
    definition, 481
    formation, 482
    horizons, 482–483, 491
    oil shale, 363
    rangelands, 503
    tropical, 123
Soil Conservation Service Land Capability
        Classification, 485, 492
Solar architecture, 285–286, 297–298, 303
Solar radiation:
    constant, 290
    flux, 285, 290–293
    global heat balance, 384–386
Solar satellite, 317
Source bed, coal, 160
Source rocks, oil formation, 155–156
Soybeans, pollution induced losses, 398,
        402
Steady state economy, 529–534
Steam, 251, 329–330
Steel:
    coal, 99–100, 244–246
    coal fired electricity, 379–380
    electric arc furnace, 328
    energy cost, 326, 335
    solar collectors, 379
    water cost, 521
Strip mining:
    coal, 240
    oil shale, 362
Structural conservation, 324
Subbituminous coal, 230
Subseabed disposal, 376–377
Subsidence, coal mining, 368–369
Subsidies:
    insulation, 340
    natural gas, 348–349
Substitution:
    agriculture, 135
    biophysical model, 46
    copper, 84
    employment, 325
    energy, 144
    energy efficiency, 54–55
    EROI, 198–199
    imported oil, 198–199
    natural resources, 79, 84–85

neoclassical model, 44–46
    prices, 44–46
Suess effect, 385–386
Sulfur:
    acid main drainage, 368
    acid rain, 390–391
    coal, 160, 229–232, 239, 250
    community structure, 400
    emission levels, 392
    environmental impact, 398–401
    fluidized bed combustion, 254
    fossil fuel, 384, 418
    haze, 396–397
    human health, 419
    industrial equivalent, 402
    secondary product, 394–395
    soil organic matter, 483
    synergistic environment, 401
    uranium milling, 260
Surface Mine Control Reclamation Act, 240,
        366, 369
Surface mining:
    coal, 233–237, 240, 257
    environmental impact, 368
    uranium, 371
Swamps, 354
Sweden, 321–323
Swidden, 123
Symbiosis, 123
Synergistic health impact, 400, 411, 417,
        427
System boundaries:
    conservation, 326–327, 331
    entrainment, 409
    human health impacts, 413–414
Systems:
    economic, 36–40
    feedback loops, 75–76
    open vs. closed, 10, 13–15, 36–40
Systems approach:
    community stress, 400
    conservation, 332–340
    environmental management, 352
    fisheries, 453–455

Taconite, 99–100
Tailings, 371, 374
Tar sands, 223
Taylor Grazing Act, 500–502
Technology:
    agriculture, 120, 135–139
    biophysical perspective, 42–44
    coal extraction, 238–239
    coke, 232
    corn production, 136
    economic role, 34
    energy efficiency, 476
    energy storage, 313
    energy supply, 31, 529–530
    EROI, 254
    erosion, 492
    fishing, 437, 451, 459
    forest products, 467
    hardwood substitution, 478

land, 135, 481, 492, 494–495
    location, 484
    lumber industry, 469, 475–476
    maximum power, 66
    neoclassical model, 42
    oil exploration, 165, 174, 177–178,
        180–181
    oil recovery, 165, 346–347
    oil shale, 221–222, 224–227
    production function, 42
    resource quality, 79–80, 85, 87, 91,
        99–100, 527
    structural conservation, 328
    uranium enrichment, 277
Texas railroad commission, 206
Thermal pollution, 404–415, 432
Thermocline, 387, 438
Thorium, 263
Three Mile Island, 260, 279, 281–283
Tidal energy, 438
Tilth, 483
Tokamak, 266
Torrey Canyon, 359–360
Trace elements, environmental impact
        395–396. See also Individual
        elements
Transportation:
    beef, 500
    forest products, 467, 477–478
    health impact, 417, 424
Trap rocks, 157
Tree plantations, 310–311
Trigger price:
    imported oil, 198–199
    oil shale, 225
    unconventional gas, 215
Tritium, 265
Trophic efficiency, fisheries, 444
Tropics:
    agriculture, 123
    deforestation, 388
    soil, 123
Tuna, 444, 455
Turbidity:
    biomass, 380
    grazing, 503
    power plant construction, 404

Underground mining:
    coal, 233–238, 255
    drift, 237
    environmental impact, 368
    fires, 369
    groundwater, 368
    oil shale, 362
    shaft, 236
    slope, 236
    uranium, 273, 371
Unions, 318, 414–415
Unit train, 248–250
Upwelling, 438, 445
Uranium, 260–265, 282
    coal, 394, 418
    concentration, 277

conversion process, 373–374
copper by-product, 271
enrichment, 260–265, 275–276, 374
fuel fabrication, 374
half life, 370
mining, 371
reprocessing, 280
tailings, 371–374
yellowcake, 371
Uranium Mill Tailings Act, 371
Urbanization, 489–490, 493
Utility, 72–73, 147

WASH-1400, *see* Rasmussen report
Waste heat boiler, 329–330
Waste isolation pilot plant, 377
Water:
coal, 249, 367–368
cooling, 515
energy production, 357, 520
land quality, 485
oil shale, 363–364
prior appropriation doctrine, 364
radium, 371
rangelands, 505–506
supply, 515–517
synthetic fuels, 253–254

tertiary waste treatment, 354
timber products, 469
U.S. use pattern, 518
watershed, 352
Watt, James, 287
Wet gas, 202
Wetlands, 14–15, 360–362
Wheat:
dollar costs, 137
energy cost, 195, 197
exports, 197
gasahol, 311–312
irrigation, 519
nitrogen fertilizer, 138
Wildcatting, 165, 171, 174
Wilderness areas, 396
Work:
economic, 28, 43, 50
everyday activities, 22–23
maximum power, 63–66
physical definition, 3
physical *vs.* economic, 46, 50–53
power, 3
tractors, 489
World War II:
catch per unit effort, 447–449
coal liquifaction, 253
energy efficiency, 50, 325

inflation, 60–61
nuclear weapons, 266
Wyodak coal bed, 233

Yellowcake, *see* Milling
Yield per effort:
agricultural production, 489
beef production, 507
coal, 254–257
crop yields, 133–134
fisheries, 446–451, 456
land, 493–496
natural gas exploration, 211, 214–215
oil exploration, 176, 178–180
petroleum exploration, 172–173, 175
petroleum extraction, 166, 344–349
rangeland management, 497
stocking reductions, 509–513
uranium, 274
wood products, 475

Zapp hypothesis, 344–349
Zooplankton:
chlorine, 410
entrainment, 409
fecal pellets, 358, 438–439
fish production, 452–453